Selected Papers on

Discrete Mathematics

Selected Papers on Discrete Mathematics

DONALD E. KNUTH

Copyright ©2003
Center for the Study of Language and Information
Leland Stanford Junior University
09 08 07 06 05 04 03   5 4 3 2 1

Library of Congress Cataloging-in-Publication Data

Knuth, Donald Ervin, 1938-
    Selected papers in discrete mathematics / Donald E. Knuth.
    xvi,812 p. 23 cm. -- (CSLI lecture notes ; no. 106)
    Includes bibliographical references and index.
    ISBN 1-57586-249-2 (cloth : alk. paper) --
ISBN 1-57586-248-4 (paper : alk. paper)
    1. Mathematics. 2. Computer science--Mathematics. I. Title.
II. Series.
    QA39.3 .K59 2001
    510--dc21

                                              2001025325

Internet page
    http://www-cs-faculty.stanford.edu/~knuth/dm.html
contains further information and links to related books.

to Professor Marshall Hall, Jr. (1910–1990)
the grand thesis advisor
of all my students

Contents

1 Combinatorial Analysis and Computers 1

2 Two Notes on Notation 15

3 Bracket Notation for the 'Coefficient of' Operator 45

4 Johann Faulhaber and Sums of Powers 61

5 Notes on Thomas Harriot 85

6 A Permanent Inequality 89

7 Overlapping Pfaffians 105

8 The Sandwich Theorem 123

9 Combinatorial Matrices 177

10 Aztec Diamonds, Checkerboard Graphs, and Spanning Trees 187

11 Partitioned Tensor Products and Their Spectra 193

12 Oriented Subtrees of an Arc Digraph 203

13 Another Enumeration of Trees 209

14 Abel Identities and Inverse Relations 221

15 Convolution Polynomials 225

16 Polynomials Involving the Floor Function 257

17 Construction of a Random Sequence 265

18 An Imaginary Number System 271

19 Tables of Finite Fields 277

20 Finite Semifields and Projective Planes 305

21 A Class of Projective Planes 345

22 Notes on Central Groupoids 357

23 Huffman's Algorithm via Algebra 377

24 Wheels Within Wheels 387

25 Complements and Transitive Closures 393

26 Random Matroids **405**

27 The Asymptotic Number of Geometries **425**

28 Permutations with Nonnegative Partial Sums **429**

29 Efficient Balanced Codes **433**

30 The Knowlton–Graham Partition Problem **439**

31 Permutations, Matrices, and Generalized Young Tableaux **445**

32 Enumeration of Plane Partitions **465**

33 A Note on Solid Partitions **483**

34 Identities from Partition Involutions **493**

35 Subspaces, Subsets, and Partitions **511**

36 The Power of a Prime That Divides a Generalized Coefficient **515**

37 An Almost Linear Recurrence **525**

38 Recurrence Relations Based on Minimization **537**

39 A Recurrence Related to Trees **565**

40 The First Cycles in an Evolving Graph **585**

41 The Birth of the Giant Component **643**

Index **793**

Preface

This book brings together almost everything that I've written about mathematical topics during the past four decades. I'm grateful for this opportunity to put the materials into a consistent format, and to correct errors in the original publications that have come to my attention. If any of this work deserves to be remembered, it is now in the form that I most wish people to remember it.

Some of these papers were written purely as expositions in which I tried to introduce important work of other people and to explain it as well as I could. Other papers contain ideas that were novel when I wrote them down; but even in such cases I did not address the papers to specialists. I tried to make every paper self-contained, so that readers with a general mathematical background would be able to understand the details. My tendency has always been not to emphasize the theorems on the "bottom line," but rather to tell a story of how various results fit together in harmonious patterns.

This book is dedicated to Marshall Hall, my mathematical mentor in graduate school and later when I joined the Caltech faculty. Although I was never able to think straight in his presence, because I knew that his knowledge vastly exceeded my own, I took notes of our discussions and they always made sense afterwards. Several chapters in this book are directly related to Hall's famous paper of 1943 on finite projective planes, one of the most frequently cited works in all of mathematics. I also spent two years trying to help him study relations on commutators so that he could extend his well-known theorem about the Burnside problem in group theory; but alas, nothing came out of that joint work, except for the experience I gained in writing computer programs for algebraic manipulation, because the subject matter resisted all of our attempts at simplification. Above all I thank him for introducing me to the ways in which world-class mathematical research is done.

Although the present book is by no means a text on discrete math, readers will find just about every basic aspect of that subject in here somewhere: permutations, partitions, identities, recurrences, and combinatorial designs; matrix theory, number theory, graph theory, probability theory, and a bit of algebra. Therefore I hope this book will be of interest to students and teachers of mathematics, as well as to everyone else who has become seduced by the deep pleasures of this subject.

Chapter 1 surveys the initial impact that computers had upon combinatorial theory, as viewed in 1965. Many mathematicians had been warning me in those days not to waste my time playing with computing machines; thank goodness Marshall Hall was not one of them.

Chapter 2 discusses two aspects of mathematical notation that have proved to be quite important in my own work. This chapter also gave me an opportunity to present the interesting history of Stirling numbers and some of their most significant properties.

Chapter 3 is about another key innovation in notation, by which coefficients are extracted from generating functions.

I enjoyed writing all of these chapters, but some of them gave me special pleasure. Chapter 4, which deals with mathematics of the early 17th century, is one of the special ones, because it involves the solution to a 350-year-old riddle. Chapter 5, similarly, reports on some fascinating manuscripts, written about 1600, that I was able to study briefly at the British Museum.

One of the most famous problems in discrete mathematics, the conjecture by B. L. van der Waerden in 1926 about the permanent of a doubly stochastic matrix, was resolved in 1980. When I learned of this work I couldn't resist writing Chapter 6, which explains "from scratch" the remarkable combination of methods that finally led to this breakthrough.

Another important operation on arrays of numbers, called the Pfaffian because it was introduced by J. F. Pfaff in 1814, is explored in Chapter 7, which argues that Pfaffians deserve to be much better known because they are more fundamental than determinants. (A determinant is just the "bipartite" special case of a Pfaffian.)

Chapter 8 is a tutorial about a fascinating concept called the Lovász number of a graph. This number $\vartheta(G)$ is especially interesting because it can be computed efficiently although it lies between two other fundamental numbers whose evaluation is NP-hard.

Chapter 9, which I wrote at the same time as Chapter 8 but never submitted for publication, discusses a family of matrices inspired by graph theory and set theory that I propose to call "combinatorial matrices." Relations between matrices, graphs, and trees are explored further in Chapters 10–14, culminating in some far-reaching extensions of Abel's generalized binomial theorem.

Chapter 15 is one of my favorite papers, yet few have seen it because it was published in a journal that was not yet well-established. It discusses a unified way to understand a large number of formulas that I encountered repeatedly during the first 30 years of my mathematical life.

One of the things that distinguishes discrete mathematics from continuous mathematics is an emphasis on integers as opposed to real values. Therefore it is almost inevitable that a book about discrete mathematics should include a chapter about the "floor function" $\lfloor x \rfloor$, which denotes the integer part of x. Chapter 16 discusses some nonobvious identities satisfied by this function. (Incidentally, the floor function was commonly written '$[x]$' until Iverson's proposal to write '$\lfloor x \rfloor$' became almost universally adopted during the 1970s. I have silently changed from $[x]$ to $\lfloor x \rfloor$ when reprinting papers that were written in the 60s. Similarly, I have changed the notation for intervals of numbers, from oldstyle '$(0, 1)$' and '$[0, 1]$' to the more distinctive '$(0 \mathbin{..} 1)$' and '$[0 \mathbin{..} 1]$', in order to be consistent with the current editions of my other books.)

Chapter 17 presents a combinatorial construction by which we can conclude that some theoretical definitions of randomness have absolutely no relevance to practical concerns.

Chapter 18 is a lightweight and slightly fallacious work from my student days, included in this collection only as an antidote to the heavier going in some of the other chapters — and perhaps also to prove to myself that I did learn a thing or two over the years.

Finite fields surely rank among the most astonishing structures in all of mathematics. The operations of addition and multiplication can be defined on sets of q elements, in such a way that the usual properties of algebra are valid, if and only if q is a power of a prime number; and for every such q, there is essentially only one way to do it. Many codes and other combinatorial constructions have been based on finite fields. Yet these addition and multiplication operations do not fit any known pattern that works for arbitrary q; we know that a unique field always does exist, but we don't know how to construct it except by trial and error. Chapter 19 presents tables by which fields can be computed and explored when q is not too large, and demonstrates some of these mysteries. Chapters 20 and 21 go beyond fields to a wider class of algebraic systems called "semifields."

Chapter 22 is another of my favorites, in which I was able to bask in the pure pleasure of deriving amazing consequences from a single axiom. It consists of eight sections, in which eight different approaches are used to explore a simple problem in abstract algebra. Chapter 23 uses yet another kind of algebra to explain the properties of Huffman's famous algorithm for optimum coding.

Directed graphs are the focus of Chapters 24 and 25, in which we see how such graphs are built up from "strong components" and "weak components" in a fundamental and natural way.

Chapter 26 turns to the theory of matroids, which are the most important structures that arise in algorithms for combinatorial optimization. All matroids are constructed here in a purely combinatorial fashion. Chapter 27 gives a simple construction by which we can estimate the total number of matroids that are definable on a given set.

Discrete mathematics abounds with intriguing, simply stated problems that sometimes have elegant solutions. Chapters 28–30 present short answers to three questions that once were making the rounds.

More substantial, yet equally elementary, concepts are treated in Chapter 31, which discusses a nonobvious correspondence between matrices of nonnegative integers and pairs of so-called "tableaux." In this case I had the good luck to hit on some ideas that have subsequently been cited hundreds of times and carried much further than I could ever have imagined. A first application of these ideas, by which the celebrated generating function for plane partitions found by P. A. MacMahon in 1912 could be given a simple proof for the first time, appears in Chapter 32. Chapter 33 considers the next case, three-dimensional partitions, together with P-partitions when P is any partial ordering.

Classical 19th-century methods for manipulating partitions are shown to have remarkable consequences in Chapter 34. Then Chapter 35 relates partitions to finite vector spaces and to Gauss's generalization of binomial coefficients. Chapter 36 considers the number-theoretical implications when such coefficients are generalized even further.

Recurrence relations, by which a sequence of numbers is defined in terms of itself, arise frequently in discrete mathematics. Chapter 37 discusses an amusing recurrence for a slowly growing sequence related to partitions into powers of 2, which I encountered early in my career. Later I learned that it is a very special case of an important class of recurrences that arise in dynamic programming problems; this class is investigated in Chapter 38. Still another family of recurrences, highly relevant to the study of trees and random graphs, is the subject of Chapter 39.

The final chapters feature a new paradigm for the study of random graphs, based on ideas of Philippe Flajolet, by which much of 19th-century mathematics can be brought to bear on problems of great interest as we enter the 21st century. Therefore I've tried my best to make those chapters expository and to take a rather leisurely path through the ideas. As a result, Chapters 40 and 41 are quite lengthy — they comprise more than 25% of this book, and an entire issue of the journal *Random Structures and Algorithms* was devoted to Chapter 41 when that paper first came out. But I hope a few readers will have the courage to look carefully at this material and see why I was captivated by the subject.

I've been blessed with the opportunity to carry out this work in collaboration with outstanding coauthors: Jack D. Alanen, Edward A. Bender, Philippe Flajolet, Michael L. Fredman, Ronald L. Graham, Inger Johanne Håland, Svante Janson, Tomasz Łuczak, Theodor S. Motzkin, Michael S. Paterson, Boris Pittel, Herbert S. Wilf, and of course Marshall Hall. Without their help I would have been unable to make much progress on the majority of the problems whose solutions are presented here.

The present book has also been made possible by the skills of Kimberly Lewis Brown, Jim C. Chou, William J. Croft, Max Etchemendy, Tony Gee, Lauri Kanerva, William E. McMechan, and Christine Sosa, who helped to convert hundreds of pages published long, long ago into a modern electronic format. Dan Eilers, Mark Ward, and Udo Wermuth gave invaluable help with the nitty-gritty details of proofreading. I've used John Hobby's marvelous METAPOST to redraw all of the illustrations.

I have included an Addendum at the close of most chapters, commenting on subsequent developments. And I have tried to put all of the bibliographies into a consistent format, so that they will be as useful as possible to everyone who pursues the subject. I have also seized this opportunity to improve the wording of the chapters that were originally written before my publishers' copy editors taught me a thing or two about writing.

This book is number six in a series of eight that Stanford's Center for the Study of Language and Information (CSLI) plans to publish, containing archival forms of the papers I have written. The first volume, *Literate Programming*, appeared in 1992; the second, *Selected Papers on Computer Science*, appeared in 1996; the third, *Digital Typography*, in 1999; the fourth, *Selected Papers on Analysis of Algorithms*, in 2000; and the fifth, *Selected Papers on Computer Languages*, in 2003. Still in preparation are *Selected Papers on Design of Algorithms*; *Selected Papers on Fun and Games*.

Donald E. Knuth
Stanford, California
May 2003

Acknowledgments

"Combinatorial analysis and computers" originally appeared in *Computers and Computing*, American Mathematical Monthly Slaught Memorial Papers **10** (February 1965), pp. 21–28. Copyright ©1965 by the Mathematical Association of America (Incorporated). Reprinted by permission.

"Two notes on notation" originally appeared in *American Mathematical Monthly* **99** (1992), pp. 403–422; addendum, **102** (1995), p. 562. Copyright ©1992 and Copyright ©1995 by the Mathematical Association of America (Incorporated). Reprinted by permission.

"Bracket notation for the 'coefficient-of' operator" originally appeared in *A Classical Mind*, essays in honour of C. A. R. Hoare, edited by A. W. Roscoe (1994), pp. 247–258. Copyright ©1994 Prentice–Hall. Reprinted by permission.

"Johann Faulhaber and sums of powers" originally appeared in *Mathematics of Computation* **61** (1993), pp. 277–294. Copyright ©1993 American Mathematical Society. Reprinted by permission.

"Notes on Thomas Harriot" originally appeared in *Historia Mathematica* **10** (1983), as part of a review of *History of Binary and other Nondecimal Numeration* by Anton Glaser, pp. 236–243. Copyright ©1983 Academic Press, Inc. Reprinted by permissionof Elsevier Science.

"A permanent inequality" originally appeared in *American Mathematical Monthly* **88** (1981), pp. 731–740, 798. Copyright ©1981 by the Mathematical Association of America (Incorporated). Reprinted by permission.

"Overlapping Pfaffians" originally appeared in *Electronic Journal of Combinatorics* $3, 2$, the Foata Festschrift (1996), paper R5, 13 pp. Copyright ©1996 by Donald E. Knuth.

"The Sandwich Theorem" originally appeared in *Electronic Journal of Combinatorics* **1** (1994), article 1, 48 pp. Copyright ©1994 by Donald E. Knuth.

"Aztec diamonds, checkerboard graphs, and spanning trees" originally appeared in *Journal of Algebraic Combinatorics* **6** (1997), pp. 253–257. Copyright ©1997 by Kluwer Academic/Plenum Publishers. Reprinted by permission.

"Partitioned tensor products and their spectra" originally appeared in *Journal of Algebraic Combinatorics* **6** (1997), pp. 259–267. Copyright ©1997 by Kluwer Academic/Plenum Publishers. Reprinted by permission.

"Oriented subtrees of an arc digraph" originally appeared in *Journal of Combinatorial Theory* **3** (1967), pp. 309–314. Copyright ©1967 Academic Press, Inc. Reprinted by permission of Elsevier Science.

"Another enumeration of trees" originally appeared in *Canadian Journal of Mathematics* **20** (1968), pp. 1077–1086. Copyright ©1968 Canadian Mathematical Society. Reprinted by permission.

"Abel identities and inverse relations" originally appeared in *Combinatorial Mathematics and Its Applications*, R. C. Bose and T. A. Dowling, eds., University of North Carolina Monograph Series in Probability and Statistics **4** (Chapel Hill, North Carolina, 1969), 91–94. Copyright ©1969 University of North Carolina Press. Reprinted by permission.

"Convolution polynomials" originally appeared in *The Mathematica Journal* **2**, 4 (Fall 1992), pp. 67–78. Copyright ©1992 Miller Freeman Incorporated. Reprinted by permission.

"Polynomials involvinging the floor function" originally appeared in *Mathematica Scandinavica* **76** (1995), pp. 194–200. Copyright ©1995 Mathematica Scandinavica. Reprinted by permission.

"Construction of a random sequence" originally appeared in *BIT* **5** (1965), pp. 246–250. Copyright ©1965 Nordisk Tidskrift for Informationsbehandling. Reprinted by permission.

"An imaginary number system" originally appeared in *Communications of the ACM* **3** (April 1960), pp. 245–247. Errata, *Communications of the ACM* **4** (August 1961), pp. 355. Copyright presently held by the author.

"Tables of finite fields" originally appeared in *Sankhyā*, series A, **26** (1964), pp. 305–328. Copyright ©1964 Sankhyā. Reprinted by permission.

"Finite semifields and projective planes" originally appeared in *Journal of Algebra* **2** (1965), pp. 182–217. Copyright ©1965 Academic Press, Inc. Reprinted by permission of Elsevier Science.

"A class of projective planes" originally appeared in *Transactions of the American Mathematical Society* **115** (1965), pp. 541–549. Copyright ©1965 American Mathematical Society. Reprinted by permission.

"Notes on central groupoids" originally appeared in *Journal of Combinatorial Theory* **8** (1970), pp. 376–390. Copyright ©1970 Academic Press, Inc. Reprinted by permission of Elsevier Science.

"Huffman's algorithm via algebra" originally appeared in *Journal of Combinatorial Theory* (A) **32** (1982), pp. 216–224. Copyright ©1982 Academic Press, Inc. Reprinted by permission of Elsevier Science.

"Wheels within wheels" originally appeared in *Journal of Combinatorial Theory* (B) **16** (1974), pp. 42–46. Copyright ©1974 Academic Press, Inc. Reprinted by permission of Elsevier Science.

"Complements and transitive closures" originally appeared in *Discrete Mathematics* **2** (1972), pp. 17–30. Copyright ©1972 by North-Holland Publishing Company. Reprinted by permission of Elsevier Science.

"Random matroids" originally appeared in *Discrete Mathematics* **12** (1975), pp. 341–358. Copyright ©1975 by North-Holland Publishing Company. Reprinted by permission of Elsevier Science.

"The asymptotic number of geometries" originally appeared in *Journal of Combinatorial Theory* (B) **16** (1974), pp. 398–400. Copyright ©1974 Academic Press, Inc. Reprinted by permission of Elsevier Science.

Chapter 1

Combinatorial Analysis and Computers

[Written with Marshall Hall, Jr. Originally published in Computers and Computing, Number 10 of the Herbert Ellsworth Slaught Memorial Papers (February 1965), 21–28, a supplement to American Mathematical Monthly 72.]

The "logical decision-making" characteristics of computers enable them to attack many problems that are not basically numerical. Combinatorial constructions and searches are applications of this kind in which computers have been very successful. The two principal ways in which computers have been used in these problems are:

1) generation of a sequence of combinatorial patterns as part of a larger problem; and

2) constructions and searches for which hand calculations are not feasible.

When the size of a combinatorial problem grows very rapidly, we can usually expect the computer to do only about one case larger than can be done by hand. For example, the problem of constructing a finite projective plane of order n is equivalent to choosing, in a restricted way, $(n-1)^2$ of the $n!$ permutations of n letters; thus, changing n to $n+1$ will make the problem many orders of magnitude larger.

Combinatorial Sequences

Many computer applications call for the generation of combinatorial sequences such as the set of all permutations of n elements, the set of all combinations of m things taken n at a time, and so on.

The simplest combinatorial sequence is the set of all ordered n-tuples $(x_1, x_2, \ldots, x_n)$ such that $0 \le x_i < m$ for $1 \le i \le n$. This set can, of course, be generated by repeatedly adding 1 in the m-ary number system; in applications, however, the results are usually more useful if the n-tuples are generated in such a way that only one x_i changes at

1

each step [6, 8]. The simplest method for doing this is the following: At the kth step, $1 \le k < m^n$, add 1 to x_{n-i} (modulo m) if m^i is the highest power of m dividing k. Whenever k is a multiple of m, it is not difficult to prove that this will be the digit immediately left of the rightmost nonzero digit, assuming that $(0, 0, \ldots, 0)$ was the starting n-tuple. The case $m = 2$ gives rise to the so-called Gray binary numbers, which have found many engineering applications, and the sequence is the well-known procedure for solving the classical Chinese Ring Puzzle [2].

The problem of generating all permutations has stimulated much interest [1, 29, 42]. The following classical method [7] is still being rediscovered fairly often: Permutations can be generated in lexicographic order if we obtain the successor of $(x_1, x_2, \ldots, x_n)$ by (1) finding r and s such that

$$x_r < x_{r+1} \ge x_{r+2} \ge \cdots \ge x_n, \quad x_{r+s} > x_r \ge x_{r+s+1};$$

then (2) interchanging x_r and x_{r+s}; and finally (3) replacing the subsequence $(x_{r+1}, x_{r+2}, \ldots, x_n)$ by its reflection $(x_n, \ldots, x_{r+2}, x_{r+1})$. This method is valid even when the elements permuted are not distinct.

As above, however, it is usually advantageous to generate the permutations in such a way that successive permutations are "near" each other, namely that only two elements have been interchanged. The first method of this kind was due to Wells [47], but an improved algorithm was later discovered almost simultaneously by Trotter [44] and by Johnson [24]. In the latter method, all permutations of n distinct objects are obtained by the successive interchange of two adjacent elements, by using essentially the following recursive rule: Let $(y_1, y_2, \ldots, y_{n-1})$ be the $(k + 1)$st permutation of the sequence for all permutations on $(1, 2, \ldots, n - 1)$; then the $(kn + 1)$st, $(kn + 2)$nd, $\ldots$, $(kn + n)$th permutations on $(1, 2, \ldots, n)$ are

$$(y_1, y_2, \ldots, y_{n-2}, y_{n-1}, n), \ (y_1, y_2, \ldots, y_{n-2}, n, y_{n-1}),$$
$$\ldots, \ (y_1, n, y_2, \ldots, y_{n-2}, y_{n-1}), \ (n, y_1, y_2, \ldots, y_{n-2}, y_{n-1})$$

if k is even, and the same sequence in reverse order if k is odd. For example, here is the set of all 24 permutations on 4 elements obtained by repeated interchanges of adjacent elements:

1234	4132	3124	4321	2314	4213
1243	1432	3142	3421	2341	2413
1423	1342	3412	3241	2431	2143
4123	1324	4312	3214	4231	2134

This technique is the fastest known method for generating the complete set of permutations on a computer.

Similarly, an algorithm is known for successively choosing all combinations of m things, n at a time, by changing the choice of only one object each time. Algorithms for obtaining all partitions of an integer n are given in [40]; the partitions of a set are produced by the algorithm in [21]; and many similar techniques are known.

Backtrack

The basic method [10, 45] used for most combinatorial searches has been christened the "backtrack" technique by D. H. Lehmer. A great many combinatorial problems can be stated in the form, "Find all vectors $(x_1, x_2, \ldots, x_n)$ that satisfy p_n," where $x_1, x_2, \ldots, x_n$ are to be chosen from some finite set of m distinct objects, and p_n is some property. The combinatorial sequences of the preceding section all fit into this framework; for example, if we are to generate all *permutations* of distinct objects, the property p_n states simply that $x_i \neq x_j$ if $i \neq j$. If we wish to generate all *combinations* of the numbers $(1, 2, \ldots, m)$ taken n at a time, p_n is the property

$$1 \leq x_1 < x_2 < \cdots < x_n \leq m.$$

The backtrack method consists of defining intermediate properties p_k for $1 \leq k \leq n$ in such a way that whenever $(x_1, x_2, \ldots, x_k)$ satisfies p_k, then $(x_1, \ldots, x_{k-1})$ necessarily satisfies p_{k-1}. The computer is programmed in a straightforward manner to consider only those partial solutions $(x_1, \ldots, x_k)$ that satisfy p_k; for if p_k is not satisfied, the m^{n-k} vectors $(x_1, \ldots, x_k, x_{k+1}, \ldots, x_n)$ need not be examined. If the properties p_k can be chosen in an efficient way, comparatively few cases need to be considered.

Although the backtrack approach frequently gives satisfactory results, there are equally many cases on record for which it has failed; in fact, it is not uncommon (see [33]) to find situations in which thousands of centuries would be required for the algorithm to be completed! Therefore it is desirable to have some convenient means for estimating the computational time before putting the algorithm onto a computer. One can proceed by playing the following little game: Let a_1 be the number of ways to choose x_1 so that (x_1) satisfies p_1; then let y_1 be one such solution, chosen at random with probability $1/a_1$. Similarly let a_k be the number of ways to choose x_k so that $(y_1, \ldots, y_{k-1}, x_k)$ satisfies p_k; let y_k be one such solution, chosen at random with probability $1/a_k$. If

$a_k = 0$ or $k = n$, the game terminates. It is not difficult to verify that the expected value of

$$a_1 + a_1 a_2 + a_1 a_2 a_3 + \cdots + a_1 a_2 \ldots a_n$$

will be precisely the number of cases that are examined by the backtrack algorithm, and therefore by playing this game several times one obtains a good idea of the size of this number. When the number is greater than, say, 10 million, the backtrack method is probably unfeasible with contemporary equipment.

This random process can be combined with backtracking to give "random" solutions $(x_1, x_2, \ldots, x_n)$ to the combinatorial problem. The solutions are not, however, truly random in general; that is, they are not necessarily obtained with equal probability. For example, Fig. 1 shows several 10×10 latin squares, which are arrays whose rows and columns are permutations of the indices. If we assume that the first row and first column are fixed, square B [33, 34] would be obtained with probability $2^{-9}3^{-11}4^{-12}5^{-6}6^{-3}7^{-4}9^{-1} \approx 9.0 \times 10^{-27}$. But square A, which certainly appears to be "less random," is actually produced with the much higher probability $9^{-1}8^{-3}7^{-5}6^{-7}5^{-6}4^{-4}3^{-2} \approx 1.3 \times 10^{-21}$.

A	*B*	*C*
0 1 2 3 4 5 6 7 8 9	0 1 2 3 4 5 6 7 8 9	0 1 2 3 4 5 6 7 8 9
1 2 3 4 5 6 7 8 9 0	1 8 3 2 5 4 7 6 9 0	9 6 5 4 7 8 3 0 1 2
2 3 4 5 6 7 8 9 0 1	2 9 5 6 3 0 8 4 7 1	1 3 8 5 2 6 0 9 4 7
3 4 5 6 7 8 9 0 1 2	3 7 0 9 8 6 1 5 2 4	7 8 4 0 5 3 2 1 9 6
4 5 6 7 8 9 0 1 2 3	4 6 7 5 2 9 0 8 1 3	5 2 9 6 3 7 8 4 0 1
5 6 7 8 9 0 1 2 3 4	5 0 9 4 7 8 3 1 6 2	3 9 6 2 0 1 4 5 7 8
6 7 8 9 0 1 2 3 4 5	6 5 4 7 1 3 2 9 0 8	8 4 0 1 6 9 7 2 5 3
7 8 9 0 1 2 3 4 5 6	7 4 1 8 0 2 9 3 5 6	6 7 3 9 1 0 5 8 2 4
8 9 0 1 2 3 4 5 6 7	8 3 6 0 9 1 5 2 4 7	2 0 1 7 8 4 9 6 3 5
9 0 1 2 3 4 5 6 7 8	9 2 8 1 6 7 4 0 3 5	4 5 7 8 9 2 1 3 6 0

FIGURE 1. Three latin squares of order 10. Squares B and C are *orthogonal*; that is, the 100 pairs (b_{ij}, c_{ij}) are distinct.

(If we number the rows and columns from 0 to 9, the probability of obtaining a given latin square X is $1/\prod_{i=1}^{9}\prod_{j=1}^{9} a_{ij}$, where a_{ij} is the number of ways to select an element in row i and column j such that there exists at least one latin square Y with $y_{kl} = x_{kl}$ for all indices (k, l) that are lexicographically less than (i, j). This probability is not difficult to calculate, because any $m \times n$ latin rectangle for $m < n$ can be completed to an $n \times n$ latin square [32].)

The backtrack algorithm usually produces its results in lexicographic order, and this fact can often be utilized to shorten the computational time greatly by eliminating isomorphic solutions. If it can be established that any completion of the partial solution $(x_1, x_2, \ldots, x_k)$ must be isomorphic to a solution $(y_1, y_2, \ldots, y_n)$ with $(y_1, \ldots, y_k) < (x_1, \ldots, x_k)$ in the lexicographic order, we may omit consideration of $(x_1, x_2, \ldots, x_k)$. (This principle is to be incorporated into the property p_k.) This and other methods for rejecting isomorphic solutions are discussed by Swift [41]. A technique for possibly speeding up backtrack solutions by preliminary identification of objects has been suggested by Tompkins [43].

Backtrack has been used in most of the applications discussed later in this paper; it has also been used to determine all 80 Steiner triple systems of order fifteen [15], to discover a Hadamard matrix of order 92 [9], to construct certain codes [22, 48], to study the four-color problem [49], and to do problems with a more "recreational" flavor such as the fitting together of geometric shapes [38]. The techniques of dynamic programming [3] have proved useful as an alternative to backtrack in some combinatorial applications.

Latin Squares

E. T. Parker [34] has discovered an ingenious way to determine all latin squares orthogonal to a given one. A straightforward use of backtracking would fill each cell of the potential orthogonal mate successively until all cells have values. Parker's method, however, proceeds in two steps: He first applies backtrack to obtain the set of all *transversals* of the given square; a transversal is a set of cells with exactly one cell in each row, one in each column, and one containing each symbol. The problem of finding an orthogonal mate to the square is equivalent to finding a set of *disjoint* transversals to cover the square; so Parker's second stage is to use backtrack again in order to find such coverings by transversals. With 10×10 latin squares, this reduces the number of cases to be considered from approximately 10^{17} to approximately 5×10^5, a factor of some 200 billion! The total time required, when the problem is split into two parts in this way, may be thought of as the sum of the time for two separate steps rather than the product.

Parker used his program to show that square C of Fig. 1 is the only latin square orthogonal to square B. On the other hand, square A is known to have no transversals, hence no orthogonal mate, by theorems proved in [14]. Parker's program has been applied to nearly one hundred 10×10 latin squares and the majority of them have possessed orthogonal mates. This fact is quite remarkable, since Euler's conjecture that no such squares exist had been believed for so many years. Yet no triple of mutually orthogonal 10×10 squares has ever been found, after much computer searching; further theoretical advances will be necessary before this problem can be reduced to a size that computers can handle.

A set of five mutually orthogonal 12×12 latin squares has been found by computer [4, 23]. Starting with the multiplication table of a group whose elements are $(x_1, x_2, \ldots, x_n)$, the square formed by permuting the rows so that the first column has the form $(y_1, y_2, \ldots, y_n)^T$ will be orthogonal to it if and only if $(x_1y_1^{-1}, x_2y_2^{-1}, \ldots, x_ny_n^{-1})$ is a permutation of the group elements. We may assume that y_1 is the identity. If we start with the Abelian group having elements x_i for $0 \leq x < 6$ and $0 \leq i < 2$, with product $x_iy_j = ((x + y) \bmod 6)_{(i+j) \bmod 2}$, the following first columns will yield five mutually orthogonal latin squares:

$$
\begin{array}{cccccccccccc}
0_0 & 1_0 & 2_0 & 3_0 & 4_0 & 5_0 & 0_1 & 1_1 & 2_1 & 3_1 & 4_1 & 5_1 \\
0_0 & 2_0 & 1_0 & 4_1 & 3_1 & 5_1 & 3_0 & 5_0 & 4_0 & 1_1 & 0_1 & 2_1 \\
0_0 & 4_1 & 4_0 & 2_1 & 2_0 & 0_1 & 3_1 & 1_0 & 1_1 & 5_0 & 5_1 & 3_0 \\
0_0 & 5_0 & 3_1 & 2_0 & 4_1 & 1_1 & 2_1 & 4_0 & 5_1 & 1_0 & 3_0 & 0_1 \\
0_0 & 0_1 & 5_0 & 1_1 & 5_1 & 4_0 & 2_0 & 2_1 & 4_1 & 3_0 & 1_0 & 3_1
\end{array}
$$

It has been shown that no set of six mutually orthogonal latin squares of this kind exist. Similarly, if we start with the non-Abelian group having elements x_i, where x is a permutation of $\{1, 2, 3\}$ and $0 \leq i < 2$, and with multiplication defined by multiplying the permutations and adding the subscripts (mod 2), we can get three mutually orthogonal squares:

$$
\begin{array}{cccccccccccc}
123_0 & 132_0 & 213_0 & 231_0 & 312_0 & 321_0 & 123_1 & 132_1 & 213_1 & 231_1 & 312_1 & 321_1 \\
123_0 & 213_1 & 231_1 & 132_1 & 321_0 & 213_0 & 321_1 & 312_1 & 132_0 & 123_1 & 312_0 & 231_0 \\
123_0 & 321_1 & 312_1 & 132_0 & 132_1 & 231_0 & 312_0 & 213_1 & 231_1 & 213_0 & 123_1 & 321_0
\end{array}
$$

(The multiplication rule for permutations implied here is, for example, $132 \times 321 = 312$, because we have

$$
\binom{123}{132} \times \binom{123}{321} = \binom{123}{132} \times \binom{132}{312} = \binom{123}{312}
$$

when the permutations are expressed in two-line form.)

Starting with the alternating group on 4 elements, there are 3840 pairs of orthogonal squares but no three mutually orthogonal. From this data it appears that as the group gets less Abelian, fewer orthogonal squares are present, although no theoretical grounds for such behavior are known.

Projective Planes and Symmetric Block Designs

At the present time, more computer studies have been made relating to finite projective geometries than to any other branch of combinatorial analysis. A survey of this work appears in [18] and we mention only the principal results here.

It is easy to show by hand computation that the projective planes of orders 2, 3, 4, and 5 are unique. The Bruck–Ryser theorem [5] shows that there is no plane of order 6. There is a unique plane of order 7; this result was first established by Norton [31], being incidental to his listing of the 147 nonisomorphic latin squares of order 7. Norton found only 146; an omission was found by Sade [36] who confirmed the completeness of the list of 147. A more direct proof of the uniqueness of the plane of order 7 has been given by Hall [12] and Pierce [35].

In a plane of order 8, the multiplication table of nonzero elements forms a 7×7 latin square. Using the list of 147 such squares and a few further simplifications, a computer was programmed to help establish the uniqueness of the plane of order 8 [13]; this proof is a good example of the effective interplay of machine and hand methods.

There are four known projective planes of order 9, and work is under way to determine whether these are in fact the complete set. This problem appears to lie at the borderline of computational feasibility with respect to an exhaustive search for planes of a small order.

Since planes exist for every order that is a prime or prime power, the question as to the existence of any plane of non-prime-power order is of considerable interest, and the smallest plane of this kind permitted by the Bruck–Ryser theorem would have order 10. A search is under way for a plane of order 10 with a nontrivial automorphism, but even this is a large undertaking. A plane of order 10 is equivalent to a set of nine mutually orthogonal 10×10 latin squares, but as stated above not even three such squares are known.

If we can find residues $b_1, b_2, \ldots, b_k$ (modulo v) where every difference $d \not\equiv 0 \pmod{v}$ has exactly λ representations $d \equiv b_i - b_j$ (modulo v) we call the b's a difference set modulo v. The relation $k(k-1) = \lambda(v-1)$ must necessarily hold. If $\{b_1, \ldots, b_k\}$ is a difference set modulo v, then the "blocks" $B_j = \{b_1 + j, b_2 + j, \ldots, b_k + j\} \bmod v$ for $j = 0, \ldots, v-1$ form a symmetric block design. In the special case that $\lambda = 1$, $k = n+1$, and $v = n^2 + n + 1$, the block design is a finite projective plane of order n, and the blocks are the lines of the plane.

For example, the following difference sets may be used to construct the planes of small orders:

$$n = 2: \quad \{1, 2, 4\} \pmod 7;$$
$$n = 3: \quad \{0, 1, 3, 9\} \pmod{13};$$
$$n = 4: \quad \{3, 6, 7, 12, 14\} \pmod{21}.$$

By a theorem of one of the authors [16], we may assume under certain conditions that the difference set $\{b_1, \ldots, b_k\}$ is fixed by "multipliers," where a multiplier is a prime p dividing $k - \lambda$. If p is such a multiplier then the numbers $\{pb_1 \bmod v, \ldots, pb_k \bmod v\}$ are the same as $\{b_1, \ldots, b_k\}$ in some order. This fact considerably simplifies a computer search for difference sets, and those of small order are discussed in [16, 19]. In particular, the SWAC computer was used to show that all difference sets with $\lambda = 13$, $k = 40$, and $v = 121$ are isomorphic to one of the following four:

$1, 3, 4, 7, 9, 11, 12, 13, 21, 25, 27, 33, 34, 36, 39, 44, 55, 63, 64, 67, 68, 70,$
$71, 75, 80, 81, 82, 83, 85, 89, 92, 99, 102, 103, 104, 108, 109, 115, 117, 119;$

$1, 3, 4, 5, 9, 12, 13, 14, 15, 16, 17, 22, 23, 27, 32, 34, 36, 39, 42, 45, 46, 48,$
$51, 64, 66, 69, 71, 77, 81, 82, 85, 86, 88, 92, 96, 102, 108, 109, 110, 117;$

$1, 3, 4, 7, 8, 9, 12, 21, 24, 25, 26, 27, 34, 36, 40, 43, 49, 63, 64, 68, 70, 71,$
$72, 75, 78, 81, 82, 83, 89, 92, 94, 95, 97, 102, 104, 108, 112, 113, 118, 120;$

$1, 3, 4, 5, 7, 9, 12, 14, 15, 17, 21, 27, 32, 36, 38, 42, 45, 46, 51, 53, 58, 63,$
$67, 68, 76, 79, 80, 81, 82, 83, 96, 100, 103, 106, 107, 108, 114, 115, 116, 119.$

Parker has recently shown that these four lead to nonisomorphic designs, by having a computer determine the intersection patterns of three blocks at a time.

When $\lambda = 1$, the difference set will always give a projective plane, and all primes p dividing $n = k - \lambda$ must be multipliers. Singer [39] proved that all finite Desarguesian planes can be constructed by using difference sets; Mann and Evans [30], by hand computation, have shown that for $n \leq 1600$ no projective planes can be constructed from a difference set unless n is the power of a prime.

Other searches have been based on the algebraic properties of the planes. Killgrove [25] showed that no plane of order 9 has a cyclic additive group, and it was determined in [17] that the only planes with an elementary Abelian additive group are the four known ones. Therefore addition in a further plane of order 9, if there is one, cannot be a group.

A projective plane of order n is called *Desarguesian* if it is obtained from a finite field of n elements by representing the points and lines as equivalence classes of nonzero triples (x, y, z), where $(x, y, z) \equiv (ax, ay, az)$ when $a \neq 0$; the number of such equivalence classes is clearly $(n^3 - 1)/(n - 1) = n^2 + n + 1$. In terms of these coordinates, point

(x, y, z) lies on line (X, Y, Z) if and only if $xX + yY + zZ = 0$. Non-Desarguesian planes can be given coordinates in a similar way [11], but the corresponding algebraic system is not a field.

Kleinfeld [26] found all Veblen–Wedderburn systems of order 16 that have a left nucleus of order 4. These systems lead to four planes: the Desarguesian plane, the Hall plane, and two new planes coordinatized by semifields (nonassociative division rings). The two latter planes have inspired constructions of several new types of projective planes [20, 27, 37].

Non-Desarguesian planes of all orders p^m, where p is prime and $m > 1$, have been known for many years except in the cases 2^p when p is prime. As mentioned above, there are no non-Desarguesian planes of order 2^2 or 2^3; hence there was interest in the next smallest case 2^5. A search was undertaken for all semifields of order 32, independently by Walker [46] and by one of the authors, and the complete set of 2502 nonisomorphic systems was found. These semifields yield five non-Desarguesian planes whose structure was also determined by computer [27]. Fortunately, it was possible to observe a pattern in one of the semifields, and this observation led to a construction [28] of semifields of all orders 2^{mn} where $mn > 3$ and $n > 1$ is odd. In this way the solution for all remaining cases 2^p was obtained as a byproduct of a computer examination of the smallest case; such results are the main goals of combinatorial computer explorations.

The preparation of this paper was supported, in part, by National Science Foundation grant number GP-212.

References

[1] "Index by subject to algorithms, 1960–1963," *Communications of the ACM* **7** (1964), 146–148.

[2] W. W. Rouse Ball, *Mathematical Recreations and Essays*, 11th edition, revised by H. S. M. Coxeter (New York: Macmillan, 1947), 305–310.

[3] Richard Bellman, "Dynamic programming treatment of the traveling salesman problem," *Journal of the Association for Computing Machinery* **9** (1962), 61–63.

[4] R. C. Bose, I. M. Chakravarti, and D. E. Knuth, "On methods of constructing sets of mutually orthogonal latin squares using a computer," *Technometrics* **2** (1960), 507–516; **3** (1961), 111–117.

[5] R. H. Bruck and H. J. Ryser, "The nonexistence of certain projective planes," *Canadian Journal of Mathematics* **1** (1949), 88–93.

[6] Martin Cohn, "Affine m-ary Gray codes," *Information and Control* **6** (1963), 70–78.

[7] Ludwig Joseph Fischer and K. Chr. Fr. Krause, *Lehrbuch der Combinationlehre und der Arithmetik* (Dresden: 1812).

[8] Ivan Flores, "Reflected number systems," *IRE Transactions on Electronic Computers* **EC-5** (1956), 79–82.

[9] Solomon W. Golomb and Leonard D. Baumert, "The search for Hadamard matrices," *American Mathematical Monthly* **70** (1963), 12–17.

[10] Solomon W. Golomb and Leonard D. Baumert, "Backtrack programming," *Journal of the Association for Computing Machinery* **12** (1965), 516–524.

[11] Marshall Hall, "Projective planes," *Transactions of the American Mathematical Society* **54** (1943), 229–277; **65** (1949), 473–474. See also Marshall Hall, Jr., *The Theory of Groups* (New York: Macmillan, 1959), 346–420.

[12] Marshall Hall, Jr., "Uniqueness of the projective plane with 57 points," *Proceedings of the American Mathematical Society* **4** (1953), 912–916; **5** (1954), 994–997.

[13] Marshall Hall, Jr., J. Dean Swift, and Robert J. Walker, "Uniqueness of the projective plane of order eight," *Mathematical Tables and Other Aids to Computation* **10** (1956), 186–194.

[14] Marshall Hall, Jr., and L. J. Paige, "Complete mappings of finite groups," *Pacific Journal of Mathematics* **5** (1955), 541–549.

[15] Marshall Hall, Jr., and J. D. Swift, "Determination of Steiner triple systems of order 15," *Mathematical Tables and Other Aids to Computation* **9** (1955), 146–152.

[16] Marshall Hall, Jr., "A survey of difference sets," *Proceedings of the American Mathematical Society* **7** (1956), 975–986.

[17] Marshall Hall, Jr., J. Dean Swift, and Raymond Killgrove, "On projective planes of order nine," *Mathematical Tables and Other Aids to Computation* **13** (1959), 233–246.

[18] Marshall Hall, Jr., "Numerical analysis of finite geometries," *IBM Scientific Computing Symposium on Combinatorial Problems*, Yorktown Heights, New York, 16–18 March 1964 (White Plains, New York: IBM Data Processing Division, 1966), 11–22.

[19] Harry S. Hayashi, "Computer investigation of difference sets," *Mathematics of Computation* **19** (1965), 73–78.

[20] D. R. Hughes and Erwin Kleinfeld, "Semi-nuclear extensions of Galois fields," *American Journal of Mathematics* **82** (1960), 389–392.

[21] G. Hutchinson, "Partitioning algorithms for finite sets," *Communications of the ACM* **6** (1963), 613–614.

[22] B. H. Jiggs, "Recent results in comma-free codes," *Canadian Journal of Mathematics* **15** (1963), 178–187.

[23] Diane M. Johnson, A. L. Dulmage, and N. S. Mendelsohn, "Orthomorphisms of groups and orthogonal latin squares, I," *Canadian Journal of Mathematics* **13** (1961), 356–372.

[24] Selmer M. Johnson, "Generation of permutations by adjacent transposition," *Mathematics of Computation* **17** (1963), 282–285.

[25] Raymond B. Killgrove, "A note on the nonexistence of certain projective planes of order nine," *Mathematics of Computation* **14** (1960), 70–71.

[26] Erwin Kleinfeld, "Techniques for enumerating Veblen–Wedderburn systems," *Journal of the Association for Computing Machinery* **7** (1960), 330–337.

[27] Donald E. Knuth, "Finite semifields and projective planes," *Journal of Algebra* **2** (1965), 182–217. [Reprinted as Chapter 20 of the present volume.]

[28] Donald E. Knuth, "A class of projective planes," *Transactions of the American Mathematical Society* **115** (1965), 541–549. [Reprinted as Chapter 21 of the present volume.]

[29] D. H. Lehmer, "Teaching combinatorial tricks to a computer," *Proceedings of Symposia in Applied Mathematics* **10** (Providence, Rhode Island: American Mathematical Society, 1960), 179–193.

[30] H. B. Mann and T. A. Evans, "On simple difference sets," *Sankhyā* **11** (1951), 357–364.

[31] H. W. Norton, "The 7 × 7 squares," *Annals of Eugenics* **9** (1939), 269–307.

[32] R. T. Ostrowski and K. D. Van Duren, "On a theorem of Mann on latin squares," *Mathematics of Computation* **15** (1961), 293–295.

[33] L. J. Paige and C. B. Tompkins, "The size of the 10 × 10 orthogonal latin square problem," *Proceedings of Symposia in Applied Mathematics* **10** (Providence, Rhode Island: American Mathematical Society, 1960), 71–84.

[34] E. T. Parker, "Computer investigation of orthogonal latin squares of order ten," *Proceedings of Symposia in Applied Mathematics* **15**

(Providence, Rhode Island: American Mathematical Society, 1963), 73–82.

[35] William A. Pierce, "The impossibility of Fano's configuration in a projective plane with eight points per line," *Proceedings of the American Mathematical Society* **4** (1953), 908–912.

[36] Albert Sade, "An omission in Norton's list of 7×7 squares," *Annals of Mathematical Statistics* **22** (1951), 306–307.

[37] Reuben Sandler, "Autotopism groups of some finite non-associative algebras," *American Journal of Mathematics* **84** (1962), 239–264.

[38] Dana S. Scott, *Programming a combinatorial puzzle*, Technical Report No. 1 (Princeton, New Jersey: Princeton University Department of Electrical Engineering, 10 June 1958), ii + 14 + 5 pages.

[39] James Singer, "A theorem in finite projective geometry and some applications to number theory," *Transactions of the American Mathematical Society* **43** (1938), 377–385.

[40] Frank Stockmal, "Algorithms 95 and 114," *Communications of the ACM* **5** (1962), 344, 434.

[41] J. D. Swift, "Isomorph rejection in exhaustive search techniques," *Proceedings of Symposia in Applied Mathematics* **10** (Providence, Rhode Island: American Mathematical Society, 1960), 195–200.

[42] C. Tompkins, "Machine attacks on problems whose variables are permutations," *Proceedings of Symposia in Applied Mathematics* **6** (Providence, Rhode Island: American Mathematical Society, 1956), 195–212.

[43] C. B. Tompkins, "Methods of successive restrictions in computational problems involving discrete variables," *Proceedings of Symposia in Applied Mathematics* **15** (Providence, Rhode Island: American Mathematical Society, 1963), 95–106.

[44] H. F. Trotter, "Algorithm 115," *Communications of the ACM* **5** (1962), 434–435.

[45] R. J. Walker, "An enumerative technique for a class of combinatorial problems," *Proceedings of Symposia in Applied Mathematics* **10** (Providence, Rhode Island: American Mathematical Society, 1960), 91–94.

[46] R. J. Walker, "Determination of division algebras with 32 elements," *Proceedings of Symposia in Applied Mathematics* **15** (Providence, Rhode Island: American Mathematical Society, 1963), 83–85.

[47] Mark B. Wells, "Generation of permutations by transposition," *Mathematics of Computation* **15** (1961), 192–195.

[48] John M. Wozencraft and Barney Reiffen, *Sequential Decoding* (Cambridge, Massachusetts: MIT Press, 1961).

[49] Hidehiko Yamabe and David Pope, "A computational approach to the four-color problem," *Mathematics of Computation* **15** (1961), 250–253.

Addendum

Further study has turned up better information about the early history of permutation generation. Instead of reference [7], the classical algorithm for lexicographic permutations should be ascribed to C. F. Hindenburg in his preface to *Specimen Analyticum de Lineis Curvis Secundi Ordinis* by C. F. Rüdiger (Leipzig: 1784), xlvi–xlvii. And the "Trotter–Johnson" algorithm for permutations by adjacent interchanges actually goes back much further, to a famous book called *Tintinnalogia*, published anonymously in 1668 but now known to have been written by Richard Duckworth and Fabian Stedman (London: Thomas Archer, 1671), 1–60. Faster methods are also known; see Donald E. Knuth, *The Art of Computer Programming*, Section 7.2.1.2.

Indeed, there has been enormous improvement since 1965 on virtually every subject mentioned in the survey paper above. The fact that no projective plane of order 10 exists was finally established in 1989, after theoretical advances by many mathematicians and massive computer searches conducted primarily by J. L. Carter, C. W. H. Lam, L. Thiel, and S. Swiercz; Lam's account of this work, "The search for a projective plane of order 10," *American Mathematical Monthly* **98** (1991), 305–318, is particularly interesting. A combination of coding theory and computer coding has, similarly, established the fact that there are only four distinct projective planes of order 9 [C. W. H. Lam, G. Kolesova, and L. Thiel, "A computer search for finite projective planes of order 9," *Discrete Mathematics* **92** (1991), 187–195].

Chapter 2

Two Notes on Notation

[Originally published in American Mathematical Monthly **99** *(1992), 403–422;* **102** *(1995), 562.]*

Mathematical notation evolves like all languages do. As new experiments are made, we sometimes witness the survival of the fittest, sometimes the survival of the most familiar. A healthy conservatism keeps things from changing too rapidly; a healthy radicalism keeps things in tune with new theoretical emphases. Our mathematical language continues to improve, just as "the d-ism of Leibniz overtook the dotage of Newton" in past centuries [4, Chapter 4].

In 1970 I began teaching a class at Stanford University entitled Concrete Mathematics. The students and I studied how to manipulate formulas in continuous and discrete mathematics, and the problems we investigated were often inspired by new developments in computer science. As the years went by we began to see that a few changes in notational traditions would greatly facilitate our work.

The notes from that class have recently been published as a textbook [15], and as I wrote the final drafts of that book I learned to my surprise that two of the notations we had been using were considerably more useful than I had previously realized. The ideas "clicked" so well, in fact, that I've decided to write this article, blatantly attempting to promote these notations among the mathematicians who have no use for [15]. I hope that within five years everybody will be able to use these notations in published papers without needing to explain what they mean.

The notations I'm talking about are (1) Iverson's convention for characteristic functions; and (2) the "right" notation for Stirling numbers, at last.

15

1. Iverson's Convention

The first notational development I want to discuss was introduced by Kenneth E. Iverson in the early 60s, on page 11 of the pioneering book [21] that led to his well known APL.

"If α and β are arbitrary entities and $\mathcal{R}$ is any relation defined on them, the *relational statement* $(a\mathcal{R}b)$ is a logical variable which is true (equal to 1) if and only if α stands in the relation $\mathcal{R}$ to β. For example, if x is any real number, then the function

$$(x > 0) - (x < 0)$$

(commonly called the *sign function* or sgn x) assumes the values 1, 0, or -1 according as x is strictly positive, 0, or strictly negative."

When I read that definition, long ago, I found it mildly interesting but not especially significant. I began using his convention informally but infrequently, in class discussions and in private notes. I allowed it to slip, undefined, into an obscure corner of one of my books (see page 117 of [16]). But when I prepared the final manuscript of [15], I began to notice that Iverson's idea led to substantial improvements in exposition and in technique.

Before I can explain why the notation now works so well for me, I need to say a few words about the manipulation of sums and summands. I realized long ago that "boundary conditions" on indices of summation are often a handicap and a waste of time. Instead of writing

$$(1 + z)^n = \sum_{k=0}^{n} \binom{n}{k} z^k , \tag{1.1}$$

it is much better to write

$$(1 + z)^n = \sum_{k} \binom{n}{k} z^k ; \tag{1.2}$$

the sum now extends over all integers k, but only finitely many terms are nonzero. The second formula (1.2) is instantly converted to other forms:

$$(1+z)^n = \sum_{k} \binom{n}{k} z^k = \sum_{k} \binom{n}{k+1} z^{k+1} = \sum_{k} \binom{n}{\lfloor n/2 \rfloor - k} z^{\lfloor n/2 \rfloor - k} ; \tag{1.3}$$

by contrast, we must work harder when dealing with (1.1), because we have to think about the limits:

$$(1+z)^n = \sum_{k=0}^{n} \binom{n}{k} z^k = \sum_{k=-1}^{n-1} \binom{n}{k+1} z^{k+1}$$

$$= \sum_{k=-\lceil n/2 \rceil}^{\lfloor n/2 \rfloor} \binom{n}{\lfloor n/2 \rfloor - k} z^{\lfloor n/2 \rfloor - k}. \quad (1.4)$$

Furthermore, (1.2) and (1.3) make sense also when n is not a positive integer.

Even when limits are necessary, it is best to keep them as simple as possible. For example, it's almost always a mistake to write

$$\sum_{k=2}^{n-1} k(k-1)(n-k) \qquad \text{instead of} \qquad \sum_{k=0}^{n} k(k-1)(n-k); \quad (1.5)$$

the additional zero terms are more helpful than harmful (and the former sum is problematical when $n = 0$, 1, or 2).

Finally it dawned on me that Iverson's convention allows us to write *any* sum as an infinite sum without limits: If $P(k)$ is any property of the integer k, we have

$$\sum_{P(k)} f(k) = \sum_{k} f(k) \, [P(k)] . \quad (1.6)$$

For example, the sums in (1.5) become

$$\sum_{k} k(k-1)(n-k) \, [0 \le k \le n] = \sum_{k} k(k-1)(n-k) \, [k \ge 0] \, [k \le n] . \quad (1.7)$$

(At the time I made this observation, I had forgotten that Iverson originally defined his convention only for single relational operators enclosed in parentheses; I began to put *arbitrary* logical statements in square brackets, and to assume that this would produce the value 0 or 1.) In this particular case nothing much has been gained when passing from (1.5) to (1.7), although we might be able to make use of identities like

$$k \, [k \ge 0] \; = \; k \, [k \ge 1] . \quad (1.8)$$

But in general, the ability to manipulate "on the line" instead of "below the line" turns out to be a great advantage.

For example, in my first book [25] I had found it necessary to include the rule

$$\sum_{k \in A} f(k) + \sum_{k \in B} f(k) = \sum_{k \in A \cup B} f(k) + \sum_{k \in A \cap B} f(k) \qquad (1.9)$$

as a separate axiom for $\sum$ manipulation. But this axiom is unnecessary in [15], because it can be derived easily from other basic laws: The left-hand side is

$$\sum_{k \in A} f(k) + \sum_{k \in B} f(k) = \sum_{k} f(k) \, [k \in A] + \sum_{k} f(k) \, [k \in B]$$

$$= \sum_{k} f(k) \, ([k \in A] + [k \in B])$$

and the right-hand side is the same, because we have

$$[k \in A] + [k \in B] = [k \in A \cup B] + [k \in A \cap B]. \qquad (1.10)$$

The interchange of summation order in multiple sums also comes out simpler now. I used to have trouble understanding and/or explaining why

$$\sum_{j=1}^{n} \sum_{k=1}^{j} f(j, k) = \sum_{k=1}^{n} \sum_{j=k}^{n} f(j, k); \qquad (1.11)$$

but now it's easy for me to see that the left-hand sum is

$$\sum_{j,k} f(j, k) \, [1 \leq j \leq n] \, [1 \leq k \leq j] = \sum_{j,k} f(j, k) \, [1 \leq k \leq j \leq n]$$

$$= \sum_{j,k} f(j, k) \, [1 \leq k \leq n] \, [k \leq j \leq n],$$

and this is the right-hand sum.

Here's another example: We have

$$[k \text{ even}] = \sum_{m} [k = 2m] \qquad \text{and} \qquad [k \text{ odd}] = \sum_{m} [k = 2m + 1]; \quad (1.12)$$

therefore

$$\sum_{k} f(k) = \sum_{k} f(k) \, ([k \text{ even}] + [k \text{ odd}])$$

$$= \sum_{k,m} f(k) \, [k = 2m] + \sum_{k,m} f(k) \, [k = 2m + 1]$$

$$= \sum_{m} f(2m) + \sum_{m} f(2m + 1). \qquad (1.13)$$

The result in (1.13) is hardly surprising; but I like to have mechanical operations like this available so that I can do manipulations reliably, without thinking. Then I'm less apt to make mistakes.

Let lg stand for logarithms to base 2. Then we have

$$\sum_{k\geq 1}\binom{n}{\lfloor \lg k\rfloor} = \sum_{k\geq 1}\sum_{m}\binom{n}{m}\,[m = \lfloor \lg k\rfloor]$$

$$= \sum_{k,m}\binom{n}{m}\,[m \leq \lg k < m+1]\,[k \geq 1]$$

$$= \sum_{m,k}\binom{n}{m}\,[2^m \leq k < 2^{m+1}]\,[k \geq 1]$$

$$= \sum_{m}\binom{n}{m}(2^{m+1} - 2^m)\,[m \geq 0]$$

$$= \sum_{m}\binom{n}{m}2^m = 3^n. \tag{1.14}$$

If we are doing infinite products we can use Iversonian brackets as exponents:

$$\prod_{P(k)} f(k) = \prod_{k} f(k)^{[P(k)]}. \tag{1.15}$$

For example, the largest squarefree divisor of n is

$$\prod_{p} p^{[p\text{ prime}]\,[p\text{ divides }n]}.$$

Everybody is familiar with one special case of an Iverson-like convention, the "Kronecker delta" symbol

$$\delta_{ik} = \begin{cases} 1, & i = k; \\ 0, & i \neq k. \end{cases} \tag{1.16}$$

Leopold Kronecker introduced this notation in his work on bilinear forms [30, page 276] and in his lectures on determinants (see [31, page 316]); it soon became widespread. Many of his followers wrote δ_j^k, which is a bit more ambiguous because it conflicts with ordinary exponentiation. I now prefer to write $[j = k]$ instead of δ_{jk}, because Iverson's convention is much more general. Although '$[j = k]$' involves five written characters instead of the three in 'δ_{jk}', we lose nothing in common cases when '$[j = k + 1]$' takes the place of '$\delta_{j(k+1)}$'.

Another familiar example of a 0–1 function, this time from continuous mathematics, is Oliver Heaviside's unit step function $[x \geq 0]$. (See [44] and [37] for expositions of Heaviside's methods.) It is clear that Iverson's convention will be as useful with integration as it is with summation, perhaps even more so. I have not yet explored this in detail, because [15] deals mostly with sums.

It's interesting to look back into the history of mathematics and see how there was a craving for such notations before they existed. For example, an Italian count named Guglielmo Libri published several papers in the 1830s concerning properties of the function 0^{0^x}. He noted [32] that 0^x is either 0 (if $x > 0$) or 1 (if $x = 0$) or ∞ (if $x < 0$), hence

$$0^{0^x} = [x > 0]. \tag{1.17}$$

But of course he didn't have Iverson's convention to work with; he was pleased to discover a way to denote the discontinuous function $[x > 0]$ without leaving the realm of operations acceptable in his day. He believed that "la fonction $0^{0^{x-n}}$ est d'un grand usage dans l'analyse mathématique." And he noted in [33] that his formulas "ne renferment aucune notation nouvelle. . . . Les formules qu'on obtient de cette manière sont très simples, et rentrent dans l'algèbre ordinaire."

Libri wrote, for example,

$$(1 - 0^{0^{-x}})(1 - 0^{0^{x-a}})$$

for the function $[0 \leq x \leq a]$, and he gave the integral formula

$$\frac{2}{\pi} \int_0^\infty \frac{dq \cos qx}{1 + q^2} = e^x \cdot 0^{0^{-x}} + e^{-x}\left(1 - 0^{0^{-x}}\right) = \frac{e^x}{0^{-x} + 1} + \frac{e^{-x}}{0^x + 1}.$$

(Of course, we would now write the value of that integral as $e^{-|x|}$, but a simple notation for absolute value wasn't introduced until many years later. I believe that the first appearance of '$|z|$' for absolute value in Crelle's journal — the journal containing Libri's papers [32] and [33] — occurred on page 227 of [56] in 1881. Karl Weierstrass was the inventor of this notation, which was applied at first only to complex numbers; Weierstrass seems to have published it first in 1876 [55].)

Libri applied his 0^{0^x} function to number theory by exhibiting a complicated way to describe the fact that x is a divisor of m. In essence, he gave the following recursive formulation: Let $P_0(x) = 1$ and for $k > 0$ let

$$P_k(x) = 0^{0^{x-k}} P_0(x) - 0^{0^{x-k+1}} P_1(x) - \cdots - 0^{0^{x-1}} P_{k-1}(x) .$$

Then the quantity

$$\frac{1 - m \cdot 0^{0^{x-m}} P_0(x) - (m-1) \cdot 0^{0^{x-m+1}} P_1(x) - \cdots - (1) \cdot 0^{0^{x-1}} P_{m-1}(x)}{x}$$

turns out to equal 1 if x divides m, otherwise it is 0. (One way to prove this, Iverson-wise, is to replace $0^{0^{x-k}}$ in Libri's formulas by $[x > k]$, and to show first by induction that $P_k(x) = [x$ divides $k] - [x$ divides $k - 1]$ for all $k > 0$. Then if $a_k(x) = k \, [x > k]$, we have

$$\sum_{k=0}^{m-1} a_{m-k}(x) P_k(x) = \sum_{k=0}^{m-1} a_{m-k}(x) \left([x \text{ divides } k] - [x \text{ divides } k - 1]\right)$$

$$= \sum_{k=0}^{m-1} [x \text{ divides } k] \left(a_{m-k}(x) - a_{m-k-1}(x)\right) .$$

If the positive integer x is not a divisor of m, the terms of this new sum are zero except when $m - k = m \bmod x$, when we have $a_{m-k}(x) - a_{m-k-1}(x) = 1$. On the other hand if x is a divisor of m, the only nonvanishing term occurs for $m - k = x$, when we have $a_{m-k}(x) - a_{m-k-1}(x) = 0 - (x-1)$. Hence the sum is $1 - x \, [x \text{ divides } m]$. Libri obtained his complicated formula by a less direct method, applying Newton's identities to compute the sum of the mth powers of the roots of the equation $t^{x-1} + t^{x-2} + \cdots + 1 = 0$.)

Evidently Libri's main purpose was to show that unlikely functions can be expressed in algebraic terms, somewhat as we might wish to show that some complex functions can be computed by a Turing Machine. "Give me the function 0^{0^x}, and I'll give you an expression for $[x \text{ divides } m]$." But our goal with Iverson's notation is, by contrast, to find a simple and natural way to express quantities that help us solve problems. If we need a function that is 1 if and only if x divides m, we can now write $[x \text{ divides } m]$.

Some of Libri's papers are still well remembered, but [32] and [33] are not. I found no mention of them in *Science Citation Index*, after searching through all years of that index available in our library (1955 to date). However, the paper [33] did produce several ripples in mathematical waters when it originally appeared, because it stirred up a controversy about whether 0^0 is defined. Most mathematicians agreed that $0^0 = 1$, but Cauchy [5, page 70] had listed 0^0 together with other expressions like $0/0$ and $\infty - \infty$ in a table of undefined forms. Libri's justification for the equation $0^0 = 1$ was far from convincing, and a commentator who signed his name simply "S" rose to the attack [45]. August Möbius [36] defended Libri, by presenting his former professor's reason for believing that $0^0 = 1$ (basically a proof that $\lim_{x \to 0+} x^x = 1$). Möbius also went further and presented a supposed proof that

$$\lim_{x \to 0+} f(x)^{g(x)} = 1$$

whenever $\lim_{x \to 0+} f(x) = \lim_{x \to 0+} g(x) = 0$. Of course "S" then asked [3] whether Möbius knew about functions such as $f(x) = e^{-1/x}$ and $g(x) = x$. (And paper [36] was quietly omitted from the historical record when the collected works of Möbius were ultimately published.) The debate stopped there, apparently with the conclusion that 0^0 should be undefined.

But no, no, ten thousand times no! Anybody who wants the binomial theorem

$$(x + y)^n = \sum_{k=0}^{n} \binom{n}{k} x^k y^{n-k} \tag{1.18}$$

to hold for at least one nonnegative integer n *must* believe that $0^0 = 1$, for we can plug in $x = 0$ and $y = 1$ to get 1 on the left and 0^0 on the right.

The number of mappings from the empty set to the empty set is 0^0. It *has* to be 1.

On the other hand, Cauchy had good reason to consider 0^0 as an undefined *limiting form*, in the sense that the limiting value of $f(x)^{g(x)}$ is not known *a priori* when $f(x)$ and $g(x)$ approach 0 independently. In this much stronger sense, the value of 0^0 is less defined than, say, the value of $0 + 0$. Both Cauchy and Libri were right, but Libri and his defenders did not understand why truth was on their side.

Well, it's instructive to study mathematical history and to observe how tastes change as progress is made. But let's come closer to the present, to see how Iverson's convention might be useful nowadays. Today's mathematical literature is, in fact, filled with instances where analogs of Iversonian brackets are being used — but the concepts must be expressed in a roundabout way, because his convention is not yet established. Here are two examples that I happened to notice just before writing this paper:

1) Hardy and Wright, in the course of proving the Staudt–Clausen theorem about the denominators of Bernoulli numbers [20, §7.9], consider the sum

$$\sum_{p-1 \text{ divides } k} \frac{1}{p}$$

where p runs through primes. They define $\epsilon_k(p)$ to be 1 if $p-1$ divides k, otherwise $\epsilon_k(p) = 0$; then the sum becomes

$$\sum_p \frac{\epsilon_k(p)}{p} .$$

They proceed to show that $\sum_{m=1}^{p-1} m^k \equiv -\epsilon_k(p)$ (modulo p) whenever p is prime, and the theorem follows with a bit more manipulation.

2) Mark Kac, introducing the relevance of ergodic theory to continued fractions [24, §5.4], says: "Let now $P_0 \in \Omega$ and $g(P)$ the characteristic function of the measurable set A; i.e.,

$$g(P) = \begin{cases} 1, & p \in A, \\ 0, & p \overline{\in} A. \end{cases}$$

It is now clear that $t(\tau, P_0, A)$ is given by the formula

$$t(\tau, P_0, A) = \int_0^\tau g\big(T_t(P_0)\big) \, dt \,,$$

and … ".

I hope it is now clear why my students and I would find it quite natural to say directly that

$$t(\tau, P_0, A) = \int_0^\tau [T_t(P_0) \in A] \, dt \,.$$

Also, in the context of Hardy and Wright, we would evaluate the sum $\left(\sum_{m=1}^{p-1} m^k\right)$ mod p and discover that it is $(p - 1) [p - 1 \text{ divides } k]$.

If you are a typical hard-working, conscientious mathematician, interested in clear exposition and sound reasoning — and I like to include myself as a member of that set — then your experiences with Iverson's convention may well go through several stages, just as mine did. First, I learned about the idea, and it certainly seemed straightforward enough. Second, I decided to use it informally while solving problems. At this stage it seemed too easy to write just '$[k \geq 0]$'; my natural tendency was to write something like '$\delta(k \geq 0)$', giving an implicit bow to Kronecker, or '$\tau(k \geq 0)$' where τ stands for truth. Adriano Garsia, similarly, decided to write '$\chi(k \geq 0)$', knowing that χ often denotes a characteristic function; he has used χ notation effectively in dozens of papers, beginning with [10], and quite a few other mathematicians have begun to follow his lead. (Garsia was one of my professors in graduate school, and I recently showed him the first draft of this note. He replied, "My definition from the very start was

$$\chi(\mathcal{A}) = \begin{cases} 1 & \text{if } \mathcal{A} \text{ is true} \\ 0 & \text{if } \mathcal{A} \text{ is false} \end{cases}$$

where $\mathcal{A}$ is any statement whatever. But just like you, I got it by generalizing from Iverson's APL. ...I don't have to tell you the magic that the use of the χ notation can do.")

If you go through the stages I did, however, you'll soon tire of writing δ, τ, or χ, when you recognize that the notation is quite unambiguous without an additional symbol. Then you will have arrived at the philosophical position adopted by Iverson when he wrote [21]. And I had also reached that stage when I completed the first edition of [15]; I adopted Iverson's original suggestion to enclose logical statements in ordinary parentheses, not square brackets.

Unfortunately, not all was well with that first edition. Students found cases where I had parenthesized a complicated logical statement for clarity, for example when I wrote something of the form 'α and (β or γ)'; they pointed out that the simple act of putting parentheses around 'β or γ' automatically caused it to be evaluated as either 0 or 1, according to a strict interpretation of Iverson's rule as I had extended it.

Worse yet, as I began to read the first edition of [15] with fresh eyes, I found that the formulas involved too many parentheses. It was hard for me to perceive the structure of complex expressions that involved Iversonian statements; the statements had been clear to me when I wrote them down, but they looked confusing when I came back to them several months later. A computer could readily parse each expression, but good notation must be engineered for human beings.

Therefore in the second and subsequent printings of [15], my co-authors and I now use square brackets instead of parentheses, whenever we wish to transform logical statements into the values 0 or 1. This resolves both problems, and we now believe that the notation has proved itself well enough to be thrust upon the world. Square brackets are used also for other purposes, but not in a conflicting way, and not so often that the multiple uses become confusing.

One small glitch remains: We want to be able to write things like

$$\sum_p [p \text{ prime}] \, [p \le x]/p \tag{1.19}$$

to denote the sum of all reciprocals of primes $\le x$. But this summand unfortunately reduces to $0/0$ when $p = 0$. In general, when an Iverson-bracketed statement is false, we want it to evaluate into a "very strong 0," namely a zero so strong that it annihilates anything it is multiplied by — even if that other factor is undefined.

Similarly, in formulas like (1.2) it is convenient to regard $\binom{n}{k}$ as strongly zero when k is negative, so that, for example, $\binom{n}{-10} z^{-10} = 0$ when $z = 0$.

The strong-zero convention is enough to handle 99% of the difficult situations, but we may also be using $1 - [P(k)]$ to stand for the quantity $[\text{not } P(k)]$; then we want $[P(k)]$ to give a "strong 1." And paradoxes can still arise, whenever irresistible forces meet immovable objects. (What happens if a strong zero appears in the denominator? And so on.)

In spite of these potential problems in extreme cases, Iverson's convention works beautifully in the vast majority of applications. It is, in fact, far less dangerous than most of the other notations of mathematics, whose dark corners we have learned to avoid long ago. The safe use of Iverson's simple and convenient idea is quite easy to learn.

2. Stirling Numbers

The second plea I wish to make for perspicuous notation concerns the famous coefficients introduced by James Stirling at the beginning of his *Methodus Differentialis* in 1730 [52]. The lack of a widely accepted way to refer to those numbers has become almost scandalous. For example, Goldberg, Newman, and Haynsworth begin their chapter on Combinatorial Analysis in the NBS Handbook [1] by remarking that notations for Stirling numbers "have never been standardized ... We feel that a capital S is natural for Stirling numbers of the first kind; it is infrequently used for other notation in this context. But once it is used we have difficulty finding a suitable symbol for Stirling numbers of the second kind. The numbers are sufficiently important to warrant a special and easily recognizable symbol, and yet that symbol must be easy to write. We have settled on a script capital $\mathcal{S}$ without any certainty that we have settled this question permanently."

The present predicament came about because Stirling numbers are indeed important enough to have arisen in a wide variety of applications, yet they are not quite important enough to have deserved a prominent place in the most influential textbooks of mathematics. Therefore they have been rediscovered many times, and each author has chosen a notation that was optimized for one particular application.

The great utility of Stirling numbers has become clearer and clearer with time, and mathematicians have now reached a stage where we can intelligently choose a notation that will serve us well in the whole range of applications.

I came into the picture rather late, having never heard of Stirling numbers until after receiving my Ph.D. in mathematics. But I soon encountered them as I was beginning to analyze the performance of algorithms and to write the manuscript for my books on *The Art of Computer Programming*. I quickly realized the truth of Imanuel Marx's comment that "these numbers have similarities with the binomial coefficients $\binom{n}{k}$; indeed, formulas similar to those known for the binomial coefficients are easily established" [35]. In order to emphasize those similarities and to facilitate pattern recognition when manipulating formulas, Marx recommended using bracket symbols $\left[\begin{smallmatrix} n \\ k \end{smallmatrix}\right]$ for Stirling numbers of the first kind and brace symbols $\left\{\begin{smallmatrix} n \\ k \end{smallmatrix}\right\}$ for Stirling numbers of the second kind. A similar proposal was being made at about the same time in Italy by Antonio Salmeri [46].

I was strongly motivated by Charles Jordan's book, *Calculus of Finite Differences* [23], which introduced me to the important analogies between sums of factorial powers and integrals of ordinary powers. But I kept getting mixed up when I tried to use Stirling numbers as he defined them, because half of his "first kind" numbers were negative and the other half were positive. I had similar problems with Marx's suggestions in [35]; he made all Stirling numbers of the first kind positive, but then he attached a minus sign to half the numbers of the *second* kind. I decided that I'd never be able to keep my head above water unless I worked with Stirling numbers that were entirely signless.

And I soon learned that the signless Stirling numbers have important combinatorial significance. So I decided to try a definition that combined the best qualities of the other notations I'd seen; I defined the quantities $\left[\begin{smallmatrix} n \\ k \end{smallmatrix}\right]$ and $\left\{\begin{smallmatrix} n \\ k \end{smallmatrix}\right\}$ as follows:

$\left[\begin{smallmatrix} n \\ k \end{smallmatrix}\right]$ = the number of permutations of n objects having k cycles;

$\left\{\begin{smallmatrix} n \\ k \end{smallmatrix}\right\}$ = the number of partitions of n objects into k nonempty subsets.

For example, $\left[\begin{smallmatrix} 4 \\ 2 \end{smallmatrix}\right] = 11$, because there are eleven different ways to arrange four elements into two cycles:

$$[1, 2, 3]\,[4] \qquad [1, 2, 4]\,[3] \qquad [1, 3, 4]\,[2] \qquad [2, 3, 4]\,[1]$$
$$[1, 3, 2]\,[4] \qquad [1, 4, 2]\,[3] \qquad [1, 4, 3]\,[2] \qquad [2, 4, 3]\,[1]$$
$$[1, 2]\,[3, 4] \qquad [1, 3]\,[2, 4] \qquad [1, 4]\,[2, 3].$$

And $\left\{\begin{smallmatrix} 4 \\ 2 \end{smallmatrix}\right\} = 7$, because the partitions of $\{1, 2, 3, 4\}$ into two subsets are

$$\{1, 2, 3\}\{4\} \qquad \{1, 2, 4\}\{3\} \qquad \{1, 3, 4\}\{2\} \qquad \{2, 3, 4\}\{1\}$$
$$\{1, 2\}\{3, 4\} \qquad \{1, 3\}\{2, 4\} \qquad \{1, 4\}\{2, 3\}.$$

Notice that this notation is mnemonic: The meaning of $\left\{\begin{smallmatrix} n \\ k \end{smallmatrix}\right\}$ is easily remembered, because braces { } are commonly used to denote sets and subsets. We could also adopt the convention of writing cycles in brackets, as in my examples above, where $[1, 2, 3] = [2, 3, 1] = [3, 1, 2]$ is a typical three-cycle; that would make the notation $\left[\begin{smallmatrix} n \\ k \end{smallmatrix}\right]$ equally mnemonic. But I don't insist on this.

I have never decided how to pronounce '$\left[\begin{smallmatrix} n \\ k \end{smallmatrix}\right]$' and '$\left\{\begin{smallmatrix} n \\ k \end{smallmatrix}\right\}$' when I'm reading formulas aloud in class. Many people have begun to verbalize '$\binom{n}{k}$' as "n choose k"; hence I've been saying "n cycle k" for $\left[\begin{smallmatrix} n \\ k \end{smallmatrix}\right]$ and "n subset k" for $\left\{\begin{smallmatrix} n \\ k \end{smallmatrix}\right\}$. But I have also caught myself calling them "n bracket k" and "n brace k."

One of the advantages of these notational conventions is that binomial coefficients and Stirling numbers can be defined by very simple recurrence relations having a nice pattern:

$$\binom{n+1}{k} = \binom{n}{k} + \binom{n}{k-1} ; \tag{2.1}$$

$$\left[\begin{matrix} n+1 \\ k \end{matrix}\right] = n \left[\begin{matrix} n \\ k \end{matrix}\right] + \left[\begin{matrix} n \\ k-1 \end{matrix}\right] ; \tag{2.2}$$

$$\left\{\begin{matrix} n+1 \\ k \end{matrix}\right\} = k \left\{\begin{matrix} n \\ k \end{matrix}\right\} + \left\{\begin{matrix} n \\ k-1 \end{matrix}\right\} . \tag{2.3}$$

Moreover — and this is extremely important — all three identities hold for all integers n and k, whether positive, negative, or zero. Therefore we can apply them in the midst of any formula (for example, to "absorb" an n or a k that appears in the context $n \left[\begin{smallmatrix} n \\ k \end{smallmatrix}\right]$ or $k \left\{\begin{smallmatrix} n \\ k \end{smallmatrix}\right\}$), without worrying about exceptional circumstances of any kind.

I introduced these notations in the first edition of my first book [25], and by now my students and I have accumulated some 25 years of experience with them; the conventions have served us well. However, such brackets and braces have still not become widely enough adopted that they could be considered "standard." For example, Stanley's magnificent book on *Enumerative Combinatorics* [51] uses $c(n, k)$ for $\begin{bmatrix} n \\ k \end{bmatrix}$ and $S(n, k)$ for $\begin{Bmatrix} n \\ k \end{Bmatrix}$. His notation conveys combinatorial significance, but it fails to suggest the analogies to binomial coefficients that prove helpful in manipulations. Such analogies were evidently not important enough in his mind to warrant an extravagant two-line notation — although he does use $\left(\binom{n}{k}\right)$ to denote $\binom{n+k-1}{k} = (-1)^k \binom{-n}{k}$, the number of combinations with repetitions permitted. (In a sense, Stanley's $\left(\binom{n}{k}\right)$ is a signless version of the numbers $\binom{-n}{k}$.)

When I wrote *Concrete Mathematics* in 1988, I explored Stirling numbers more carefully than I had ever done before, and I learned two things that really clinch the argument for $\begin{bmatrix} n \\ k \end{bmatrix}$ and $\begin{Bmatrix} n \\ k \end{Bmatrix}$ as the best possible Stirling number notations. Ron Graham sent me a preview copy of a memorandum by B. F. Logan [34], which presented a number of interesting connections between Stirling numbers and other mathematical quantities. One of the first things that caught my attention was Logan's Table 1, a two-dimensional array that contained the numbers $\begin{bmatrix} n \\ k \end{bmatrix}$ and $\begin{Bmatrix} n \\ k \end{Bmatrix}$ simultaneously — implying that there really is only one "kind" of Stirling number. Indeed, when I translated Logan's results into my own favorite notation, I was astonished to find that his arrangement of numbers was equivalent to a beautiful and easily remembered law of duality,

$$\begin{Bmatrix} n \\ k \end{Bmatrix} = \begin{bmatrix} -k \\ -n \end{bmatrix}. \tag{2.4}$$

Once I had this clue, it was easy to check that the recurrence relations (2.2) and (2.3) are equivalent to each other. And the boundary conditions

$$\begin{bmatrix} 0 \\ k \end{bmatrix} = \begin{Bmatrix} 0 \\ k \end{Bmatrix} = [k = 0] \quad \text{and} \quad \begin{bmatrix} n \\ 0 \end{bmatrix} = \begin{Bmatrix} n \\ 0 \end{Bmatrix} = [n = 0] \tag{2.5}$$

yield unique solutions to (2.2) and (2.3) for all integers k and n, when we run the recurrences forward and backward; the "negative" region for Stirling numbers of one kind turns out to contain precisely the numbers of the other kind. For example, the following subset of Logan's table gives the values of $\left[{n \atop k}\right]$ when $|n|$ and $|k|$ are at most 4:

	$\left[{n \atop -4}\right]$	$\left[{n \atop -3}\right]$	$\left[{n \atop -2}\right]$	$\left[{n \atop -1}\right]$	$\left[{n \atop 0}\right]$	$\left[{n \atop 1}\right]$	$\left[{n \atop 2}\right]$	$\left[{n \atop 3}\right]$	$\left[{n \atop 4}\right]$
$n = -4$	1	0	0	0	0	0	0	0	0
$n = -3$	6	1	0	0	0	0	0	0	0
$n = -2$	7	3	1	0	0	0	0	0	0
$n = -1$	1	1	1	1	0	0	0	0	0
$n = 0$	0	0	0	0	1	0	0	0	0
$n = 1$	0	0	0	0	0	1	0	0	0
$n = 2$	0	0	0	0	0	1	1	0	0
$n = 3$	0	0	0	0	0	2	3	1	0
$n = 4$	0	0	0	0	0	6	11	6	1

The reflection of this matrix about a 45° diagonal gives the values of $\left\{{n \atop k}\right\} = \left[{-k \atop -n}\right]$.

Naturally I wondered how I could have been working with Stirling numbers for so many years without having been aware of such a basic fact. Surely it must have been known before? After several hours of searching in the library, I learned that identity (2.4) had indeed been known, but largely forgotten by succeeding generations of mathematicians, primarily because previous notations for Stirling numbers made it impossible to state the identity in such a memorable form. These investigations also turned up several things about the history of Stirling numbers that I had not previously realized.

During the nineteenth century, Stirling's connection with these numbers had been almost entirely forgotten. The numbers themselves were studied, in the role of "sums of products of combinations of the numbers $\{1, 2, \ldots, n\}$ taken k at a time." Let $C_k(n)$ and $\Gamma_k(n)$ denote those sums, when the combinations are respectively without or with repetitions; thus, for example,

$$C_3(4) = 1 \cdot 2 \cdot 3 + 1 \cdot 2 \cdot 4 + 1 \cdot 3 \cdot 4 + 2 \cdot 3 \cdot 4 = 50 \, ;$$

$$\Gamma_3(3) = 1 \cdot 1 \cdot 1 + 1 \cdot 1 \cdot 2 + 1 \cdot 1 \cdot 3 + 1 \cdot 2 \cdot 2 + 1 \cdot 2 \cdot 3$$
$$+ 1 \cdot 3 \cdot 3 + 2 \cdot 2 \cdot 2 + 2 \cdot 2 \cdot 3 + 2 \cdot 3 \cdot 3 + 3 \cdot 3 \cdot 3 = 90 \, .$$

It turns out that

$$C_k(n) = \left[{n+1 \atop n+1-k}\right] \quad \text{and} \quad \Gamma_k(n) = \left\{{n+k \atop n}\right\}. \tag{2.6}$$

Christian Kramp [28] proved near the end of the eighteenth century that

$$C_k(n) = \sum \binom{n+1}{k+l} \frac{(k+l)!}{j_1! \, 2^{j_1} \, j_2! \, 3^{j_2} \, j_3! \, 4^{j_3} \, \cdots}, \qquad (2.7)$$

$$\Gamma_k(n) = \sum \binom{n+k}{k+l} \frac{(k+l)!}{j_1! \, 2!^{j_1} \, j_2! \, 3!^{j_2} \, j_3! \, 4!^{j_3} \, \cdots}, \qquad (2.8)$$

where the sums are taken over all sequences of nonnegative integers $\langle j_1, j_2, j_3, \ldots \rangle$ such that we have $j_1 + 2j_2 + 3j_3 + \cdots = k$ (that is, over all partitions of k), and where $l = j_1 + j_2 + j_3 + \cdots$. For example,

$$C_2(n) = \binom{n+1}{4} \frac{1}{8} + \binom{n+1}{3} \frac{1}{3}; \qquad \Gamma_2(n) = \binom{n+2}{4} \frac{1}{8} + \binom{n+2}{3} \frac{1}{6}.$$

Notice that $C_k(n)$ and $\Gamma_k(n)$ are polynomials in n, of degree $2k$. The duality law (2.4) and the notational transformations of (2.6) are equivalent to the amazing polynomial identity

$$C_k(n-1) = \Gamma_k(-n); \qquad (2.9)$$

but hardly anybody was aware of this surprising fact, otherwise we would almost certainly find it mentioned explicitly in the comprehensive surveys compiled in the 1890s [19, 38].

On the other hand, a rereading of Stirling's original treatment [52] makes it clear that Stirling himself would not have found the duality law (2.4) at all surprising. From the very beginning, he thought of the numbers as two triangles hooked together in tandem. Indeed, his entire motivation for studying them was the general identity

$$z^n = \sum_k \left\{ {n \atop k} \right\} z^{\underline{k}}, \qquad (2.10)$$

which expresses ordinary powers in terms of falling factorial powers. When n is positive, the nonzero terms in this sum occur for positive values of $k \leq n$; but when n is negative, the nonzero terms occur for negative $k \leq n$. Stirling presented his tables by displaying $\left\{ {n \atop k} \right\}$ with k as the row index and $\left[{n \atop k} \right]$ with k as the column index; thus, he visualized a tandem arrangement exactly as in the matrix of numbers above, with each column containing a sequence of coefficients for (2.10).

I need to digress a bit about factorial powers. If n is a positive integer and z is a complex number, I like to write

$$z^{\underline{n}} = z(z-1) \ldots (z-n+1), \qquad (2.11)$$

which I call "z to the n falling," and

$$z^{\overline{n}} = z(z+1)\ldots(z+n-1)\,, \tag{2.12}$$

which is "z to the n rising." More generally, if α is any complex number, factorial powers are defined by

$$z^{\underline{\alpha}} = z!/(z-\alpha)! \qquad \text{and} \qquad z^{\overline{\alpha}} = \Gamma(z+\alpha)/\Gamma(z)\,, \tag{2.13}$$

unless these formulas reduce to ∞/∞, when limiting values are used. My use of underlined and overlined exponents is still controversial, but I cannot resist mentioning a curious fact: Many people (for example, specialists in hypergeometric series) have become accustomed to the notation $(z)_n$ for rising factorial powers, while many other people (for example, statisticians) use that same notation for *falling* powers. The curious fact is that this notation is called "Pochhammer's symbol," but Pochhammer himself [43] used $(z)_n$ to stand for the binomial coefficient $\binom{z}{n}$. I prefer the underline/overline notation because it is unambiguous and mnemonic, especially when I'm doing work that involves factorial powers of both kinds. (Moreover, I know that $z^{\underline{n}}$ and $z^{\overline{n}}$ are easy to typeset, using macros available in the file gkpmac.tex in the standard UNIX distribution of TeX.)

In the special case $n=3$, Stirling's formula (2.10) gives

$$z^3 = \left\{ {3 \atop 3} \right\} z^{\underline{3}} + \left\{ {3 \atop 2} \right\} z^{\underline{2}} + \left\{ {3 \atop 1} \right\} z^{\underline{1}} = z(z-1)(z-2) + 3z(z-1) + z\,.$$

And in the special case $n=-1$, it reduces to the infinite sum

$$\frac{1}{z} = \sum_k \left\{ {-1 \atop k} \right\} z^{\underline{k}}$$

$$= \sum_k \left[{k \atop 1} \right] z^{-\underline{k}}$$

$$= \frac{0!}{z+1} + \frac{1!}{(z+1)(z+2)} + \frac{2!}{(z+1)(z+2)(z+3)} + \cdots\,, \tag{2.14}$$

because

$$\left[{n \atop 1} \right] = (n-1)!\,[n>0]\,. \tag{2.15}$$

(Stirling did not discuss convergence; he was, after all, writing in 1730. We have the partial sum

$$\frac{1}{z} = \sum_{k=1}^{n} \frac{(k-1)!}{(z+1)\ldots(z+k)} + \frac{n!}{z(z+1)\ldots(z+n)};$$

this is a special case of the general identity

$$\frac{1}{z} = \sum_{k=1}^{n} \frac{z_1 \ldots z_{k-1}}{(z+z_1)\ldots(z+z_k)} + \frac{z_1 \ldots z_n}{z(z+z_1)\ldots(z+z_n)} \qquad (2.16)$$

discovered by François Nicole [39] a few years before Stirling's treatise appeared. Therefore the infinite series (2.14) converges if and only if the real part of z, $\Re(z)$, is positive. By induction on n, the same condition is necessary and sufficient for (2.10) when n is any negative integer. See [41, §30] for further discussion of (2.10).)

We noted above that the numbers $\left[{n \atop k}\right]$ can be regarded as sums of products of combinations. The first identity in (2.6) is equivalent to the formula

$$z^{\overline{n}} = \sum_{k} \left[{n \atop k}\right] z^k, \qquad (2.17)$$

when n is a nonnegative integer, if we expand the product $z^{\overline{n}}$ and sum the coefficients of each power of z. Similarly, we have

$$z^{\underline{n}} = \sum_{k} \left[{n \atop k}\right] (-1)^{n-k} z^k. \qquad (2.18)$$

These equations are valid also when n is a negative integer; in that case both infinite series converge for $|z| > |n|$. Notice that (2.10) and (2.18) tell us how to convert back and forth between ordinary powers and factorial powers.

Let's turn now to the nineteenth century. Kramp [29] decided to explore a slightly generalized type of factorial power, for which he used the notations

$$a^{n|r} = a(a+r)\ldots\big(a+(n-1)r\big) \qquad (2.19)$$

$$a^{-n|r} = 1/\big((a-r)(a-2r)\ldots(a-nr)\big) \qquad (2.20)$$

when n is a positive integer. Then he considered the expansion

$$a^{n|r} = a^n + n\,\ell\,1.\,a^{n-1}r + n\,\ell\,2.\,a^{n-2}r^2 + \cdots, \qquad (2.21)$$

where the coefficients $n\,\ell\,m$ are independent of a and r [29, §539 and §540]; thus, $n\,\ell\,m$ was his notation for $\left[\begin{smallmatrix} n \\ n-m \end{smallmatrix}\right]$. He obtained [29, §557] a series of formulas equivalent to

$$m\begin{bmatrix} n \\ n-m \end{bmatrix} = \sum_{k=0}^{m-1} \binom{n-k}{m+1-k}\begin{bmatrix} n \\ n-k \end{bmatrix}, \qquad (2.22)$$

thereby giving a new proof that $\left[\begin{smallmatrix} n \\ n-m \end{smallmatrix}\right]$ is a polynomial in n of degree $2m$. This proof, independent of his earlier formulas (2.7) and (2.8), works for both positive and negative values of n.

Kramp implicitly understood the duality principle (2.4), in the sense that he regarded the coefficients $\left[\begin{smallmatrix} n \\ k \end{smallmatrix}\right]$ and $\left\{\begin{smallmatrix} n \\ k \end{smallmatrix}\right\}$ as the positive and negative portions of a doubly infinite array of numbers. In fact, he assumed that equation (2.21) would hold for arbitrary real values of n. He differentiated $a^{x|r}$ with respect to x and gave formal derivations of several interesting series. However, his expansion (2.21) is equivalent to

$$z^{\overline{n}} = \sum_k \begin{bmatrix} n \\ n-k \end{bmatrix} z^{n-k} \qquad (2.23)$$

(a slight variation of (2.17)), and this series is not always convergent for noninteger n. We can show, for example, that

$$\left| \begin{bmatrix} 1/2 \\ 1/2 - k \end{bmatrix} \right| > k!/7^k \qquad \text{for infinitely many } k\,; \qquad (2.24)$$

hence (2.23) diverges for all z when $n = 1/2$. Kramp lived before the days when convergence of infinite series was understood. (See [29, §574], where he says that the divergent series $\sum_{k>0} B_k y^k / k$ is "très convergente pour peu que y soit une petite fraction"!)

Several other nineteenth-century authors developed the theory of factorial powers, notably Andreas von Ettingshausen [6], Ludwig Schläfli [47, 48], and Oskar Schlömilch [49], who used the respective notations

$$\overset{n}{\mathrm{F}}_m, \qquad \overset{n}{A}_m, \qquad \text{and} \qquad \overset{n}{C}_m$$

for the coefficients $\left[\begin{smallmatrix} n \\ n-m \end{smallmatrix}\right]$. All of these authors considered both positive and negative integers n. Thus, for example, Ettingshausen's notation for a Stirling number such as $\left\{\begin{smallmatrix} n+m \\ n \end{smallmatrix}\right\} = \left[\begin{smallmatrix} -n \\ -n-m \end{smallmatrix}\right]$ was

$$\overset{-n}{\mathrm{F}}_m$$

(see [6, §151]).

Incidentally, these works of Kramp and Ettingshausen proved to be important in the history of mathematical notations. Kramp's book introduced the notation $n!$ for factorials [29, pages V and 219], and Ettingshausen's book introduced the notation $\binom{n}{k}$ for binomial coefficients [6, page 30]. Ettingshausen wrote his book shortly after Fourier [8] had invented $\sum$-notation for sums; Ettingshausen tried a German variation, writing $\mathfrak{S}^k_{a,b}$ for what has evolved into $\sum_{k=a}^{b}$. He also wrote $(a,r)^n$ for Kramp's $a^{n|r}$; thus, for example, Ettingshausen [6, §153 and §156] gave the equations

$$(a,d)^n = \overset{w}{\underset{0}{\mathfrak{S}}}\, \overset{n}{F_w}\, a^{n-w}\, d^w \qquad \text{and} \qquad a^n = \overset{r}{\underset{0}{\mathfrak{S}}}\,(-1)^r \,\overset{-n+r}{F_r}\,(a,d)^{n-r}\, d^r$$

as equivalents of Kramp's (2.21) and Stirling's (2.10). He presented Kramp's (2.22) in the form

$$v\,\overset{n}{F_v} = \overset{w}{\underset{0,v-1}{\mathfrak{S}}}\,\binom{n-w}{v+1-w}\,\overset{n}{F_w}\,,$$

and remarked [6, §154] that this holds for both negative and positive n. Ettingshausen had related the F coefficients to sums of products of combinations with and without repetition; thus he implicitly confirmed (2.9).

The first person to attach Stirling's name to the numbers we now call Stirling numbers was Niels Nielsen in 1904 [40]; he said that this new nomenclature had been suggested to him by T. N. Thiele. (The numbers may have been studied before Stirling's time; for example, I once found the values of $\left[{n \atop k}\right]$ for $1 \leq n \leq 7$ in some unpublished manuscripts of Thomas Harriot, dating from about 1600, in the British Museum [26, page 241]. But Stirling almost surely deserves the credit for being first to deduce nontrivial facts about $\left[{n \atop k}\right]$ and $\left\{{n \atop k}\right\}$.)

Nielsen wrote C_n^k for $\left[{n \atop n-k}\right]$, which he called a "Stirling number of rank n"; and he wrote $\mathfrak{C}_n^k$ for $\left\{{n+k-1 \atop n-1}\right\}$, which he called a "Stirling number of rank $-n$." (He should really have defined its rank to be $1-n$.) In equation (41) of his paper, Nielsen obtained a rigorous proof of the duality law (2.4); but he had to state it in a peculiar way, because he had defined C_n^k and $\mathfrak{C}_n^k$ only for nonnegative n and k. Thus, he could not write the formula $C_n^k = \mathfrak{C}_{1-n}^k$; he had to say instead that $f_k(n) = g_k(1-n)$, where $f_k(n)$ and $g_k(n)$ were the polynomials defined by C_n^k and $\mathfrak{C}_n^k$. Tweedie [54] expressed (2.4) with similar circumlocutions.

When Jordan took up Stirling numbers [22], he wrote S_n^k for $(-1)^{n-k}\left[{n \atop k}\right]$ and $\mathfrak{S}_n^k$ for $\left\{{n \atop k}\right\}$. He does not seem to have known the duality law (2.4), probably because he had learned about Stirling numbers from Nielsen's book [41], which omitted some of the details in Nielsen's paper [40]. And as far as I know, the duality law largely disappeared from mathematicians' collective consciousness during most of the twentieth century; it seems to have been mentioned explicitly only in a few scattered places: (1) Hansraj Gupta, "working in a small township away from what was then the only University in the Panjab" [18, page 5], rediscovered Stirling numbers and Stirling duality by himself, in the early 1930s. This became part of his Ph.D. dissertation [17], and he included it in a book on number theory prepared many years later [18, Chapter 5]. (2) H. W. Gould [12] was probably the first twentieth-century mathematician to observe that we can use the polynomials $\left[{n \atop n-k} \right]$ and $\left\{ {n \atop n-k} \right\}$ to extend the domain of Stirling numbers to negative values of n. Gould's way of writing (2.4) was $S_1(-n-1,k) = S_2(n,k)$; and shortly thereafter [13], he mentioned the equivalent formula

$$S_{-k}^{-n} = (-1)^{n-k} \mathfrak{S}_n^k \,,$$

in Jordan's notation. (3) R. V. Parker [42], like Gupta, displayed both of Stirling's triangles in tandem, presenting them in a single table as Logan did later. (4) In 1976, Ira Gessel and Richard Stanley investigated some of the deeper structure underlying the Stirling polynomials $f_k(n) = \left\{ {n+k \atop n} \right\}$ and $g_k(n) = \left[{n \atop n-k} \right]$. They noted in particular [11, equation (3)] that $f_k(-n) = g_k(n)$. This fact is equivalent to the duality law (2.4).

Stanley had discovered a beautiful theorem in his Ph.D. thesis a few years earlier [50, Proposition 13.2(i)], now called the reciprocity theorem for order polynomials: If P is any finite partially ordered set, let $\Omega(P,n)$ be the number of order-preserving mappings from P into the totally ordered set $\{1,2,\ldots,n\}$; and let $\overline{\Omega}(P,n)$ be the number of such mappings that are *strictly* order-preserving. Thus, if $x \prec y$ in P, the mappings f enumerated by $\Omega(P,n)$ must satisfy $f(x) \leq f(y)$, and the mappings g enumerated by $\overline{\Omega}(P,n)$ must satisfy $g(x) < g(y)$. Stanley's theorem states that, in general, we have $f(-n) = (-1)^p g(n)$, where p is the number of elements of P. For example, if P consists of p isolated points with no order constraints whatever, we have $\Omega(P,n) = \overline{\Omega}(P,n) = n^p$. And if the points of P are themselves totally ordered, then $\Omega(P,n)$ is $\binom{n+p-1}{p}$, the number of combinations of n things p at a time with repetitions permitted, while $\overline{\Omega}(P,n)$ is $\binom{n}{p}$, the combinations without repetition. In both cases Stanley's law $\Omega(P,-n) = (-1)^p \overline{\Omega}(P,n)$ holds.

I showed Stanley the first draft of this note and asked him whether the Stirling duality formula (2.4) could be derived as a special case of his general reciprocity law. Sure enough, he replied that Gessel had noticed a simple way to do exactly that, shortly after the paper [11] was written. Let P_k be the partial order on $2k$ points typified by

$$P_4 = \text{\quad}\quad ;$$

then

$$\Omega(P_k, n) = \sum_{1 \le x_1, \dots, x_k, y_1, \dots, y_k \le n} [x_1 \le \dots \le x_k][x_1 \ge y_1] \dots [x_k \ge y_k]$$

$$= \sum_{1 \le x_1, \dots, x_k \le n} [x_1 \le \dots \le x_k] \, x_1 \dots x_k \, ,$$

and

$$\overline{\Omega}(P_k, n) = \sum_{1 \le x_1, \dots, x_k, y_1, \dots, y_k \le n} [x_1 < \dots < x_k][x_1 > y_1] \dots [x_k > y_k]$$

$$= \sum_{2 \le x_1, \dots, x_k \le n} [x_1 < \dots < x_k](x_1 - 1) \dots (x_k - 1)$$

$$= \sum_{1 \le x_1, \dots, x_k \le n-1} [x_1 < \dots < x_k] \, x_1 \dots x_k \, .$$

Thus the sums are respectively $\Gamma_k(n)$ and $C_k(n-1)$; by (2.6) we have $\Omega(P_k, n) = \left\{ {n+k \atop n} \right\}$ and $\overline{\Omega}(P_k, n) = \left[{n \atop n-k} \right]$, hence (2.4) is indeed an instance of Stanley's theorem.

Now we are ready to discuss the second reason why I became convinced that $\left[{n \atop k} \right]$ is the right symbolism for these coefficients after I had translated Logan's memo [34] into that notation: We know that $\left[{n \atop n-k} \right]$ is a polynomial in n, when k is an integer; hence, as Kramp knew, we can sensibly define the quantity $\left[{\alpha \atop \alpha-k} \right]$ for arbitrary complex α and integer k, using that same polynomial. Then — and here comes the punch line — Logan noticed that the fundamental equations (2.17) and (2.18) generalize to *asymptotic formulas*, valid for arbitrary exponents α: If $z \to \infty$ and if m is any nonnegative integer, we have

$$z^{\overline{\alpha}} = \sum_{k=0}^{m} \left[{\alpha \atop \alpha - k} \right] z^{\alpha - k} + O(z^{\alpha - m - 1}); \tag{2.25}$$

$$z^{\underline{\alpha}} = \sum_{k=0}^{m} \left[{\alpha \atop \alpha - k} \right] (-1)^k \, z^{\alpha - k} + O(z^{\alpha - m - 1}). \tag{2.26}$$

(See [15, exercise 9.44]; equation (2.25) is a correct way to formulate Kramp's divergent series (2.23). These equations are special cases of a still more general result proved by Tricomi and Erdélyi [53, 9].) The easily remembered expansions in (2.25) and (2.26) were quite a revelation to me. I had often spent time laboriously calculating approximations to ratios such as $z^{\overline{1/2}} = \Gamma(z+1/2)/\Gamma(z)$, the hard way: I took logarithms, then used Stirling's approximation, and then took exponentials. But equations (2.25) and (2.26) produce the answer directly.

Moreover, Stirling's original identity (2.10) can be generalized in a similar way: If α is any complex number, we have

$$z^\alpha = \sum_k \left\{ {\alpha \atop \alpha - k} \right\} z^{\overline{\alpha-k}}, \qquad \Re(z) > 0. \qquad (2.27)$$

When I wrote the first draft of this note, I knew only that the series (2.27) was convergent, and that it was asymptotically correct as $z \to \infty$; so I conjectured that equality might hold. Soon afterward, B. F. Logan found the following proof (although he naturally stated it in his own notation): Suppose first that $\Re(\alpha) < 1$. Then we have the well known identity

$$z^{\alpha-1} = \frac{1}{\Gamma(1-\alpha)} \int_0^\infty e^{-zt} t^{-\alpha} \, dt, \qquad \Re(z) > 0, \qquad (2.28)$$

and we can substitute $e^{-t} = 1 - u$ to get

$$z^{\alpha-1} = \frac{1}{\Gamma(1-\alpha)} \int_0^1 (1-u)^{z-1} u^{-\alpha} \left(\frac{1}{u} \ln \frac{1}{1-u} \right)^{-\alpha} du.$$

Now it turns out that the powers of $\frac{1}{u} \ln \frac{1}{1-u}$ generate the Stirling numbers $\left\{ {\alpha \atop \alpha-k} \right\} = \left[{k-\alpha \atop -\alpha} \right]$, in the sense that

$$\left(\frac{1}{u} \ln \frac{1}{1-u} \right)^{-\alpha} = \sum_k \left\{ {\alpha \atop \alpha-k} \right\} \frac{u^k}{(k-\alpha) \dots (1-\alpha)}, \qquad (2.29)$$

a series that converges for $|u| < 1$ $\bigl($see [15, equations (6.45), (6.53), (7.50)]$\bigr)$. Therefore

$$z^\alpha = \sum_k \left\{ {\alpha \atop \alpha-k} \right\} \frac{z}{\Gamma(k+1-\alpha)} \int_0^1 (1-u)^{z-1} u^{k-\alpha} \, du$$

$$= \sum_k \left\{ {\alpha \atop \alpha-k} \right\} \frac{\Gamma(z+1)}{\Gamma(z+1+k-\alpha)} = \sum_k \left\{ {\alpha \atop \alpha-k} \right\} \frac{z!}{(z+k-\alpha)!},$$

and (2.27) is verified when $\Re(\alpha) < 1$. To complete the proof, we need only show that (2.27) holds for $\alpha + 1$ if it holds for α; but this is easy, because

$$
\begin{aligned}
z^{\alpha+1} &= \sum_k \left\{ {\alpha \atop \alpha - k} \right\} z \cdot z^{\underline{\alpha-k}} \\
&= \sum_k \left\{ {\alpha \atop \alpha - k} \right\} \left(z^{\underline{\alpha+1-k}} + (\alpha - k) z^{\underline{\alpha-k}} \right) \\
&= \sum_k \left\{ {\alpha \atop \alpha - k} \right\} z^{\underline{\alpha+1-k}} + \sum_k \left\{ {\alpha \atop \alpha + 1 - k} \right\} (\alpha + 1 - k) z^{\underline{\alpha+1-k}} \\
&= \sum_k \left\{ {\alpha + 1 \atop \alpha + 1 - k} \right\} z^{\underline{\alpha+1-k}}
\end{aligned}
$$

by the basic recurrence equation (2.3).

Notice that in all of the general identities (2.25)–(2.27), as in the original formulas (2.10), (2.17), and (2.18) that inspired them, the lower index within the braces or brackets is the same as the exponent of z. This makes the relations easy to remember, by analogy with the binomial theorem

$$
(1 + z)^\alpha = \sum_k \binom{\alpha}{k} z^k, \qquad \text{when } |z| < 1. \tag{2.30}
$$

Some readers will have been thinking, "This all looks fairly plausible, but unfortunately Knuth is overlooking a key point that ruins the whole proposal: We *can't* use the notation $\left[{n \atop k} \right]$ for Stirling numbers, because it has already been used for more than a century as the standard notation for Gauss's generalized binomial coefficients."

Well, there is a down side to every good idea, but this objection is not really severe. For one thing, the standard notation for Gaussian binomial coefficients involves a hidden parameter q, and it's not unusual for modern researchers to make transformations that change q. Therefore Gauss's notation is incomplete, and Andrews (for example) has used the notation $\left[{n \atop k} \right]_{q^2}$ for the Gaussian coefficient with q^2 as the hidden parameter [2, page 49]. Such examples suggest that it is appropriate to denote Gaussian binomials as $\binom{n}{k}_q$, especially since they reduce to ordinary binomials when $q = 1$. This notation also generalizes nicely to such things as Fibonomial coefficients $\binom{n}{k}_{\mathcal{F}}$; see [27]. We can then reserve the notation $\left[{n \atop k} \right]_q$ for a q–generalization of $\left[{n \atop k} \right]$. (The reverse strategy was unfortunately adopted in [14].)

Secondly, I do not believe that any existing mathematical works, including books like [2] which use Gaussian coefficients extensively, would become seriously cluttered if the Gaussian $\begin{bmatrix} n \\ k \end{bmatrix}$ were changed everywhere to $\binom{n}{k}_q$. Even so, such changes are not necessary; there is obviously no harm in beginning a mathematical paper or a book chapter or an entire book with a statement to the effect that "$\begin{bmatrix} n \\ k \end{bmatrix}$ will denote a Gaussian binomial coefficient with parameter q in what follows." All notation can be redefined for special purposes. Therefore Stirling number enthusiasts are not encroaching on Gaussian territory when they write $\begin{bmatrix} n \\ k \end{bmatrix}$, if they also mumble something about Stirling in order to set the context.

One further point is worth noting in conclusion: As soon as the notations $\begin{bmatrix} n \\ k \end{bmatrix}$ and/or $\left\{ \begin{matrix} n \\ k \end{matrix} \right\}$ are adopted, there will no longer be a need to speak about Stirling numbers "of the first and second kind," except as a concession to history. Nielsen wrote a superb book [41], but he did the world a disservice by originating the *Erster Art* and *Zweiter Art* terminology, because that terminology has no mnemonic value and is historically inaccurate. Stirling introduced the numbers $\left\{ \begin{matrix} n \\ k \end{matrix} \right\}$ first and brought in $\begin{bmatrix} n \\ k \end{bmatrix}$ second. Indeed, practical applications have always tended to involve the numbers $\left\{ \begin{matrix} n \\ k \end{matrix} \right\}$ much more often than their $\begin{bmatrix} n \\ k \end{bmatrix}$ counterparts. It seems far better to speak of $\left\{ \begin{matrix} n \\ k \end{matrix} \right\}$ as a Stirling subset number, and to call $\begin{bmatrix} n \\ k \end{bmatrix}$ a Stirling cycle number. Then the names are tied to intuitive, student-friendly concepts, not to arbitrary and offputting concepts of the kth kind.

Acknowledgments

I am extremely grateful for comments received from John Ewing, Philippe Flajolet, Adriano Garsia, B. F. Logan, Andrew Odlyzko, Richard Stanley, and H. S. Wilf, without which these notes would have been substantially poorer.

This research was supported in part by National Science Foundation grant number CCR-86-10181.

References

[1] Milton Abramowitz and Irene A. Stegun, editors, *Handbook of Mathematical Functions* (Washington: National Bureau of Standards, 1964).

[2] George E. Andrews, *The Theory of Partitions* (Reading, Massachusetts: Addison–Wesley, 1977).

[3] Anonymous and S..., "Bemerkungen zu dem Aufsatze überschrieben 'Beweis der Gleichung $0^0 = 1$ nach J. F. Pfaff,' im zweiten Hefte dieses Bandes, S. 134," *Journal für die reine und angewandte Mathematik* **12** (1834), 292–294.

[4] Charles Babbage, *Passages from the Life of a Philosopher* (London: 1864). Reprinted in *Charles Babbage and his Calculating Engines*, edited by Philip Morrison and Emily Morrison (New York: Dover, 1961).

[5] Augustin-Louis Cauchy, *Cours d'Analyse de l'Ecole Royale Polytechnique* (1821). Reprinted in his *Œuvres Complètes*, Series 2, Volume 3.

[6] Andreas v. Ettingshausen, *Die combinatorische Analysis* (Vienna: 1826).

[7] Philippe Flajolet and Andrew Odlyzko, "Singularity analysis of generating functions," *SIAM Journal on Discrete Mathematics* **3** (1990), 216–240.

[8] J. Fourier, "Refroidissement séculaire du globe terrestre," *Bulletin des Sciences par la Société philomathique de Paris* (3) **7** (1820), 58–70. Reprinted in *Œuvres de Fourier* **2**, 271–288.

[9] C. L. Frenzen, "Error bounds for asymptotic expansions of the ratio of two gamma functions," *SIAM Journal on Mathematical Analysis* **18** (1987), 890–896.

[10] Adriano M. Garsia, "On the 'maj' and 'inv' q-analogues of Eulerian polynomials," *Linear and Multilinear Algebra* **8** (1979), 21–34.

[11] Ira Gessel and Richard P. Stanley, "Stirling polynomials," *Journal of Combinatorial Theory* **A24** (1978), 24–33.

[12] H. W. Gould, "Stirling number representation problems," *Proceedings of the American Mathematical Society* **11** (1960), 447–451. For subsequent work, see his review of [42] in *Mathematical Reviews* **49** (1975), 885–886.

[13] H. W. Gould, "Note on a paper of Klamkin concerning Stirling numbers," *American Mathematical Monthly* **68** (1961), 477–479.

[14] H. W. Gould, "The q-Stirling numbers of first and second kinds," *Duke Mathematical Journal* **28** (1961), 281–289.

[15] Ronald L. Graham, Donald E. Knuth, and Oren Patashnik, *Concrete Mathematics* (Reading, Massachusetts: Addison–Wesley, 1989).

[16] Daniel H. Greene and Donald E. Knuth, *Mathematics for the Analysis of Algorithms*, second edition (Boston: Birkhäuser, 1981).

[17] H. Gupta, *Symmetric Functions in the Theory of Integral Numbers*, Lucknow University Studies **14** (Allahabad: Allahabad Law Journal Press, 1940).

[18] Hansraj Gupta, *Selected Topics in Number Theory* (Tunbridge Wells, England: Abacus Press, 1980).

[19] Johann G. Hagen, *Synopsis der Höheren Mathematik* **1** (Berlin: 1891).

[20] G. H. Hardy and E. M. Wright, *An Introduction to the Theory of Numbers* (Oxford, Clarendon Press, 1938). Fifth edition, 1979.

[21] Kenneth E. Iverson, *A Programming Language* (New York: Wiley, 1962).

[22] Charles Jordan, "On Stirling's numbers," *Tôhoku Mathematical Journal* **37** (1933), 254–278.

[23] Charles Jordan, *Calculus of Finite Differences* (Budapest: 1939). Third edition, 1965.

[24] Mark Kac, *Statistical Independence in Probability, Analysis and Number Theory*, Carus Mathematical Monographs **12** (Mathematical Association of America, 1959).

[25] Donald E. Knuth, *Fundamental Algorithms*, Volume 1 of *The Art of Computer Programming* (Reading, Massachusetts: Addison-Wesley, 1968).

[26] Donald E. Knuth, review of *History of Binary and Other Nondecimal Numeration* by Anton Glaser, *Historia Mathematica* **10** (1983), 236–243. [Reprinted, in part, as Chapter 5 of the present volume.]

[27] Donald E. Knuth and Herbert S. Wilf, "The power of a prime that divides a generalized binomial coefficient," *Journal für die reine und angewandte Mathematik* **396** (1989), 212–219. [Reprinted as Chapter 36 of the present volume.]

[28] Christian Kramp, "Coefficient des allgemeinen Gliedes jeder willkührlichen Potenz eines Infinitinomiums; Verhalten zwischen Coefficienten der Gleichungen und Summen der Produkte und der Potenzen ihrer Wurzeln; Transformation und Substitution der Reihen durch einander," in *Der polynomische Lehrsatz das wichtigste Theorem der ganzen Analysis nebst einigen verwandten und andern Sätzen*, edited by Carl Friedrich Hindenburg (Leipzig: Gerhard Fleischer dem Jüngern, 1796), 91–122.

[29] C. Kramp, *Élémens d'arithmétique universelle* (Cologne: 1808).

[30] Leopold Kronecker, "Ueber bilineare Formen," *Journal für die reine und angewandte Mathematik* **68** (1868), 273–285.

[31] Leopold Kronecker, *Vorlesungen über die Theorie der Determinanten*, edited by Kurt Hensel, Volume 1 (Leipzig: Teubner, 1903).

[32] Guillaume Libri, "Note sur les valeurs de la fonction 0^{0^x}," *Journal für die reine und angewandte Mathematik* **6** (1830), 67–72.

[33] Guillaume Libri, "Mémoire sur les fonctions discontinues," *Journal für die reine und angewandte Mathematik* **10** (1833), 303–316.

[34] B. F. Logan, "Polynomials related to the Stirling numbers," AT&T Bell Labs internal technical memorandum (10 August 1987).

[35] Imanuel Marx, "Transformation of series by a variant of Stirling numbers," *American Mathematical Monthly* **69** (1962), 530–532. His $\begin{bmatrix} n \\ k \end{bmatrix}$ is my $\begin{bmatrix} n+1 \\ k+1 \end{bmatrix}$; his $\begin{Bmatrix} n \\ k \end{Bmatrix}$ is my $(-1)^{n-k} \begin{Bmatrix} n+1 \\ k+1 \end{Bmatrix}$.

[36] A. F. Möbius, "Beweis der Gleichung $0^0 = 1$, nach J. F. Pfaff," *Journal für die reine und angewandte Mathematik* **12** (1834), 134–136.

[37] Douglas H. Moore, *Heaviside Operational Calculus: An Elementary Foundation* (New York: American Elsevier, 1971).

[38] Eugen Netto, *Lehrbuch der Combinatorik* (Leipzig: B. G. Teubner, 1901). Second edition, with additions by Thoralf Skolem and Viggo Brun, 1927.

[39] Nicole, "Méthode pour sommer une infinité de Suites nouvelles, dont on ne peut trouver les Sommes par les Méthodes connuës," *Mémoires de l'Academie Royale des Sciences* (Paris: 1727), 257–268.

[40] Niels Nielsen, "Recherches sur les polynomes et les nombres de Stirling," *Annali di Matematica pura ed applicata* (3) **10** (1904), 287–318.

[41] Niels Nielsen, *Handbuch der Theorie der Gammafunktion* (Leipzig: Teubner, 1906).

[42] R. V. Parker, "The complete polynomial grid," *Matematički Vesnik* **25** (1973), 181–203.

[43] L. Pochhammer, "Ueber hypergeometrische Functionen n^{ter} Ordnung," *Journal für die reine und angewandte Mathematik* **71** (1870), 316–352.

[44] Hillel Poritsky, "Heaviside's operational calculus — its applications and foundations," *American Mathematical Monthly* **43** (1936), 331–344.

[45] S..., "Sur la valeur de 0^0," *Journal für die reine und angewandte Mathematik* **11** (1834), 272–273.

[46] Antonio Salmeri, "Introduzione alla teoria dei coefficienti fattoriali," *Giornale di Matematiche di Battaglini* **90** (1962), 44–54. His $\left[{n \atop k}\right]$ is my $\left[{n+1 \atop n+1-k}\right]$.

[47] Schlaeffli [sic], "Sur les coëfficients du développement du produit $1.(1+x)(1+2x) \ldots (1+(n-1)x)$ suivant les puissances ascendantes de x," *Journal für die reine und angewandte Mathematik* **43** (1852), 1–22.

[48] Schläfli, "Ergänzung der Abhandlung über die Entwickelung des Products $1.(1 + x)(1 + 2x)(1 + 3x) \ldots (1 + (n - 1)x) = \overset{n}{\Pi}(x)$ in Band XLIII dieses Journals," *Journal für die reine und angewandte Mathematik* **67** (1867), 179–182.

[49] O. Schlömilch, "Recherches sur les coefficients des facultés analytiques," *Journal für die reine und angewandte Mathematik* **44** (1852), 344–355.

[50] Richard P. Stanley, *Ordered Structures and Partitions*, Memoirs of the American Mathematical Society **119** (1972).

[51] Richard P. Stanley, *Enumerative Combinatorics*, Volume 1 (Belmont, California: Wadsworth, 1986).

[52] James Stirling, *Methodus Differentialis* (London: 1730). English translation, *The Differential Method* (1749).

[53] F. G. Tricomi and A. Erdélyi, "The asymptotic expansion of a ratio of gamma functions," *Pacific Journal of Mathematics* **1** (1951), 133–142.

[54] Charles Tweedie, "The Stirling numbers and polynomials," *Proceedings of the Edinburgh Mathematical Society* **37** (1918), 2–25.

[55] Karl Weierstrass, "Zur Theorie der eindeutigen analytischen Functionen," *Mathematische Abhandlungen der Akademie der Wissenschaften zu Berlin* (1876), 11–60. Reprinted in his *Mathematische Werke* **2**, 77–124. (Florian Cajori, in *History of Mathematical Notations* **2**, cites unpublished papers of 1841 and 1859 as the first occurrences of the notation $|z|$; however, those papers were not edited for publication until 1894, and they use the notation without defining it, so their published form may differ from Weierstrass's original.)

[56] Christian Wiener, "Geometrische und analytische Untersuchung der *Weierstrass*schen Function," *Journal für die reine und angewandte Mathematik* **90** (1881), 221–252.

Addendum

I received two most pleasant surprises in June, 1993, several months after this article had appeared in print. First, I learned that the article had been selected as one of the recipients of the Lester R. Ford Award for mathematical exposition, and I wrote the following response:

> When I first mentioned to editor Herb Wilf that I planned to submit an article about notation to the MONTHLY, he replied that "the topic of notation is (intrinsically) about as exciting as IRS form 6378." He said he would rather have an article about mathematics; but he promised to look at whatever I wrote — even if I chose to write "on the sex life of the Antarctic newt." Therefore I eventually sent him a draft entitled Two Newts on Newtation, by Ursula N. Gnewt. He decided it might be a silk purse after all, and forwarded it to the new editor John Ewing.
>
> After the article was published, several correspondents informed me that an Iverson-like notation had independently been championed by Bruno de Finetti in his *Teoria delle Probababilità* **1** (Turin: 1970); English translation, *Theory of Probability* **1** (London: Wiley, 1974). De Finetti wrote, "I am often surprised by new, important applications" — and that remark is so true, it evaluates to 2 instead of 1! (Readers of my paper will understand.)

The second surprise came a few days later when I learned from Philippe Flajolet that my favorite notations $\begin{bmatrix} n \\ k \end{bmatrix}$ and $\begin{Bmatrix} n \\ k \end{Bmatrix}$ for Stirling numbers were used already in 1932 by J. Karamata of Belgrade! This astonishing news seemed too good to be true, and I suspected a clever hoax; but no, a trip to Stanford's library confirmed that Karamata's paper did indeed define $\begin{bmatrix} n \\ k \end{bmatrix}$ and $\begin{Bmatrix} n \\ k \end{Bmatrix}$ precisely as I had decided to do more than 30 years later, "puisque, par cette notation, les formules deviennent plus symétriques." On my 64th birthday in 2002, Flajolet presented me with a rare document that he had been able to acquire in Paris: an original reprint of Karamata's paper, "Théorèmes sur la sommabilité exponentielle et d'autres sommabilités s'y rattachant," *Mathematica* (Cluj) **9** (1935), 164–178, from the proceedings of a congress of Romanian mathematicians held in May 1932.

Chapter 3

Bracket Notation for the 'Coefficient of' Operator

[This chapter is a slightly extended version of a paper originally published in A Classical Mind, Essays in Honour of C. A. R. Hoare, edited by A. W. Roscoe (London: Prentice–Hall, 1994), 247–258.]

When $G(z)$ is a power series in z, many authors now write '$[z^n] G(z)$' for the coefficient of z^n in $G(z)$, using a notation introduced by Goulden and Jackson in [5, page 1]. More controversial, however, is the proposal of the same authors [5, page 160] to let '$[z^n/n!] G(z)$' denote the coefficient of $z^n/n!$, that is, $n!$ times the coefficient of z^n. An alternative generalization of $[z^n] G(z)$, in which we define $[F(z)] G(z)$ to be a linear function of both F and G, seems to be more useful because it facilitates algebraic manipulations. The purpose of this paper is to explore some of the properties of such a definition. The remarks are dedicated to Tony Hoare because of his lifelong interest in the improvement of notations that facilitate manipulation.

Informal Introduction

In this paper '$[z^2+2z^3] G(z)$' will stand for the coefficient of z^2 plus twice the coefficient of z^3 in $G(z)$, when $G(z)$ is a function of z for which such coefficients are well defined. More generally, if $F(z) = f_0 + f_1 z + f_2 z^2 + \cdots$ and $G(z) = g_0 + g_1 z + g_2 z^2 + \cdots$, we will let

$$[F(z)] G(z) = f_0 g_0 + f_1 g_1 + f_2 g_2 + \cdots$$

be the "dot product" of the vectors $(f_0, f_1, f_2, \ldots)$ and $(g_0, g_1, g_2, \ldots)$, assuming that the infinite sum exists. Still more generally, if $F(z) = \cdots + f_{-2} z^{-2} + f_{-1} z^- + f_0 + f_1 z + f_2 z^2 + \cdots$ and $G(z) = \cdots + g_{-2} z^{-2} + g_{-1} z^- + g_0 + g_1 z + g_2 z^2 + \cdots$ are doubly infinite series, we will write

$$[F(z)] G(z) = \cdots + f_{-2} g_{-2} + f_{-1} g_{-1} + f_0 g_0 + f_1 g_1 + f_2 g_2 + \cdots , \quad (1)$$

45

again assuming convergence. (It is convenient to write 'z^-' for $1/z$, as in [8].) The right side of (1) is symmetric in F and G, so we have a commutative law:

$$[F(z)]\, G(z) = [G(z)]\, F(z)\,. \tag{2}$$

There also is symmetry between positive and negative powers:

$$[F(z)]\, G(z) = [F(z^-)]\, G(z^-)\,. \tag{3}$$

In particular, we will write $[1]\, G(z)$ for the constant term g_0 of a given doubly infinite power series $G(z) = \sum_n g_n z^n$. Notice that $[z^n]\, G(z) = [1]\, z^{-n} G(z)$ and in fact

$$[F(z)]\, G(z) = [1]\, F(z^-) G(z)\,, \tag{4}$$

when the product of series is defined in the usual way:

$$\sum_n h_n z^n = \left(\sum_n f_n z^n \right) \left(\sum_n g_n z^n \right) \quad \Longleftrightarrow \quad h_n = \sum_{j+k=n} f_j g_k\,. \tag{5}$$

Relation (4) gives us a useful rule for moving factors in and out of brackets:

$$[F(z)]\, G(z) H(z) = [F(z) G(z^-)]\, H(z)\,. \tag{6}$$

Both sides reduce to $[1]\, F(z^-) G(z) H(z)$, so they must be equal. This rule is most often applied in a simple form such as

$$[z^n]\, z^3 H(z) = [z^{n-3}]\, H(z)\,,$$

but it is helpful to remember the general principle (6). Similarly,

$$[F(z) G(z)]\, H(z) = [F(z)]\, G(z^-) H(z)\,. \tag{7}$$

A Paradox

So far the extended bracket notation seems straightforward and innocuous, but if we start to play with it in an undisciplined fashion we can easily get into trouble. For example, one of the first uses we might wish to make of relation (1) is

$$\left[\frac{z^n}{1-z} \right] G(z) = g_n + g_{n+1} + g_{n+2} + \cdots\,, \tag{8}$$

because $z^n/(1-z) = z^n + z^{n+1} + z^{n+2} + \cdots$. This operation, unfortunately, turns out to be dangerous, if not outright fallacious.

The danger is sometimes muted and we might be lucky. For example, if we try combining (8) with (7) in the case $G(z) = 1/(1-z)$ and $H(z) = (1-z)^2 = 1 - 2z + z^2$, we get

$$\left[\frac{z^n}{1-z}\right] (1-z)^2 = [z^n] \frac{(1-z)^2}{1-z^-} = [z^n] (z^2 - z). \tag{9}$$

Sure enough, the sum $h_n + h_{n+1} + h_{n+2} + \cdots$ is nonzero in this case only when $n = 2$ and $n = 1$, and (9) gives the correct answer. So far so good.

But (7) and (8) lead to a contradiction when we apply them to the trivial case $F(z) = H(z) = 1$ and $G(z) = 1/(1-z)$:

$$1 = \left[\frac{1}{1-z}\right] 1 = [1] \frac{1}{1-z^-} = [1] \frac{-z}{1-z} = 0. \tag{10}$$

What went wrong?

Formal Analysis

To understand the root of the paradox (10), and to learn when (6) and (7) are indeed valid rules of transformation, we need to know the basic properties of doubly infinite power series $\sum_n g_n z^n$. The general theory can be found in Henrici [6, §4.4]; we will merely sketch it here.

If $G(z)$ is analytic in an annulus $\alpha < |z| < \beta$, it has a unique doubly infinite series representation $G(z) = \sum_n g_n z^n$. Conversely, *every* doubly infinite power series that converges in an annulus defines an analytic function there. The proof is based on the contour integral formula

$$G(z) = \frac{1}{2\pi i} \oint_{|t|=\beta'} \frac{G(t)\,dt}{(t-z)} - \frac{1}{2\pi i} \oint_{|t|=\alpha'} \frac{G(t)\,dt}{(t-z)}, \tag{11}$$

where α' is between $|z|$ and α while β' is between $|z|$ and β. The quantity $1/(t-z)$ can be expanded as $t^-(1 + z/t + z^2/t^2 + \cdots)$ when $|t| > |z|$ and as $-z^-(1 + t/z + t^2/z^2 + \cdots)$ when $|t| < |z|$.

If $F(z)$ and $G(z)$ are both analytic for $\alpha < |z| < \beta$, their product $H(z)$ is an analytic function whose coefficients are given by (5). Moreover, the infinite sum over all j and k with $j + k = n$ in (5) is absolutely convergent: The terms are $O((\alpha'/\beta')^k)$ as $k \to +\infty$ and $O((\beta'/\alpha')^k)$ as $k \to -\infty$.

The coefficients of $G(z)$ in its doubly infinite power series depend on α and β. For example, suppose $G(z) = 1/(2 - z)$; we have

$$\frac{1}{2 - z} = \begin{cases} \frac{1}{2} + \frac{1}{4}z + \frac{1}{8}z^2 + \cdots, & \text{when } |z| < 2; \\ -z^- - 2z^{-2} - 4z^{-3} - \cdots, & \text{when } |z| > 2. \end{cases} \tag{12}$$

Thus if $F(z) = 1/(2 - z) + 1/(2 - z^-)$, there are three expansions

$$F(z) = \begin{cases} \frac{1}{2} + \left(\frac{1}{4} - 1\right)z + \left(\frac{1}{8} - 2\right)z^2 + \left(\frac{1}{16} - 4\right)z^3 + \cdots, & |z| < \frac{1}{2}; \\ \cdots + \frac{1}{8}z^{-2} + \frac{1}{4}z^- + 1 + \frac{1}{4}z + \frac{1}{8}z^2 + \cdots, & \frac{1}{2} < |z| < 2; \\ \cdots + \left(\frac{1}{16} - 4\right)z^{-3} + \left(\frac{1}{8} - 2\right)z^{-2} + \left(\frac{1}{4} - 1\right)z^- + \frac{1}{2}, & |z| > 2. \end{cases} \tag{13}$$

Here's another example, this time involving a function that has an essential singularity instead of a pole:

$$e^{z/(1-z)} = \begin{cases} 1 + z + \frac{3}{2}z^2 + \frac{13}{6}z^3 + \frac{73}{24}z^4 + \cdots, & |z| < 1; \\ e^- - e^- z^- - \frac{1}{2}e^- z^{-2} - \frac{1}{6}e^- z^{-3} + \frac{1}{24}e^- z^{-4} + \cdots, & |z| > 1. \end{cases} \tag{14}$$

The coefficients when $|z| < 1$ are $P_n/n!$, where P_n is the number of "sets of lists" of order n (see Motzkin [9]).

Explaining the Paradox

The dependency of coefficients on α and β makes the bracket notation $[F(z)]\,G(z)$ ambiguous; that is why we ran into trouble in the paradoxical "equation" (10). We can legitimately use bracket notation only when the context specifies a family of "safe" functions — functions with well defined coefficients.

The basic definition of $[F(z)]\,G(z)$ in (4) should be used only if the product $F(z^-)G(z)$ is safe. Operation (6), which moves a factor $G(z)$ into the bracket, should be used only if $F(z^-)G(z)H(z)$ is safe. Operation (7), which removes a factor $G(z)$ from the bracket, should be used only if $F(z^-)G(z^-)H(z)$ is safe.

The root of our problem in (10) begins in (8), where we used the expansion $F(z) = z^n/(1 - z) = z^n + z^{n+1} + z^{n+2} + \cdots$; in other words, $F(z^-) = z^{-n}/(1 - z^-) = z^{-n} + z^{-n-1} + z^{-n-2} + \cdots$. The latter expansion is valid only when $|z| > 1$, so the bracket notation of (8) refers to coefficients in the region $1 < |z| < \infty$. In the last step of (10), however, we said that $[1]\left(-z/(1 - z)\right) = 0$, using coefficients from the region $|z| < 1$. The correct result for $|z| > 1$ is

$$[1]\,\frac{-z}{1 - z} = [1]\,\frac{1}{1 - z^-} = [1]\left(\cdots + z^{-2} + z^- + 1\right) = 1.$$

Bracket notation is most often used when $|z|$ is small, so we should actually forget the "rightward sum" appearing in equation (8); it hardly ever yields the formula we want. The "leftward sum" rule

$$\left[\frac{z^n}{1 - z^-}\right] G(z) = \cdots + g_{n-2} + g_{n-1} + g_n \tag{15}$$

should be used instead, because $z^n/(1 - z^-) = \cdots + z^{n-2} + z^{n-1} + z^n$ is valid for $|z^-| < 1$. *When the bracket notation* $[F(z)] G(z)$ *is being used in the annulus* $(\alpha \mathbin{..} \beta)$, *the functions* $F(z^-)$ *and* $G(z)$ *should be analytic in* $(\alpha \mathbin{..} \beta)$. Note that $f(z^-)$ is analytic in $(\alpha \mathbin{..} \beta)$ if and only if $f(z)$ is analytic in $(\beta^- \mathbin{..} \alpha^-)$.

Formal Series

Manipulations of generating functions are often done on formal power series, when the coefficients are arbitrary and convergence is disregarded. However, formal power series are not allowed to be infinite in both directions; a formal series $G(z) = \sum_n g_n z^n$ is generally required to be a "formal Laurent series" — a series in which $g_n = 0$ for all sufficiently negative values of n. We shall call such series *L-series* for short. Similarly, we shall say that a reverse formal Laurent series, in which $g_n = 0$ for all sufficiently *positive* values of n, is an *R-series*. A power series is both an *L*-series and an *R*-series if and only if it is a polynomial in z and z^-.

Henrici [6, §1.2–1.8] shows that the normal operations on power series — addition, subtraction, multiplication, division by nonzero, differentiation, composition — can all be done rigorously on *L*-series without regard to convergence. Thus *L*-series are "safe" functions: *We can define bracket notation* $[F(z)] G(z)$ *by rule (4) whenever* $F(z)$ *is an R-series and* $G(z)$ *is an L-series.* Convergence is not then an issue. This definition provides the default meaning of bracket notation, whenever no other context is specified. The transformations in (7) and (8) are valid when the functions inside brackets are *R*-series and the functions outside brackets are *L*-series. Equations (2) and (3) should not be used unless F and G are both *L*-series and *R*-series.

In such cases paradoxes do not rear their ugly heads. The ill-fated equation (8) may fail, but equation (15) is always true.

Additional Properties

The bracket notation satisfies several identities in addition to (2), (3), (6), and (7), hence we can often transform formulas in which it appears.

In the first place, the operation is linear in both operands:

$$[aF(z) + bG(z)]\, H(z) = a[F(z)]\, H(z) + b[G(z)]\, H(z)\,; \quad (16)$$

$$[F(z)]\bigl(aG(z) + bH(z)\bigr) = a[F(z)]\, G(z) + b[F(z)]\, H(z)\,. \quad (17)$$

In the second place, there is a general multiplication law

$$[F_1(z)\, F_2(z)]\, G_1(z)\, G_2(z) = \sum_k \bigl([F_1(z)z^k]\, G_1(z)\bigr)\bigl([F_2(z)z^{-k}]\, G_2(z)\bigr)\,. \tag{18}$$

If $F_1(z) = F_2(z) = 1$, this equation is simply the special case $n = 0$ of (5), and for general F_1 and F_2 it follows from the special case because we can replace $G_1(z)$ and $G_2(z)$ by $F_1(z^-)G_1(z)$ and $F_2(z^-)G_2(z)$ using (7).

We also have

$$[F(z^m)]\, G(z^m) = [F(z)]\, G(z) \tag{19}$$

for any nonzero integer m; this equation, which includes (3) as the special case $m = -1$, follows immediately from (4) because $[1]\, H(z) = [1]\, H(z^m)$. Equation (19) suggests that we generalize bracket notation to functions that are sums over nonintegral powers, in which case m would not need to be an integer. Then we could write (19) as

$$[F(z)]\, G(z^m) = [F(z^{1/m})]\, G(z)\,, \quad m \neq 0\,. \tag{19'}$$

Such generalizations, extending perhaps to integrals as well as to sums, may prove to be quite interesting, but they will not be pursued further here.

If a is any nonzero constant, we have $[1]\, H(az) = [1]\, H(z)$. This rule implies that $[1]\, F(z^-)G(az) = [1]\, F(az^-)G(z)$, and (4) yields

$$[F(z)]\, G(az) = [F(az)]\, G(z)\,. \tag{20}$$

The special case where $F(z)$ is simply z^m is, of course, already familiar:

$$[z^m]\, G(az) = [(az)^m]\, G(z) = a^m[z^m]\, G(z)\,.$$

Bracket notation also interacts with differentiation in interesting ways. We have, for instance,

$$[z^-]\, G'(z) = 0 \tag{21}$$

for any function $G(z) = \sum_{n=-\infty}^{\infty} g_n z^n$. More significantly,

$$[F(z)] \, z \, G'(z) = [z \, F'(z)] \, G(z) \,. \tag{22}$$

Equation (21) is essentially the special case $F(z) = 1$ of (22), but we can also derive (22) from (21): Let $H(z) = F(z^-)G(z)$; then we have $0 = [1] \, z \, H'(z) = [1] \, z \left(F(z^-)G'(z) - z^{-2}F'(z^-)G(z) \right)$, hence $[1] \, F(z^-) \, z \, G'(z) = [1] \, z^- F'(z^-)G(z)$, which is (22).

Let ϑ be the operator $z \frac{d}{dz}$. Then (22) implies by induction on m that

$$[F(z)] \, \vartheta^m G(z) = [\vartheta^m F(z)] \, G(z)$$

for all integers $m \geq 0$, and we have

$$[F(z)] \, P(\vartheta) G(z) = [P(\vartheta) F(z)] \, G(z) \tag{23}$$

for any polynomial P. If $F(z) = \sum_n f_n z^n$ and $G(z) = \sum_n g_n z^n$, both sides of (23) evaluate to $\sum_n P(n) f_n g_n$.

Additional Variables

When $G(w, z)$ is a bivariate generating function we also wish to write $[w^m z^n]$ for the coefficient of $w^m z^n$ in G. In general we can define

$$[F(w, z)] \, G(w, z) = [w^0 z^0] \, F(w^-, z^-) G(w, z) \,, \tag{24}$$

extending (4).

Variables must be clearly distinguished from constants. If w and z are both variables, we have for instance $[z] \, wz = 0$, while if w is constant we have $[z] \, wz = w$. If the set of variables is not clear from the context, we can specify it by letting each relevant variable appear explicitly in the brackets, as suggested by Renzo Sprugnoli. According to this convention we have, for example, $[w^0 z] \, wz = 0$ but $[z] \, wz = w$. All variables mentioned in brackets are essentially "bound" by the notation; that is why '$[w^0 z^0]$' appears in (24) instead of simply '$[1]$'.

We have

$$[F(w) \, G(z)] \, H(w, z) = [F(w)] \, ([G(z)] \, H(w, z)) \tag{25}$$

because the former is $[w^0 z^0] \, F(w^-) G(z^-) H(w, z)$ while the latter is

$$[w^0] \left(F(w^-) [z^0] \left(G(z^-) H(w, z) \right) \right) = [w^0] [z^0] \, F(w^-) G(z^-) H(w, z)$$

and the operator $[w^0 z^0]$ is the same as $[w^0] [z^0]$.

After we have evaluated the parenthesized expression on the right side of (25), the ambiguity disappears, because w is no longer present. For example, if $m \geq 0$ we have

$$[w^m z^n] \frac{1}{1 - wF(z)} = [z^n] \left([w^m] \frac{1}{1 - wF(z)} \right) = [z^n] F(z)^m . \qquad (26)$$

Similarly

$$[w^m z^n] e^{wF(z)} = [z^n] \frac{F(z)^m}{m!} ; \qquad (27)$$

$$[w^m z^n] G(wF(z)) H(z) = [z^n] F(z)^m H(z) [w^m] G(w) . \qquad (28)$$

Suppose w and z are variables. Then laws (19) and (20) extend to

$$[F(w, z)] G(aw, z) = [F(aw, z)] G(w, z), \qquad a \neq 0; \qquad (29)$$

$$[F(w^m, z)] G(w^m, z) = [F(w, z)] G(w, z), \quad \text{integer } m \neq 0; \ (30)$$

$$[F(w, w^m z)] G(w, w^m z) = [F(w, z)] G(w, z); \qquad (31)$$

and we have indeed the general rule

$$[F(a^- w^k z^l, b^- w^m z^n)] G(aw^k z^l, bw^m z^n) = [F(w, z)] G(w, z) \qquad (32)$$

when $a \neq 0$, $b \neq 0$, and $\begin{vmatrix} k & l \\ m & n \end{vmatrix} \neq 0$, that is, $kn \neq lm$. A similar formula applies with respect to any number of variables.

The following example from the theory of random graphs [3, (10.10) and (10.14)] illustrates how these rules are typically applied. Suppose we want to evaluate the coefficient of $[w^m z^n]$ in the expression $e^{U(wz)/w + V(wz)}$, where U and V are known functions with $U(0) = 0$. The two-variable problem is reduced to a one-variable problem as follows:

$$[w^m z^n] e^{U(wz)/w + V(wz)} = [(w^-)^{n-m} (wz)^n] e^{U(wz)w^- + V(wz)}$$

$$= [w^{n-m} z^n] e^{U(z)w + V(z)}$$

$$= \frac{1}{(n-m)!} [z^n] U(z)^{n-m} e^{V(z)} , \qquad (33)$$

by (32) with $F(w, z) = w^{n-m} z^n$, $G(w, z) = e^{U(z)w + V(z)}$, $a = b = 1$, $k = -1$, $l = 0$, $m = n = 1$. The final step uses (28) with $F(z) = U(z)$, $G(w) = e^w$, and $H(z) = e^{V(z)}$.

As before, we need to check that the functions are safe before we can guarantee that such manipulations are legitimate. For formal power series, the functions inside brackets should be R-series and the functions outside should be L-series. This condition holds in each step of (33) because $U(0) = 0$.

Additional Identities

The bracket notation also obeys more complex laws that deserve further
study. For example, Gessel and Stanton [4, Eq. (3)] have shown among
other things that

$$[F(w,z)] \frac{G(w,z)}{1 - wz} = [F(w(1+z^-), z(1+w^-))] \, G\left(\frac{w}{1+z}, \frac{z}{1+w}\right). \quad (34)$$

If we set $F(w,z) = w^k z^l$ and $G(w,z) = (1+w)^m (1+z)^n / (1-wz)^{m+n}$,
Gessel and Stanton observe that we obtain Saalschütz's identity after
some remarkable cancellation:

$$\sum_r \binom{m}{k-r} \binom{n}{l-r} \binom{m+n+r}{r} = [w^k z^l] \, (1+w)^m (1+z)^n / (1-wz)^{m+n+1}$$

$$= [w^k(1+z^-)^k z^l (1+w^-)^l] \, (1+w)^m (1+z)^n$$

$$= [w^k z^l] \, (1+w)^{m+l} (1+z)^{n+k}$$

$$= \binom{m+l}{k} \binom{n+k}{l}. \quad (35)$$

And if we set $F(w,z) = w^{l+n} z^{m+n}$, $G(w,z) = (w-z)^{l+m}/(1-wz)^{l+m}$,
the left side of (34) reduces to

$$[w^{l+n} z^{m+n}] \frac{(w-z)^{l+m}}{(1-wz)^{l+m+1}} = (-1)^m \frac{(l+m+n)!}{l! \, m! \, n!}; \quad (36)$$

the right side is

$$[w^{l+n}(1+z^-)^{l+n} z^{m+n}(1+w^-)^{m+n}] \, (w-z)^{l+m}$$

$$= [w^{l+n} z^{m+n}] \, (w-z)^{l+m} (1+w)^{m+n} (1+z)^{l+n}$$

$$= \sum_k (-1)^{k+m} \binom{l+m}{k+m} \binom{m+n}{k+n} \binom{n+l}{k+l}. \quad (37)$$

The fact that (36) = (37) is Dixon's identity [7, exercise 1.2.6–62].

Equation (34) can be generalized to n variables, and we can replace
the '1' on the right by any nonzero constant a:

$$[F(z_1, \ldots, z_n)] \frac{G(z_1, \ldots, z_n)}{1 - z_1 \ldots z_n}$$

$$= [F(z_1(a+z_2^-), \ldots, z_n(a+z_1^-))] \, G\left(\frac{z_1}{a+z_2}, \ldots, \frac{z_n}{a+z_1}\right). \quad (38)$$

We need to prove this only when $F(z_1, \ldots, z_n) = 1$ and $G(z_1, \ldots, z_n) = z_1^{m_1} \ldots z_n^{m_n}$; and in that case both sides are 0 unless $m_1 = \cdots = m_n \leq 0$, when both sides are 1. Equation (38) holds in particular when $n = 1$:

$$[F(z)]\frac{G(z)}{1-z} = [F(1+az)]G\left(\frac{z}{a+z}\right), \qquad a \neq 0. \qquad (39)$$

Returning to the case of a single variable, we should also state the general rule for composition of series:

$$G(F(z)) = \sum_n F(z)^n [z^n] G(z). \qquad (40)$$

Special conditions are needed to ensure that this infinite sum is well defined.

Lagrange's Inversion Formula

Let $F(z) = f_1 z + f_2 z^2 + f_3 z^3 + \cdots$, with $f_1 \neq 0$, and let $G(z)$ be the inverse function so that

$$F(G(z)) = G(F(z)) = z. \qquad (41)$$

Lagrange's celebrated formula for the coefficients of G can be expressed in bracket notation in several ways; for example, we have

$$n[z^n] G(z)^m = m[z^{-m}] F(z)^{-n}, \qquad (42)$$

for all integers m and n.

One way to derive (42), following Paule [10], is to note first that (40) implies

$$z^m = G(F(z))^m = \sum_k F(z)^k [z^k] G(z)^m. \qquad (43)$$

Differentiating with the ϑ operator and dividing by $F(z)^n$ yields

$$\frac{mz^m}{F(z)^n} = \sum_k kF(z)^{k-1-n}\vartheta F(z) [z^k] G(z)^m. \qquad (44)$$

Now we will study the constant terms of (44). If $k \neq n$,

$$[1] F(z)^{k-1-n}\vartheta F(z) = [1]\frac{\vartheta(F(z)^{k-n})}{k-n} = 0, \qquad (45)$$

by (22). And if $k = n$,

$$[1]\frac{\vartheta F(z)}{F(z)} = [1]\frac{f_1 + 2f_2 z + 3f_3 z^2 + \cdots}{f_1 + f_2 z + f_3 z^2 + \cdots} = 1, \qquad (46)$$

because $f_1 \neq 0$. Therefore the constant terms of (44) are

$$[1]\frac{mz^m}{F(z)^n} = n[z^n] G(z)^m;$$

this is Lagrange's formula (42).

Composition of Functions

If $G(z) = g_0 + g_1 z + g_2 z^2 + \cdots$, Eqs. (26) and (40) yield

$$[z^n] G(F(z)) = \left[[z^n] \frac{1}{1 - wF(z)} \right] G(w) \qquad (47)$$

because both sides are $[z^n] \sum_{m \geq 0} g_m F(z)^m$. Thus, for example, if $F(z) = a + bz$ we obtain

$$\sum_{m=0}^{\infty} \binom{m}{n} a^{m-n} b^n g_m = \left[\frac{b^n w^n}{(1 - aw)^{n+1}} \right] G(w). \qquad (48)$$

And if $F(z) = 1/(1 - z)$ we have

$$[z^n] G\left(\frac{1}{1-z} \right) = \left[\frac{w}{(1-w)^{n+1}} \right] G(w) = [G(w)] \frac{w}{(1-w)^{n+1}} \qquad (49)$$

for $n > 0$. Setting $a = b = 1$ in (48) and replacing $G(z)$ by $z^{n-1}G(z)$ also tells us that

$$[z^n] (1 + z)^{n-1} G(1 + z) = [w^n/(1 - w)^{n+1}] w^{n-1} G(w)$$
$$= [w^{n-1} G(w)] w^n/(1 - w)^{n+1}$$
$$= [G(w)] w/(1 - w)^{n+1};$$

therefore by (49) we obtain the curious relation

$$[z^n] G\left(\frac{1}{1-z} \right) = [z^n] (1 + z)^{n-1} G(1 + z), \qquad (50)$$

which holds also when $n = 0$.

Equation (50) can be rewritten in the form

$$[1] \frac{1}{1-z} H\left(\frac{z}{1-z} \right) = [1] H(z) \qquad (51)$$

if we set $H(z) = (1 + z)^{n-1} G(1 + z)/z^n$. And indeed, we can prove the general formula

$$[1] \frac{z F'(z)}{F(z)} H(F(z)) = [1] H(z) \qquad (52)$$

whenever $F(z) = f_1 z + f_2 z^2 + f_3 z^3 + \cdots$ with $f_1 \neq 0$ and H is arbitrary. The reason is that (52) holds when $H(z) = z^m$ for any integer m; the case $m \neq 0$ gives 0 on both sides, and the case $m = 0$ is (46).

Conclusions

Many years of experience have confirmed the great importance of generating functions in the analysis of algorithms, and we can reasonably expect that some fluency in manipulating the "coefficient-of" operator will therefore be rewarding.

If, for example, we are faced with the task of simplifying a formula such as

$$\sum_k \binom{m}{k} [z^{n-k}] F(z)^k \,,$$

a rudimentary acquaintance with the properties of brackets will tell us that it can be written as $\sum_k \binom{m}{k}[z^n] z^k F(z)^k = [z^n] \sum_k \binom{m}{k} z^k F(z)^k$ and then summed to yield

$$[z^n]\big(1 + zF(z)\big)^m \,.$$

We have seen several examples above in which formulas that are far less obvious can be derived rapidly by bracket manipulation, when we use quantities more general than monomials inside the brackets.

In most applications we use bracket notation in connection with formal Laurent series, in which case it is important to remember that our identities for $[F(z)]\,G(z)$ require $G(z)$ to have only finitely many *negative* powers of z while $F(z)$ must have only finitely many *positive* powers. If we write, for example,

$$\left[\frac{z^n}{z-1} \right] G(z)\,, \tag{53}$$

we should think of the quantity in brackets as an infinite series

$$z^{n-1} + z^{n-2} + z^{n-3} + \cdots$$

that descends to arbitrarily *negative* powers of z; the bracket notation then denotes the sum $g_{n-1} + g_{n-2} + g_{n-3} + \cdots$, which will be finite. We have seen that other interpretations of bracket notation are possible for functions analytic in an annulus; but great care must be taken to avoid paradoxes in such cases, hence the extra effort might not be worthwhile.

Bracket notation, like all notations, is "dispensable," in the sense that we can prove the same theorems without it as with it. But the use of a good notation can shorten proofs and help us see patterns that would otherwise be difficult to perceive.

Let us close with one more example, illustrating that the notation (53) helps to simplify some of the formulas in [2]. The *coupon collector's problem* asks for the expected number of trials needed to obtain n distinct coupons from a set C of m given coupons, where each trial independently produces coupon c with probability $p(c)$. Theorem 2 of [2] says, when rewritten in the notation discussed above, that this expected number is

$$\int_0^\infty \left[\frac{z^n}{z-1}\right] \prod_{c\in C} \left(1 + z(e^{p(c)t} - 1)\right) e^{-t}\,dt\,. \qquad (54)$$

We can evaluate (54) by expanding the integrand as follows:

$$\left[\frac{z^n}{z-1}\right] \prod_{c\in C} \left(1 + z(e^{p(c)t}-1)\right) = \sum_{\substack{B\subseteq C \\ |B|<n}} \prod_{c\in B} (e^{p(c)t} - 1)$$

$$= \sum_{\substack{A\subseteq B\subseteq C \\ |B|<n}} (-1)^{|B|-|A|} e^{p(A)t}$$

$$= \sum_{\substack{A\subseteq C \\ |A|<n}} e^{p(A)t} \sum_{|A|\leq k<n} (-1)^{k-|A|} \binom{|C|-|A|}{k-|A|}$$

$$= \sum_{\substack{A\subseteq C \\ |A|<n}} e^{p(A)t}(-1)^{n-1-|A|} \binom{|C|-|A|-1}{|C|-n}\,,$$

where $p(A)$ denotes $\sum_{a\in A} p(a)$. The integral (54) therefore is

$$\sum_{\substack{A\subseteq C \\ |A|<n}} (-1)^{n-1-|A|} \binom{|C|-|A|-1}{|C|-n} \bigg/ \left(1 - p(A)\right)\,. \qquad (55)$$

(This is Corollary 3 of [2], which was stated without proof.)

Related Work

Steven Roman's book on umbral calculus [11] develops extensive properties of his notation $\langle G(t) \mid F(x)\rangle$, which equals $\sum_{n\geq 0} f_n g_n$ when $F(x) = \sum_{n=0}^\infty f_n x^n$ is a polynomial and $G(t) = \sum_{n=0}^\infty g_n t^n/n!$. Thus, if

D is the operator d/dx, Roman's expression $\langle G(t) \mid F(x) \rangle$ is the constant term of the polynomial $G(D)F(x)$. Chapter 6 of [11] considers generalizations in which $\langle G(t) \mid F(x) \rangle$ is defined to be $\sum_{n \geq 0} f_n g_n$ when $G(t) = \sum_{n \geq 0} g_n t^n / c_n$ and c_n is an arbitrary sequence of constants; the case $c_n = 1$ corresponds to the special case of bracket notation $[F(z)]\,G(z)$ when F and G involve no negative powers of z. Roman traces the theory back to a paper by Morgan Ward [12].

G. P. Egorychev's book [1] includes a great many examples that demonstrate the value of coefficient extraction in the midst of formulas.

Open Problems

One reason formal power series are usually restricted to L-series is that certain doubly infinite power series are divisors of zero. For example, $\sum_{n=-\infty}^{\infty} z^n$ is a divisor of zero because multiplication by $1 - z$ annihilates it. (This series causes no problem in the theory of non-formal power series because it does not converge for any value of z.) All doubly infinite series having the form $\sum_n n^m \alpha^n z^n$ for $\alpha \neq 0$ and integer $m \geq 0$ can also be shown to be divisors of zero. Question: Do there exist divisors of zero besides finite linear combinations of the doubly infinite series just mentioned? Conjecture: There is no formal, nonzero, doubly infinite power series $F(z)$ such that $e^z F(z) = 0$. (A counterexample would necessarily be divergent.)

It may be possible and interesting to extend the theory of formal Laurent series to arbitrary functions of the form $F(z) \sum_n g_n z^n$, where g_n is zero for all sufficiently negative n and where $F(z)$ is analytic for $0 < |z| < \infty$.

Acknowledgments

I wish to thank Edsger and Ria Dijkstra for the splendid opportunity to write this paper in the guest room of their Texas home; also Peter Paule for his penetrating comments on the first draft; and Ira Gessel for his comments on (50) received in November 1994.

Bibliography

[1] G. P. Egorychev, *Integral Representation and the Computation of Combinatorial Sums* (Providence, Rhode Island: American Mathematical Society, 1984).

[2] Philippe Flajolet, Danièle Gardy, and Loÿs Thimonier, "Birthday paradox, coupon collectors, caching algorithms and self-organizing search," *Discrete Applied Mathematics* **39** (1992), 207–229.

[3] Philippe Flajolet, Donald E. Knuth, and Boris Pittel, "The first cycles in an evolving graph," *Discrete Mathematics* **75** (1989), 167–215. [Reprinted as Chapter 40 of the present volume.]

[4] Ira Gessel and Dennis Stanton, "Short proofs of Saalschütz's and Dixon's theorems," *Journal of Combinatorial Theory* **A38** (1985), 87–90.

[5] I. P. Goulden and D. M. Jackson, *Combinatorial Enumeration* (New York: Wiley, 1983).

[6] Peter Henrici, *Applied and Computational Complex Analysis*, Volume 1 (New York: Wiley, 1974).

[7] Donald E. Knuth, *Fundamental Algorithms*, Volume 1 of *The Art of Computer Programming* (Reading, Massachusetts: Addison–Wesley, 1968).

[8] Donald E. Knuth, "Efficient representation of perm groups," *Combinatorica* **11** (1991), 33–43.

[9] T. S. Motzkin, "Sorting numbers for cylinders and other classification numbers," *Proceedings of Symposia in Pure Mathematics* **19** (Providence, Rhode Island: American Mathematical Society, 1971), 167–176.

[10] Peter Paule, "Ein neuer Weg zur q-Lagrange Inversion," *Bayreuther Mathematische Schriften* **18** (1985), 1–37.

[11] Steven Roman, *The Umbral Calculus* (Orlando, Florida: Academic Press, 1984).

[12] Morgan Ward, "A calculus of sequences," *American Journal of Mathematics* **58** (1936), 255–266.

Addendum

Philippe Jacquet and Philippe Flajolet found infinitely many counterexamples to my conjecture that e^z is not a divisor of zero in the ring of doubly infinite power series. For example, we have $e^z F(z) = 0$ if

$$F(z) = \sum_{m=0}^{\infty} m!\, [t^m]\, f(t)\, z^{-m} \;-\; z \int_0^1 e^{-zt} f(t)\, dt, \qquad f(t) = e^{-1/\sqrt{1-t}}.$$

Chapter 4

Johann Faulhaber and Sums of Powers

The early 17th-century mathematical publications of Johann Faulhaber contain some remarkable theorems, such as the fact that the r-times-repeated summation of 1^m, 2^m, ..., n^m is a polynomial in $n(n + r)$ times the r-fold sum of 1, 2, ..., n, when m is a positive odd number. The present paper explores a computation-based approach by which Faulhaber may well have discovered such results, and solves a 360-year-old riddle that Faulhaber presented to his readers. It also shows that similar results hold when we express the sums in terms of central factorial powers instead of ordinary powers. Faulhaber's coefficients can moreover be generalized to factorial powers of noninteger exponents, obtaining asymptotic series for $1^\alpha + 2^\alpha + \cdots + n^\alpha$ in powers of $n^{-1}(n + 1)^{-1}$.

[Dedicated to the memory of Derrick H. Lehmer. Originally published in Mathematics of Computation **61** (1993), 277–294, the Lehmer Memorial Issue.]

1. Introduction

Johann Faulhaber of Ulm (1580–1635), the founder of a school for engineers early in the 17th century, loved numbers. His passion for arithmetic and algebra led him to devote a considerable portion of his life to the computation of formulas for the sums of powers, significantly extending all previously known results. He may well have carried out more computing than anybody else in Europe during the first half of the 17th century.

Faulhaber's greatest mathematical achievements appear in a booklet entitled *Academia Algebræ* (written in German in spite of its latin title), published in Augsburg, 1631 [2]. Here we find, for example, the following formulas for sums of odd powers:

$$1^1 + 2^1 + \cdots + n^1 = N, \qquad N = (n^2 + n)/2;$$

$$1^3 + 2^3 + \cdots + n^3 = N^2;$$

$$1^5 + 2^5 + \cdots + n^5 = (4N^3 - N^2)/3;$$

$$1^7 + 2^7 + \cdots + n^7 = (12N^4 - 8N^3 + 2N^2)/6;$$

$$1^9 + 2^9 + \cdots + n^9 = (16N^5 - 20N^4 + 12N^3 - 3N^2)/5;$$

$$1^{11} + 2^{11} + \cdots + n^{11} = (32N^6 - 64N^5 + 68N^4 - 40N^3 + 10N^2)/6;$$

$$1^{13} + 2^{13} + \cdots + n^{13} = (960N^7 - 2800N^6 + 4592N^5 - 4720N^4 \\ + 2764N^3 - 691N^2)/105;$$

$$1^{15} + 2^{15} + \cdots + n^{15} = (192N^8 - 768N^7 + 1792N^6 - 2816N^5 \\ + 2872N^4 - 1680N^3 + 420N^2)/12;$$

$$1^{17} + 2^{17} + \cdots + n^{17} = (1280N^9 - 6720N^8 + 21120N^7 - 46880N^6 \\ + 72912N^5 - 74220N^4 \\ + 43404N^3 - 10851N^2)/45.$$

Other mathematicians had studied Σn^1, Σn^2, ..., Σn^7, and Faulhaber himself had gotten as far as Σn^{12}; but the sums had always previously been expressed as polynomials in n, not N.

Faulhaber began his book by simply stating these novel formulas and proceeding to expand them into the corresponding polynomials in n. Then he verified the results when $n = 4$, $N = 10$. But he gave no clues about how he derived the expressions; he stated only that the leading coefficient in Σn^{2m-1} is $2^{m-1}/m$, and that the trailing coefficients will have the form $4\alpha_m N^3 - \alpha_m N^2$ when $m \geq 3$.

Faulhaber believed that similar polynomials in N, with alternating signs, would continue to exist for all m, but he may not really have known how to prove such a theorem. In his day, mathematics was treated like all other sciences; the goal was merely to amass a large body of evidence for an observed phenomenon. No rigorous proof of Faulhaber's assertion was published until Jacobi treated the subject in 1834 [6]. A. W. F. Edwards showed recently how to obtain Faulhaber's coefficients by matrix inversion [1], based on another proof that was given by L. Tits in 1923 [8]; but none of these derivations use methods that are very close to those known in 1631.

Faulhaber went on to consider sums of sums. Let us write $\Sigma^r n^m$ for the r-fold summation of mth powers from 1 to n; thus,

$$\Sigma^0 n^m = n^m; \qquad \Sigma^{r+1} n^m = \Sigma^r 1^m + \Sigma^r 2^m + \cdots + \Sigma^r n^m.$$

Johann Faulhaber in 1630. (Engraving by Sebastian Furck, in Stadtarchiv Ulm.)

He discovered that $\Sigma^r n^{2m}$ can be written as a polynomial in the quantity

$$N_r = (n^2 + rn)/2,$$

times $\Sigma^r n^2$. For example, he gave the formulas

$$\Sigma^2 n^4 = (4N_2 - 1)\,\Sigma^2 n^2/5\,;$$
$$\Sigma^3 n^4 = (4N_3 - 1)\,\Sigma^3 n^2/7\,;$$
$$\Sigma^4 n^4 = (6N_4 - 1)\,\Sigma^4 n^2/14\,;$$
$$\Sigma^6 n^4 = (4N_6 + 1)\,\Sigma^6 n^2/15\,;$$
$$\Sigma^2 n^6 = (6N_2^2 - 5N_2 + 1)\,\Sigma^2 n^2/7\,;$$
$$\Sigma^3 n^6 = (10N_3^2 - 10N_3 + 1)\,\Sigma^3 n^2/21\,;$$
$$\Sigma^4 n^6 = (4N_4^2 - 4N_4 - 1)\,\Sigma^4 n^2/14\,;$$
$$\Sigma^2 n^8 = (16N_2^3 - 28N_2^2 + 18N_2 - 3)\,\Sigma^2 n^2/15\,.$$

He also gave similar formulas for odd exponents, factoring out $\Sigma^r n^1$ instead of $\Sigma^r n^2$:

$$\Sigma^2 n^5 = (8N_2^2 - 2N_2 - 1)\,\Sigma^2 n^1/14\,;$$
$$\Sigma^2 n^7 = (40N_2^3 - 40N_2^2 + 6N_2 + 6)\,\Sigma^2 n^1/60\,.$$

And he claimed that, in general, $\Sigma^r n^m$ can be expressed as a polynomial in N_r times either $\Sigma^r n^2$ or $\Sigma^r n^1$, depending on whether m is even or odd.

Faulhaber had probably verified this remarkable theorem in many cases including $\Sigma^{11} n^6$, because he exhibited a polynomial in n for $\Sigma^{11} n^6$ that would have been quite difficult to obtain by repeated summation. His polynomial, which has the form

$$\frac{6n^{17} + 561n^{16} + \cdots + 1021675563656n^5 + \cdots - 96598656000n}{2964061900800},$$

turns out to be absolutely correct, according to calculations with a modern computer. (The denominator is 17!/120. One cannot help thinking that nobody has ever checked these numbers since Faulhaber himself wrote them down, until today.)

Did he, however, know how to prove his claim, in the sense that 20th century mathematicians would regard his argument as conclusive? He may in fact have known how to do so, because there is an extremely simple way to verify the result using only methods that he would have found natural.

2. Reflective Functions

Let us begin by studying an elementary property of integer functions. We will say that the function $f(x)$ is *r-reflective* if

$$f(x) = f(y) \quad \text{whenever} \quad x + y + r = 0\,;$$

and it is *anti-r-reflective* if

$$f(x) = -f(y) \quad \text{whenever} \quad x + y + r = 0\,.$$

The values of x, y, r will be assumed to be integers for simplicity. When $r = 0$, reflective functions are even, and anti-reflective functions are odd. Notice that r-reflective functions are closed under addition and multiplication; the product of two anti-r-reflective functions is r-reflective.

Given a function f, we define its backward difference ∇f in the usual way:

$$\nabla f(x) = f(x) - f(x - 1)\,.$$

It is now easy to verify a simple basic fact.

Lemma 1. *If f is r-reflective then ∇f is anti-$(r-1)$-reflective. If f is anti-r-reflective then ∇f is $(r-1)$-reflective.*

Proof. If $x+y+(r-1) = 0$ then $x+(y-1)+r = 0$ and $(x-1)+y+r = 0$. Thus $f(x) = \pm f(y-1)$ and $f(x-1) = \pm f(y)$ when f is r-reflective or anti-r-reflective. □

Faulhaber almost certainly knew this lemma, because he published a table of $n^8, \nabla n^8, \ldots, \nabla^8 n^8$ in which the reflection phenomenon is clearly apparent [2, folio D.iii recto]. He stated that he had constructed "grosse Tafeln," but that this example should be "alles gnugsam vor Augen sehen und auf höhere quantiteten [exponents] continuiren könde."

The converse of Lemma 1 is also true, if we are careful. Let us define Σ as an inverse to the ∇ operator:

$$\Sigma f(n) = \begin{cases} C + f(1) + \cdots + f(n), & \text{if } n \geq 0; \\ C - f(0) - \cdots - f(n+1), & \text{if } n < 0. \end{cases}$$

Here C is an unspecified constant, which we will choose later; whatever its value, we have

$$\nabla \Sigma f(n) = \Sigma f(n) - \Sigma f(n-1) = f(n)$$

for all n.

Lemma 2. *If f is r-reflective, there is a unique C such that Σf is anti-$(r+1)$-reflective. If f is anti-r-reflective, then Σf is $(r+1)$-reflective for all C.*

Proof. If r is odd, Σf can be anti-$(r+1)$-reflective only if C is chosen so that we have $\Sigma f\big(-(r+1)/2\big) = 0$. If r is even, Σf can be anti-$(r+1)$-reflective only if $\Sigma f(-r/2) = -\Sigma f(-r/2-1) = -\big(\Sigma f(-r/2) - f(-r/2)\big)$, that is, only if $\Sigma f(-r/2) = \frac{1}{2} f(-r/2)$.

Once we have found x and y such that $x + y + r + 1 = 0$ and $\Sigma f(x) = -\Sigma f(y)$, it is easy to see that we will also have $\Sigma f(x-1) = -\Sigma f(y+1)$, if f is r-reflective, since $\Sigma f(x) - \Sigma f(x-1) = f(x) = f(y+1) = \Sigma f(y+1) - \Sigma f(y)$.

Suppose on the other hand that f is anti-r-reflective. If r is odd, clearly $\Sigma f(x) = \Sigma f(y)$ if $x = y = -(r+1)/2$. If r is even, then $f(-r/2) = 0$; so $\Sigma f(x) = \Sigma f(y)$ when $x = -r/2$ and $y = -r/2-1$. Once we have found x and y such that $x + y + r + 1 = 0$ and $\Sigma f(x) = \Sigma f(y)$, it is easy to verify as above that $\Sigma f(x-1) = \Sigma f(y+1)$. □

Lemma 3. *If f is any even function with $f(0) = 0$, the r-fold repeated sum $\Sigma^r f$ is r-reflective for all even r and anti-r-reflective for all odd r, if we choose the constant $C = 0$ in each summation. If f is any odd function, the r-fold repeated sum $\Sigma^r f$ is r-reflective for all odd r and anti-r-reflective for all even r, if we choose the constant $C = 0$ in each summation.*

Proof. Note that $f(0) = 0$ if f is odd. If $f(0) = 0$ and if we always choose $C = 0$, it is easy to verify by induction on r that $\Sigma^r f(x) = 0$ for $-r \le x \le 0$. Therefore the choice $C = 0$ always agrees with the unique choice stipulated in the proof of Lemma 2, whenever a specific value of C is necessary in that lemma.   □

When m is a positive integer, the function $f(x) = x^m$ obviously satisfies the condition of Lemma 3. Therefore we have proved that each function $\Sigma^r n^m$ is either r-reflective or anti-r-reflective, for all $r > 0$ and $m > 0$. And Faulhaber presumably knew this too. His theorem can now be proved if we supply one small additional fact, specializing from arbitrary functions to polynomials:

Lemma 4. *A polynomial $f(x)$ is r-reflective if and only if it can be written as a polynomial in $x(x + r)$; it is anti-r-reflective if and only if it can be written as $(x + r/2)$ times a polynomial in $x(x + r)$.*

Proof. The second statement follows from the first, because we have already observed that an anti-r-reflective function must have $f(-r/2) = 0$ and because the function $x + r/2$ is obviously anti-r-reflective. Furthermore, any polynomial in $x(x + r)$ is r-reflective, because $x(x + r) = y(y + r)$ when $x + y + r = 0$. Conversely, if $f(x)$ is r-reflective we have $f(x - r/2) = f(-x - r/2)$, so $g(x) = f(x - r/2)$ is an even function of x; hence $g(x) = h(x^2)$ for some polynomial h. Then $f(x) = g(x + r/2) = h\big(x(x + r) + r^2/4\big)$ is a polynomial in $x(x + r)$.   □

Theorem (Faulhaber). *There exist polynomials $g_{r,m}$ for all positive integers r and m such that*

$$\Sigma^r n^{2m-1} = g_{r,2m+1}\big(n(n+r)\big)\Sigma^r n^1, \qquad \Sigma^r n^{2m} = g_{r,2m}\big(n(n+r)\big)\Sigma^r n^2.$$

Proof. Lemma 3 tells us that $\Sigma^r n^m$ is r-reflective if $m + r$ is even and anti-r-reflective if $m + r$ is odd.

Note that $\Sigma^r n^1 = \binom{n+r}{r+1}$. Therefore a polynomial in n is a multiple of $\Sigma^r n^1$ if and only if it vanishes at $-r, \dots, -1, 0$. We have shown in the proof of Lemma 3 that $\Sigma^r n^m$ has this property for all m; therefore

$\Sigma^r n^m / \Sigma^r n^1$ is an r-reflective polynomial when m is odd, an anti-r-reflective polynomial when m is even. In the former case, we are done, by Lemma 4. In the latter case, Lemma 4 establishes the existence of a polynomial g such that $\Sigma^r n^m / \Sigma^r n^1 = (n + r/2)g\big(n(n+r)\big)$. Again, we are done, because the identity

$$\Sigma^r n^2 = \frac{2n + r}{r + 2} \Sigma^r n^1$$

is readily verified.   □

3. A Plausible Derivation

Faulhaber probably didn't think about r-reflective and anti-r-reflective functions in exactly the way we have described them, but his book [2] certainly indicates that he was quite familiar with the territory encompassed by that theory.

In fact, he could have found his formulas for power sums without knowing the theory in detail. A simple approach, illustrated here for Σn^{13}, would suffice: Suppose

$$14\Sigma n^{13} = n^7(n + 1)^7 - S(n),$$

where $S(n)$ is a 1-reflective function to be determined. Then

$$14n^{13} = n^7(n + 1)^7 - (n - 1)^7 n^7 - \nabla S(n)$$

$$= 14n^{13} + 70n^{11} + 42n^9 + 2n^7 - \nabla S(n),$$

and we have

$$S(n) = 70\Sigma n^{11} + 42\Sigma n^9 + 2\Sigma n^7.$$

In other words

$$\Sigma n^{13} = \frac{64}{7}N^7 - 5\Sigma n^{11} - 3\Sigma n^9 - \frac{1}{7}\Sigma n^7$$

when $N = n(n+1)/2$, and we can complete the calculation by subtracting multiples of previously computed results.

The great advantage of using polynomials in N rather than n is that the new formulas are considerably shorter. The method Faulhaber and others had used before making this discovery was most likely equivalent to the laborious calculation

$$\Sigma n^{13} = \tfrac{1}{14}n^{14} + \tfrac{13}{2}\Sigma n^{12} - 26\Sigma n^{11} + \tfrac{143}{2}\Sigma n^{10} - 143\Sigma n^9$$

$$+ \tfrac{429}{2}\Sigma n^8 - \tfrac{1716}{7}\Sigma n^7 + \tfrac{429}{2}\Sigma n^6 - 143\Sigma n^5$$

$$+ \tfrac{143}{2}\Sigma n^4 - 26\Sigma n^3 + \tfrac{13}{2}\Sigma n^2 - \Sigma n^1 + \tfrac{1}{14}n;$$

the coefficients of Σn^{12}, Σn^{11}, ... here are $\tfrac{1}{14}\binom{14}{12}$, $-\tfrac{1}{14}\binom{14}{11}$, ..., $\tfrac{1}{14}\binom{14}{0}$.

To handle sums of even exponents, Faulhaber knew that

$$\Sigma n^{2m} = \frac{n + 1/2}{2m + 1}(a_1 N + a_2 N^2 + \cdots + a_m N^m)$$

holds if and only if

$$\Sigma n^{2m+1} = \frac{a_1}{2}N^2 + \frac{a_2}{3}N^3 + \cdots + \frac{a_m}{m+1}N^{m+1}.$$

Therefore he could get two sums for the price of one [2, folios C.iv verso and D.i recto]. It is not difficult to prove this relation by establishing an isomorphism between the calculations of Σn^{2m+1} and the calculations of the quantities $S_{2m} = \big((2m + 1)\Sigma n^{2m}\big)/(n + 1/2)$; for example, the recurrence for Σn^{13} above corresponds to the formula

$$S_{12} = 64N^6 - 5S_{10} - 3S_8 - \tfrac{1}{7}S_6,$$

which can be derived in essentially the same way. Since the recurrences are essentially identical, we obtain a correct formula for Σn^{2m+1} from the formula for S_{2m} if we replace N^k by $N^{k+1}/(k + 1)$.

4. Faulhaber's Cryptomath

Mathematicians of Faulhaber's day tended to conceal their methods and hide results in secret code. Faulhaber ends his book [2] with a curious exercise of this kind, evidently intended to prove to posterity that he had in fact computed the formulas for sums of powers as far as Σn^{25} although he published the results only up to Σn^{17}.

His puzzle can be translated into modern notation as follows: Let

$$\Sigma^9 n^8 = \frac{a_{17}n^{17} + \cdots + a_2 n^2 + a_1 n}{d},$$

where the a's are integers having no common factor and where $d = a_{17} + \cdots + a_2 + a_1$. Let

$$\Sigma n^{25} = \frac{A_{26}n^{26} + \cdots + A_2 n^2 + A_1 n}{D}$$

be the analogous formula for Σn^{25}. Also let

$$\Sigma n^{22} = \frac{(b_{10}N^{10} - b_9 N^9 + \cdots + b_0)}{b_{10} - b_9 + \cdots + b_0}\Sigma n^2,$$

$$\Sigma n^{23} = \frac{(c_{10}N^{10} - c_9 N^9 + \cdots + c_0)}{c_{10} - c_9 + \cdots + c_0}\Sigma n^3,$$

$$\Sigma n^{24} = \frac{(d_{11}N^{11} - d_{10}N^{10} + \cdots - d_0)}{d_{11} - d_{10} + \cdots - d_0}\Sigma n^2,$$

$$\Sigma n^{25} = \frac{(e_{11}N^{11} - e_{10}N^{10} + \cdots - e_0)}{e_{11} - e_{10} + \cdots - e_0}\Sigma n^3,$$

where the integers b_k, c_k, d_k, e_k are as small as possible so that b_k, c_k, d_k, and e_k are multiples of 2^k. (He wants them to be multiples of 2^k so that $b_k N^k$, $c_k N^k$, $d_k N^k$, $e_k N^k$ are polynomials in n with integer coefficients; that is why he wrote, for example, $\Sigma n^7 = (12N^2 - 8N + 2)N^2/6$ instead of $(6N^2 - 4N + 1)N^2/3$. See [2, folio D.i verso].) Then compute

$$x_1 = (c_3 - a_{12})/7924252\,;$$

$$x_2 = (b_5 + a_{10})/112499648\,;$$

$$x_3 = (a_{11} - b_9 - c_1)/2945002\,;$$

$$x_4 = (a_{14} + c_7)/120964\,;$$

$$x_5 = (A_{26}a_{11} - D + a_{13} + d_{11} + e_{11})/199444\,.$$

These values $(x_1, x_2, x_3, x_4, x_5)$ specify the five letters of a "hochgerühmte Nam," if we use five designated alphabets [2, folio F.i recto].

It is doubtful whether anybody solved this puzzle during the first 360 years after its publication, but the task is relatively easy with modern computers. We have

$$a_{10} = 532797408\,, \quad a_{11} = 104421616\,, \quad a_{12} = 14869764\,,$$
$$a_{13} = 1526532\,, \quad a_{14} = 110160\,;$$

$$b_5 = 29700832\,, \quad b_9 = 140800\,;$$

$$c_1 = 205083120\,, \quad c_3 = 344752128\,, \quad c_7 = 9236480\,;$$

$$d_{11} = 559104\,; \quad e_{11} = 86016\,; \quad A_{26} = 42\,; \quad D = 1092\,.$$

The fact that $x_2 = (29700832 + 532797408)/112499648 = 5$ is an integer is reassuring: We must be on the right track! But alas, the other values are not integral.

A bit of experimentation soon reveals that we do obtain good results if we divide all the c_k by 4. Then, for example, $x_1 = (344752128/4 - 14869764)/7924252 = 9$, and we also find $x_3 = 18$, $x_4 = 20$. It appears that Faulhaber calculated $\Sigma^9 n^8$ and Σn^{22} correctly, and that he also had a correct expression for Σn^{23} as a polynomial in N; but he probably never went on to express Σn^{23} as a polynomial in n, because he would then have multiplied his coefficients by 4 in order to compute $c_6 N^6$ with integer coefficients.

The values of (x_1, x_2, x_3, x_4) correspond to the letters I E S U, so the concealed name in Faulhaber's riddle is undoubtedly I E S U S (Jesus).

But his formula for x_5 does not check out at all; it is way out of range and not an integer. This is the only formula that relates to Σn^{24} and Σn^{25}, and it involves only the simplest elements of those sums — the leading coefficients A_{26}, D, d_{11}, e_{11}. Therefore we have no evidence that Faulhaber's calculations beyond Σn^{23} were reliable. It is tempting to imagine that he meant to say '$A_{26}a_{11}/D$' instead of '$A_{26}a_{11} - D$' in his formula for x_5, but even then major corrections are needed to the other terms and it is unclear what he intended.

5. All-Integer Formulas

Faulhaber's theorem allows us to express the power sum Σn^m in terms of about $\frac{1}{2}m$ coefficients. The elementary theory above also suggests another approach that produces a similar effect: We can write, for example,

$$n = \binom{n}{1};$$

$$n^3 = 6\binom{n+1}{3} + \binom{n}{1};$$

$$n^5 = 120\binom{n+2}{5} + 30\binom{n+1}{3} + \binom{n}{1}.$$

(It is easy to see that any odd function $g(n)$ of the integer n can be expressed uniquely as a linear combination

$$g(n) = a_1\binom{n}{1} + a_3\binom{n+1}{3} + a_5\binom{n+2}{5} + \cdots$$

of the odd functions $\binom{n}{1}$, $\binom{n+1}{3}$, $\binom{n+2}{5}$, $\ldots$, because we can determine the coefficients $a_1, a_3, a_5, \ldots$ successively by plugging in the values $n = 1, 2, 3, \ldots$. The coefficients a_k will be integers if and only if $g(n)$ is an integer for all n.) Once $g(n)$ has been expressed in this way, we clearly have

$$\Sigma g(n) = a_1\binom{n+1}{2} + a_3\binom{n+2}{4} + a_5\binom{n+3}{6} + \cdots.$$

This approach therefore yields the following identities for sums of odd powers:

$$\Sigma n^1 = \binom{n+1}{2};$$

$$\Sigma n^3 = 6\binom{n+2}{4} + \binom{n+1}{2};$$

$$\Sigma n^5 = 120\binom{n+3}{6} + 30\binom{n+2}{4} + \binom{n+1}{2};$$

$$\Sigma n^7 = 5040\binom{n+4}{8} + 1680\binom{n+3}{6} + 126\binom{n+2}{4} + \binom{n+1}{2};$$

$$\Sigma n^9 = 362880\binom{n+5}{10} + 151200\binom{n+4}{8} + 17640\binom{n+3}{6}$$
$$+ 510\binom{n+2}{4} + \binom{n+1}{2};$$

$$\Sigma n^{11} = 39916800\binom{n+6}{12} + 19958400\binom{n+5}{10} + 3160080\binom{n+4}{8}$$
$$+ 168960\binom{n+3}{6} + 2046\binom{n+2}{4} + \binom{n+1}{2};$$
$$\Sigma n^{13} = 6227020800\binom{n+7}{14} + 3632428800\binom{n+6}{12} + 726485760\binom{n+5}{10}$$
$$+ 57657600\binom{n+4}{8} + 1561560\binom{n+3}{6} + 8190\binom{n+2}{4} + \binom{n+1}{2}.$$

And repeated sums are equally easy; we have

$$\Sigma^r n^1 = \binom{n+r}{1+r}, \qquad \Sigma^r n^3 = 6\binom{n+1+r}{3+r} + \binom{n+r}{1+r}, \qquad \text{etc.}$$

The coefficients in these formulas are related to what Riordan [7, page 213] has called *central factorial numbers* of the second kind. In his notation

$$x^m = \sum_{k=1}^{m} T(m,k)x^{[k]}, \quad x^{[k]} = x\left(x+\tfrac{k}{2}-1\right)\left(x+\tfrac{k}{2}-2\right) \cdots \left(x-\tfrac{k}{2}+1\right),$$

when $m > 0$, and $T(m,k) = 0$ when $m - k$ is odd; hence

$$n^{2m-1} = \sum_{k=1}^{m} (2k-1)!\, T(2m,2k)\binom{n+k-1}{2k-1},$$

$$\Sigma n^{2m-1} = \sum_{k=1}^{m} (2k-1)!\, T(2m,2k)\binom{n+k}{2k}.$$

The coefficients $T(2m,2k)$ are always integers, because the identity $x^{[k+2]} = x^{[k]}(x^2 - k^2/4)$ implies the recurrence

$$T(2m+2,2k) = k^2 T(2m,2k) + T(2m,2k-2).$$

The generating function for these numbers turns out to be

$$\cosh\left(2x\sinh(y/2)\right) = \sum_{m=0}^{\infty}\left(\sum_{k=0}^{m} T(2m,2k)x^{2k}\right)\frac{y^{2m}}{(2m)!}.$$

Notice that the power-sum formulas obtained in this way are more "efficient" than the well-known formulas based on Stirling numbers (see [5, (6.12)]):

$$\Sigma n^m = \sum_{k} k!\begin{Bmatrix}m\\k\end{Bmatrix}\binom{n+1}{k+1} = \sum_{k} k!\begin{Bmatrix}m\\k\end{Bmatrix}(-1)^{m-k}\binom{n+k}{k+1}.$$

The latter formulas give, for example,

$$\Sigma n^7 = 5040\binom{n+1}{8} + 15120\binom{n+1}{7} + 16800\binom{n+1}{6} + 8400\binom{n+1}{5}$$
$$+ 1806\binom{n+1}{4} + 126\binom{n+1}{3} + \binom{n+1}{2}$$
$$= 5040\binom{n+7}{8} - 15120\binom{n+6}{7} + 16800\binom{n+5}{6} - 8400\binom{n+4}{5}$$
$$+ 1806\binom{n+3}{4} - 126\binom{n+2}{3} + \binom{n+1}{2}.$$

There are about twice as many terms, and the coefficients are larger. (The Faulhaberian expression $\Sigma n^7 = (6N^4 - 4N^3 + N^2)/3$ is, of course, better yet.)

Similar formulas for even powers can be obtained as follows. We have

$$n^2 = n\binom{n}{1} \qquad\qquad\qquad = U_1(n),$$
$$n^4 = 6n\binom{n+1}{3} + n\binom{n}{1} \qquad\qquad = 12U_2(n) + U_1(n),$$
$$n^6 = 120n\binom{n+2}{5} + 30n\binom{n+1}{3} + n\binom{n}{1} = 360U_3(n) + 60U_2(n) + U_1(n),$$

and so on, where

$$U_k(n) = \frac{n}{k}\binom{n+k-1}{2k-1} = \binom{n+k}{2k} + \binom{n+k-1}{2k}.$$

Hence

$$\Sigma n^2 = T_1(n),$$
$$\Sigma n^4 = 12T_2(n) + T_1(n),$$
$$\Sigma n^6 = 360T_3(n) + 60T_2(n) + T_1(n),$$
$$\Sigma n^8 = 20160T_4(n) + 5040T_3(n) + 252T_2(n) + T_1(n),$$
$$\Sigma n^{10} = 1814400T_5(n) + 604800T_4(n) + 52920T_3(n)$$
$$+ 1020T_2(n) + T_1(n),$$
$$\Sigma n^{12} = 239500800T_6(n) + 99792000T_5(n) + 12640320T_4(n)$$
$$+ 506880T_3(n) + 4092T_2(n) + T_1(n),$$

et cetera, where

$$T_k(n) = \binom{n+k+1}{2k+1} + \binom{n+k}{2k+1} = \frac{2n+1}{2k+1}\binom{n+k}{2k}.$$

Curiously, we have found a relation here between Σn^{2m} and Σn^{2m-1}, somewhat analogous to Faulhaber's relation between Σn^{2m} and Σn^{2m+1}: The formula

$$\frac{\Sigma n^{2m}}{2n+1} = a_1 \binom{n+1}{2} + a_2 \binom{n+2}{4} + \cdots + a_m \binom{n+m}{2m}$$

holds for all n if and only if

$$\Sigma n^{2m-1} = \frac{3}{1} a_1 \binom{n+1}{2} + \frac{5}{2} a_2 \binom{n+2}{4} + \cdots + \frac{2m+1}{m} a_m \binom{n+m}{2m}.$$

6. Reflective Decomposition

The forms of the expressions in the previous section lead naturally to useful representations of arbitrary functions $f(n)$ defined on the integers. It is easy to see that any $f(n)$ can be written uniquely in the form

$$f(n) = \sum_{k \geq 0} a_k \binom{n + \lfloor k/2 \rfloor}{k},$$

for some coefficients a_k; indeed, we have

$$a_k = \nabla^k f(\lfloor k/2 \rfloor).$$

(Thus $a_0 = f(0)$, $a_1 = f(0) - f(-1)$, $a_2 = f(1) - 2f(0) + f(-1)$, etc.) The a_k are integers if and only if $f(n)$ is always an integer. The a_k are eventually zero if and only if f is a polynomial. The a_{2k} are all zero if and only if f is odd. The a_{2k+1} are all zero if and only if f is 1-reflective.

Similarly, there is a unique expansion

$$f(n) = b_0 T_0(n) + b_1 U_1(n) + b_2 T_1(n) + b_3 U_2(n) + b_4 T_2(n) + \cdots,$$

in which the b_k are integers if and only if $f(n)$ is always an integer. The b_{2k} are all zero if and only if f is even and $f(0) = 0$. The b_{2k+1} are all zero if and only if f is anti-1-reflective. Using the recurrence relations

$$\nabla T_k(n) = U_k(n), \qquad \nabla U_k(n) = T_{k-1}(n-1),$$

we find

$$a_k = \nabla^k f(\lfloor k/2 \rfloor) = 2b_{k-1} + (-1)^k b_k$$

and therefore

$$b_k = \sum_{j=0}^{k} (-1)^{\lceil j/2 \rceil + \lfloor k/2 \rfloor} 2^{k-j} a_j.$$

In particular, when $f(n) = 1$ for all n, we have $b_k = (-1)^{\lfloor k/2 \rfloor} 2^k$. The infinite series is finite for each n.

Theorem. *If f is any function defined on the integers, and if r and s are arbitrary integers, we can always express f in the form*

$$f(n) = g(n) + h(n)$$

where $g(n)$ is r-reflective and $h(n)$ is anti-s-reflective. This representation is unique, except when r is even and s is odd; in the latter case the representation is unique if we specify the value of g or h at any point.

Proof. It suffices to consider $0 \le r, s \le 1$, because $f(x)$ is (anti)-r-reflective if and only if $f(x + a)$ is (anti)-$(r + 2a)$-reflective.

When $r = s = 0$, the result is just the well known decomposition of a function into even and odd parts,

$$g(n) = \tfrac{1}{2}\big(f(n) + f(-n)\big), \qquad h(n) = \tfrac{1}{2}\big(f(n) - f(-n)\big).$$

When $r = s = 1$, we have similarly

$$g(n) = \tfrac{1}{2}\big(f(n) + f(-1 - n)\big), \qquad h(n) = \tfrac{1}{2}\big(f(n) - f(-1 - n)\big).$$

When $r = 1$ and $s = 0$, it is easy to deduce that $h(0) = 0$, $g(0) = f(0)$, $h(1) = f(0) - f(-1)$, $g(1) = f(1) - f(0) + f(-1)$, $h(2) = f(1) - f(0) + f(-1) - f(-2)$, $g(2) = f(2) - f(1) + f(0) - f(-1) + f(-2)$, and so on.

And when $r = 0$ and $s = 1$, the general solution is $g(0) = f(0) - C$, $h(0) = C$, $g(1) = f(-1) + C$, $h(1) = f(1) - f(-1) - C$, $g(2) = f(1) - f(-1) + f(-2) - C$, $h(2) = f(2) - f(1) + f(-1) - f(-2) + C$, and so forth. □

When $f(n) = \sum_{k \ge 0} a_k \binom{n + \lfloor k/2 \rfloor}{k}$, the case $r = 1$ and $s = 0$ corresponds to the decomposition

$$g(n) = \sum_{k=0}^{\infty} a_{2k} \binom{n + k}{2k}, \qquad h(n) = \sum_{k=0}^{\infty} a_{2k+1} \binom{n + k}{2k + 1}.$$

The representation $f(n) = \sum_{k \ge 0} b_{2k} T_k(n) + \sum_{k \ge 0} b_{2k+1} U_{k+1}(n)$ corresponds similarly to the case $r = 0$, $s = 1$, $C = f(0)$.

7. Back to Faulhaber's Form

Let us now return to representations of Σn^m as polynomials in $n(n+1)$. Setting $u = 2N = n^2 + n$, we have

$$\Sigma n = \tfrac{1}{2}u \qquad\qquad = \tfrac{1}{2}A_0^{(1)}u\,;$$

$$\Sigma n^3 = \tfrac{1}{4}u^2 \qquad\qquad = \tfrac{1}{4}\big(A_0^{(2)}u^2 + A_1^{(2)}u\big)\,;$$

$$\Sigma n^5 = \tfrac{1}{6}\big(u^3 - \tfrac{1}{2}u^2\big) \qquad = \tfrac{1}{6}\big(A_0^{(3)}u^3 + A_1^{(3)}u^2 + A_2^{(3)}u\big)\,;$$

$$\Sigma n^7 = \tfrac{1}{8}\big(u^4 - \tfrac{4}{3}u^3 + \tfrac{2}{3}u^2\big) = \tfrac{1}{8}\big(A_0^{(4)}u^4 + A_1^{(4)}u^3 + A_2^{(4)}u^2 + A_3^{(4)}u\big)\,;$$

and so on, for certain coefficients $A_k^{(m)}$.

Faulhaber never discovered the Bernoulli numbers; that is, he never realized that a single sequence of constants $B_0, B_1, B_2, \ldots$ would provide a uniform formula

$$\Sigma n^m = \tfrac{1}{m+1}\Big(B_0 n^{m+1} - \tbinom{m+1}{1} B_1 n^m + \tbinom{m+1}{2} B_2 n^{m-1} - \cdots$$
$$+ (-1)^m \tbinom{m+1}{m} B_m n\Big)$$

for all sums of powers. He never mentioned, for example, the fact that almost half of the coefficients turned out to be zero after he had converted his formulas for Σn^m from polynomials in N to polynomials in n. (He did notice that the coefficient of n was zero when $m > 1$ was odd.)

However, we know now that Bernoulli numbers exist, and we know that $B_3 = B_5 = B_7 = \cdots = 0$. This is a strong condition. Indeed, it completely defines the constants $A_k^{(m)}$ in the Faulhaber polynomials above, given that $A_0^{(m)} = 1$.

For example, let's consider the case $m = 4$, namely the formula for Σn^7: We need to find coefficients $a = A_1^{(4)}$, $b = A_2^{(4)}$, $c = A_3^{(4)}$ such that the polynomial

$$n^4(n+1)^4 + an^3(n+1)^3 + bn^2(n+1)^2 + cn(n+1)$$

has vanishing coefficients of n^5, n^3, and n. The polynomial is

$$
\begin{aligned}
n^8 + 4n^7 &+ 6n^6 + 4n^5 + n^4 \\
&+ an^6 + 3an^5 + 3an^4 + an^3 \\
&\quad\;\; + bn^4 + 2bn^3 + bn^2 \\
&\qquad\qquad\qquad + cn^2 + cn\,;
\end{aligned}
$$

so we must have $3a + 4 = 2b + a = c = 0$. In general the coefficient of, say, n^{2m-5} in the polynomial for $2m\Sigma n^{2m-1}$ is easily seen to be

$$\tbinom{m}{5} A_0^{(m)} + \tbinom{m-1}{3} A_1^{(m)} + \tbinom{m-2}{1} A_2^{(m)}\,.$$

Thus the Faulhaber coefficients can be defined by the rules

$$A_0^{(w)} = 1;\qquad \sum_{j=0}^{k} \binom{w-j}{2k+1-2j} A_j^{(w)} = 0,\quad k > 0.\qquad (*)$$

(The upper parameter will often be called w instead of m, in the following discussion, because we will want to generalize to noninteger values.)

Notice that (∗) defines the coefficients for each exponent w without reference to other exponents; for every integer $k \geq 0$, the quantity $A_k^{(w)}$ is a certain rational function of w. For example, we have

$$-A_1^{(w)} = w(w-2)/6,$$

$$A_2^{(w)} = w(w-1)(w-3)(7w-8)/360,$$

$$-A_3^{(w)} = w(w-1)(w-2)(w-4)(31w^2 - 89w + 48)/15120,$$

$$A_4^{(w)} = w(w-1)(w-2)(w-3)(w-5)$$
$$\times (127w^3 - 691w^2 + 1038w - 384)/6048000,$$

and in general $A_k^{(w)}$ is $w^{\underline{k}} = w(w-1) \ldots (w-k+1)$ times a polynomial of degree k, with leading coefficient equal to $(2-2^{2k})B_{2k}/(2k)!$; if $k > 0$, that polynomial vanishes when $w = k+1$.

Jacobi mentioned these coefficients $A_k^{(m)}$ in his paper [6], and tabulated them for $m \leq 6$, although he did not consider the recurrence (∗). He observed that the derivative of Σn^m with respect to n is $m \Sigma n^{m-1} + B_m$; this follows because power sums can be expressed in terms of Bernoulli polynomials,

$$\Sigma n^m = \tfrac{1}{m+1}\big(B_{m+1}(n+1) - B_{m+1}(0)\big),$$

and because $B_m'(x) = m B_{m-1}(x)$. Thus Jacobi obtained a new proof of Faulhaber's formulas for even exponents:

$$\Sigma n^2 = \tfrac{1}{3}\big(\tfrac{2}{4}A_0^{(2)}u + \tfrac{1}{4}A_1^{(2)}\big)(2n+1),$$

$$\Sigma n^4 = \tfrac{1}{5}\big(\tfrac{3}{6}A_0^{(3)}u^2 + \tfrac{2}{6}A_1^{(3)}u + \tfrac{1}{6}A_2^{(3)}\big)(2n+1),$$

$$\Sigma n^6 = \tfrac{1}{7}\big(\tfrac{4}{8}A_0^{(4)}u^3 + \tfrac{3}{8}A_1^{(4)}u^2 + \tfrac{2}{8}A_2^{(4)}u + \tfrac{1}{8}A_3^{(4)}\big)(2n+1),$$

etc. (The constant terms are zero, but they are shown explicitly here so that the pattern is plain.) Differentiating again gives, for example,

$$\Sigma n^5 = \frac{1}{6 \cdot 7 \cdot 8}\big((4 \cdot 3\, A_0^{(4)}u^2 + 3 \cdot 2\, A_1^{(4)}u + 2 \cdot 1\, A_2^{(4)})(2n+1)^2$$
$$+ 2(4A_0^{(4)}u^3 + 3A_1^{(4)}u^2 + 2A_2^{(4)}u + 1A_3^{(4)})\big) - \tfrac{1}{6}B_6$$

$$= \frac{1}{6 \cdot 7 \cdot 8}\big(8 \cdot 7\, A_0^{(4)}u^3 + (6 \cdot 5\, A_1^{(4)} + 4 \cdot 3\, A_0^{(4)})u^2$$
$$+ (4 \cdot 3\, A_2^{(4)} + 3 \cdot 2\, A_1^{(4)})u$$
$$+ (2 \cdot 1\, A_3^{(4)} + 2 \cdot 1\, A_2^{(4)})\big) - \tfrac{1}{6}B_6.$$

This approach yields Jacobi's recurrence

$$(2w - 2k)(2w - 2k - 1)A_k^{(w)} + (w - k + 1)(w - k)A_{k-1}^{(w)}$$
$$= 2w(2w - 1)A_k^{(w-1)}, \qquad (**)$$

which is valid for all integers $w > k + 1$ so it must be valid for all w. Our derivation of $(**)$ also allows us to conclude that

$$A_{m-2}^{(m)} = \binom{2m}{2} B_{2m-2}, \quad m \geq 2,$$

by considering the constant term of the second derivative of Σn^{2m-1}.

Recurrence $(*)$ does not define $A_m^{(m)}$, except as the limit of $A_m^{(w)}$ when $w \to m$. But we can compute this value by setting $w = m + 1$ and $k = m$ in $(**)$, which reduces to

$$2A_{m-1}^{(m+1)} = (2m + 2)(2m + 1)A_m^{(m)}$$

because $A_m^{(m+1)} = 0$. Thus

$$A_m^{(m)} = B_{2m}, \quad \text{integer } m \geq 0.$$

8. Solution to the Recurrence

An explicit formula for $A_k^{(m)}$ can be found as follows: We have

$$\Sigma n^{2m-1} = \frac{1}{2m}\left(B_{2m}(n + 1) - B_{2m}\right) = \frac{1}{2m}(A_0^{(m)}u^m + \cdots + A_{m-1}^{(m)}u),$$

and $n + 1 = (\sqrt{1 + 4u} + 1)/2$; hence, using the known values of $A_m^{(m)}$, we obtain

$$\sum_{k=0}^{\infty} A_k^{(m)} u^{m-k} = B_{2m}\left(\frac{\sqrt{1 + 4u} + 1}{2}\right) = B_{2m}\left(\frac{1 - \sqrt{1 + 4u}}{2}\right),$$

a closed form in terms of Bernoulli polynomials. (We have used the fact that $A_{m+1}^{(m)} = A_{m+2}^{(m)} = \cdots = 0$, together with the identity $B_n(x + 1) = (-1)^n B_n(-x)$.) Expanding the right side in powers of u gives

$$\sum_l \binom{2m}{l}\left(\frac{1 - \sqrt{1 + 4u}}{2}\right)^l B_{2m-l}$$

$$= \sum_{j,l} \binom{2m}{l}\binom{2j + l}{j}\frac{l}{2j + l}(-u)^{j+l}B_{2m-l},$$

using equation (5.70) of [5]. Setting $j + l = m - k$ finally yields

$$A_k^{(m)} = (-1)^{m-k} \sum_j \binom{2m}{m-k-j} \binom{m-k+j}{j} \frac{m-k-j}{m-k+j} B_{m+k+j},$$

for $0 \leq k < m$. This formula, which was first obtained by Gessel and Viennot [4], makes it easy to confirm that $A_{m-1}^{(m)} = 0$ and $A_{m-2}^{(m)} = \binom{2m}{2} B_{2m-2}$, and to derive additional values such as

$$A_{m-3}^{(m)} = -2 \binom{2m}{2} B_{2m-2} = -2 A_{m-2}^{(m)}, \quad m \geq 3;$$

$$A_{m-4}^{(m)} = \binom{2m}{4} B_{2m-4} + 5 \binom{2m}{2} B_{2m-2}, \quad m \geq 4.$$

The author's interest in Faulhaber polynomials was inspired by the work of Edwards [1], who resurrected Faulhaber's work after it had been long forgotten and undervalued by historians of mathematics. Ira Gessel responded to the same stimulus by submitting problem E3204 to the *Math Monthly* [3] regarding a bivariate generating function for Faulhaber's coefficients. Such a function is obtainable from the univariate generating function above, using the standard generating function for Bernoulli polynomials: Since

$$\sum B_{2m}\left(\frac{x+1}{2}\right) \frac{z^{2m}}{(2m)!} = \frac{1}{2} \sum B_m\left(\frac{x+1}{2}\right) \frac{z^m}{m!}$$

$$+ \frac{1}{2} \sum B_m\left(\frac{x+1}{2}\right) \frac{(-z)^m}{m!}$$

$$= \frac{z\, e^{(x+1)z/2}}{2(e^z - 1)} - \frac{z\, e^{-(x+1)z/2}}{2(e^{-z} - 1)} = \frac{z \cosh(xz/2)}{2 \sinh(z/2)},$$

we have

$$\sum_{k,m} A_k^{(m)} u^{m-k} \frac{z^{2m}}{(2m)!} = \sum_m B_{2m}\left(\frac{\sqrt{1+4u}+1}{2}\right) \frac{z^{2m}}{(2m)!}$$

$$= \frac{z}{2} \frac{\cosh\left(\sqrt{1+4u}\, z/2\right)}{\sinh(z/2)};$$

$$\sum_{k,m} A_k^{(m)} u^k \frac{z^{2m}}{(2m)!} = \frac{z\sqrt{u}\, \cosh(\sqrt{u+4}\, z/2)}{2 \sinh(z\sqrt{u}/2)}.$$

The numbers $A_k^{(m)}$ are obtainable by inverting a lower triangular matrix, as Edwards showed; indeed, recurrence $(*)$ defines such a matrix. Gessel and Viennot [4] observed that we can therefore express them in terms of a $k \times k$ determinant,

$$A_k^{(w)} = \frac{1}{(1-w) \ldots (k-w)} \begin{vmatrix} \binom{w-k+1}{3} & \binom{w-k+1}{1} & 0 & \cdots & 0 \\ \binom{w-k+2}{5} & \binom{w-k+2}{3} & \binom{w-k+2}{1} & \cdots & 0 \\ \vdots & \vdots & \vdots & & \vdots \\ \binom{w-1}{2k-1} & \binom{w-1}{2k-3} & \binom{w-1}{2k-5} & \cdots & \binom{w-1}{1} \\ \binom{w}{2k+1} & \binom{w}{2k-1} & \binom{w}{2k-3} & \cdots & \binom{w}{3} \end{vmatrix}.$$

When w and k are positive integers, Gessel and Viennot proved that this determinant is the number of sequences of positive integers $a_1 a_2 a_3 \ldots a_{3k}$ such that

$$a_{3j-2} < a_{3j-1} < a_{3j} \le w - k + j, \quad \text{for} \quad 1 \le j \le k;$$

$$a_{3j-2} < a_{3j+1}, \quad a_{3j-1} < a_{3j+3}, \quad \text{for} \quad 1 \le j < k.$$

In other words, it is the number of ways to put positive integers into a k-rowed triple staircase such as

with all rows and all columns strictly increasing from left to right and from top to bottom, and with all entries in row j at most $w - k + j$. This theorem provides a surprising combinatorial interpretation of the Bernoulli number B_{2m} when $w = m + 1$ and $k = m - 1$ (in which case the top row of the staircase is forced to contain $1, 2, 3$).

The combinatorial interpretation proves in particular that

$$(-1)^k A_k^{(m)} \ge 0 \qquad \text{for all } k \ge 0.$$

Faulhaber stated this, but he may not have known how to prove it.

Denoting the determinant by $D(w, k)$, Jacobi's recurrence $(**)$ implies that we have

$$(w - k)^2 (w - k + 1)(w - k - 1)D(w, k - 1)$$
$$= (2w - 2k)(2w - 2k - 1)(w - k - 1)D(w, k)$$
$$- 2w(2w - 1)(w - 1)D(w - 1, k);$$

this formula can also be written in a slightly tidier form, using a special case of the "integer basis" polynomials discussed above:

$$U_2(w-k)D(w,k-1) = T_1(w-k-1)D(w,k) - T_1(w-1)D(w-1,k).$$

It does not appear obvious that the determinant satisfies such a recurrence, nor that the solution to the recurrence should have integer values when w and k are integers. But, identities are not always obvious.

9. Generalization to Noninteger Powers

Recurrence $(*)$ does not require w to be a positive integer, and we can in fact solve it in closed form when $w = 3/2$:

$$\sum_{k\geq 0} A_k^{(3/2)} u^{3/2-k} = B_3\left(\frac{\sqrt{1+4u}+1}{2}\right)$$

$$= \frac{u}{2}\sqrt{1+4u} = u^{3/2}\sum_{k\geq 0}\binom{1/2}{k}(4u)^{-k}.$$

Therefore $A_k^{(3/2)} = \binom{1/2}{k}4^{-k}$ is related to the kth Catalan number. A similar closed form exists for the coefficient $A_k^{(m+1/2)}$ when m is any nonnegative integer.

For other cases of w, our generating function for $A_k^{(w)}$ involves $B_n(x)$ with noninteger subscripts. The Bernoulli polynomials can be generalized to a family of functions $B_z(x)$, for arbitrary z, in several ways; the best generalization for our present purposes seems to arise when we define

$$B_z(x) = x^z\sum_{k\geq 0}\binom{z}{k}x^{-k}B_k,$$

choosing a suitable branch of the function x^z. With this definition we can develop the right-hand side of

$$\sum_{k\geq 0}A_k^{(w)}u^{-k} = B_{2w}\left(\frac{\sqrt{1+4u}+1}{2}\right)u^{-w}$$

$$= \left(\frac{\sqrt{1+4u}+1}{2\sqrt{u}}\right)^{2w}\sum_{k\geq 0}\binom{2w}{k}\left(\frac{\sqrt{1+4u}+1}{2}\right)^{-k}B_k \quad (***)$$

as a power series in u^{-1}, letting $u \to \infty$.

The factor outside the $\sum$ sign is rather nice; we have

$$\left(\frac{\sqrt{1+4u}+1}{2\sqrt{u}}\right)^{2w} = \sum_{j\geq 0} \frac{w}{w+j/2}\binom{w+j/2}{j}u^{-j/2},$$

because the generalized binomial series $\mathcal{B}_{1/2}(u^{-1/2})$ [5, equation (5.58)] is the solution to

$$f(u)^{1/2} - f(u)^{-1/2} = u^{-1/2}, \qquad f(\infty) = 1,$$

namely

$$f(u) = \left(\frac{\sqrt{1+4u}+1}{2\sqrt{u}}\right)^2.$$

Similarly we find

$$\left(\frac{\sqrt{1+4u}+1}{2}\right)^{-k} = \sum_j \frac{-k}{j-k}\binom{j/2-k/2}{j}u^{-k/2-j/2}$$

$$= u^{-k/2} - \sum_{j\geq 1}\frac{k}{2j}\binom{j/2-k/2-1}{j-1}u^{-k/2-j/2}.$$

So we can indeed expand the right-hand side as a power series with coefficients that are polynomials in w. It is actually a power series in $u^{-1/2}$, not u; but since the coefficients of odd powers of $u^{-1/2}$ vanish when w is a positive integer, they must be identically zero. Sure enough, a check with computer algebra on formal power series yields $1 + A_1^{(w)}u^{-1} + A_2^{(w)}u^{-2} + A_3^{(w)}u^{-3} + O(u^{-4})$, where the values of $A_k^{(w)}$ for $k \leq 3$ agree perfectly with those obtained directly from (∗). Therefore this approach allows us to express $A_k^{(w)}$ as a polynomial in w, using ordinary Bernoulli number coefficients:

$$A_k^{(w)} = \sum_{l=0}^{2k} \frac{w}{w+l/2}\binom{w+l/2}{l}$$

$$\times \left(\binom{2w}{2k-l}B_{2k-l} - \frac{1}{2}\sum_{j=1}^{2k-l-1}\binom{2w}{j}\frac{j}{2k-l-j}\binom{k-l/2-j-1}{2k-l-j-1}B_j\right).$$

The power series (∗∗∗) we have used in this successful derivation is actually divergent for all u unless $2w$ is a nonnegative integer, because B_k grows superexponentially while the factor

$$\binom{2w}{k} = (-1)^k\binom{k-2w-1}{k} = \frac{(-1)^k\,\Gamma(k-2w)}{\Gamma(k+1)\,\Gamma(-2w)} \sim \frac{(-1)^k}{\Gamma(-2w)}k^{-2w-1}$$

does not decrease very rapidly as $k \to \infty$.

Still, (∗∗∗) is easily seen to be a valid asymptotic series as $u \to \infty$, because asymptotic series multiply like formal power series. This means that, for any positive integer p, we have

$$\sum_{k=0}^{2p} \binom{2w}{k} \left(\frac{\sqrt{1+4u}+1}{2} \right)^{2w-k} B_k = \sum_{k=0}^{p} A_k^{(w)} u^{w-k} + O(u^{w-p-1}).$$

We can now apply these results to obtain sums of noninteger powers, as asymptotic series of Faulhaber's type. Suppose, for example, that we are interested in the sum

$$H_n^{(1/3)} = \sum_{k=1}^{n} \frac{1}{k^{1/3}}.$$

Euler's summation formula [5, exercise 9.27] tells us that

$$H_n^{(1/3)} - \zeta(\tfrac{1}{3}) \sim \tfrac{3}{2} n^{2/3} + \tfrac{1}{2} n^{-1/3} - \tfrac{1}{36} n^{-4/3} + \tfrac{7}{4860} n^{-10/3} - \cdots$$

$$= \frac{3}{2} \left(\sum_{k \geq 0} \binom{2/3}{k} n^{2/3-k} B_k + n^{-1/3} \right),$$

where the parenthesized quantity is what we have called $B_{2/3}(n+1)$. And when $u = n^2 + n$ we have $B_{2/3}(n+1) = B_{2/3}\big((\sqrt{1+4u}+1)/2\big)$; hence

$$H_n^{(1/3)} - \zeta(\tfrac{1}{3}) \sim \tfrac{3}{2} \sum_{k \geq 0} A_k^{(1/3)} u^{1/3-k}$$

$$= \tfrac{3}{2} u^{1/3} + \tfrac{5}{36} u^{-2/3} - \tfrac{17}{1215} u^{-5/3} + \tfrac{77}{26244} u^{-8/3} - \cdots$$

as $n \to \infty$. (We can't claim that this series converges twice as fast as the usual one, because both series diverge! Nor do we get twice as much precision in a fixed number of terms, because half of the Bernoulli numbers are zero. But the new series gets off to a faster start.)

In general, the same argument establishes the asymptotic series

$$\sum_{k=1}^{n} k^{\alpha} - \zeta(-\alpha) \sim \frac{1}{\alpha+1} \sum_{k \geq 0} A_k^{((\alpha+1)/2)} u^{(\alpha+1)/2-k},$$

whenever $\alpha \neq -1$. The series on the right is finite when α is a positive odd integer; it is convergent (for sufficiently large n) if and only if α is a nonnegative integer.

The special case $\alpha = -2$ has historic interest, so it deserves a special look:

$$\sum_{k=1}^{n} \frac{1}{k^2} \sim \frac{\pi^2}{6} - A_0^{(-1/2)} u^{-1/2} - A_1^{(-1/2)} u^{-3/2} - \cdots$$

$$= \frac{\pi^2}{6} - u^{-1/2} + \frac{5}{24} u^{-3/2} - \frac{161}{1920} u^{-5/2}$$

$$+ \frac{401}{7168} u^{-7/2} - \frac{32021}{491520} u^{-9/2} + \cdots.$$

These coefficients do not seem to have a simple closed form; the prime factorization $32021 = 11 \cdot 41 \cdot 71$ is no doubt just a quirky coincidence.

Acknowledgments

This paper could not have been written without the help provided by several correspondents. Anthony Edwards kindly sent me a photocopy of Faulhaber's *Academia Algebræ*, a book that is evidently extremely rare: An extensive search of printed indexes and electronic indexes indicates that no copies have ever been recorded to exist in America, in the British Library, or the Bibliothèque Nationale. Edwards found it at Cambridge University Library, where the volume once owned by Jacobi now resides. (I have annotated the photocopy and deposited it in the Mathematical Sciences Library at Stanford, so that other interested scholars can take a look.) Ivo Schneider, who is currently preparing a book about Faulhaber and his work, helped me understand some of the archaic German phrases. Herb Wilf gave me a vital insight by discovering the first half of Lemma 4, in the case $r = 1$. And Ira Gessel pointed out that the coefficients in the expansion $n^{2m+1} = \sum a_k \binom{n+k}{2k+1}$ are central factorial numbers in slight disguise.

References

[1] A. W. F. Edwards, "A quick route to sums of powers," *American Mathematical Monthly* **93** (1986), 451–455.

[2] Johann Faulhaber, *Academia Algebræ*, Darinnen die miraculosische Inventiones zu den höchsten Cossen weiters *continuirt* und *profitiert* werden (Augspurg [sic]: Johann Ulrich Schönigs, 1631). Call number QA154.8 F3 1631a f MATH at Stanford University Libraries.

[3] Ira Gessel and University of South Alabama Problem Group, "A formula for power sums," *American Mathematical Monthly* **95** (1988), 961–962.

[4] Ira M. Gessel and Gérard Viennot, "Determinants, paths, and plane partitions," preprint (28 July 1989).

[5] Ronald L. Graham, Donald E. Knuth, and Oren Patashnik, *Concrete Mathematics* (Reading, Massachusetts: Addison–Wesley, 1989). [Some of the equations cited above were numbered incorrectly in the original printing of 1989.]

[6] C. G. J. Jacobi, "De usu legitimo formulae summatoriae Maclaurinianae," *Journal für die reine und angewandte Mathematik* **12** (1834), 263–272.

[7] John Riordan, *Combinatorial Identities* (New York: John Wiley & Sons, 1968).

[8] L. Tits, "Sur la sommation des puissances numériques," *Mathesis* **37** (1923), 353–355.

Addendum

Two biographies of Faulhaber were published shortly after this paper was written:

Ivo Schneider, *Johannes Faulhaber 1580–1635: Rechenmeister in einer Welt des Umbruchs* (Basel: Birkhäuser Verlag, 1993), xiv + 271 pages.

Kurt Hawlitschek, *Johann Faulhaber 1580–1635: Eine Blütezeit der mathematischen Wissenschaften in Ulm* (Ulm: Stadtbibliothek Ulm, 1995), 376 pages.

Chapter 5

Notes on Thomas Harriot

Thomas Harriot (1560–1621) was a truly remarkable man. After graduating from Oxford in 1580 he was employed by Sir Walter Raleigh, on whose behalf he went to America in 1585 as part of an unsuccessful attempt at colonization. He returned to England a year later and wrote the famous book A Briefe and True Report of the New Found Land of Virginia. *He was interested in all the sciences, especially chemistry, metallurgy, and astronomy; he discovered sunspots, he made the first maps of the moon, and he investigated a great many other things. But he did not publish his discoveries, probably because free-thinkers in those days were likely to be imprisoned or killed as heretics. His pioneering work on algebra, in which among other things he invented signs for the relations "less than" and "greater than," became known only because some of his associates published portions of it in 1631, ten years after his death.*

Many interesting stories can be told about him — for example, that he introduced tobacco to Britain, and that inhaling tobacco eventually led to his death via cancer of the nose — but my interest in Harriot was sparked chiefly by reports that he had invented binary arithmetic. The following notes were originally published in Historia Mathematica **10** *(1983), 240–243, as part of a book review about the history of binary and other nondecimal notations.*

Thomas Harriot's manuscripts contain numerous things that deserve to be better known, as Jon Pepper remarked in 1967 [2]. It is true that they remained unpublished, and therefore they probably failed to influence any subsequent work. Yet I believe Leibniz was right when he said that "the art of making discoveries should be extended by considering noteworthy examples of it" [1, page 22].

In 1970 I had the chance to spend about 12 hours glancing through thousands of Harriot's manuscripts in the British Museum, looking for appearances of binary arithmetic. To my surprise, I stumbled across a variety of other interesting items as well. I hope that the following notes will whet the appetite of future researchers and encourage others to undertake a more careful search.

1. Harriot made an extensive investigation of permutations and combinations; but this purely combinatorial work is not connected with his explorations of binary arithmetic. For example, on folio 2.38r* he shows constructively that $\sum_k \binom{n}{k} = 2^n$ and says "The practise is playne. The theorem is to be demonstrated out of Boethius or Maurolicus." He tabulated $n!$ for $n \leq 25$ on folio 2.332r; and he played with interesting ways to arrange the sixteen 4-bit binary patterns on a wheel, with $\bar{x}$ opposite x (folio 6.378v). But in this general area of mathematics Harriot's researches fell far short of results known centuries earlier in India.

2. Binary arithmetic appears on folios 2.247r, 6.347r, 6.516v, and 8.244v. Examples of conversion to and from binary notation are on 6.243v and 6.346v. Harriot began to explore ternary notation on 6.516v, but only in a trivial way; there are simple instances of octal notation on 2.1r and 2.247r; and there is a table of powers of 2 in radix 60 on 6.305r. I found no other instances of nondecimal numeration. (I did not have time to view the other collection of Harriot manuscripts in Petworth House.)

3. Of all these folios, 8.244v was the most revealing about Harriot's process of discovery. The preceding and following pages are filled with laboratory measurements, like those on the top of 8.244v, where we find, for example,

Glasse & water	$11^{\text{lb}} \cdot 0 \cdot 0 \cdot \frac{1}{8} \cdot 0 \cdot + 28^{\text{gr}}$
Rounde measuring glasse weyeth	$3^{\text{lb}} \cdot \frac{1}{2} \cdot 0 \cdot \frac{1}{8} \cdot \frac{1}{16} \cdot + 21^{\text{gr}}$
Water	$7^{\text{lb}} \cdot 0 \cdot \frac{1}{4} \cdot \frac{1}{8} \cdot \frac{1}{16} \cdot + \ 7^{\text{gr}}$

Clearly he was using a balance scale with half-pound, quarter-pound, etc., weights; such a subtraction was undoubtedly a natural thing to do. Now comes the flash of insight: He realized that he was essentially doing a calculation with radix 2, and he abstracted the situation. At the bottom of that same folio, 8.244v, the binary numbers from 1 to 16 appear; then Harriot tried out the system by doing three multiplications. It is plausible to believe that his discovery was made at exactly this time. And since the preceding folio 8.243 is dated June 1605, while the next one (8.245) is dated July 1604, we can conjecture that the undated 8.244 was written at about the same time.

---

* In these notes I write '2.38r' for the recto side of folio 38 in the box "Additional 6782" from the British Museum collection. Similarly, '6.378v' stands for the verso side of folio 378 in "Additional 6786."

Of course it must be admitted that the Harriot folios are often out of order. He may also have written on the bottom of 8.244 some years after writing on the top. All kinds of hypotheses are possible: Perhaps he even visited Napier, who told him about a binary abacus, etc.! But I believe the most plausible explanation is that Harriot invented binary arithmetic one day in 1604 or 1605 while doing a chemical experiment.

4. On 2.11r he tabulated the Stirling numbers of the first kind. (Actually he computed the polynomials $\binom{n}{k}$ for $k \leq 7$, writing

$$\frac{nnnn - 6nnn + 11nn - 6n}{24}$$

for $\binom{n}{4}$.) I know of no other appearance of Stirling numbers before Stirling's time.

5. On folios 5.333–338 Harriot calculated the transitive closure of a binary relation. To my knowledge, mathematicians did not rediscover this important closure operation until the late 19th century. He also considered a relation and its inverse on folios 5.363–364. The relations studied are based on the dependencies among theorems in Euclid.

6. Folio 6.408v considers the number of ways that three circles can be put together; for example,

etc. I have never seen this problem treated anywhere else to this day.

7. On folio 2.31r he concluded that at most 424,905,528,000,000 people could inhabit the whole world. On 8.537r, the number of descendants born to one man and one woman in 240 years is calculated to be 10,068,606,874, assuming that they have one offspring each year, that each child starts producing at age 20, and that nobody dies. This calculation is accompanied by an examination of genealogies in Genesis.

8. On 6.235v he tried out various mnemonic ways of remembering the digits of π, finally hitting on the following:

cadaueribus fero centũ humi gummi cubos: cum humo do
3141 5 92　65　35　　8　　9 7　　9 3 2　　3　　8　　4

　　fumum bufonum, donum cuculorum hoc buglossum inter K.
　　6　　　2 6　　　4　　　3 3　　　8 3 2 7　　　9 5 0

["To corpses I bring a hundred cubes of gum on the ground ..."] He assigned the digits 1 through 9 to the letters a through i, and represented 0 by k; other letters were ignored. (I thank Cecily Tanner for helping me to decipher Harriot's writing.)

9. Finally, there is an extremely interesting application of binary numeration, together with what is now called interval arithmetic, on folio 6.243v. For some reason Harriot wanted to know the approximate value of 2^{28262}, and he proceeded as follows:

[As Harriot wrote it]   [In modern notation]

$$\begin{array}{l} 27.\ \ 13421 + 4 \\ 27.\ \ 13422 + 4 \end{array} \qquad 13421 \times 10^4 < \ 2^{27}\ < 13422 \times 10^4$$

$$\begin{array}{l} 54.\ \ 18012 + 12 \\ 54.\ \ 1801501 + 10 \end{array} \qquad 18012 \times 10^{12} < \ 2^{54}\ < 1801501 \times 10^{10}$$

$$\begin{array}{l} 55.\ \ 36024 + 12 \\ 55.\ \ 3603002 + 10 \end{array} \qquad 36024 \times 10^{12} < \ 2^{55}\ < 3603002 \times 10^{10}$$

$$\begin{array}{l} 110.\ \ 12977 + 29 \\ 110.\ \ 12983 + 29 \end{array} \qquad 12977 \times 10^{29} < \ 2^{110}\ < 12983 \times 10^{29}$$

$\cdots$

$$\begin{array}{l} 28262.\ \ 46 + 8506 \\ 28262.\ \ 57 + 8506 \end{array} \qquad 46 \times 10^{8506} < 2^{28262} < 57 \times 10^{8506}$$

Exponentiation by repeated squaring is, of course, a well-known technique going back at least to al-Uqlīdisī in the 10th century. But to combine it with floating-point arithmetic and upper/lower bounds may well be another first for Harriot.

(Jon Pepper points out that Harriot also calculated $.9997092387^{675}$ in a similar way. See [3, page 371].)

References

[1] J. M. Child, *The Early Mathematical Manuscripts of Leibniz* (London: Open Court, 1920).

[2] Jon V. Pepper, "The study of Thomas Harriot's manuscripts. II. Harriot's unpublished papers," *History of Science* **6** (1967), 17–40.

[3] Jon V. Pepper, "Harriot's calculation of the meridional parts as logarithmic tangents," *Archive for History of Exact Sciences* **4** (1967–1968), 359–413.

Chapter 6

A Permanent Inequality

*[Originally published in American Mathematical Monthly **88** (1981), 731–740, 798.]*

A famous problem that had resisted the attacks of many of the world's greatest mathematicians was resolved in 1980 by G. P. Egorychev [5], who proved a conjecture that B. L. van der Waerden had made in 1926 [15]. The conjecture, which is now a theorem, states that the permanent of an $n \times n$ doubly stochastic matrix is never less than $n!/n^n$; the latter value, which is obtained when all entries of the matrix are equal to $1/n$, is therefore the minimum. The purpose of this note is to give an essentially self-contained exposition of Egorychev's proof and of the auxiliary results that preceded it, using only elementary concepts of mathematics (except at one point).

1. Introduction to Quadratic Forms

Our discussion will be based mostly on facts about matrices and quadratic forms that we will prove "from scratch." A *quadratic form* $f(x_1, \ldots, x_n)$ of n variables is an expression

$$f(x_1, \ldots, x_n) = \sum_{i,j} f_{ij} x_i x_j \qquad (1.1)$$

defined by some $n \times n$ matrix of real coefficients f_{ij}. The sum is over $1 \le i, j \le n$, which we abbreviate to 'i, j'. Since the coefficient of $x_i x_j$ in $f(x_1, \ldots, x_n)$ is $f_{ij} + f_{ji}$ when $i \ne j$, we can assume without loss of generality that $f_{ij} = f_{ji} =$ one half of the coefficient of $x_i x_j$; then the matrix of coefficients is symmetric about its diagonal. For example, the quadratic form $x_1^2 + x_2^2 + x_3^2 - 2x_1x_2 - 2x_1x_3 - 2x_2x_3$ can be specified either by the triangular matrix $\begin{pmatrix} 1 & -2 & -2 \\ 0 & 1 & -2 \\ 0 & 0 & 1 \end{pmatrix}$ or by the symmetric matrix $\begin{pmatrix} 1 & -1 & -1 \\ -1 & 1 & -1 \\ -1 & -1 & 1 \end{pmatrix}$; we will always assume that it corresponds to the latter.

89

The most successful way to deal with quadratic forms seems to be to consider what happens when we do linear transformations of the variables, namely to replace x_i by $\sum_j t_{ij} y_j$ for $1 \le i \le n$; then $f(x_1, \dots, x_n)$ is transformed into another quadratic form $g(y_1, \dots, y_n)$, whose matrix of coefficients has $g_{ij} = \sum_{k,l} f_{kl} t_{ki} t_{lj}$. If the transformation matrix (t_{ij}) is nonsingular, so that we can invert it and express the y's in terms of the x's, the quadratic forms f and g are called *equivalent*.

We shall now prove that every quadratic form is equivalent in this sense to a quadratic form of a very simple type. The idea is to use a transformation that can be thought of as a highly generalized version of "completing the square":

Lemma 1.1. *Let $f(x_1, \dots, x_n) = \sum_{i,j} f_{ij} x_i x_j$ be a quadratic form and let $(a_1, \dots, a_n)$ be a vector of real numbers for which we have $a_1 \ne 0$ and $f(a_1, \dots, a_n) = c \ne 0$. Then the nonsingular transformation defined by*

$$x_i = a_i \left(y_1 - \sum_{j \ge 2} y_j \sum_k f_{jk} a_k / c \right) + y_i \cdot [i \ge 2], \qquad (1.2)$$

$$y_1 = \sum_{i,j} f_{ij} a_j x_i / c, \quad y_i = x_i - x_1 a_i / a_1 \quad \text{for } i \ge 2, \qquad (1.3)$$

makes $f(x_1, \dots, x_n) = c y_1^2 + g(y_2, \dots, y_n)$, where g is a quadratic form in $n - 1$ variables.   (The notation '$[i \ge 2]$' in (1.2) denotes 1 if $i \ge 2$ .and 0 otherwise; it is often convenient to use relations as expressions in this way.)

Proof. It is not difficult to verify that (1.2) and (1.3) are inverses of each other. Let t_{ij} denote the coefficient of y_j in x_i. The coefficient of y_1^2 in $f(x_1, \dots, x_n)$ is $\sum_{i,j} f_{ij} t_{i1} t_{j1} = \sum f_{ij} a_i a_j = c$; and the coefficient of $y_1 y_k$ for $k \ge 2$ is

$$\sum_{i,j} f_{ij} (t_{i1} t_{jk} + t_{j1} t_{ik})$$

$$= \sum_{i,j} f_{ij} \left(a_i \left([j = k] - \frac{a_j}{c} \sum_l f_{kl} a_l \right) + a_j \left([i = k] - \frac{a_i}{c} \sum_l f_{kl} a_l \right) \right),$$

which nicely cancels to zero because $f_{ij} = f_{ji}$. $\square$

Lemma 1.2. *Every quadratic form $f(x_1, \dots, x_n)$ is equivalent to a simple quadratic form*

$$g(y_1, \dots, y_n) = y_1^2 + \cdots + y_p^2 - y_{p+1}^2 - \cdots - y_r^2 \qquad (1.4)$$

for some indices p and r, where $0 \leq p \leq r \leq n$.

Proof. If $f(a_1, \ldots, a_n)$ is zero for all $(a_1, \ldots, a_n)$, we have form (1.4) with $p = r = 0$. If $f(a_1, \ldots, a_n)$ is nonzero for at least one vector $(a_1, \ldots, a_n)$, then some $a_i \neq 0$; by permuting variables if necessary we can assume that $a_1 \neq 0$. Now we use the construction of Lemma 1.1, but with y_1 replaced by $z = y_1 |c|^{1/2}$, thereby obtaining $f(x_1, \ldots, x_n) = \pm z^2 + g(y_2, \ldots, y_n)$. The result follows by induction. $\square$

As an example of the process used in these proofs, let us find a reduced form equivalent to $f(x_1, x_2, x_3) = x_1^2 + x_2^2 + x_3^2 - 2x_1x_2 - 2x_1x_3 - 2x_2x_3$. First, $f(1, 0, 0) = 1$, so the transformation of Lemma 1.1 applies with $x_1 = y_1 + y_2 + y_3$, $x_2 = y_2$, $x_3 = y_3$; we have $f(x_1, x_2, x_3) = y_1^2 - 4y_2y_3$. Let $g(y_2, y_3) = -4y_2y_3$; since $g(1, 1) = -4$, we can let $y_2 = \frac{1}{2}z_1 - \frac{1}{2}z_2$, $y_3 = \frac{1}{2}z_1 + \frac{1}{2}z_2$, so $-4y_2y_3 = -z_1^2 + z_2^2$. Thus we have $f(x_1, x_2, x_3) = y_1^2 + z_2^2 - z_1^2 = (x_1 - x_2 - x_3)^2 + (x_3 - x_2)^2 - (x_2 + x_3)^2$.

The method of reduction in Lemma 1.1 was essentially discovered by Lagrange in 1759. The next basic fact that we need is "Sylvester's law of inertia," which dates from 1852.

Lemma 1.3. *The numbers p and r of Lemma 1.2 are unique. In other words, if the quadratic forms*

$$y_1^2 + \cdots + y_p^2 - y_{p+1}^2 - \cdots - y_r^2 = z_1^2 + \cdots + z_q^2 - z_{q+1}^2 - \cdots - z_s^2 \quad (1.5)$$

are equivalent, then $p = q$ and $r = s$.

Proof. Let $y_i = \sum t_{ij}z_j$, and suppose that $p < q$. Then we can find values $z_1^*, \ldots, z_q^*$ not all zero such that $\sum_{j=1}^{q} t_{ij}z_j^* = 0$ for $1 \leq i \leq p$, since this is a system of p homogeneous linear equations in $q > p$ unknowns z_j^*. Let $z_{q+1}^* = \cdots = z_n^* = 0$ and $y_i^* = \sum_{j=1}^{n} t_{ij}z_j^*$ for all i. Then (1.5) reduces to

$$-y_{p+1}^{*2} - \cdots - y_r^{*2} = z_1^{*2} + \cdots + z_q^{*2},$$

which can hold only if $z_1^* = \cdots = z_q^* = 0$, a contradiction. Hence $p = q$. Now if $r < s$, we can similarly find $z_{q+1}^*, \ldots, z_s^*$ not all zero such that $\sum_{j=q+1}^{s} t_{ij}z_j^* = 0$ for $p < i \leq r$; setting $z_1^* = \cdots = z_q^* = z_{s+1}^* = \cdots = z_n^* = 0$, we get

$$y_1^{*2} + \cdots + y_p^{*2} = -z_{q+1}^{*2} - \cdots - z_s^{*2},$$

another contradiction. $\square$

As a consequence of Lemma 1.3 we can refer to $p(f)$ and $r(f)$ as well-defined invariants of the quadratic form f.

We will need a somewhat more subtle fact later:

Lemma 1.4. Let $f_\theta(x_1, \ldots, x_n)$ be the quadratic form

$$f_\theta(x_1, \ldots, x_n) = (1 - \theta)f_0(x_1, \ldots, x_n) + \theta f_1(x_1, \ldots, x_n) \qquad (1.6)$$

that changes from f_0 to f_1 as θ varies from 0 to 1. If $r(f_\theta) = n$ for $0 \le \theta \le 1$, then $p(f_0) = p(f_1)$.

Proof. This lemma can be proved by using well-known facts of analysis, since $p(f_\theta)$ is the number of positive eigenvalues of the real symmetric matrix f_θ, and since $r(f_\theta) = n$ is equivalent to saying that f_θ has no zeros as eigenvalues; the result follows because the roots of a polynomial are continuous functions of their coefficients, and the eigenvalues of f_θ are roots of polynomials whose coefficients are continuous functions of θ. A proof closer to first principles can be formulated by showing that for all θ_0 in $[0 \mathinner{.\,.} 1]$ there exists $\epsilon > 0$ such that $p(f_\theta)$ is constant for $|\theta - \theta_0| < \epsilon$; by compactness, we can then cover $[0 \mathinner{.\,.} 1]$ with a finite number of open intervals of constant p. The existence of ϵ follows from a proof like that of Lemma 1.3; details are left to the interested reader. □

2. Quadratic Forms and Permanents

The *permanent* of an $n \times n$ matrix $A = (a_{ij})$ is the sum

$$\operatorname{per}(A) = \sum_\pi a_{1\pi(1)} \cdots a_{n\pi(n)} , \qquad (2.1)$$

taken over all permutations $\pi = \pi(1) \ldots \pi(n)$ of $\{1, \ldots, n\}$. We will write a_i for the ith row $(a_{i1}, \ldots, a_{in})$ of the matrix A, and $\operatorname{per}(A) = \operatorname{per}(a_1, \ldots, a_n)$ if we are enumerating its rows.

The next part of the theory is a proof of two lemmas, which are proved simultaneously by induction on n. That is, we prove both of them for $n = 2$ first, then both for $n = 3$, and so on.

Lemma 2.1. Let $a_1, \ldots, a_{n-1}$ be vectors of nonnegative numbers in which at least $n + 1 - i$ elements of a_i are positive, and suppose that $b = (b_1, \ldots, b_n)$ is any vector of real numbers such that

$$\operatorname{per}(a_1, \ldots, a_{n-1}, b) = 0 . \qquad (2.2)$$

Then

$$\operatorname{per}(a_1, \ldots, a_{n-2}, b, b) \le 0 ; \qquad (2.3)$$

furthermore, $\operatorname{per}(a_1, \ldots, a_{n-2}, b, b) = 0$ if and only if $b_1 = \cdots = b_n = 0$.

Lemma 2.2. *Let $a_1, \ldots, a_{n-2}$ be as in Lemma 2.1, and let f be the quadratic form*

$$f(x_1, \ldots, x_n) = \operatorname{per}(a_1, \ldots, a_{n-2}, x, x) \qquad (2.4)$$

where x stands for the vector $(x_1, \ldots, x_n)$. Then $r(f) = n$ and $p(f) = 1$.

Proof. Both results are clear for $n = 2$: If $a_{11} > 0$ and $a_{12} > 0$ and $a_{11}b_2 + a_{12}b_1 = 0$ then $2b_1b_2 \le 0$; and in this case we have $2b_1b_2 = 0$ if and only if $b_1 = b_2 = 0$. Furthermore the quadratic form $2x_1x_2$ is $\big((x_1 + x_2)/\sqrt{2}\big)^2 - \big((x_1 - x_2)/\sqrt{2}\big)^2$. So we shall assume that $n \ge 3$ and that both lemmas have been proved for $n - 1$.

In the quadratic form (2.4), the value of f_{ij} is the permanent of the $(n-2) \times (n-2)$ matrix obtained by removing columns i and j of the matrix $(a_1, \ldots, a_{n-2})$, if $i \neq j$; also $f_{ii} = 0$. If $r(f)$ is less than n, the matrix f is singular, so there is a nonzero vector $(b_1, \ldots, b_n)$ such that $\sum_j f_{ij}b_j = 0$ for all i. This condition is equivalent to saying that $\operatorname{per}(a_1, \ldots, a_{n-2}, b, x) = 0$ for all x, because of our interpretation of f_{ij}; therefore in particular, $\operatorname{per}(a_1, \ldots, a_{n-2}, b, b) = 0$. Furthermore we have $\operatorname{per}_j(a_1, \ldots, a_{n-2}, b) = 0$ for all j, where per_j denotes the permanent obtained by removing column j. By induction, $\operatorname{per}_j(a_1, \ldots, a_{n-3}, b, b) \le 0$ for all j. Now

$$0 = \operatorname{per}(a_1, \ldots, a_{n-2}, b, b) = \sum_i a_{(n-2)j} \operatorname{per}_j(a_1, \ldots, a_{n-3}, b, b) \le 0 \, ;$$

hence we have $\operatorname{per}_j(a_1, \ldots, b, b) = 0$ whenever $a_{(n-2)j} > 0$. This inequality occurs for at least two values of j, in fact for at least three; hence $b_1 = \cdots = b_n = 0$, a contradiction.

We have proved half of Lemma 2.2, the fact that $r(f) = n$. For the other half, it suffices by Lemma 1.4 to compute $p(f)$ in the special case that $a_1 = \cdots = a_{n-2} = (1, 1, \ldots, 1)$, since we can transform the rows one by one from this case into (2.4), and all the intermediate quadratic forms have rank n by what we have already proved. In this special case the quadratic form is $(n-2)!$ times the special quadratic form defined by $f_{ij} = [i \neq j]$. When $n = 4$, for example, the special form is $2(x_1x_2 + x_1x_3 + x_1x_4 + x_2x_3 + x_2x_4 + x_3x_4)$. The general transformation method of Lemma 1.2 shows that this form can be represented for example as

$$\sum_{i \neq j} x_i x_j = (n-1)s^2 - (x_2 - s)^2 - \cdots - (x_n - s)^2 \, , \qquad (2.5)$$

where $s = x_1 + \frac{1}{2}(x_2 + \cdots + x_n)$. Hence $p(f) = 1$.

We turn now to the proof of Lemma 2.1. The hypotheses on a_1, ..., a_{n-1} imply that $\mathrm{per}(a_1, \ldots, a_{n-1}, a_{n-1}) > 0$; hence we must have $f(a_{(n-1)1}, \ldots, a_{(n-1)n}) = c > 0$, in terms of the quadratic form (2.4). We may assume without loss of generality that $a_{(n-1)1} > 0$, by permutation of columns. Therefore if we apply the construction of Lemma 1.1 we obtain

$$f(x_1, \ldots, x_n) = cy_1^2 + g(y_2, \ldots, y_n).$$

By Lemma 2.2, $g(y_2, \ldots, y_n) \le 0$ for all $(y_2, \ldots, y_n)$, and it is zero if and only if $y_2 = \cdots = y_n$. We have

$$y_1 = \sum_{i=1}^{n} \sum_{j=1}^{n} f_{ij} a_{(n-1)i} x_j / c = \mathrm{per}(a_1, \ldots, a_{n-1}, x)/c$$

by (1.2); this fact, together with our hypothesis (2.2), implies that $f(b_1, \ldots, b_n) \le 0$. Finally, $f(b_1, \ldots, b_n) = 0$ implies that we have $y_1 = \cdots = y_n = 0$ when $(x_1, \ldots, x_n) = (b_1, \ldots, b_n)$; hence $b_1 = \cdots = b_n = 0$. □

Now we come to the main theorem on which Egorychev's proof of van der Waerden's conjecture rests. This theorem is essentially due to A. D. Aleksandrov, who published it in 1938 [1], using a more general framework that was not obviously related to permanents. Aleksandrov's motivation was the study of the volume of n-dimensional convex sets, for which a similar geometric inequality had been deduced by W. Fenchel in 1936 [7]. A search through *Science Citation Index* for 1960–1980 reveals that the geometric aspects of Aleksandrov's work became well known, but the algebraic aspects were rarely cited in the western world. The only exceptions are Buseman's book [3], which paraphrases part of Aleksandrov's proof, and Schneider's paper [12], which presents an alternative derivation. (See also the recent work by Teissier [14] and Stanley [13].) Egorychev discovered that Aleksandrov's work was relevant to permanents after studying a variety of new formulas for the permanent function [6].

Theorem 2.3. *Let a_1, ..., a_{n-1} be nonnegative vectors such that a_i contains at least $n + 1 - i$ positive entries, and let a_n be any vector of real numbers. Then*

$$\mathrm{per}(a_1, \ldots, a_{n-1}, a_n)^2 \\ \ge \mathrm{per}(a_1, \ldots, a_{n-2}, a_{n-1}, a_{n-1}) \, \mathrm{per}(a_1, \ldots, a_{n-2}, a_n, a_n); \tag{2.6}$$

and equality holds if and only if $a_n = \lambda a_{n-1}$ for some real number λ.

Proof. Let $\mathrm{per}(a_1, \ldots, a_{n-1}, a_n) = \lambda \, \mathrm{per}(a_1, \ldots, a_{n-1}, a_{n-1})$. The value of λ is well defined since $\mathrm{per}(a_1, \ldots, a_{n-1}, a_{n-1}) > 0$ by Lemma 2.1. If we set $b = a_n - \lambda a_{n-1}$, we have (2.2), since the permanent is a linear function of each of its rows. Hence Lemma 2.1 tells us that

$$
\begin{aligned}
0 &\geq \mathrm{per}(a_1, \ldots, a_{n-2}, b, b) \\
&= \mathrm{per}(a_1, \ldots, a_{n-2}, b, a_n) - \lambda \, \mathrm{per}(a_1, \ldots, a_{n-2}, b, a_{n-1}) \\
&= \mathrm{per}(a_1, \ldots, a_{n-2}, a_n, a_n) - 2\lambda \, \mathrm{per}(a_1, \ldots, a_n) \\
&\qquad + \lambda^2 \, \mathrm{per}(a_1, \ldots, a_{n-2}, a_{n-1}, a_{n-1}) \\
&= \mathrm{per}(a_1, \ldots, a_{n-2}, a_n, a_n) - \lambda^2 \, \mathrm{per}(a_1, \ldots, a_{n-2}, a_{n-1}, a_{n-1}) \\
&= \mathrm{per}(a_1, \ldots, a_{n-2}, a_n, a_n) - \frac{\mathrm{per}(a_1, \ldots, a_n)^2}{\mathrm{per}(a_1, \ldots, a_{n-2}, a_{n-1}, a_{n-1})} .
\end{aligned}
$$

Equality holds if and only if $b = 0$, that is, $a_n = \lambda a_{n-1}$.   □

Corollary 2.4. *Let $a_1, \ldots, a_{n-1}$ be nonnegative vectors and let a_n be arbitrary. Then the inequality (2.6) holds.*

Proof. Consider the vectors $a_i + (\epsilon, \ldots, \epsilon)$ and take the limit as ϵ approaches zero through positive values.   □

Incidentally, the example

$$
\left(\mathrm{per} \begin{pmatrix} 1 & 1 & 0 \\ 0 & 1 & 1 \\ 1 & 0 & 1 \end{pmatrix} \right)^2 = \mathrm{per} \begin{pmatrix} 1 & 1 & 0 \\ 0 & 1 & 1 \\ 0 & 1 & 1 \end{pmatrix} \mathrm{per} \begin{pmatrix} 1 & 1 & 0 \\ 1 & 0 & 1 \\ 1 & 0 & 1 \end{pmatrix}
$$

shows that some condition on positivity is necessary to obtain Theorem 2.3's strong statement about the conditions of equality in (2.6).

3. Doubly Stochastic Matrices

The matrix (a_{ij}) is called *doubly stochastic* if its elements are nonnegative and if all row sums and column sums are equal to 1. Thus, each row can be regarded as a probability distribution, and so can each column; it is for this reason that the matrix is "stochastic" in a double sense.

Doubly stochastic matrices have many pleasant properties that make them important in applications. For example, it is easy to verify that the product AB of two doubly stochastic matrices A and B is doubly stochastic; and if $0 \leq \theta \leq 1$, the convex combination $(1 - \theta)A + \theta B$ is also doubly stochastic. If we imagine a network of n points with a_{ij} units of flow proceeding from point i to point j, the total flow in and out of each point is equal to unity.

The simplest kind of doubly stochastic matrix is a *permutation matrix*, in which all a_{ij} are 0 or 1; there is exactly one 1 in every row and every column. If $\pi(1) \ldots \pi(n)$ is a permutation of $\{1, \ldots, n\}$, we let P_π be the corresponding permutation matrix (p_{ij}), where $p_{ij} = [j = \pi(i)]$. Garrett Birkhoff proved in 1946 [2] that every doubly stochastic matrix is a convex combination of permutation matrices:

Lemma 3.1. *The $n \times n$ matrix is doubly stochastic if and only if there exist nonnegative numbers t_π such that*

$$A = \sum_\pi t_\pi P_\pi \quad \text{and} \quad \sum_\pi t_\pi = 1, \tag{3.1}$$

where the sums are taken over all $n!$ permutations $\pi = \pi(1) \ldots \pi(n)$ of $\{1, \ldots, n\}$.

Proof. It is clear that every matrix of the form (3.1) is doubly stochastic, so the problem is to show that every doubly stochastic matrix can be represented in terms of permutation matrices. Such a decomposition relies on an important lemma published in 1935 by Philip Hall [8] (see also Egerváry's work of 1931 [4]), which we can formulate as follows:

Lemma 3.2. *Consider n men and n women such that each man–woman pair is either compatible or incompatible. If there is no way to match the men and women into n compatible marriages, then for some $k > 0$ there is a set of k men who are compatible with only $k - 1$ women.*

Proof. Suppose there is a way to obtain m compatible marriages but no way to obtain $m+1$, for some $m < n$, and let x be an unmarried man in one of these maximum matchings. Consider all chains of relationships of the form

$$i_1 \to j_1 \Rightarrow i_2 \to j_2 \Rightarrow \cdots \Rightarrow i_r \to j_r \tag{3.2}$$

where $x = i_1$, and where the relation $i \to j$ means "man i is compatible with woman j" while $j \Rightarrow i$ means "woman j is married to man i." In every such chain, the woman j_r must be married; for if she were not, there would be a way to have $m+1$ compatible marriages, by performing $r - 1$ divorces and then marrying i_l with j_l for $1 \leq l \leq r$. Consider now the set S of all men i_l appearing in chains (3.2), and the set T of all women j_l that appear. Then each woman in T is married to a man in S, and each man in S (except for x) is married to a woman in T. Therefore S contains k elements while T contains only $k - 1$. □

Returning now to the proof of Lemma 3.1, let A be doubly stochastic and let us imagine n men and women such that man i is compatible with woman j if and only if $a_{ij} > 0$. In these circumstances a set of n compatible marriages is possible. For if S were a set of k men that are compatible with only $k - 1$ women, and if we let T be the set of those women, the sum $\sum\{a_{ij} \mid i \in S, j \in T\}$ would equal k because it includes all the nonzero elements of k rows; yet it involves only $k - 1$ columns, so it cannot exceed $k - 1$. This contradiction shows that there is a permutation $\pi(1)\ldots\pi(n)$ such that $a_{i\pi(i)} > 0$ for $1 \leq i \leq n$.

Let $t_\pi = \min(a_{1\pi(1)}, \ldots, a_{n\pi(n)})$. If $t_\pi = 1$, A is a permutation matrix, so it clearly has the form (3.1). Otherwise

$$A = t_\pi P_\pi + (1 - t_\pi)B$$

for some doubly stochastic matrix B, where B contains at least one more zero than A. By induction on the number of nonzero entries, the matrix B can be expressed in the form (3.1), and we obtain the desired representation of A. $\quad\square$

A doubly stochastic matrix $A = (a_{ij})$ in which a_{ij} represents the probability of going from state j to state i is said to have an *equilibrium vector* $x = (x_1, \ldots, x_n)^T$ if $Ax = x$, that is, if $\sum_j a_{ij}x_j = x_i$ for all i. The matrix A is called *decomposable* if the set $\{1, \ldots, n\}$ can be partitioned into disjoint nonempty subsets S and T such that $a_{ij} = 0$ whenever $i \in S$ and $j \in T$. These concepts are related to each other by the following simple observation:

Lemma 3.3. *If $x = (x_1, \ldots, x_n)^T$ is an equilibrium vector for A having some components unequal, then A is decomposable.*

Proof. If $(x_1, \ldots, x_n)^T$ is an equilibrium vector, so is the shifted vector $(x_1 + c, \ldots, x_n + c)^T$; hence we can assume that all components of x are nonnegative, and that at least one component is zero. Let $S = \{i \mid x_i = 0\}$ and $T = \{i \mid x_i > 0\}$; then $\sum_j a_{ij}x_j = x_i$ implies that $a_{ij} = 0$ whenever $i \in S$ and $j \in T$. $\quad\square$

The proof of Lemma 3.3 uses only the fact that A satisfies the "singly stochastic" property $\sum_j a_{ij} = 1$ for $1 \leq i \leq n$. Doubly stochastic matrices satisfy a much stronger condition:

Lemma 3.4. *If A is doubly stochastic and if S and T are sets of decomposability for A, then $a_{ij} = 0$ unless $i, j \in S$ or $i, j \in T$.*

Proof. We have $\sum_{i,j \in S} a_{ij} = \sum_{i \in S} \sum_{j=1}^n a_{ij} = \sum_{i \in S} 1 = \sum_{j \in S} 1 = \sum_{j \in S} \sum_{i=1}^n a_{ij} = \sum_{i,j \in S} a_{ij} + \sum_{i \in T, j \in S} a_{ij}$. Therefore a_{ij} must be zero when $i \in T$ and $j \in S$. $\quad\square$

4. Minimal Matrices

For purposes of this discussion we shall say that an $n \times n$ matrix $A = (a_{ij})$ is *minimal* if it is doubly stochastic and if it has the smallest permanent among all $n \times n$ doubly stochastic matrices. There is at least one minimal matrix; that is, the smallest value is actually achieved. This fact follows because the permanent is a continuous function of the matrix elements, and because the set of doubly stochastic matrices is a closed and bounded subset of n^2-dimensional space. Lemma 3.1 implies that the permanent of A is at least $\sum_\pi t_\pi^n$ for some nonnegative numbers with $\sum_\pi t_\pi = 1$, so the minimal permanent cannot be zero.

According to van der Waerden's conjecture [15] there is only one minimal matrix of order n, namely the matrix with $a_{ij} = 1/n$ for all i and j. We are not ready to prove the conjecture yet, but we shall see that standard methods of minimization give us rather strict conditions that minimal matrices must satisfy.

From now on we shall use the notation A_{ij} to stand for the "(i,j) minor of A," namely the $(n-1) \times (n-1)$ matrix obtained from A by removing row i and column j. Clearly

$$\operatorname{per}(A) = \sum_j a_{ij} \operatorname{per}(A_{ij}) = \sum_i a_{ij} \operatorname{per}(A_{ij}); \qquad (4.1)$$

in fact either of these equations can be used to provide an alternative definition of the permanent.

Another basic fact about permanents of minors is the formula

$$\operatorname{per}(A + \epsilon B) = \operatorname{per}(A) + \epsilon \sum_{i,j} b_{ij} \operatorname{per}(A_{ij}) + O(\epsilon^2), \qquad (4.2)$$

where A and B are arbitrary matrices and where $O(\epsilon^2)$ refers to ϵ^2 times a polynomial in ϵ and the elements of A and B.

If A is doubly stochastic let us say that B is a *valid modification* for A if the row sums and column sums of B are zero and if $b_{ij} \geq 0$ whenever $a_{ij} = 0$; then $A + \epsilon B$ is doubly stochastic for all sufficiently small $\epsilon > 0$.

Lemma 4.1. *If A is a minimal matrix and if B is a valid modification for A, then*

$$\sum_{i,j} b_{ij} \operatorname{per}(A_{ij}) \geq 0. \qquad (4.3)$$

Proof. This inequality follows immediately from identity (4.2) and the definition of minimality.   □

Lemma 4.2. *A minimal matrix is indecomposable.*

Proof. Suppose that A is doubly stochastic and that $a_{ij} > 0$ only when $i, j \in S$ or $i, j \in T$, where S and T are nonempty and $S \cup T = \{1, \ldots, n\}$. We know that $\text{per}(A) > 0$, so there is a permutation $\pi(1) \ldots \pi(n)$ with $a_{i\pi(i)} > 0$ for all i. Let s and t be elements of S and T, respectively, and let $B = (b_{ij})$ be a matrix that is entirely zero except that $b_{s\pi(s)} = b_{t\pi(t)} = -1$ and $b_{s\pi(t)} = b_{t\pi(s)} = +1$. Since $\pi(s) \in S$ and $\pi(t) \in T$, the matrix B is a valid modification for A; therefore

$$\text{per}(A_{s\pi(s)}) + \text{per}(A_{t\pi(t)}) \leq \text{per}(A_{s\pi(t)}) + \text{per}(A_{t\pi(s)})$$

by Lemma 4.1. But this cannot happen, because both $\text{per}(A_{s\pi(s)})$ and $\text{per}(A_{t\pi(t)})$ are positive, while $\text{per}(A_{s\pi(t)})$ and $\text{per}(A_{t\pi(s)})$ are zero. For example, $\text{per}(A_{t\pi(s)})$ is zero because the matrix $A_{t\pi(s)}$ has k rows corresponding to S in which all nonzero entries occur in only $k - 1$ columns corresponding to $S \setminus \{\pi(s)\}$. $\square$

Lemma 4.3. *If A is a minimal matrix, then $\text{per}(A_{ij}) > 0$ for all i and j.*

Proof. If $\text{per}(A_{ij}) = 0$, it has some set S of $k > 0$ rows in which all nonzero entries occur in some $k - 1$ columns, by Lemma 3.2. Let $T = \{1, \ldots, n\} \setminus S$; note that T is nonempty, since $i \in T$. We can now permute the columns of A to obtain a matrix A' in which all of the nonzero entries for the rows of S appear in the columns of S. Clearly A' is also a minimal matrix. But A' is decomposable by definition, contradicting Lemma 4.2. $\square$

The next property of minimal matrices is the key to everything that follows; it was first proved by Marcus and Newman in 1959 [10].

Theorem 4.4. *If A is a minimal matrix and $a_{ij} > 0$, then $\text{per}(A_{ij}) = \text{per}(A)$.*

Proof. The basic idea is to consider the set of all B that are valid modifications for A and such that $b_{ij} = 0$ when $a_{ij} = 0$. The other values of b_{ij} are unconstrained except for the fact that row sums and column sums are zero; therefore the inequality (4.3) can be used with the technique of Lagrange multipliers to prove that there exist constants $\lambda_1, \ldots, \lambda_n, \mu_1, \ldots, \mu_n$ such that

$$\text{per}(A_{ij}) = \lambda_i + \mu_j \qquad \text{if } a_{ij} > 0. \tag{4.4}$$

However, we shall present a purely combinatorial proof, because it is based on elementary principles showing that a very simple class of matrices B is sufficient to prove (4.4).

By permuting the columns, we can assume without loss of generality that $a_{ii} > 0$ for all i. Let us write $i \to j$ if $a_{ij} > 0$; thus $i \to i$. Since A is indecomposable there is a "path"

$$1 = j_0 \to j_1 \to \cdots \to j_l = j \qquad (4.5)$$

from 1 to j for all j. We say that j is at distance l if the shortest such path is of length l. For every $j > 1$ at distance l, let $p(j)$ be the smallest index at distance $l-1$ such that $p(j) \to j$. Then the path (4.5) is unique if we insist that $j_{k-1} = p(j_k)$ for $1 \le k \le l$. The arcs $\{\, p(j) \to j \mid j > 1 \,\}$ may be regarded as an oriented tree emanating from point 1. We say that i is an ancestor of j (and we write $i \prec j$) if we have $i = p(j)$ or $i = p(p(j))$ or $\ldots$; the notation $i \preceq j$ means that $i = j$ or $i \prec j$.

Now we can establish (4.4). First we set $\lambda_1 = 0$ and $\mu_1 = \operatorname{per}(A_{11})$. Then for the indices $j \ge 2$ in increasing order of their distance from index 1, we can define μ_j so that (4.4) holds when $i = p(j)$, since the value of $\lambda_{p(j)}$ was previously defined; after μ_j has been set, we immediately define λ_j so that (4.4) holds when $i = j$.

The construction above assigns values to $\lambda_1, \ldots, \lambda_n$ and $\mu_1, \ldots, \mu_n$; we still must prove (4.4) for the pairs $(\hat{\imath}, \hat{\jmath})$ such that $a_{\hat{\imath}\hat{\jmath}} > 0$ and $\hat{\imath} \ne \hat{\jmath}$ and $\hat{\imath} \ne p(\hat{\jmath})$. Consider the matrix B whose entries are all zero, except that $b_{\hat{\imath}\hat{\jmath}} = 1$ and that

$$\begin{aligned} b_{jj} &= [j \prec \hat{\jmath}] - [j \preceq \hat{\imath}], \quad \text{for } 1 \le j \le n; \\ b_{p(j)j} &= [j \preceq \hat{\imath}] - [j \preceq \hat{\jmath}], \quad \text{for } 1 \le j \le n. \end{aligned} \qquad (4.6)$$

The sum $\sum_i b_{ij} = [j = \hat{\jmath}] + ([j \prec \hat{\jmath}] - [j \preceq \hat{\imath}]) + ([j \preceq \hat{\imath}] - [j \preceq \hat{\jmath}]) = 0$. And for fixed i the sum of all b_{ij} such that $i = p(j)$ is $[i \prec \hat{\jmath}] - [i \prec \hat{\jmath}]$, hence $\sum_j b_{ij} = [i = \hat{\imath}] + ([i \prec \hat{\jmath}] - [i \preceq \hat{\imath}]) + ([i \prec \hat{\imath}] - [i \prec \hat{\jmath}]) = 0$. (Consider the paths from 1 to $\hat{\imath}$ and from 1 to $\hat{\jmath}$; at most one such b_{ij} is positive and at most one is negative.) Therefore B is a valid modification for A, and so is $-B$. According to Lemma 4.1, the sum $\sum_{i,j} b_{ij} \operatorname{per}(A_{ij})$ must be zero. The sum actually turns out to be

$$\sum_{i,j} b_{ij} \left(\operatorname{per}(A_{ij}) - \lambda_i - \mu_j \right) = \operatorname{per}(A_{\hat{\imath}\hat{\jmath}}) - \lambda_{\hat{\imath}} - \mu_{\hat{\jmath}},$$

because $\lambda_i \sum_j b_{ij} = \mu_j \sum_i b_{ij} = 0$, and because (4.4) holds for all pairs (i, j) such that $b_{ij} \ne 0$ except possibly for $(\hat{\imath}, \hat{\jmath})$. Thus (4.4) holds in general.

Now we are ready to complete the proof. We have $a_{ij} \operatorname{per}(A_{ij}) = a_{ij}(\lambda_i + \mu_j)$ for all i, j; hence by (4.1) we have

$$\operatorname{per}(A) = \lambda_i + \sum_j a_{ij}\mu_j = \mu_j + \sum_i a_{ij}\lambda_i$$

for all i, j. In matrix notation, $\lambda + A\mu = \mu + A^T\lambda = \operatorname{per}(A)e$, where $\lambda = (\lambda_1, \ldots, \lambda_n)^T$, $\mu = (\mu_1, \ldots, \mu_n)^T$, and e is a column vector of all 1s. Since $Ae = A^Te = e$, we have $A^T\lambda + A^TA\mu = \operatorname{per}(A)e$ and $A\mu + AA^T\lambda = \operatorname{per}(A)e$; hence

$$\mu = A^TA\mu \quad \text{and} \quad \lambda = AA^T\lambda.$$

Now A is indecomposable and has nonzero elements on the diagonal, so the matrices A^TA and AA^T are even less decomposable. Consequently Lemma 3.3 implies that $\lambda_1 = \cdots = \lambda_n$ and $\mu_1 = \cdots = \mu_n$. We originally chose $\lambda_1 = 0$; hence $\lambda_1 = \cdots = \lambda_n = 0$ and $\mu_1 = \cdots = \mu_n = \operatorname{per}(A)$.   □

David London proved in 1971 [9] that something also can be said about $\operatorname{per}(A_{ij})$ when $a_{ij} = 0$.

Lemma 4.5. *If A is a minimal matrix, then $\operatorname{per}(A_{ij}) \geq \operatorname{per}(A)$ for all i and j.*

Proof. (The following proof is based on an idea of Henryk Minc [11].) Because of Lemma 4.4, we need only consider (i, j) such that $a_{ij} = 0$; without loss of generality, assume that $i = j = 1$ and $a_{11} = 0$. By Lemma 4.3, $\operatorname{per}(A_{11}) > 0$; hence we can assume that $a_{jj} > 0$ for $2 \leq j \leq n$.

Let $B = I - A$. All elements of B such that $a_{ij} = 0$ are nonnegative, and the row and column sums are zero, so B is a valid modification of A. By Lemma 4.1 and equation (4.1) we have

$$0 \leq \sum_{i,j} b_{ij} \operatorname{per}(A_{ij}) = \sum_j \operatorname{per}(A_{jj}) - \sum_{i,j} a_{ij} \operatorname{per}(A_{ij})$$

$$= \operatorname{per}(A_{11}) + (n-1)\operatorname{per}(A) - n\operatorname{per}(A)$$

$$= \operatorname{per}(A_{11}) - \operatorname{per}(A),$$

since $\operatorname{per}(A_{jj}) = \operatorname{per}(A)$ for $j > 1$ by Lemma 4.4.   □

5. Egorychev's Theorem

The results of Section 2 can now be combined with those of Section 4 to complete the analysis.

Lemma 5.1. *If A is a minimal matrix, then* $\operatorname{per}(A_{ij}) = \operatorname{per}(A)$ *for all* i *and* j.

Proof. Without loss of generality, assume that $i = j = n$; we may assume further that $\operatorname{per}(A_{nn}) > \operatorname{per}(A)$ and $a_{(n-1)n} > 0$. Then $n > 1$, and Corollary 2.4 implies that

$$\operatorname{per}(A)^2 = \operatorname{per}(a_1, \ldots, a_n)^2$$
$$\geq \operatorname{per}(a_1, \ldots, a_{n-2}, a_{n-1}, a_{n-1}) \operatorname{per}(a_1, \ldots, a_{n-2}, a_n, a_n).$$

However,

$$\operatorname{per}(a_1, \ldots, a_{n-2}, a_{n-1}, a_{n-1}) = \sum_j a_{(n-1)j} \operatorname{per}(A_{nj})$$
$$> \sum_j a_{(n-1)j} \operatorname{per}(A) = \operatorname{per}(A);$$
$$\operatorname{per}(a_1, \ldots, a_{n-2}, a_n, a_n) = \sum_j a_{nj} \operatorname{per}(A_{(n-1)j})$$
$$\geq \sum_j a_{nj} \operatorname{per}(A) = \operatorname{per}(A);$$

so we have $\operatorname{per}(A)^2 > \operatorname{per}(A)^2$, a sort of contradiction.   □

Lemma 5.2. *If A is a minimal matrix of order n, with $a_{ij} > 0$ for all i and j except possibly when $i = n$, then $a_{ij} = 1/n$ for all i and j.*

Proof. We have $\operatorname{per}(a_1, \ldots, a_{n-2}, a_n, a_n) = \sum_j a_{nj} \operatorname{per}(A) = \operatorname{per}(A)$ by (4.1) and Lemma 5.1, and similarly $\operatorname{per}(a_1, \ldots, a_{n-2}, a_{n-1}, a_{n-1}) = \operatorname{per}(A)$. Therefore equality holds in (2.6), and Theorem 2.3 implies that $a_n = \lambda a_{n-1}$ for some λ. Clearly $\lambda = 1$ since A is doubly stochastic; hence $a_n = a_{n-1}$. Similarly, all rows of A are equal. Therefore all columns of A consist of identical elements. Therefore all elements of A are equal to $1/n$.   □

The pieces of the puzzle are almost all in place, and we only need to dispense with the hypothesis that $a_{ij} > 0$.

Theorem 5.3. *If A is a minimal matrix of order n, then $a_{ij} = 1/n$ for all i and j; hence*

$$\operatorname{per}(A) = n!/n^n. \tag{5.1}$$

Proof. Let B be obtained from A by replacing some row a_i by some other row a_k; then $\operatorname{per}(B) = \sum_j a_{kj} \operatorname{per}(A_{ij}) = \operatorname{per}(A)$. Similarly, let C be obtained from A by replacing the row a_k by a_i; again $\operatorname{per}(C) = \operatorname{per}(A)$. Of course, we do not know that B and C are doubly stochastic; but the matrix $D = \frac{1}{2}(B + C)$ certainly is doubly stochastic, and $\operatorname{per}(D) = \frac{1}{4}\big(\operatorname{per}(B) + 2\operatorname{per}(A) + \operatorname{per}(C)\big)$, so D is a minimal matrix.

By a finite number of "averaging" steps like the ones that formed D from A, we obtain a minimal matrix E having the same bottom row as A, but with $e_{ij} = 0$ only if $i = n$ or $a_{1j} = \cdots = a_{(n-1)j} = 0$. Since A is indecomposable, we cannot have $a_{1j} = \cdots = a_{(n-1)j} = 0$; hence E is a minimal matrix satisfying the condition of Lemma 5.2. It follows that its bottom row a_n is $(1/n, \ldots, 1/n)$. Similarly, all rows of A have this value. $\square$

The preparation of this report was supported in part by National Science Foundation grant MCS-77-23738 and in part by Office of Naval Research contract N00014-76-C-0330.

References

[1] A. D. Aleksandrov, "K teorii smeshannykh ob'ëmov vypuklykh tel," *Matematicheskiĭ Sbornik* (new series) **3** (1938), 227–251.

[2] Garrett Birkhoff, "Tres observaciones sobre al algebra lineal," *Universidad Nacional de Tucumán, Revista* **A5** (1946), 147–151.

[3] Herbert Buseman, *Convex Surfaces* (New York: Interscience, 1958), 51–56.

[4] E. Egerváry, "Matrixok kombinatorius tulajfonságairól," *Matematikai és Fizikai Lapok* **38** (1931), 16–28.

[5] G. P. Egorychev, "Reshenie problemy van-der-Wardena dli͡a permanentov," preprint IFSO-13M (Krasnoi͡arsk: USSR Academy of Sciences, Siberian branch, Institut Fiziki imeni L. V. Kirenskogo, 1980); *Sibirskiĭ Matematicheskiĭ Zhurnal* **22,** 6 (1981), 65–71, 225. [English translation, "The solution of van der Waerden's problem for permanents," *Advances in Mathematics* **42** (1981), 299–305.]

[6] G. P. Egorychev, "Novye formuly dli͡a permanenta," *Doklady Akademii͡a Nauk SSSR* **254** (1980), 784–787.

[7] Werner Fenchel, "Inégalités quadratiques entre les volumes mixtes des corps convexes," *Comptes Rendus hebdomadaires des séances de l'Académie des Sciences* **203** (Paris: 1936), 647–650.

[8] Philip Hall, "On representatives of subsets," *Journal of the London Mathematical Society* **10** (1935), 26–30.

[9] David London, "Some notes on the van der Waerden conjecture," *Linear Algebra and Its Applications* **4** (1971), 155–160.

[10] Marvin Marcus and Morris Newman, "On the minimum of the permanent of a doubly stochastic matrix," *Duke Mathematical Journal* **26** (1959), 61–72.

[11] Henryk Minc, "Doubly stochastic matrices with minimal permanents," *Pacific Journal of Mathematics* **58** (1975), 155–157.

[12] Rolf Schneider, "On A. D. Aleksandrov's inequalities for mixed discriminants," *Journal of Mathematics and Mechanics* **15** (1966), 285–290.

[13] Richard P. Stanley, "Two combinatorial applications of the Aleksandrov–Fenchel inequalities," *Journal of Combinatorial Theory* **A17** (1981), 56–65.

[14] Bernard Teissier, "Du théorème de l'index de Hodge aux inégalités isopérimétriques," *Comptes Rendus hebdomadaires des séances de l'Académie des Sciences* **A288** (Paris: 1979), 287–289.

[15] B. L. van der Waerden, "Problem," *Jahresbericht der Deutschen Mathematiker-Vereinigung* **25** (1926), 117.

Addendum

Part of Egorychev's result was anticipated a year earlier by D. I. Falikman, whose elegant proof appears in *Matematicheskie Zametki* **19** (1981), 931–948. Falikman's paper, which was received for publication on 14 May 1979, established the minimum value of doubly stochastic permanents but did not show that the value is uniquely attained. The following references contain further information:

D. I. Falikman, "Proof of the van der Waerden conjecture regarding the permanent of a doubly stochastic matrix," *Mathematical Notes of the Academy of Sciences of the USSR* **29** (1981), 475–479. [An English translation of his Russian article.]

J. C. Lagarias, "The van der Waerden conjecture: Two Soviet solutions," *Notices of the American Mathematical Society* **29**, 2 (February 1982), 130–133.

J. H. van Lint, "The van der Waerden conjecture: Two proofs in one year," *Mathematical Intelligencer* **4** (1982), 72–77. [Includes remarks by van der Waerden on the genesis of the conjecture, and how he had not known of its fame.]

A. Schrijver, "Bounds on permanents and the number of 1-factors and 1-factorizations of a bipartite graph," in *Surveys in Combinatorics*, edited by E. Keith Lloyd, London Mathematical Society Lecture Note Series, Number 82 (Cambridge: Cambridge University Press, 1983), 107–134.

Chapter 7

Overlapping Pfaffians

[To Dominique Cyprien Foata on his 60th birthday, 12 October 1994. Originally published in The Foata Festschrift, edited by Jacques Désarménien, Adalbert Kerber, and Volker Strehl (Gap, France: Imprimerie Louie-Jean, 1996), 151–163 = The Electronic Journal of Combinatorics 3, 2 (1996), #R5.]

Abstract

A combinatorial construction proves an identity for the product of the Pfaffian of a skew-symmetric matrix by the Pfaffian of one of its submatrices. Several applications of this identity are followed by a brief history of Pfaffians.

0. Definitions

Let X be a possibly infinite index set. We consider quantities $f[xy]$ defined on ordered pairs of elements of X, satisfying the law of skew symmetry

$$f[xy] = -f[yx], \qquad \text{for} \quad x, y \in X. \tag{0.0}$$

This notation is extended to $f[\alpha]$ for arbitrary words $\alpha = x_1 \ldots x_{2n}$ of even length over X by defining the *Pfaffian*

$$f[x_1 \ldots x_{2n}] = \sum s(x_1 \ldots x_{2n}, y_1 \ldots y_{2n}) f[y_1 y_2] \ldots f[y_{2n-1} y_{2n}], \tag{0.1}$$

where the sum is over all $(2n-1)(2n-3)\ldots(1)$ ways to write the set $\{x_1, \ldots, x_{2n}\}$ as a union of pairs $\{y_1, y_2\} \cup \cdots \cup \{y_{2n-1}, y_{2n}\}$, and where the coefficient $s(x_1 \ldots x_{2n}, y_1 \ldots y_{2n})$ is the sign of the permutation that takes $x_1 \ldots x_{2n}$ into $y_1 \ldots y_{2n}$.

The Pfaffian is well defined, even though there are $2^n n!$ different permutations $y_1 \ldots y_{2n}$ that yield the same partition $\{y_1, y_2\} \cup \cdots \cup \{y_{2n-1}, y_{2n}\}$ into pairs. For if we interchange y_{2j-1} with y_{2j}, we change

105

the sign of both $s(x_1 \ldots x_{2n}, y_1 \ldots y_{2n})$ and $f[y_1 y_2] \ldots f[y_{2n-1} y_n]$, by (0.0); if we interchange y_{2i-1} with y_{2j-1} and y_{2i} with y_{2j}, both factors stay the same. Thus, for example,

$$
\begin{aligned}
f[wxyz] &= f[wx]f[yz] - f[wy]f[xz] + f[wz]f[xy] \\
&= f[wx]f[yz] + f[wy]f[zx] + f[wz]f[xy] \,.
\end{aligned} \tag{0.2}
$$

A partition into pairs is commonly called a *perfect matching*. Therefore it is convenient to abbreviate (0.1) in the form

$$
f[\alpha] = \sum_{\mu \in M(\alpha)} s(\alpha, \mu) \, \Pi f[\mu] \tag{0.3}
$$

where $M(\alpha)$ is the set of perfect matchings of α represented as words $y_1 \ldots y_{2n}$ in some canonical way, and $\Pi f[y_1 \ldots y_{2n}]$ denotes the product $f[y_1 y_2] \ldots f[y_{2n-1} y_{2n}]$.

Notice that we have

$$
f[wxyz] = -f[xyzw] \,. \tag{0.4}
$$

In general, an odd permutation of α will reverse the sign of $f[\alpha]$, because every term in (0.3) changes sign.

Pfaffians can also be defined recursively, starting with the null word ϵ and proceeding to words of greater length:

$$
f[\epsilon] = 1 \,;
$$

$$
f[x_1 \ldots x_{2n}] = \sum_{j=2}^{2n} f[x_1 x_j] f[x_{j+1} \ldots x_{2n} x_2 \ldots x_{j-1}] \,, \quad n > 0 \,. \tag{0.5}
$$

This recurrence (see Jacobi [9]) corresponds to a procedure that constructs all perfect matchings by starting with $\{x_1, x_2\} \cup \cdots \cup \{x_{2n-1}, x_{2n}\}$ and making cyclic permutations of the indices in positions $\{2, \ldots, 2n\}$, $\{4, \ldots, 2n\}$, $\{6, \ldots, 2n\}$, $\ldots$; each of those permutations is even.

It will be convenient in the sequel to extend the sign function s to $s(\alpha, \beta)$ for arbitrary words $\alpha, \beta \in X^*$. We define $s(\alpha, \beta) = 0$ if either α or β has a repeated letter, or if β contains a letter not in α. Otherwise $s(\alpha, \beta)$ is the sign of the permutation that takes α into the word

$$
\beta \, (\alpha \backslash \beta) \,,
$$

where $\alpha\backslash\beta$ is the word that remains when the elements of β are removed from α. Thus, for example,

$$s(\alpha\beta\gamma,\beta) = \begin{cases} 0, & \text{if } \alpha\beta\gamma \text{ contains a repeated letter;} \\ (-1)^{|\alpha|\,|\beta|}, & \text{otherwise.} \end{cases} \qquad (0.6)$$

(Here, as usual, $|\alpha|$ denotes the length of α.) We also have

$$s(\alpha,\beta\gamma) = s(\alpha,\beta)s(\alpha\backslash\beta,\gamma), \qquad (0.7)$$

since both sides vanish unless the letters of $\beta\gamma$ are distinct and contained in the distinct letters of α, and in the latter case $s(\alpha,\beta\gamma)$ is the parity of the number of transpositions needed to bring β to the left of α and γ to the left of the remaining word $\alpha\backslash\beta$.

If α has repeated letters, the Pfaffian $f[\alpha]$ is zero, because $f[\alpha] = -f[\alpha]$ when we transpose two identical letters. Therefore our convention that $s(\alpha,\beta) = 0$ when α or β has repeated letters does not invalidate definition (0.1), which used a different convention for $s(x_1\ldots x_{2n},y_1\ldots y_{2n})$. One consequence of the new convention is the identity

$$f[\alpha] = \sum_{x_1<\cdots<x_n} \sum_{y_1>x_1} \cdots \sum_{y_n>x_n} s(\alpha,x_1y_1\ldots x_ny_n)\, f[x_1y_1]\ldots f[x_ny_n] \qquad (0.8)$$

for any word α of length $2n$, assuming that X is an ordered set; the sum is over all conceivable perfect matchings $\mu = x_1y_1\ldots x_ny_n$, but $s(\alpha,\mu)$ is zero unless μ is a perfect matching of α.

1. The Basic Identity

The following identity due to H. W. L. Tanner [24] can now be proved:

$$f[\alpha]\,f[\alpha\beta] = \sum_y s(\beta,xy)\,f[\alpha xy]\,f[\alpha\beta\backslash xy], \qquad \text{for all } x \in \beta. \qquad (1.0)$$

This formula is vacuous when $|\beta| = 0$ and trivial when $|\beta| = 2$, but when $|\beta| = 4$ it says in particular that

$$\begin{aligned} f[\alpha]\,f[\alpha wxyz] &= f[\alpha wx]\,f[\alpha yz] - f[\alpha wy]\,f[\alpha xz] + f[\alpha wz]\,f[\alpha xy] \\ &= f[\alpha wx]\,f[\alpha yz] + f[\alpha wy]\,f[\alpha zx] + f[\alpha wz]\,f[\alpha xy]. \end{aligned} \qquad (1.1)$$

We will demonstrate (1.0) by giving a combinatorial interpretation to each term on the left and right sides of the equation, when the Pfaffians are expanded as sums over perfect matchings.

A typical term on the right of (1.0) is

$$s(\beta, xy)\, s(\alpha xy, \mu)\, s(\alpha\beta\backslash xy, \nu)\, \Pi f[\mu]\, \Pi f[\nu]\,, \qquad (1.2)$$

where x and y are distinct elements of β, and μ is a perfect matching of αxy while ν is a perfect matching of $\alpha\beta\backslash xy$. Ignoring the sign for the moment, we can construct a graph by superimposing the matchings μ and ν. In this graph all vertices of α have degree 2 because they are matched in both μ and ν; all vertices of β have degree 1.

There is a unique maximal path that starts at y and uses edges from μ and ν alternately. This path ends at some element of β, call it z. Let μ_1 and ν_1 be the edges of μ and ν on this path; let μ_0 and ν_0 be the other edges. Then we define corresponding matchings

$$\mu' = \mu_0 \cup \nu_1\,, \qquad \nu' = \nu_0 \cup \mu_1\,, \qquad (1.3)$$

which will be the key to establishing (1.0).

Case 1, $z \neq x$. In this case $|\mu_1| = |\nu_1|$, since the path from y starts with an element of μ and ends with an element of ν. Thus the matchings μ' and ν' correspond to another term on the right side of (1.0); we will prove that this other term cancels with (1.2). Since $\mu'' = \mu$ and $\nu'' = \nu$, this will set up an involution between cancelling terms.

We have

$$\Pi f[\mu]\, \Pi f[\nu] = \Pi f[\mu_0]\, \Pi f[\mu_1]\, \Pi f[\nu_0]\, \Pi f[\nu_1] = \Pi f[\mu']\, \Pi f[\nu']\,, \qquad (1.4)$$

so (1.2) will cancel with its counterpart if the signs differ. The sign of (1.2) is

$$s(\alpha xyz, \mu_0\mu_1 z)\, s(\alpha\beta, xy\nu_0\nu_1)\,, \qquad (1.5)$$

because $s(\beta, xy) = s(\alpha\beta, xy)$ and $s(\alpha\beta, xy)\, s(\alpha\beta\backslash xy, \nu) = s(\alpha\beta, xy\nu)$ by (0.7). The sign of the permutation that takes $\mu_1 z$ into $\nu_1 y$ is the same as the sign of the permutation that takes $y\nu_0\nu_1$ into $z\nu_0\mu_1$, hence (1.5) equals

$$s(\alpha xyz, \mu_0\nu_1 y)\, s(\alpha\beta, xz\nu_0\mu_1)\,.$$

But this is the negative of $s(\alpha xzy, \mu_0\nu_1 y)\, s(\alpha\beta, xz\nu_0\mu_1)$, the sign of the term that corresponds to μ' and ν'.

Case 2, $z = x$. In this case we have $|\mu_1| = |\nu_1| + 2$, since μ_1 includes both x and y while ν_1 is contained in α. It follows that μ' and ν' are perfect matchings of α and $\alpha\beta$, respectively, so they define a typical term

$$s(\alpha, \mu') \, s(\alpha\beta, \nu') \, \Pi f[\mu'] \, \Pi f[\nu'] \qquad (1.6)$$

from the left side of (1.0). Conversely, every such term corresponds to matchings μ and ν for a uniquely defined term (1.2) on the right. The sign of this term,

$$s(\alpha xy, \mu_0 \mu_1) \, s(\alpha\beta, xy\nu_0\nu_1) \,,$$

agrees with $s(\alpha, \mu') \, s(\alpha\beta, \nu') = s(\alpha xy, \mu_0\nu_1 xy) \, s(\alpha\beta, \nu_0\mu_1)$, because the permutation that takes μ_1 into $\nu_1 xy$ has the same sign as the permutation that takes $xy\nu_0\nu_1$ into $\nu_0\mu_1$.

2. Basic Applications

The special case $\alpha = \epsilon$ of (1.0) reads

$$f[\beta] = \sum_y s(\beta, xy) \, f[xy] \, f[\beta \backslash xy] \,, \qquad \text{for all } x \in \beta. \qquad (2.0)$$

This is a mild generalization of the recurrence (0.5); it tells us how to expand $f[\beta]$ with respect to any element of β. We can get rid of the constraint $x \in \beta$ by summing over all x:

$$f[\beta] = \frac{1}{|\beta|} \sum_x \sum_y s(\beta, xy) \, f[xy] \, f[\beta \backslash xy] \,. \qquad (2.1)$$

Applying this rule to $f[\beta \backslash xy]$ and repeating until words of length 2 are reached yields a $|\beta|$-fold sum,

$$f[\beta] = \frac{1}{(2n)(2n-2)\ldots 2} \sum_{x_1} \cdots \sum_{x_{2n}} s(\beta, x_1 \ldots x_{2n}) \, f[x_1 x_2] \ldots f[x_{2n-1} x_{2n}] \,, \qquad (2.2)$$

when $|\beta| = 2n$; this is, of course, the same as (0.8) when we collect equal terms.

Now let α be a fixed word such that $f[\alpha] \neq 0$, and consider the function

$$g(\beta) = f[\alpha\beta]/f[\alpha] \qquad (2.3)$$

on the words of X. Tanner's identity (1.0) tells us that

$$g(\beta) = \sum_y s(\beta, xy) \, g(xy) \, g(\beta \backslash xy) \,, \qquad \text{for all } x \in \beta. \qquad (2.4)$$

But this is the same relation as (2.0); so g satisfies the Pfaffian recurrence (0.5). Therefore any identity for Pfaffians leads *a fortiori* to an identity for g. In particular, (0.3) tells us that

$$g(\beta) = \sum_{\mu \in M(\beta)} s(\beta, \mu) \, \Pi g(\mu) \,,$$

which is equivalent to

$$f[\alpha]^{n-1} f[\alpha\beta] = \sum_{M(\beta)} s(\beta, x_1 y_1 \ldots x_n y_n) \, f[\alpha x_1 y_1] \ldots f[\alpha x_n y_n] \quad (2.5)$$

when $|\beta| = 2n$, where the sum is over all perfect matchings $x_1 y_1 \ldots x_n y_n$ of β. The special case $n = 2$ appears in (1.1).

We can also construct a dual formula by starting with a fixed word $\alpha\beta$ such that $f[\alpha\beta] \neq 0$ and defining

$$h(\gamma) = s(\alpha\beta, \gamma) \, f[\alpha\beta \backslash \gamma] / f[\alpha\beta] \quad (2.6)$$

on the words γ contained in $\alpha\beta$. Then (1.0) yields

$$h(\beta) = \sum_y s(\beta, xy) \, h(\beta \backslash xy) \, h(xy) \,, \qquad \text{for all } x \in \beta; \quad (2.7)$$

so we can derive a companion to (2.5) in a similar fashion:

$$f[\alpha] \, f[\alpha\beta]^{n-1} = \sum_{M(\beta)} s(\beta, x_1 y_1 \ldots x_n y_n) \, f[\alpha\beta \backslash x_1 y_1] \ldots f[\alpha\beta \backslash x_n y_n] \,.$$
$$(2.8)$$

Identities (2.4) and (2.7) are the Pfaffian analogs of theorems about determinants that Muir called the Law of Extensible Minors and the Law of Complementaries. (See [15], §179 and §98 in the original edition; §187 and §179 in Metzler's revision.)

3. Applications to Determinants

Determinants are the special case of Pfaffians in which the index set is bipartite with respect to f, in the sense that $f[xy] = 0$ when x and y belong to the same part. It is convenient to imagine that the set of indices consists of two disjoint parts X and $\bar{X}$, so that x belongs to X if and only if $\bar{x}$ belongs to $\bar{X}$, and $f[xy] = f[\bar{x}\bar{y}] = 0$ for all $x, y \in X$. The independent quantities are now $f[x\bar{y}] = -f[\bar{y}x]$; we can regard X as a set of "rows" and $\bar{X}$ as a set of "columns," so that $f[x\bar{y}]$ is essentially an element of the matrix f. We use $f[x,y]$ as an alternative notation for $f[x\bar{y}]$. In fact, when α and β are arbitrary words of X we write

$$f[\alpha, \beta] = f[\alpha\bar{\beta}^R] \tag{3.0}$$

for the determinant formed from rows α and columns β. Here $\bar{\beta}^R$ stands for the reverse complement of β:

$$\overline{y_1 y_2 \cdots y_n}^R = \bar{y}_n \cdots \bar{y}_2 \bar{y}_1 . \tag{3.1}$$

Definition (3.0) agrees with the usual definition of determinants, when $|\alpha| = |\beta| = n$, since the perfect matchings of $\alpha\bar{\beta}^R$ that do not have vanishing products correspond to the products

$$f[x_1\bar{y}_1] \cdots f[x_n\bar{y}_n] = f[x_1, y_1] \cdots f[x_n, y_n] , \tag{3.2}$$

where $\alpha = x_1 \ldots x_n$ and $y_1 \ldots y_n$ is a permutation of β; the corresponding sign $s(\alpha\bar{\beta}^R, x_1\bar{y}_1 \ldots x_n\bar{y}_n)$ is just $s(\beta, y_1 \ldots y_n)$, because the permutation that takes $x_1 \ldots x_n \bar{y}_n \ldots \bar{y}_1$ to $x_1\bar{y}_1 \ldots x_n\bar{y}_n$ is even. For example, we have

$$\begin{aligned}
f[wx, yz] &= f[wx\bar{z}\bar{y}] \\
&= f[wx]\,f[\bar{z}\bar{y}] - f[w\bar{z}]\,f[x\bar{y}] + f[w\bar{y}]\,f[x\bar{z}] \\
&= 0 - f[w,z]\,f[x,y] + f[w,y]\,f[x,z] ,
\end{aligned}$$

and this is the usual 2×2 determinant

$$\begin{vmatrix} f[w,y] & f[w,z] \\ f[x,y] & f[x,z] \end{vmatrix} .$$

Theorem (1.0) immediately yields a corresponding identity for determinants, when we apply these definitions:

$$f[\alpha, \beta]\,f[\alpha\gamma, \beta\delta] = \sum_y s(\gamma, x)\,s(\delta, y)\,f[\alpha x, \beta y]\,f[\alpha\gamma\backslash x, \beta\delta\backslash y] , \tag{3.3}$$

for all $x \in \gamma$. When $|\gamma| = |\delta|$ is 2 or 3, this identity reads

$$\begin{aligned}
f[\alpha, \beta]\,f[\alpha wx, \beta yz] = \ &f[\alpha w, \beta y]\,f[\alpha x, \beta z] \\
&- f[\alpha w, \beta z]\,f[\alpha x, \beta y] ; \tag{3.4}
\end{aligned}$$

$$\begin{aligned}
f[\alpha, \beta]\,f[\alpha uvw, \beta xyz] = \ &f[\alpha u, \beta x]\,f[\alpha vw, \beta yz] \\
&- f[\alpha u, \beta y]\,f[\alpha vw, \beta xz] \\
&+ f[\alpha u, \beta z]\,f[\alpha vw, \beta xy] . \tag{3.5}
\end{aligned}$$

Here are some small examples written in more conventional notation:

$$a_{11}\begin{vmatrix} a_{11} & a_{12} & a_{13} \\ a_{21} & a_{22} & a_{23} \\ a_{31} & a_{32} & a_{33} \end{vmatrix} = \begin{vmatrix} a_{11} & a_{12} \\ a_{21} & a_{22} \end{vmatrix}\begin{vmatrix} a_{11} & a_{13} \\ a_{31} & a_{33} \end{vmatrix} \\ - \begin{vmatrix} a_{11} & a_{13} \\ a_{21} & a_{23} \end{vmatrix}\begin{vmatrix} a_{11} & a_{12} \\ a_{31} & a_{32} \end{vmatrix} ; \tag{3.6}$$

$$\begin{vmatrix} a_{11} & a_{12} \\ a_{21} & a_{22} \end{vmatrix}\begin{vmatrix} a_{11} & a_{12} & a_{13} & a_{14} \\ a_{21} & a_{22} & a_{23} & a_{24} \\ a_{31} & a_{32} & a_{33} & a_{34} \\ a_{41} & a_{42} & a_{43} & a_{44} \end{vmatrix} = \begin{vmatrix} a_{11} & a_{12} & a_{13} \\ a_{21} & a_{22} & a_{23} \\ a_{31} & a_{32} & a_{33} \end{vmatrix}\begin{vmatrix} a_{11} & a_{12} & a_{14} \\ a_{21} & a_{22} & a_{24} \\ a_{41} & a_{42} & a_{44} \end{vmatrix} \\ - \begin{vmatrix} a_{11} & a_{12} & a_{14} \\ a_{21} & a_{22} & a_{24} \\ a_{31} & a_{32} & a_{34} \end{vmatrix}\begin{vmatrix} a_{11} & a_{12} & a_{13} \\ a_{21} & a_{22} & a_{23} \\ a_{41} & a_{42} & a_{43} \end{vmatrix} ; \tag{3.7}$$

$$a_{11}\begin{vmatrix} a_{11} & a_{12} & a_{13} & a_{14} \\ a_{21} & a_{22} & a_{23} & a_{24} \\ a_{31} & a_{32} & a_{33} & a_{34} \\ a_{41} & a_{42} & a_{43} & a_{44} \end{vmatrix} = \begin{vmatrix} a_{11} & a_{12} \\ a_{21} & a_{22} \end{vmatrix}\begin{vmatrix} a_{11} & a_{13} & a_{14} \\ a_{31} & a_{33} & a_{34} \\ a_{41} & a_{43} & a_{44} \end{vmatrix} \\ - \begin{vmatrix} a_{11} & a_{13} \\ a_{21} & a_{23} \end{vmatrix}\begin{vmatrix} a_{11} & a_{12} & a_{14} \\ a_{31} & a_{32} & a_{34} \\ a_{41} & a_{42} & a_{44} \end{vmatrix} \\ + \begin{vmatrix} a_{11} & a_{14} \\ a_{21} & a_{24} \end{vmatrix}\begin{vmatrix} a_{11} & a_{12} & a_{13} \\ a_{31} & a_{32} & a_{33} \\ a_{41} & a_{42} & a_{43} \end{vmatrix} . \tag{3.8}$$

Of course determinants have been investigated rather thoroughly for nearly 250 years, so it would be surprising indeed if these identities were new. Equation (3.6) was, for instance, noted by Lagrange in 1773 [16, page 39]; (3.7) and higher examples of (3.4) were discussed by Desnanot in 1819 [16, page 142].

One particularly interesting case in which (3.4) played a crucial role is C. L. Dodgson's elegant "condensation method" for determinant evaluation [7], discovered between the times when he wrote *Alice in Wonderland* and *Through the Looking Glass*: Suppose the index set X is the integers, and let $f_0[x, y] = 1$ for all x and y, while $f_1[x, y]$ is the entry in row x and column y of a given matrix. Then for $k \geq 1$ let

$$f_{k+1}[x, y] = \begin{vmatrix} f_k[x, y] & f_k[x, y + 1] \\ f_k[x + 1, y] & f_k[x + 1, y + 1] \end{vmatrix} / f_{k-1}[x + 1, y + 1]. \tag{3.9}$$

It follows that

$$f_k[x, y] = f_1[x(x + 1) \ldots (x + k - 1), y(y + 1) \ldots (y + k - 1)] \tag{3.10}$$

for all $k \geq 0$, by induction on k using (3.4). To evaluate the $n \times n$ determinant $f[1\,2\,\ldots\,n,\,1\,2\,\ldots\,n]$, we may therefore simply compute $f_k[x, y]$ for $1 \leq x, y \leq n + 1 - k$ and $k = 2, \ldots, n$, hoping that it will not be necessary to divide by zero. Dodgson's condensation method provided the original motivation for Robbins and Rumsey's recent work on alternating sign matrices [19].

The earliest known identity involving products of determinants is

$$f[ab, 12]\, f[ab, 34] - f[ab, 13]\, f[ab, 24] + f[ab, 14]\, f[ab, 23] = 0\,, \quad (3.11)$$

which Alexis Fontaine des Bertins proudly wrote out 126 times for different choices of the indices and then said "et cetera." He submitted this and other memoirs to the French academy in 1748, but the works remained unpublished until 1764 [16, pages 10–11]. From (1.0) we can now recognize that the left-hand side of (3.11) is actually a Pfaffian product

$$f[ab]\, f[ab\bar{1}\bar{2}\bar{3}\bar{4}]\,,$$

which is indeed zero in the bipartite case. Bezout, in 1779, gave the similar formula

$$\begin{aligned}
f[abc, 123]\, f[abc, 456] &- f[abc, 124]\, f[abc, 356] \\
&+ f[abc, 125]\, f[abc, 346] - f[abc, 126]\, f[abc, 345] = 0\,, \quad (3.12)
\end{aligned}$$

and said "on voit qu'il y a une infinité d'autres combinaisons à faire" [16, page 51]; the left-hand side in this case is

$$f[abc\bar{1}\bar{2}]\, f[abc\bar{1}\bar{2}\bar{3}\bar{4}\bar{5}\bar{6}]$$

when we replace determinants by Pfaffians.

Another instance of (1.0) yields

$$\begin{aligned}
f[ab]\, f[abc\bar{1}\bar{2}\bar{3}\bar{4}\bar{5}] = f[ab\bar{1}\bar{2}]\, f[abc\bar{3}\bar{4}\bar{5}] &- f[ab\bar{1}\bar{3}]\, f[abc\bar{2}\bar{4}\bar{5}] \\
&+ f[ab\bar{1}\bar{4}]\, f[abc\bar{2}\bar{3}\bar{5}] - f[ab\bar{1}\bar{5}]\, f[abc\bar{2}\bar{3}\bar{4}] \\
&- f[ab\bar{1}c]\, f[ab\bar{2}\bar{3}\bar{4}\bar{5}]\,. \quad (3.13)
\end{aligned}$$

Under bipartite restrictions this formula becomes an identity in determinants,

$$\begin{aligned}
f[ab, 12]\, f[abc, 345] &- f[ab, 13]\, f[abc, 245] \\
&+ f[ab, 14]\, f[abc, 235] - f[ab, 15]\, f[abc, 234] = 0\,, \quad (3.14)
\end{aligned}$$

which Desnanot [6] seems to have known only in the special case where column 1 = column 5,

$$f[ab, 12] f[abc, 134] - f[ab, 13] f[abc, 124]$$
$$+ f[ab, 14] f[abc, 123] = 0, \qquad (3.15)$$

although he knew the general result (3.3) [16, page 145].

Thus we see that the single Pfaffian identity (1.0) unifies a variety of different-appearing determinant identities that arise when the indices have been assigned a bipartite structure in different ways.

When identity (2.8) is specialized to determinants, it gives a formula for minors of the adjugate of a matrix (determinants of cofactors):

$$f[\alpha, \beta] f[\alpha x_1 \ldots x_n, \beta y_1 \ldots y_n]^{n-1}$$
$$= \begin{vmatrix} f[\alpha x_2 \ldots x_n, \beta y_2 \ldots y_n] & \cdots & f[\alpha x_2 \ldots x_n, \beta y_1 \ldots y_{n-1}] \\ \vdots & & \vdots \\ f[\alpha x_1 \ldots x_{n-1}, \beta y_2 \ldots y_n] & \cdots & f[\alpha x_1 \ldots x_{n-1}, \beta y_1 \ldots y_{n-1}] \end{vmatrix}.$$
$$(3.16)$$

This general formula was first published by Jacobi in 1834, although special cases had been found by Lagrange in 1773 and Minding in 1829 [16, pages 39, 197, 208–209]. The formula that corresponds to (2.5),

$$f[\alpha, \beta]^{n-1} f[\alpha x_1, \ldots x_n, \beta y_1 \ldots y_n] = \begin{vmatrix} f[\alpha x_1, \beta y_1] & \cdots & f[\alpha x_1, \beta y_n] \\ \vdots & & \vdots \\ f[\alpha x_n, \beta y_1] & \cdots & f[\alpha x_n, \beta y_n] \end{vmatrix},$$
$$(3.17)$$

is simpler but was not discovered until Sylvester introduced a new viewpoint in 1851 [17, pages 60–61].

4. Applications to Closed Forms

Let g be the skew-symmetric Blaschke operator

$$g[xy] = \frac{x - y}{1 - xy}. \qquad (4.0)$$

Laksov, Lascoux, and Thorup [10, (A.12.3)] and John R. Stembridge [23, Proposition 2.3(e)] independently discovered the remarkable identity

$$g[x_1 x_2 \ldots x_n] = \prod_{1 \leq i < j \leq n} \frac{x_i - x_j}{1 - x_i x_j}, \qquad n \text{ even}, \qquad (4.1)$$

for which they gave ingenious but rather special-purpose proofs.

We can, however, prove (4.1) as a special case of a more general theorem that follows from a special case of (1.0):

Theorem. *The identity*

$$f[x_1 \dots x_n] = \prod_{1 \le i < j \le n} f[x_i x_j] \tag{4.2}$$

holds for all even n if and only if it holds for $n = 4$.

Proof. If $n > 4$ and the identity holds for smaller even values of n, let α be any word of length $n - 4$. Then

$$
\begin{aligned}
f[\alpha]\,f[\alpha wxyz] &= f[\alpha wx]\,f[\alpha yz] - f[\alpha wy]\,f[\alpha xz] + f[\alpha wz]\,f[\alpha xy] \\
&= R(f[wx]\,f[yz] - f[wy]\,f[xz] + f[wz]\,f[xy]) \\
&= R\,f[wxyz] \\
&= R\,f[wx]\,f[wy]\,f[wz]\,f[xy]\,f[xz]\,f[yz] \,,
\end{aligned}
$$

where if $\alpha = x_1 \dots x_{n-4}$ the common factor R is

$$
\left(\prod_{1 \le i < j \le n-4} f[x_i x_j]^2 \right) \left(\prod_{1 \le i \le n-4} f[x_i w]\,f[x_i x]\,f[x_i y]\,f[x_i z] \right).
$$

Therefore

$$f[x_1 \dots x_{n-4}]\,f[x_1 \dots x_n] = f[x_1 \dots x_{n-4}] \prod_{1 \le i < j \le n} f[x_i x_j].$$

Equation (4.2) follows unless $f[x_1 \dots x_{n-4}] = 0$.

If $f[y_1 \dots y_{n-4}] = 0$ for all subwords $y_1 \dots y_{n-4}$ of $x_1 \dots x_n$, then $f[x_1 \dots x_n] = 0$ and again (4.2) holds. Finally, if $y_1 \dots y_{n-4}$ is a subword such that $f[y_1 \dots y_{n-4}] \ne 0$, there is a permutation $y_1 \dots y_n$ of $x_1 \dots x_n$ for which our argument proves $f[y_1 \dots y_n] = \prod_{1 \le i < j \le n} f[y_i y_j]$. This establishes (4.2), because permutations of the indices change the signs of both sides in the same manner. □

The theorem is of interest because it applies not only to (4.0) but also to the simpler function

$$f[x_i x_j] = \frac{x_i - x_j}{c + x_i + x_j} \tag{4.3}$$

when c is any complex constant. Thus we obtain a more-or-less "closed form" (4.2) for the Pfaffian of a new kind of matrix. (The special case $c = 0$ was previously noted by Schur [22, §36].)

In fact, the general function

$$f[x_i x_j] = \frac{x_i - x_j}{c + b(x_i + x_j) + a\,x_i x_j}\,, \qquad b^2 = ac \pm 1\,, \qquad (4.4)$$

also satisfies the necessary conditions; this expression includes both (4.0) and (4.3).

Are there other skew-symmetric rational functions of two variables that satisfy

$$f[wx]\,f[yz] + f[wy]\,f[zx] + f[wz]\,f[xy]$$
$$= f[wx]\,f[wy]\,f[wz]\,f[xy]\,f[xz]\,f[yz]\,? \qquad (4.5)$$

One can, of course, replace $f[xy]$ by $f[r(x)r(y)]$ for any rational function r, so any solution of (4.5) implies an infinite class of equivalent solutions. Alain Lascoux [11] has recently found strong reasons for believing that there are no other solutions, up to changes of variables.

When $f[xy]$ is a polynomial, an amusing closed form of a similar type was noticed by G. Torelli [25]: Let $f_k[xy] = (x - y)^k$; then

$$f_{n-1}[x_1 \ldots x_n] = (-1)^{\binom{n/2}{2}} \left(\prod_{k=0}^{n/2-1} \binom{n-1}{k} \right) \prod_{1 \le i < j \le n} (x_i - x_j) \quad (4.6)$$

when n is even. It is easy to prove this identity, as well as the fact that $f_{2m-1}[x_1 \ldots x_n] = 0$ for $2m < n$, by observing that the Pfaffian must vanish when $x_i = x_j$.

5. Generalization of the Basic Identity

Equation (1.0), which gives an expression for $f[\alpha]\,f[\alpha\beta]$ when α is a proper subword of $\alpha\beta$, leads to a similar identity that is useful when two words have an odd number of letters in common. Suppose $\alpha\beta\gamma$ has no repeated letters, and let $x \in \beta$. Then

$$f[\alpha\beta]\,f[\alpha\gamma] = \sum_y s(\beta, xy)\,f[\alpha\beta\backslash xy]\,f[\alpha\gamma xy]$$
$$+ \sum_y s(\beta, x)\,s(\gamma, y)\,f[\alpha y\beta\backslash x]\,f[\alpha x\gamma\backslash y]\,. \qquad (5.0)$$

For example, when $|\alpha|$ is odd we have

$$f[\alpha xyz]\,f[\alpha uvw] = f[\alpha z]\,f[\alpha uvwxy] - f[\alpha y]\,f[\alpha uvwxz]$$
$$+ f[\alpha uyz]\,f[\alpha xvw] - f[\alpha vyz]\,f[\alpha xuw]$$
$$+ f[\alpha wyz]\,f[\alpha xuv]\,. \qquad (5.1)$$

To prove (5.0), let $\gamma = x_1 \ldots x_k$. We will construct a "cancelling" word $\gamma' = x'_k \ldots x'_1$ on new indices, by defining

$$f[yx'_j] = 0 \quad \text{if} \quad y \neq x_j; \qquad f[x_j x'_j] = 1. \tag{5.2}$$

Then $f[\alpha\beta] = f[\alpha\gamma\gamma'\beta]$, and we can use (1.0) to conclude that

$$f[\alpha\beta]\,f[\alpha\gamma] = \sum_y s(\gamma'\beta, xy)\,f[\alpha\gamma\gamma'\beta\backslash xy]\,f[\alpha\gamma xy]\,. \tag{5.3}$$

Now if $y \in \beta$ we have $s(\gamma'\beta, xy) = s(\beta, xy)$, and $f[\alpha\gamma\gamma'\beta\backslash xy] = f[\alpha\beta\backslash xy]$. But if $y = x'_j$ we have $s(\gamma'\beta, xy) = (-1)^j s(\beta, x)$, while $f[\alpha\gamma\gamma'\beta\backslash xy] = (-1)^{j-1} f[\alpha y \beta\backslash x]$, $f[\alpha\gamma xy] = (-1)^j f[\alpha x\gamma\backslash y]$, and $s(\gamma, y) = (-1)^{j-1}$.

6. A Brief History of Pfaffians

Johann Friedrich Pfaff introduced the functions that now bear his name in 1815 [18] [16, pages 396–401], while studying a general way to solve systems of first-order partial differential equations. He gave two procedures for listing all perfect matchings, and observed that when the matchings are ordered lexicographically the corresponding signs are strictly alternating $+, -, +, \ldots, +$.

Jacobi developed Pfaff's method further in 1827 [9], and discovered an analog of "Cramer's rule" for the solution of general systems of skew-symmetric linear equations

$$\sum_{j=1}^{2n} f[ij]\,z_j = f[i0]\,, \qquad n \text{ even}; \tag{6.0}$$

namely,

$$z_j = \frac{f[1\ldots(j-1)\,0\,(j+1)\ldots n]}{f[1\ldots n]}\,. \tag{6.1}$$

This implicitly proves that the Pfaffian $f[1\ldots n]$ is a factor of the general skew-symmetric determinant

$$\begin{vmatrix} f[11] & \cdots & f[1n] \\ \vdots & & \vdots \\ f[n1] & \cdots & f[nn] \end{vmatrix}, \qquad n \text{ even}. \tag{6.2}$$

Cayley demonstrated in 1849 [3] that this determinant is in fact equal to the *square* of $f[1\ldots n]$.

An elegant graph-theoretic proof of Cayley's theorem, somewhat analogous to the derivation of (1.0) above, was found by Veltmann in 1871 [26] and independently by Mertens in 1877 [14]. Their proof anticipated 20th-century studies on the superposition of two matchings, and the ideas have frequently been rediscovered. Cayley himself had claimed that such a proof would be possible, after doing the calculations for $n = 4$ on the final page of a paper he wrote in 1861 [5]. But we should note that his original method was simpler. In fact, Cayley originally [3] gave a short inductive proof of the more general formula

$$\begin{vmatrix} f[xy] & f[x2] & f[x3] & \dots & f[xn] \\ f[2y] & f[22] & f[23] & \dots & f[2n] \\ f[3y] & f[32] & f[33] & \dots & f[3n] \\ \vdots & \vdots & \vdots & & \vdots \\ f[ny] & f[n2] & f[n3] & \dots & f[nn] \end{vmatrix} = f[x23\dots n]\, f[y23\dots n], \quad (6.3)$$

for arbitrary x and y when n is even. And he proved several years later [17, pages 269, 278] that the determinant on the left of (6.3) is $f[xy23\dots n]\, f[23\dots n]$ when n is odd. (This determinant is incidentally *not* the same as $f[x23\dots n, y23\dots n]$; the elements of the latter are $f[x, y]$, $f[x, 2]$, $f[x, 3]$, $\dots = f[x\bar{y}]$, $f[x\bar{2}]$, $f[x\bar{3}]$, $\dots$, not $f[xy]$, $f[x2]$, $f[x3]$, $\dots$, according to our conventions. Moreover, we generally use the notation $f[x, y]$ only when we assume that $f[xy] = 0$.)

It was Cayley who introduced the name *Pfaffian*, because of its "connexion with the researches of Pfaff on differential equations" [4]. Another name *semideterminant* (German *Halbdeterminant*) was proposed by Wilhelm Scheibner [21], but it did not gain many adherents.

Theorem (1.0) was discovered by Henry William Lloyd Tanner in 1878 [24], who gave inductive proofs for the cases $|\beta| = 4$ and $|\beta| = 6$ from which proof schemata for higher cases could be inferred. Władysław Zajaczkowski found another proof shortly afterward [28] [29] based on Jacobi's determinant theorem (3.16). The theorem was independently rediscovered in 1901 by J. Brill [1], who found a still better proof. He first established the identity

$$\binom{n-1}{k} f[x_1 \dots x_{2n}] = \sum_{1 \le j_1 < \dots < j_{2k} \le 2n} s(x_1 \dots x_{2n}, x_{j_1} \dots x_{j_{2k}})$$
$$\times f[x_{j_1} \dots x_{j_{2k}}]\, f[x_1 \dots x_{2n} \backslash x_{j_1} \dots x_{j_{2k}}] \quad (6.4)$$

by induction on k; then he made the left side zero by setting $x_{2n} = x_1$. A series of further steps led him to (1.0). But the combinatorial proof in Section 1 above seems preferable to all three of these early approaches.

Identity (5.0) was recently discovered by Wenzel [27, Proposition 2.3], and demonstrated via exterior algebra by Dress and Wenzel [8].

The fact that Pfaffians are more fundamental than determinants, in the sense that determinants are merely the bipartite special case of a general sum over matchings, went unnoticed for a long time. The first person to observe that every $n \times n$ determinant is a Pfaffian was apparently Louis Saalschütz in 1908 [20], but the implicitly bipartite nature of his construction was not stated in his paper; a modern reader sees it only with hindsight. Brioschi had found a complicated way to express a $2n \times 2n$ determinant as a Pfaffian, in 1856 [2]: If A is any $2n \times 2n$ matrix and if $Q = I_n \otimes \left(\begin{smallmatrix} 0 & 1 \\ -1 & 0 \end{smallmatrix} \right)$, the determinant of A is the Pfaffian of $A^T Q A$.

Pfaffians continue to find numerous applications, for example in matching theory [13] and in the enumeration of plane partitions [23]. It should prove interesting to extend Leclerc's combinatorics of relations for determinants [12] to the analogous rules for Pfaffians.

Acknowledgments

Discussions with Lyle Ramshaw helped greatly to clarify my proof of (1.0). Paul Algoet kindly corrected several typographical errors in my preprint. Alain Lascoux referred me to [10] and [12], John Stembridge told me about [22], and an anonymous referee called my attention to [8]. I also thank the editors for their patience.

References

[1] J. Brill, "Note on the algebraic properties of Pfaffians," *Proceedings of the London Mathematical Society* **34** (1901), 143–151.

[2] F. Brioschi, "Sur l'analogie entre une classe de déterminants d'order pair; et sur les déterminants binaires," *Journal für die reine und angewandte Mathematik* **52** (1856), 133–141.

[3] A. Cayley, "Sur les déterminants gauches," *Journal für die reine und angewandte Mathematik* **38** (1849), 93–96. Reprinted in his *Collected Mathematical Papers* **1**, 410–413.

[4] A. Cayley, "On the theory of permutants," *Cambridge and Dublin Mathematical Journal* **7** (1852), 40–51. Reprinted in his *Collected Mathematical Papers* **2**, 16–26.

[5] A. Cayley, "Note on the theory of determinants," *Philosophical Magazine* **21** (1861), 180–185. Reprinted in his *Collected Mathematical Papers* **5**, 45–49.

[6] P. Desnanot, *Complément de la Théorie des Équations du Premier Degré* (Paris: 1819).

[7] C. L. Dodgson, "Condensation of determinants, being a new and brief method for computing their arithmetical values," *Proceedings of the Royal Society* **84** (1866), 150–155. Reprinted in *The Mathematical Pamphlets of Charles Lutwidge Dodgson and Related Pieces*, edited by Francine F. Abeles (Charlottesville, Virginia: The University Press of Virginia, 1994), 170–180.

[8] Andreas W. M. Dress and Walter Wenzel, "A simple proof of an identity concerning Pfaffians of skew symmetric matrices," *Advances in Mathematics* **112** (1995), 120–134.

[9] C. G. J. Jacobi, "Ueber die *Pfaff*sche Methode, eine gewöhnliche lineäre Differentialgleichung zwischen $2n$ Variabeln durch ein System von n Gleichungen zu integriren," *Journal für die reine und angewandte Mathematik* **2** (1827), 347–357. Reprinted in *C. G. J. Jacobi's Gesammelte Werke* **4** (1886), 17–29.

[10] D. Laksov, A. Lascoux, and A. Thorup, "On Giambelli's theorem on complete correlations," *Acta Mathematica* **162** (1989), 143–199.

[11] Alain Lascoux, personal communication (10 April 1995).

[12] Bernard Leclerc, "On identities satisfied by minors of a matrix," *Advances in Mathematics* **100** (1993), 101–132.

[13] László Lovász and Michael D. Plummer, *Matching Theory* (Budapest: Akadémiai Kiadó, 1986) = North-Holland Mathematics Studies **121**.

[14] F. Mertens, "Über die Determinanten, deren correspondirende Elemente a_{pq} und a_{qp} entgegengesetzt gleich sind," *Journal für die reine und angewandte Mathematik* **82** (1877), 207–211.

[15] Thomas Muir, *A Treatise on the Theory of Determinants* (London: Macmillan, 1882). Revised and enlarged by William H. Metzler (London: Longmans, Green, 1933; New York: Dover, 1960).

[16] Thomas Muir, *The Theory of Determinants in the Historical Order of Development* (London: MacMillan, 1906).

[17] Thomas Muir, *The Theory of Determinants in the Historical Order of Development*, Volume 2 (London: MacMillan, 1911).

[18] J. F. Pfaff, "Methodus generalis, aequationes differentiarum partialium, nec non aequationes differentiales vulgares, utrasque primi ordinis, inter quotcunque variabiles, completi integrandi," *Abhandlungen der Königlich-Preußischen Akademie der Wissenschaften zu Berlin*, Mathematische Klasse (1814–1815), 76–136.

[19] David P. Robbins and Howard Rumsey, Jr., "Determinants and alternating sign matrices," *Advances in Mathematics* **62** (1986), 169–184.

[20] Louis Saalschütz, "Zur Determinanten-Lehre," *Journal für die reine und angewandte Mathematik* **134** (1908), 187–197.

[21] W. Scheibner, "Über Halbdeterminanten," *Berichte über die Verhandlungen der Königlich Sächsischen Gesellschaft der Wissenschaften zu Leipzig* **11** (1859), 151–159.

[22] J. [sic] Schur, "Über die Darstellung der symmetrischen und der alternierenden Gruppe durch gebrochene lineare Substitutionen," *Journal für die reine und angewandte Mathematik* **139** (1910–1911), 155–250. Reprinted in Issai Schur, *Gesammelte Abhandlungen* **1** (1973), 346–441.

[23] John R. Stembridge, "Nonintersecting paths, Pfaffians, and plane partitions," *Advances in Mathematics* **83** (1990), 96–131.

[24] H. W. Lloyd Tanner, "A theorem relating to Pfaffians," *Messenger of Mathematics* **8** (1878), 56–59.

[25] Gabriele Torelli, "Quistione 64," *Giornale di Matematiche* **24** (1886), 377.

[26] W. Veltmann, "Beiträge zur Theorie der Determinanten," *Zeitschrift für Mathematik und Physik* **16** (1871), 516–525.

[27] Walter Wenzel, "Pfaffian forms and Δ-matroids," *Discrete Mathematics* **115** (1993), 253–266.

[28] W. Zajaczkowski, "A theorem relating to Pfaffians," *Messenger of Mathematics* **10** (1880), 36–37.

[29] W. Zajączkowski, "O pewnéj własności pfafianu," *Rozprawy i Sprawozdania z Posiedzeń*, Wydziału Matematyczno-Przyrodniczego Akademii Umiejętności **7** (Krakow: 1880), 67–74.

Addendum

The approach used here to prove (1.0) has been developed further by A. M. Hamel, "Pfaffian identities: A combinatorial approach," *Journal of Combinatorial Theory* **A94** (2001), 205–217.

Chapter 8

The Sandwich Theorem

This report contains expository notes about a function $\vartheta(G)$ that is popularly known as the Lovász number of a graph G. There are many ways to define $\vartheta(G)$, and the surprising variety of different characterizations indicates in itself that $\vartheta(G)$ should be interesting. But the most interesting property of $\vartheta(G)$ is probably the fact that it can be computed efficiently, although it lies "sandwiched" between other classic graph numbers whose computation is NP-hard. I have tried to make these notes self-contained so that they might serve as an elementary introduction to the growing literature on Lovász's fascinating function.

[Originally published in The Electronic Journal of Combinatorics 1 (1994), #A1.]

It is NP-complete to compute $\omega(G)$, the size of the largest clique in a graph G, and it is NP-complete to compute $\chi(G)$, the minimum number of colors needed to color the vertices of G. But Grötschel, Lovász, and Schrijver [5] proved that we can compute in polynomial time a real number that is "sandwiched" between these hard-to-compute integers:

$$\omega(G) \leq \vartheta(\overline{G}) \leq \chi(G). \qquad (*)$$

(As usual, $\overline{G}$ denotes the complement of G.) Lovász [13] has called this a "sandwich theorem." The book [7] develops further facts about the function $\vartheta(G)$ and shows that it possesses many interesting properties. Therefore I think it's worthwhile to study $\vartheta(G)$ closely, in hopes of getting acquainted with it and finding faster ways to compute it.

Caution: The function called $\vartheta(G)$ in [13] is called $\vartheta(\overline{G})$ in [7] and [12]. I am following the latter convention because it is more likely to be adopted by other researchers; [7] is a classic book that contains complete proofs, while [13] is simply an extended abstract.

In these notes I am mostly following [7] and [12] with minor simplifications and a few additions. I mention several natural problems that
I was not able to solve immediately although I expect (and fondly hope)
that they will be resolved before I get to writing this portion of my
forthcoming book on Combinatorial Algorithms. I'm grateful to many
people — especially to Martin Grötschel and László Lovász — for their
comments on my first drafts of this material.

The following notes have been organized into numbered sections,
and there is at most one Lemma, Theorem, Corollary, or Example in
each section. Thus, "Lemma 2" will mean "the lemma in Section 2."

0. Preliminaries

Let's begin slowly by defining some notational conventions and by
stating some basic things that will be assumed without proof. All vectors in these notes will be regarded as column vectors, indexed either by
the vertices of a graph or by integers. The notation $x \geq y$, when x and y
are vectors, will mean that $x_v \geq y_v$ for all v. If A is a matrix, A_v will
denote column v, and A_{uv} will be the element in row u of column v.
The zero vector and the zero matrix and zero itself will all be denoted
by 0.

We will use several properties of matrices and vectors of real numbers
that are familiar to everyone who works with linear algebra but not to
everyone who studies graph theory, so it seems wise to list them here:

(i) The *dot product* of (column) vectors a and b is

$$a \cdot b = a^T b; \qquad (0.1)$$

the vectors are *orthogonal* (also called perpendicular) if $a \cdot b = 0$. The
length of vector a is

$$\|a\| = \sqrt{a \cdot a}. \qquad (0.2)$$

Cauchy's inequality asserts that

$$a \cdot b \leq \|a\| \, \|b\|; \qquad (0.3)$$

equality holds if and only if a is a scalar multiple of b or $b = 0$. Notice
that if A is any matrix we have

$$(A^T A)_{uv} = \sum_{k=1}^{n} (A^T)_{uk} A_{kv} = \sum_{k=1}^{n} A_{ku} A_{kv} = A_u \cdot A_v; \qquad (0.4)$$

in other words, the elements of $A^T A$ represent all dot products of the
columns of A.

(ii) An *orthogonal matrix* is a square matrix Q such that $Q^T Q$ is the identity matrix I. Thus, by (0.4), Q is orthogonal if and only if its columns are unit vectors perpendicular to each other. The transpose of an orthogonal matrix is orthogonal, because the condition $Q^T Q = I$ implies that Q^T is the inverse of Q, hence $QQ^T = I$.

(iii) A given matrix A is *symmetric* (that is, $A = A^T$) if and only if it can be expressed in the form

$$A = QDQ^T \tag{0.5}$$

where Q is orthogonal and D is a diagonal matrix. Notice that (0.5) is equivalent to the matrix equation

$$AQ = QD, \tag{0.6}$$

which is equivalent to the equations

$$AQ_v = \lambda_v Q_v$$

for all v, where $\lambda_v = D_{vv}$. Hence the diagonal elements of D are the eigenvalues of A and the columns of Q are the corresponding eigenvectors.

Properties (i), (ii), and (iii) are proved in any textbook of linear algebra. We can get some practice using these concepts by giving a constructive proof of another well known fact:

Lemma. *Given k mutually perpendicular unit vectors, there is an orthogonal matrix having these vectors as the first k columns.*

Proof. Suppose first that $k = 1$ and that x is a d-dimensional vector with $\|x\| = 1$. If $x_1 = 1$ we have $x_2 = \cdots = x_d = 0$, so the orthogonal matrix $Q = I$ satisfies the desired condition. Otherwise we let

$$y_1 = \sqrt{(1 - x_1)/2}, \qquad y_j = -x_j/(2y_1) \quad \text{for } 1 < j \le d. \tag{0.7}$$

Then

$$y^T y = \|y\|^2 = y_1^2 + \frac{x_2^2 + \cdots + x_d^2}{4y_1^2} = \frac{1 - x_1}{2} + \frac{1 - x_1^2}{2(1 - x_1)} = 1.$$

And x is the first column of the Householder [8] matrix

$$Q = I - 2yy^T, \tag{0.8}$$

which is easily seen to be orthogonal because

$$Q^T Q = Q^2 = I - 4yy^T + 4yy^T yy^T = I .$$

Now suppose the lemma has been proved for some $k \geq 1$; we will show how to increase k by 1. Let Q be an orthogonal matrix and let x be a unit vector perpendicular to its first k columns. We want to construct an orthogonal matrix Q' agreeing with Q in columns 1 to k and having x in column $k + 1$. Notice that we have

$$Q^T x = \begin{pmatrix} 0 \\ \vdots \\ 0 \\ y \end{pmatrix}$$

by hypothesis, where there are 0s in the first k rows. The $(d - k)$-dimensional vector y has squared length

$$\|y\|^2 = Q^T x \cdot Q^T x = x^T Q Q^T x = x^T x = 1 ,$$

so it is a unit vector. (In particular, $y \neq 0$, so we must have $k < d$.) Using the construction above, we can find a $(d - k) \times (d - k)$ orthogonal matrix R with y in its first column. Then the matrix

$$Q' = Q \begin{pmatrix} 1 & & & \\ & \ddots & & 0 \\ & & 1 & \\ 0 & & & R \end{pmatrix}$$

does what we want.   □

1. Orthogonal Labelings

Let G be a graph on the vertices V. If u and v are distinct elements of V, the notation $u - v$ means that they are adjacent in G; $u \not\!- v$ means that they are not (hence they are adjacent in $\overline{G}$).

An assignment of vectors a_v to each vertex v is called an *orthogonal labeling* of G if $a_u \cdot a_v = 0$ whenever $u \not\!- v$. In other words, whenever a_u is not perpendicular to a_v in the labeling, we must have $u - v$ in the graph. The vectors may have any desired dimension d; the components of a_v are a_{jv} for $1 \leq j \leq d$. *Example:* The labeling $a_v = 0$ for all v always works trivially.

The *cost* $c(a_v)$ of a vector a_v in an orthogonal labeling is defined to be 0 if $a_v = 0$, otherwise

$$c(a_v) = \frac{a_{1v}^2}{\|a_v\|^2} = \frac{a_{1v}^2}{a_{1v}^2 + \cdots + a_{dv}^2}.$$

Notice that we can multiply any vector a_v by a nonzero scalar t_v without changing its cost, and without violating the orthogonal labeling property. We can also get rid of a zero vector by increasing d by 1 and adding a new component 0 to each vector, except that the zero vector gets the new component 1. In particular, we can if we like assume that all vectors have unit length. Then the cost of a_v will be a_{1v}^2.

Lemma. *If $S \subseteq V$ is a stable set of vertices (that is, no two vertices of S are adjacent) and if a is an orthogonal labeling then*

$$\sum_{v \in S} c(a_v) \le 1. \tag{1.1}$$

Proof. We can assume that $\|a_v\| = 1$ for all v. Then the vectors a_v for $v \in S$ must be mutually orthogonal, and Lemma 0 tells us we can find a $d \times d$ orthogonal matrix Q with these vectors as its leftmost columns. The sum of the costs will then be at most $q_{11}^2 + q_{12}^2 + \cdots + q_{1d}^2 = 1$. $\square$

Relation (1.1) makes it possible for us to study stable sets geometrically.

2. Convex Labelings

An assignment x of real numbers x_v to the vertices v of G is called a *real labeling* of G. Several families of such labelings will be of importance to us:

The *characteristic labeling* for $U \subseteq V$ has $x_v = \begin{cases} 1 & \text{if } v \in U; \\ 0 & \text{if } v \notin U. \end{cases}$

A *stable labeling* is a characteristic labeling for a stable set.

A *clique labeling* is a characteristic labeling for a clique (a set of mutually adjacent vertices).

STAB(G) is the smallest convex set containing all stable labelings; that is, STAB(G) = convex hull$\{\, x \mid x \text{ is a stable labeling of } G \,\}$.

QSTAB(G) = $\{\, x \ge 0 \mid \sum_{v \in Q} x_v \le 1 \text{ for all cliques } Q \text{ of } G \,\}$.

TH(G) = $\{\, x \ge 0 \mid \sum_{v \in V} c(a_v) x_v \le 1 \text{ for all orthogonal labelings } a \text{ of } G \,\}$.

Lemma. TH *is sandwiched between* STAB *and* QSTAB:

$$\text{STAB}(G) \subseteq \text{TH}(G) \subseteq \text{QSTAB}(G). \tag{2.1}$$

Proof. Relation (1.1) tells that every stable labeling belongs to TH(G). Since TH(G) is obviously convex, it must contain the convex hull STAB(G). On the other hand, every clique labeling is an orthogonal labeling of dimension 1. Therefore every constraint of QSTAB(G) is one of the constraints of TH(G). □

Note: QSTAB was first defined by Shannon [19], and the first systematic study of STAB was undertaken by Padberg [18]. TH was first defined by Grötschel, Lovász, and Schrijver in [6].

3. Monotonicity

Suppose G and G' are graphs on the same vertex set V, with

$$G \subseteq G'$$

(that is, $u - v$ in G implies $u - v$ in G'). Then

every stable set in G' is stable in G, hence STAB$(G) \supseteq$ STAB(G');

every clique in G is a clique in G', hence QSTAB$(G) \supseteq$ QSTAB(G');

every orthogonal labeling of G is an orthogonal labeling of G', hence TH$(G) \supseteq$ TH(G').

In particular, if G is the empty graph $\overline{K}_n$ on $|V| = n$ vertices, all sets are stable and all cliques have size ≤ 1. Hence

STAB$(\overline{K}_n)$ = TH$(\overline{K}_n)$ = QSTAB$(\overline{K}_n)$ = $\{\, x \mid 0 \leq x_v \leq 1 \text{ for all } v \,\}$ is the n-cube.

If G is the complete graph K_n, all stable sets have size ≤ 1 and there is an n-clique; so

STAB(K_n) = TH(K_n) = QSTAB(K_n) = $\{\, x \geq 0 \mid \sum_v x_v \leq 1 \,\}$ is the n-simplex.

Thus the convex sets STAB(G), TH(G), and QSTAB(G) all lie between the n-simplex and the n-cube when G has n vertices.

Consider, for example, the case $n = 3$. Then there are three coordinates, so we can visualize the sets in 3-space (although there aren't many interesting graphs). The QSTAB of $\overset{x}{\bullet}\!-\!\overset{y}{\bullet}\!-\!\overset{z}{\bullet}$ is obtained from the unit cube by restricting the coordinates to $x + y \leq 1$ and $y + z \leq 1$; we can think of making two cuts in a piece of cheese:

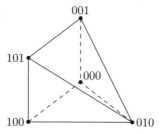

The vertices $\{000, 100, 010, 001, 101\}$ correspond to the stable labelings, so once again we have $\texttt{STAB}(G) = \texttt{TH}(G) = \texttt{QSTAB}(G)$.

4. The Theta Function

The function $\vartheta(G)$ mentioned in the introduction is a special case of a two-parameter function $\vartheta(G, w)$, where w is a nonnegative real labeling:

$$\vartheta(G, w) = \max\{\, w \cdot x \mid x \in \texttt{TH}(G) \,\}; \tag{4.1}$$

$$\vartheta(G) = \vartheta(G, \mathbb{1}) \text{ where } \mathbb{1} \text{ is the labeling } w_v = 1 \text{ for all } v. \tag{4.2}$$

This function, called the *Lovász number* of G (or the *weighted Lovász number* when $w \neq \mathbb{1}$), tells us about 1-dimensional projections of the n-dimensional convex set $\texttt{TH}(G)$.

Notice, for example, that the monotonicity properties of §3 tell us

$$G \subseteq G' \;\Rightarrow\; \vartheta(G, w) \geq \vartheta(G', w) \tag{4.3}$$

for all $w \geq 0$. It is also obvious that ϑ is monotone in its other parameter:

$$w \leq w' \;\Rightarrow\; \vartheta(G, w) \leq \vartheta(G, w'). \tag{4.4}$$

The smallest possible value of ϑ is

$$\vartheta(K_n, w) = \max\{w_1, \ldots, w_n\}; \quad \vartheta(K_n) = 1. \tag{4.5}$$

The largest possible value is

$$\vartheta(\overline{K}_n, w) = w_1 + \cdots + w_n; \quad \vartheta(\overline{K}_n) = n. \tag{4.6}$$

Similar definitions can be given for $\texttt{STAB}$ and $\texttt{QSTAB}$:

$$\alpha(G, w) = \max\{\, w \cdot x \mid x \in \texttt{STAB}(G) \,\}, \quad \alpha(G) = \alpha(G, \mathbb{1}); \tag{4.7}$$

$$\kappa(G, w) = \max\{\, w \cdot x \mid x \in \texttt{QSTAB}(G) \,\}, \quad \kappa(G) = \kappa(G, \mathbb{1}). \tag{4.8}$$

Clearly $\alpha(G)$ is the size of the largest stable set in G, because every stable labeling x corresponds to a stable set with $\mathbb{1} \cdot x$ vertices. It is also easy to see that $\kappa(G)$ is at most $\overline{\chi}(G)$, the smallest number of cliques that cover the vertices of G. For if the vertices can be partitioned into k cliques $Q_1, \ldots, Q_k$ and if $x \in \mathtt{QSTAB}(G)$, we have

$$\mathbb{1} \cdot x = \sum_{v \in Q_1} x_v + \cdots + \sum_{v \in Q_k} x_v \leq k.$$

Sometimes $\kappa(G)$ is less than $\overline{\chi}(G)$. For example, consider the cyclic graph C_n, with vertices $\{0, 1, \ldots, n-1\}$ and $u \,-\, v$ if and only if $u \equiv v \pm 1 \pmod{n}$. Adding up the inequalities $x_0 + x_1 \leq 1, \ldots,$ $x_{n-2} + x_{n-1} \leq 1$, $x_{n-1} + x_0 \leq 1$ of QSTAB gives $2(x_0 + \cdots + x_{n-1}) \leq n$, and this upper bound is achieved when all x's are $1/2$; hence $\kappa(C_n) = n/2$, if $n > 3$. But $\overline{\chi}(G)$ is always an integer, and $\overline{\chi}(C_n) = \lceil n/2 \rceil$ is greater than $\kappa(C_n)$ when n is odd.

Incidentally, these remarks establish the "sandwich inequality" (∗) stated in the introduction, because

$$\alpha(G) \leq \vartheta(G) \leq \kappa(G) \leq \overline{\chi}(G) \tag{4.9}$$

and $\omega(G) = \alpha(\overline{G})$, $\overline{\chi}(G) = \chi(\overline{G})$.

5. Alternative Definitions of ϑ

Four additional functions ϑ_1, ϑ_2, ϑ_3, ϑ_4 are defined in [7], and they all turn out to be identical to ϑ. Thus, we can understand ϑ in many different ways; this may help us to compute it.

We will show, following [7], that if w is any fixed nonnegative real labeling of G, the inequalities

$$\vartheta(G, w) \leq \vartheta_1(G, w) \leq \vartheta_2(G, w) \leq \vartheta_3(G, w) \leq \vartheta_4(G, w) \leq \vartheta(G, w) \tag{5.1}$$

can be proved. Thus we will establish the theorem of [7], and all inequalities in our proofs will turn out to be equalities.

We will introduce the alternative definitions ϑ_k one at a time; any one of these definitions could have been taken as the starting point. First,

$$\vartheta_1(G, w) = \min_a \max_v \frac{w_v}{c(a_v)}, \quad \text{over all orthogonal labelings } a. \tag{5.2}$$

Here we regard $w_v / c(a_v) = 0$ when $w_v = c(a_v) = 0$; but the max is ∞ if some $w_v > 0$ has $c(a_v) = 0$.

Lemma. $\vartheta(G, w) \leq \vartheta_1(G, w)$.

Proof. Suppose $x \in \text{TH}(G)$ maximizes $w \cdot x$, and suppose a is an orthogonal labeling that achieves the minimum value $\vartheta_1(G, w)$. Then

$$\vartheta(G, w) = w \cdot x = \sum_v w_v x_v \leq$$

$$\left(\max_v \frac{w_v}{c(a_v)} \right) \sum_v c(a_v) x_v \leq \max_v \frac{w_v}{c(a_v)} = \vartheta_1(G, w). \quad \square$$

Incidentally, the fact that all inequalities are exact will imply later that every nonzero weight vector w has an orthogonal labeling a such that

$$c(a_v) = \frac{w_v}{\vartheta(G, w)} \quad \text{for all } v. \tag{5.3}$$

We will restate such consequences of (5.1) later, but it may be helpful to keep that future goal in mind.

6. Characterization via Eigenvalues

The second variant of ϑ is rather different; this is the only one Lovász chose to mention in [13].

We say that A is a *feasible matrix* for G and w if A is indexed by vertices and if

A is real and symmetric;

$A_{vv} = w_v$ for all $v \in V$;

$A_{uv} = \sqrt{w_u w_v}$ whenever $u \not\sim v$ in G. $\tag{6.1}$

The other elements of A are unconstrained (that is, they can be anything between $-\infty$ and $+\infty$).

If A is any real, symmetric matrix, let $\Lambda(A)$ be its maximum eigenvalue. This quantity is well defined because all eigenvalues of A are real. Suppose A has eigenvalues $\{\lambda_1, \ldots, \lambda_n\}$; then $A = Q \, \text{diag}(\lambda_1, \ldots, \lambda_n) Q^T$ for some orthogonal Q, and $\|Qx\| = \|x\|$ for all vectors x, so there is a nice way to characterize $\Lambda(A)$:

$$\Lambda(A) = \max\{ x^T A x \mid \|x\| = 1 \}. \tag{6.2}$$

Notice that $\Lambda(A)$ might not be the largest eigenvalue in absolute value. We now let

$$\vartheta_2(G, w) = \min\{ \Lambda(A) \mid A \text{ is a feasible matrix for } G \text{ and } w \}. \tag{6.3}$$

Lemma. $\vartheta_1(G, w) \le \vartheta_2(G, w)$.

Proof. Note first that the trace $\operatorname{tr} A = \sum_v w_v$ is nonnegative for any feasible matrix A. The trace is also well-known to be the sum of the eigenvalues; this fact is an easy consequence of the identity

$$\operatorname{tr} XY = \sum_{j=1}^{m} \sum_{k=1}^{n} x_{jk} y_{kj} = \operatorname{tr} YX \qquad (6.4)$$

valid for any matrices X and Y of respective sizes $m \times n$ and $n \times m$. In particular, $\vartheta_2(G, w)$ is always nonnegative, and it is zero if and only if $w = 0$ $\big($when also $\vartheta_1(G, w) = 0\big)$.

So suppose $w \ne 0$ and let A be a feasible matrix that attains the minimum value $\Lambda(A) = \vartheta_2(G, w) = \lambda > 0$. Let

$$B = \lambda I - A. \qquad (6.5)$$

The eigenvalues of B are λ minus the eigenvalues of A. $\big($For if $A = Q \operatorname{diag}(\lambda_1, \ldots, \lambda_n) Q^T$ then $B = Q \operatorname{diag}(\lambda - \lambda_1, \ldots, \lambda - \lambda_n) Q^T$.$\big)$ Thus they are all nonnegative; such a matrix B is called *positive semidefinite*. We can write

$$B = X^T X, \qquad \text{that is,} \quad B_{uv} = x_u \cdot x_v, \qquad (6.6)$$

where $X = \operatorname{diag}(\sqrt{\lambda - \lambda_1}, \ldots, \sqrt{\lambda - \lambda_n}) Q^T$.

Let a_v be the vector $(\sqrt{w_v}, x_{1v}, \ldots, x_{rv})^T$. Then we have $c(a_v) = w_v / \|a_v\|^2 = w_v / (w_v + x_{1v}^2 + \cdots + x_{rv}^2)$ and $x_{1v}^2 + \cdots + x_{rv}^2 = B_{vv} = \lambda - w_v$, hence $c(a_v) = w_v / \lambda$. Also if $u \not\sim v$ we have $a_u \cdot a_v = \sqrt{w_u w_v} + x_u \cdot x_v = \sqrt{w_u w_v} + B_{uv} = \sqrt{w_u w_v} - A_{uv} = 0$. Therefore a is an orthogonal labeling and $\max_v w_v / c(a_v) = \lambda \ge \vartheta_1(G, w)$.   □

7. A Complementary Characterization

Still another variation is based on orthogonal labelings of the complementary graph $\overline{G}$.

In this case we let b be an orthogonal labeling of $\overline{G}$, normalized so that $\sum_v \|b_v\|^2 = 1$, and we let

$$\vartheta_3(G, w) = \max\Big\{ \Big| \sum_{u,v} (\sqrt{w_u}\, b_u) \cdot (\sqrt{w_v}\, b_v) \Big|$$

$$b \text{ is a normalized orthogonal labeling of } \overline{G} \Big\}. \qquad (7.1)$$

A normalized orthogonal labeling b is equivalent to an $n \times n$ symmetric positive semidefinite matrix B, where $B_{uv} = b_u \cdot b_v$ is zero when $u - v$, and where $\operatorname{tr} B = 1$.

Lemma. $\vartheta_2(G, w) \le \vartheta_3(G, w)$.

This lemma is the "heart" of the proof that all ϑs are equivalent, according to [7]. It relies on a fact about positive semidefinite matrices that we will prove in §9.

Fact. *If A is a symmetric matrix such that $A \cdot B \ge 0$ for all symmetric positive semidefinite B with $B_{uv} = 0$ for u — v, then $A = X + Y$ where X is symmetric positive semidefinite and Y is symmetric and $Y_{vv} = 0$ for all v and $Y_{uv} = 0$ for $u \not{-} v$.*

Here $C \cdot B$ stands for the dot product of matrices, that is, the sum $\sum_{u,v} C_{uv} B_{uv}$, which can also be written $\text{tr}\, C^T B$. The stated fact is a duality principle for quadratic programming.

Proof of the lemma. Assuming the Fact, let W be the matrix with $W_{uv} = \sqrt{w_u w_v}$, and let $\vartheta_3 = \vartheta_3(G, w)$. By definition (7.1), if b is any nonzero orthogonal labeling of $\overline{G}$ (not necessarily normalized), we have

$$\sum_{u,v} (\sqrt{w_u}\, b_u) \cdot (\sqrt{w_v}\, b_v) \le \vartheta_3 \sum_v \|b_v\|^2 . \tag{7.2}$$

In matrix terms this says $W \cdot B \le (\vartheta_3 I) \cdot B$ for all symmetric positive semidefinite B with $B_{uv} = 0$ for u — v. The Fact now tells us that we can write

$$\vartheta_3 I - W = X + Y \tag{7.3}$$

where X is symmetric positive semidefinite, Y is symmetric and diagonally zero, and $Y_{uv} = 0$ when $u \not{-} v$. Therefore the matrix A defined by

$$A = W + Y = \vartheta_3 I - X$$

is a feasible matrix for G, and $\Lambda(A) \le \vartheta_3$. This completes the proof that $\vartheta_2(G, w) \le \vartheta_3(G, w)$, because $\Lambda(A)$ is an upper bound on ϑ_2 by definition of ϑ_2.   □

8. Elementary Facts About Cones

A *cone* in N-dimensional space is a set of vectors closed under addition and under multiplication by nonnegative scalars. (In particular, it is convex: If c and c' are in cone C and $0 < t < 1$ then tc and $(1 - t)c'$ are in C, hence $tc + (1 - t)c' \in C$.) A *closed cone* is a cone that is also closed under taking limits.

F1. *If C is a closed convex set and $x \notin C$, there is a hyperplane separating x from C. This means there is a vector y and a number b such that $c \cdot y \le b$ for all $c \in C$ but $x \cdot y > b$.*

Proof. Let d be the greatest lower bound of $\|x - c\|^2$ for all $c \in C$. Then there's a sequence of vectors c_k with $\|x - c_k\|^2 < d + 1/k$; this infinite set of vectors contained in the sphere $\{ y \mid \|x - y\|^2 \le d + 1 \}$ must have a limit point c_∞, and $c_\infty \in C$ since C is closed. Therefore $\|x - c_\infty\|^2 \ge d$; in fact $\|x - c_\infty\|^2 = d$, since $\|x - c_\infty\| \le \|x - c_k\| + \|c_k - c_\infty\|$ and the right-hand side can be made arbitrarily close to d. Since $x \notin C$, we must have $d > 0$. Now let $y = x - c_\infty$ and $b = c_\infty \cdot y$. Clearly $x \cdot y = y \cdot y + b > b$. And if c is any element of C and ϵ is any small positive number, the vector $\epsilon c + (1 - \epsilon) c_\infty$ is in C, hence $\left\| x - \left(\epsilon c + (1 - \epsilon) c_\infty \right) \right\|^2 \ge d$. But

$$\left\| x - \left(\epsilon c + (1 - \epsilon) c_\infty \right) \right\|^2 - d = \|x - c_\infty - \epsilon(c - c_\infty)\|^2 - d$$
$$= -2\epsilon y \cdot (c - c_\infty) + \epsilon^2 \|c - c_\infty\|^2$$

can be nonnegative for all small ϵ only if $y \cdot (c - c_\infty) \le 0$, that is, only if $c \cdot y \le b$. $\square$

If A is any set of vectors, let $A^* = \{ b \mid a \cdot b \ge 0 \text{ for all } a \in A \}$. The following facts are immediate:

F2. $A \subseteq A'$ *implies* $A^* \supseteq A'^*$.

F3. $A \subseteq A^{**}$.

F4. A^* *is a closed cone.*

From F1 we also get a result that, in the special case that $C = \{ Ax \mid x \ge 0 \}$ for a matrix A, is called Farkas's Lemma:

F5. *If C is a closed cone, $C = C^{**}$.*

Proof. Suppose $x \in C^{**}$ and $x \notin C$, and let (y, b) be a separating hyperplane as in F1. Then $(y, 0)$ is also a separating hyperplane; for we have $x \cdot y > b \ge 0$ because $0 \in C$, and we cannot have $c \cdot y > 0$ for $c \in C$ because $(\lambda c) \cdot y$ would then be unbounded. But then $c \cdot (-y) \ge 0$ for all $c \in C$, so $-y \in C^*$; hence $x \cdot (-y) \ge 0$, a contradiction. $\square$

If A and B are sets of vectors, we define

$$A + B = \{ a + b \mid a \in A \text{ and } b \in B \}.$$

F6. *If C and C' are closed cones, $(C \cap C')^* = C^* + C'^*$.*

Proof. If A and B are arbitrary sets we have $A^* + B^* \subseteq (A \cap B)^*$, for if $x \in A^* + B^*$ and $y \in A \cap B$ then $x \cdot y = a \cdot y + b \cdot y \geq 0$. If A and B are arbitrary sets including 0 then $(A + B)^* \subseteq A^* \cap B^*$ by F2, because $A + B \supseteq A$ and $A + B \supseteq B$. Thus for arbitrary A and B we have $(A^* + B^*)^* \subseteq A^{**} \cap B^{**}$, hence

$$(A^* + B^*)^{**} \supseteq (A^{**} \cap B^{**})^*.$$

Now let A and B be closed cones; then $A^* + B^* \supseteq (A \cap B)^*$ by F5. □

F7. *If C and C' are closed cones, $(C + C')^* = C^* \cap C'^*$.*

Proof. F6 says $(C^* \cap C'^*)^* = C^{**} + C'^{**}$; apply F5 and $*$ again. □

F8. *Let S be any set of indices and let $\overline{S}$ be all the indices not in S. Also let $A_S = \{ a \mid a_s = 0 \text{ for all } s \in S \}$. Then*

$$A_S^* = A_{\overline{S}}.$$

Proof. If $b_s = 0$ for all $s \notin S$ and $a_s = 0$ for all $s \in S$, obviously $a \cdot b = 0$; so $A_{\overline{S}} \subseteq A_S^*$. If $b_s \neq 0$ for some $s \notin S$ and $a_t = 0$ for all $t \neq s$ and $a_s = -b_s$, then $a \in A_S$ and $a \cdot b < 0$; so $b \notin A_S^*$, hence $A_{\overline{S}} \supseteq A_S^*$. □

9. Definite Proof of a Semidefinite Fact

Now we are almost ready to prove the result needed in the proof of Lemma 7.

Let D be the set of real symmetric positive semidefinite matrices (called "spuds" henceforth for brevity), considered as vectors in N-dimensional space, where $N = (n + 1)n/2$. We use the inner product $A \cdot B = \operatorname{tr} A^T B$; this is justified if we divide off-diagonal elements by $\sqrt{2}$. For example, if $n = 3$ the correspondence between 6-dimensional vectors and 3×3 symmetric matrices is

$$(a, b, c, d, e, f) \leftrightarrow \begin{pmatrix} a & d/\sqrt{2} & e/\sqrt{2} \\ d/\sqrt{2} & b & f/\sqrt{2} \\ e/\sqrt{2} & f/\sqrt{2} & c \end{pmatrix}$$

preserving sum, scalar product, and dot product. Clearly D is a closed cone.

F9. $D^* = D$.

Proof. If A and B are spuds then $A = X^T X$ and $B = Y^T Y$ and $A \cdot B = \operatorname{tr} X^T X Y^T Y = \operatorname{tr} X Y^T Y X^T = (Y X^T) \cdot (Y X^T) \geq 0$; hence $D \subseteq D^*$. (In fact, this argument shows that $A \cdot B = 0$ if and only if $AB = 0$, for any spuds A and B, since $A = A^T$.)

If A is symmetric but has a negative eigenvalue λ we can write

$$A = Q \operatorname{diag}(\lambda, \lambda_2, \ldots, \lambda_n) Q^T$$

for some orthogonal matrix Q. Let $B = Q \operatorname{diag}(1, 0, \ldots, 0) Q^T$; then B is a spud, and

$$A \cdot B = \operatorname{tr} A^T B = \operatorname{tr} Q \operatorname{diag}(\lambda, 0, \ldots, 0) Q^T = \lambda < 0.$$

So A is not in D^*; this proves $D \supseteq D^*$. $\square$

Let E be the set of all real symmetric matrices such that $E_{uv} = 0$ when $u - v$ in a graph G; let F be the set of all real symmetric matrices such that $F_{uv} = 0$ when $u = v$ or $u \not\!\sim v$. The Fact stated in Section 7 is now equivalent in our new notation to

Fact. $(D \cap E)^* \subseteq D + F$.

Proof.   And we know now that

$$\begin{aligned}
(D \cap E)^* &= D^* + E^* &&\text{by F6} \\
&= D + F &&\text{by F9 and F8.}   \square
\end{aligned}$$

10. Another Characterization

Remember ϑ, ϑ_1, ϑ_2, and ϑ_3? We are now going to introduce yet another function

$$\vartheta_4(G, w) = \max \Big\{ \sum_v c(b_v) w_v \ \Big|$$

$$b \text{ is an orthogonal labeling of } \overline{G} \Big\}. \qquad (10.1)$$

Lemma. $\vartheta_3(G, w) \leq \vartheta_4(G, w)$.

Proof.   Suppose b is a normalized orthogonal labeling of $\overline{G}$ that achieves the maximum ϑ_3; and suppose the vectors of this labeling have dimension d. Let

$$x_k = \sum_v b_{kv} \sqrt{w_v}, \qquad \text{for } 1 \leq k \leq d; \qquad (10.2)$$

then

$$\vartheta_3(G, w) = \sum_{u,v} \sqrt{w_u}\, b_u \cdot b_v \sqrt{w_v} = \sum_{u,v,k} \sqrt{w_u w_v}\, b_{ku} b_{kv} = \sum_k x_k^2 \,.$$

Let Q be an orthogonal $d \times d$ matrix whose first row is

$$(x_1/\sqrt{\vartheta_3}, \ldots, x_d/\sqrt{\vartheta_3}),$$

and let $b_v' = Qb_v$. Then $b_u' \cdot b_v' = b_u^T Q^T Q b_v = b_u^T b_v = b_u \cdot b_v$, so b' is a normalized orthogonal labeling of $\overline{G}$. Also

$$x_k' = \sum_v b_{kv}' \sqrt{w_v} = \sum_{v,j} Q_{kj} b_{jv} \sqrt{w_v}$$

$$= \sum_j Q_{kj} x_j = \begin{cases} \sqrt{\vartheta_3}\,, & k = 1; \\ 0\,, & k > 1. \end{cases} \qquad (10.3)$$

Hence by Cauchy's inequality

$$\vartheta_3(G, w) = \left(\sum_v b_{1v}' \sqrt{w_v}\right)^2 \le \left(\sum_v \|b_v'\|^2\right)\left(\sum_{\substack{v \\ b_v' \ne 0}} \frac{b_{1v}'^2}{\|b_v'\|^2}\, w_v\right)$$

$$= \sum_v c(b_v') w_v \le \vartheta_4(G, w) \qquad (10.4)$$

because $\sum_v \|b_v'\|^2 = \sum_v \|b_v\|^2 = 1$.   □

11. The Final Link

Now we can close the loop:

Lemma. $\vartheta_4(G, w) \le \vartheta(G, w)$.

Proof. If b is an orthogonal labeling of $\overline{G}$ that achieves the maximum ϑ_4, we will show that the real labeling x defined by $x_v = c(b_v)$ is in TH(G). Therefore $\vartheta_4(G, w) = w \cdot x$ is $\le \vartheta(G, w)$.

We will prove that if a is any orthogonal labeling of G, and if b is any orthogonal labeling of $\overline{G}$, then

$$\sum_v c(a_v) c(b_v) \le 1\,. \qquad (11.1)$$

Suppose a is a labeling of dimension d and b is of dimension d'. Then consider the $d \times d'$ matrices

$$A_v = a_v b_v^T \tag{11.2}$$

as elements of a vector space of dimension dd'. If $u \neq v$ we have

$$A_u \cdot A_v = \operatorname{tr} A_u^T A_v = \operatorname{tr} b_u a_u^T a_v b_v^T = \operatorname{tr} a_u^T a_v b_v^T b_u = 0, \tag{11.3}$$

because $a_u^T a_v = 0$ when $u \not\!\!\frac{\ }{\ } v$ and $b_v^T b_u = 0$ when $u - v$. If $u = v$ we have

$$A_v \cdot A_v = \|a_v\|^2 \|b_v\|^2.$$

The upper left corner element of A_v is $a_{1v} b_{1v}$, hence the "cost" of A_v is $(a_{1v} b_{1v})^2 / \|A_v\|^2 = c(a_v) c(b_v)$. This, with (11.3), proves (11.1). (See the proof of Lemma 1.) □

12. The Main Theorem

Lemmas 5, 6, 7, 10, and 11 establish the five inequalities claimed in (5.1); hence all five variants of ϑ are the same function of G and w. Moreover, all the inequalities in those five proofs are equalities (with the exception of (11.1)). We can summarize the results as follows.

Theorem. *For all graphs G and any nonnegative real labeling w of G we have*

$$\vartheta(G, w) = \vartheta_1(G, w) = \vartheta_2(G, w) = \vartheta_3(G, w) = \vartheta_4(G, w). \tag{12.1}$$

Moreover, if $w \neq 0$, there exist orthogonal labelings a and b of G and $\overline{G}$, respectively, such that

$$c(a_v) = w_v / \vartheta; \tag{12.2}$$

$$\sum c(a_v) c(b_v) = 1. \tag{12.3}$$

Proof. Relation (12.1) is, of course, (5.1); and (12.2) is (5.3). The desired labeling b is what we called b' in the proof of Lemma 10. The fact that the application of Cauchy's inequality in (10.4) is actually an equality,

$$\vartheta = \left(\sum_v b_{1v} \sqrt{w_v} \right)^2 = \left(\sum_v \|b_v\|^2 \right) \left(\sum_{\substack{v \\ b_v \neq 0}} \frac{b_{1v}^2}{\|b_v\|^2} w_v \right), \tag{12.4}$$

tells us that the vectors whose dot product has been squared are proportional: There is a number t such that

$$\|b_v\| = t\,\frac{b_{1v}\sqrt{w_v}}{\|b_v\|}\,, \quad \text{if } b_v \neq 0; \quad b_v = 0 \iff b_{1v}\sqrt{w_v} = 0. \quad (12.5)$$

The labeling in the proof of Lemma 10 also satisfies

$$\sum_v \|b_v\|^2 = 1\,; \quad (12.6)$$

hence $t = \pm 1/\sqrt{\vartheta}\,$.

We can now show

$$c(b_v) = \|b_v\|^2\,\vartheta/w_v\,, \quad \text{when } w_v \neq 0. \quad (12.7)$$

This relation is obvious if $\|b_v\| = 0$; otherwise we have

$$c(b_v) = \frac{b_{1v}^2}{\|b_v\|^2} = \frac{\|b_v\|^2}{t^2 w_v}$$

by (12.5). Summing the product of (12.2) and (12.7) over v gives the desired result (12.3).   □

13. The Main Converse

The nice thing about Theorem 12 is that conditions (12.2) and (12.3) also provide a *certificate* that a given value ϑ is the minimum or maximum stated in the definitions of ϑ, ϑ_1, ϑ_2, ϑ_3, and ϑ_4.

Theorem. *If a is an orthogonal labeling of G and b is an orthogonal labeling of $\overline{G}$ such that relations (12.2) and (12.3) hold for some ϑ and w, then ϑ is the value of $\vartheta(G, w)$.*

Proof. Plugging (12.2) into (12.3) gives $\sum w_v c(b_v) = \vartheta$, hence $\vartheta \leq \vartheta_4(G, w)$ by definition of ϑ_4. Also,

$$\max_v \frac{w_v}{c(a_v)} = \vartheta\,,$$

hence $\vartheta \geq \vartheta_1(G, w)$ by definition of ϑ_1.   □

14. Another Look at TH

We originally defined $\vartheta(G, w)$ in (4.1) in terms of the convex set TH defined in Section 2:

$$\vartheta(G, w) = \max\{\, w \cdot x \mid x \in \mathrm{TH}(G)\,\}, \qquad \text{when } w \geq 0. \qquad (14.1)$$

We can also go the other way, defining TH in terms of ϑ:

$$\mathrm{TH}(G) = \{\, x \geq 0 \mid w \cdot x \leq \vartheta(G, w) \text{ for all } w \geq 0 \,\}. \qquad (14.2)$$

Every $x \in \mathrm{TH}(G)$ belongs to the right-hand set, by (14.1). Conversely, if x belongs to the right-hand set and if a is any orthogonal labeling of G, not entirely zero, let $w_v = c(a_v)$, so that $w \cdot x = \sum_v c(a_v)x_v$. Then

$$\vartheta_1(G, w) \leq \max_v \big(w_v/c(a_v)\big) = 1$$

by definition (5.2), so we know by Lemma 5 that $\sum c(a_v)x_v \leq 1$. This proves that x belongs to $\mathrm{TH}(G)$.

Theorem 12 tells us even more.

Lemma. $\mathrm{TH}(G) = \{\, x \geq 0 \mid \vartheta(\overline{G}, x) \leq 1 \,\}.$

Proof. By definition (10.1),

$$\vartheta_4(\overline{G}, w) = \max\Big\{ \sum_v c(a_v)w_v \,\Big|$$
$$a \text{ is an orthogonal labeling of } G \Big\}. \qquad (14.3)$$

Thus $x \in \mathrm{TH}(G)$ if and only if $\vartheta_4(\overline{G}, x) \leq 1$, when $x \geq 0$.   □

Theorem. $\mathrm{TH}(G) = \{\, x \mid x_v = c(b_v) \text{ for some orthogonal labeling } b \text{ of } \overline{G} \,\}.$

Proof. We already proved in (11.1) that the right side is contained in the left.

Let $x \in \mathrm{TH}(G)$ and let $\vartheta = \vartheta(\overline{G}, x)$. By the lemma, $\vartheta \leq 1$. Therefore, by (12.2), there is an orthogonal labeling b of $\overline{G}$ such that $c(b_v) = x_v/\vartheta \geq x_v$ for all v. It is easy to reduce the cost of any vector in an orthogonal labeling to any desired value, simply by increasing the dimension and giving this vector an appropriate nonzero value in the new component while all other vectors remain zero there. The dot products are unchanged, so the new labeling is still orthogonal. Repeating this construction for each v produces a labeling with $c(b_v) = x_v$.   □

This theorem makes the definition of ϑ_4 in (10.1) identical to the definition of ϑ in (4.1).

15. Zero Weights

Our next result shows that when a weight is zero, the corresponding vertex might as well be absent from the graph.

Lemma. *Let U be a subset of the vertices V of a graph G, and let $G' = G|U$ be the graph induced by U (that is, the graph on vertices U with $u - v$ in G' if and only if $u - v$ in G). Then if w and w' are nonnegative labelings of G and G' such that*

$$w_v = w'_v \quad \text{when} \quad v \in U, \qquad w_v = 0 \quad \text{when} \quad v \notin U, \qquad (15.1)$$

we have

$$\vartheta(G, w) = \vartheta(G', w'). \qquad (15.2)$$

Proof. Let a and b satisfy (12.2) and (12.3) for G and w. Then $c(a_v) = 0$ for $v \notin U$, so $a|U$ and $b|U$ satisfy (12.2) and (12.3) for G' and w'. (Here $a|U$ means the vectors whose components are a_v for $v \in U$.) By Theorem 13, they determine the same ϑ. $\square$

16. Nonzero Weights

We can also get some insight into the significance of nonzero weights by "splitting" vertices instead of removing them.

Lemma. *Let v be a vertex of G and let G' be a graph obtained from G by adding a new vertex v' and new edges*

$$u - v' \quad \text{if and only if} \quad u - v. \qquad (16.1)$$

Let w and w' be nonnegative labelings of G and G' such that

$$w_u = w'_u, \qquad \text{when } u \neq v; \qquad (16.2)$$

$$w_v = w'_v + w'_{v'}. \qquad (16.3)$$

Then

$$\vartheta(G, w) = \vartheta(G', w'). \qquad (16.4)$$

Proof. By Theorem 12 there are labelings a and b of G and $\overline{G}$ satisfying (12.2) and (12.3). We can modify them to obtain labelings a' and b' of G' and $\overline{G'}$ as follows, with the vectors of a' having one more component than the vectors of a:

$$a'_u = \begin{pmatrix} a_u \\ 0 \end{pmatrix}, \qquad b'_u = b_u, \qquad \text{when } u \neq v; \qquad (16.5)$$

$$a'_v = \begin{pmatrix} a_v \\ \alpha \end{pmatrix}, \quad a'_{v'} = \begin{pmatrix} a_v \\ -\beta \end{pmatrix}, \quad \alpha = \sqrt{\frac{w'_{v'}}{w'_v}} \, \|a_v\|, \quad \beta = \sqrt{\frac{w'_v}{w'_{v'}}} \, \|a_v\|; \quad (16.6)$$

$$b'_v = b'_{v'} = b_v. \quad (16.7)$$

(We can assume by Lemma 15 that w'_v and $w'_{v'}$ are nonzero.) All orthogonality relations are preserved; and since $v \not\sim v'$ in G', we also need to verify

$$a'_v \cdot a'_{v'} = \|a_v\|^2 - \alpha\beta = 0.$$

We have

$$c(a'_v) = \frac{c(a_v) \, \|a_v\|^2}{\|a_v\|^2 + \alpha^2} = \frac{c(a_v)}{1 + w'_{v'}/w'_v} = \frac{c(a_v) w'_v}{w_v} = \frac{w'_v}{\vartheta},$$

and similarly $c(a'_{v'}) = w'_{v'}/\vartheta$; thus (12.2) and (12.3) are satisfied by a' and b' for G' and w'. □

Notice that if all the weights are integers we can apply this lemma repeatedly to establish that

$$\vartheta(G, w) = \vartheta(G'), \quad (16.8)$$

where G' is obtained from G by replacing each vertex v by a cluster of w_v mutually nonadjacent vertices that are adjacent to each of v's neighbors. (Recall that $\vartheta(G') = \vartheta(G', \mathbb{1})$, by definition (4.2).) In particular, if G is the trivial graph K_2 and if we assign the weights M and N, we have $\vartheta\big(K_2, (M, N)^T\big) = \vartheta(K_{M,N})$ where $K_{M,N}$ denotes the complete bipartite graph on M and N vertices.

A similar operation called "duplicating" a vertex has a similarly simple effect:

Corollary. *Let G' be constructed from G as in Lemma 16 but with an additional edge between v and v'. Then $\vartheta(G, w) = \vartheta(G', w')$ if w' is defined by (16.2) and*

$$w_v = \max(w'_v, w'_{v'}). \quad (16.9)$$

Proof. We may assume that $w_v = w'_v$ and $w'_{v'} \neq 0$. Most of the construction (16.5)–(16.7) can be used again, but we set $\alpha = 0$ and $b'_{v'} = 0$ and

$$\beta = \sqrt{\frac{w_v - w'_{v'}}{w'_{v'}}} \, \|a_v\|.$$

Again the necessary and sufficient conditions are readily verified.   □

If the corollary is applied repeatedly, it tells us that $\vartheta(G)$ is unchanged when we replace the vertices of G by cliques.

17. Simple Examples

We observed in Section 4 that $\vartheta(G, w)$ always is at least

$$\vartheta_{\min} = \vartheta(K_n, w) = \max\{w_1, \ldots, w_n\} \qquad (17.1)$$

and at most

$$\vartheta_{\max} = (\overline{K}_n, w) = w_1 + \cdots + w_n . \qquad (17.2)$$

What are the corresponding orthogonal labelings?

For K_n the vectors of a have no orthogonal constraints, while the vectors of b must satisfy $b_u \cdot b_v = 0$ for all $u \neq v$. We can let a be the two-dimensional labeling

$$a_v = \begin{pmatrix} \sqrt{w_v} \\ \sqrt{\vartheta - w_v} \end{pmatrix}, \qquad \vartheta = \vartheta_{\min} \qquad (17.3)$$

so that $\|a_v\|^2 = \vartheta$ and $c(a_v) = w_v/\vartheta$ as desired; and b can be one-dimensional,

$$b_v = \begin{cases} (1), & \text{if } v = v_{\max}, \\ (0), & \text{if } v \neq v_{\max}, \end{cases} \qquad (17.4)$$

where $v_{\max}$ is any particular vertex that maximizes w_v. Clearly

$$\sum_v c(a_v)c(b_v) = \frac{c(a_{v_{\max}})}{\vartheta} = \frac{w_{v_{\max}}}{\vartheta} = 1 .$$

For $\overline{K}_n$ the vectors of a must be mutually orthogonal while the vectors of b are unrestricted. We can let the vectors a be the columns of any orthogonal matrix whose top row contains the element

$$\sqrt{w_v/\vartheta}, \qquad \vartheta = \vartheta_{\max} \qquad (17.5)$$

in column v. Then $\|a_v\|^2 = 1$ and $c(a_v) = w_v/\vartheta$. Once again a one-dimensional labeling suffices for b; we can let $b_v = (1)$ for all v.

18. The Direct Sum of Graphs

Let $G = G' + G''$ be the graph on vertices

$$V = V' \cup V'' \qquad (18.1)$$

where the vertex sets V' and V'' of G' and G'' are disjoint, and where $u - v$ in G if and only if $u, v \in V'$ and $u - v$ in G', or $u, v \in V''$ and $u - v$ in G''. In this case

$$\vartheta(G, w) = \vartheta(G', w') + \vartheta(G'', w''), \qquad (18.2)$$

where w' and w'' are the sublabelings of w on vertices of V' and V''. We can prove (18.2) by constructing orthogonal labelings (a, b) satisfying (12.2) and (12.3).

Suppose a' is an orthogonal labeling of G' such that

$$\|a'_v\|^2 = \vartheta' \qquad a'_{1v} = \sqrt{w'_v}\,, \tag{18.3}$$

and suppose a'' is a similar orthogonal labeling of G''. If a' has dimension d' and a'' has dimension d'', we construct a new labeling a of dimension $d = d' + d''$ as follows, where j' runs from 2 to d' and j'' runs from 2 to d'':

if $v \in V'$	if $v \in V''$

$$
\begin{aligned}
a_{1v} &= \sqrt{w'_v} = a'_{1v}\,, & a_{1v} &= \sqrt{w''_v} = a''_{1v}\,, \\
a_{j'v} &= \sqrt{\vartheta/\vartheta'}\,a'_{j'v}\,, & a_{j'v} &= 0\,, \\
a_{(d'+1)v} &= \sqrt{\vartheta''w'_v/\vartheta'}\,, & a_{(d'+1)v} &= -\sqrt{\vartheta'w''_v/\vartheta'}\,, \\
a_{(d'+j'')v} &= 0\,, & a_{(d'+j'')v} &= \sqrt{\vartheta/\vartheta''}\,a''_{j''v}\,.
\end{aligned}
\tag{18.4}
$$

Now if $u, v \in V'$ we have

$$
\begin{aligned}
a_u \cdot a_v &= \sqrt{w'_u w'_v} + \frac{\vartheta}{\vartheta'}\left(a'_u \cdot a'_v - \sqrt{w'_u w'_v}\right) + \frac{\vartheta''}{\vartheta'}\sqrt{w'_u w'_v} \\
&= \frac{\vartheta}{\vartheta'}\,a'_u \cdot a'_v\,;
\end{aligned}
\tag{18.5}
$$

thus $a_u \cdot a_v = 0$ when $a'_u \cdot a'_v = 0$, and

$$\|a_v\|^2 = \frac{\vartheta}{\vartheta'}\,\|a'_v\|^2 = \vartheta\,. \tag{18.6}$$

It follows that $c(a_v) = w_v/\vartheta$ as desired. A similar derivation holds for $u, v \in V''$. And if $u \in V'$, $v \in V''$, then

$$a_u \cdot a_v = \sqrt{w'_u w''_v} - \sqrt{w'_u w''_v} = 0\,. \tag{18.7}$$

The orthogonal labeling b of $\overline{G' + G''}$ is much simpler; we just let $b_v = b'_v$ for $v \in V'$ and $b_v = b''_v$ for $v \in V''$. Then (12.2) and (12.3) are clearly preserved. This proves (18.2).

There is a close relation between the construction (18.4) and the construction (16.6), suggesting that we might be able to define another operation on graphs that generalizes both the splitting and direct sum operation.

19. The Direct Cosum of Graphs

If G' and G'' are graphs on disjoint vertex sets V' and V'' as in Section 18, we can also define

$$G = G' \mp G'' \iff \overline{G} = \overline{G'} + \overline{G''}. \tag{19.1}$$

This means $u - v$ in G if and only if either $u - v$ in G' or $u - v$ in G'' or u and v belong to opposite vertex sets. In this case

$$\vartheta(G, w) = \max\big(\vartheta(G', w'),\ \vartheta(G'', w'')\big) \tag{19.2}$$

and again there is an easy way to construct (a, b) from (a', b') and (a'', b'') to prove (19.2). Assume "without lots of generality" that

$$\vartheta(G', w') \geq \vartheta(G'', w'') \tag{19.3}$$

and suppose again that we have (18.3) and its counterpart for a''. Then we can define

if $v \in V'$	if $v \in V''$	
$a_{1v} = \sqrt{w_{v'}} = a'_{1v},$	$a_{1v} = \sqrt{w_{v''}} = a''_{1v},$	
$a_{j'v} = a'_{j'v},$	$a_{j'v} = 0,$	(19.4)
$a_{(d'+1)v} = 0,$	$a_{(d'+1)v} = \sqrt{(\vartheta' - \vartheta'')w''_v/\vartheta''},$	
$a_{(d'+j'')v} = 0,$	$a_{(d'+j'')v} = \sqrt{\vartheta'/\vartheta''}\, a''_{j''v}.$	

Now a_v is essentially unchanged when $v \in V'$; and when $u, v \in V''$ we have

$$
\begin{aligned}
a_u \cdot a_v &= \sqrt{w''_u w''_v} + \left(\frac{\vartheta'}{\vartheta''} - 1\right)\sqrt{w''_u w''_v} + \frac{\vartheta'}{\vartheta''}\left(a''_u \cdot a''_v - \sqrt{w''_u w''_v}\right) \\
&= \frac{\vartheta'}{\vartheta''}\, a''_u \cdot a''_v.
\end{aligned} \tag{19.5}
$$

Again we retain the necessary orthogonality, and we have $c(a_v) = w_v/\vartheta$ for all v.

For the b's, we let $b_v = b'_v$ when $v \in V'$ and $b_v = 0$ when $v \in V''$.

20. A Product of Graphs

Now let G' and G'' be graphs on vertices V' and V'' and let V be the $n = n'n''$ ordered pairs

$$V = V' \times V''. \tag{20.1}$$

We define the "strong product"

$$G = G' * G'' \tag{20.2}$$

on V by the rule

$(u', u'') - (v', v'')$ or $(u', u'') = (v', v'')$ in G $\iff$
$(u' - v'$ or $u' = v'$ in $G')$ and $(u'' - v''$ or $u'' = v''$ in $G'')$. $\quad$ (20.3)

In this case we have, for example, $K_{n'} * K_{n''} = K_{n'n''}$ and $\overline{K}_{n'} * \overline{K}_{n''} = \overline{K}_{n'n''}$. More generally, if G' is regular of degree r' and G'' is regular of degree r'', then $G' * G''$ is regular of degree $(r' + 1)(r'' + 1) - 1 = r'r'' + r' + r''$.

The value of $\vartheta(G, w)$ seems to be complicated for arbitrary w, but simple in the special case

$$w_{(v',v'')} = w'_{v'} \, w''_{v''} \,. \tag{20.4}$$

Lemma. *If G and w are given by (20.2) and (20.4), then*

$$\vartheta(G, w) = \vartheta(G', w') \, \vartheta(G'', w'') \,. \tag{20.5}$$

Proof. (The following proof is in [12].) Given orthogonal labelings (a', b') and (a'', b'') of G' and G'', we let a be the Kronecker product

$$a_{(j',j'')(v',v'')} = a'_{j'v'} a''_{j''v''} \,, \qquad 1 \le j' \le d', \quad 1 \le j'' \le d'', \tag{20.6}$$

where d' and d'' are the respective dimensions of the vectors in a' and a''. Then

$$a_{(u',u'')} \cdot a_{(v',v'')} = \sum_{j',j''} a'_{j'u'} a''_{j''u''} a'_{j'v'} a''_{j''v''}$$

$$= (a'_{u'} \cdot a'_{v'})(a''_{u''} \cdot a''_{v''}) \,. \tag{20.7}$$

Thus $\|a_{(v',v'')}\|^2 = \|a'_{v'}\|^2 \|a''_{v''}\|^2$ and

$$c(a_{(v',v'')}) = c(a'_{v'}) c(a''_{v''}) \,. \tag{20.8}$$

The same construction is used for b in terms of b' and b''.

All necessary orthogonalities are preserved, because we have

$(u', u'') — (v', v'')$ and $(u', u'') \neq (v', v'')$ in G

$\implies (u' — v'$ and $u' \neq v'$ in $G')$ or $(u'' — v''$ and $u'' \neq v''$ in $G'')$

$\implies b_{(u',u'')} \cdot b_{(v',v'')} = 0$;

$(u', u'') \not\!\!— (v', v'')$ and $(u', u'') \neq (v', v'')$ in G

$\implies (u' \not\!\!— v'$ and $u' \neq v'$ in $G')$ or $(u'' \not\!\!— v''$ and $u'' \neq v''$ in $G'')$

$\implies a_{(u',u'')} \cdot a_{(v',v'')} = 0$.

(In fact one of these relations is $\iff$, but we need only $\implies$ to make (20.7) zero when it needs to be zero.) Therefore a and b are orthogonal labelings of G that satisfy (12.2) and (12.3).   □

21. A Coproduct of Graphs

Guess what? We also define

$$G = G' \barstar G'' \iff \overline{G} = \overline{G'} * \overline{G''}. \tag{21.1}$$

This graph tends to be "richer" than $G' * G''$; we have

$(u', u'') — (v', v'')$ and $(u', u'') \neq (v', v'')$ in G $\iff$

$(u' — v'$ and $u' \neq v'$ in $G')$ or $(u'' — v''$ and $u'' \neq v''$ in $G'')$.   (21.2)

Now, for instance, if G' is regular of degree r' and G'' is regular of degree r'', then $G' \barstar G''$ is regular of degree

$$n'n'' - (n' - r')(n'' - r'') = r'n'' + r''n' - r'r''.$$

(This quantity is always $\geq r'r'' + r' + r''$, because $r'(n''-1-r'') + r''(n'-1-r') \geq 0$.) Indeed, $G' \barstar G'' \supseteq G' * G''$ for all graphs G' and G''. The Kronecker product construction used in Section 20 can be applied word-for-word to prove that

$$\vartheta(G, w) = \vartheta(G', w') \vartheta(G'', w'') \tag{21.3}$$

when G satisfies (21.1) and w has the special factored form (20.4).

It follows that many graphs have identical ϑ's:

Corollary. If $G' * G'' \subseteq G \subseteq G' \mathbin{\overline{\ast}} G''$ and w satisfies (20.4), then (21.3) holds.

Proof. This is just the monotonicity relation (4.3). The reason it works is that we have $a_{(u',v')} \cdot a_{(u'',v'')} = b_{(u',v')} \cdot b_{(u'',v'')} = 0$ for all pairs of vertices (u', u'') and (v', v'') whose adjacency differs in $G' * G''$ and $G' \mathbin{\overline{\ast}} G''$. $\square$

Some small examples will help clarify the results of the past few sections. Let P_3 be the path of length 2 on 3 vertices, ●—●—● , and consider the four graphs we get by taking its strong product and coproduct with $\overline{K}_2$ and K_2:

$$\overline{K}_2 * P_3 = \qquad \vartheta = \max(u+w, v) + \max(x+z, y)$$

(Since P_3 may be regarded as $\overline{K}_2 \mathbin{\overline{\mp}} K_1$ and $\overline{K}_2$ is $K_1 + K_1$, this graph is

$$\big((K_1 + K_1) \mathbin{\overline{\mp}} K_1\big) + \big((K_1 + K_1) \mathbin{\overline{\mp}} K_1\big)$$

and the formula for ϑ follows from (18.2) and (19.2).)

$$\overline{K}_2 \mathbin{\overline{\ast}} P_3 = \qquad \vartheta = \max(u+w+x+z, v+y)$$

(This graph is $\overline{K}_2 \mathbin{\overline{\mp}} \overline{K}_4$; we could also obtain it by applying Lemma 16 three times to P_3.)

$$K_2 * P_3 = \qquad \vartheta = \max\big(\max(u,x) + \max(w,z), \max(v,y)\big)$$

$$K_2 \mathbin{\overline{\ast}} P_3 = \qquad \vartheta = \max\big(\max(u+w, x+z), \max(v,y)\big)$$

If the weights satisfy $u = \lambda x$, $v = \lambda y$, $w = \lambda z$ for some parameter λ, the first two formulas for ϑ both reduce to $(1 + \lambda) \max(u + w, v)$, in agreement with (20.5) and (21.3). Similarly, the last two formulas for ϑ reduce to $\max(1, \lambda) \max(u + w, v)$ in such a case.

22. Odd Cycles

Now let $G = C_n$ be the graph with vertices $0, 1, \ldots, n-1$ and

$$u \text{—} v \iff u - v \equiv \pm 1 \pmod{n}, \tag{22.1}$$

where n is an *odd* number. A general formula for $\vartheta(C_n, w)$ appears to be very difficult; but we can compute $\vartheta(C_n)$ without too much labor when all weights are 1, because of the cyclic symmetry.

It is easier to construct orthogonal labelings of $\overline{C}_n$ than of C_n, so we begin with that. Given a vertex v, $0 \le v < n$, let b_v be the three-dimensional vector

$$b_v = \begin{pmatrix} \alpha \\ \cos v\varphi \\ \sin v\varphi \end{pmatrix}, \tag{22.2}$$

where α and φ remain to be determined. We have

$$\begin{aligned} b_u \cdot b_v &= \alpha^2 + \cos u\varphi \, \cos v\varphi + \sin u\varphi \, \sin v\varphi \\ &= \alpha^2 + \cos(u - v)\varphi. \end{aligned} \tag{22.3}$$

Therefore we can make $b_u \cdot b_v = 0$ when $u \equiv v \pm 1$ by setting

$$\alpha^2 = -\cos\varphi, \qquad \varphi = \frac{\pi(n-1)}{n}. \tag{22.4}$$

This choice of φ makes $n\varphi$ a multiple of 2π, because n is odd. We have found an orthogonal labeling b of $\overline{C}_n$ such that

$$c(b_v) = \frac{\alpha^2}{1 + \alpha^2} = \frac{\cos \pi/n}{1 + \cos \pi/n}. \tag{22.5}$$

Turning now to orthogonal labelings of C_n, we can use $(2n - 1)$-dimensional vectors

$$a_v = \begin{pmatrix} \alpha_0 \\ \alpha_1 \cos v\varphi \\ \alpha_1 \sin v\varphi \\ \alpha_2 \cos 2v\varphi \\ \alpha_2 \sin 2v\varphi \\ \vdots \\ \alpha_{n-1} \cos(n-1)v\varphi \\ \alpha_{n-1} \sin(n-1)v\varphi \end{pmatrix}, \tag{22.6}$$

with $\varphi = \pi(n-1)/n$ as before. As in (22.3), we find

$$a_u \cdot a_v = \sum_{k=0}^{n-1} \alpha_k^2 \cos k(u-v)\varphi \, ; \qquad (22.7)$$

so the result depends only on $(u-v) \bmod n$. Let $\omega = e^{i\varphi}$. We can find values of α_k such that $a_u \cdot a_v = x_{(u-v) \bmod n}$ by solving the equations

$$x_j = \sum_{k=0}^{n-1} \alpha_k^2 \omega^{jk} \, . \qquad (22.8)$$

Now ω is a primitive nth root of unity; that is, $\omega^k = 1$ if and only if k is a multiple of n. So (22.8) is just a finite Fourier transform, and we can easily invert it: For $0 \le m < n$ we have

$$\sum_{j=0}^{n-1} \omega^{-mj} x_j = \sum_{k=0}^{n-1} \alpha_k^2 \sum_{j=0}^{n-1} \omega^{j(k-m)} = n\alpha_m^2 \, .$$

In our case we want a solution with $x_2 = x_3 = \cdots = x_{n-2} = 0$, and we can set $x_0 = 1$, $x_{n-1} = x_1 = x$, so we find

$$n\alpha_k^2 = x_0 + \omega^{-k} x_1 + \omega^k x_{n-1} = 1 + 2x \cos k\varphi \, .$$

We must choose x so that these values are nonnegative; this means $2x \le -1/\cos\varphi$, since $\cos k\varphi$ is most negative when $k = 1$. Setting x to this maximum value yields

$$c(a_v) = \alpha_0^2 = \frac{1}{n}\left(1 - \frac{1}{\cos\varphi}\right) = \frac{1 + \cos\pi/n}{n\cos\pi/n} \, . \qquad (22.9)$$

So (22.5) and (22.9) give

$$\sum_v c(a_v)c(b_v) = \sum_v \frac{1}{n} = 1 \, . \qquad (22.10)$$

This is (12.3), hence from (12.2) we know that $\vartheta(C_n) = \lambda$. We have proved, in fact, that

$$\vartheta(C_n, \mathbb{1}) = \frac{n\cos\pi/n}{1 + \cos\pi/n} \, ; \qquad (22.11)$$

$$\vartheta(\overline{C}_n, \mathbb{1}) = \frac{1 + \cos\pi/n}{\cos\pi/n} \, . \qquad (22.12)$$

When $n = 3$, $C_n = K_n$ and these values agree with our previous formulas $\vartheta(K_3) = 1$, $\vartheta(\overline{K}_3) = 3$; when $n = 5$, $\overline{C}_5$ is isomorphic to C_5, so we have $\vartheta(C_5) = \sqrt{5}$; when n is large,

$$\vartheta(C_n) = \frac{n}{2} - \frac{\pi^2}{8n} + O(n^{-3}); \qquad \vartheta(\overline{C}_n) = 2 + \frac{\pi^2}{2n^2} + O(n^{-4}). \quad (22.13)$$

Instead of an explicit construction of vectors a_v as in (22.6), we could also find $\vartheta(C_n)$ by using the matrix characterization ϑ_2 of Section 6. When all weights are 1, a feasible A has 1 everywhere except on the superdiagonal, the subdiagonal, and the corners. This suggests that we look at "circulant" matrices; for example, when $n = 5$,

$$A = \begin{pmatrix} 1 & 1+x & 1 & 1 & 1+x \\ 1+x & 1 & 1+x & 1 & 1 \\ 1 & 1+x & 1 & 1+x & 1 \\ 1 & 1 & 1+x & 1 & 1+x \\ 1+x & 1 & 1 & 1+x & 1 \end{pmatrix} = J + xP + xP^{-1}, \quad (22.14)$$

where J is all 1's and P is the permutation matrix that takes j into $(j + 1) \bmod n$. It is well known and not difficult to prove that the eigenvalues of the circulant matrix $a_0 I + a_1 P + \cdots + a_{n-1} P^{n-1}$ are

$$\sum_{0 \le j < n}^{n-1} \omega^{jk} a_j, \qquad 0 \le k < n, \quad (22.15)$$

where $\omega = e^{2\pi i/n}$. (Indeed, it suffices to find the eigenvalues of P itself. We could also use the primitive root ω from (22.8).) Hence the eigenvalues of (22.14) are

$$n + 2x, \quad x(\omega + \omega^{-1}), \quad x(\omega^2 + \omega^{-2}), \quad \ldots, \quad x(\omega^{n-1} + \omega^{1-n}). \quad (22.16)$$

We minimize the maximum of these values if we choose x so that

$$n + 2x = -2x \cos \pi/n ;$$

then

$$\Lambda(A) = -2x \cos \pi/n = \frac{n \cos \pi/n}{1 + \cos \pi/n} \quad (22.17)$$

is the value of $\vartheta(G)$.

If n is even, the graph C_n is bipartite. We will prove later that bipartite graphs are perfect, hence $\vartheta(C_n) = n/2$ and $\vartheta(\overline{C}_n) = 2$ in the even case.

23. Comments on the Previous Example

The cycles C_n provide us with infinitely many graphs G for which $\vartheta(G)\vartheta(\overline{G}) = n$, and it is natural to wonder whether this identity holds in general. Of course it doesn't: If $G = \overline{K}_m + K_{n-m}$ then we have $\overline{G} = K_m \mp \overline{K}_{n-m}$, hence we know from (18.2) and (19.2) that

$$\vartheta(G) = m + 1, \qquad \vartheta(\overline{G}) = \max(1, n - m). \tag{23.1}$$

In particular, we can make the product $\vartheta(G)\vartheta(\overline{G})$ as high as $n^2/4 + n/2$ when $m = \lfloor n/2 \rfloor$.

We can, however, prove without difficulty that $\vartheta(G)\vartheta(\overline{G}) \geq n$:

Lemma. *If w and w' are any weight vectors for G we have*

$$\vartheta(G, w)\vartheta(\overline{G}, w') \geq w \cdot w'. \tag{23.2}$$

Proof. By Theorem 12 there is an orthogonal labeling a of G and an orthogonal labeling b of $\overline{G}$ such that

$$c(a_v) = w_v/\vartheta(G, w), \qquad c(b_v) = w'_v/\vartheta(\overline{G}, w'). \tag{23.3}$$

By (11.1) we have

$$\sum_v c(a_v)c(b_v) \leq 1. \tag{23.4}$$

QED.   □

24. Regular Graphs

When each vertex of G has exactly r neighbors, Lovász and Hoffman observed that the construction in (22.14) can be generalized. (See [12, Theorem 9].) Let B be the adjacency matrix of G, that is, the $n \times n$ matrix with

$$B_{uv} = \begin{cases} 1, & \text{if } u - v; \\ 0, & \text{if } u = v \text{ or } u \not\!- v. \end{cases} \tag{24.1}$$

Lemma. *If G is a regular graph,*

$$\vartheta(G) \leq \frac{n\Lambda(-B)}{\Lambda(B) + \Lambda(-B)}. \tag{24.2}$$

Proof. Let A be a matrix analogous to (22.14),

$$A = J + xB. \tag{24.3}$$

Since G is regular, the all-1's vector $\mathbb{1}$ is an eigenvector of B, and the other eigenvectors are orthogonal to $\mathbb{1}$ so they are eigenvectors also of A. Thus if the eigenvalues of B are

$$r = \Lambda(B) = \lambda_1 \geq \lambda_2 \geq \cdots \geq \lambda_n = -\Lambda(-B), \qquad (24.4)$$

the eigenvalues of A are

$$n + rx, \; x\lambda_2, \; \ldots, \; x\lambda_n. \qquad (24.5)$$

(The Perron–Frobenius theorem tells us that $\lambda_1 = r$.) We have $\lambda_1 + \cdots + \lambda_n = \operatorname{tr} B = 0$, so $\lambda_n < 0$, and we minimize the maximum of (24.5) by choosing $n + rx = x\lambda_n$; thus

$$\Lambda(A) = x\lambda_n = \frac{-n\lambda_n}{r - \lambda_n},$$

which is the right-hand side of (24.2). By (6.3) and Theorem 12 this quantity is an upper bound on ϑ. $\Box$

Incidentally, we need to be a little careful in (24.2): The denominator can be zero, but only when $G = \overline{K}_n$.

25. Automorphisms

An automorphism of a graph G is a permutation p of the vertices such that

$$p(u) \;\text{—}\; p(v) \qquad \text{if and only if} \qquad u \;\text{—}\; v. \qquad (25.1)$$

Such permutations are closed under multiplication, so they form a group.

We call G *vertex-symmetric* if its automorphism group is vertex-transitive, that is, if given u and v there is an automorphism p such that $p(u) = v$. We call G *edge-symmetric* if its automorphism group is edge-transitive, that is, if given $u \;\text{—}\; v$ and $u' \;\text{—}\; v'$ there is an automorphism p such that either $p(u) = u'$ and $p(v) = v'$ or $p(u) = v'$ and $p(v) = u'$.

Any vertex-symmetric graph is regular, but edge-symmetric graphs need not be regular. For example,

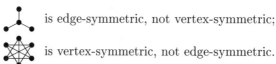 is edge-symmetric, not vertex-symmetric;

is vertex-symmetric, not edge-symmetric.

The graph $\overline{C}_n$ is not edge-symmetric for $n > 7$ because it has more edges than automorphisms. Also, $\overline{C}_7$ has no automorphism that takes $0 \;\text{—}\; 2$ into $0 \;\text{—}\; 3$.

Lemma. *If the graph G is edge-symmetric and regular, equality holds in Lemma 24.*

Proof. Say that A is an optimum feasible matrix for G if it is a feasible matrix with

$$\Lambda(A) = \vartheta(G)$$

as in Section 6. We can prove that optimum feasible matrices form a convex set, as follows. First, $tA + (1 - t)B$ is clearly feasible when A and B are feasible. Second,

$$\Lambda\big(tA + (1 - t)B\big) \leq t\Lambda(A) + (1 - t)\Lambda(B), \qquad 0 \leq t \leq 1 \qquad (25.2)$$

holds for all symmetric matrices A and B, by (6.2); this relation follows because there is a unit vector x such that $\Lambda\big(tA + (1 - t)B\big) = x^T\big(tA + (1 - t)B\big)x = tx^TAx + (1 - t)x^TBx \leq t\Lambda(A) + (1 - t)\Lambda(B)$. Third, if A and B are optimum feasible matrices, the right side of (25.2) is $\vartheta(G)$ while the left side is $\geq \vartheta(G)$ by (6.3). Therefore equality holds.

If A is an optimum feasible matrix for G, so is $p(A)$, the matrix obtained by permuting rows and columns by an automorphism p. (Namely, $p(A)_{uv} = A_{p(u)p(v)}$.) Therefore the average, $\bar{A}$, over all p is also an optimal feasible matrix. Since $p(\bar{A}) = \bar{A}$ for all automorphisms p, and since G is edge-symmetric, $\bar{A}$ has the form $J + xB$ where B is the adjacency matrix of G. The bound in Lemma 24 is therefore tight.   □

(*Note:* If p is a permutation, let $P_{uv} = 1$ if $u = p(v)$, otherwise 0. Then $(P^TAP)_{uv} = \sum_{j,k}(P^T)_{uj}A_{jk}P_{kv} = A_{p(u)p(v)}$, so $p(A) = P^TAP$.)

The argument in this proof shows that the set of all optimum feasible matrices A for G has a common eigenvector x such that $Ax = \vartheta(G)x$. The argument also shows that, if G has an edge automorphism taking $u — v$ into $u' — v'$, we can assume without loss of generality that $A_{uv} = A_{u'v'}$ in an optimum feasible matrix. This simplifies the computation of $\Lambda(A)$, and justifies our restriction to circulant matrices (22.14) in the case of cyclic graphs.

Theorem. *If G is vertex-symmetric, we have $\vartheta(G)\,\vartheta(\overline{G}) = n$.*

Proof. Say that b is an optimum normalized labeling of $\overline{G}$ if it is a normalized orthogonal labeling of $\overline{G}$ achieving equality in (7.1) when all weights are 1:

$$\vartheta = \sum_{u,v} b_u \cdot b_v, \qquad \sum_v \|b_v\|^2 = 1, \qquad b_u \cdot b_v = 0 \text{ when } u — v. \qquad (25.3)$$

Let B be the corresponding spud; that is, let $B_{uv} = b_u \cdot b_v$ and $\vartheta = \sum_{u,v} B_{uv}$. Then $p(B)$ is also equivalent to an optimum normalized labeling, whenever p is an automorphism; and such matrices B form a convex set, so we can assume as in the lemma that $B = p(B)$ for all automorphisms p. Since G is vertex-symmetric, we must have $B_{vv} = 1/n$ for all vertices v. Thus there is an optimum normalized labeling b with $\|b_v\|^2 = 1/n$, and the arguments of Lemma 10 and Theorem 12 establish the existence of such a b with

$$c(b_v) = \vartheta(G)/n \tag{25.4}$$

for all v. But b is an orthogonal labeling of $\overline{G}$, hence

$$\vartheta_1(\overline{G}, \mathbb{1}) \leq n/\vartheta(G)$$

by the definition (5.2) of ϑ_1. Thus $\vartheta(\overline{G})\vartheta(G) \leq n$. And we have already proved the reverse inequality in Lemma 23.   □

26. Consequence for Eigenvalues

A curious corollary of the results just proved is the following fact about eigenvalues.

Corollary. *If the graphs G and $\overline{G}$ are vertex-symmetric and edge-symmetric, and if the adjacency matrix of G has eigenvalues*

$$\lambda_1 \geq \lambda_2 \geq \cdots \geq \lambda_n, \tag{26.1}$$

then

$$(\lambda_1 - \lambda_n)(n - \lambda_1 + \lambda_2) = -\lambda_n(\lambda_2 + 1)n. \tag{26.2}$$

Proof. By Lemma 25 and Theorem 25,

$$\frac{n\Lambda(-B)}{\Lambda(B) + \Lambda(-B)} \frac{n\Lambda(-\overline{B})}{\Lambda(\overline{B}) + \Lambda(-\overline{B})} = n, \tag{26.3}$$

where B and $\overline{B}$ are the adjacency matrices of G and $\overline{G}$, and where we interpret $0/0$ as 1. We have

$$\overline{B} = J - I - B. \tag{26.4}$$

If the eigenvalues of B are given by (26.1), the eigenvalues of $\overline{B}$ are therefore

$$n - 1 - \lambda_1 \geq -1 - \lambda_n \geq \cdots \geq -1 - \lambda_2. \tag{26.5}$$

(We use the fact that G is regular of degree λ_1.) Formula (26.2) follows if we plug the values $\Lambda(B) = \lambda_1$, $\Lambda(-B) = -\lambda_n$, $\Lambda(\overline{B}) = n - 1 - \lambda_1$, $\Lambda(-\overline{B}) = 1 + \lambda_2$ into (26.3).   □

27. Further Examples of Symmetric Graphs

Consider the graph $P(m,t,q)$ whose vertices are all $\binom{m}{t}$ subsets of cardinality t of some given set S of cardinality m, where

$$u - v \iff |u \cap v| = q. \tag{27.1}$$

We want $0 \le q < t$ and $m \ge 2t - q$, so that the graph isn't empty. In fact, we can assume that $m \ge 2t$, because $P(m,r,q)$ is isomorphic to $P(m, m-t, m-2t+q)$ if we map each subset u into the set difference $S \setminus u$:

$$|(S \setminus u) \cap (S \setminus v)| = |S| - |u \cup v| = |S| - |u| - |v| + |u \cap v|. \tag{27.2}$$

The letter P stands for Petersen, because $P(5,2,0)$ is the well known "Petersen graph" on 10 vertices,

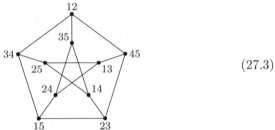

$$\text{(27.3)}$$

The P-graphs are clearly vertex-symmetric and edge-symmetric, because every permutation of S induces an automorphism. For example, to find an automorphism that maps $u - v$ into $u' - v'$, let $u = (u \cap v) \cup \bar{u}$, $v = (u \cap v) \cup \bar{v}$, $u' = (u' \cap v') \cup \bar{u}'$, $v' = (u' \cap v') \cup \bar{v}'$, and apply any permutation that takes the q elements of $u \cap v$ into the q elements of $u' \cap v'$, the $t - q$ elements of $\bar{u}$ into the $t - q$ elements of u', and the $t - q$ elements of $\bar{v}$ into $\bar{v}'$. Thus we can determine $\vartheta\bigl(P(m,t,q)\bigr)$ from the eigenvalues of the adjacency matrix. Lovász [12] discusses the case $q = 0$, and his discussion readily generalizes to other values of q. It turns out that $\vartheta\bigl(P(m,t,0)\bigr) = \binom{m-1}{t-1}$. This is also the value of $\alpha\bigl(P(m,t,0)\bigr)$, because the $\binom{m-1}{t-1}$ vertices containing any given point form a stable set.

The special case $t = 2$, $q = 0$ is especially interesting because those graphs also satisfy the condition of Corollary 26. We have

$$n = \binom{m}{2}, \quad \lambda_1 = \binom{m-2}{2}, \quad \lambda_2 = 1, \quad \lambda_n = 3 - m, \tag{27.4}$$

and (26.2) does indeed hold (but not "trivially"). It is possible to cover $P(m,2,0)$ with disjoint maximum cliques; hence $\kappa\bigl(P(m,2,0)\bigr) = \binom{m}{2}/\lfloor \frac{m}{2} \rfloor = 2\lceil \frac{m}{2} \rceil - 1$. In particular, when G is the Petersen graph we have $\alpha(G) = \vartheta(G) = 4$, $\kappa(G) = 5$; also $\alpha(\overline{G}) = 2$, $\vartheta(\overline{G}) = \kappa(\overline{G}) = \frac{5}{2}$.

28. A Bound on ϑ

Lovász's paper [12] contains one more result about ϑ that is not in [7], so we will wrap up our discussion of [12] by describing [12, Theorem 11].

Theorem. *If G has an orthogonal labeling of dimension d with no zero vectors, we have $\vartheta(G) \leq d$.*

Proof. Given a nonzero orthogonal labeling a of dimension d, we can assume that $\|a_v\|^2 = 1$ for all v. (The hypothesis about zeros is important, since there is trivially an orthogonal labeling of any desired dimension if we allow zeros. The labeling needn't be optimum.) Then we construct an orthogonal labeling a'' of dimension d^2, with $c(a''_v) = 1/d$ for all v, as follows:

Let a'_v have d^2 components where the (j, k) component is $a_{jv}a_{kv}$. Then

$$a'_u \cdot a'_v = (a_u \cdot a_v)^2 \tag{28.1}$$

as in (20.7). Let Q be any orthogonal matrix with d^2 rows and columns, such that the (j, k) entry in row $(1, 1)$ is $1/\sqrt{d}$ for $j = k$, 0 otherwise. Then we define

$$a''_v = Qa'_v. \tag{28.2}$$

Once again $a''_u \cdot a''_v = (a_u \cdot a_v)^2$, so a'' is an orthogonal labeling. And the first component of a''_v is

$$a''_{(1,1)v} = \sum_{j,k} \frac{[j = k]}{\sqrt{d}} a'_{(j,k)v} = \sum_k \frac{a^2_{kv}}{\sqrt{d}} = \frac{1}{\sqrt{d}}; \tag{28.3}$$

hence $c(a''_v) = 1/d$. This proves $\vartheta(G) \leq d$, by definition of ϑ_1.   □

This theorem improves the obvious lower bound $\alpha(G)$ on the dimension of an optimum labeling.

29. Compatible Matrices

There's another way to formulate the theory we've been developing, by looking at things from a somewhat higher level, following ideas developed by Lovász and Schrijver [15] a few years after the book [7] was written. Let us say that the matrix A is λ-*compatible with G and w* if A is an $(n + 1) \times (n + 1)$ spud indexed by the vertices of G and by a special value 0, having the following properties:

- $A_{00} = \lambda$;
- $A_{vv} = A_{0v} = w_v$ for all vertices v;
- $A_{uv} = 0$ whenever $u \not\sim v$ in G.

Lemma. *There exists an orthogonal labeling a for G with costs $c(a_v) = w_v/\lambda$ if and only if there exists a matrix A that is λ-compatible with G and w.*

Proof. Given such an orthogonal labeling, we can normalize each vector so that $\|a_v\|^2 = w_v$. Then when $w_v \neq 0$ we have

$$\frac{w_v}{\lambda} = c(a_v) = \frac{a_{1v}^2}{w_v},$$

so we can assume that $a_{1v} = w_v/\sqrt{\lambda}$ for all v. Add a new vector a_0, having $a_{10} = \sqrt{\lambda}$ and $a_{j0} = 0$ for all $j > 1$. Then the matrix A with $A_{uv} = a_u \cdot a_v$ is easily seen to be λ-compatible with G and w.

Conversely, if such a matrix A exists, there are $n + 1$ vectors a_0, ..., a_n such that $A_{uv} = a_u \cdot a_v$; in particular, $\|a_0\|^2 = \lambda$. Let Q be an orthogonal matrix with first row $a_0^T/\sqrt{\lambda}$, and define $a_v' = Qa_v$ for all v. Then $a_{10}' = \sqrt{\lambda}$ and $a_{j0}' = 0$ for all $j > 1$. Also $a_u' \cdot a_v' = a_u \cdot a_v = A_{uv}$ for all u and v. Hence $\sqrt{\lambda} a_{1v}' = a_0' \cdot a_v' = A_{0v} = w_v$ and $\|a_v'\|^2 = a_v' \cdot a_v' = A_{vv} = w_v$, for all $v \in G$, proving that $c(a_v') = w_v/\lambda$. Finally a' is an orthogonal labeling, since $a_u' \cdot a_v' = A_{uv} = 0$ whenever $u \not\sim v$. $\square$

Corollary. $x \in \mathrm{TH}(G)$ *if and only if there exists a matrix 1-compatible with $\overline{G}$ and x.*

Proof. Set $\lambda = 1$ in the lemma and apply Theorem 14. $\square$

The corollary and definition (4.1) tell us that $\vartheta(G, w)$ is the maximum of $w_1 x_1 + \cdots + w_n x_n$ over all x that appear in matrices that are 1-compatible for $\overline{G}$ and x. Theorem 12 tells us that $\vartheta(G, w)$ is also the minimum λ such that there exists a λ-compatible matrix for G and w. The "certificate" property of Theorem 13 has an even stronger formulation in matrix terms:

Theorem. *Given a nonnegative weight vector $w = (w_1, \ldots, w_n)^T$, let A be λ-compatible with G and w, where λ is as small as possible, and let B be 1-compatible with $\overline{G}$ and x, where $w_1 x_1 + \cdots + w_n x_n$ is as large as possible. Then*

$$ADB = 0, \tag{29.1}$$

where D is the diagonal matrix with $D_{00} = -1$ and $D_{vv} = +1$ for all $v \neq 0$. Conversely, if A is λ-compatible with G and w and if B is 1-compatible with $\overline{G}$ and x, then (29.1) implies that $\lambda = w_1 x_1 + \cdots + w_n x_n = \vartheta(G, w)$.

Proof. Assume that A is λ-compatible with G and w, and B is 1-compatible with $\overline{G}$ and x. Let $B' = DBD$, so that B' is a spud with

$B'_{00} = 1$, $B'_{0v} = B'_{v0} = -x_v$, and $B'_{uv} = B_{uv}$ when u and v are nonzero. Then the dot product $A \cdot B'$ is

$$\lambda - w_1 x_1 - \cdots - w_n x_n - w_1 x_1 - \cdots - w_n x_n + w_1 x_1 + \cdots + w_n x_n$$
$$= \lambda - (w_1 x_1 + \cdots + w_n x_n),$$

because $A_{uv} B_{uv} = 0$ when u and v are vertices of G. We showed in the proof of F9 in Section 9 that the dot product of spuds is nonnegative; in fact, that proof implies that the dot product is zero if and only if the ordinary matrix product is zero. So $\lambda = w_1 x_1 + \cdots + w_n x_n = \vartheta(G, w)$ if and only if $AB' = 0$, and this is equivalent to (29.1). $\quad\Box$

Equation (29.1) gives us further information about the orthogonal labelings a and b that appear in Theorems 12 and 13. Normalize those labelings so that $\|a\|^2 = w_v$ and $\|b\|^2 = x_v$. Then we have

$$\sum_{t \in G} w_t (b_t \cdot b_v) = \vartheta x_v, \tag{29.2}$$

$$\sum_{t \in G} x_t (a_t \cdot a_v) = w_v, \tag{29.3}$$

$$\sum_{t \in G} (a_t \cdot a_u)(b_t \cdot b_v) = w_u x_v, \tag{29.4}$$

for all vertices u and v of G. (Indeed, (29.2) and (29.4) are respectively equivalent to $(AB')_{0v} = 0$ and $(AB')_{vv} = 0$; (29.3) is equivalent to $(B'A)_{0v} = 0$.) Notice that if $\widehat{A}$ and $\widehat{B}$ are the $n \times n$ spuds obtained by deleting row 0 and column 0 from optimum matrices A and B, these equations are equivalent to

$$\widehat{B} w = \vartheta x, \qquad \widehat{A} x = w, \qquad \widehat{A}\widehat{B} = w x^T. \tag{29.5}$$

Equation (29.1) is equivalent to (29.5) together with the condition $w_1 x_1 + \cdots + w_n x_n = \vartheta$.

Since $AB' = 0$ if and only if $B'A = 0$ when A and B' are symmetric matrices, the optimum matrices A and B' commute. This implies that they have common eigenvectors: There is an orthogonal matrix Q such that

$$A = Q \operatorname{diag}(\lambda_0, \ldots, \lambda_n) Q^T, \qquad B' = Q \operatorname{diag}(\mu_0, \ldots, \mu_n) Q^T. \tag{29.6}$$

Moreover, the product is zero, so

$$\lambda_0 \mu_0 = \cdots = \lambda_n \mu_n = 0. \tag{29.7}$$

The number of zero eigenvalues λ_k is $n + 1 - d$, where d is the smallest dimension for which there is an orthogonal labeling a with $A_{uv} = a_u \cdot a_v$. A similar statement holds for B', since the eigenvalues of B and B' are the same; y is an eigenvector for B if and only if Dy is an eigenvector for B'. In the case $G = C_n$, studied in Section 22, we constructed an orthogonal labeling (22.3) with only three dimensions, so all but 3 of the eigenvalues μ_k were zero. When all the weights w_v are nonzero and $\vartheta(G)$ is large, Theorem 28 implies that a large number of λ_k must be nonzero, hence a large number of μ_k must be zero.

The "optimum feasible matrices" A studied in Section 6 are related to the matrices $\widehat{A}$ of (29.5) by the formula

$$\vartheta \widehat{A} = ww^T - \vartheta \operatorname{diag}(w_1, \ldots, w_n)$$
$$-\operatorname{diag}(\sqrt{w_1}, \ldots, \sqrt{w_n}) \, A \operatorname{diag}(\sqrt{w_1}, \ldots, \sqrt{w_n}), \qquad (29.8)$$

because of the construction following (6.6). If the largest eigenvalue $\Lambda(A) = \vartheta$ of A occurs with multiplicity r, the rank of $\vartheta I - A$ will be $n - r$, hence $\widehat{A}$ will have rank $n - r$ or $n - r + 1$, and the number of zero eigenvalues λ_k in (29.6) will be $r + 1$ or r.

30. Antiblockers

The convex sets STAB, TH, and QSTAB defined in Section 2 have many special properties. For example, they are always nonempty, closed, convex, and nonnegative; they also satisfy the condition

$$0 \leq y \leq x \quad \text{and} \quad x \in X \;\Rightarrow\; y \in X. \qquad (30.1)$$

A set X of vectors satisfying all five of these properties is called a *convex corner*.

If X is any set of nonnegative vectors we define its *antiblocker* by the condition

$$\operatorname{abl} X = \{\, y \geq 0 \mid x \cdot y \leq 1 \text{ for all } x \in X \,\}. \qquad (30.2)$$

Clearly $\operatorname{abl} X$ is a convex corner, and $\operatorname{abl} X \supseteq \operatorname{abl} X'$ when $X \subseteq X'$.

Lemma. *If X is a convex corner we have* $\operatorname{abl}\operatorname{abl} X = X$.

Proof. (Compare with the proof of F5 in Section 8.) The relation $X \subseteq \operatorname{abl}\operatorname{abl} X$ is obvious by definition (30.2), so the lemma can fail only if there is some $z \in \operatorname{abl}\operatorname{abl} X$ with $z \notin X$. Then there is a hyperplane separating z from X, by F1; that is, there is a vector y and a number b

such that $x \cdot y \leq b$ for all $x \in X$ but $z \cdot y > b$. Let y' be the same as y but with all negative components changed to zero. Then (y', b) is also a separating hyperplane. [*Proof:* If $x \in X$, let x' be the same as x but with all components changed to zero where y has a negative entry; then $x' \in X$, and $x \cdot y' = x' \cdot y \leq b$. Furthermore $z \cdot y' \geq z \cdot y > b$.] If $b = 0$, we have $\lambda y' \in \text{abl } X$ for all $\lambda > 0$; this contradicts $z \cdot \lambda y' \leq 1$. We cannot have $b < 0$, since $0 \in X$. Hence $b > 0$, and the vector $y'/b \in \text{abl } X$. But then $z \cdot (y'/b)$ must be ≤ 1, a contradiction.   □

Corollary. *If G is any graph we have*

$$\text{STAB}(\overline{G}) = \text{abl QSTAB}(G), \tag{30.3}$$

$$\text{TH}(\overline{G}) = \text{abl TH}(G), \tag{30.4}$$

$$\text{QSTAB}(\overline{G}) = \text{abl STAB}(G). \tag{30.5}$$

Proof. First we show that

$$\text{abl } X = \text{abl}(\text{convex hull } X). \tag{30.6}$$

The left side surely contains the right. And any element $y \in \text{abl } X$ will satisfy

$$(\alpha_1 x^{(1)} + \cdots + \alpha_k x^{(k)}) \cdot y \leq 1$$

when the α's are nonnegative scalars summing to 1 and the $x^{(j)}$ are in X. This proves (30.6), because the convex hull of X is the set of all such vectors $\alpha_1 x^{(1)} + \cdots + \alpha_k x^{(k)}$.

Now (30.6) implies (30.5), because the definitions in Section 2 say that

$$\text{QSTAB}(\overline{G}) = \text{abl} \{ x \mid x \text{ is a clique labeling of } \overline{G} \}$$

$$= \text{abl} \{ x \mid x \text{ is a stable labeling of } G \},$$

$$\text{STAB}(G) = \text{convex hull} \{ x \mid x \text{ is a stable labeling of } G \}.$$

And (30.5) is equivalent to (30.3) by the lemma, because $\text{STAB}(G)$ is a convex corner. (We must prove (30.1), and it suffices to do this when y equals x in all but one component; and in fact by convexity we may assume that y is 0 in that component; and then we can easily prove it, because any subset of a stable set is stable.)

Finally, (30.4) is equivalent to Theorem 14, because $\text{TH}(G) = \text{abl} \{ x \mid x_v = c(a_v) \text{ for some orthogonal labeling of } G \}$.   □

The sets STAB and QSTAB are polytopes, that is, they are bounded and can be defined by a finite number of inequalities. But the antiblocker concept applies also to sets with curved boundaries. For example, let

$$X = \{\, x \geq 0 \mid \|x\| \leq 1 \,\} \tag{30.7}$$

be the intersection of the unit ball with the nonnegative orthant. Cauchy's inequality implies that $x \cdot y \leq 1$ whenever $\|x\| \leq 1$ and $\|y\| \leq 1$, hence $X \subseteq \text{abl}\, X$. And if $y \in \text{abl}\, X$ we have $y \in X$, since $y \neq 0$ implies $\|y\| = y \cdot (y/\|y\|) \leq 1$. Therefore $X = \text{abl}\, X$.

In fact, the set X in (30.7) is the only set that equals its own antiblocker. If $Y = \text{abl}\, Y$ and $y \in Y$ we have $y \cdot y \leq 1$, hence $Y \subseteq X$; this implies $\text{abl}\, Y \supseteq X$.

31. Perfect Graphs

Let $\omega(G)$ be the size of a largest clique in G. The graph G is called *perfect* if every induced subgraph G' of G can be colored with $\omega(G')$ colors. (See Section 15 for the notion of induced subgraph. This definition of perfection was introduced by Claude Berge in 1961.)

Let G^+ be G with vertex v duplicated, as described in Section 16. This means we add a new vertex v' with the same neighbors as v and with $v - v'$.

Lemma. *If G is perfect, so is G^+.*

Proof. Any induced subgraph of G^+ that is not G^+ itself is either an induced subgraph of G (if it omits v or v' or both), or has the form G'^+ for some induced subgraph G' of G (if it retains v and v'). Therefore it suffices to prove that G^+ can be colored with $\omega(G^+)$ colors.

Color G with $\omega(G)$ colors and suppose v is red. Let G' be the subgraph induced from G by leaving out all red vertices except v. Recolor G' with $\omega(G')$ colors, and assign a new color to the set $G^+ \setminus G'$, which is stable in G^+. This colors G^+ with $\omega(G') + 1$ colors, hence $\omega(G^+) \leq \omega(G') + 1$.

We complete the proof by showing that $\omega(G^+) = \omega(G') + 1$. Let Q be a clique of size $\omega(G')$ in G'.

Case 1. $v \in Q$. Then $Q \cup \{v'\}$ is a clique of G^+.

Case 2. $v \notin Q$. Then Q contains no red element, so $\omega(G') < \omega(G)$. In both cases we can conclude that $\omega(G^+) \geq \omega(G') + 1$. $\square$

Theorem. *If G is perfect,* STAB$(G) = $QSTAB$(G)$.

Proof. It suffices to prove that every $x \in$ QSTAB(G) with *rational* coordinates is a member of STAB(G), because STAB(G) is a closed set.

Suppose $x \in \text{QSTAB}(G)$ and qx has integer coordinates. Let G^+ be the graph obtained from G by repeatedly duplicating vertices until each original vertex v of G has been replaced by a clique of size qx_v. Call the vertices of that clique the *clones* of v.

By definition of $\text{QSTAB}(G)$, if Q is any clique of G we have

$$\sum_{v \in Q} x_v \leq 1 \, .$$

Every clique Q' of G^+ is contained in a clique of size $\sum_{v \in Q} qx_v$ for some clique Q of G. (Including all clones of each element yields this possibly larger clique.) Thus $\omega(G^+) \leq q$, and the lemma tells us that G^+ can be colored with q colors because G^+ is perfect.

For each color k, where $1 \leq k \leq q$, let $x_v^{(k)} = 1$ if some clone of v is colored k, otherwise $x_v^{(k)} = 0$. Then $x^{(k)}$ is a stable labeling. Hence

$$\frac{1}{q} \sum_{k=1}^{q} x^{(k)} \in \text{STAB}(G) \, .$$

But every vertex of G^+ is colored, so $\sum_{k=1}^{q} x_v^{(k)} = qx_v$ for all v, so $q^{-1} \sum_{k=1}^{q} x^{(k)} = x$.   □

32. A Characterization of Perfection

The converse of Theorem 31 is also true; but before we prove it we need another fact about convex polyhedra.

Lemma. *Suppose P is the set $\{ x \geq 0 \mid x \cdot z \leq 1 \text{ for all } z \in Z \} = \text{abl } Z$ for some finite set Z and suppose $y \in \text{abl } P$; that is, y is a nonnegative vector such that $x \cdot y \leq 1$ for all $x \in P$. Then the set*

$$Q = \{ x \in P \mid x \cdot y = 1 \} \tag{32.1}$$

is contained in the set $\{ x \mid x \cdot z = 1 \}$ for some $z \in Z$ (unless Q and Z are both empty).

Proof. This lemma is "geometrically obvious" — it says that every vertex, edge, etc., of a convex polyhedron is contained in some "facet" — but we ought also to prove it. The proof is by induction on $|Z|$. If Z is empty, the result holds because P is the set of all nonnegative x, hence y must be 0 and Q must be empty.

Suppose z is an element of Z that does not satisfy the condition; namely, there is an element $x \in P$ with $x \cdot y = 1$ and $x \cdot z \neq 1$. Then

$x \cdot z < 1$. Let $Z' = Z \setminus \{z\}$ and $P' = \text{abl}\, Z'$. It follows that $x' \cdot y \leq 1$ for all $x' \in P'$. For if $x' \cdot y > 1$, a convex combination $x'' = \epsilon x + (1 - \epsilon) x'$ will lie in P for sufficiently small ϵ, but $x'' \cdot y > 1$.

Therefore by induction, $Q' = \{\, x \in P' \mid x \cdot y = 1 \,\}$ is contained in $\{\, x \mid x \cdot z' = 1 \,\}$ for some $z' \in Z'$, unless Q' is empty, when we can take $z' = z$. And $Q \subseteq Q'$, since $P \subseteq P'$. $\square$

Theorem. *G is perfect if and only if $\text{STAB}(G) = \text{QSTAB}(G)$.*

Proof. As in Section 15, let $G|U$ be the graph induced from G by restriction to vertices U. If X is a set of vectors indexed by V and if $U \subseteq V$, let $X|U$ be the set of all vectors indexed by U that arise from the vectors of X when we suppress all components x_v with $v \notin U$. Then it is clear that

$$\text{QSTAB}(G|U) = \text{QSTAB}(G)|U\,, \tag{32.2}$$

because every $x \in \text{QSTAB}(G|U)$ belongs to $\text{QSTAB}(G)$ if we set $x_v = 0$ for $v \notin U$, and every $x \in \text{QSTAB}(G)$ satisfies $\sum_{v \in Q} x_v \leq 1$ for every clique $Q \subseteq U$. Also

$$\text{STAB}(G|U) = \text{STAB}(G)|U\,, \tag{32.3}$$

because every stable labeling of $G|U$ is a stable labeling of G if we extend it with zeros, and every stable labeling of G is stable for $G|U$ if we ignore components not in U.

Therefore $\text{STAB}(G) = \text{QSTAB}(G)$ if and only if $\text{STAB}(G') = \text{QSTAB}(G')$ for all induced graphs. By Theorem 31 we need only prove that $\text{STAB}(G) = \text{QSTAB}(G)$ implies that G can be colored with $\omega(G)$ colors.

Suppose $\text{STAB}(G) = \text{QSTAB}(G)$. Then by Corollary 30,

$$\text{STAB}(\overline{G}) = \text{QSTAB}(\overline{G})\,. \tag{32.4}$$

Let $P = \text{STAB}(\overline{G})$, and let $y = \mathbb{1}/\omega(G)$. Then $x \cdot y \leq 1$ whenever x is a clique labeling of G, that is, whenever x is a stable labeling of $\overline{G}$; so $x \cdot y \leq 1$ for all $x \in P$. Let Z be the set of all stable labelings of G, that is, all clique labelings of $\overline{G}$. Then $P = \text{QSTAB}(\overline{G}) = \text{abl}\, Z$ and Z is nonempty. So the lemma applies, and it tells us that the set Q defined in (32.1) is contained in $\{\, x \mid x \cdot z = 1 \,\}$ for some stable labeling z of G. Therefore every maximum clique labeling x satisfies $x \cdot z = 1$; that is, every clique of size $\omega(G)$ intersects the stable set S corresponding to z. So $\omega(G') = \omega(G) - 1$, where

$$G' = G|(V \setminus S)\,. \tag{32.5}$$

By induction on $|V|$ we can color the vertices of G' with $\omega(G')$ colors, then we can use a new color for the vertices of S; this colors G with $\omega(G)$ colors.   □

Lovász states in [13] that he knows no polynomial time algorithm to test if G is perfect; but he conjectures ("guesses") that such an algorithm exists, because the results we are going to discuss next suggest that much more might be provable.

33. Another Definition of ϑ

The following result generalizes Lemma 9.3.21 of [7].

Lemma. *Let a and b be orthogonal labelings of G and $\overline{G}$ that satisfy the conditions of Theorem 12, normalized so that*

$$\|a_v\|^2\|b_v\|^2 = w_v c(b_v), \qquad a_{1v} \geq 0, \qquad \text{and} \qquad b_{1v} \geq 0, \qquad (33.1)$$

for all v. Then

$$\sum_v a_{jv} b_{kv} = \begin{cases} \sqrt{\vartheta(G, w)}, & \text{if } j = k = 1; \\ 0, & \text{otherwise.} \end{cases} \qquad (33.2)$$

Proof. Let $a_0 = (\sqrt{\vartheta}, 0, \ldots, 0)^T$ and $b_0 = (-1, 0, \ldots, 0)^T$. Then the $(n+1) \times (n+1)$ matrices $A = a^T a$ and $B = b^T b$ are spuds, and $A \cdot B = 0$. (In the special case $\|a_v\|^2 = w_v$ and $\|b_v\|^2 = c(b_v)$, matrix B is what we called B' in the proof of Theorem 29.) Therefore $0 = \operatorname{tr} A^T B = \operatorname{tr} a^T a b^T b = \operatorname{tr} b a^T a b^T = (ab^T) \cdot (ab^T)$, and we have $ab^T = 0$. In other words

$$a_{j0} b_{k0} + \sum_v a_{jv} b_{kv} = 0$$

for all j and k.   □

We now can show that ϑ has yet another definition, in some ways nicer than the one we considered in Section 6. (Perhaps the reader can find a simpler way to derive all these facts.) Call the matrix B *dual feasible* for G and w if it is indexed by vertices and

$$B \text{ is real and symmetric;}$$

$$B_{vv} = w_v \text{ for all } v \in V;$$

$$B_{uv} = 0 \text{ whenever } u \not\sim v \text{ in } G; \qquad (33.3)$$

and define

$$\vartheta_6(G, w) = \max\{ \Lambda(B) \mid B \text{ is positive semidefinite}$$
$$\text{and dual feasible for } G \text{ and } w \}. \qquad (33.4)$$

(Compare with the analogous definitions in (6.1) and (6.3).)

Theorem. $\vartheta(G,w) = \vartheta_6(G,w)$.

Proof. If B is positive semidefinite and dual feasible, and if λ is any eigenvalue of B, we can write $B = QDQ^T$ where Q is orthogonal and D is diagonal, with $D_{11} = \lambda$. Let $b = \sqrt{D}\,Q^T$; then b is an orthogonal labeling of $\overline{G}$ with $\|b_v\|^2 = w_v$ for all v. Furthermore $c(b_v) = b_{1v}^2/w_v = \lambda\, q_{v1}^2/w_v$, where $(q_{11},\ldots,q_{n1})$ is the first column of Q. Therefore $\sum_v c(b_v)w_v = \lambda \sum_v q_{v1}^2 = \lambda$, and we have $\lambda \leq \vartheta_4(G,w)$ by (10.1). This proves that $\vartheta_6 \leq \vartheta$.

Conversely, let a and b be orthogonal labelings of G and $\overline{G}$ that satisfy the conditions of Theorem 12. Normalize them so that $\|a_v\|^2 = c(b_v)$ and $\|b_v\|^2 = w_v$. Then $a_{1v}^2 = c(a_v)c(b_v) = w_v c(b_v)/\vartheta = b_{1v}^2/\vartheta$. The lemma now implies that $(b_{11},\ldots,b_{1n})^T$ is an eigenvector of $b^T b$, with eigenvalue ϑ. This proves that $\vartheta \leq \vartheta_6$. $\square$

Corollary. *If G is any graph,*

$$\vartheta(G) = 1 + \max\{\, \Lambda(B)/\Lambda(-B) \mid B \text{ is dual feasible for } G \text{ and } 0 \,\}.$$

Proof. If B is dual feasible for G and 0, its eigenvalues are $\lambda_1 \geq \cdots \geq \lambda_n$ where $\lambda_1 = \Lambda(B)$ and $\lambda_n = -\Lambda(-B)$. Then $B' = I + B/\Lambda(-B)$ has eigenvalues $1 - \lambda_1/\lambda_n, \ldots, 1 - \lambda_n/\lambda_n = 0$. Consequently B' is positive semidefinite and dual feasible for G and $\mathbb{1}$, and $1 + \Lambda(B)/\Lambda(-B) = \Lambda(B') \leq \vartheta_6(G)$.

Conversely, suppose B' is positive semidefinite and dual feasible for G and $\mathbb{1}$, with $\Lambda(B') = \vartheta = \vartheta(G)$. Let $B = B' - I$. Then B is dual feasible for G and 0, and $0 \leq \Lambda(-B) \leq 1$ since the sum of the eigenvalues of B is $\operatorname{tr} B = 0$. Hence $\vartheta - 1 = \Lambda(B) \leq \Lambda(B)/\Lambda(-B)$. $\square$

34. Facets of TH

We know that $\mathrm{TH}(G)$ is a convex corner set in n-dimensional space, so it is natural to ask whether it might have $(n-1)$-dimensional facets on its nontrivial boundary — for example, a straight line segment in two dimensions, or a region of a plane in three dimensions. This means it would have n linearly independent vectors $x^{(k)}$ such that

$$\sum_v x_v^{(k)} c(a_v) = 1 \tag{34.1}$$

for some orthogonal labeling a of G.

Theorem. *If G has n vertices and an orthogonal labeling such that* TH(G) *contains linearly independent solutions $x^{(1)}, \ldots, x^{(n)}$ of (34.1), then there is a maximal clique Q of G such that*

$$c(a_v) = \begin{cases} 1, & v \in Q; \\ 0, & v \notin Q. \end{cases} \tag{34.2}$$

Proof. Theorem 14 tells us that every $x^{(k)} \in$ TH(G) has $x_v^{(k)} = c(b_v^{(k)})$ for some orthogonal labeling of $\overline{G}$. Set $w_v = c(a_v)$; then $\vartheta(G, w) = 1$, by Theorem 13. We can normalize the labelings so that $\|a_v\|^2 = a_{1v} = w_v$ and $\|b_v^{(k)}\|^2 = b_{1v}^{(k)} = x_v^{(k)}$. Hence, by Lemma 33,

$$\sum_v x_v^{(k)} a_v = \begin{pmatrix} 1 \\ 0 \\ \vdots \\ 0 \end{pmatrix} = e_1. \tag{34.3}$$

Let

$$Q = \{ v \mid a_{1v} \neq 0 \} = \{ v \mid c(a_v) \neq 0 \} \tag{34.4}$$

and suppose Q has m elements. Then (34.3) is equivalent to the matrix equation

$$A x^{(k)} = e_1 \tag{34.5}$$

where A is a $d \times m$ matrix and $x^{(k)}$ has m components $x_v^{(k)}$, one for each $v \in Q$. By hypothesis there are m linearly independent solutions to (34.5), because there are n linearly independent solutions to (34.3). But then there are $m - 1$ linearly independent solutions to $Ax = 0$, and it follows that A has rank 1: Every row of A must be a multiple of the top row (which is nonzero). And then (34.5) tells us that all rows but the top row are zero. We have proved that

$$c(a_v) \neq 0 \quad \Longrightarrow \quad c(a_v) = 1. \tag{34.6}$$

Therefore if u and v are elements of Q we have $a_u \cdot a_v \neq 0$, hence $u \mathrel{-\!\!-} v$; Q is a clique.

Moreover, Q is maximal. For if $v \notin Q$ is adjacent to all elements of Q, there is a k such that $x_v^{(k)} > 0$. But the characteristic labeling of $Q \cup \{v\}$ is an orthogonal labeling a' such that $\sum_u x_u^{(k)} c(a'_u) = 1 + x_v^{(k)} > 1$, hence $x^{(k)} \notin$ TH(G).   □

Conversely, it is easy to see that the characteristic labeling of any maximal clique Q does have n linearly independent vectors satisfying

(34.1), so it does define a facet. For each vertex u we let $x_u^{(u)} = 1$, and $x_v^{(u)} = 0$ for all $v \neq u$ except for one vertex $v \in Q$ with $v \not\sim u$ (when $u \notin Q$). Then $x^{(u)}$ is a stable labeling so it is in $\mathrm{TH}(G)$. The point of the theorem is that a constraint $\sum_v x_v c(a_v) \leq 1$ of $\mathrm{TH}(G)$ that is not satisfied by all $x \in \mathrm{QSTAB}(G)$ cannot correspond to a facet of $\mathrm{TH}(G)$.

Corollary. *If G is any graph,*

$$\mathrm{TH}(G) \text{ is a polytope} \iff \mathrm{TH}(G) = \mathrm{QSTAB}(G)$$
$$\iff \mathrm{TH}(G) = \mathrm{STAB}(G) \iff G \text{ is perfect.}$$

Proof. If $\mathrm{TH}(G)$ is a polytope it is defined by facets as in the theorem, which are nothing more than the constraints of $\mathrm{QSTAB}(G)$; hence $\mathrm{TH}(G) = \mathrm{QSTAB}(G)$. Also the antiblocker of a convex corner polytope is a polytope, so $\mathrm{TH}(\overline{G})$ is a polytope by (30.4); it must be equal to $\mathrm{QSTAB}(\overline{G})$. Taking antiblockers, we have $\mathrm{TH}(G) = \mathrm{STAB}(G)$ by (30.3). The converses are easy since STAB and QSTAB are always polytopes. The connection to perfection is an immediate consequence of Theorem 32 and Lemma 2.   □

We cannot strengthen the corollary to say that $\vartheta(G) = \alpha(G)$ holds if and only if $\vartheta(G) = \kappa(G)$; the Petersen graph (Section 27) is a counterexample.

35. Orthogonal Labelings in a Perfect Graph

A perfect graph has

$$\vartheta(G, w) = \alpha(G, w) = \max\{\, x \cdot w \mid x \text{ is a stable labeling of } G \,\}, \quad (35.1)$$

and Theorem 12 tells us there exist orthogonal labelings of G and $\overline{G}$ such that (12.2) and (12.3) hold. But it isn't obvious what those labelings might be; the proof was not constructive.

The problem is to find vectors a_v such that $a_u \cdot a_v = 0$ when $u \not\sim v$ and such that (12.2) holds; then it is easy to satisfy (12.3) by simply letting b be a stable labeling where the maximum occurs in (35.1).

The following general construction gives an orthogonal labeling (not necessarily optimum) in any graph: Let $g(Q)$ be a nonnegative number for every clique Q, chosen so that

$$\sum_{v \in Q} g(Q) = w_v, \qquad \text{for all } v. \tag{35.2}$$

Furthermore, for each clique Q, let

$$a_{Qv} = \begin{cases} \sqrt{g(Q)}, & \text{if } v \in Q; \\ 0, & \text{otherwise.} \end{cases} \tag{35.3}$$

Then

$$a_u \cdot a_v = \sum_{\{u,v\} \subseteq Q} g(Q),$$

hence $a_u \cdot a_v = 0$ when $u \not\sim v$. If we also let $a_{Q0} = \sqrt{g(Q)}$ for all Q, $a_{00} = 0$, we find

$$a_0 \cdot a_v = a_v \cdot a_v = \sum_{v \in Q} g(Q) = w_v.$$

We have constructed a matrix A that is λ-compatible with G and w, in the sense of Section 29, where

$$\lambda = a_0 \cdot a_0 = \sum_Q g(Q). \tag{35.4}$$

An orthogonal labeling with costs $c(a_v') = w_v/\lambda$ can now be found as in the proof of Lemma 29.

The duality theorem of linear programming tells us that the minimum of (35.4) subject to the constraints (35.2) is equal to the maximum value of $w \cdot x$ over all x with $\sum_{v \in Q} x_v \leq 1$ for all Q. When x maximizes $w \cdot x$, we can assume that $x \geq 0$, because a negative x_v can be replaced by 0 without decreasing $w \cdot x$ or violating a constraint. (Every subset of a clique is a clique.) Thus, we are maximizing $w \cdot x$ over QSTAB(G); the construction in the previous paragraph allows us to reduce λ as low as $\kappa(G, w)$. But $\kappa(G, w) = \vartheta(G, w)$ in a perfect graph, so this construction solves our problem, once we have computed $g(Q)$.

The special case of a bipartite graph is especially interesting, because its cliques have only one or two vertices. Suppose all edges of G have the form $u \longrightarrow v$ where $u \in U$ and $v \in V$, and consider the network defined as follows: There is a special source vertex s connected to all $u \in U$ by a directed arc of capacity w_u, and a special sink vertex t connected from all $v \in V$ by a directed arc of capacity w_v. The edges $u \longrightarrow v$ of G are also present, directed from u to v with infinite capacity. Any flow from s to t in this network defines a suitable function g, if we let

$$g(\{u, v\}) = \text{the flow in } u \to v,$$
$$g(\{u\}) = w_u \text{ minus the flow in } s \to u,$$
$$g(\{v\}) = w_v \text{ minus the flow in } v \to t,$$

for all $u \in U$ and $v \in V$. Let S be a subset of $U \cup V$. If we cut the edges that connect s or t with vertices not in S, we cut off all paths from s to t if and only if S is a stable set. The minimum cut (that is, the minimum sum of capacities of cut edges) is equal to the maximum flow; and it is also equal to

$$\sum_{u \in U} w_u + \sum_{v \in V} w_v - \max\{\, w \cdot x \mid x \text{ is a stable labeling}\,\}$$
$$= \sum_{u \in U} w_u + \sum_{v \in V} w_v - \alpha(G, w).$$

Thus the value of $\lambda = \sum_Q g(Q)$ is

$$\sum_{u \in U} w_u - \{\text{flow from } s\} + \sum_{v \in V} w_v$$
$$+ \{\text{flow in } u \to v \text{ arcs}\} - \{\text{flow to } t\} = \alpha(G, w) = \vartheta(G, w)$$

as desired.

For general perfect graphs G, a solution to (35.4) with $\lambda = \vartheta(G, w)$ can be found in polynomial time as shown in equation (9.4.6) of [7]. However, the methods described in [7] are not efficient enough for practical calculation, even on small graphs.

36. The Smallest Non-Perfect Graph

The cyclic graph C_5 is of particular interest because it is the smallest graph that isn't perfect, and the smallest case where the function $\vartheta(G, w)$ is not completely known.

The discussion following Theorem 34 points out that $\mathtt{TH}(G)$ always has facets in common with $\mathtt{QSTAB}(G)$, when those facets belong also to $\mathtt{STAB}(G)$. It is not hard to see that $\mathtt{QSTAB}(C_5)$ has ten facets, defined by $x_j = 0$ and $x_j + x_{(j+1) \bmod 5} = 1$ for $0 \le j < 5$; and $\mathtt{STAB}(C_5)$ has an additional facet defined by $x_0 + x_1 + x_2 + x_3 + x_4 = 2$. The weighted functions α and κ of Section 4 are evaluated by considering the vertices of $\mathtt{STAB}$ and $\mathtt{QSTAB}$:

$$\alpha(C_5, (w_0, \ldots, w_4)^T)$$
$$= \max(w_0 + w_2, w_1 + w_3, w_2 + w_4, w_3 + w_0, w_4 + w_1); \quad (36.1)$$

$$\kappa(C_5, (w_0, \ldots, w_4)^T)$$
$$= \max\big(\alpha(C_5, (w_0, \ldots, w_4)^T), (w_0 + \cdots + w_4)/2\big). \quad (36.2)$$

Where these functions agree, they tell us also the value of ϑ.

For example, let $f(x) = \vartheta(C_5, (x, 1, 1, 1, 1)^T)$. Relations (36.1) and (36.2) imply that $f(x) = x + 1$ when $x \geq 2$. Clearly $f(0) = 2$, and Section 22 tells us that $f(1) = \sqrt{5}$. Other values of $f(x)$ are not yet known. Equation (23.2) gives the lower bound $f(x)^2 \geq x^2 + 4$. Incidentally, the a vectors

$$\begin{pmatrix} \sqrt{x} \\ 1 \\ 0 \\ 0 \\ 0 \end{pmatrix} \begin{pmatrix} \sqrt{x} \\ 1 \\ \sqrt{x+1} \\ \sqrt{(x-2)(x+1)} \\ 0 \end{pmatrix} \begin{pmatrix} 1 \\ -\sqrt{x} \\ 0 \\ 0 \\ 0 \end{pmatrix} \begin{pmatrix} 1 \\ -\sqrt{x} \\ 0 \\ 0 \\ 0 \end{pmatrix} \begin{pmatrix} \sqrt{x} \\ 1 \\ -\sqrt{x+1} \\ 0 \\ \sqrt{(x-2)(x+1)} \end{pmatrix}$$

and $b = (1)(0)(0)(0)(0)$ establish $f(x)$ for $x \geq 2$ in the fashion of Theorems 12 and 13.

Let $\phi = (1 + \sqrt{5})/2$ be the golden ratio. The matrices A and B' of Theorem 29, when $G = C_5$ and $w = \mathbb{1}$, are

$$A = \begin{pmatrix} \sqrt{5} & 1 & 1 & 1 & 1 & 1 \\ 1 & 1 & \phi-1 & 0 & 0 & \phi-1 \\ 1 & \phi-1 & 1 & \phi-1 & 0 & 0 \\ 1 & 0 & \phi-1 & 1 & \phi-1 & 0 \\ 1 & 0 & 0 & \phi-1 & 1 & \phi-1 \\ 1 & \phi-1 & 0 & 0 & \phi-1 & 1 \end{pmatrix} ;$$

$$B' = \frac{1}{\sqrt{5}} \begin{pmatrix} \sqrt{5} & -1 & -1 & -1 & -1 & -1 \\ -1 & 1 & 0 & \phi-1 & \phi-1 & 0 \\ -1 & 0 & 1 & 0 & \phi-1 & \phi-1 \\ -1 & \phi-1 & 0 & 1 & 0 & \phi-1 \\ -1 & \phi-1 & \phi-1 & 0 & 1 & 0 \\ -1 & 0 & \phi-1 & \phi-1 & 0 & 1 \end{pmatrix} .$$

They have the common eigenvectors

$$\begin{pmatrix} \sqrt{5} \\ 1 \\ 1 \\ 1 \\ 1 \\ 1 \end{pmatrix} \begin{pmatrix} \sqrt{5} \\ -1 \\ -1 \\ -1 \\ -1 \\ -1 \end{pmatrix} \begin{pmatrix} 0 \\ \phi \\ 1 \\ -1 \\ -\phi \\ 0 \end{pmatrix} \begin{pmatrix} 0 \\ 0 \\ \phi \\ 1 \\ -1 \\ -\phi \end{pmatrix} \begin{pmatrix} 0 \\ 1 \\ -\phi \\ \phi \\ -1 \\ 0 \end{pmatrix} \begin{pmatrix} 0 \\ 0 \\ 1 \\ -\phi \\ \phi \\ -1 \end{pmatrix} ,$$

with respective eigenvalues $(\lambda_0, \ldots, \lambda_5) = (2\sqrt{5}, 0, \sqrt{5}/\phi, \sqrt{5}/\phi, 0, 0)$ and $(\mu_0, \ldots, \mu_5) = (0, 2, 0, 0, 1/\phi, 1/\phi)$. (See (29.6) and (29.7).)

37. Perplexing Questions

The book [7] explains how to compute $\vartheta(G, w)$ with given tolerance ϵ, in polynomial time using an ellipsoid method. But that algorithm is too slow and numerically unstable to deal with graphs that have more than 10 or so vertices. Fortunately, however, new "interior-point methods" have been developed for this purpose, especially by Alizadeh [1,2], who has computed $\vartheta(G)$ when G has hundreds of vertices and thousands of edges. He has also shown how to find large stable sets, as a byproduct of evaluating $\vartheta(G, w)$ when w has integer coordinates. Calculations on somewhat smaller cyclically symmetric graphs have also been reported by Overton [17]. Further computational experience with such programs should prove to be very interesting.

Solutions to the following four concrete problems may also help shed light on the subject:

P1.   Describe $\mathrm{TH}(C_5)$ geometrically. This convex corner is isomorphic to its own antiblocker. (Namely, if $(x_0, x_1, x_2, x_3, x_4) \in \mathrm{TH}(C_5)$, then so are its cyclic permutations $(x_1, x_2, x_3, x_4, x_0)$, etc., as well as the cyclic permutations of $(x_0, x_4, x_3, x_2, x_1)$; $\mathrm{TH}(\overline{C}_5)$ contains the cyclic permutations of $(x_0, x_2, x_4, x_1, x_3)$ and $(x_0, x_3, x_1, x_4, x_2)$.) Can the values $f(x) = \vartheta(C_5, (x, 1, 1, 1, 1)^T)$, discussed in Section 36, be expressed in closed form when $0 < x < 2$, using familiar functions?

P2.   What is the probable value of $\vartheta(G, w)$ when G is a random graph on n vertices, where each of the $\binom{n}{2}$ possible edges is independently present with some fixed probability p? (Juhász [9] has solved this problem in the case $w = \mathbb{1}$, showing that $\vartheta(G)/\sqrt{(1-p)n/p}$ lies between $\frac{1}{2}$ and 2 with probability approaching 1 as $n \to \infty$.)

P3.   What is the minimum d for which G almost surely has an orthogonal labeling of dimension d with no zero vectors, when G is a random graph as in Problem P2? (Theorem 28 and the theorem of Juhász [9] show that d must be at least of order $\sqrt{n}$. But Lovász tells me that he suspects the correct answer is near n. Theorem 29 and its consequences might be helpful here.)

P4.   Is there a constant c such that $\vartheta(G) \leq c\sqrt{n}\,\alpha(G)$ for all n-vertex graphs G? (This conjecture was suggested by Lovász in a recent letter. He knows no infinite family of graphs where $\vartheta(G)/\alpha(G)$ grows faster than $O(\sqrt{n}/\log n)$. The latter behavior occurs for random graphs, which have $\alpha(G) = \log_{1/p} n$ with high probability [4, Chapter XI].)

Another, more general, question is to ask whether it is feasible to study two- or three-dimensional projections of $\mathrm{TH}(G)$, and whether they have combinatorial significance. The function $\vartheta(G, w)$ gives just a one-dimensional glimpse.

Lovász and Schrijver have recently generalized the topics treated here to a wide variety of more powerful techniques for studying 0–1 vectors associated with graphs [15]. In particular, one of their methods can be described as follows: Let us say that a *strong orthogonal labeling* is a vector labeling such that $\|a_v\|^2 = c(a_v)$ and $a_u \cdot a_v \geq 0$, also satisfying the relation

$$c(a_u) + c(a_v) + c(a_w) - 1 \leq a_u \cdot a_v + a_v \cdot a_w \leq c(a_v) \qquad (37.1)$$

whenever $u \not\!\frown w$. In particular, when $w = v$ this relation implies that $a_u \cdot a_v = 0$, so the labeling is orthogonal in the former sense.

Notice that every stable labeling is a strong orthogonal labeling of $\overline{G}$. Let S be a stable set and let u and w be vertices such that $u \frown w$. If u and w are not in S, condition (37.1) just says that $0 \leq c(a_v) \leq 1$, which surely holds. If u is in S, then $w \notin S$ and (37.1) reduces to $c(a_v) \leq c(a_v) \leq c(a_v)$; this holds even more surely.

Let

$$\texttt{TH}_-(G) = \{\, x \mid x_v = c(b_v) \text{ for some}$$
$$\text{strong orthogonal labeling of } \overline{G} \,\}. \qquad (37.2)$$

(This set is called $N_+(\texttt{FR}(G))$ in [15].) We also define

$$\vartheta_-(G, w) = \max\{\, w \cdot x \mid x \in \texttt{TH}_-(G) \,\}. \qquad (37.3)$$

The argument in the two previous paragraphs implies that

$$\texttt{STAB}(G) \subseteq \texttt{TH}_-(G) \subseteq \texttt{TH}(G),$$

hence

$$\alpha(G, w) \leq \vartheta_-(G, w) \leq \vartheta(G, w). \qquad (37.4)$$

The authors of [15] prove that $\vartheta_-(G, w)$ can be computed in polynomial time, about as easily as $\vartheta(G, w)$; moreover, $\vartheta_-(G, w)$ can be a significantly better approximation to $\alpha(G, w)$. They show, for example, that $\texttt{TH}_-(G) = \texttt{STAB}(G)$ when G is any cyclic graph C_n. In fact, they prove that if $x \in \texttt{TH}_-(G)$ and if $v_0 \frown v_1$, $v_1 \frown v_2$, ..., $v_{2n} \frown v_0$ is any circuit or multicircuit of G, then $x_{v_0} + x_{v_1} + \cdots + x_{v_{2n}} \leq n$. This work suggests additional research problems:

P5. What is the smallest graph such that $\texttt{STAB}(G) \neq \texttt{TH}_-(G)$?

P6. What is the probable value of $\vartheta_-(G)$ when G is a random graph as in Problem P2?

A recent theorem by Arora, Lund, Motwani, Sudan, and Szegedy [3] proves that there is an $\epsilon > 0$ such that no polynomial algorithm can compute a number between $\alpha(G)$ and $n^\epsilon \alpha(G)$ for all n-vertex graphs G, unless $P = NP$. Therefore it would be surprising if the answer to P6 turns out to be that $\vartheta_-(G)$ is, say, $O(\log n)^2$ with probability $\to 1$ for random G. Still, such a result would not be inconsistent with [3], because the graphs for which $\alpha(G)$ is hard to approximate might be decidedly nonrandom.

Lovász has called the author's attention to papers by Kashin and Konîagin [10, 11], which prove (in a very disguised form, related to (6.2) and Theorem 33) that if G has no stable set with 3 elements we have

$$\vartheta(G) \leq 2^{2/3} n^{1/3} \, ; \tag{37.5}$$

moreover, such graphs exist with

$$\vartheta(G) = \Omega(n^{1/3}/\sqrt{\log n}) \, . \tag{37.6}$$

Further study of methods like those in [15] promises to be exciting indeed. Lovász has sketched yet another approach in [14].

The preparation of this paper was supported in part by the Mittag-Leffler Institute in Djursholm, Sweden.

References

[1] Farid Alizadeh, "A sublinear-time randomized parallel algorithm for the maximum clique problem in perfect graphs," *ACM–SIAM Symposium on Discrete Algorithms* **2** (1991), 188–194.

[2] Farid Alizadeh, "Interior point methods in semidefinite programming with applications to combinatorial optimization," *SIAM Journal on Optimization* **5** (1995), 13–51.

[3] Sanjeev Arora, Carsten Lund, Rajeev Motwani, Madhu Sudan, and Mario Szegedy, "Proof verification and intractability of approximation problems," *Proceedings of the 33rd IEEE Symposium on Foundations of Computer Science* (1992), 14–23. [See also "Proof verification and the hardness of approximation problems," *Journal of the ACM* **45** (1998), 501–555.]

[4] Béla Bollobás, *Random Graphs* (London: Academic Press, 1985).

[5] Martin Grötschel, L. Lovász, and A. Schrijver, "The ellipsoid method and its consequences in combinatorial optimization," *Combinatorica* **1** (1981), 169–197.

[6] M. Grötschel, L. Lovász, and A. Schrijver, "Relaxations of vertex packing," *Journal of Combinatorial Theory* **B40** (1986), 330–343.

[7] Martin Grötschel, László Lovász, and Alexander Schrijver, *Geometric Algorithms and Combinatorial Optimization* (Berlin: Springer–Verlag, 1988), §9.3.

[8] A. S. Householder, "Unitary triangularization of a nonsymmetric matrix," *Journal of the Association for Computing Machinery* **5** (1958), 339–342.

[9] Ferenc Juhász, "The asymptotic behaviour of Lovász' ϑ function for random graphs," *Combinatorica* **2** (1982), 153–155.

[10] B. S. Kashin and S. V. Konîagin, "O sistemakh vektorov v Gil'bertovom prostranstve," *Trudy Matematicheskogo Instituta imeni V. A. Steklova* **157** (1981), 64–67. English translation, "On systems of vectors in a Hilbert space," *Proceedings of the Steklov Institute of Mathematics* (American Mathematical Society, 1983, issue 3), 67–70.

[11] S. V. Konîagin, "O sistemakh vektorov v Evclidovom prostranstve i odnoĭ ékstremal'noĭ zadache dlîa mnogochlenov," *Matematicheskie Zametki* **29** (1981), 63–74. English translation, "Systems of vectors in Euclidean space and an extremal problem for polynomials," *Mathematical Notes of the Academy of Sciences of the USSR* **29** (1981), 33–39.

[12] L. Lovász, "On the Shannon capacity of a graph," *IEEE Transactions on Information Theory* **IT-25** (1979), 1–7.

[13] L. Lovász, *An Algorithmic Theory of Numbers, Graphs, and Convexity*, CBMS Regional Conference Series in Applied Mathematics (SIAM, 1986), §3.2.

[14] L. Lovász, "Stable sets and polynomials," *Discrete Mathematics* **124** (1994), 137–153.

[15] L. Lovász and A. Schrijver, "Cones of matrices and set functions and 0–1 optimization," *SIAM Journal on Optimization* **1** (1991), 166–190.

[16] R. J. McEliece, E. R. Rodemich, and H. C. Rumsey, Jr., "The Lovász bound and some generalizations," *Journal of Combinatorics, Information and System Sciences* **3** (1978), 134–152.

[17] Michael Overton, "Large-scale optimization of eigenvalues," *SIAM Journal on Optimization* **2** (1992), 88–120.

[18] M. W. Padberg, "On the facial structure of set packing polyhedra," *Mathematical Programming* **5** (1973), 199–215.

[19] Claude E. Shannon, "The zero error capacity of a channel," *IRE Transactions on Information Theory* **2**, 3 (September 1956), 8–19.

Addendum

Further information can be found in A. Galtman, "Spectral characterizations of the Lovász number and the Delsarte number of a graph," *Journal of Algebraic Combinatorics* **12** (2000), 131–143.

Chapter 9

Combinatorial Matrices

*[The following previously unpublished notes were composed at the same
time as I wrote the preceding chapter. After I posted them informally
on the Internet, several people have referred to them in print; so I have
decided to include them in the present book — with apologies for not
having had time to polish them to the point where they meet the standards of ordinary publications.]*

Exercises 1.2.3–36, 39, and 42 of *Fundamental Algorithms* discuss the
determinants and inverses of matrices of the form

$$
\begin{pmatrix}
a & b & b & b & \dots & b \\
b & a & b & b & \dots & b \\
b & b & a & b & \dots & b \\
b & b & b & a & \dots & b \\
\vdots & \vdots & \vdots & \vdots & & \vdots \\
b & b & b & b & \dots & a
\end{pmatrix}
= bJ + (a - b)I, \tag{0.1}
$$

which I called "combinatorial matrices" because they arise for example
in the study of block designs. Those matrices are the special case $t = 1$
of what I now believe are more properly called combinatorial matrices
of type (m, t).

In general, a *combinatorial matrix of type* (m, t) has $n = \binom{m}{t}$ rows
and columns indexed by the t-element subsets of an m-element set S.
The value in row u and column v should depend only on the cardinality
$|u \cap v|$.

We can assume without loss of generality that $m \geq 2t$, because subsets of cardinality t are isomorphic (under complementation) to subsets
of cardinality $m - t$.

A general combinatorial matrix of type (m, t) has at most $t + 1$
distinct entries, because two t-sets can intersect in $0, 1, \dots, t$ elements.

177

For example, the most general combinatorial matrices when $t = 2$ and when $m = 4$ or 5 are

	12	13	14	23	24	34
12	a	b	b	b	b	c
13	b	a	b	b	c	b
14	b	b	a	c	b	b
23	b	b	c	a	b	b
24	b	c	b	b	a	b
34	c	b	b	b	b	a

	12	13	14	15	23	24	25	34	35	45
12	a	b	b	b	b	b	b	c	c	c
13	b	a	b	b	b	c	c	b	b	c
14	b	b	a	b	c	b	c	b	c	b
15	b	b	b	a	c	c	b	c	b	b
23	b	b	c	c	a	b	b	b	b	c
24	b	c	b	c	b	a	b	b	c	b
25	b	c	c	b	b	b	a	c	b	b
34	c	b	b	c	b	b	c	a	b	b
35	c	b	c	b	b	c	b	b	a	b
45	c	c	b	b	c	b	b	b	b	a

1. Basic Combinatorial Matrices

Let $J_{m,t,q}$ be the combinatorial matrix of type (m,t) having 1 in row u and column v when $|u \cap v| = q$, and 0 elsewhere. When (m,t) is understood, we write simply J_q for $J_{m,t,q}$. Thus the general combinatorial matrix of type (m,t) is

$$a_0 J_0 + a_1 J_1 + \cdots + a_t J_t . \tag{1.1}$$

Clearly $J_t = I$ and $J_0 + J_1 + \cdots + J_t = J$.

Lemma. *Combinatorial matrices of type (m,t) are closed under multiplication.*

Proof. It suffices to show that $J_p J_q$ is a combinatorial matrix of type (m,t). And if $|u \cap v| = r$ we have

$$(J_p J_q)_{uv} = \sum_w [\,|u \cap w| = p \text{ and } |v \cap w| = q\,]$$

$$= \sum_k \binom{r}{k}\binom{t-r}{p-k}\binom{t-r}{q-k}\binom{m-2t+r}{t-p-q+k} = c_{pqr} \tag{1.2}$$

where k represents $|u \cap v \cap w|$, hence

$$J_p J_q = c_{pq0} J_0 + c_{pq1} J_1 + \cdots + c_{pqt} J_t . \quad \square \tag{1.3}$$

For example, when $t = 1$ we have $J_0 J_0 = (m-2)J_0 + (m-1)J_1$; $J_1 = I$. When $t = 2$ the multiplication table is

$$J_0 J_0 = \binom{m-4}{2} J_0 + \binom{m-3}{2} J_1 + \binom{m-2}{2} J_2 \,;$$

$$J_0 J_1 = 2(m-4)J_0 + (m-3)J_1 \,; \qquad\qquad (1.4)$$

$$J_1 J_1 = 4J_0 + (m-2)J_1 + 2(m-2)J_2 \,.$$

One can check that $(J_0 + J_1 + J_2)^2 = \binom{m}{2}(J_0 + J_1 + J_2)$.

2. Basic (m, t) Sets

We call the integers $\{a_1, a_2, \ldots, a_t\}$ an "(m, t) set" if

$$1 \le a_1 < a_2 < \cdots < a_t \le m \,, \qquad\qquad (2.1)$$

and a *basic* (m, t) set if in addition

$$a_k \ge 2k \quad \text{for} \quad 1 \le k \le t \,. \qquad\qquad (2.2)$$

For example, the basic $(6, 3)$ sets are 246, 256, 346, 356, 456 [omitting braces and commas].

Lemma. *There are exactly* $\binom{m}{t} - \binom{m}{t-1}$ *basic* (m, t) *sets, when* $m \ge 2t - 1 > 0$.

Proof. By induction on t, the result being clear when $t = 1$. For fixed $t > 1$, the proof is by induction on m, the result being clear when $m = 2t - 1$. If $m \ge 2t$ and $t > 1$, the basic (m, t) sets not containing m are basic $(m-1, t)$ sets; those containing m are obtained by appending m to basic $(m - 1, t - 1)$ sets. Hence the total number is

$$\binom{m-1}{t} - \binom{m-1}{t-1} + \binom{m-1}{t-1} - \binom{m-1}{t-2} = \binom{m}{t} - \binom{m}{t-1} \,. \quad \square$$

Alternative proof. We can show that there are exactly $\binom{m}{t-1}$ nonbasic (m, t) sets by exhibiting a one-to-one correspondence between arbitrary $(m, t - 1)$ sets and nonbasic (m, t) sets. Given $1 \le a_1 < a_2 < \cdots < a_{t-1} \le m$, let j be maximum such that $j = 0$ or $a_j \le 2j + 1$. The corresponding nonbasic set is $\{1, \ldots, 2j+1\} \setminus \{a_1, \ldots, a_j\} \cup \{a_{j+1}, \ldots, a_{t-1}\}$.

3. A Linear Covering Relation

We say set u covers set v if $u \supseteq v$ and $|u| = |v| + 1$. Let's denote this relation by $u \succ v$.

Consider the system of equations

$$\sum_{u \succ v} x_u = y_v \tag{3.1}$$

where x_u is defined for all t-element subsets u of $\{1, \ldots, m\}$ and y_v is defined for all $(t-1)$-element subsets. If the values of y_v are given, this is a system of $\binom{m}{t-1}$ equations in $\binom{m}{t}$ unknown variables x_u. The following lemma asserts that the variables x_u for basic (m, t) sets are independent.

Lemma. *Given values of y_v for all $(m, t-1)$ sets v and of x_u for all basic (m, t) sets u, we can solve system (3.1) for the remaining (m, t) sets x_u, when $m \geq 2t - 1 > 0$.*

Proof. Again we use induction on t, the result being obvious when $t = 1$.

Suppose $m = 2t - 1$. Then there are no basic (m, t) sets, and (3.1) has as many x's as y's. Fix any (m, t) set w and consider the sum

$$a_0 \sum_{|w \cap v| = 0} y_v + a_1 \sum_{|w \cap v| = 1} y_v + \cdots + a_{t-1} \sum_{|w \cap v| = t-1} y_v . \tag{3.2}$$

If $|u \cap w| = q$, the coefficient of x_u in (3.2) is

$$q\, a_{q-1} + (t - q) a_q , \tag{3.3}$$

because u covers q sets v that have $q - 1$ elements in common with w and $t - q$ sets v that have q elements in common. Furthermore $q > 0$, because $m < 2t$. We can choose $a_0, \ldots, a_{t-1}$ so that (3.3) vanishes for $1 \leq q < t$ and equals 1 for $q = t$. Then (3.2) gives the value of x_w. Indeed, the solution is

$$a_k = (-1)^{t-k-1}(t - k)^{-1}\binom{t}{k}^{-1} , \qquad 0 \leq k < t . \tag{3.4}$$

Now if $m \geq 2t$ we use induction on m. There are $\binom{m-1}{t-2}$ equations of (3.1) in which $m \in v$; these involve only variables x_u such that $m \in u$. There are $\binom{m-1}{t-1} - \binom{m-1}{t-2}$ basic (m, t) sets of that kind, so we can use induction to find x_u for all u containing m. The remaining $\binom{m-1}{t-1}$ equations of (3.1) now allow us to determine the remaining x_u, again by induction. (We transpose the term in which $u = v \cup \{m\}$ to the right-hand side, combining it with y_v.) $\square$

4. Eigenvectors

Let's say that the set of values x_u for (m, s) sets u is an (m, s) *kernel system* if (3.1) is satisfied with $y_v = 0$ for all $(m, s - 1)$ sets v.

Lemma. *If $\{x_u\}$ is any (m, s) kernel system and if w is any (m, t) set, for $t \geq s$, then*

$$\sum_{|u \cap w| = q} x_u = (-1)^{s-q} \binom{s}{q} z_w, \tag{4.1}$$

where

$$z_w = \sum_{u \subseteq w} x_u. \tag{4.2}$$

Proof. Equation (4.1) surely holds for $q = s$. And an argument like our derivation of (3.3) shows that

$$\sum_{|v \cap w| = q} y_v = (q + 1) \sum_{|u \cap w| = q+1} x_u + (s - q) \sum_{|u \cap w| = q} x_u. \tag{4.3}$$

The left side of (4.3) is zero in a kernel system, hence (4.1) follows by induction on $s - q$. $\square$

Corollary. *If $\{x_u\}$ is an (m, s) kernel system and if we define z_w by equation (4.2) for all (m, t) sets w, where $t \geq s$, then z is an eigenvector for any combinatorial matrix of type (m, t).*

Proof. It suffices to show that z is an eigenvector for J_q when $0 \leq q \leq t$. Component w of the vector $J_q z$ is

$$\sum_{|w \cap v| = q} z_v = \sum_{|w \cap v| = q} \sum_{u \subseteq v} x_u$$

$$= \sum_r \sum_{|u \cap w| = r} x_u \sum_{v \supseteq u} [\,|w \cap v| = q\,]$$

$$= \sum_r \sum_{|u \cap w| = r} x_u \binom{t - r}{q - r} \binom{m - t - s + r}{t - q - s + r}$$

$$= \sum_r (-1)^{s-r} \binom{s}{r} \binom{t - r}{q - r} \binom{m - t - s + r}{t - q - s + r} z_w. \quad \square \tag{4.4}$$

The coefficient of z_w in (4.4) is the eigenvalue, which does not appear to simplify. (It is the sth difference of a polynomial in r of degree $(t - q) + (m - 2t + q) = m - t$.) When $q = 0$ it is

$$(-1)^s \binom{m - t - s}{t - s}; \tag{4.5}$$

when $q = 1$ it can be written

$$(-1)^s \left((t-s)\binom{m-t-s}{t-s-1} - s\binom{m-t-s}{t-s} \right) ; \qquad (4.6)$$

when $t = s$ it reduces to the term for $r = q$,

$$(-1)^{t-q}\binom{t}{q}. \qquad (4.7)$$

5. Eigenvalues

The all-ones vector is clearly an eigenvector of J_q, with eigenvalue

$$\binom{t}{q}\binom{m-t}{t-q}. \qquad (5.1)$$

This is the coefficient of z_w in (4.4) when $s = 0$. The other $\binom{m}{t} - 1$ eigenvalues are all obtained by the construction of the previous section when $s > 0$:

Theorem. *The coefficient of z_w in (4.4) occurs as an eigenvalue of J_q exactly*

$$\binom{m}{s} - \binom{m}{s-1} \qquad (5.2)$$

times.

[Summing this quantity over $1 \le s \le t$ gives $\binom{m}{t} - \binom{m}{0}$, as desired.]

Proof. The eigenvectors for different values of s are orthogonal, because they give different eigenvalues when $q = 0$ by (4.5). (When $m = 2t$ we need to consider also the case $q = 1$.) Therefore we must show only that (5.2) linearly independent eigenvectors z are constructed by the method of Corollary 4.

We know (5.2) is the number of basic (m, s) sets, by Lemma 2. Lemma 3 tells us that there is an (m, s) kernel system with $x_u = 1$ for any desired basic (m, s) set and $x_u = 0$ for all the other basic sets; thus we have (5.2) linearly independent (m, s) kernel systems.

It remains to show that linearly independent kernel systems lead to linearly independent eigenvectors z. For this it suffices to show that we can calculate each x_u from the values of z_w, if the linear system (4.2) holds.

Consider the expression

$$a_0 \sum_{|w \cap u|=0} z_w + a_1 \sum_{|w \cap u|=1} z_w + \cdots + a_s \sum_{|w \cap u|=s} z_w . \tag{5.3}$$

The coefficient of x_v, if $|u \cap v| = q$, is

$$\sum_r a_r \binom{s-q}{r-q} \binom{m-2s+q}{t-r-s+q} . \tag{5.4}$$

We can therefore choose a_s, a_{s-1}, ..., a_0 so that (5.4) is 1 when $q = s$ and 0 when $q < s$. Then (5.3) reduces to x_u. □

6. Integer Kernels

When the eigenvectors z are actually computed according to the method implicit in the proofs above, a surprising simplification occurs. We are told to construct (m, s) kernel systems by setting $x_u = 1$ for some basic (m, s) set u, and $x_u = 0$ for all other basic u; then we are supposed to invert the linear system (3.1) to obtain the values of x_u for nonbasic u. This inversion, as we have described it, involves the coefficients a_k in (3.4), which are never integers when $t > 1$. The surprising fact is that the final values x_u all turn out to be integers. (The reader may wish to try the case $m = 6$, $t = 3$, in order to appreciate the surprise.)

Lemma. *If x_u is an integer for all basic u in an (m, s) kernel system, it is an integer for all (m, s) sets u.*

Proof. It suffices to exhibit, for each basic (m, s) set v, an integer-valued (m, s) kernel system such that $x_v = 1$ and $x_u = 0$ for all basic sets u that are greater than v in some total ordering of the (m, s) sets.

Let the elements of v be $\{a_1, \ldots, a_s\}$, where

$$1 \le a_1 < a_2 < \cdots < a_s \le m \tag{6.1}$$

and $a_k \ge 2k$ for $1 \le k \le s$, and let the elements of the complementary set $\{1, \ldots, m\} \setminus u$ be

$$1 \le b_1 < b_2 < \cdots < b_{m-s} \le m . \tag{6.2}$$

Then we have

$$b_k < a_k \quad \text{for} \quad 1 \le k \le s . \tag{6.3}$$

Indeed, this property characterizes basic sets — it is *equivalent* to the condition $a_k \geq 2k$, when $m \geq 2s$.

Now consider all 2^s sets that contain either a_k or b_k, for $1 \leq k \leq s$, and no other elements. If u is such a set, let $x_u = (-1)^l$ where l is the number of b's. Let $x_u = 0$ for all other u. This rule defines an (m, s) kernel system. For if v is any $(m, s - 1)$ set, containing both a_k and b_k for some k, it is covered by no sets u with $x_u \neq 0$, hence (3.1) surely holds with $y_v = 0$. And if v contains no pairs $\{a_k, b_k\}$, there is at least one k such that v contains neither a_k nor b_k; this k is unique if $v \subseteq \{a_1, \ldots, a_s, b_1, \ldots, b_s\}$. The only sets u that cover v and have $x_u \neq 0$ are $v \cup \{a_k\}$ and $v \cup \{b_k\}$, and these x_u have opposite signs.

This completes the proof, because we can use, say, lexicographic order on $(a_1, \ldots, a_s)$ to totally order the (m, s) sets in the requisite fashion. □

As an example of the construction in this lemma, here is an easy way to find the $(8, 3)$ kernel system in which $x_{468} = 1$ and $x_u = 0$ for all other basic u. We start with

$$468 - 168 - 248 - 346 + 128 + 136 + 234 - 123 \,, \qquad (6.4)$$

which is a shorthand for $x_{468} = 1$, $x_{168} = -1$, $\ldots$, $x_{123} = -1$, and all other $x_u = 0$; this is the kernel system described above when $\{a_1, \ldots, a_s\} = \{4, 6, 8\}$, since $\{b_1, \ldots, b_{m-s}\} = \{1, 2, 3, 5, 7\}$. The basic sets with nonzero coefficients are 468, 248, and 346, so we want to eliminate 248 and 346. Adding

$$248 - 148 - 238 - 245 + 138 + 145 + 235 - 135$$

and

$$346 - 146 - 236 - 345 + 126 + 145 + 235 - 125$$

yields

$$468 - 345 - 245 - 238 - 236 + 2 \cdot 235 + 234 - 168 - 148$$
$$- 146 + 2 \cdot 145 + 138 + 136 - 135 + 128 + 126 - 125 - 123 \,. \quad (6.5)$$

Incidentally, this system has $x_{235} = x_{145} = 2$, illustrating the fact that nonzero coefficients of the fully diagonalized systems need not be ± 1. But if we stick to a triangularized (not diagonalized) set of coefficients, we can avoid large coefficients in general:

Theorem. *A combinatorial matrix of type* (m, t) *has* $\binom{m}{t}$ *linearly independent eigenvectors whose components are all* $0, +1,$ *or* -1.

Proof. For $0 \leq s \leq t$ we obtain $\binom{m}{s} - \binom{m}{s-1}$ eigenvectors as in Corollary 4 and Theorem 5, using kernel systems with 2^s nonzero coefficients as in (6.4). The values of z_w, defined by (4.2), will then be 0 or ± 1. For we have $z_w = 0$ unless w contains either a_k or b_k for $1 \leq k \leq s$, where $\{a_1, \ldots, a_s\}$ and $\{b_1, \ldots, b_{m-s}\}$ are the complementary sets in the proof above; also $z_w = 0$ if both a_k and b_k appear in w, for some k, because terms x_u with opposite signs will cancel out. Otherwise $z_w = (-1)^l$, where $l = |w \cap \{b_1, \ldots, b_s\}|$. $\square$

7. Examples

The eigenvalues and a set of linearly independent for the two combinatorial matrices in the introduction can now be read off from these constructions. The 6×6 matrix of type $(4, 2)$ has the following eigenvectors z:

z_{12}	z_{13}	z_{14}	z_{23}	z_{24}	z_{34}	s	basic set	eigenvalue
1	1	1	1	1	1	0	$\emptyset$	$a + 4b + c$
0	-1	-1	1	1	0	1	$\{2\}$	$a - c$
-1	0	-1	1	0	1	1	$\{3\}$	$a - c$
-1	-1	0	0	1	1	1	$\{4\}$	$a - c$
0	1	-1	-1	1	0	2	$\{2, 4\}$	$a - 2b + c$
1	0	-1	-1	0	1	2	$\{3, 4\}$	$a - 2b + c$

And for type $(5, 2)$ we have

z_{12}	z_{13}	z_{14}	z_{15}	z_{23}	z_{24}	z_{25}	z_{34}	z_{35}	z_{45}	s	basic set	eigenvalue
1	1	1	1	1	1	1	1	1	1	0	$\emptyset$	$a + 6b + 3c$
0	-1	-1	-1	1	1	1	0	0	0	1	$\{2\}$	$a + b - 2c$
-1	0	-1	-1	1	0	0	1	1	0	1	$\{3\}$	$a + b - 2c$
-1	-1	0	-1	0	1	0	1	0	1	1	$\{4\}$	$a + b - 2c$
-1	-1	-1	0	0	0	1	0	1	1	1	$\{5\}$	$a + b - 2c$
0	1	-1	0	-1	1	0	0	0	0	2	$\{2, 4\}$	$a - 2b + c$
0	1	0	-1	-1	0	1	0	0	0	2	$\{2, 5\}$	$a - 2b + c$
1	0	-1	0	-1	0	0	1	0	0	2	$\{3, 4\}$	$a - 2b + c$
1	0	0	-1	-1	0	0	0	1	0	2	$\{3, 5\}$	$a - 2b + c$
1	0	0	-1	0	-1	0	0	0	1	2	$\{4, 5\}$	$a - 2b + c$

8. Remarks

Lovász [*IEEE Transactions on Information Theory* **IT-25** (1979), page 6] says it is "well known" that the system (4.2) can be solved for the x's, if equality holds for all w of fixed cardinality t, but he does not give a reference. I imagine I will find all this in, say, papers of Gill Williamson when I get around to studying them, and probably the whole construction is also generalized to arbitrary partial ordered sets (or at least distributive lattices) instead of the n-cube. However, the n-cube is interesting enough in itself to deserve a study of the eigenvalues in fairly explicit form as shown above. Lovász presented only the case $q = 0$, but I'm sure he would not be surprised to hear that the eigenvectors for that case work also for general q. I would be pleased to see a reference to any prior work in which the determinant of matrices like those in the introduction was shown to have a simple form.

Herb Wilf points out that Eqs. (1.2) and (1.3) imply immediately the identity $J_p J_q = J_q J_p$; hence combinatorial matrices of the same type are commutative. This guarantees the existence of linearly independent eigenvectors common to all of J_0, J_1, ..., and it may simplify the computations of Section 4.

I thank Howard Karloff for correcting an error in my original statement of the theorem. He has pointed out that it is sometimes impossible to find a complete set of eigenvectors that are mutually orthogonal unless we use components outside the set $\{-1, 0, 1\}$.

Postscript: Richard Stanley tells me that much of the above follows from the theory of association schemes, in particular the so-called Johnson scheme due to Selmer Johnson. Stanley's paper, "Variations on differential posets," in *Invariant Theory and Tableaux*, edited by Dennis Stanton, *IMA Volumes in Mathematics and Its Applications* **19** (1990), 145–165, is also closely connected, because a Boolean algebra is a sequentially differential poset. I have not yet had time to follow up on these leads and explore the situation further.

These notes were drafted during a visit to the Mittag-Leffler Institute in Djursholm, Sweden.

Chapter 10

Aztec Diamonds, Checkerboard Graphs, and Spanning Trees

This note derives the characteristic polynomial of a graph that represents nonjump moves in a generalized game of checkers. The number of spanning trees is also determined.

*[Originally published in Journal of Algebraic Combinatorics **6** (1997), 253–257.]*

Consider the graph on mn vertices $\{\,(x,y) \mid 1 \leq x \leq m,\ 1 \leq y \leq n\,\}$, with (x,y) adjacent to (x',y') if and only if $|x - x'| = |y - y'| = 1$. This graph consists of disjoint subgraphs

$$\mathrm{EC}_{m,n} = \{\,(x,y) \mid x + y \text{ is even}\,\},$$
$$\mathrm{OC}_{m,n} = \{\,(x,y) \mid x + y \text{ is odd}\,\},$$

having respectively $\lceil mn/2 \rceil$ and $\lfloor mn/2 \rfloor$ vertices. When mn is even, $\mathrm{EC}_{m,n}$ and $\mathrm{OC}_{m,n}$ are isomorphic. The special case $\mathrm{OC}_{2n+1,2n+1}$ has been called an *Aztec diamond of order n* by Elkies, Kuperberg, Larsen, and Propp [6], who gave several interesting proofs that it contains exactly $2^{n(n+1)/2}$ perfect matchings.

Richard Stanley recently conjectured [11] that $\mathrm{OC}_{2n+1,2n+1}$ contains exactly 4 times as many spanning trees as $\mathrm{EC}_{2n+1,2n+1}$, and it was his conjecture that motivated the present note. We will see that Stanley's conjecture follows from some even more remarkable properties of these graphs.

In general, if G and H are arbitrary bipartite graphs having parts of respective sizes (p, q) and (r, s), their *weak direct product* $G \times H$ has $(p + q)(r + s)$ vertices (u, v), with (u, v) adjacent to (u', v') if and only if u is adjacent to u' and v is adjacent to v'. This graph $G \times H$ divides naturally into even and odd components

$$E(G, H) = \{\,(u, v) \mid u \in G \text{ and } v \in H \text{ are in corresponding parts}\,\},$$
$$O(G, H) = \{\,(u, v) \mid u \in G \text{ and } v \in H \text{ are in opposite parts}\,\},$$

187

which are disjoint. Notice that $E(G, H)$ and $O(G, H)$ have $pr + qs$ and $ps + qr$ vertices, respectively. Our graphs $EC_{m,n}$ and $OC_{m,n}$ are just $E(P_m, P_n)$ and $O(P_m, P_n)$, where P_n denotes a simple path on n points.

Let $P(G; x)$ be the characteristic polynomial of the adjacency matrix of a graph G. The eigenvalues of $E(G, H)$ and $O(G, H)$ turn out to have a simple relation to the eigenvalues of G and H:

Theorem 1. If $P(G; x) = \prod_{j=1}^{p+q}(x - \mu_j)$ and $P(H; x) = \prod_{k=1}^{r+s}(x - \lambda_k)$, *the characteristic polynomials* $P(E(G, H); x)$ *and* $P(O(G, H); x)$ *satisfy*

$$P(E(G, H); x)P(O(G, H); x) = \prod_{j=1}^{p+q}\prod_{k=1}^{r+s}(x - \mu_j\lambda_k);\tag{1}$$

$$P(E(G, H); x) = x^{(p-q)(r-s)}P(O(G, H); x).\tag{2}$$

Proof. This theorem is a consequence of more general results proved in [7], as remarked at the top of page 67 in that paper, but for our purposes a direct proof is preferable.

Let A and B be the adjacency matrices of G and H. It is well known [2; 12] that the adjacency matrix of $G \times H$ is the Kronecker product $A \otimes B$, and that the eigenvalues of $A \otimes B$ are $\mu_j\lambda_k$ when A and B are square matrices having eigenvalues μ_j and λ_k, respectively [10, page 24]. Since the left side of (1) is just $P(G \times H; x)$, equation (1) is therefore clear.

Equation (2) is more surprising, because the graphs $E(G, H)$ and $O(G, H)$ often look completely different from each other. But we can express A and B in the form

$$A = \begin{pmatrix} O_p & C \\ C^T & O_q \end{pmatrix}, \qquad B = \begin{pmatrix} O_r & D \\ D^T & O_s \end{pmatrix},\tag{3}$$

where C and D have respective sizes $p \times q$ and $r \times s$, and where O_k denotes a $k \times k$ matrix of zeros. It follows that the adjacency matrices of $E(G, H)$ and $O(G, H)$ are respectively

$$\begin{pmatrix} O_{pr} & C \otimes D \\ C^T \otimes D^T & O_{qs} \end{pmatrix} \quad \text{and} \quad \begin{pmatrix} O_{ps} & C \otimes D^T \\ C^T \otimes D & O_{qr} \end{pmatrix}.\tag{4}$$

We want to show that these matrices have the same eigenvalues, except for the multiplicity of 0.

One way to complete the proof is to observe that the kth powers of both matrices have the same trace, for all k. When $k = 2l$ is even, both matrix powers have trace

$$\left(\operatorname{tr}(CC^T)^l + \operatorname{tr}(C^T C)^l\right)\left(\operatorname{tr}(DD^T)^l + \operatorname{tr}(D^T D)^l\right),$$

by [10, pages 8, 18]; and when k is odd the traces are zero. The coefficients a_1, a_2, ... of $P(G; x) = x^{|G|}(1 - a_1 x^{-1} + a_2 x^{-2} - \cdots)$ are completely determined by the traces of powers of the adjacency matrix of any graph G, via Newton's identities; therefore (2) holds. □

Corollary 1. *For all positive integers m and n, the characteristic polynomials $P(\mathrm{EC}_{m,n}; x)$ and $P(\mathrm{OC}_{m,n}; x)$ satisfy*

$$P(\mathrm{EC}_{m,n}; x)\, P(\mathrm{OC}_{m,n}; x) = \prod_{j=1}^{m} \prod_{k=1}^{n} \left(x - 4\cos\frac{j\pi}{m+1}\cos\frac{k\pi}{n+1}\right); \quad (5)$$

$$P(\mathrm{EC}_{m,n}; x) = x^{mn \bmod 2}\, P(\mathrm{OC}_{m,n}; x). \quad (6)$$

Proof. It is well known [9, problem 1.29; or 4, page 73] that the eigenvalues of the path graph P_m are

$$\left\{2\cos\frac{\pi}{m+1}, \; 2\cos\frac{2\pi}{m+1}, \; \ldots, \; 2\cos\frac{m\pi}{m+1}\right\}. \quad (7)$$

Therefore (1) and (2) reduce to (5) and (6). □

Theorem 2. *If $m \geq 2$ and $n \geq 2$, the number of spanning trees of $\mathrm{EC}_{m,n}$ is $P(\mathrm{OC}_{m-2,n-2}; 4)$, and the number of spanning trees of $\mathrm{OC}_{m,n}$ is $P(\mathrm{EC}_{m-2,n-2}; 4)$.*

Proof. Both $\mathrm{EC}_{m,n}$ and $\mathrm{OC}_{m,n}$ are connected planar graphs, so they have exactly as many spanning trees as their duals [9, problem 5.23]. The dual graph $\mathrm{EC}^*_{m,n}$ has vertices (x, y) where $1 < x < m$ and $1 < y < n$ and $x + y$ is odd, corresponding to the face centered at (x, y); it also has an additional vertex ∞ corresponding to the exterior face. All its non-infinite vertices have degree 4, and when $\mathrm{EC}^*_{m,n}$ is restricted to those vertices it is just $\mathrm{OC}_{m-2,n-2}$. Therefore the submatrix of the Laplacian of $\mathrm{EC}^*_{m,n}$ that we obtain by omitting row ∞ and column ∞ is just $4I - M$, where M is the adjacency matrix of $\mathrm{OC}_{m-2,n-2}$. And the number of spanning trees of $\mathrm{EC}^*_{m,n}$ is just the determinant of this matrix, according to the Matrix Tree Theorem [1; 9, problem 4.9; 4, page 38].

A similar argument enumerates the spanning trees of $\mathrm{OC}_{m,n}$. The basic idea of this proof is due to Cvetković and Gutman [5]; see also [3, pages 85–88]. □

Combining Theorem 2 with equation (6) now yields a generalization of Stanley's conjecture [11]:

Corollary 2. *When m and n are both odd, $OC_{m,n}$ contains exactly 4 times as many spanning trees as $EC_{m,n}$.* □

Another corollary that does not appear to be obvious *a priori* follows from Theorem 2 and equation (5):

Corollary 3. *When m and n are both even, $EC_{m,n}$ contains an odd number of spanning trees.*

Proof. The adjacency matrix B_m of P_m is nonsingular mod 2 when m is even. Hence the adjacency matrix $B_m \otimes B_n$ of $EC_{m,n} \cup OC_{m,n}$ is nonsingular mod 2. Hence $P(EC_{m,n}; 4) \equiv 1$ (modulo 2). □

Stanley [11] tabulated the number of spanning trees in $OC_{2n+1,2n+1}$ for $n \leq 6$ and observed that the numbers consisted entirely of small prime factors. For example, the Aztec diamond graph $OC_{13,13}$ has exactly $2^{32} \cdot 3^7 \cdot 5^5 \cdot 7^3 \cdot 11^3 \cdot 13^2 \cdot 73^2 \cdot 193^2$ spanning trees. One way to account for this phenomenon is to note that the number of spanning trees in $OC_{2n+1,2n+1}$ is

$$4^{2n-1} \prod_{j=1}^{n-1} \prod_{k=1}^{n-1} \left(4 - 4\cos\frac{j\pi}{2n}\cos\frac{k\pi}{2n}\right)\left(4 + 4\cos\frac{j\pi}{2n}\cos\frac{k\pi}{2n}\right)$$

$$= 4^{2n-1} \prod_{j=1}^{n-1} \prod_{k=1}^{n-1} \left(4 - (\omega^j+\omega^{-j})(\omega^k+\omega^{-k})\right)\left(4 + (\omega^j+\omega^{-j})(\omega^k+\omega^{-k})\right),$$

where $\omega = e^{\pi i/2n}$ is a primitive $4n$th root of unity. Thus each factor such as $4 - (\omega^j + \omega^{-j})(\omega^k + \omega^{-k})$ is an algebraic integer in a cyclotomic number field, and all of its conjugates $4 - (\omega^{jt} + \omega^{-jt})(\omega^{kt} + \omega^{-kt})$ appear. Each product of conjugate factors is therefore an integer factor of the overall product.

Let us say that the edge from (x, y) to (x', y') in the graph is positive or negative, according as $(x - x')(y - y')$ is $+1$ or -1. The authors of [6] showed that the generating function for perfect matchings in $OC_{2n+1,2n+1}$ is $(u^2 + v^2)^{n(n+1)/2}$, in the sense that the coefficient of $u^k v^l$ in this function is the number of perfect matchings with k positive edges and l negative ones. It is natural to consider the analogous question for spanning trees: What is the generating function for spanning trees of $EC_{m,n}$ and $OC_{m,n}$ that use a given number of positive and negative edges? A careful analysis of the proof of Theorem 2 shows that the generating function for cotrees (the complements of spanning trees) in $OC_{m,n}$ is $P(EC_{m-2,n-2}; 2u + 2v)$, where P now represents the characteristic polynomial of the weighted adjacency matrix with positive and negative edges represented respectively by u and v. There are $\lceil (m - 1)(n - 1)/2 \rceil$ positive edges and $\lfloor (m - 1)(n - 1)/2 \rfloor$ negative edges altogether, so we get the generating function for trees instead of cotrees by replacing u and v by u^{-1} and v^{-1}, then multiplying by $u^{\lceil (m-1)(n-1)/2 \rceil} v^{\lfloor (m-1)(n-1)/2 \rfloor}$. A similar approach works for $EC_{m,n}$.

Unfortunately, the polynomial P does not seem to simplify nicely for general u and v, as it does when $u = v = 1$. In the case $m = n = 3$, the results look reasonably encouraging because we have

$$P(EC_{3,3}; x) = x^3 \left(x^2 - 2(u^2 + v^2) \right),$$

$$P(OC_{3,3}; x) = (x + u + v)(x - u - v)(x + u - v)(x - u + v).$$

When n increases to 5, however, we get

$$P(EC_{3,5}; x) = x^4 \left(x^2 - 2(u^2 + uv + v^2) \right) \left(x^2 - 2(u^2 - uv - v^2) \right),$$

$$P(OC_{3,5}; x) = x \left(x^2 - (u^2 + v^2) \right) \left(x^4 - 3(u^2 + v^2)x^2 + 2(u^2 - v^2) \right).$$

The quartic factor of $P(OC_{3,5}; x)$ cannot be decomposed into quadratics having the general form

$$\left(x^2 - (\alpha u^2 + \beta uv + \gamma v^2) \right) \left(x^2 - (\alpha' u^2 + \beta' uv + \gamma' v^2) \right),$$

so it is unclear how to proceed. Additional factors do appear when we set $x = 2u + 2v$; for example,

$$P(EC_{3,5}; 2u + 2v) = 64(u + v)^4 (u^2 + 3uv + v^2)(u^2 + 5uv + v^2).$$

But these factors are explained by the symmetries of $EC_{m,n}$ and $OC_{m,n}$.

Acknowledgments

I thank an anonymous referee for suggesting the present form of Theorem 1, which is considerably more general (and easier to prove) than the special case I had observed in the first draft of this note. Noam Elkies gave me the insight about algebraic conjugates, when I was studying the related problem of spanning trees in grids [8].

References

[1] C. W. Borchardt, "Ueber eine der Interpolation entsprechende Darstellung der Eliminations-Resultante," *Journal für die reine und angewandte Mathematik* **57** (1860), 111–121.

[2] Karel Čulik, "Zur Theorie der Graphen," *Časopis pro Pěstování Matematiky* **83** (1958), 133–155.

[3] Dragoš M. Cvetković, Michael Doob, Ivan Gutman, and Aleksandar Torgašev, *Recent Results in the Theory of Graph Spectra*, Annals of Discrete Mathematics **36** (1988).

[4] Dragoš M. Cvetković, Michael Doob, and Horst Sachs, *Spectra of Graphs* (New York: Academic Press, 1980).

[5] D. Cvetković and I. Gutman, "A new spectral method for determining the number of spanning trees," *Publications de l'Institut Mathématique* **29** (43) (Beograd, 1981), 49–52.

[6] Noam Elkies, Greg Kuperberg, Michael Larsen, and James Propp, "Alternating-sign matrices and domino tilings," *Journal of Algebraic Combinatorics* **1** (1992), 111–132, 219–234.

[7] C. Godsil and B. McKay, "Products of graphs and their spectra," in *Combinatorial Mathematics IV*, edited by A. Dold and B. Eckmann, *Lecture Notes in Mathematics* **560** (1975), 61–72.

[8] Germain Kreweras, "Complexité et circuits eulériens dans les sommes tensorielles de graphes," *Journal of Combinatorial Theory* **B24** (1978), 202–212.

[9] László Lovász, *Combinatorial Problems and Exercises*, 2nd edition (Budapest: Akadémiai Kiadó, 1993).

[10] Marvin Marcus and Henrik Minc, *A Survey of Matrix Theory and Matrix Inequalities* (Boston: Allyn and Bacon, 1964).

[11] Richard P. Stanley, "Problem 251. Spanning trees of Aztec diamonds," *Discrete Mathematics* **157** (1996), 383–385.

[12] Paul M. Weichsel, "The Kronecker product of graphs," *Proceedings of the American Mathematical Society* **13** (1962), 47–52.

Addendum

The following papers develop the results further:

Timothy Y. Chow, "The Q-spectrum and spanning trees of tensor products of bipartite graphs," *Proceedings of the American Mathematical Society* **125** (1997), 3155–3161.

Mihai Ciucu, "A complementation theorem for perfect matchings of graphs having a cellular completion," *Journal of Combinatorial Theory* **A81** (1998), 34–68.

Chapter 11

Partitioned Tensor Products and Their Spectra

A pleasant family of graphs defined by Godsil and McKay is shown to have easily computed eigenvalues in many cases.

[Originally published in Journal of Algebraic Combinatorics **6** (1997), 259–267.]

Let G and H be directed graphs on the respective vertices U and V, and suppose that the vertex sets have each been partitioned into disjoint subsets $U = U_0 \cup U_1$ and $V = V_0 \cup V_1$. The *partitioned tensor product* $G \underline{\times} H$ of G and H with respect to this partitioning is defined as follows:

a) Each vertex of U_0 is replaced by a copy of $H|V_0$, the subgraph of H induced by V_0;

b) Each vertex of U_1 is replaced by a copy of $H|V_1$;

c) Each arc of G that runs from U_0 to U_1 is replaced by a copy of the arcs of H that run from V_0 to V_1;

d) Each arc of G that runs from U_1 to U_0 is replaced by a copy of the arcs of H that run from V_1 to V_0.

For example, Figure 1 shows two partitioned tensor products. The example in Figure 1b is undirected; this is the special case of a directed graph where each undirected edge corresponds to a pair of arcs in opposite directions. Arcs of G that stay within U_0 or U_1 do not contribute to $G \underline{\times} H$, so we may assume that no such arcs exist; in other words, we may assume that G is bipartite.

Figure 2 shows what happens if we interchange the roles of U_0 and U_1 in G but leave everything else intact. (Equivalently, we could interchange the roles of V_0 and V_1.) These graphs, which may be denoted by $G^R \underline{\times} H$ to distinguish them from the graphs $G \underline{\times} H$ in Figure 1, typically look quite different from their left-right duals, yet it turns out that the characteristic polynomials of $G \underline{\times} H$ and $G^R \underline{\times} H$ are strongly related.

$G =$

$H =$

$G \underline{\times} H =$

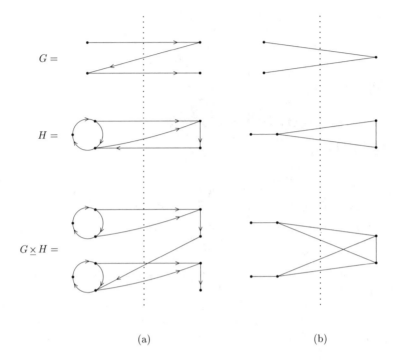

(a)   (b)

FIGURE 1. Partitioned tensor products, directed and undirected.

$G^R \underline{\times} H =$

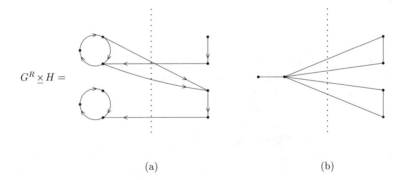

(a)   (b)

FIGURE 2. Dual products after right-left reflection of G.

Let E_{ij} be the arcs from U_i to U_j in G, and F_{ij} the arcs from V_i to V_j in H; multiple arcs are allowed, so E_{ij} and F_{ij} are multisets. It follows that $G \times H$ has $|U_0||V_0| + |U_1||V_1|$ vertices and $|U_0||F_{00}| + |U_1||F_{11}| + |E_{01}||F_{01}| + |E_{10}||F_{10}|$ arcs. Similarly, $G^R \times H$ has $|U_1||V_0| + |U_0||V_1|$ vertices and $|U_1||F_{00}| + |U_0||F_{11}| + |E_{10}||F_{01}| + |E_{01}||F_{10}|$ arcs.

The definition of partitioned tensor product is due to Godsil and McKay [3], who proved the remarkable fact that

$$p(G \underline{\times} H)\, p(H|V_0)^{|U_1|-|U_0|} = p(G^R \underline{\times} H)\, p(H|V_1)^{|U_1|-|U_0|} ,$$

where p denotes the characteristic polynomial of a graph. They also observed [4] that Figures 1b and 2b represent the smallest pair of connected undirected graphs having the same spectrum (the same p). The purpose of the present note is to refine their results by showing how to calculate $p(G \underline{\times} H)$ explicitly in terms of G and H.

We can use the symbols G and H to stand for the adjacency matrices as well as for the graphs themselves. Thus we have

$$G = \begin{pmatrix} G_{00} & G_{01} \\ G_{10} & G_{11} \end{pmatrix} \quad \text{and} \quad H = \begin{pmatrix} H_{00} & H_{01} \\ H_{10} & H_{11} \end{pmatrix}$$

in partitioned form, where G_{ij} and H_{ij} denote the respective adjacency matrices corresponding to the arcs E_{ij} and F_{ij}. (These submatrices are not necessarily square; G_{ij} has size $|U_i| \times |U_j|$ and H_{ij} has size $|V_i| \times |V_j|$.) It follows by definition that

$$G \underline{\times} H = \begin{pmatrix} I_{|U_0|} \otimes H_{00} & G_{01} \otimes H_{01} \\ G_{10} \otimes H_{10} & I_{|U_1|} \otimes H_{11} \end{pmatrix}$$

where $\otimes$ denotes the Kronecker product or tensor product [7, page 8] and I_k denotes an identity matrix of size $k \times k$.

Let $H \uparrow \sigma$ denote the graph obtained from H by σ-fold repetition of each arc that joins V_0 to V_1. In matrix form

$$H \uparrow \sigma = \begin{pmatrix} H_{00} & \sigma H_{01} \\ \sigma H_{10} & H_{11} \end{pmatrix} .$$

This definition applies to the adjacency matrix when σ is any complex number, but of course $H \uparrow \sigma$ is difficult to "draw" unless σ is a nonnegative integer. We will show that the characteristic polynomial of $G \underline{\times} H$ factors into characteristic polynomials of graphs $H \uparrow \sigma$, times a power of the characteristic polynomials of H_{00} or H_{11}. The proof is simplest when G is undirected.

Theorem 1. *Let G be an undirected graph, and let $(\sigma_1, \ldots, \sigma_l)$ be the singular values of $G_{01} = G_{10}^T$, where $l = \min(|U_0|, |U_1|)$. Then*

$$
p(G \underline{\times} H) =
\begin{cases}
\left(\prod_{j=1}^{l} p(H \uparrow \sigma_j)\right) p(H_{00})^{|U_0| - |U_1|}, & \text{if } |U_0| \geq |U_1|; \\
\left(\prod_{j=1}^{l} p(H \uparrow \sigma_j)\right) p(H_{11})^{|U_1| - |U_0|}, & \text{if } |U_1| \geq |U_0|.
\end{cases}
$$

Proof. Any real $m \times n$ matrix A has a singular value decomposition

$$A = QSR^T$$

where Q is an $m \times m$ orthogonal matrix, R is an $n \times n$ orthogonal matrix, and S is an $m \times n$ matrix with $S_{jj} = \sigma_j \geq 0$ for $1 \leq j \leq \min(m, n)$ and $S_{ij} = 0$ for $i \neq j$ [6, page 16]. The numbers $\sigma_1, \ldots, \sigma_{\min(m,n)}$ are called the singular values of A.

Let $m = |U_0|$ and $n = |U_1|$, and suppose that QSR^T is the singular value decomposition of G_{01}. Then $(\sigma_1, \ldots, \sigma_l)$ are the nonnegative eigenvalues of the bipartite graph G, and we have

$$
\begin{pmatrix} Q^T \otimes I_{|V_0|} & O \\ O & R^T \otimes I_{|V_1|} \end{pmatrix} G \underline{\times} H \begin{pmatrix} Q \otimes I_{|V_0|} & O \\ O & R \otimes I_{|V_1|} \end{pmatrix}
$$

$$
= \begin{pmatrix} I_{|U_0|} \otimes H_{00} & S \otimes H_{01} \\ S^T \otimes H_{10} & I_{|U_1|} \otimes H_{11} \end{pmatrix}
$$

because $G_{10} = RS^T Q^T$. Row and column permutations of this matrix transform it into the block diagonal form

$$
\begin{pmatrix} H \uparrow \sigma_1 & & & \\ & \ddots & & \\ & & H \uparrow \sigma_l & \\ & & & D \end{pmatrix},
$$

where D consists of $m - n$ copies of H_{00} if $m \geq n$, or $n - m$ copies of H_{11} if $n \geq m$. □

A similar result holds when G is directed, but we cannot use the singular value decomposition because the eigenvalues of G might not be real and the elementary divisors of $\lambda I - G$ might not be linear. The following lemma can be used in place of the singular value decomposition in such cases.

Lemma. *Let A and B be arbitrary matrices of complex numbers, where A is $m \times n$ and B is $n \times m$. Then we can write*

$$A = QSR^{-1}, \qquad B = RTQ^{-1},$$

where Q is a nonsingular $m \times m$ matrix, R is a nonsingular $n \times n$ matrix, S is an $m \times n$ matrix, T is an $n \times m$ matrix, and the matrices (S,T) are triangular with consistent diagonals:

$$S_{ij} = T_{ij} = 0, \qquad\qquad \text{for } i > j;$$
$$S_{jj} = T_{jj} \quad \text{or} \quad S_{jj}T_{jj} = 0, \qquad \text{for } 1 \le j \le \min(m,n).$$

Proof. We may assume that $m \le n$. If AB has a nonzero eigenvalue λ, let σ be any square root of λ and let x be a nonzero m-vector such that $ABx = \sigma^2 x$. Then the n-vector $y = Bx/\sigma$ is nonzero, and we have

$$Ay = \sigma x, \qquad Bx = \sigma y.$$

On the other hand, if all eigenvalues of AB are zero, let x be a nonzero vector such that $ABx = 0$. Then if $Bx \ne 0$, let $y = Bx$. If $Bx = 0$, let y be any nonzero vector such that $Ay = 0$; this is possible unless all n columns of A are linearly independent, in which case we must have $m = n$ and we can find y such that $Ay = x$. In all cases we have therefore demonstrated the existence of nonzero vectors x and y such that

$$Ay = \sigma x, \qquad Bx = \tau y, \qquad \sigma = \tau \quad \text{or} \quad \sigma\tau = 0.$$

Let X be a nonsingular $m \times m$ matrix whose first column is x, and let Y be a nonsingular $n \times n$ matrix whose first column is y. Then

$$X^{-1}AY = \begin{pmatrix} \sigma & a \\ 0 & A_1 \end{pmatrix}, \qquad Y^{-1}BX = \begin{pmatrix} \tau & b \\ 0 & B_1 \end{pmatrix}$$

where A_1 is $(m-1) \times (n-1)$ and B_1 is $(n-1) \times (m-1)$. If $m = 1$, let $Q = X$, $R = Y$, $S = (\sigma a)$, and $T = \binom{\tau}{0}$. Otherwise we have $A_1 = Q_1 S_1 R_1^{-1}$ and $B_1 = R_1 T_1 Q_1^{-1}$ by induction, and we can let

$$Q = X\begin{pmatrix} 1 & 0 \\ 0 & Q_1 \end{pmatrix}, \quad R = Y\begin{pmatrix} 1 & 0 \\ 0 & R_1 \end{pmatrix}, \quad S = \begin{pmatrix} \sigma & aR_1 \\ 0 & S_1 \end{pmatrix}, \quad T = \begin{pmatrix} \tau & bQ_1 \\ 0 & T_1 \end{pmatrix}.$$

All conditions are now fulfilled. □

Theorem 2. *Let G be an arbitrary graph, and let $(\sigma_1, \ldots, \sigma_l)$ be such that $\sigma_j = S_{jj} = T_{jj}$ or $\sigma_j = 0 = S_{jj}T_{jj}$ when $G_{01} = QSR^{-1}$ and $G_{10} = RTQ^{-1}$ as in the lemma, where $l = \min(|U_0|, |U_1|)$. Then $p(G \times H)$ satisfies the identities of Theorem 1.*

Proof. Proceeding as in the proof of Theorem 1, we have

$$
\begin{pmatrix} Q^{-1} \otimes I_{|V_0|} & O \\ O & R^{-1} \otimes I_{|V_1|} \end{pmatrix} G \underline{\times} H \begin{pmatrix} Q \otimes I_{|V_0|} & O \\ O & R \otimes I_{|V_1|} \end{pmatrix}
$$

$$
= \begin{pmatrix} I_{|U_0|} \otimes H_{00} & S \otimes H_{01} \\ T \otimes H_{10} & I_{|U_1|} \otimes H_{11} \end{pmatrix}.
$$

This time a row and column permutation converts the right-hand matrix to a block *triangular* form, with zeros below the diagonal blocks. Each block on the diagonal is either $H{\uparrow}\sigma_j$ or H_{00} or H_{11}, or of the form

$$
\begin{pmatrix} H_{00} & \sigma H_{01} \\ \tau H_{10} & H_{11} \end{pmatrix}, \qquad \sigma\tau = 0.
$$

In the latter case the characteristic polynomial is clearly $p(H_{00})p(H_{11}) = p(H{\uparrow}0)$, so the remainder of the proof of Theorem 1 carries over in general. □

The proof of the lemma shows that the numbers $\sigma_1^2, \ldots, \sigma_p^2$ are the characteristic roots of $G_{01}G_{10}$, when $|U_0| \le |U_1|$; otherwise they are the characteristic roots of $G_{10}G_{01}$. Either square root of σ_j^2 can be chosen, since the matrix $H{\uparrow}\sigma$ is similar to $H{\uparrow}(-\sigma)$.

We have now reduced the problem of computing $p(G \times H)$ to the problem of computing the characteristic polynomial of the graphs $H{\uparrow}\sigma$. The latter is easy when $\sigma = 0$, and some graphs G have only a few nonzero singular values. For example, if G is the complete bipartite graph having parts U_0 and U_1 of sizes m and n, all singular values vanish except for $\sigma = \sqrt{mn}$.

If H is small, and if only a few nonzero σ need to be considered, the computation of $p(H{\uparrow}\sigma)$ can be carried out directly. For example, suppose

$$
H = \begin{pmatrix} 0 & 1 & 1 & 0 & 0 \\ 1 & 0 & 0 & 0 & 1 \\ 1 & 0 & 0 & 1 & 0 \\ 0 & 0 & 1 & 0 & 1 \\ 0 & 1 & 0 & 1 & 0 \end{pmatrix}.
$$

It turns out that

$$\det \begin{pmatrix} \lambda & -1 & -\sigma & 0 & 0 \\ -1 & \lambda & 0 & 0 & -\sigma \\ -\sigma & 0 & \lambda & -1 & 0 \\ 0 & 0 & -1 & \lambda & -1 \\ 0 & -\sigma & 0 & -1 & \lambda \end{pmatrix} = (\lambda^2 + \lambda - \sigma^2)(\lambda^3 - \lambda^2 - (2+\sigma^2)\lambda + 2);$$

so we can compute the spectrum of $G \underline{\times} H$ by solving a few quadratic and cubic equations, when H is this particular 5-vertex graph (a partitioned 5-cycle). But it is interesting to look for large families of graphs in which simple formulas yield $p(H \uparrow \sigma)$ as a function of σ.

One such family consists of graphs that have only one edge crossing the partition. Let H_{00} and H_{11} be graphs on V_0 and V_1, and form the graph $H = H_{00} \bullet\!-\!\bullet H_{11}$ by adding a single edge between designated vertices $x_0 \in V_0$ and $x_1 \in V_1$. Then a glance at the adjacency matrix of H shows that

$$p(H \uparrow \sigma) = p(H_{00})p(H_{11}) - \sigma^2 p(H_{00} \,|\, V_0 \backslash \{x_0\}) p(H_{11} \,|\, V_1 \backslash \{x_1\}).$$

(The special case $\sigma = 1$ of this formula is Theorem 4.2(ii) of [5].)

Another case where $p(H \uparrow \sigma)$ has a simple form arises when the matrices

$$H_0 = \begin{pmatrix} H_{00} & 0 \\ 0 & H_{11} \end{pmatrix} \quad \text{and} \quad H_1 = \begin{pmatrix} 0 & H_{01} \\ H_{10} & 0 \end{pmatrix}$$

commute with each other. Then it is well known [2] that the eigenvalues of $H_0 + \sigma H_1$ are $\lambda_j + \sigma\mu_j$, for some ordering of the eigenvalues λ_j of H_0 and μ_j of H_1. Let us say that (V_0, V_1) is a *compatible partition* of H if $H_0 H_1 = H_1 H_0$; that is, if

$$H_{00}H_{01} = H_{01}H_{11} \quad \text{and} \quad H_{11}H_{10} = H_{10}H_{00}.$$

When H is undirected, so that $H_{00} = H_{00}^T$ and $H_{11} = H_{11}^T$ and $H_{10} = H_{01}^T$, the compatibility condition boils down to the single relation

$$H_{00}H_{01} = H_{01}H_{11}. \tag{$*$}$$

Let $m = |V_0|$ and $n = |V_1|$, so that H_{00} is $m \times m$, H_{01} is $m \times n$, and H_{11} is $n \times n$. One obvious way to satisfy $(*)$ is to let H_{00} and H_{11} both be zero, so that H is bipartite as well as G. Then $H \uparrow \sigma$ is simply σH, the σ-fold repetition of the arcs of H, and its eigenvalues are just those of H multiplied by σ. For example, if G is the M-cube P_2^M and H is a path P_N on N points, and if U_0 consists of the vertices of even parity in G while V_0 is one of H's bipartite parts, the characteristic polynomial of $G \times H$ is

$$\prod_{\substack{1 \leq j \leq M \\ 1 \leq k \leq N}} \left(\lambda - (2N - 4j) \cos \frac{k\pi}{N+1} \right)^{\binom{M}{j}/2},$$

because of the well-known eigenvalues of G and H [1]. Figure 3 illustrates this construction in the special case $M = N = 3$. The smallest pair of cospectral graphs, ✕ and ▪, is obtained in a similar way by considering the eigenvalues of $P_3 \times P_3$ and $P_3^T \times P_3$ [4].

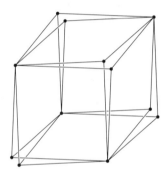

FIGURE 3. $P_2^3 \times P_3$.

Another simple way to satisfy the compatibility condition $(*)$ with symmetric matrices H_{00} and H_{11} is to let H_{01} consist entirely of 1s, and to let H_{00} and H_{11} both be regular graphs of the same degree d. Then the eigenvalues of H_0 are $(\lambda_1, \ldots, \lambda_m, \lambda'_1, \ldots, \lambda'_n)$, where $(\lambda_1, \ldots, \lambda_m)$ belong to H_{00} and $(\lambda'_2, \ldots, \lambda'_n)$ belong to H_{11} and $\lambda_1 = \lambda'_1 = d$. The eigenvalues of H_1 are $(\sqrt{mn}, -\sqrt{mn}, 0, \ldots, 0)$. We can match the eigenvalues of H_0 properly with those of H_1 by looking at the common eigenvectors $(1, \ldots, 1)^T$ and $(1, \ldots, 1, -1, \ldots, -1)^T$ that correspond to d in H_0 and $\pm\sqrt{mn}$ in H_1; the eigenvalues of $H \uparrow \sigma$ are therefore

$$(d + \sigma\sqrt{mn}, \lambda_2, \ldots, \lambda_m, d - \sigma\sqrt{mn}, \lambda'_2, \ldots, \lambda'_n).$$

Yet another easy way to satisfy $(*)$ is to assume that $m = n$ and to let $H_{00} = H_{11}$ commute with H_{01}. One general construction of this kind arises when the vertices of V_0 and V_1 are the elements of a group, and when $H_{00} = H_{11}$ is a Cayley graph on that group. In other words, two elements α and β are adjacent in H_{00} if and only if $\alpha\beta^{-1} \in X$, where X is an arbitrary set of group elements closed under inverses. And we can let $\alpha \in V_0$ be adjacent to $\beta \in V_1$ if and only if $\alpha\beta^{-1} \in Y$, where Y is any normal subgroup. Then H_{00} commutes with H_{01}. The effect is to make the cosets of Y fully interconnected between V_0 and V_1, while retaining a more interesting Cayley graph structure inside V_0 and V_1. If Y is the trivial subgroup, so that H_{01} is simply the identity matrix, our partitioned tensor product $G\times H$ becomes simply the ordinary Cartesian product $I_{|U|} \otimes H + G \otimes I_{|V|}$. But in many other cases this construction gives something more general.

A fourth family of compatible partitions is illustrated by the following graph H in which $m = 6$ and $n = 12$:

$$\left(\begin{array}{cccccc|cccccccccccc}
0 & 0 & 1 & 1 & 1 & 0 & 1 & 0 & 0 & 0 & 0 & 0 & 0 & 1 & 0 & 0 & 0 & 0 \\
0 & 0 & 0 & 1 & 1 & 1 & 0 & 1 & 0 & 0 & 0 & 0 & 0 & 0 & 1 & 0 & 0 & 0 \\
1 & 0 & 0 & 0 & 1 & 1 & 0 & 0 & 1 & 0 & 0 & 0 & 0 & 0 & 0 & 1 & 0 & 0 \\
1 & 1 & 0 & 0 & 0 & 1 & 0 & 0 & 0 & 1 & 0 & 0 & 0 & 0 & 0 & 0 & 1 & 0 \\
1 & 1 & 1 & 0 & 0 & 0 & 0 & 0 & 0 & 0 & 1 & 0 & 0 & 0 & 0 & 0 & 0 & 1 \\
0 & 1 & 1 & 1 & 0 & 0 & 0 & 0 & 0 & 0 & 0 & 1 & 1 & 0 & 0 & 0 & 0 & 0 \\
\hline
1 & 0 & 0 & 0 & 0 & 0 & 0 & 0 & 1 & 0 & 1 & 0 & 0 & 0 & 0 & 0 & 1 & 0 \\
0 & 1 & 0 & 0 & 0 & 0 & 0 & 0 & 0 & 1 & 0 & 1 & 0 & 0 & 0 & 0 & 0 & 1 \\
0 & 0 & 1 & 0 & 0 & 0 & 1 & 0 & 0 & 0 & 1 & 0 & 1 & 0 & 0 & 0 & 0 & 0 \\
0 & 0 & 0 & 1 & 0 & 0 & 0 & 1 & 0 & 0 & 0 & 1 & 0 & 1 & 0 & 0 & 0 & 0 \\
0 & 0 & 0 & 0 & 1 & 0 & 1 & 0 & 1 & 0 & 0 & 0 & 0 & 0 & 1 & 0 & 0 & 0 \\
0 & 0 & 0 & 0 & 0 & 1 & 0 & 1 & 0 & 1 & 0 & 0 & 0 & 0 & 0 & 1 & 0 & 0 \\
0 & 0 & 0 & 0 & 0 & 1 & 0 & 0 & 1 & 0 & 0 & 0 & 0 & 0 & 1 & 0 & 1 & 0 \\
1 & 0 & 0 & 0 & 0 & 0 & 0 & 0 & 0 & 1 & 0 & 0 & 0 & 0 & 0 & 1 & 0 & 1 \\
0 & 1 & 0 & 0 & 0 & 0 & 0 & 0 & 0 & 0 & 1 & 0 & 1 & 0 & 0 & 0 & 1 & 0 \\
0 & 0 & 1 & 0 & 0 & 0 & 0 & 0 & 0 & 0 & 0 & 1 & 0 & 1 & 0 & 0 & 0 & 1 \\
0 & 0 & 0 & 1 & 0 & 0 & 1 & 0 & 0 & 0 & 0 & 0 & 1 & 0 & 1 & 0 & 0 & 0 \\
0 & 0 & 0 & 0 & 1 & 0 & 0 & 1 & 0 & 0 & 0 & 0 & 0 & 1 & 0 & 1 & 0 & 0
\end{array}\right)$$

In general, let C_{2k} be the matrix of a cyclic permutation on $2k$ elements, and let $m = 2k$, $n = 4k$. Then we obtain a compatible partition if

$$H_{00} = \left(C_{2k}^j + C_{2k}^k + C_{2k}^{-j}\right), \quad H_{01} = \left(I_{2k} \quad C_{2k}\right),$$

$$H_{10} = \begin{pmatrix} I_{2k} \\ C_{2k}^{-1} \end{pmatrix}, \quad H_{11} = \begin{pmatrix} C_{2k}^j + C_{2k}^{-j} & C_{2k}^{k+1} \\ C_{2k}^{k-1} & C_{2k}^j + C_{2k}^{-j} \end{pmatrix}.$$

The 18×18 example matrix is the special case $j = 2$, $k = 3$. The eigenvalues of $H{\uparrow}\sigma$ in general are

$$\omega^{jl} + \omega^{-jl} + 1, \qquad \omega^{jl} + \omega^{-jl} - 1 + \sqrt{2}\,\sigma, \qquad \omega^{jl} + \omega^{-jl} - 1 - \sqrt{2}\,\sigma$$

for $0 \le l < 2k$, where $\omega = e^{\pi i/k}$.

Compatible partitionings of digraphs are not difficult to construct. But it would be interesting to find further examples of undirected graphs, without multiple edges, that have a compatible partition.

References

[1] Dragoš M. Cvetković, Michael Doob, and Horst Sachs, *Spectra of Graphs* (New York: Academic Press, 1980).

[2] G. Frobenius, "Über vertauschbare Matrizen," *Sitzungsberichte der Königlich Preußischen Akademie der Wissenschaften zu Berlin* (1896), 601–614. Reprinted in his *Gesammelte Abhandlungen* **2** (Berlin: Springer, 1968), 705–718.

[3] C. Godsil and B. McKay, "Products of graphs and their spectra," in *Combinatorial Mathematics IV*, edited by A. Dold and B. Eckmann, *Lecture Notes in Mathematics* **560** (1975), 61–72.

[4] C. Godsil and B. McKay, "Some computational results on the spectra of graphs," in *Combinatorial Mathematics IV*, edited by A. Dold and B. Eckmann, *Lecture Notes in Mathematics* **560** (1975), 73–82.

[5] C. D. Godsil and B. D. McKay, "Constructing cospectral graphs," *Æquationes Mathematicæ* **25** (1982), 257–268.

[6] Gene H. Golub and Charles F. Van Loan, *Matrix Computations* (Baltimore: Johns Hopkins University Press, 1983).

[7] Marvin Marcus and Henrik Minc, *A Survey of Matrix Theory and Matrix Inequalities* (Boston: Allyn and Bacon, 1964).

Chapter 12

Oriented Subtrees of an Arc Digraph

A simple formula can be given for the number of subtrees of the arc digraph D^* of a directed graph D, in terms of the number of subtrees of D, thereby generalizing certain results of van Aardenne-Ehrenfest and de Bruijn [1].

[Originally published in Journal of Combinatorial Theory **3** (1967), 309–314.]

Let D be a directed graph with vertices v_1, v_2, ..., v_n, and with a_{jk} arcs leading from v_j to v_k. We write

$$\sigma_j = \sum_{1 \leq k \leq n} a_{jk} = \text{out-degree of } v_j; \tag{1}$$

$$\tau_k = \sum_{1 \leq j \leq n} a_{jk} = \text{in-degree of } v_k. \tag{2}$$

The *arc digraph* D^* is a directed graph with $\sigma_1 + \sigma_2 + \cdots + \sigma_n$ vertices, one for each arc of D; a vertex of D^* that corresponds to an arc from v_j to v_k in D will be denoted A_{jk}. (There may be several vertices of D^* with the same name A_{jk}, but they will be regarded as distinct.) In this directed graph, exactly 0 or 1 arcs lead from A_{jk} to $A_{j'k'}$ according as $k \neq j'$ or $k = j'$. Hence D^* has $\sigma_1\tau_1 + \cdots + \sigma_n\tau_n$ arcs.

(The concept of an arc digraph seems to have appeared first for special digraphs in a paper by de Bruijn [2], and for the general case in a paper by Harary and Norman [3].)

A vertex v_k is called a *source* or "transmitter" if $\tau_k = 0$; the arc digraph D^* contains no vertices A_{jk} when v_k is a source. We will assume that vertices v_1, ..., v_m of D are non-sources, and vertices v_{m+1}, ..., v_n are sources. The *derived graph* D' is the graph on vertices v_1, ..., v_m obtained from D by deleting all sources (and of course all arcs leading from sources). The number τ_k will always refer to the number of arcs whose final vertex is v_k in D, not in D'.

The *matrix* $C = (C_{jk})$ of D is defined by the rule

$$C_{jk} = \sigma_j \delta_{jk} - a_{jk}. \tag{3}$$

This matrix clearly has the partitioned form

$$\begin{pmatrix} C' & O \\ * & * \end{pmatrix}$$

where C' is the matrix of D' and O is all zeros. (As usual we use '$*$' to stand for part of a matrix whose entries are not being specified in detail.)

An *oriented subtree* of D with root v_j is a set of $n-1$ arcs $e_2, \ldots, e_n$ such that, for $1 \le k \le n$, there is an oriented path along some subset of these arcs from v_k to v_j. A well-known theorem due to Tutte [4] states that *the number of oriented subtrees of D with root v_j is the cofactor of C_{jj} in the matrix of D.*

These concepts are all illustrated in Figure 1, which shows a digraph D with 4 vertices, v_4 being a source. The arc digraph D^* has 8 vertices and 16 arcs. One of the four oriented subtrees of D with root v_1 is shown. (Subtrees using different arcs from v_2 to v_1 are considered distinct.) The matrices of D, D', and D^* are

$$C = \begin{pmatrix} 1 & 0 & -1 & 0 \\ -2 & 3 & -1 & 0 \\ 0 & -1 & 1 & 0 \\ -1 & -1 & 0 & 2 \end{pmatrix}, \quad C' = \begin{pmatrix} 1 & 0 & -1 \\ -2 & 3 & -1 \\ 0 & -1 & 1 \end{pmatrix},$$

$$C^* = \left(\begin{array}{cccc|cc|cc} 1 & 0 & 0 & 0 & 0 & 0 & -1 & 0 \\ -1 & 2 & 0 & 0 & 0 & 0 & -1 & 0 \\ -1 & 0 & 2 & 0 & 0 & 0 & -1 & 0 \\ -1 & 0 & 0 & 2 & 0 & 0 & -1 & 0 \\ \hline 0 & -1 & -1 & 0 & 3 & 0 & 0 & -1 \\ 0 & -1 & -1 & 0 & 0 & 3 & 0 & -1 \\ \hline 0 & 0 & 0 & 0 & -1 & 0 & 1 & 0 \\ 0 & 0 & 0 & 0 & -1 & 0 & 0 & 1 \end{array} \right). \tag{4}$$

Notice that we have partitioned C^* into a 3×3 array of submatrices; in general, it is possible to partition C^* into an $m \times m$ array of submatrices $B_{kk'}$ for $1 \le k, k' \le m$, if we group together vertices A_{jk} with equal k.

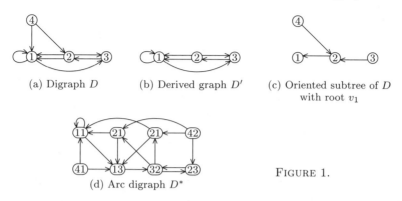

(a) Digraph D (b) Derived graph D' (c) Oriented subtree of D
                                                              with root v_1

(d) Arc digraph D^* FIGURE 1.

The submatrix $B_{kk'}$ then represents the relations between vertices of the forms A_{jk} and $A_{j'k'}$ in D^*.

We wish to count the number of oriented subtrees of D^* having a given root, for a general directed graph D. The following theorem gives the desired formula:

Theorem. *The number of oriented subtrees of D^* with root $A_{j_0 k_0}$ is*

$$\sigma_1^{\tau_1 - 1} \sigma_2^{\tau_2 - 1} \ldots \sigma_m^{\tau_m - 1} \left(t_{k_0} - \sum t_j / \sigma_{k_0} \right) \tag{5}$$

where t_j is the number of oriented subtrees of D' with root v_j, and where the sum is over all j such that $A_{j k_0}$ is a vertex of D^* distinct from $A_{j_0 k_0}$. (If v_j is a source, $t_j = 0$.)

Proof. There is no loss of generality in assuming that $k_0 = 1$, since $k_0 \leq m$. We construct the matrix C^*, partitioned into submatrices $B_{kk'}$ as above, with its first row and column corresponding to $A_{j_0 1}$. Let C_λ^* be the matrix C^* with the formal variable λ added to the upper left corner element. Since $\det C^* = 0$, Tutte's theorem implies that $\det C_\lambda^* = \lambda t$ where t is the desired number of subtrees.

The determinant of C_λ^* may be evaluated as follows: Add columns together so that the first column of each submatrix contains the sum of the columns. Then subtract the first row of each submatrix from the other rows. This transforms C_λ^* into submatrices where

$$B_{kk'} = \begin{pmatrix} C_{kk'} & * & \cdots & * \\ 0 & 0 & \cdots & 0 \\ \vdots & \vdots & & \vdots \\ 0 & 0 & \cdots & 0 \end{pmatrix}, \qquad \text{if } k \neq k',$$

since $B_{kk'}$ originally contained columns of the form

$$\begin{pmatrix} 0 \\ \vdots \\ 0 \end{pmatrix} \quad \text{or} \quad \begin{pmatrix} -1 \\ \vdots \\ -1 \end{pmatrix},$$

and the number of columns of the latter type is $-C_{kk'}$. Similarly for $k = k'$ we have the square submatrix

$$B_{kk} = \begin{pmatrix} C_{kk} + \lambda\delta_{k1} & * & * & \cdots & * \\ -\lambda\delta_{k1} & \sigma_k & 0 & \cdots & 0 \\ -\lambda\delta_{k1} & 0 & \sigma_k & \cdots & 0 \\ \vdots & \vdots & \vdots & \ddots & \vdots \\ -\lambda\delta_{k1} & 0 & 0 & \cdots & \sigma_k \end{pmatrix}.$$

For example the matrix C^* of (4) would be transformed into

$$\left(\begin{array}{cccc|cc|cc} \lambda+1 & 0 & 0 & 0 & 0 & 0 & -1 & 0 \\ -\lambda & 2 & 0 & 0 & 0 & 0 & 0 & 0 \\ -\lambda & 0 & 2 & 0 & 0 & 0 & 0 & 0 \\ -\lambda & 0 & 0 & 2 & 0 & 0 & 0 & 0 \\ \hline -2 & -1 & -1 & 0 & 3 & 0 & -1 & 0 \\ 0 & 0 & 0 & 0 & 0 & 3 & 0 & 0 \\ \hline 0 & 0 & 0 & 0 & -1 & 0 & 1 & 0 \\ 0 & 0 & 0 & 0 & 0 & 0 & 0 & 1 \end{array} \right).$$

Consequently

$$\det C^*_\lambda = \sigma_2^{\tau_2-1}\ldots\sigma_m^{\tau_m-1} \det \begin{pmatrix} \lambda & * & * & \cdots & * & C_{12} & \cdots & C_{1m} \\ -\lambda & \sigma_1 & 0 & \cdots & 0 & 0 & \cdots & 0 \\ -\lambda & 0 & \sigma_1 & \cdots & 0 & 0 & \cdots & 0 \\ \vdots & \vdots & \vdots & \ddots & \vdots & \vdots & & \vdots \\ -\lambda & 0 & 0 & \cdots & \sigma_1 & 0 & \cdots & 0 \\ 0 & * & * & \cdots & * & C_{22} & \cdots & C_{2m} \\ \vdots & \vdots & \vdots & & \vdots & \vdots & \ddots & \vdots \\ 0 & * & * & \cdots & * & C_{m2} & \cdots & C_{mm} \end{pmatrix} \quad (6)$$

where we have factored out and removed rows and columns when the row contains σ_k on the diagonal as its only nonzero element, and we have then added the last $m - 1$ columns to the first column.

The $*$'s remaining in (6) consist of at most one -1 in each column, appearing in the row containing $C_{j2} \ldots C_{jm}$ if the column corresponds to A_{j1}. Hence the value of the determinant in (6) comes to

$$\lambda \sigma_1^{\tau_1 - 1} \operatorname{cofactor}(C'_{11}) - \lambda \sum \sigma_1^{\tau_1 - 2} \operatorname{cofactor}(C'_{j1})$$

summed over the same values of j as in (5). But $\operatorname{cofactor}(C'_{j1}) = \operatorname{cofactor}(C'_{jj})$, so an application of Tutte's theorem completes the proof. $\square$

Let us now consider important special cases of the theorem. The directed graph D is *balanced* if $\sigma_k = \tau_k$ for all k; this means that the row and column sums of its matrix C both vanish, hence all cofactors are identical (and the number of subtrees is independent of the root). A "source" vertex of a balanced digraph has both σ_k and $\tau_k = 0$, so it is completely isolated and D contains no subtrees (unless $n = 1$). However, notice that it is possible for D^* to have subtrees even though D has isolated vertices; this fact accounts for the appearance of D' in the statement of the theorem. If D is balanced, D' is just D minus its isolated vertices, and the theorem takes the following simple form in this special case:

Corollary. *If D is a balanced digraph, the number of oriented subtrees of D^* with root $A_{j_0 k_0}$ is*

$$\sigma_1^{\sigma_1 - 1} \sigma_2^{\sigma_2 - 1} \ldots \sigma_m^{\sigma_m - 1} t / \sigma_{k_0}, \tag{7}$$

where t is the number of oriented subtrees of D' with a given root.

Proof. There are $\sigma_{k_0} - 1$ equal terms in the sum within (5). $\square$

The arc digraph D^* is balanced if D is *regular*, that is, if

$$\tau_1 = \cdots = \tau_n = \sigma_1 = \cdots = \sigma_n = \sigma.$$

In this case formula (7) becomes

$$\sigma^{n(\sigma - 1) - 1} t. \tag{8}$$

A well-known theorem of van Aardenne-Ehrenfest and de Bruijn [1] states that the number of Eulerian circuits in a balanced digraph D is $(\sigma_1 - 1)! \ldots (\sigma_m - 1)! \, t$, where t is the number of oriented subtrees of D' having a given root. The special case (8) of our theorem therefore implies that when D is regular, *the number of Eulerian circuits of D^* is $\sigma^{-1}(\sigma!)^{n(\sigma-1)}$ times the number of Eulerian circuits of D.* This corollary was proved by van Aardenne-Ehrenfest and de Bruijn [1] using an interesting but considerably more complicated argument.

The relative simplicity of equation (5) suggests that the theorem above could be established by some combinatorial construction relating subtrees of D^* with subtrees of D, but there seems to be no obvious way to do this.

This work was supported by the National Science Foundation. The author wishes to thank F. Harary for several improvements to the terminology and notation used in an earlier draft of this paper.

References

[1] T. van Aardenne-Ehrenfest and N. G. de Bruijn, "Circuits and trees in oriented linear graphs," *Simon Stevin* **28** (1951), 203–217.

[2] N. G. de Bruijn, "A combinatorial problem," *Indagationes Mathematicæ* **8** (1946), 461–467.

[3] F. Harary and R. Z. Norman, "Some properties of line digraphs," *Rendiconti del Circolo Matematico de Palermo* (2) **9** (1960), 161–168.

[4] W. T. Tutte, "The dissection of equilateral triangles into equilateral triangles," *Proceedings of the Cambridge Philosophical Society* **44** (1948), 463–482.

Chapter 13

Another Enumeration of Trees

Given a set of vertices that have each been assigned one of the colors C_1, C_2, ..., C_m, with n_j vertices of color C_j, we can derive a formula for the number of oriented trees on these vertices, having a designated root, and subject to any number of restrictions of the form "no arc goes from a vertex of color C_i to a vertex of color C_j." The formula is based on a combinatorial construction that defines a correspondence between such trees and certain sequences.

[Originally published in Canadian Journal of Mathematics **20** (1968), 1077–1086.]

In 1889, A. Cayley [2] found that the number of oriented trees that can be constructed on n vertices, having a specified root, is exactly n^{n-2}. Cayley's formula has been generalized in several interesting ways; see, for example, Raney [8], Riordan [9], Knuth [5], Good [3], Moon [6]. In this paper we present a combinatorial construction that leads to another rather pleasant generalization of Cayley's formula.

The term "oriented tree" is used in this paper to distinguish the trees discussed here from "free trees" (which have no root and no orientation specified for the arcs) and from "ordered trees" (in which the relative order of the vertices pointing to a vertex is significant as well as the orientation of the arcs).

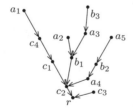

FIGURE 1. An oriented tree.

Figure 1 shows an oriented tree on the vertices a_1, a_2, a_3, a_4, a_5, b_1, b_2, b_3, c_1, c_2, c_3, c_4, r, in which all arcs go from an 'a' to a 'b' or a 'c', or from a 'b' to an 'a' or a 'c', or from a 'c' to a 'c'

209

or an 'r'. The admissible kinds of arcs just described are represented graphically in Figure 2. It is natural to ask: "How many ways are there to draw arcs on the specified vertices so that an oriented tree of this type is obtained?" In general, we will find that if there are $|A|$, $|B|$, and $|C|$ vertices of the corresponding types, then the total number of oriented trees subject to the restrictions of Figure 2 is exactly $(|B|+|C|)^{|A|-1}(|A|+|C|)^{|B|-1}(|C|+1)^{|C|-1}(|A||C|+|B||C|+|C|^2)$. The theorem below shows that similar formulas may be obtained when any diagram of "chromatic constraints" is considered in place of Figure 2.

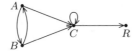

FIGURE 2. Chromatic constraints.

In the following discussion, the notation $|X|$ stands for the number of elements in the (finite) set X. Furthermore, if f is a function, we write $f^0(x) = x$ and $f^{r+1}(x) = f(f^r(x))$ when the latter is defined.

1. The Basic Construction

Let us say (U, V, f) is a *T-graph* if V is a finite set of vertices, $U \subseteq V$, and f is a function from U into V such that there are no "cycles," that is, no vertices x with $f^m(x) = x$ for some $m > 0$. It follows that for all $x \in V$ there is a least integer $m \geq 0$ such that $f^m(x) \notin U$, and in this case we write $f^\infty(x) = f^m(x)$. In terms of this notation, an *oriented tree* with root r is a T-graph of the form $(U, U \cup \{r\}, f)$.

The enumeration formulas to be derived rest essentially on the following construction, which generalizes a theorem due to Prüfer [7].

Lemma. *Let U, V, and W be sets of vertices, with W disjoint from $U \cup V$. Let f be a function from $V \setminus U$ into U. The number of functions h from U into $V \cup W$, such that $(U \cup V, U \cup V \cup W, f \cup h)$ is a T-graph, is*

$$|V \cup W|^{|U|-1}|W|.$$

Proof. Let $n = |U|$. We will prove the more interesting result that there is a one-to-one correspondence between such functions h and sequences of vertices $(a_1, a_2, \ldots, a_n)$ such that $a_k \in V \cup W$ for $1 \leq k < n$ and $a_n \in W$. For this purpose, assume that the set $U \cup V \cup W$ has been linearly ordered in some fashion.

First, suppose such a function h is given, and consider the directed graph G with vertices $U \cup V \cup W$, with arcs from v to $f(v)$ for $v \in V \setminus U$, and with arcs from u to $h(u)$ for $u \in U$. Let us say that the vertex $u \in U$ is "free" with respect to G if there is no oriented path from u' to u for any other $u' \in U$. Since there are no oriented cycles in G, there is at least one free vertex. Let u_1 be the lowest free vertex (in the assumed linear ordering). Once $u_1, \ldots, u_t$ have been defined, let u_{t+1} be the lowest free vertex in the directed graph obtained from G by removing u_k and the arc from u_k to $h(u_k)$ for $1 \leq k \leq t$. This rule defines a sequence $(u_1, u_2, \ldots, u_n)$ that contains each of the n vertices of U. Now let $a_k = h(u_k)$ for $1 \leq k \leq n$. Clearly, $a_k \in V \cup W$ for $1 \leq k < n$ and $a_n \in W$.

Conversely, assume that such a sequence $(a_1, a_2, \ldots, a_n)$ is given. Let us now call a vertex $u \in U$ "free" with respect to the sequence if there is no j for which $a_j = u$ or for which $a_j \in V \setminus U$ and $f(a_j) = u$. Since $a_n \in W$, there must be at least one free vertex. Let u_1 be the lowest free vertex (in the assumed linear ordering). Once $u_1, \ldots, u_t$ have been defined, let u_{t+1} be the lowest free vertex with respect to $a_{t+1}, \ldots, a_n$ that is different from $u_1, \ldots, u_t$. This rule defines a sequence $u_1, u_2, \ldots, u_n$ containing each of the n vertices of U. Now let $h(u_k) = a_k$ for $1 \leq k \leq n$. Then $(U \cup V, U \cup V \cup W, f \cup h)$ is a T-graph, since the relation $(f \cup h)^m(u_k) = u_r$ for $m > 0$ implies that $r > k$.

The two constructions just given are obviously inverse to each other, so the stated one-to-one correspondence has been established. (Prüfer essentially published the special case in which $U = V$ and $|W| = 1$.) We may note also from the construction that if u is the highest vertex of U, in the assumed linear ordering, then

$$(f \cup h)^\infty(u) = a_n;$$

for in the rule that determines the sequence $(u_1, u_2, \ldots, u_n)$, vertex u becomes free only when there is an oriented path from u to all remaining vertices. $\square$

2. The Main Construction

Let $\mathcal{C}$ be a family of nonempty disjoint sets, and let $V = \bigcup \mathcal{C}$, that is, $V = \bigcup_{C \in \mathcal{C}} C$. We will assume that V is a finite set of vertices, partitioned into the classes represented by $\mathcal{C}$. Let $\mathcal{G}$ be a directed graph on the elements of $\mathcal{C}$ (like Figure 2); and for $C \in \mathcal{C}$, let $\mathcal{G}(C)$ be the set of all $C' \in \mathcal{C}$ such that there is an arc from C to C' in $\mathcal{G}$. The family $\mathcal{C}$ and the directed graph $\mathcal{G}$ will be fixed throughout this section.

If $\mathcal{K}$ is a subset of $\mathcal{C}$, we say a $\mathcal{K}$-structure is a T-graph (U, V, f) for which $U = \bigcup \mathcal{K}$, $V = \bigcup \mathcal{C}$, and if $u \in C \in \mathcal{K}$, then $f(u) \in \bigcup \mathcal{G}(C)$. In other words, we are considering the set of T-graphs on V satisfying the "chromatic constraints" of $\mathcal{G}$, where we think of $\mathcal{C}$ as a set of colors. Our goal is to enumerate the number of possible $\mathcal{K}$-structures, that is, the number of functions f satisfying the restrictions just mentioned.

Theorem. *The number of possible* $\mathcal{K}$*-structures is equal to*

$$\sum_g \left(\prod_{C \in \mathcal{K}} \left| \bigcup \mathcal{G}(C) \right|^{|C|-1} |g(C)| \right),$$

where the sum is over all functions g *such that* $(\mathcal{K}, \mathcal{C}, g)$ *is a* T*-graph and* $g(C) \in \mathcal{G}(C)$ *for all* $C \in \mathcal{K}$.

(*Note:* Using a theorem of Tutte [11], this formula can also be written

$$\det A \cdot \prod_{C \in \mathcal{K}} \left| \bigcup \mathcal{G}(C) \right|^{|C|-1},$$

where A is a matrix whose rows and columns are indexed by the elements of $\mathcal{K}$; $A_{CC'} = -|C' \cap \bigcup \mathcal{G}(C)|$ when $C \neq C'$, and $A_{CC} = |\bigcup \mathcal{G}(C) \setminus C|$.)

Proof. Let $n = |\mathcal{K}|$. We will prove in fact that there is a one-to-one correspondence between structures and sets of n sequences of the form

$$(a_{C1}, a_{C2}, \ldots, a_{Cr}), \quad r = |C|, \tag{*}$$

where $a_{Ck} \in \bigcup \mathcal{G}(C)$ for $1 \leq k < r$ and $a_{Cr} \in g(C)$, for all $C \in \mathcal{K}$, where $(\mathcal{K}, \mathcal{C}, g)$ is a T-graph contained in $\mathcal{G}$. We may assume that the vertices V are linearly ordered, and so are the "colors" $\mathcal{C}$.

First suppose that a $\mathcal{K}$-structure (U, V, f) is given. We will define a function g as required, and a set of sequences $(*)$, and a sequence $(C_1, C_2, \ldots, C_n)$ representing the colors of $\mathcal{K}$, and also a sequence $(u_1, u_2, \ldots, u_n)$ with $u_k \in C_k$. Start with C_1, the lowest color in the assumed linear order, and u_1 the highest vertex of C_1. If $t < n$ and if C_t and u_t have been chosen, we define C_{t+1} and u_{t+1} as follows: Let m be maximal such that $f^m(u_t) \in C_t$, and $f^k(u_t) \notin \{C_1, \ldots, C_{t-1}\}$ for $0 \leq k \leq m$. Let $v_t = f^{m+1}(u_t)$ and let $g(C_t)$ be the class in $\mathcal{G}(C_t)$ such that $v_t \in g(C_t)$. Now if $g(C_t) = C_k$ for some k, $1 \leq k < t$, or if $g(C_t) \notin \mathcal{K}$, choose C_{t+1} to be the lowest color of $\mathcal{K} \setminus \{C_1, \ldots, C_t\}$ and let u_{t+1} be the highest vertex of that color. Otherwise, let $C_{t+1} = g(C_t)$, $u_{t+1} = v_t$.

We wish to prove that $(\mathcal{K}, \mathcal{C}, g)$ is a T-graph. Note that if $g(C_t) = C_k$ and $t \leq k \leq n$, then $k = t + 1$. If $(\mathcal{K}, \mathcal{C}, g)$ is not a T-graph, there is some t such that $g^r(C_t) = C_t$ and $g(C_t) = C_k$ for some $r > 0$ and $k < t$. We can find $s \leq t$ such that $g(C_j) = C_{j+1}$ for $s \leq j < t$ but $C_s \neq g(C_j)$ for $j < s$; it follows that $g(C_t) = C_k$ for some k such that $s \leq k < t$. Consider the values

$$u_s, f(u_s), f^2(u_s), \ldots, v_s = u_{s+1}, f(u_{s+1}), \ldots, u_t, f(u_t), \ldots, v_t, \ldots, u,$$

where u is the first element encountered that is in

$$\bigcup (\mathcal{C} \setminus \mathcal{K}) \cup C_1 \cup \cdots \cup C_{s-1}.$$

By construction, none of the elements of this sequence after u_{s+1} are in C_s; none of the elements after u_{s+2} are in C_{s+1}; and so on. It is therefore impossible for v_t to be an element of C_k for $s \le k < t$. This contradiction proves that $(\mathcal{K}, \mathcal{C}, g)$ is a T-graph.

Finally, for $t = n, n - 1, \ldots, 1$, we successively construct the sequence $(*)$ for $C = C_t$. Consider the T-graph (U_t, V, f_t), where $U_t = C_{t+1} \cup \cdots \cup C_n$ and f_t is f restricted to U_t. Reorder the elements of C_t, if necessary, so that u_t is the highest element, and apply the construction of the lemma with U, V, W, and f replaced, respectively, by C_t, V_t, $\bigcup \mathcal{G}(C_t) \setminus V_t$, and ϕ_t, where $V_t = \{ v \in \bigcup \mathcal{G}(C_t) \mid f_t^\infty(v) \in C_t \}$, and ϕ_t is f_t^∞ restricted to V_t. The values of f restricted to C_t now correspond to a function h as stated in the lemma, so we obtain a sequence $(*)$ in which the last element is v_t.

Conversely, let us suppose we are given a set of sequences $(*)$ for each $C \in \mathcal{K}$, defining a function g of the required type. We will define a function f such that (U, V, f) is a T-graph of the required type, and we will also define a sequence $(C_1, C_2, \ldots, C_n)$ representing the colors of $\mathcal{K}$, and a sequence $(u_1, u_2, \ldots, u_n)$ with $u_k \in C_k$. Start with C_1, the lowest color in the assumed linear order, and u_1, the highest vertex of C_1. If $t < n$ and if C_t and u_t have been chosen, let v_t be the last element of the sequence $(*)$ for C_t. Now if $g(C_t) = C_k$ for some k, $1 \le k < t$, or if $g(C_t) \notin \mathcal{K}$, choose C_{t+1} to be the lowest color of $\mathcal{K} \setminus \{C_1, \ldots, C_t\}$ and let u_{t+1} be the highest vertex of that color. Otherwise, let $C_{t+1} = g(C_t)$ and $u_{t+1} = v_t$.

Now for $t = n, n-1, \ldots, 1$, we successively define f on the elements of C_t so that no cycles are introduced. Suppose f has already been defined on $U_t = C_{t+1} \cup \cdots \cup C_n$ and let f_t be this function. Reorder the elements of C_t, if necessary, so that u_t is the highest element, and apply the construction of the lemma with U, V, W, and f replaced, respectively, by C_t, V_t, $\bigcup \mathcal{G}(C_t) \setminus V_t$, and ϕ_t (as above). The construction has been carried out so that $v_t \notin V_t$, since, if $g(C_t) = C_{t+1}$, we have $f_t^\infty(v_t) = f_t^\infty(u_{t+1}) = f_t^\infty(v_{t+1})$ and, continuing in this manner, it is clear that $f_t^\infty(v_t) \notin C_t$ when $(\mathcal{K}, C, g)$ is a T-graph. Therefore the lemma applies, and it determines a function h that can be used to define $f_{t-1} = f_t \cup h$.

The two constructions just described are inverses of each other, so the theorem has been proved. $\quad\square$

It is possible to give a much simpler proof based directly on the theorem of Tutte [11], which shows that the number of subtrees of a directed graph having a given root can be expressed as a determinant. We consider the directed graph having $\bigcup \mathcal{C} \cup \{r_0\}$ as vertices, where r_0 is a new symbol; there is an arc in this graph from v to v' if and only if either $v \in C \in \mathcal{K}$ and $v' \in \bigcup \mathcal{G}(C)$, or if $v \in C \in \mathcal{C}\backslash\mathcal{K}$ and $v' = r_0$. The number of $\mathcal{K}$-structures is obviously the number of subtrees of this directed graph having root r_0. The corresponding determinant is easily evaluated by using elementary row and column operations. For example, given the situation in Figures 1 and 2, we have $\mathcal{C} = \mathcal{K} \cup \{R\}$, $\mathcal{K} = \{A, B, C\}$, $A = \{a_1, a_2, a_3, a_4, a_5\}$, $B = \{b_1, b_2, b_3\}$, and $C = \{c_1, c_2, c_3, c_4\}$. By Tutte's theorem, the number of $\mathcal{K}$-structures is

$$\det \left(\begin{array}{ccccc|ccc|cccc|c}
\multicolumn{5}{c}{A} & \multicolumn{3}{c}{B} & \multicolumn{4}{c}{C} & R \\
7 & 0 & 0 & 0 & 0 & -1 & -1 & -1 & -1 & -1 & -1 & -1 & 0 \\
0 & 7 & 0 & 0 & 0 & -1 & -1 & -1 & -1 & -1 & -1 & -1 & 0 \\
0 & 0 & 7 & 0 & 0 & -1 & -1 & -1 & -1 & -1 & -1 & -1 & 0 \\
0 & 0 & 0 & 7 & 0 & -1 & -1 & -1 & -1 & -1 & -1 & -1 & 0 \\
0 & 0 & 0 & 0 & 7 & -1 & -1 & -1 & -1 & -1 & -1 & -1 & 0 \\
\hline
-1 & -1 & -1 & -1 & -1 & 9 & 0 & 0 & -1 & -1 & -1 & -1 & 0 \\
-1 & -1 & -1 & -1 & -1 & 0 & 9 & 0 & -1 & -1 & -1 & -1 & 0 \\
-1 & -1 & -1 & -1 & -1 & 0 & 0 & 9 & -1 & -1 & -1 & -1 & 0 \\
\hline
0 & 0 & 0 & 0 & 0 & 0 & 0 & 0 & 4 & -1 & -1 & -1 & -1 \\
0 & 0 & 0 & 0 & 0 & 0 & 0 & 0 & -1 & 4 & -1 & -1 & -1 \\
0 & 0 & 0 & 0 & 0 & 0 & 0 & 0 & -1 & -1 & 4 & -1 & -1 \\
0 & 0 & 0 & 0 & 0 & 0 & 0 & 0 & -1 & -1 & -1 & 4 & -1 \\
\hline
0 & 0 & 0 & 0 & 0 & 0 & 0 & 0 & 0 & 0 & 0 & 0 & 1
\end{array} \right)$$

$$= \det \left(\begin{array}{ccccc|ccc|cccc|c}
7 & 0 & 0 & 0 & 0 & -1 & -1 & -1 & -1 & -1 & -1 & -1 & 0 \\
-7 & 7 & 0 & 0 & 0 & 0 & 0 & 0 & 0 & 0 & 0 & 0 & 0 \\
-7 & 0 & 7 & 0 & 0 & 0 & 0 & 0 & 0 & 0 & 0 & 0 & 0 \\
-7 & 0 & 0 & 7 & 0 & 0 & 0 & 0 & 0 & 0 & 0 & 0 & 0 \\
-7 & 0 & 0 & 0 & 7 & 0 & 0 & 0 & 0 & 0 & 0 & 0 & 0 \\
\hline
-1 & -1 & -1 & -1 & -1 & 9 & 0 & 0 & -1 & -1 & -1 & -1 & 0 \\
0 & 0 & 0 & 0 & 0 & -9 & 9 & 0 & 0 & 0 & 0 & 0 & 0 \\
0 & 0 & 0 & 0 & 0 & -9 & 0 & 9 & 0 & 0 & 0 & 0 & 0 \\
\hline
0 & 0 & 0 & 0 & 0 & 0 & 0 & 0 & 4 & -1 & -1 & -1 & -1 \\
0 & 0 & 0 & 0 & 0 & 0 & 0 & 0 & -5 & 5 & 0 & 0 & 0 \\
0 & 0 & 0 & 0 & 0 & 0 & 0 & 0 & -5 & 0 & 5 & 0 & 0 \\
0 & 0 & 0 & 0 & 0 & 0 & 0 & 0 & -5 & 0 & 0 & 5 & 0 \\
\hline
0 & 0 & 0 & 0 & 0 & 0 & 0 & 0 & 0 & 0 & 0 & 0 & 1
\end{array} \right)$$

$$= \det \left(\begin{array}{ccccc|ccc|cccc|c} 7 & 0 & 0 & 0 & 0 & -3 & -1 & -1 & -4 & -1 & -1 & -1 & 0 \\ 0 & 7 & 0 & 0 & 0 & 0 & 0 & 0 & 0 & 0 & 0 & 0 & 0 \\ 0 & 0 & 7 & 0 & 0 & 0 & 0 & 0 & 0 & 0 & 0 & 0 & 0 \\ 0 & 0 & 0 & 7 & 0 & 0 & 0 & 0 & 0 & 0 & 0 & 0 & 0 \\ 0 & 0 & 0 & 0 & 7 & 0 & 0 & 0 & 0 & 0 & 0 & 0 & 0 \\ \hline -5 & -1 & -1 & -1 & -1 & 9 & 0 & 0 & -4 & -1 & -1 & -1 & 0 \\ 0 & 0 & 0 & 0 & 0 & 0 & 9 & 0 & 0 & 0 & 0 & 0 & 0 \\ 0 & 0 & 0 & 0 & 0 & 0 & 0 & 9 & 0 & 0 & 0 & 0 & 0 \\ \hline 0 & 0 & 0 & 0 & 0 & 0 & 0 & 0 & 1 & -1 & -1 & -1 & -1 \\ 0 & 0 & 0 & 0 & 0 & 0 & 0 & 0 & 0 & 5 & 0 & 0 & 0 \\ 0 & 0 & 0 & 0 & 0 & 0 & 0 & 0 & 0 & 0 & 5 & 0 & 0 \\ 0 & 0 & 0 & 0 & 0 & 0 & 0 & 0 & 0 & 0 & 0 & 5 & 0 \\ \hline 0 & 0 & 0 & 0 & 0 & 0 & 0 & 0 & 0 & 0 & 0 & 0 & 1 \end{array}\right)$$

$$= (3+4)^{5-1}(5+4)^{3-1}(4+1)^{4-1} \det \begin{pmatrix} 7 & -3 & -4 \\ -5 & 9 & -4 \\ 0 & 0 & 1 \end{pmatrix}.$$

The formula in the theorem can also be obtained by means of multivariate generating functions and a generalization of Lagrange's inversion formula, as shown by Good [3, page 512; several misprints on that page need to be corrected].

Even though there are alternative ways to prove the theorem, the proof given above has several advantages, since it establishes a useful correspondence with sequences. The sequences make it possible to enumerate such oriented trees with a given number of vertices of in-degree 2, etc., as in Riordan [9], since the in-degree of each vertex is the number of times it appears in $(*)$.

As an example of the construction in the sequence-based proof, consider the tree in Figure 1 and suppose we order the colors

$$A < B < C < R.$$

Figure 1 is an $\{A, B, C\}$-structure. The construction selects $C_1 = A$, and we may take $a_1 < a_2 < a_3 < a_4 < a_5$ as an ordering of the elements of A; therefore $u_1 = a_5$. Since $f^2(a_5) \in A$ but $f^{3+k}(a_5)$ is not, we set $v_1 = f^3(a_5) = c_2$, $g(A) = C$, $C_2 = C$, and $u_2 = c_2$. The elements of C must be ordered so that c_2 is highest; therefore let $c_1 < c_3 < c_4 < c_2$. Now $f(c_2) = r$ and thus we let $v_2 = r$ and $g(C) = R$. Finally, we take $C_3 = B$ and $b_1 < b_2 < b_3 = u_3$. In this case, $f^2(b_3) = b_1 \in B$, but since $f(b_3)$ is in $C_1 = A$ we take $v_3 = a_3$, not $v_3 = c_2$; hence $g(B) = A$.

The construction of the lemma is now used, starting with a sequence for $C_3 = B$. Here, all vertices are free since V_3 is vacuous, and the sequence is simply $\big(f(b_1), f(b_2), f(b_3)\big)$, namely

$$B : (c_2, a_4, a_3).$$

The C_2 ($= C$) sequence is constructed next (remembering that $c_1 < c_3 < c_4 < c_2$); V_2 is vacuous and the sequence is $\big(f(c_3), f(c_4), f(c_1), f(c_2)\big)$, namely

$$C : (r, c_1, c_2, r).$$

Finally, the C_1 ($= A$) sequence is constructed, with $V_1 = \{b_2, b_3\}$:

$$A : (c_4, b_1, b_1, b_2, c_2).$$

The original tree is reconstructible from these three sequences. Conversely from any sequences of this type (that is, the A sequence contains five elements of B and C; the B sequence contains three elements of A and C, and if the last element is in A, the last element of A is not in B; and the C sequence contains four elements of C and R, the last in R) we can construct an oriented tree that will lead to these sequences.

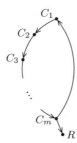

FIGURE 3. The cyclic case.

3. Examples and Applications

Consider a cyclic directed graph like that in Figure 3; for oriented trees suppose $|R| = 1$. The theorem tells us that the number of oriented trees in which all arcs go from color C_j to C_{j+1} or from C_m to C_1 or from C_m to R is

$$n_2^{n_1} n_3^{n_2} \ldots n_m^{n_{m-1}} (n_1 + 1)^{n_m - 1}; \quad n_j = |C_j|.$$

If we like, we may merge together colors R and C_1; then we find

$$n_2^{n_1 - 1} n_3^{n_2} \ldots n_m^{n_{m-1}} n_1^{n_m}.$$

is the number of oriented trees on $C_1 \cup C_2 \cup \cdots \cup C_m$ in which all arcs go from color C_j to $C_{(j \bmod m)+1}$ and the root is in C_1. The case $m = 1$ is Cayley's theorem; the case $m = 2$ was proved by Scoins [10] and it also follows from a more general result due to Austin [1].

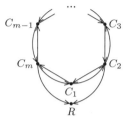
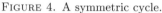

FIGURE 4. A symmetric cycle.

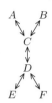

FIGURE 5. A free tree.

If the arcs are allowed to go in either direction between color C_j and color C_{j+1}, we get a situation like Figure 4. We may consider the vertex R of Figure 4 to be essentially a specified element of C_1, which has temporarily been given a new name. Since the arcs in the remaining graph are symmetric, we may consider free trees instead of oriented trees, namely, connected graphs without cycles: *The number of free trees on $C_1 \cup C_2 \cup \cdots \cup C_m$, with a vertex of color C_j adjacent to a vertex of color C_k only if $j \equiv (k \pm 1)$ (modulo m), is*

$$(n_m + n_2)^{n_1 - 1}(n_1 + n_3)^{n_2 - 1}(n_2 + n_4)^{n_3 - 1} \ldots (n_{m-1} + n_1)^{n_m - 1} n_1 n_2 \ldots n_m$$

$$\times \left(\frac{1}{n_m n_1} + \frac{1}{n_1 n_2} + \cdots + \frac{1}{n_{m-1} n_m} \right), \quad n_j = |C_j|, \quad m \geq 3.$$

In general, enumeration formulas for free trees can be obtained in this way whenever the graph of chromatic constraints has a symmetric incidence matrix. Another interesting case occurs when the directed graph $\mathcal{G}$ is itself a free tree with two-way arcs. Thus, for example, *the number of free trees on $A \cup B \cup C \cup D \cup E \cup F$, such that adjacent vertices have adjacent colors in the diagram of Figure 5, is*

$$|C|^{|A|-1}|C|^{|B|-1}|A \cup B \cup D|^{|C|-1}|C \cup E \cup F|^{|D|-1}|D|^{|E|-1}|D|^{|F|-1}|C|^2|D|^2.$$

In general, the number of such free trees is

$$\prod_{C \in \mathcal{C}} \left| \bigcup \{ C' \mid C' \text{ adjacent to } C \} \right|^{|C|-1} |C|^{\text{degree}(C)-1}$$

when the chromatic constraints themselves form a free tree.

The formulas above can also be used to derive nonobvious summation identities. Let $\mathcal{G}$ be a directed graph on $\{C_1, C_2, \ldots, C_m, R\}$ and let $p(n_1, n_2, \ldots, n_m, x)$ be the formula for the number of $(C_1, C_2, \ldots, C_m)$-structures according to the theorem in §2, where $n_k = |C_k|$ and $x = |R|$. Then we have the convolution formula

$$\sum_{k_1, k_2, \ldots, k_m} \binom{n_1}{k_1} \binom{n_2}{k_2} \cdots \binom{n_m}{k_m} p(k_1, k_2, \ldots, k_m, x)$$

$$\times\, p(n_1 - k_1, n_2 - k_2, \ldots, n_m - k_m, y)$$

$$= p(n_1, n_2, \ldots, n_m, x + y).$$

For in every T-graph $(C_1 \cup \cdots \cup C_m, C_1 \cup \cdots \cup C_m \cup R, f)$ we can partition the vertices v of $C_1 \cup \cdots \cup C_m$ according to the values of $f^\infty(v) \in R$; the left-hand side expresses the number of ways in which colors $C_1, \ldots,$ C_m can be split into $k_1, \ldots, k_m$ and $n_1 - k_1, \ldots, n_m - k_m$ respective elements so that the first group falls into x specified elements of R and the second group falls into the other y elements of R.

As an example, the simple graph

$$\circlearrowleft A \longrightarrow R$$

yields the identity

$$\sum_k \binom{n}{k} x(x+k)^{k-1} y(y+n-k)^{n-k-1} = (x+y)(x+y+n)^{n-1},$$

which is directly related to Abel's generalization of the binomial theorem (see [4]). From the slightly more complex graph

$$\circlearrowleft B \longleftrightarrow A \longrightarrow R$$

we get the following identity in integers m, n, x, and y:

$$\sum_{j,k} \binom{m}{j}\binom{n}{k} xj(x+k)^{j-1}(j+k)^{k-1}$$

$$\times\, y(m-j)(y+n-k)^{m-j-1}(m+n-j-k)^{n-k-1}$$

$$= (x+y)m(x+y+n)^{m-1}(m+n)^{n-1}.$$

(Suitable conventions are assumed when $0/0$ appears.) This identity appears to be very difficult to derive by any other means, and more complicated graphs will give still more intricate formulas of this type.

This research was supported in part by National Science Foundation grant GP-3909.

References

[1] T. L. Austin, "The enumeration of point labelled chromatic graphs and trees," *Canadian Journal of Mathematics* **12** (1960), 535–545.

[2] A. Cayley, "A theorem on trees," *Quarterly Journal of Pure and Applied Mathematics* **23** (1889), 376–378. Reprinted in his *Collected Mathematical Papers* **13**, 26–28.

[3] I. J. Good, "The generalization of Lagrange's expansion and the enumeration of trees," *Proceedings of the Cambridge Philosophical Society* **61** (1965), 499–517.

[4] H. W. Gould, "Note on problems 4960 and 4984," *American Mathematical Monthly* **69** (1962), 572.

[5] Donald E. Knuth, "Oriented subtrees of an arc digraph," *Journal of Combinatorial Theory* **3** (1967), 309–314. [Reprinted as Chapter 12 of the present volume.]

[6] J. W. Moon, "Various proofs of Cayley's formula for counting trees," in *A Seminar on Graph Theory*, edited by F. Harary (New York: Holt, Rinehart, and Winston, 1967), 70–78.

[7] H. Prüfer, "Neuer Beweis eines Satzes über Permutationen," *Archiv der Mathematik und Physik* (3) **27** (1918), 142–144.

[8] George N. Raney, "A formal solution of $\sum_{i=1}^{\infty} A_i \exp(B_i X) = X$," *Canadian Journal of Mathematics* **16** (1964), 755–762.

[9] John Riordan, "The enumeration of labeled trees by degrees," *Bulletin of the American Mathematical Society* **72** (1966), 110–112.

[10] H. I. Scoins, "The number of trees with nodes of alternate parity," *Proceedings of the Cambridge Philosophical Society* **58** (1962), 12–16.

[11] W. T. Tutte, "The dissection of equilateral triangles into equilateral triangles," *Proceedings of the Cambridge Philosophical Society* **44** (1948), 463–482.

Addendum

Soon after writing this paper I learned that I should really have attributed the formula n^{n-2} to Borchardt, as Cayley himself did. [C. W. Borchardt, "Ueber eine der Interpolation entsprechende Darstellung der Eliminations-Resultante," *Journal für die reine und angewandte Mathematik* **57** (1860), 111–121.]

Chapter 14

Abel Identities and Inverse Relations

[The organizers of a conference on combinatorial mathematics, held in April 1967 at the University of North Carolina, decided to have a rather unusual format in which "discussants" were invited to comment on each of the major papers that would be presented. Thus I was provided in advance with a copy of John Riordan's paper, "Abel identities and inverse relations," and asked to prepare a few remarks that would complement its main points. My remarks were subsequently published in the conference proceedings: Combinatorial Mathematics and Its Applications, edited by R. C. Bose and T. A. Dowling, The University of North Carolina Monograph Series in Probability and Statistics, Number 4 (Chapel Hill, North Carolina: University of North Carolina Press, 1969), 91–94.]

I would like to begin this discussion by making some presumptuous comments about notation, in the fond hope that the distinguished audience gathered here may help to overcome a rather unfortunate situation that exists today. Mr. Riordan has used the symbolism '$(x)_n$' to stand for the falling factorial power $x(x-1)\ldots(x-n+1)$. This is one of several commonly used notations for this function, which is also often written '$x^{(n)}$', or '$x_{(n)}$', or '$x^{[n]}$', etc.

The problem is that the same notations are also commonly used for the *rising* factorial powers $x(x+1)\ldots(x+n-1)$, especially in the literature concerning hypergeometric functions.

Since the history of mathematics clearly indicates the importance of good notations, I believe it necessary for me to speak up in behalf of a notational convention that offers a good solution to this dilemma: Let us agree to write

$$x^{\underline{n}} = x(x-1)\ldots(x-n+1),$$
$$x^{\overline{n}} = x(x+1)\ldots(x+n-1), \qquad \text{when } n \geq 0.$$

Such a symbolism appears to possess all of the properties of a good notation. For example, it is mnemonic, unambiguous, brief, and relatively

221

easy to set in type. Moreover, it strongly suggests analogies with the usual notation for non-factorial powers, as in the Abel-like sums

$$(x + y + na)^{\underline{n}} = \sum_k \binom{n}{k} x(x + ka - 1)^{\underline{k-1}}(y + (n - k)a)^{\underline{n-k}};$$

$$(x + y + na)^{\overline{n}} = \sum_k \binom{n}{k} x(x + ka + 1)^{\overline{k-1}}(y + (n - k)a)^{\overline{n-k}}.$$

Note that another notational convention, the use of "umbral variables," is the key reason why Mr. Riordan has been able to compile the formulas shown in Table 1 of his paper in such a suggestive form.

Abel published his short paper on the extended binomial formula in the first volume of Crelle's *Journal für die reine und angewandte Mathematik*, together with two of his most famous other memoirs dealing with the unsolvability of the quintic and with the beginnings of a rigorous theoretical treatment of infinite series, in connection with the convergence of $\sum \binom{n}{k} x^k$ when n is not an integer. In this connection it is interesting to raise a problem that is still apparently unsolved: For what complex values of x, y, and n does the sum $\sum_{k \geq 0} \binom{n}{k} x(x + k)^{k-1}(y + n - k)^{n-k}$ converge to $(x + y + n)^n$?

I believe Mr. Riordan's references to Adolf Hurwitz's paper give a misleading impression of the contents of that paper. The main point of Hurwitz's work, at least as I understand it, was not to give a multinomial generalization of Abel's identity; that follows easily from the binomial case, and it is discussed (in a much more general setting) only in the final paragraphs. His point was rather to exhibit and prove a much more interesting generalization of Abel's formula, namely

$$(x + y)(x + y + z_1 + z_2 + \cdots + z_n)^{n-1} \tag{1}$$
$$= \sum x(x + \epsilon_1 z_1 + \cdots + \epsilon_n z_n)^{\epsilon_1 + \cdots + \epsilon_n - 1}$$
$$\times y(y + (1 - \epsilon_1)z_1 + \cdots + (1 - \epsilon_n)z_n)^{(1 - \epsilon_1) + \cdots + (1 - \epsilon_n) - 1}$$

summed over all 2^n choices of $\epsilon_1, \ldots, \epsilon_n$ independently taking the values 0 and 1. This is an identity in $2n + 2$ variables, and Abel's binomial formula is the special case $z_1 = z_2 = \cdots = z_n$. Hurwitz proved this identity and some related results by using techniques quite similar to those of the present paper, inserting an arbitrary p and q in place of the -1's that occur in the exponents and deriving appropriate recurrence relations.

I have mentioned Hurwitz's general identity (1) here primarily because it has an interesting combinatorial significance. To see this connection, we need to know an auxiliary result about cycle-free directed graphs: Let $r \geq 1$ and $1 \leq p \leq q$, and consider a directed graph with $q + r$ vertices $U_1, \ldots, U_q, V_1, \ldots, V_r$, and with $q - p$ arcs from U_t to $V_{f(t)}$ for $p < t \leq q$, where f is any function from $\{p + 1, \ldots, q\}$ into $\{1, 2, \ldots, r\}$. The number of ways to add r additional arcs, from V_t to $U_{g(t)}$ for $1 \leq t \leq r$ where g takes $\{1, \ldots, r\}$ into $\{1, \ldots, q\}$, in such a way that the resulting directed graph contains no cycles, is exactly $q^{r-1}p$. In fact there is a one-to-one correspondence between such functions g and the sequences of r integers $(a_1, a_2, \ldots, a_r)$ where $1 \leq a_j \leq q$ and $a_r \leq p$; this correspondence may be obtained by generalizing a construction of Prüfer [2] in a straightforward manner.

Now consider a directed graph with $x + y + n + z_1 + z_2 + \cdots + z_n$ vertices $U_1, U_2, \ldots, U_{x+y}, V_1, V_2, \ldots, V_n$, and W_{ij} for $1 \leq i \leq n$, $1 \leq j \leq z_i$; let it have $z_1 + z_2 + \cdots + z_n$ arcs, from W_{ij} to V_i for $1 \leq i \leq n$ and $1 \leq j \leq z_i$. By the theorem just quoted, the number of ways to add n additional arcs, from V_j to one of the U's or W's for $1 \leq j \leq n$, without introducing a cycle, is exactly $(x + y + z_1 + z_2 + \cdots + z_n)^{n-1}(x + y)$.

In any graph that results from this construction, there is a unique path from each of the vertices $V_1, \ldots, V_n$ to one of the U's. If we set $\epsilon_j = 1$ when there is a path from V_j to U_k for some $k \leq x$, but $\epsilon_j = 0$ when the path from V_j goes to U_k for some $k > x$, the formula just derived yields Hurwitz's equation (1) whenever $x, y, z_1, \ldots, z_n$ are positive integers. Furthermore, both sides of (1) are polynomials, so the identity must hold in general.

Finally I would like to point out that the relations between trees and Abel's identities that appear in this paper have been discussed in a much more general setting in two important works by George N. Raney [3, 4].

References

[1] N. H. Abel, "Beweis eines Ausdruckes, von welchem die Binomial-Formel ein einzelner Fall ist," *Journal für die reine und angewandte Mathematik* **1** (1826), 159–160. French translation, "Démonstration d'une expression de laquelle la formule binôme est un cas particulier," in Abel's *Œuvres Complètes* **1** (Christiania: Grøndahl, 1881), 102–103.

[2] H. Prüfer, "Neuer Beweis eines Satzes über Permutationen," *Archiv der Mathematik und Physik* (3) **27** (1918), 142–144.

[3] George N. Raney, "Functional composition patterns and power series reversion," *Transactions of the American Mathematical Society* **94** (1960), 441–451.

[4] George N. Raney, "A formal solution of $\sum_{i=1}^{\infty} A_i \exp(B_i X) = X$," *Canadian Journal of Mathematics* **16** (1964), 755–762.

Addendum

Partial information about the apparent non-validity of Abel's binomial theorem when n is not a nonnegative integer can be found in the answer to exercise 1.2.6–52 in my book *Fundamental Algorithms*. But the general question stated above has not yet been resolved, as far as I know.

Chapter 15

Convolution Polynomials

The polynomials that arise as coefficients when a power series is raised to the power x include many important special cases, which have surprising properties that are not widely known. This paper explains how to recognize and use such properties, and it closes with a general result about approximating such polynomials asymptotically.

[Originally published in The Mathematica Journal **2**, 4 *(Fall 1992), 67–78.]*

A family of polynomials $F_0(x), F_1(x), F_2(x), \ldots$ forms a *convolution family* if $F_n(x)$ has degree $\leq n$ and if the convolution condition

$$F_n(x+y) = F_n(x)F_0(y) + F_{n-1}(x)F_1(y) + \cdots + F_1(x)F_{n-1}(y) + F_0(x)F_n(y)$$

holds for all x and y and for all $n \geq 0$. Many such families are known, and they appear frequently in applications. For example, we can let $F_n(x) = x^n/n!$; the condition

$$\frac{(x+y)^n}{n!} = \sum_{k=0}^{n} \frac{x^k}{k!} \frac{y^{n-k}}{(n-k)!}$$

is equivalent to the binomial theorem for integer exponents. Or we can let $F_n(x)$ be the binomial coefficient $\binom{x}{n}$; the corresponding identity

$$\binom{x+y}{n} = \sum_{k=0}^{n} \binom{x}{k}\binom{y}{n-k}$$

is commonly called *Vandermonde's convolution*.

225

How special is the convolution condition? *Mathematica* will readily find all sequences of polynomials that work for, say, $0 \leq n \leq 4$:

```
F[n_,x_] := Sum[f[n,j]x^j,{j,0,n}]/n!
conv[n_] := LogicalExpand[Series[F[n,x+y],{x,0,n},{y,0,n}]
     == Series[Sum[F[k,x]F[n-k,y],{k,0,n}],{x,0,n},{y,0,n}]]
Solve[Table[conv[n],{n,0,4}],
          Flatten[Table[f[i,j],{i,0,4},{j,0,4}]]]
```

Mathematica replies that the F's are either identically zero or that the coefficients of $F_n(x) = \left(f_{n0} + f_{n1}x + f_{n2}x^2 + \cdots + f_{nn}x^n\right)/n!$ satisfy

$$f_{00} = 1, \quad f_{10} = f_{20} = f_{30} = f_{40} = 0,$$

$$f_{22} = f_{11}^2, \quad f_{32} = 3f_{11}f_{21}, \quad f_{33} = f_{11}^3,$$

$$f_{42} = 4f_{11}f_{31} + 3f_{21}^2, \quad f_{43} = 6f_{11}^2 f_{21}, \quad f_{44} = f_{11}^4.$$

Thus we are allowed to choose f_{11}, f_{21}, f_{31}, and f_{41} freely.

Suppose we weaken the requirements by asking only that the convolution condition hold when $x = y$. The definition of conv then becomes simply

```
conv[n_] := LogicalExpand[Series[F[n,2x],{x,0,n}]
          == Series[Sum[F[k,x]F[n-k,x],{k,0,n}],{x,0,n}]]
```

and we discover that exactly the same solutions occur. In other words, the weaker requirements imply that the strong requirements are fulfilled as well, at least when $n \leq 4$.

In fact, it is not difficult to discover a simple rule that characterizes all "convolution families." *Let*

$$F(z) = 1 + F_1 z + F_2 z^2 + F_3 z^3 + \cdots$$

be any power series with $F(0) = 1$. Then the polynomials

$$F_n(x) = [z^n] F(z)^x$$

form a convolution family. Conversely, every convolution family arises in this way or is identically zero. (Here the notation '$[z^n]$ *expr*' stands for what *Mathematica* calls Coefficient[*expr*,z,n].)

Proof. Let $f(z) = \ln F(z) = f_1 z + f_2 z^2/2! + f_3 z^3/3! + \cdots$. It is easy to verify that the coefficient of z^n in $F(z)^x$ is indeed a polynomial in x of degree $\leq n$, because

$$F(z)^x = e^{xf(z)} = \exp(xf_1 z + xf_2 z^2/2! + xf_3 z^3/3! + \cdots)$$

expands to the power series

$$\sum_{k_1, k_2, k_3, \ldots \geq 0} x^{k_1 + k_2 + k_3 + \cdots} \frac{f_1^{k_1} f_2^{k_2} f_3^{k_3} \cdots}{1!^{k_1} k_1! \, 2!^{k_2} k_2! \, 3!^{k_3} k_3! \, \ldots} z^{k_1 + 2k_2 + 3k_3 + \cdots};$$

when $k_1 + 2k_2 + 3k_3 + \cdots = n$, the coefficient of z^n is a polynomial in x with terms of degree $k_1 + k_2 + k_3 + \cdots \leq n$. This construction produces a convolution family because of the rule for forming coefficients of the product $F(z)^{x+y} = F(z)^x F(z)^y$.

Conversely, suppose that the polynomials $F_n(x)$ form a convolution family. The condition $F_0(0) = F_0(0)^2$ can hold only if $F_0(x) = 0$ or $F_0(x) = 1$. In the former case it is easy to prove by induction that $F_n(x) = 0$ for all n. Otherwise, the condition $F_n(0) = 2F_n(0)$ for $n > 0$ implies that $F_n(0) = 0$ for $n > 0$. If we equate coefficients of x^k on both sides of

$$F_n(2x) = F_n(x)F_0(x) + F_{n-1}(x)F_1(x) + \cdots + F_1(x)F_{n-1}(x) + F_0(x)F_n(x),$$

we now find that the coefficient f_{nk} of x^k in $n! \, F_n(x)$ is forced to have certain values based on the coefficients of $F_1(x), \ldots, F_{n-1}(x)$, when $k > 1$, because $2^k f_{nk}$ occurs on the left and $2f_{nk}$ on the right. The coefficient f_{n1} can, however, be chosen freely. Any such choice must make $F_n(x) = [z^n] \exp(xf_{11} z + xf_{21} z^2/2! + xf_{31} z^3/3! + \cdots)$, by induction on n. $\square$

Examples

The first example mentioned above, $F_n(x) = x^n/n!$, comes from the power series $F(z) = e^z$; the second example, $F_n(x) = \binom{x}{n}$, comes from $F(z) = 1 + z$. Several other power series are also known to have simple coefficients when we raise them to the power x. If $F(z) = 1/(1-z)$, for instance, we find

$$[z^n] (1-z)^{-x} = \binom{-x}{n}(-1)^n = \binom{x+n-1}{n}.$$

It is convenient to use the notations

$$x^{\underline{n}} = x(x-1) \ldots (x-n+1) = x!/(x-n)!$$

$$x^{\overline{n}} = x(x+1) \ldots (x+n-1) = \Gamma(x+n)/\Gamma(x)$$

for falling factorial powers and rising factorial powers. Since $\binom{x}{n} = x^{\underline{n}}/n!$ and $\binom{x+n-1}{n} = x^{\overline{n}}/n!$, our last two examples have demonstrated that the polynomials $x^{\underline{n}}/n!$ and $x^{\overline{n}}/n!$ form convolution families, corresponding to $F(z) = 1 + z$ and $F(z) = 1/(1-z)$. Similarly, the polynomials

$$F_n(x) = \frac{x(x-s)(x-2s) \ldots (x-(n-1)s)}{n!}$$

form a convolution family corresponding to $(1 + sz)^{1/s}$ when $s \neq 0$.

The cases $F(z) = 1 + z$ and $F(z) = 1/(1-z)$ are in fact simply the cases $t = 0$ and $t = 1$ of a general formula for the binomial power series $\mathcal{B}_t(z)$, which satisfies

$$\mathcal{B}_t(z) = 1 + z\,\mathcal{B}_t(z)^t \, .$$

When t is any real or complex number, exponentiation of this series is known to yield

$$[z^n]\,\mathcal{B}_t(z)^x = \binom{x+tn}{n} \frac{x}{x+tn} = \frac{x(x+tn-1)\ldots(x+tn-n+1)}{n!} \, ;$$

see, for example, [10, §7.5, Example 5], where a combinatorial proof is given.

The special cases $t = 2$ and $t = -1$,

$$\mathcal{B}_2(z) = \frac{1 - \sqrt{1-4z}}{2z} = 1 + z + 2z^2 + 5z^3 + 14z^4 + 42z^5 + \cdots \, ,$$

$$\mathcal{B}_{-1}(z) = \frac{1 + \sqrt{1+4z}}{2} = 1 + z - z^2 + 2z^3 - 5z^4 + 14z^5 - \cdots \, ,$$

in which the coefficients are the Catalan numbers, arise in numerous applications. For example, $\mathcal{B}_2(z)$ is the generating function for binary trees, and $\mathcal{B}_{-1}(-z)$ is the reciprocal of $\mathcal{B}_2(z)$. We can get identities in trigonometry by noting that $\mathcal{B}_2\left((\frac{1}{2}\sin\theta)^2\right) = \sec^2(\theta/2)$. Furthermore, if p and q are probabilities with $p + q = 1$, it turns out that $\mathcal{B}_2(pq) = 1/\max(p, q)$. The case $t = 1/2$,

$$\mathcal{B}_{1/2}(z) = \left(\frac{z+\sqrt{4+z^2}}{2}\right)^2 = 1 + z + \frac{z^2}{2} + \frac{z^3}{2^3} - \frac{z^5}{2^7} + \frac{2z^7}{2^{11}} - \frac{5z^9}{2^{15}} + \frac{14z^{11}}{2^{19}} - \cdots,$$

is another interesting series in which the Catalan numbers can be seen. The convolution polynomials in this case are the "central factorials" $x(x + \frac{n}{2} - 1)^{\underline{n-1}}/n!$ (see Riordan [24, §6.5]), also called Steffensen polynomials [27, Example 6].

The convolution formula corresponding to $\mathcal{B}_t(z)$,

$$\binom{x + y + tn}{n} \frac{x + y}{x + y + tn}$$
$$= \sum_{k=0}^{n} \binom{x + tk}{k} \frac{x}{x + tk} \binom{y + t(n - k)}{n - k} \frac{y}{y + t(n - k)}$$

is a rather startling generalization of Vandermonde's convolution; it is an identity for all x, y, t, and n.

The limit of $\mathcal{B}_t(z/t)^t$ as $t \to \infty$ is another important function $T(z)/z$; here

$$T(z) = \sum_{n \geq 1} \frac{n^{n-1}}{n!} z^n = z + z^2 + \frac{3z^3}{2} + \frac{8z^4}{3} + \frac{125z^5}{24} + \cdots$$

is called the *tree function* because n^{n-1} is the number of labeled, rooted trees. The tree function satisfies

$$T(z) = ze^{T(z)},$$

and we have the corresponding convolution family

$$[z^n]\left(\frac{T(z)}{z}\right)^x = [z^n]e^{xT(z)} = \frac{x(x + n)^{n-1}}{n!}.$$

The related power series

$$1 + zT'(z) = \frac{1}{1 - T(z)} = \sum_{n \geq 0} \frac{n^n z^n}{n!}$$

$$= 1 + z + 2z^2 + \frac{9z^3}{2} + \frac{32z^4}{3} + \frac{625z^5}{24} + \cdots$$

defines yet another convolution family of importance: We have

$$[z^n] \frac{1}{\left(1 - T(z)\right)^x} = \frac{t_n(x)}{n!},$$

where $t_n(x)$ is called the *tree polynomial* of order n [18]. The coefficients of $t_n(x) = t_{n1}x + t_{n2}x^2 + \cdots + t_{nn}x^n$ are integers with combinatorial significance; namely, t_{nk} is the number of mappings of an n-element set into itself having exactly k cycles.

A similar but simpler sequence arises from the coefficients of powers of e^{ze^z}:

$$n! \, [z^n] \, e^{xze^z} = \sum_{k=0}^{n} \binom{n}{k} k^{n-k} x^k \, .$$

The coefficient of x^k is the number of *idempotent* mappings of an n-element set into itself, having exactly k cycles [11].

If the reader still isn't convinced that convolution families are worthy of detailed study, well, there's not much hope, although another example or two might clinch the argument. Consider the power series

$$e^{e^z - 1} = \sum \frac{\varpi_n z^n}{n!} = 1 + \frac{z}{1!} + \frac{2z^2}{2!} + \frac{5z^3}{3!} + \frac{15z^4}{4!} + \frac{52z^5}{5!} + \cdots \, ;$$

these coefficients ϖ_n are the so-called *Bell numbers*, the number of ways to partition sets of size n into subsets. For example, the five partitions that make $\varpi_3 = 5$ are

$$\{1, 2, 3\}, \quad \{1\}\{2, 3\}, \quad \{1, 2\}\{3\}, \quad \{1, 3\}\{2\}, \quad \{1\}\{2\}\{3\} \, .$$

The corresponding convolution family is

$$[z^n] \, e^{(e^z - 1)x} = \frac{\left\{{n \atop 0}\right\} + \left\{{n \atop 1}\right\}x + \left\{{n \atop 2}\right\}x^2 + \cdots + \left\{{n \atop n}\right\}x^n}{n!},$$

where the Stirling number $\left\{{n \atop k}\right\}$ is the number of partitions into exactly k subsets.

Need more examples? If the coefficients of $F(z)$ are arbitrary nonnegative numbers with a finite sum S, then $F(z)/S$ defines a discrete probability distribution, and the convolution polynomial $F_n(x)$ is S^x times the probability of obtaining the value n as the sum of x independent random variables having that distribution.

A Derived Convolution

Every convolution family $\{F_n(x)\}$ satisfies another general convolution formula in addition to the one we began with:

$$(x + y) \sum_{k=0}^{n} k\, F_k(x)\, F_{n-k}(y) = x\, n\, F_n(x + y)\,.$$

For example, if $F_n(x)$ is the convolution family corresponding to powers of $\mathcal{B}_t(z)$, this formula says that

$$(x + y) \sum_{k=0}^{n} k \binom{x + tk}{k} \frac{x}{x + tk} \binom{y + t(n - k)}{n - k} \frac{y}{y + t(n - k)}$$

$$= xn \binom{x + y + tn}{n} \frac{x + y}{x + y + tn}\,;$$

it looks messy, but it simplifies to another amazing identity in four parameters,

$$\sum_{k=0}^{n} \binom{x + t(n - k)}{n - k} \binom{y + tk}{k} \frac{y}{y + tk} = \binom{x + y + tn}{n}$$

if we replace n by $n+1$, k by $n+1-k$, and x by $x-t+1$. This identity has an interesting history going back to Rothe in 1793 (see [9]).

The alternative convolution formula is proved by differentiating the basic identity $F(z)^x = \sum_{n \geq 0} F_n(x) z^n$ with respect to z and multiplying by z:

$$xz F'(z)\, F(z)^{x-1} = \sum_{n \geq 0} n\, F_n(x)\, z^n\,.$$

Now $\sum_{k=0}^{n} k\, F_k(x)\, F_{n-k}(y)$ is the coefficient of z^n in $xz F'(z)\, F(z)^{x+y-1}$, while $n F_n(x + y)$ is the coefficient of z^n in $(x + y)z F'(z)\, F(z)^{x+y-1}$. Q.E.D. ☐

Convolution and Composition

Once upon a time I was trying to remember the form of a general convolution family, so I gave *Mathematica* the following command:

```
Simplify[Series[(1+Sum[A[k]z^k,{k,4}])^x,{z,0,4}]]
```

The result was a surprise. Instead of presenting the coefficient of z^n as a polynomial in x, *Mathematica* chose another form: The coefficient of z^2, for example, was $\frac{1}{2}A_1^2 x(x-1) + A_2 x$. In the notation of falling factorial powers, *Mathematica*'s response took the form

$$1 + A_1 xz + \left(\tfrac{1}{2}A_1^2 x^{\underline{2}} + A_2 x\right) z^2 + \left(\tfrac{1}{6}A_1^3 x^{\underline{3}} + A_1 A_2 x^{\underline{2}} + A_3 x\right) z^3$$
$$+ \left(\tfrac{1}{24}A_1^4 x^{\underline{4}} + \tfrac{1}{2}A_1^2 A_2 x^{\underline{3}} + \left(A_1 A_3 + \tfrac{1}{2}A_2^2\right) x^{\underline{2}} + A_4 x\right) z^4 + O(z)^5 .$$

I wasn't prepared to work with factorial powers, so I tried another tack:

```
Simplify[Series[Exp[Sum[a[k]z^k,{k,4}]x],{z,0,4}]]
```

This time I got ordinary polynomials in x, but — lo and behold — they were

$$1 + a_1 xz + \left(\tfrac{1}{2}a_1^2 x^2 + a_2 x\right) z^2 + \left(\tfrac{1}{6}a_1^3 x^3 + a_1 a_2 x^2 + a_3 x\right) z^3$$
$$+ \left(\tfrac{1}{24}a_1^4 x^4 + \tfrac{1}{2}a_1^2 a_2 x^3 + \left(a_1 a_3 + \tfrac{1}{2}a_2^2\right) x^2 + a_4 x\right) z^4 + O(z)^5 .$$

The result was exactly the same as before, but with a's in place of A's, and with normal powers in place of the factorials!

So I learned a curious phenomenon: *If we take any convolution family and replace each power x^k by $x^{\underline{k}}$, we get another convolution family.* (By the way, the replacement can be done in *Mathematica* by saying

```
Expand[F[n,x]]/.Power[x,k_]->k!Binomial[x,k];
```

expansion is necessary in case $F_n(x)$ has been factored.)

The proof was not difficult to find, once I psyched out how *Mathematica* might have come up with its factorial-based formula: We have

$$e^{xf(z)} = 1 + f(z)\,x + \frac{f(z)^2}{2!}\,x^2 + \frac{f(z)^3}{3!}\,x^3 + \cdots ,$$

and furthermore

$$\left(1 + f(z)\right)^x = 1 + f(z)\,x + \frac{f(z)^2}{2!}\,x^{\underline{2}} + \frac{f(z)^3}{3!}\,x^{\underline{3}} + \cdots .$$

Therefore if we start with the convolution family $F_n(x)$ corresponding to $F(z) = e^{f(z)}$, and replace each x^k by $x^{\underline{k}}$, we get the convolution family corresponding to $1 + f(z) = 1 + \ln F(z)$.

A similar derivation shows that if we replace x^k by the *rising* factorial power $x^{\overline{k}}$ instead, we get the convolution family corresponding to

$1/\big(1 - f(z)\big) = 1/\big(1 - \ln F(z)\big)$. In particular, if we begin with the family $F_n(x) = x(x+n)^{n-1}/n!$ corresponding to $T(z)/z = e^{T(z)}$, and if we replace x^k by $x^{\overline{k}}$ to get

$$\frac{1}{n!} \sum_{k=0}^{n-1} \binom{n-1}{k} x^{\overline{k+1}} n^{n-1-k} \, ,$$

this must be $[z^n]\big(1 - T(z)\big)^{-x} = t_n(x)/n!$, the tree polynomial.

Indeed, we can replace each x^k by $k!\,G_k(x)$, where $\{G_k(x)\}$ is *any* convolution family whatever! The previous examples, $x^{\underline{k}}$ and $x^{\overline{k}}$, are merely the special cases $k!\binom{x}{k}$ and $k!\binom{x+k-1}{k}$ corresponding to two of the simplest and most basic families we have considered. In general we get

$$1 + f(z)\,G_1(x) + \frac{f(z)^2}{2!}\,2!\,G_2(x) + \frac{f(z)^3}{3!}\,3!\,G_3(x) + \cdots \, ,$$

which is none other than $G\big(f(z)\big)^x = G\big(\ln F(z)\big)^x$.

For example,

$$G_k(x) = \binom{x+2k}{k}\frac{x}{x+2k} = x(x+2k-1)^{\underline{k-1}}/k!$$

is the family corresponding to $\mathcal{B}_2(z)$. If we know the family $F_n(x)$ corresponding to $e^{f(z)}$ we can replace x^k by $x(x+2k-1)^{\underline{k-1}}$, thereby obtaining the family that corresponds to

$$\mathcal{B}_2\big(f(z)\big) = \big(1 + \sqrt{1 - 4f(z)}\,\big)/\big(2f(z)\big).$$

Convolution Matrices

I knew that such remarkable facts must have been discovered before, although they were new to me at the time. And indeed, it was not difficult to find them in books, once I knew what to look for. (Special cases of general theorems are not always easy to recognize, because any particular formula is a special case of infinitely many generalizations, almost all of which are false.)

In the special case that each polynomial $F_n(x)$ has degree exactly n, namely when $f_1 \neq 0$, the polynomials $n!\,F_n(x)$ are said to be of *binomial type* [22]. An extensive theory of such polynomial sequences has been developed [28, 6, 27], based on the theory of linear operators, and the reader will find it quite interesting to compare the instructive treatment in those papers to the related but rather different directions explored in the present work. A comprehensive exposition of the operator approach appears in [26]. Actually, Steffensen had defined a concept called *poweroids*, many years earlier [29], and poweroids are almost exactly the same as sequences of binomial type; but Steffensen apparently did not realize that his poweroids satisfy the convolution property, which we can readily deduce (with hindsight) from equations (6) and (7) of his paper.

Eri Jabotinsky [13] introduced a nice way to understand the phenomena of convolution polynomials, by considering the infinite matrix of coefficients f_{nk}. Let us recapitulate the notation that was introduced informally above:

$$e^{xf(z)} = F(z)^x = 1 + F_1(x)\,z + F_2(x)\,z^2 + \cdots\,;$$

$$F_n(x) = (f_{n1}x + f_{n2}x^2 + \cdots + f_{nn}x^n)/n!\,;$$

$$f(z) = f_1 z + f_2 z^2/2! + f_3 z^3/3! + \cdots\,.$$

Then Jabotinsky's matrix $F = (f_{nk})$ is a lower triangular matrix containing the coefficients of $n!\,F_n(x)$ in the nth row. The first few rows are

$$
\begin{array}{llll}
f_1 & & & \\
f_2 & f_1^2 & & \\
f_3 & 3f_1 f_2 & f_1^3 & \\
f_4 & 4f_1 f_3 + 3f_2^2 & 6f_1^2 f_2 & f_1^4\,,
\end{array}
$$

as we saw earlier. In general,

$$f_{nk} = \sum \frac{n!}{1!^{k_1}\,k_1!\,2!^{k_2}\,k_2!\,3!^{k_3}\,k_3!\,\ldots}\,f_1^{k_1} f_2^{k_2} f_3^{k_3} \cdots\,,$$

summed over all $k_1, k_2, k_3, \ldots \geq 0$ with

$$k_1 + k_2 + k_3 + \cdots = k\,, \qquad k_1 + 2k_2 + 3k_3 + \cdots = n\,.$$

(The summation is over all partitions of the integer n into k parts, where k_j of the parts are equal to j.) The matrix entry f_{nk} is often called a "partial Bell polynomial" in the variables f_j (see [2]). We will call such an array a *convolution matrix*.

If each original coefficient f_j is an integer, all entries of the corresponding convolution matrix will be integers, because the complicated quotient of factorials in the sum is an integer — it is the number of ways to partition a set of n elements into k subsets with exactly k_j of the subsets having size j. Given the first column we can compute the other columns from left to right and from top to bottom by using the recurrence

$$f_{nk} = \sum_{j=1}^{n-k+1} \binom{n-1}{j-1} f_j\, f_{(n-j)(k-1)}.$$

This recurrence is based on set partitions in which the element n occurs in a subset of size j: There are $\binom{n-1}{j-1}$ ways to choose the other $j-1$ elements of the subset, and the factor $f_{(n-j)(k-1)}$ corresponds to partitioning the remaining $n-j$ elements into $k-1$ parts.

For example, if each $f_j = 1$, the convolution matrix begins

$$
\begin{array}{ccccc}
1 \\
1 & 1 \\
1 & 3 & 1 \\
1 & 7 & 6 & 1 \\
1 & 15 & 25 & 10 & 1
\end{array}
$$

These are the numbers $\{{n \atop k}\}$ that *Mathematica* calls StirlingS2[n,k]; they arose in our example of Bell numbers when $f(z) = e^z - 1$. Similarly, if each $f_j = (j-1)!$, the first five rows are

$$
\begin{array}{ccccc}
1 \\
1 & 1 \\
2 & 3 & 1 \\
6 & 11 & 6 & 1 \\
24 & 50 & 35 & 10 & 1
\end{array} \quad ;
$$

Mathematica calls these numbers (-1)^(n-k)StirlingS1[n,k]. In this case $f(z) = \ln\big(1/(1-z)\big)$, and $F_n(x) = \binom{x+n-1}{n}$. The signed numbers StirlingS1[n,k],

$$
\begin{array}{ccccc}
1 \\
-1 & 1 \\
2 & -3 & 1 \\
-6 & 11 & -6 & 1 \\
24 & -50 & 35 & -10 & 1
\end{array}
$$

correspond to $f(z) = \ln(1+z)$ and $F_n(x) = \binom{x}{n}$. In general if we replace z by αz and x by βx, the effect is to multiply row n of the matrix by α^n and to multiply column k by β^k. Thus when $\beta = \alpha^{-1}$, the net effect is to multiply f_{nk} by α^{n-k}. Transforming the signs by a factor $(-1)^{n-k}$ corresponds to changing $F(z)$ to $1/F(-z)$ and $f(z)$ to $-f(-z)$. Therefore the matrix that begins

$$
\begin{array}{rrrrr}
1 & & & & \\
-1 & 1 & & & \\
1 & -3 & 1 & & \\
-1 & 7 & -6 & 1 & \\
1 & -15 & 25 & -10 & 1
\end{array}
$$

corresponds to $f(z) = 1 - e^{-z}$.

Let's look briefly at some of our other examples in matrix form. When $F(z) = \mathcal{B}_t(z)$, we have $f_j = (tj - 1)^{\underline{j-1}}$, which is an integer when t is an integer. In particular, the Catalan case $t = 2$ produces a matrix that begins

$$
\begin{array}{rrrrr}
1 & & & & \\
3 & 1 & & & \\
20 & 9 & 1 & & \\
210 & 107 & 18 & 1 & \\
3024 & 1650 & 335 & 30 & 1
\end{array} \quad .
$$

When $t = 1/2$, we can remain in an all-integer realm by replacing z by $2z$ and x by $x/2$. Then $f_j = 0$ when j is even, while $f_{2j+1} = (-1)^j (2j-1)!!^2$:

$$
\begin{array}{rrrrr}
1 & & & & \\
0 & 1 & & & \\
-1 & 0 & 1 & & \\
0 & -4 & 0 & 1 & \\
9 & 0 & -10 & 0 & 1
\end{array} \quad .
$$

If we now replace z by iz and x by x/i to eliminate the minus signs, we find that $f(z) = \arcsin z$, because $\ln\left(iz + \sqrt{1 - z^2}\,\right) = i\theta$ when $z = \sin\theta$. Thus we can deduce a closed form for the coefficients of $e^{x \arcsin z} = \mathcal{B}_{1/2}(2iz)^{x/(2i)}$:

$$
n!\,[z^n]\,e^{x \arcsin z} = (2i)^{n-1} x \left(\frac{x}{2i} + \frac{n}{2} - 1\right) \cdots \left(\frac{x}{2i} - \frac{n}{2} + 1\right)
$$

$$
= \begin{cases}
x^2(x^2 + 2^2) \cdots \left(x^2 + (n-2)^2\right), & n \text{ even;} \\
x(x^2 + 1^2)(x^2 + 3^2) \cdots \left(x^2 + (n-2)^2\right), & n \text{ odd.}
\end{cases}
$$

This remarkable formula is equivalent to the theorem of Gomes Teixeira [7].

If $f_j = 2^{1-j}$ when j is odd but $f_j = 0$ when j is even, we get the convolution matrix corresponding to $e^{2x\sinh(z/2)}$:

$$
\begin{array}{ccccccccc}
1 \\
0 & 1 \\
\frac{1}{4} & 0 & 1 \\
0 & 1 & 0 & 1 \\
\frac{1}{16} & 0 & \frac{5}{2} & 0 & 1 \\
0 & 1 & 0 & 5 & 0 & 1 \\
\frac{1}{64} & 0 & \frac{91}{16} & 0 & \frac{35}{4} & 0 & 1 \\
0 & 1 & 0 & 21 & 0 & 14 & 0 & 1
\end{array}\,.
$$

Again we could stay in an all-integer realm if we replaced z by $2z$ and x by $x/2$; but the surprising thing in this case is that the entries in even-numbered rows and columns are all integers *before* we make any such replacement. The reason is that the entries satisfy $f_{nk} = k^2 f_{(n-2)k}/4 + f_{(n-2)(k-2)}$. (See Riordan [24, pages 213–217], where the notation $T(n,k)$ is used for these "central factorial numbers" f_{nk} of the second kind.)

We can complete our listing of noteworthy examples by setting $f_j = \sum_{k=1}^{n} n^{n-k-1} n^{\underline{k}}$; then we get the coefficients of the tree polynomials:

$$
\begin{array}{ccccc}
1 \\
3 & 1 \\
17 & 9 & 1 \\
142 & 95 & 18 & 1 \\
1569 & 1220 & 305 & 30 & 1
\end{array}\,.
$$

The sum of the entries in row n is n^n.

Composition and Iteration

Jabotinsky's main reason for defining things as he did was his observation that *the product of convolution matrices is a convolution matrix.* Indeed, if F and G are the convolution matrices corresponding to the functions $e^{xf(z)}$ and $e^{xg(z)}$ we have the vector/matrix identities

$$
e^{xf(z)} - 1 = (z, z^2/2!, z^3/3!, \dots)\, F\, (x, x^2, x^3, \dots)^T;
$$

$$
e^{xg(z)} - 1 = (z, z^2/2!, z^3/3!, \dots)\, G\, (x, x^2, x^3, \dots)^T.
$$

If we now replace x^k in $e^{xf(z)}$ by $k!\,G_k(x)$, as in our earlier discussion, we get

$$(z, z^2/2!, z^3/3!, \dots)\,F\,\big(G_1(x),\,2!\,G_2(x),\,3!\,G_3(x),\,\dots\big)^T$$

$$= (z, z^2/2!, z^3/3!, \dots)\,FG\,(x, x^2, x^3, \dots)^T$$

$$= \big(f(z), f(z)^2/2!, f(z)^3/3!, \dots\big)\,G\,(x, x^2, x^3, \dots)^T$$

$$= e^{xg(f(z))} - 1.$$

Multiplication of convolution matrices corresponds to composition of the functions in the exponent.

Why did the function corresponding to FG turn out to be $g\big(f(z)\big)$ instead of $f\big(g(z)\big)$? Jabotinsky, in fact, defined his matrices as the transposes of those given here. The rows of his (upper triangular) matrices were the power series $f(z)^k$, while the columns were the polynomials $F_n(x) = [z^n]\,e^{xf(z)}$; with those conventions the product of his matrices $F^T G^T$ corresponded to $f\big(g(z)\big)$. (In fact, he defined a considerably more general representation, in which the matrix F could be $U^{-1}FU$ for any nonsingular matrix U.) However, when our interest is focused on the polynomials $n!\,F_n(x)$, as when we study Stirling numbers or tree polynomials or the Stirling polynomials to be discussed below, it is more natural to work with lower triangular matrices and to insert factorial coefficients, as Comtet did [5, §3.7]. The two conventions are isomorphic. Without the factorials, convolution matrices are sometimes called *renewal arrays* [25]. We would get a non-reversed order if we had been accustomed to using postfix notation $(z)f$ for functions, as we do for operations such as squaring or taking transposes or factorials; then $g\big(f(z)\big)$ would be $\big((z)f\big)g$.

Recall that the Stirling numbers $\left\{{n\atop k}\right\}$ correspond to $f(z) = e^z - 1$, and the other Stirling numbers $\left[{n\atop k}\right]$ correspond to $g(z) = \ln\big(1/(1-z)\big)$. Therefore the product of Stirling's triangles gives us the convolution matrix

$$FG = \begin{matrix}
1 & & & & \\
2 & 1 & & & \\
6 & 6 & 1 & & \\
26 & 36 & 12 & 1 & \\
150 & 250 & 120 & 20 & 1
\end{matrix}\,,$$

which corresponds to $g\big(f(z)\big) = \ln\big(1/(2 - e^z)\big)$. Voila! These convolution polynomials represent the coefficients of $(2 - e^z)^{-x}$. Cayley [4] showed that $(2 - e^z)^{-1}$ is the exponential generating function for the sequence 1, 3, 13, 75, 541, ..., which counts *preferential arrangements* of

n objects, that is, different outcomes of sorting when equality is possible as well as inequality. The kth element of row n in FG is the number of preferential arrangements in which the "current minimum" changes k times when we examine the elements one by one in some fixed order. (See [10, exercise 7.44].)

Similarly, the reverse matrix product of Stirling's triangles yields the so-called Lah numbers [20],

$$
GF = \begin{matrix}
1 \\
2 & 1 \\
6 & 6 & 1 \\
24 & 36 & 12 & 1 \\
120 & 240 & 120 & 20 & 1
\end{matrix} \quad ;
$$

here $f_j = j!$ and the rows represent the coefficients of $\exp\bigl(xf(g(z))\bigr) = \exp(xz + xz^2 + xz^3 + \cdots)$. Indeed, the convolution polynomials in this case are the generalized Laguerre polynomials

$$
L_n^{(-1)}(-x),
$$

which *Mathematica* calls LaguerreL[n,-1,-x]. These polynomials can also be expressed as $L_n(-x) - L_{n-1}(-x)$; or as LaguerreL[n,-x]-LaguerreL[n-1,-x] if we say

    Unprotect[LaguerreL];
    LaguerreL[-1,x_]:=0; Protect[LaguerreL]

first. The row sums $1, 3, 13, 73, 501, \ldots$ of GF enumerate what Motzkin [21] called "sets of lists"; the coefficients are

$$
(GF)_{nk} = n!\,[z^n]\,f\bigl(g(z)\bigr)^k/k! = \binom{n}{k}\binom{n-1}{k-1}(n-k)!
$$

(see [24, exercise 5.7]).

Since convolution matrices are closed under multiplication, they are also closed under exponentiation, that is, under taking of powers. The qth power F^q of a convolution matrix then corresponds to q-fold *iteration* of the function $\ln F = f$. Let us denote $f\bigl(f(z)\bigr)$ by $f^{[2]}(z)$; in general, the qth iterate $f^{[q]}(z)$ is defined to be $f\bigl(f^{[q-1]}(z)\bigr)$, where $f^{[0]}(z) = z$. This is *Mathematica*'s Nest[f,z,q].

The qth iterate can be obtained by doing $O(\log q)$ matrix multiplications, but in the interesting case $f'(0) = f_1 = 1$ we can also compute the coefficients of $f^{[q]}(z)$ by using formulas in which q is simply a numerical parameter. Namely, as suggested by Jabotinsky [13], we can express the matrix power F^q as

$$\left(I + (F-I)\right)^q = I + \binom{q}{1}(F-I) + \binom{q}{2}(F-I)^2 + \binom{q}{3}(F-I)^3 + \cdots .$$

This infinite series converges, because the entry in row n and column k of $(F-I)^j$ is zero for all $j > n - k$. When q is any positive integer, the result defined in this way is a convolution matrix. Furthermore, the matrix entries are all polynomials in q. Therefore the matrix obtained by this infinite series is a convolution matrix for *all* values of q.

Jabotinsky presented another formula for the entries of F^q in [15]. Let $f_{nk}^{(q)}$ be the element in row n and column k of F^q; then

$$f_{nk}^{(q)} = \sum_{l=0}^{m} \binom{q}{l} (F - I)_{nk}^{l}$$

$$= \sum_{j=0}^{m} f_{nk}^{(j)} \sum_{l=j}^{m} \binom{q}{l}\binom{l}{j}(-1)^{l-j}$$

$$= \sum_{j=0}^{m} f_{nk}^{(j)} \binom{q}{j} \sum_{l=j}^{m} \binom{q-j}{l-j}(-1)^{l-j}$$

$$= \sum_{j=0}^{m} f_{nk}^{(j)} \binom{q}{j} \binom{q-j-1}{m-j}(-1)^{m-j} ,$$

for any $m \geq n-k$. Indeed, we have $p(q) = \sum_{j=0}^{m} p(j)\binom{q}{j}\binom{q-j-1}{m-j}(-1)^{m-j}$ whenever p is a polynomial of degree $\leq m$; this is a special case of Lagrange interpolation.

It is interesting to set $q = 1/2$ and compute convolution square roots of the Stirling number matrices. We have

$$\begin{pmatrix} 1 & & & & \\ 1/2 & 1 & & & \\ 1/8 & 3/2 & 1 & & \\ 0 & 5/4 & 3 & 1 & \\ 1/32 & 5/8 & 5 & 5 & 1 \end{pmatrix}^2 = \begin{pmatrix} 1 & & & & \\ 1 & 1 & & & \\ 1 & 3 & 1 & & \\ 1 & 7 & 6 & 1 & \\ 1 & 15 & 25 & 10 & 1 \end{pmatrix} ;$$

$$
\begin{pmatrix}
1 & & & & \\
1/2 & 1 & & & \\
5/8 & 3/2 & 1 & & \\
5/4 & 13/4 & 3 & 1 & \\
109/32 & 75/8 & 10 & 5 & 1
\end{pmatrix}^2
=
\begin{pmatrix}
1 & & & & \\
1 & 1 & & & \\
2 & 3 & 1 & & \\
6 & 11 & 6 & 1 & \\
24 & 50 & 35 & 10 & 1
\end{pmatrix}.
$$

The function $z + z^2/4 + z^3/48 + z^5/3840 - 7z^6/92160 + \cdots$ therefore lies "halfway" between z and $e^z - 1 = z + z^2/2! + z^3/3! + \cdots$, and the function $z + z^2/4 + 5z^3/48 + 5z^4/96 + 109z^5/3840 + 497z^6/30720 + \cdots$ lies halfway between z and $\ln(1/(1-z)) = z + z^2/2 + z^3/3 + \cdots$. These half-iterates are unfamiliar functions; but it is not difficult to prove that $z/(1 - z/2) = z + z^2/2 + z^3/4 + \cdots$ is halfway between z and $z/(1-z) = z + z^2 + z^3 + \cdots$. In general when $f(z) = z/(1 - cz^k)^{1/k}$ we have $f^{[q]}(z) = z/(1 - qcz^k)^{1/k}$.

It seems natural to conjecture that the coefficients of $f^{[q]}(z)$ are positive for $q > 0$ when $f(z) = \ln(1/(1-z))$; but this conjecture turns out to be false, because *Mathematica* reports that $[z^8] f^{[q]}(z) = -11q/241920 + O(q^2)$. Is there a simple necessary and sufficient condition on f that characterizes when all coefficients of $f^{[q]}$ are nonnegative for nonnegative q? This will happen if and only if the entries in the first column of

$$
\ln F = (F - I) - \tfrac{1}{2}(F - I)^2 + \tfrac{1}{3}(F - I)^3 - \cdots
$$

are nonnegative. (See Kuczma [19] for iteration theory and an extensive bibliography.)

Reversion

The case $q = -1$ of iteration is often called reversion of series, although *Mathematica* uses the more proper name `InverseSeries`. Given $f(z) = f_1 z + f_2 z^2/2! + \cdots$, we seek $g(z) = f^{[-1]}(z)$ such that $g(f(z)) = z$. This is clearly equivalent to finding the first column of the inverse of the convolution matrix.

The inverse does not exist when $f_1 = 0$, because the diagonal of F is zero in that case. Otherwise we can assume that $f_1 = 1$, because $h(z) = g(z/f_1)$ reverts the power series $f(z) = f_1 z + \cdots$ when $g(z)$ reverts $f(z)/f_1$.

When $f_1 = 1$ we can obtain the inverse by setting $q = -1$ in our general formula for iteration. But Lagrange's celebrated inversion theorem for power series tells us that there is another, more informative, way to compute the function $g = f^{[-1]}$. Let us set $\widehat{F}(z) = f(z)/z =$

$1 + f_2 z/2! + f_3 z^2/3! + \cdots$. Then Lagrange's theorem states that the elements of the matrix $G = F^{-1}$ are

$$g_{nk} = \frac{(n-1)!}{(k-1)!} \, \widehat{F}_{n-k}(-n) \,,$$

where $\widehat{F}_n(x)$ denotes the convolution family corresponding to $\widehat{F}(z)$.

There is a surprisingly simple way to prove Lagrange's theorem, using our knowledge of convolution families. Note first that

$$f_{nk} = n! \, [z^n x^k] \, e^{x f(z)} = \frac{n!}{k!} \, [z^n] \, f(z)^k = \frac{n!}{k!} \, [z^{n-k}] \, \widehat{F}(z)^k \,;$$

therefore

$$f_{nk} = \frac{n!}{k!} \, \widehat{F}_{n-k}(k) \,.$$

Now we need only verify that the matrix product GF is the identity, by computing its element in row n and column m:

$$\sum_{k=m}^{n} g_{nk} f_{km} = \sum_{k=m}^{n} \frac{(n-1)!}{(k-1)!} \, \widehat{F}_{n-k}(-n) \, \frac{k!}{m!} \, \widehat{F}_{k-m}(m) \,.$$

When $m = n$ the sum is obviously 1. When $m = n - p$ for $p > 0$ it is $(n-1)!/(n-p)!$ times

$$\sum_{k=n-p}^{n} k \, \widehat{F}_{n-k}(-n) \, \widehat{F}_{k-n+p}(n-p) = \sum_{k=0}^{p} (n-k) \, \widehat{F}_k(-n) \, \widehat{F}_{p-k}(n-p)$$

$$= n \sum_{k=0}^{p} \widehat{F}_k(-n) \, \widehat{F}_{p-k}(n-p)$$

$$- \sum_{k=0}^{p} k \, \widehat{F}_k(-n) \, \widehat{F}_{p-k}(n-p)$$

$$= n \, \widehat{F}_p(-p) - n \, \widehat{F}_p(-p) = 0$$

by the original convolution formula and the one we derived from it. The proof is complete.

Extending the Matrix

The simple formula for f_{nk} that we used to prove Lagrange's theorem when $f_1 = 1$ can be written in another suggestive form, if we replace k by $n - k$:

$$f_{n(n-k)} = n^{\underline{k}} \widehat{F}_k(n - k).$$

For every fixed k, this is a polynomial in n, of degree $\leq 2k$. Therefore we can define the quantity $f_{y(y-k)}$ for all real or complex y to be $y^{\underline{k}} \widehat{F}_k(y-k)$; and in particular we can define f_{nk} in this manner for all integers n and k, letting $f_{nk} = 0$ when $k > n$. For example, in the case of Stirling numbers this analysis establishes the well-known fact that $\left\{ {y \atop y-k} \right\}$ and $\left[{y \atop y-k} \right]$ are polynomials in y of degree $2k$, and that these polynomials are multiples of $y^{\underline{k+1}} = y(y - 1) \dots (y - k)$ when $k > 0$.

The two flavors of Stirling numbers are related in two important ways. First, their matrices are inverse to each other if we attach the signs $(-1)^{n-k}$ to the elements in one matrix:

$$\sum_{k=0}^{n} \left\{ {n \atop k} \right\} \left[{k \atop m} \right] (-1)^{n-k} = \sum_{k=0}^{m} \left[{n \atop k} \right] \left\{ {k \atop m} \right\} (-1)^{n-k} = \delta_{mn}.$$

This follows since the numbers $\left\{ {n \atop k} \right\}$ correspond to $f(z) = e^z - 1$ and the numbers $\left[{n \atop k} \right] (-1)^{n-k}$ correspond to $g(z) = \ln(1 + z)$, as mentioned earlier, and we have $g\bigl(f(z)\bigr) = z$.

The other important relationship between $\left\{ {n \atop k} \right\}$ and $\left[{n \atop k} \right]$ is the striking identity

$$\left\{ {n \atop k} \right\} = \left[{-k \atop -n} \right],$$

which holds for all integers n and k when we use the polynomial extension method. We can prove, in fact, that the analogous relation

$$f_{nk} = (-1)^{k-n} g_{(-k)(-n)}$$

holds in the extended matrices F and G that correspond to *any* pair of inverse functions $g\bigl(f(z)\bigr) = z$, when $f'(0) = 1$. For we have

$$(-1)^{n-k} g_{(-k)(-n)} = (-1)^{n-k}(-k - 1)(-k - 2) \dots (-n) \, \widehat{F}_{n-k}(k)$$

$$= \frac{n!}{k!} \, \widehat{F}_{n-k}(k) = f_{nk}$$

in the formulas above. (The interesting history of the identity $\left\{ {n \atop k} \right\} = \left[{-k \atop -n} \right]$ is traced in [17]. The fact that the analogous formula holds in any convolution matrix was pointed out by Ira Gessel after he had read a draft of that paper. See also Jabotinsky [14]; Carlitz [3]; Roman and Rota [27, §10].)

Suppose we denote the Lah numbers $\binom{n}{k}\binom{n-1}{k-1}(n-k)!$ by $\left|\begin{smallmatrix} n \\ k \end{smallmatrix}\right|$. The extended matrix in that case has a pleasantly symmetrical property

$$\left|\begin{matrix} n \\ k \end{matrix}\right| = \left|\begin{matrix} -k \\ -n \end{matrix}\right|,$$

because the corresponding function $f(z) = z/(1-z)$ is a "co-involution," satisfying $f(-f(-z)) = z$. (See Mullin and Rota [22, §9].) Near the origin $n = k = 0$, the nonzero entries look like this:

$$
\begin{array}{rrrrrrr}
\cdots & 1 & & & & & \\
\cdots & 12 & 1 & & & & \\
\cdots & 36 & 6 & 1 & & & \\
\cdots & 24 & 6 & 2 & 1 & & \\
 & & & & 1 & & \\
 & & & & 1 & & \\
 & & & & 2 & 1 & \\
 & & & & 6 & 6 & 1 \\
 & & & & 24 & 36 & 12 & 1 \\
 & & & & \vdots & \vdots & \vdots & \vdots
\end{array}
$$

Still More Convolutions

Our proof of Lagrange's theorem yields yet another corollary. Suppose $g(f(z)) = z$ and $f'(0) = 1$, and let $\widehat{F}(z) = f(z)/z$, $\widehat{G}(z) = g(z)/z$. Then the equations

$$g_{nk} = \frac{n!}{k!}\,\widehat{G}_{n-k}(k) = \frac{(n-1)!}{(k-1)!}\,\widehat{F}_{n-k}(-n)$$

tell us, after replacing n by $n + k$, that the identity

$$\frac{n+k}{k}\,\widehat{G}_n(k) = \widehat{F}_n(-n - k)$$

holds for all positive integers k. Thus the polynomials $\widehat{G}_n(x)$ and $\widehat{F}_n(x)$ must be related by the formula

$$(x + n)\,\widehat{G}_n(x) = x\widehat{F}_n(-x - n).$$

Now $\widehat{F}_n(x)$ is an arbitrary convolution family, and $\widehat{F}_n(-x)$ is another. We can conclude that *if $\{F_n(x)\}$ is any convolution family, then so is the set of polynomials $\{xF_n(x+n)/(x+n)\}$.* Indeed, if $F_n(x)$ corresponds to the coefficients of $F(z)^x$, our argument proves that the coefficients of $G(z)^x$ are $x\,F_n(x+n)/(x+n)$, where $zG(z)$ is the inverse of the power series $z/F(z)$:

$$G(z) = F\big(zG(z)\big), \qquad G\big(z/F(z)\big) = F(z).$$

The case $F(z) = 1 + z$ and $G(z) = 1/(1 - z)$ provides a simple example, where we know that the corresponding polynomials are $F_n(x) = \binom{x}{n}$ and $G_n(x) = \binom{x+n-1}{n} = xF_n(x+n)/(x+n)$.

A more interesting example arises when $F(z) = ze^z/(e^z - 1) = z + z/(e^z - 1) = 1 + z/2 + B_2 z^2/2! + B_4 z^4/4! + \cdots$; then $F(-z)$ is the exponential generating function for the Bernoulli numbers. The convolution family for $F(z)^x$ is $F_n(x) = x\sigma_n(x)$, where $\sigma_n(x)$ is called a *Stirling polynomial*. (Actually $\sigma_0(x) = 1/x$, but $\sigma_n(x)$ is a genuine polynomial when $n \geq 1$.) The function G defined by the relation $G\big(z/F(z)\big) = F(z)$ is $G(z) = z^{-1}\ln\big(1/(1 - z)\big)$; therefore the convolution family for $G(z)^x$ is $G_n(x) = xF_n(x+n)/(x+n) = x\sigma_n(x+n)$.

In this example the convolution family for $e^{xzG(z)} = (1 - z)^{-x}$ is

$$\binom{x+n-1}{n} = \frac{1}{n!}\left(\begin{bmatrix} n \\ 0 \end{bmatrix} + \begin{bmatrix} n \\ 1 \end{bmatrix} x + \cdots + \begin{bmatrix} n \\ n \end{bmatrix} x^n\right);$$

therefore

$$\begin{aligned}
\begin{bmatrix} n \\ n-k \end{bmatrix} &= \frac{n!}{(n-k)!}\, G_k(n-k) = \frac{n!}{(n-k)!}\,(n-k)\,\sigma_k(n) \\
&= n(n-1)\,\ldots\,(n-k)\,\sigma_k(n).
\end{aligned}$$

We also have

$$\begin{Bmatrix} n \\ n-k \end{Bmatrix} = \begin{bmatrix} k-n \\ -n \end{bmatrix} = (k-n)(k-1-n)\,\ldots\,(-n)\,\sigma_k(k-n).$$

These formulas, which are polynomials in n of degree $2k$ for every fixed k, explain why the σ functions are called Stirling polynomials. Notice that $\sigma_n(1) = (-1)^n B_n/n!$; it can also be shown that $\sigma_n(0) = -B_n/(n \cdot n!)$.

The process of going from $F_n(x)$ to $xF_n(x+n)/(x+n)$ can be iterated: Another replacement gives $xF_n(x+2n)/(x+2n)$, and after

t iterations we discover that the polynomials $xF_n(x + tn)/(x + tn)$ also form a convolution family. This holds for all positive integers t, and the convolution condition is expressible as a set of polynomial relations in t; therefore $\{xF_n(x + tn)/(x + tn)\}$ *is a convolution family for all complex numbers* t. If $F_n(x) = [z^n] F(z)^x$, then $xF_n(x + tn)/(x + tn) = [z^n] \mathcal{F}_t(z)^x$, where $\mathcal{F}_t(z)$ is defined implicitly by the equation

$$\mathcal{F}_t(z) = F\big(z\mathcal{F}_t(z)^t\big).$$

In particular, we could have deduced the convolution properties of the coefficients of $\mathcal{B}_t(z)^x$ in this way.

Let us restate what we have just proved, combining it with the "derived convolution formula" obtained earlier:

Theorem. *Let* $F_n(x)$ *be any family of polynomials in* x *such that* $F_n(x)$ *has degree* $\leq n$. *If the identity*

$$F_n(2x) = \sum_{k=0}^{n} F_k(x) F_{n-k}(x)$$

holds for all n *and* x, *then the following identities hold for all* n, x, y, *and* t:

$$\frac{(x + y) F_n(x + y + tn)}{x + y + tn} = \sum_{k=0}^{n} \frac{x F_k(x + tk)}{x + tk} \frac{y F_{n-k}\big(y + t(n - k)\big)}{y + t(n - k)};$$

$$\frac{n F_n(x + y + tn)}{x + y + tn} = \sum_{k=1}^{n} \frac{k F_k(x + tk)}{x + tk} \frac{y F_{n-k}\big(y + t(n - k)\big)}{y + t(n - k)}. \quad \Box$$

Additional Constructions

We have considered several ways to create new convolution families from given ones, by multiplication or exponentiation of the associated convolution matrices, or by replacing $F_n(x)$ by $x F_n(x + tn)/(x + tn)$. It is also clear that the polynomials $\alpha^n F_n(\beta x)$ form a convolution family whenever the polynomials $F_n(x)$ do.

One further operation deserves to be mentioned: If $F_n(x)$ and $G_n(x)$ are convolution families, then so is the family $H_n(x)$ defined by

$$H_n(x) = \sum_{k=0}^{n} F_k(x) G_{n-k}(x).$$

This is obvious, since $H_n(x) = [z^n] F(z)^x G(z)^x$. The corresponding operation on matrices $F = (f_{nk})$, $G = (g_{nk})$, $H = (h_{nk})$ is

$$h_{nk} = \sum_{i,j} \binom{n}{j} f_{ji}\, g_{(n-j)(k-i)}\,.$$

If we denote this binary operation by $H = F \circ G$, it is interesting to observe that the associative law holds: $(E \circ F) \circ G = E \circ (F \circ G)$ is true for all matrices E, F, G, not just for convolution matrices. A convolution matrix is characterized by the special property $F \circ F = F \operatorname{diag}(2, 4, 8, \dots)$.

The construction just mentioned is merely a special case of the one-parameter family

$$H_n^{(t)}(x) = \sum_{k=0}^{n} F_k(x)\, G_{n-k}(x + tk)\,.$$

Again, $\{H_n^{(t)}(x)\}$ turns out to be a convolution family, for arbitrary t: We have

$$\sum_{k=0}^{n} H_n^{(t)}(x) z^n = \sum_{n \geq k \geq 0} F_k(x)\, G_{n-k}(x + tk) z^n$$

$$= \sum_{n,k \geq 0} F_k(x)\, G_n(x + tk) z^{n+k}$$

$$= \sum_{k \geq 0} F_k(x) z^k G(z)^{x+tk} = G(z)^x F\big(z G(z)^t\big)^x,$$

so $H_n(x) = [z^n] \big(G(z) F\big(z G(z)^t\big)\big)^x$.

Applications

What's the use of all this? Well, we have shown that many interesting convolution families exist, and that we can deduce nonobvious facts with comparatively little effort once we know that we're dealing with a convolution family.

One moral to be drawn is therefore the following. Whenever you encounter a triangular pattern of numbers that you haven't seen before, check to see if the first three rows have the form

$$
\begin{array}{ccc}
a & & \\
b & a^2 & \\
c & 3ab & a^3
\end{array}
$$

for some a, b, c. (You may have to multiply or divide the nth row by $n!$ first, and/or reflect its entries left to right.) If so, and if the problem you are investigating is mathematically "clean," chances are good that the fourth row will look like

$$d \quad 4ac + 3b^2 \quad 6a^2b \quad a^4 \,.$$

And if so, chances are excellent that you are dealing with a convolution family. And if so, you may well be able to solve your problem.

In fact, exactly that scenario has helped the author on several occasions.

Asymptotics

Once you have identified a convolution family $F_n(x)$, you may well want to know the approximate value of $F_n(x)$ when n and x are large. The remainder of this paper discusses a remarkable general power series expansion, discovered with the help of *Mathematica*, which accounts for the behavior of $F_n(x)$ when n/x stays bounded and reasonably small as $x \to \infty$, although n may also vary as a function of x. We will assume that $F_n(x)$ is the coefficient of z^n in $F(z)^x$, where $F(0) = F'(0) = 1$.

Our starting point is the classical "saddle point method," which shows that in many cases the coefficient of z^n in a power series $P(z)$ can be approximated by considering the value of P at a point where the derivative of $P(z)/z^n$ is zero. (See Good [8].) In our case we have $P(z) = e^{xf(z)}$, where $f(z) = \ln F(z) = z + f_2 z^2/2! + \cdots$; and the derivative is zero when $x f'(z) = n/z$. Let this saddle point occur at $z = s$; thus, we have

$$s f'(s) = n/x \,.$$

Near s we have $f(z) = f(s) + (z - s)f'(s) + O\big((z - s)^2\big)$; so we will base our approximation on the assumption that the $O\big((z - s)^2\big)$ contribution is zero. The approximation to $F_n(x)$ will be $\widetilde{F}_n(x)$, where

$$\widetilde{F}_n(x) = [z^n] \exp\big(x\,f(s) + x\,(z - s)\,f'(s)\big)$$

$$= \frac{e^{x(f(s) - sf'(s))}}{n!}\, x^n f'(s)^n = \frac{F(s)^x}{n!}\left(\frac{n}{es}\right)^n \,.$$

First let's look at some examples; later we will show that the ratio $F_n(x)/\widetilde{F}_n(x)$ is well behaved as a formal power series. Throughout this discussion we will let

$$y = n/x \,;$$

our goal, remember, is to find approximations that are valid when y is not too large, as x and possibly n go to ∞.

The simplest example is, of course, $F(z) = e^z$ and $f(z) = z$; but we needn't sneeze at it because it will give us some useful calibration. In this case $f''(z) = 0$, so our approximation will be exact. We have $s = y$, hence

$$\widetilde{F}_n(x) = \frac{e^{xy}}{n!} \left(\frac{n}{ey} \right)^n = \frac{e^n}{n!} \left(\frac{x}{e} \right)^n = \frac{x^n}{n!} = F_n(x) .$$

Next let's consider the case $F(z) = T(z)/z$, $f(z) = T(z)$, when we know from properties of the tree function that $F_n(x) = x(x+n)^{n-1}/n!$. In this case $z T'(z) = T(z)/(1 - T(z))$, so we have $T(s)/(1 - T(s)) = y$ or

$$T(s) = \frac{y}{1+y} , \qquad s = \frac{y}{1+y} e^{-y/(1+y)}$$

because $T(z) = ze^{T(z)}$. Therefore

$$\widetilde{F}_n(x) = \frac{e^{xy/(1+y)}}{n!} \left(\frac{n(1+y)}{ey \, e^{-y/(1+y)}} \right)^n = \frac{(x+n)^n}{n!} ;$$

the ratio $F_n(x)/\widetilde{F}_n(x) = x/(x+n) = 1/(1+y)$ is indeed near 1 when y is small.

If $F(z) = 1 + z$ we find, similarly, $s = y/(1-y)$ and

$$n! \, \widetilde{F}_n(x) = \left(\frac{1}{1-y} \right)^x \left(\frac{n(1-y)}{ey} \right)^n = \frac{x^x e^{-n}}{(x-n)^{x-n}} ;$$

by Stirling's approximation we also have

$$n! \, F_n(x) = \frac{x!}{(x-n)!} = \frac{x^x e^{-n}}{(x-n)^{x-n}} (1-y)^{-1/2} \left(1 + O(x^{-1}) \right) .$$

Again the ratio $F_n(x)/\widetilde{F}_n(x)$ is near 1. In general if $F(z) = \mathcal{B}_t(z)$ the saddle point s turns out to be $y(1 + (t-1)y)^{t-1}/(1 + ty)^t$, and

$$n! \, \widetilde{F}_n(x) = \frac{(x+tn)^{x+tn} e^{-n}}{\left(x + (t-1)n \right)^{x+(t-1)n}} ;$$

a similar analysis shows that this approximation is quite good, for any fixed t.

We know that

$$F_n(x) = \frac{x^n}{n!} \left(1 + \frac{f_{n(n-1)}}{x} + \frac{f_{n(n-2)}}{x^2} + \cdots \right)$$

and that $f_{n(n-k)}$ is always a polynomial in n of degree $\leq 2k$. Therefore if $n^2/x \to 0$ as $x \to \infty$, we can simply use the approximation $F_n(x) = (x^n/n!)\left(1 + O(n^2/x)\right)$. But there are many applications where we need a good estimate of $F_n(x)$ when $n^2/x \to \infty$ while $n/x \to 0$; for example, x might be $n \log n$. In such cases $\widetilde{F}_n(x)$ is close to $F_n(x)$ but $x^n/n!$ is not.

We can express s/y as a power series in y by inverting the power series expression $sf'(s) = y$:

$$s/y = 1 - f_2 y + (4f_2^2 - f_3)y^2/2 + (15f_2 f_3 - 30f_2^3 - f_4)y^3/6 + \cdots .$$

From this we can get a formal series for $\widetilde{F}_n(x)$,

$$\widetilde{F}_n(x) = \frac{x^n}{n!} \frac{\exp\big(n(s/y)(1 + f_2 s/2! + f_3 s^2/3! + \cdots) - n\big)}{(s/y)^n}$$

$$= \frac{x^n}{n!} \left(1 + \frac{nf_2}{2} y + \frac{3n^2 f_2^2 - 12nf_2^2 + 4nf_3}{24} y^2 + O(n^3 y^3) \right).$$

We can also use the formula

$$f_{n(n-k)} = \sum \frac{n^{k+k_2+k_3+\cdots}}{2!^{k_2} k_2! \, 3!^{k_3} k_3! \, \cdots} f_2^{k_2} f_3^{k_3} \cdots ,$$

where the sum is over all nonnegative $k_2, k_3, \ldots$ with $k_2 + 2k_3 + \cdots = k$, to write

$$F_n(x) = \frac{x^n}{n!} \left(1 + \frac{nf_2 - f_2 + O(x^{-1})}{2} y \right.$$

$$+ \frac{3n^2 f_2^2 - 18nf_2^2 + 4nf_3 + 33f_2^2 - 12f_3 + O(x^{-1})}{24} y^2$$

$$\left. + O(n^3 y^3) \right).$$

These series are not useful asymptotically unless $ny = n^2/x$ is small. But the approximation $\widetilde{F}_n(x)$ itself is excellent, because amazing cancellations occur when we compute the ratio:

$$\frac{F_n(x)}{\widetilde{F}_n(x)} = 1 - \frac{f_2}{2} y + \frac{11f_2^2 - 4f_3}{8} y^2 + O(y^3) + O(x^{-1}).$$

Theorem. *When $F(z) = \exp(z + f_2 z^2/2! + f_3 z^3/3! + \cdots)$ and the functions $F_n(x)$ and $\widetilde{F}_n(x)$ are defined as above, the ratio $F_n(x)/\widetilde{F}_n(x)$ can be written as a formal power series $\sum_{i,j \geq 0} c_{ij} y^i x^{-j}$, where $y = n/x$ and the coefficients c_{ij} are polynomials in $f_2, f_3, \ldots$.*

The derivation just given shows that we can write $F_n(x)/\widetilde{F}_n(x)$ as a formal power series of the form $\sum_{i,j \geq 0} a_{ij} n^i x^{-j}$, where $a_{ij} = 0$ when $i > 2j$; the surprising thing is that we also have $a_{ij} = 0$ whenever $i > j$. Therefore we can let $c_{ij} = a_{i(i+j)}$.

To prove the theorem, we let $R(z) = 1 + R_1 z + R_2 z^2 + \cdots$ stand for the terms neglected in our approximation:

$$F(z)^x = e^{xf(s) - xsf'(s)} \left(1 + \frac{n}{s} \frac{z}{1!} + \frac{n^2}{s^2} \frac{z^2}{2!} + \frac{n^3}{s^3} \frac{z^3}{3!} + \cdots \right) R(z).$$

The coefficient of z^n is

$$F_n(x) = \widetilde{F}_n(x) \left(1 + R_1 s + \frac{n-1}{n} R_2 s^2 + \frac{(n-1)(n-2)}{n^2} R_3 s^3 + \cdots \right);$$

so the ratio $F_n(x)/\widetilde{F}_n(x)$ is equal to

$$\sum_{k \geq 0} \frac{n^k}{n^k} R_k s^k = \sum_{j,k \geq 0} (-n)^{-j} \begin{bmatrix} k \\ k-j \end{bmatrix} R_k s^k = \sum_j (-n)^{-j} P_j,$$

where $P_j = \sum_k \begin{bmatrix} k \\ k-j \end{bmatrix} R_k s^k$ is a certain power series in s and x. The coefficients R_k are themselves power series in s and x, because we have

$$R(z) = \exp\left(x(z-s)^2 \frac{f''(s)}{2!} + x(z-s)^3 \frac{f'''(s)}{3!} + \cdots \right).$$

We know from the discussion above that

$$\begin{bmatrix} k \\ k-j \end{bmatrix} = k(k-1) \ldots (k-j) \, \sigma_j(k)$$

is a polynomial in k. Therefore we can write

$$P_j = \begin{bmatrix} \vartheta \\ \vartheta - j \end{bmatrix} R(z) \Bigg|_{z=s},$$

where ϑ is the operator that takes $z^k \mapsto k \, z^k$ for all k; that is, $\vartheta G(z) = z \, G'(z)$ for all power series $G(z)$. The theorem will be proved if we can show that $P_j/(-n)^j$ is a formal power series in y and x^{-1}, and if the sum of these formal power series over all j is also such a series.

Consider, for example, the simplest case $P_0 = R(s)$; obviously $P_0 = 1$. The next simplest case is

$$P_1 = \begin{bmatrix} \vartheta \\ \vartheta - 1 \end{bmatrix} R(z) \bigg|_{z=s} = \frac{1}{2}\vartheta(\vartheta - 1)R(z) \bigg|_{z=s}.$$

It is easy to see that

$$\vartheta^{\underline{j}} = z^j D^j \,,$$

where D is the differentiation operator $DG(z) = G'(z)$, because $z^j D^j$ takes z^k into $k^{\underline{j}} z^k$. Therefore

$$P_1 = \tfrac{1}{2}s^2 R''(s) = \tfrac{1}{2} x s^2 f''(s)\,.$$

It follows that $P_1/n = \frac{1}{2}(s/y)sf''(s)$ is a power series in y; it begins $\frac{1}{2}f_2 y + \frac{1}{2}(f_3 - f_2^2)y^2 + \cdots$.

Now let's consider P_j in general. We will use the fact that the Stirling numbers $\begin{bmatrix} k \\ k-j \end{bmatrix}$ can be represented in the form

$$\begin{bmatrix} k \\ k - j \end{bmatrix} = p_{j1}\binom{k}{j+1} + p_{j2}\binom{k}{j+2} + \cdots + p_{jj}\binom{k}{2j},$$

where the coefficients p_{ji} are the positive integers in the following triangular array:

$$
\begin{array}{ccccc}
1 \\
2 & 3 \\
6 & 20 & 15 \\
24 & 130 & 210 & 105 \\
120 & 924 & 2380 & 2520 & 945
\end{array} \quad .
$$

(This array is clearly not a convolution matrix; but the theory developed above implies that the numbers $j!\,p_{ji}/(i+j)!$, namely

$$
\begin{array}{ccccc}
1/2 \\
2/3 & 1/4 \\
3/2 & 1 & 1/8 \\
24/5 & 13/3 & 1 & 1/16 \\
20 & 22 & 85/12 & 5/6 & 1/32
\end{array} \quad ,
$$

do form the convolution matrix for the powers of $\exp(z/2 + z^2/3 + z^3/4 + \cdots)$. The expression $\begin{bmatrix} k \\ k-j \end{bmatrix} = \sum_{i=1}^{j} p_{ji}\binom{k}{j+i}$ was independently discovered by Appell [1], Jordan [16], and Ward [30]. The number of permutations of $i + j$ elements having no fixed points and exactly i cycles is p_{ji}, an "associated Stirling number of the first kind" [23, §4.4; 5, exercise 6.7].) It follows that

$$P_j = p_{j1}s^{j+1} \frac{R^{(j+1)}(s)}{(j+1)!} + p_{j2}s^{j+2} \frac{R^{(j+2)}(s)}{(j+2)!} + \cdots + p_{jj}s^{2j} \frac{R^{(2j)}(s)}{(2j)!}.$$

Now $R(z)$ is a sum of terms having the form

$$a_{il}x^i(z - s)^l,$$

where $l \geq 2i$ and where a_{il} is a power series in s. Such a term contributes $a_{il}x^i s^l p_{j(l-j)}$ to P_j; so it contributes $a_{il}(s/y)^j s^{l-j} x^{i-j} p_{j(l-j)}$ to P_j/n^j. This contribution is nonzero only if $j < l \leq 2j$. Since $l \geq 2i$, we have $i \leq j$; so P_j/n^j is a power series in y and x^{-1}.

For a fixed value of $j - i$, the smallest power of y that can occur in P_j/n^j is $y^{2i-j} = y^{j-2(j-i)}$. Therefore only a finite number of terms of $\sum_j P_j/(-n)^j$ contribute to any given power of y and x^{-1}. This completes the proof. □

A careful analysis of the proof, and a bit of *Mathematica* hacking, yields the more precise result

$$\frac{F_n(x)}{\widetilde{F}_n(x)} = \frac{1}{(1+s^2y^{-1}d_2)^{1/2}} + \frac{(s/y)^3 A}{x(1+s^2y^{-1}d_2)^{7/2}} + O(x^{-2}),$$

where $A = \frac{1}{12}s^3y^{-1}d_2^3 - \frac{3}{4}sd_2^2 - \frac{1}{2}s^2d_2d_3 - \frac{5}{24}s^3d_3^2 + \frac{1}{3}yd_3 + \frac{1}{8}s^3d_2d_4 + \frac{1}{8}syd_4$ and $d_k = f^{(k)}(s)$.

Acknowledgments

I wish to thank Ira Gessel and Svante Janson for stimulating my interest in this subject and for their helpful comments on the first draft. Ira Gessel and Richard Brent also introduced me to several relevant references.

References

[1] P. Appell, "Développement en série entière de $(1 + ax)^{1/x}$," *Archiv der Mathematik und Physik* **65** (1880), 171–175.

[2] E. T. Bell, "Exponential polynomials," *Annals of Mathematics* **35** (1934), 258–277. See also E. T. Bell, "Exponential numbers," *American Mathematical Monthly* **41** (1934), 411–419.

[3] L. Carlitz, "Generalized Stirling and related numbers," *Rivista di Matematica della Università di Parma* (4) **4** (1978), 79–99.

[4] A. Cayley, "On the analytical forms called trees. Second part," *Philosophical Magazine* **18** (1859), 371–378. Reprinted in Cayley's *Collected Mathematical Papers* **4**, 112–115.

[5] Louis Comtet, *Analyse Combinatoire*, Tomes I et II (Paris: Presses Universitaires de France, 1970). English translation, revised and enlarged, *Advanced Combinatorics* (Dordrecht: D. Reidel, 1974).

[6] Adriano M. Garsia, "An exposé of the Mullin–Rota theory of polynomials of binomial type," *Linear and Multilinear Algebra* **1** (1973), 47–65.

[7] F. Gomes Teixeira, "Sur le développement de x^k en série ordonnée suivant les puissances du sinus de la variable," *Nouvelles Annales de Mathématiques* (3) **15** (1896), 270–274.

[8] I. J. Good, "Saddle-point methods for the multinomial distribution," *Annals of Mathematical Statistics* **28** (1957), 861–881.

[9] H. W. Gould and J. Kaucký, "Evaluation of a class of binomial coefficient summations," *Journal of Combinatorial Theory* **1** (1966), 233–247; **12** (1972), 309–310.

[10] Ronald L. Graham, Donald E. Knuth, and Oren Patashnik, *Concrete Mathematics* (Reading, Massachusetts: Addison–Wesley, 1989).

[11] Bernard Harris and Lowell Schoenfeld, "The number of idempotent elements in symmetric semigroups," *Journal of Combinatorial Theory* **3** (1967), 122–135.

[12] L. C. Hsu, "Theory and application of generalized Stirling number pairs," *Journal of Mathematical Research and Exposition* **9** (1989), 211–220.

[13] Eri Jabotinsky, "Sur la représentation de la composition de fonctions par un produit de matrices. Application à l'itération de e^x et de $e^x - 1$," *Comptes Rendus hebdomadaires des séances de l'Académie des Sciences* **224** (Paris: 1947), 323–324.

[14] Eri Jabotinsky, "Representation of functions by matrices. Application to Faber polynomials," *Proceedings of the American Mathematical Society* **4** (1953), 546–553.

[15] Eri Jabotinsky, "Analytic iteration," *Transactions of the American Mathematical Society* **108** (1963), 457–477.

[16] Charles Jordan, "On Stirling's numbers," *Tôhoku Mathematical Journal* **37** (1933), 254–278.

[17] Donald E. Knuth, "Two notes on notation," *American Mathematical Monthly* **99** (1992), 403–422. [Reprinted as Chapter 2 of the present volume.]

[18] Donald E. Knuth and Boris Pittel, "A recurrence related to trees," *Proceedings of the American Mathematical Society* **105** (1989), 335–349. [Reprinted as Chapter 39 of the present volume.]

[19] Marek Kuczma, *Functional Equations in a Single Variable* (Warsaw: Polish Scientific Publishers, 1968).

[20] I. Lah, "Eine neue Art von Zahlen, ihre Eigenschaften und Anwendung in der mathematischen Statistik," *Mitteilungsblatt für Mathematische Statistik* **7** (1955), 203–212.

[21] T. S. Motzkin, "Sorting numbers for cylinders and other classification numbers," *Proceedings of Symposia in Pure Mathematics* **19** (Providence, Rhode Island: American Mathematical Society, 1971), 167–176.

[22] Ronald Mullin and Gian-Carlo Rota, "On the foundations of combinatorial theory. III. Theory of binomial enumeration," in *Graph Theory and Its Applications*, edited by Bernard Harris (Academic Press, 1970), 167–213.

[23] John Riordan, *An Introduction to Combinatorial Analysis* (New York: John Wiley & Sons, 1958).

[24] John Riordan, *Combinatorial Identities* (New York: John Wiley & Sons, 1968).

[25] D. G. Rogers, "Pascal triangles, Catalan numbers and renewal arrays," *Discrete Mathematics* **22** (1978), 301–310.

[26] Steven Roman, *The Umbral Calculus* (Orlando, Florida: Academic Press, 1984).

[27] Steven M. Roman and Gian-Carlo Rota, "The umbral calculus," *Advances in Mathematics* **27** (1978), 95–188.

[28] Gian-Carlo Rota, D. Kahaner, and A. Odlyzko, "On the foundations of combinatorial theory. VIII. Finite operator calculus," *Journal of Mathematical Analysis and Applications* **42** (1973), 684–760. Reprinted in Gian-Carlo Rota, *Finite Operator Calculus* (Academic Press, 1975), 7–82.

[29] J. F. Steffensen, "The poweroid, an extension of the mathematical notion of power," *Acta Mathematica* **73** (1941), 333–366.

[30] Morgan Ward, "The representation of Stirling's numbers and Stirling's polynomials as sums of factorials," *American Journal of Mathematics* **56** (1934), 87–95.

Addendum

After this article was published, Rob Corless called my attention to the highly relevant paper of G. Labelle, "Sur l'Inversion et l'itération continue des séries formelles," *European Journal of Combinatorics* **1** (1980), 113–138, which anticipates several of the results I had presented and introduces other ideas that had never occurred to me. In particular, Labelle shows that if $\{F_n(x)\}$ and $\{G_n(x)\}$ are convolution families for the functions $F(z)$ and $G(z)$, then

$$\sum_k F_k(x) G_{n-k}(x+k)$$

is the convolution family for $G(z) F\big(G(z)\big)$. Furthermore he observes that the inverse series for $z\mathcal{F}_t(z)^s$ is $z\mathcal{F}_{t-s}(z)^{-s}$, when $\mathcal{F}_t$ is implicitly defined by

$$\mathcal{F}_t(z) = F\big(z\mathcal{F}_t(z)^t\big)$$

as in the section entitled "still more convolutions" above.

Incidentally, in the third edition of my book *Seminumerical Algorithms* (1997), exercise 4.7–22, I decided to use the notation $F^{\{t\}}(z)$ for this function $\mathcal{F}_t(z)$, and to call it the tth *induced function* of F, since such functions tend to arise rather often.

I learned recently from Jim Pitman that the numbers studied by Lah had been previously treated by L. Toscano, "Numeri di Stirling generalizzati, operatori differenziali e polinomi ipergeometrici," *Commentationes* **3** (Vatican City: Accademia delle Scienze, 1939), 721–757, for example in Equations 17 and 117; see also "Operatori lineari e numeri di Stirling generalizzati," *Annali di Matematica* (4) **14** (1936), 287–297, Equation (16). When I looked up Toscano's article, I was astonished to learn that he used precisely the notations $x^{\overline{n}}$ and $x^{\underline{n}}$ for factorial powers that I had independently been recommending since the 1960s. (I had previously seen $x^{\overline{n}}$ in a paper by Alfredo Capelli, vintage 1893.)

Chapter 16

Polynomials Involving the Floor Function

Some identities are presented that generalize the formula

$$x^3 = 3x\lfloor x\lfloor x\rfloor\rfloor - 3\lfloor x\rfloor\lfloor x\lfloor x\rfloor\rfloor + \lfloor x\rfloor^3 + 3\{x\}\{x\lfloor x\rfloor\} + \{x\}^3$$

to a representation of the product $x_0 x_1 \ldots x_{n-1}$.

*[Written with Inger Johanne Håland. Originally published in Mathematica Scandinavica **76** (1995), 194–200.]*

1. Introduction

Let $\lfloor x\rfloor$ be the greatest integer less than or equal to x, and let $\{x\} = x - \lfloor x\rfloor$ be the fractional part of x. The purpose of this note is to show how the formulas

$$xy = \lfloor x\rfloor y + x\lfloor y\rfloor - \lfloor x\rfloor\lfloor y\rfloor + \{x\}\{y\} \tag{1.1}$$

and

$$\begin{aligned}
xyz = {} & x\lfloor y\lfloor z\rfloor\rfloor + y\lfloor z\lfloor x\rfloor\rfloor + z\lfloor x\lfloor y\rfloor\rfloor \\
& - \lfloor x\rfloor\lfloor y\lfloor z\rfloor\rfloor - \lfloor y\rfloor\lfloor z\lfloor x\rfloor\rfloor - \lfloor z\rfloor\lfloor x\lfloor y\rfloor\rfloor \\
& + \lfloor x\rfloor\lfloor y\rfloor\lfloor z\rfloor \\
& + \{x\}\{y\lfloor z\rfloor\} + \{y\}\{z\lfloor x\rfloor\} + \{z\}\{x\lfloor y\rfloor\} \\
& + \{x\}\{y\}\{z\}
\end{aligned} \tag{1.2}$$

can be extended to higher-order products $x_0 x_1 \ldots x_{n-1}$.

These identities make it possible to answer questions about the distribution mod 1 of sequences having the form

$$\alpha_1 n\lfloor\alpha_2 n\ldots\lfloor\alpha_{k-1}n\lfloor\alpha_k n\rfloor\rfloor\ldots\rfloor, \qquad n = 1, 2, \ldots. \tag{1.3}$$

257

Such sequences are known to be uniformly distributed mod 1 if the real numbers $1, \alpha_1, \ldots, \alpha_k$ are rationally independent [1]; we will prove that (1.3) is uniformly distributed in the special case $\alpha_1 = \alpha_2 = \cdots = \alpha_k = \alpha$ if and only if α^k is irrational, when k is prime. (It is interesting to compare this result to analogous properties of the sequence

$$\alpha_0 \lfloor \alpha_1 n \rfloor \lfloor \alpha_2 n \rfloor \ldots \lfloor \alpha_k n \rfloor, \qquad n = 1, 2, \ldots, \tag{1.4}$$

where $\alpha_0, \alpha_1, \ldots, \alpha_k$ are positive real numbers. If $k \geq 3$, such sequences are uniformly distributed mod 1 if and only if α_0 is irrational [2].)

2. Formulas for the Product $x_0 x_1 \ldots x_{n-1}$

The general expression that we will derive for $x_0 x_1 \ldots x_{n-1}$ contains $2^{n+1} - n - 2$ terms. Given a sequence $X = (x_0, x_1, \ldots, x_{n-1})$ we regard x_{n+j} as equivalent to x_j, and for integers $a \leq b$ we define

$$X^{a:b} = \begin{cases} 1, & \text{if } a = b; \\ x_a \lfloor X^{(a+1):b} \rfloor, & \text{otherwise.} \end{cases} \tag{2.1}$$

Thus $X^{1:4} = x_1 \lfloor x_2 \lfloor x_3 \rfloor \rfloor$ and $X^{4:(n+1)} = x_4 \lfloor x_5 \lfloor \ldots \lfloor x_{n-1} \lfloor x_0 \rfloor \rfloor \ldots \rfloor \rfloor$. Using this notation, we obtain an expression for $x_0 x_1 \ldots x_{n-1}$ by taking the sum of

$$\{X^{s_1:s_2}\} \{X^{s_2:s_3}\} \ldots \{X^{s_k:(s_1+n)}\}$$
$$- (-1)^k \lfloor X^{s_1:s_2} \rfloor \lfloor X^{s_2:s_3} \rfloor \ldots \lfloor X^{s_k:(s_1+n)} \rfloor \tag{2.2}$$

over all nonempty subsets $S = \{s_1, \ldots, s_k\}$ of $\{0, 1, \ldots, n-1\}$, where $s_1 < \cdots < s_k$. This rule defines $2^{n+1} - 2$ terms, but in the special case $k = 1$ the two terms of (2.2) reduce to

$$\{X^{s_1:(s_1+n)}\} + \lfloor X^{s_1:(s_1+n)} \rfloor = X^{s_1:(s_1+n)} \tag{2.3}$$

so we can combine them and make the overall formula n terms shorter. The right-hand side of (1.2) illustrates this construction when $n = 3$.

To prove that the sum of all terms (2.2) equals $x_0 x_1 \ldots x_{n-1}$, we replace $\{X^{a:b}\}$ by $X^{a:b} - \lfloor X^{a:b} \rfloor$ and expand all products. One of the terms in this expansion is $x_0 x_1 \ldots x_{n-1}$; it arises only from the set $S = \{0, 1, \ldots, n-1\}$. The other terms all contain at least one occurrence of the floor operator, and they can be written

$$x_{u_1} \ldots x_{v_1-1} \lfloor X^{v_1:u_2} \rfloor x_{u_2} \ldots x_{v_2-1} \lfloor X^{v_2:u_3} \rfloor x_{u_3} \ldots x_{v_3-1} \ldots \lfloor X^{v_k:(u_1+n)} \rfloor \tag{2.4}$$

where $u_1 \le v_1 < u_2 \le v_2 < u_3 \le \cdots \le v_k < n$. We want to show that all such terms cancel out. For example, some of the terms in the expansion when $n = 9$ have the form

$$x_1 \lfloor X^{2:4} \rfloor x_4 x_5 \lfloor X^{6:7} \rfloor \lfloor X^{7:10} \rfloor = x_1 \lfloor x_2 \lfloor x_3 \rfloor \rfloor x_4 x_5 \lfloor x_6 \rfloor \lfloor x_7 \lfloor x_8 \lfloor x_0 \rfloor \rfloor \rfloor ,$$

which is (2.4) with $u_1 = 1$, $v_1 = 2$, $u_2 = 4$, $v_2 = 6$, $u_3 = v_3 = 7$. It is easy to see that this term arises from the expansion of (2.2) only when S is one of the sets $\{1, 2, 4, 5, 6, 7\}$, $\{1, 4, 5, 6, 7\}$, $\{1, 2, 4, 5, 7\}$, $\{1, 4, 5, 7\}$; in those cases it occurs with the respective signs $-, +, +, -$, so it does indeed cancel out.

In general, the only sets S leading to the term (2.4) have

$$S = \{ s \mid u_j \le s < v_j \} \cup \{ v_j \mid u_j = v_j \} \cup T,$$

where T is a subset of $U = \{ v_j \mid u_j \ne v_j \}$. If U is empty, all parts of the term (2.4) appear inside floor brackets and this term is cancelled by the second term of (2.2). If U contains $m > 0$ elements, the 2^m choices for S produce 2^{m-1} terms with a coefficient of $+1$ and 2^{m-1} terms with a coefficient of -1. This completes the proof.

Notice that we used no special properties of the floor function in this argument. The same identity holds when $\lfloor x \rfloor$ is an arbitrary function, if we define $\{x\} = x - \lfloor x \rfloor$.

The formulas become simpler, of course, when all x_j are equal. Let

$$x^{:k} = \begin{cases} 1, & \text{if } k = 0; \\ x \lfloor x^{:(k-1)} \rfloor, & \text{if } k > 0; \end{cases} \tag{2.5}$$

and let

$$a_k = \{ x^{:k} \}, \qquad b_k = \lfloor x^{:k} \rfloor . \tag{2.6}$$

Then an identity for x^n can be read off from the coefficients of z^n in the formula

$$\frac{xz}{1 - xz} = \frac{a_1 z + 2a_2 z^2 + 3a_3 z^3 + \cdots}{1 - a_1 z - a_2 z^2 - a_3 z^3 - \cdots}$$
$$+ \frac{b_1 z + 2b_2 z^2 + 3b_3 z^3 + \cdots}{1 + b_1 z + b_2 z^2 + b_3 z^3 + \cdots} , \tag{2.7}$$

which can be derived from (2.2) or proved independently as shown below. For example,

$$x^2 = a_1^2 + 2a_2 - b_1^2 + 2b_2 ;$$

$$x^3 = a_1^3 + 3a_1 a_2 + 3a_3 + b_1^3 - 3b_1 b_2 + 3b_3 ;$$

$$x^4 = a_1^4 + 4a_1^2 a_2 + 4a_1 a_3 + 2a_2^2 + 4a_4$$
$$- b_1^4 + 4b_1^2 b_2 - 4b_1 b_3 - 2b_2^2 + 4b_4 .$$

In general we have

$$x^n = p_n(a_1, a_2, \ldots, a_n) - p_n(-b_1, -b_2, \ldots, -b_n), \qquad (2.8)$$

where the polynomial $p_n(a_1, a_2, \ldots, a_n) =$

$$\sum_{k_1 + 2k_2 + \cdots + nk_n = n} \frac{(k_1 + k_2 + \cdots + k_n - 1)! \, n}{k_1! \, k_2! \, \ldots \, k_n!} a_1^{k_1} a_2^{k_2} \ldots a_n^{k_n} \qquad (2.9)$$

contains one term for each partition of n.

It is interesting to note that (2.7) can be written

$$\frac{zd}{dz} \ln \frac{1}{1 - xz} = \frac{zd}{dz} \ln \frac{1}{1 - a_1 z - a_2 z^2 - \cdots} - \frac{zd}{dz} \ln \frac{1}{1 + b_1 z + b_2 z^2 + \cdots},$$

hence we obtain the equivalent identity

$$\frac{1}{1 - xz} = \frac{1 + b_1 z + b_2 z^2 + b_3 z^3 + \cdots}{1 - a_1 z - a_2 z^2 - a_3 z^3 - \cdots}. \qquad (2.10)$$

This identity is easy to prove directly, because it says that $a_k + b_k = xb_{k-1}$ for $k \geq 1$. Therefore it provides an alternative proof of (2.7). It also yields formulas for x^n with mixed a's and b's, and with no negative coefficients. For example,

$$x^2 = a_1^2 + a_2 + a_1 b_1 + b_2\,;$$
$$x^3 = a_1^3 + 2a_1 a_2 + a_3 + (a_1^2 + a_2)b_1 + a_1 b_2 + b_3\,;$$
$$x^4 = a_1^4 + 3a_1^2 a_2 + 2a_1 a_3 + a_2^2 + a_4 + (a_1^3 + 2a_1 a_2 + a_3)b_1$$
$$+ (a_1^2 + a_2)b_2 + a_1 b_3 + b_4\,.$$

3. Application to Uniform Distribution

We can now apply the identities to a problem in number theory, as stated in the introduction. Let $[0 .. 1) = \{\, x \mid 0 \leq x < 1 \,\}$.

Lemma 1.   *For all positive integers k and l, there is a function $f_{k,l}(y_1, y_2, \ldots, y_{k-1})$ from $[0 .. 1)^{k-1}$ to $[0 .. 1)$ such that*

$$\frac{x^{:k}}{l} \equiv \frac{x^k}{kl} - f_{k,l}\left(\left\{\frac{x}{k! \, l}\right\}, \left\{\frac{x^2}{k! \, l}\right\}, \ldots, \left\{\frac{x^{k-1}}{k! \, l}\right\}\right) \quad \text{(modulo 1).} \quad (3.1)$$

Proof.   Let

$$\hat{p}_n(a_1, a_2, \ldots, a_{n-1}) = p_n(a_1, a_2, \ldots, a_n) - n\, a_n \tag{3.2}$$

be the polynomial of (2.9) without its (unique) linear term. Then

$$\frac{x^{:k}}{l} = \frac{x^k}{kl} - \frac{1}{kl}\,\hat{p}_k(a_1, \ldots, a_{k-1}) + \frac{1}{kl}\,\hat{p}_k(-b_1, \ldots, -b_{k-1}). \tag{3.3}$$

We proceed by induction on k, defining the constant $f_{1,l} = 0$ for all l. Then if $y_j = \{x^j/k!\, l\}$ and $l_j = k!\, l/j!$ we have

$$a_j = \left\{ l_j\, \frac{x^{:j}}{l_j} \right\} = \left\{ l_j\big((j-1)!\, y_j - f_{j,l_j}(y_1, \ldots, y_{j-1})\big) \right\}$$

and

$$b_j = \left\lfloor l_j\, \frac{x^{:j}}{l_j} \right\rfloor = l_j \left\lfloor \frac{x^{:j}}{l_j} \right\rfloor + \sum_{h=1}^{l_j-1} \left\lfloor \left\{ \frac{x^{:j}}{l_j} \right\} + \frac{h}{l_j} \right\rfloor$$

$$\equiv \sum_{h=1}^{l_j-1} \left\lfloor \big\{ (j-1)!\, y_j - f_{j,l_j}(y_1, \ldots, y_{j-1}) \big\} + \frac{h}{l_j} \right\rfloor \quad (\text{modulo } kl),$$

because of the well-known identities

$$\{lx\} = \{l\{x\}\}, \qquad \lfloor lx \rfloor = \sum_{h=0}^{l-1} \left\lfloor x + \frac{h}{l} \right\rfloor, \tag{3.4}$$

when l is a positive integer. Therefore (3.1) holds with

$$f_{k,l}(y_1, \ldots, y_{k-1}) =$$

$$\left\{ \frac{1}{kl}\, \hat{p}_k(\bar{a}_{1,k,l}, \ldots, \bar{a}_{k-1,k,l}) - \frac{1}{kl}\, \hat{p}_k(-\bar{b}_{1,k,l}, \ldots, -\bar{b}_{k-1,k,l}) \right\} \tag{3.5}$$

where

$$\bar{a}_{j,k,l} = \left\{ \big((j-1)!\, y_j - f_{j,k!\, l/j!}(y_1, \ldots, y_{j-1})\big) k!\, l/j! \right\}, \tag{3.6}$$

$$\bar{b}_{j,k,l} = \sum_{h=1}^{k!\, l/j!-1} \left\lfloor \big\{ (j-1)!\, y_j - f_{j,k!\, l/j!}(y_1, \ldots, y_{j-1}) \big\} + \frac{j!\, h}{k!\, l} \right\rfloor. \tag{3.7}$$

For example,

$$f_{2,3}(y) = \{(\alpha_1^2 - \beta_1^2)/6\},$$

$$f_{3,1}(y, z) = \{(3\alpha_1\alpha_2 + \alpha_1^3 - 3\beta_1\beta_2 + \beta_1^3)/3\},$$

where $\alpha_1 = \{6y\}$, $\alpha_2 = \{3z - 3f_{2,3}(y)\}$, $\beta_1 = \lfloor y + \frac{1}{6} \rfloor + \lfloor y + \frac{2}{6} \rfloor + \cdots + \lfloor y + \frac{5}{6} \rfloor$, and $\beta_2 = \lfloor \{z - f_{2,3}(y)\} + \frac{1}{3} \rfloor + \lfloor \{z - f_{2,3}(y)\} + \frac{2}{3} \rfloor$. $\square$

Lemma 2. *The function $f_{k,l}$ of Lemma 1 does not preserve Lebesgue measure, and neither does $\{klmf_{k,l}\}$ for any positive integer m.*

Proof. It suffices to prove the second statement, for if $f_{k,l}$ were measure-preserving the functions $\{mf_{k,l}\}$ would preserve Lebesgue measure for all positive integers m. Notice that $\{klmf_{k,l}\} = \{m\,\hat{p}_k(\bar{a}_{1,k,l}, \ldots, \bar{a}_{k-1,k,l})\}$, because $\hat{p}_k(-\bar{b}_{1,k,l}, \ldots, -\bar{b}_{k-1,k,l})$ is an integer. The triangular construction of (3.6) makes it clear that $\bar{a}_{1,k,l}, \ldots, \bar{a}_{k-1,k,l}$ are independent random variables defined on the probability space $[0 \ldots 1)^{k-1}$, each uniformly distributed in $[0 \ldots 1)$. Therefore it suffices to prove that $\{m\,\hat{p}_k(a_1, \ldots, a_{k-1})\}$ is not uniformly distributed when $a_1, \ldots, a_{k-1}$ are independent uniform deviates.

We can express $\hat{p}_k(a_1, \ldots, a_{k-1})$ in the form

$$k\,a_1 a_{k-1} + a_1 q_1(a_1, \ldots, a_{k-2}) + k\,a_2 a_{k-2}$$
$$+ a_2 q_2(a_2, \ldots, a_{k-3}) + \cdots + \tfrac{1}{2}k\,a_{k/2}^2 ,$$

for some polynomials $q_1, \ldots, q_{\lfloor (k-1)/2 \rfloor}$, where the final term $\tfrac{1}{2}k\,a_{k/2}^2$ is absent when k is odd. Then we can let

$$y_j = \begin{cases} a_j, & \text{if } j \le \tfrac{1}{2}k, \\ a_j - q_{k-j}(a_{k-j}, \ldots, a_{j-1})/k, & \text{if } j > \tfrac{1}{2}k, \end{cases}$$

obtaining $k - 2$ independent uniform deviates $y_1, \ldots, y_{k-2}$ for which $m\,\hat{p}_k(a_1, \ldots, a_{k-1})$ equals

$$g_k(y_1, \ldots, y_{k-1}) = mk\,y_1 y_{k-1} + mk\,y_2 y_{k-2} + \cdots$$
$$+ \left(\tfrac{m}{2}k\,y_{k/2}^2 [k \text{ even}]\right). \quad (3.8)$$

For example, $g_4(y_1, y_2, y_3) = 4y_1 y_3 + 2y_2^2$ and $g_5(y_1, y_2, y_3, y_4) = 5y_1 y_4 + 5y_2 y_3$ when $m = 1$.

The individual terms of (3.8) are independent, and they have monotone decreasing density functions mod 1. (The density function for the probability that $\{kxy\} \in [t \ldots t+dt)$ is $\sum_{j=0}^{k-1} \tfrac{1}{k} \ln \tfrac{k}{j+t}\, dt$.) Therefore they cannot possibly yield a uniform distribution. For if $f(x)$ is the density function for a random variable on $[0 \ldots 1)$, we have $E(e^{2\pi i X}) = \int_0^1 e^{2\pi i x} f(x)\, dx \ne 0$ when $f(x)$ is monotone; for example, if $f(x)$ is decreasing, the imaginary part is $\int_0^{1/2} \sin(2\pi x)\big(f(x) - f(1 - x)\big)\, dx > 0$. If Y is an independent random variable with monotone density, we have $E(e^{2\pi i \{X+Y\}}) = E(e^{2\pi i (X+Y)}) = E(e^{2\pi i X})E(e^{2\pi i Y}) \ne 0$. But

$E(e^{2\pi i U}) = 0$ when U is a uniform deviate. Therefore (3.8) cannot be uniform mod 1. □

Now we can deduce properties of sequences like

$$(\alpha n)^{:k} = \alpha n \lfloor \alpha n \lfloor \dots \lfloor \alpha n \rfloor \dots \rfloor \rfloor$$

as n runs through integer values.

Theorem. *If the powers $\alpha^2, \dots, \alpha^{k-1}$ are irrational, then the sequence $\{m(\alpha n)^k - km(\alpha n)^{:k}\}$ for $n = 1, 2, \dots$ is not uniformly distributed in $[0..1)$ for any integer m.*

Proof. This result is trivial when $k = 1$ and obvious when $k = 2$, since $\{(\alpha n)^2 - 2(\alpha n)^{:2}\} = \{\alpha n\}^2$. But for large values of k it seems to require a careful analysis. By Lemma 1 we have

$$\{m(\alpha n)^k - km(\alpha n)^{:k}\} = \left\{ km f_{k,1}\left(\left\{\frac{\alpha n}{k!}\right\}, \dots, \left\{\frac{\alpha^{k-1} n^{k-1}}{k!}\right\}\right)\right\},$$
(3.9)

and Lemma 2 tells that $\{km f_{k,1}\}$ is not measure preserving.

Let S be an interval of $[0..1)$, and T its inverse image in $[0..1)^{k-1}$ under $\{k f_{k,1}\}$, where $\mu(T) \neq \mu(S)$. It is easy to see that if $(y_1, \dots, y_{k-1})$ is in T and $y_1, \dots, y_{k-1}$ are irrational, there are values $\epsilon_1, \dots, \epsilon_{k-1}$ such that $[y_1 .. y_1 + \epsilon_1) \times \dots \times [y_{k-1} .. y_{k-1} + \epsilon_{k-1}) \subseteq T$. Therefore the irrational points of T can be covered by disjoint half-open hyperrectangles. We will show that (3.9) is not uniform by using Theorem 6.4 of [3], which implies that the sequence $(\{\alpha_1 n^{e_1}\}, \dots, \{\alpha_s n^{e_s}\})$ is uniformly distributed in $[0..1)^s$ whenever $\alpha_1, \dots, \alpha_s$ are irrational numbers and the integer exponents $e_1, \dots, e_s$ are distinct. Thus the probability that $\{(\alpha n)^k - k(\alpha n)^{:k}\} \in S$ approaches $\mu(T)$ as $n \to \infty$; the distribution is nonuniform. □

Corollary. *If the powers $\alpha^2, \dots, \alpha^{k-1}$ are irrational, the sequence $\{(\alpha n)^{:k}\}$ for $n = 1, 2, \dots$ is uniformly distributed in $[0..1)$ if and only if α^k is irrational.*

Proof. If α^k is irrational, $\{\alpha^k n^k/k\}$ is uniformly distributed in $[0..1)$ and independent of $(\{\alpha n/k!\}, \dots, \{\alpha^{k-1} n^{k-1}/k!\})$, by the theorem quoted above from [3]. Therefore the right-hand side of (3.1) is uniform, even if some of the powers $\alpha^2, \dots, \alpha^{k-1}$ are rational. (Rational components can be ignored because the sequence is periodic there.)

If α^k is rational, say $\alpha^k = p/q$, assume that $\{(\alpha n)^{:k}\}$ is uniform. Then $\{q(\alpha^k n^k - k(\alpha n)^{:k})\} = \{-qk(\alpha n)^{:k}\}$ is also uniform, contradicting what we proved. □

We conjecture that the theorem and its corollary remain true for all real α, without the hypothesis that $\alpha^2, \ldots, \alpha^{k-1}$ are irrational.

References

[1] Inger Johanne Håland, "Uniform distribution of generalized polynomials," *Journal of Number Theory* **45** (1993), 327–366.

[2] Inger Johanne Håland, "Uniform distribution of generalized polynomials of the product type," *Acta Arithmetica* **67** (1994), 13–27.

[3] L. Kuipers and H. Niederreiter, *Uniform Distribution of Sequences* (New York: Wiley, 1974).

Chapter 17

Construction of a Random Sequence

A completely equidistributed sequence of real numbers is constructed explicitly.

[Originally published in BIT **5** *(1965), 246–250.]*

It is a familiar fact that nobody has ever given a satisfactory definition of what a random sequence really is, even though we speak frequently about such sequences. If "random" means that the sequence satisfies no predictable rules, the title of this paper is contradictory. On the other hand, there are good reasons to formulate a definition of randomness in terms of simple properties satisfied by the sequence, and such a definition does not preclude the possible construction of such sequences. The sequences will be random in the sense that they possess the statistical properties common to all random sequences.

Several significant steps towards this sort of definition of a random sequence have been made in an important paper by Franklin [3]. He shows that if the sequence is completely equidistributed (in the sense defined below), a great many other statistical tests for randomness are valid. For example, the autocorrelation between X_n and X_{n+k} is zero, for all fixed k; the probability that k adjacent elements X_{n+1}, X_{n+2}, ..., X_{n+k} have any given relative order is $1/k!$; and the sequence has the proper distribution of ascending runs of any given length.

Franklin [3] proved that, if θ is a real number, the sequence of fractional parts $\{\theta^n\}$ is completely equidistributed for *almost all* $\theta > 1$, and also that it is necessary for θ to be transcendental. But there is still no known number θ for which this property has been verified, nor has anyone ever exhibited an explicit sequence that is completely equidistributed. The purpose of the present paper is to exhibit one way to construct such a sequence.

Equidistribution

If $\langle X_1, X_2, \ldots \rangle$ is any infinite sequence, and if S_n is any statement about this sequence depending on n, we define as usual

$$\Pr(S_n) = \lim_{N \to \infty} \nu_N / N, \tag{1}$$

provided the limit exists, where ν_N is the number of positive integers $n \leq N$ for which S_n is true. For example, if $\langle X_1, X_2, \ldots \rangle$ is a sequence of random real numbers, we would expect that $\Pr(X_n > X_{n+1}) = \frac{1}{2}$. Even when the limit does not exist, we define $\overline{\Pr}(S_n)$ and $\underline{\Pr}(S_n)$ to be the superior and inferior limits in (1).

The following statement is obvious by addition of limits:

Lemma 1. *If at least three of the limits exist,*

$$\Pr(S_n \text{ or } T_n) + \Pr(S_n \text{ and } T_n) = \Pr(S_n) + \Pr(T_n). \quad \square \tag{2}$$

In this note we will consider sequences of two kinds: real-valued sequences where the numbers X_n lie in the interval $[0 \mathinner{.\,.} 1)$, and discrete "b-ary" sequences where the numbers X_n are integers $0 \leq X_n < b$ for some integer $b \geq 2$.

A real-valued sequence $\langle X_1, X_2, \ldots \rangle$ is said to be *k-distributed* if

$$\Pr(u_1 \leq X_{n+1} < v_1, \ u_2 \leq X_{n+2} < v_2, \ \ldots, \ u_k \leq X_{n+k} < v_k)$$
$$= (v_1 - u_1)(v_2 - u_2) \ldots (v_k - u_k) \tag{3}$$

for all choices of u_j and v_j with $0 \leq u_j \leq v_j \leq 1$ for $1 \leq j \leq k$. In other words, the vectors $(X_{n+1}, \ldots, X_{n+k})$ should asymptotically have uniform density in the k-cube. Similarly, a b-ary sequence is said to be k-distributed if

$$\Pr(X_{n+1} = a_1, \ X_{n+2} = a_2, \ \ldots, \ X_{n+k} = a_k) = 1/b^k \tag{4}$$

for all ordered k-tuples of b-ary integers $(a_1, a_2, \ldots, a_k)$.

One obvious consequence of these definitions, in view of Lemma 1, is the fact that a k-distributed sequence is l-distributed for each $l \leq k$. A sequence is said to be ∞-distributed, or "completely equidistributed," if it is k-distributed for all k.

A b-ary sequence can be regarded as the representation $(.X_1 X_2 \ldots)_b$ of a real number in the radix b number system. Real numbers that correspond in this way to completely equidistributed b-ary sequences are well-known in mathematics (see [6, Chapter 9]), and they are customarily called "normal numbers to base b." The simplest known construction of normal b-ary numbers is due to Champernowne [1], and the method used below is related to his construction.

Lemma 2. *Let $\langle b_1, b_2, b_3, \ldots \rangle$ be an unbounded sequence of positive integers. The real sequence $\langle X_1, X_2, \ldots \rangle$ is completely equidistributed if and only if the discrete b_j-ary sequences*

$$\langle \lfloor b_j X_1 \rfloor, \lfloor b_j X_2 \rfloor, \ldots \rangle$$

are completely equidistributed for all $j \geq 1$.

Proof. If $\langle X_1, X_2, \ldots \rangle$ is ∞-distributed and real-valued, the sequence $\langle \lfloor b X_1 \rfloor, \lfloor b X_2 \rfloor, \ldots \rangle$ is clearly ∞-distributed, since (4) will follow from (3) by taking $u_j = a_j/b$ and $v_j = (a_j + 1)/b$.

Conversely if $\langle \lfloor b X_1 \rfloor, \lfloor b X_2 \rfloor, \ldots \rangle$ is an ∞-distributed b-ary sequence, it follows from Lemma 1 that (3) holds whenever all of the parameters u_j and v_j are integer multiples of $1/b$. If u'_j and v'_j are multiples of $1/b$ with $u'_j \leq u_j \leq u'_j + 1/b$ and $v'_j \leq v_j \leq v'_j + 1/b$, then

$$\begin{aligned}
\overline{\Pr}(u_j \leq X_j &< v_j \text{ for } 1 \leq j \leq k) \\
&\leq \Pr(u'_j \leq X_j < v'_j + 1/b \text{ for } 1 \leq j \leq k) \\
&= (v'_1 - u'_1 + 1/b) \ldots (v'_k - u'_k + 1/b) \\
&\leq (v_1 - u_1 + 2/b) \ldots (v_k - u_k + 2/b)
\end{aligned}$$

and similarly

$$\begin{aligned}
\underline{\Pr}(u_j \leq X_j &< v_j \text{ for } 1 \leq j \leq k) \\
&\geq (\max(v_1 - u_1 - 2/b, 0)) \ldots (\max(v_k - u_k - 2/b, 0)).
\end{aligned}$$

If these inequalities hold for arbitrarily large b, (3) is established.   □

A Completely Equidistributed Sequence

A periodic sequence of period p certainly cannot be p-distributed; but we have the following "best possible" result for periodic sequences, due to Good [5]:

Lemma 3. *Let b and k be integers ≥ 2. There is a periodic b-ary sequence $\langle X_1, X_2, \ldots \rangle$, with $X_{n+b^k} = X_n$, that is k-distributed.*   □

In other words, each of the b^k possible k-tuples $(X_{n+1}, \ldots, X_{n+k})$ occurs exactly once in the period.

A simple construction of such a sequence has been given by Ford [2]; we may choose $X_1 = X_2 = \cdots = X_k = 0$ and then choose X_{n+k} for $0 < n < b^k$ in any manner consistent with the following rule:

"$(X_{n+1}, \ldots, X_{n+k})$ does not duplicate a previous k-tuple, and $X_{n+k} = 0$ only if the k-tuples $(X_{n+1}, \ldots, X_{n+k-1}, j)$ have already appeared in the sequence for $0 < j < b$."

For example if $b = k = 3$, and if we repeatedly choose X_{n+k} to be the smallest value consistent with the stated rule, we obtain

$$000111211011221201021002220200 \tag{5}$$

and the two zeros at the right begin the repetition of the sequence. It is easy to prove that Ford's construction works (that is, to show that X_{n+k} can always be chosen and that every k-tuple $(a_1, a_2, \ldots, a_k)$ appears in the sequence) by induction on r, where $(a_1, a_2, \ldots, a_k)$ has precisely $k - r$ zeros at the right. Ford's construction can be greatly generalized (see [4]).

Now let $A(b, k)$ be the finite sequence consisting of the first b^k terms of some periodic sequence as in Lemma 3, with each term divided by b. Thus, each element of $A(b, k)$ is a real number in $[0 \, . \, . \, 1)$. Let $A(b, k)^n$ denote the sequence $A(b, k)$ repeated n times.

Theorem. *The real-valued sequence*

$$\left\langle A(2, 1)^{2^2}, A(2^2, 2)^{2 \cdot 2^4}, A(2^3, 3)^{3 \cdot 2^6}, \ldots \right\rangle = \langle X_1, X_2, \ldots \rangle \tag{6}$$

is completely equidistributed. (See Figure 1.)

Proof. By Lemma 2, it suffices to prove that the sequence $\langle \lfloor 2^m X_1 \rfloor, \lfloor 2^m X_2 \rfloor, \ldots \rangle$ is k-distributed for all m and k. This means we must look at the first m bits of the numbers X_j.

Let $(a_1, \ldots, a_k)$ be any ordered k-tuple of m-bit integers, and let r be an integer with $r \geq m$, $r \geq k$. Within $A(2^r, r)$, we will have $\left(\lfloor 2^r X_{n+1} \rfloor, \ldots, \lfloor 2^r X_{n+k} \rfloor \right)$ equal to a given k-tuple of r-bit integers $2^{r(r-k)}$ times, since each r-tuple occurs once. Since 2^{r-m} different values of the numbers X_j will give the same value of $\lfloor 2^m X_j \rfloor$ we see that $(a_1, \ldots, a_k)$ corresponds to $(2^{r-m})^k$ of these k-tuples of r-bit integers, and we will have $\left(\lfloor 2^m X_{n+1} \rfloor, \ldots, \lfloor 2^m X_{n+k} \rfloor \right) = (a_1, \ldots, a_k)$ a total of $2^{r(r-k)} 2^{(r-m)k} = 2^{r^2 - mk}$ times. These remarks must be slightly qualified, since we have ignored effects that occur at the right-hand end of the sequence $A(2^r, r)$; the true total number is

$$2^{r^2 - mk} + \epsilon, \qquad \text{where } |\epsilon| \leq k. \tag{7}$$

$$\left.\begin{array}{l} X_1 = .0 \\ X_2 = .1 \end{array}\right\} \text{repeat 3 more times, to obtain } X_3,\ X_4,\ \ldots,\ X_8$$

$$\left.\begin{array}{l} X_9 = .00 \\ X_{10} = .00 \\ X_{11} = .01 \\ X_{12} = .01 \\ X_{13} = .10 \\ X_{14} = .01 \\ X_{15} = .11 \\ X_{16} = .01 \\ X_{17} = .00 \\ X_{18} = .10 \\ X_{19} = .10 \\ X_{20} = .11 \\ X_{21} = .10 \\ X_{22} = .00 \\ X_{23} = .11 \\ X_{24} = .11 \end{array}\right\} \text{repeat 31 more times, to obtain } X_{25},\ X_{26},\ \ldots,\ X_{520}$$

$$\left.\begin{array}{l} X_{521} = .000 \\ X_{522} = .000 \\ X_{523} = .000 \\ X_{524} = .001 \\ \qquad \vdots \\ X_{1032} = .111 \end{array}\right\} \text{repeat } 3\cdot 2^6 - 1 = 191 \text{ more times, to obtain } X_{1033},\ \ldots,\ X_{98824}$$

FIGURE 1. The first few terms of the sequence, represented in the binary number system.

Assume now that N is large, and that the Nth term of the sequence (6) belongs to $A(2^r, r)$; then there are integers p and q such that

$$N = \sum_{s=1}^{r-1}(s2^{2s})2^{s^2} + q2^{r^2} + p, \quad 0 \le q < r2^{2r}, \quad 1 \le p \le 2^{r^2}.$$

The number of $n \le N$ for which the k-tuple $\left(\lfloor 2^m X_{n+1}\rfloor, \ldots, \lfloor 2^m X_{n+k}\rfloor\right)$ has a given value $(a_1, \ldots, a_k)$ is then

$$\nu_N = \sum_{s=1}^{r-1}(s2^{2s})2^{s^2-mk} + \sum_{s=1}^{r-1}(s2^{2s})O(k)$$
$$+ q2^{r^2-mk} + O(kr2^{2r}) + O(2^{r^2}) + O(2^{r^2}) + O\left((m+k)2^{(m+k)^2}\right)$$

where the O-terms account for the ϵ's of (7) and the value of p, and for the fact that formula (7) is true only for $r \geq \max(m, k)$. Therefore $\nu_N = 2^{-mk}N + O(2^{r^2})$ for any fixed values of k and m. And since $N > (r-1)2^{2(r-1)}2^{(r-1)^2}$, we have $2^{r^2}/N < 2/(r-1)$, and $\nu_N/N \to 2^{-mk}$ as $r \to \infty$. This completes the proof of the theorem.   □

The sequence constructed here would be quite useless in practice as a computer method for random number generation, since the construction is rather complicated and since convergence to the asymptotic behavior is extremely slow. By contrast, Franklin's sequence $\{\pi\}$, $\{\pi^2\}$, $\{\pi^3\}$, ... appears to be an excellent source of random numbers, although nothing is known about its equidistribution properties from a theoretical standpoint.

Note added in proof: Since this paper was written, it has come to the author's attention that the results of Good and Ford were anticipated by M. H. Martin, "A problem in arrangements," *Bulletin of the American Mathematical Society* **40** (1934), 859–864.

References

[1] D. G. Champernowne, "The construction of decimals normal in the scale of ten," *Journal of the London Mathematical Society* **8** (1933), 254–260.

[2] L. R. Ford, Jr., *A Cyclic Arrangement of m-tuples*, report P-1071 (Santa Monica, California: Rand Corporation, 1957).

[3] Joel N. Franklin, "Deterministic simulation of random processes," *Mathematics of Computation* **17** (1963), 28–57.

[4] S. W. Golomb and L. R. Welch, *Nonlinear Shift-Register Sequences*, Memorandum 20-149 (Pasadena, California: Jet Propulsion Laboratory, 1957).

[5] I. J. Good, "Normal recurring decimals," *Journal of the London Mathematical Society* **21** (1946), 167–169.

[6] G. H. Hardy and E. M. Wright, *An Introduction to the Theory of Numbers*, 5th edition (Oxford: Clarendon Press, 1979).

Chapter 18

An Imaginary Number System

*[Originally published in Communications of the ACM **3** (1960), 245–247;
4 (1961), 355. "The author's first publication about mathematics."]*

The decimal number system has reigned supreme for centuries, except,
perhaps, among the Mayan Indians, but the advent of digital computers
has brought the binary and octal number systems into the limelight.
This paper introduces another number system, which may prove useful
for manipulating complex numbers on machines.

The "quater-imaginary" system uses the imaginary number $2i$ (or,
$2j$ for electrical engineers) as its base or radix. That is, the number

$$d_5 d_4 d_3 d_2 d_1 d_0 . d_{-1} d_{-2} d_{-3}$$

is the quater-imaginary representation for

$$d_5(32i) + d_4(16) + d_3(-8i) + d_2(-4) + d_1(2i) + d_0$$
$$+ d_{-1}\left(-\frac{i}{2}\right) + d_{-2}\left(-\frac{1}{4}\right) + d_{-3}\left(\frac{i}{8}\right).$$

Every complex number can be represented in a unique way in the
quater-imaginary system in terms of the digits 0, 1, 2, and 3. Quater-
imaginary numbers are signless; for example, the number "plus unity"
is represented as

$$1.$$

and the number "minus unity" is represented as

$$103.$$

"Plus i" is the number

$$10.2$$

while "minus i" takes the form 0.2 in this system.

271

In general, let $z = x + iy$ be any complex number, and let

$$A = \pm\, a_p a_{p-1} \ldots a_0.a_{-1} \ldots a_{-q}$$

and

$$B = \pm\, b_r b_{r-1} \ldots b_0.b_{-1} \ldots b_{-s}$$

be representations of x and $\frac{1}{2}y$, respectively, in the quaternary (base 4) number system. The integers q and s may be infinite by convention. The representations of A and B are, of course, easily obtained from the binary forms of x and y. Adjust p and r by adding a leading zero, if necessary, so that they are both even; and similarly adjust q and s so that they are either infinite or odd.

If A is positive, form a new representation A' where

$$a'_k = \begin{cases} 1 + a_k, & k \text{ even}, \\ 4 - a_k, & k \text{ odd}, \end{cases}$$

for $k = p,\, p-1,\, \ldots,\, -q$. Then A' contains only the digits 1, 2, 3, and 4. Now form A'' by replacing the fours with zeros and lowering the digits immediately to their left by one. If the digit to the left of the 4 was zero, change it to a 3 and increase the preceding digit by 1; this process eventually terminates.

If A is negative, form a new representation A' where

$$a'_k = \begin{cases} 4 - a_k, & k \text{ even}, \\ 1 + a_k, & k \text{ odd}, \end{cases}$$

for $k = p,\, p-1,\, \ldots,\, -q+1$, and let $a'_{p+1} = 1$, $a'_{-q} = a_{-q}$. Then go to A'' by getting rid of the 4s that may occur, as above.

For example, suppose A is $+1.23013$. Then A' is 2.24421, and A'' is 2.23021. And for $A = -1.23013$, we would have $A' = 13.31133 = A''$. Notice that A'' is now the number x expressed in the system with radix *minus four*, because

$$-1.23013 = 3.01030 - 10.30103$$

in the quaternary system; a study of this example will reveal the motivation underlying the conversion routine.

The process for going from A to A'' is illustrated in the flowchart (Fig. 1). The same process[1] should now be applied to B to obtain B''. The final number

$$z = d_u d_{u-1} \ldots d_0.d_{-1} \ldots d_{-v},$$

---

[1] Faster methods for the conversion process may be obtained in an obvious manner by considering groups of consecutive 4's.

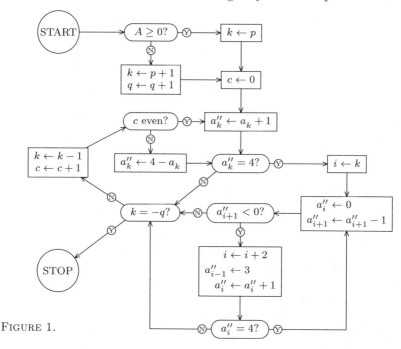

FIGURE 1.

where $u = 2\max(p + 2, r + 2)$, $v = 2\max(q, s)$, is then defined by

$$d_k = \begin{cases} a''_n, & k = 2n \\ b''_n, & k = 2n + 1 \end{cases}$$

for $k = u, u - 1, \ldots, -v$; and this is the required quater-imaginary representation of z.

Addition and Multiplication

The new system may prove useful because of the way it facilitates multiplication. The rules for addition and multiplication are identical with those of the ordinary quaternary system, with the variation that carrying is done by *subtracting* one from the column *two over*, rather than adding one to the adjacent column. "Borrowing," if necessary, is done by *adding* one in the column *two over*, rather than subtracting one from the adjacent column.

For example,

$$2 + 2 = 4 = \bar{1}00 = 10300$$

if $\bar{1}$ denotes the (temporary) digit -1.

With these ground rules, some simple multiplications are carried out as follows:

1)
$$\begin{array}{r} 10.2 \quad [i] \\ 10.2 \quad [i] \\ \hline 1.00 \\ 102 \\ \hline 103.00 \quad [-1] \end{array}$$

2)
$$\begin{array}{r} 113311 \quad [5 + 10i] \\ 10210 \quad [8 + 2i] \\ \hline 1133110 \\ 112222 \\ 1133110 \\ \hline 0000321310 \quad [20 + 90i] \end{array}$$

Clearly a floating point notation could be devised for working with quater-imaginary numbers. The biggest problem is to find a good division algorithm; most processes of division rely heavily on the order properties of the real numbers, and division in this new system appears to be fairly difficult. Examples of long division, or square-root extraction, appear to require a stroke of genius when trial divisors are being picked.

Long division: $[(1 + 5i)/(5 - i) = i]$

$$\begin{array}{r} 10.20 \\ 10301.2 \overline{)\, 000031.200} \\ 103012 \\ \hline 1123\ 20 \\ 1123\ 20 \\ \hline 0 \end{array}$$

The Bi-Imaginary System

We can just as easily consider "n-imaginary" number systems,[2] where the radix is taken to be $\sqrt{-n}$. Here n is any integer greater than 1, and the digits needed for the representation are 0, 1, ..., $n - 1$. The development above, except for the examples, can be extended merely by

---

[2] The idea of extending this theory to numbers n that are not perfect squares was suggested to the author by Dr. Arthur Wouk.

replacing '4' by 'n' and '3' by '$n-1$' wherever they occur, and by letting B be the representation of $y/\sqrt{n}$.

In particular, the flavor of the binary system can be retained in an imaginary number system by taking $n = 2$. In this "bi-imaginary" number system the only digits used are 0 and 1, and the system obtained is thus well suited to incorporation into electronic circuitry. The division problem might be more simply solved in such a system, because only two choices (whether to subtract once or not at all) need to be made as each quotient digit is being produced. Unfortunately, the number i itself must be represented in a nonterminating way in the bi-imaginary system; hence truncation and rounding errors may be more common in practice than if the quater-imaginary system were used.

Another property of the bi-imaginary system is its uniqueness: Every complex number has one and only one representation in the system. This fact may be applied, for example, to give more "elegant" proofs of mathematical theorems such as the nonenumerability of the complex numbers.

The apparent inability to divide easily may be enough of an obstacle to make the n-imaginary systems impractical to incorporate into any hardware; so here is a challenge to the reader, to find an algorithm for bi-imaginary division. If anyone can dream up such a scheme, it would certainly speed up complex number manipulations inside machines.

Errata

The article above claims repeatedly that every complex number has a *unique* representation in an n-imaginary number system. This rash statement was based simply on the fact that the author could not think of any numbers with a non-unique representation, after trying a few of the more obvious cases like integers for uniqueness. There are, however, lots of numbers with non-unique representations — they are even "dense" in the set of complex numbers — although they have only four different representations at most. But this non-uniqueness is no worse than the situation in the decimal system; for example, $1 + i$ has four representations:

$$(1.000\ldots) + (1.000\ldots)i, \qquad (0.999\ldots) + (1.000\ldots)i,$$
$$(1.000\ldots) + (0.999\ldots)i, \qquad (0.999\ldots) + (0.999\ldots)i.$$

Non-uniqueness does not hurt logical operations in any way. It merely affects mathematical proofs about the complex numbers or things of that kind.

Let us look further, in order to see exactly which numbers have non-unique representations. Any number in the n-imaginary system

$$\ldots a_3 a_2 a_1 a_0 . a_{-1} a_{-2} a_{-3} a_{-4} \ldots$$

has its real part given by $\ldots a_2 a_0 . a_{-2} a_{-4} \ldots$ and its imaginary part given by $\sqrt{n}$ times $\ldots a_3 a_1 . a_{-1} a_{-3} \ldots$, where the latter numbers are expressed the number system with radix $-n$. It suffices therefore to consider the uniqueness problem for radix $-n$.

In this system it is easy to see that the minimum value the number $(0.b_1 b_2 b_3 b_4 b_5 b_6 \ldots)_{-n}$ can have is

$$(0.m0m0m0\ldots)_{-n} \; = \; -\frac{n}{n+1},$$

where $m = n - 1$; and the maximum possible value is

$$(0.0m0m0m\ldots)_{-n} \; = \; +\frac{1}{n+1}.$$

This analysis shows that the numbers to the left of the decimal point in a representation of the real number θ are uniquely determined, unless the fractional part $\{\theta\}$ is equal to $1/(n+1)$. And if $\{\theta\} = 1/(n+1)$ there are exactly two choices; for example, $(1.m0m0m0\ldots)_{-n}$ and $(0.0m0m0m\ldots)_{-n}$ are the two ways to represent $1/(n+1)$.

Finally, therefore, it is easy to observe the following facts:

a) A real number θ has more than one representation in the radix-$(-n)$ number system if and only if there is an integer r such that $\{(-n)^r \theta\} = 1/(n+1)$.

b) A complex number has more than one representation in the n-imaginary system if and only if either its real part or $\sqrt{n}$ times its imaginary part have non-unique representation in the $(-n)$ system.

Addendum

This note was my first "mathematical" publication, composed during my senior year in college. Not only did I make an embarrassing error about uniqueness of representation, I also evidently believed that the quater-imaginary system would make multiplication faster; yet it does *not* actually decrease the number of bit operations needed by the ordinary multiplication procedure for complex numbers.

Thus the only possible merit of this paper, besides the cute coincidence that "imaginary" sort of rhymes with "binary," is that it explores a useless but curious corner of mathematics that had not previously been considered. I guess it proves most of all that I've never been able to resist rushing into print with half-baked ideas.

Chapter 19

Tables of Finite Fields

This paper contains a number of tables designed to facilitate calculations with finite fields, by hand or by digital computer. A theoretical discussion of computational techniques useful in preparing such tables is also given.

[Written with J. D. Alanen. Originally published in Sankhyā, The Indian Journal of Statistics **A26** (1964), 305–328. New material has been added to Sections 4 and 5.]

0. Introduction

Finite fields have proved to be very useful in the construction of orthogonal latin squares, block designs, factorial designs, error-correcting codes, and many other combinatorial designs. The advent of digital computers makes it possible to work in larger fields than was practicable by hand, so there is a new need for more complete tables of finite fields.

About fifty years ago, W. H. Bussey [3, 4] published extensive tables for fields with less than 1000 elements. Because his tables are rather bulky and not always easy to obtain, there is an overlap between his tables and those presented here.

The tables below are tailored for three different applications:

1) Hand computation using fields with 128 elements or less.

2) Computation by computer using fields with less than 10^9 elements.

3) Applications where *all* polynomials that generate a field are required, not merely a single polynomial, using fields with at most 1024 elements.

A summary of elementary field theory is given, followed by a description of the method by which various tables were prepared. The reader who merely wants to use the tables may omit Sections 2 and 3 of this paper. Section 4 describes the tables and demonstrates how to use them. Appendix 1 gives a simple method for obtaining all of the indexing polynomials from only one.

1. Notation and Elementary Theory

A finite set of q objects that satisfy the axioms for an algebraic field is called a Galois field, GF(q). For details of finite field theory the reader is referred to [1] where they are discussed in great detail. The following elementary facts are well known:

1) GF(q) exists if and only if q is a power of a prime number. Hereafter we will assume that $q = p^n$, where p is prime and $n \geq 1$ is an integer.

2) Two finite fields with the same number of elements are isomorphic. Therefore, we speak of *the* field GF(q).

3) The multiplication in GF(q) is cyclic; that is, there is an element π such that the q elements of GF(q) are 0, 1, π, π^2, ..., π^{q-2}. Here $\pi^{q-1} = 1$ and we call π a *primitive element* of the field.

4) When $n = 1$, the finite field GF(q) can be taken to be the set of integers 0, 1, 2, ..., $p - 1$, with addition and multiplication defined modulo p. In this case π is called a *primitive root* of the prime p.

5) GF(q) is a vector space over GF(p) with basis $\{1, \pi, \pi^2, \ldots, \pi^{n-1}\}$. That is, every element x can be written uniquely as a polynomial, $x = c_1\pi^{n-1} + c_2\pi^{n-2} + \cdots + c_n$, where $0 \leq c_i < p$ for $i = 1, \ldots, n$.

   Alternatively we can use vector notation and write

$$x = (c_1, c_2, \ldots, c_n).$$

The addition or subtraction of two vectors is done component by component, modulo p. Multiplication and division cannot be done in such a simple way, hence tables are prepared to facilitate those operations.

Multiplication by zero naturally gives zero. Multiplication and division of nonzero elements can be performed using a notation analogous to logarithms. The term "index" is used rather than logarithm.

Any nonzero vector $x = (c_1, c_2, \ldots, c_n)$ can be written as π^r for some integer r in the range $0 \leq r < q - 1$. We write $r = \operatorname{ind} x$. If $x = \pi^r$ and $y = \pi^s$, the product $xy = \pi^{r+s}$. Since $\pi^{q-1} = 1$, we obtain the basic rules for the index function:

$$\operatorname{ind} xy = (\operatorname{ind} x + \operatorname{ind} y) \bmod (q - 1); \tag{1}$$

$$\operatorname{ind} x/y = (\operatorname{ind} x - \operatorname{ind} y) \bmod (q - 1). \tag{2}$$

The function inverse to ind, corresponding to taking antilogarithms, is called the exp function, and we have

$$\exp r = \pi^r; \tag{3}$$

$$\exp(\operatorname{ind} x) = x, \qquad \operatorname{ind}(\exp r) = r. \tag{4}$$

Given a table of exp and ind functions, it is easy to carry out the four basic operations of arithmetic. Addition and subtraction are done by using vector notation; multiplication and division are done by using exponential notation; exp and ind provide the conversion from one notation to the other. Examples are given in Section 4.

2. Indexing Polynomials

Since every element of GF(q) can be expressed in terms of the basis $\{1, \pi, \pi^2, \ldots, \pi^{n-1}\}$, we have in particular

$$-\pi^n = a_1 \pi^{n-1} + a_2 \pi^{n-2} + \cdots + a_n, \tag{5}$$

for some integers a_i with $0 \le a_1, \ldots, a_n < p$. This relationship completely determines the multiplication of all elements of the field, for it allows us to calculate the exp function:

$$
\begin{aligned}
\text{If} \qquad \pi^r &= c_1 \pi^{n-1} + \cdots + c_{n-1} \pi + c_n, & (6)\\
\text{then} \qquad \pi^{r+1} &= c_1 \pi^n + \cdots + c_{n-1} \pi^2 + c_n \pi \\
&= (c_2 - c_1 a_1)\pi^{n-1} + \cdots + (c_n - c_1 a_{n-1})\pi - c_1 a_n. & (7)
\end{aligned}
$$

Let $f(x) = x^n + a_1 x^{n-1} + \cdots + a_n$. Then π is a root of $f(x)$ in GF(q), and the polynomial f must be irreducible, modulo p. For if we have $f(x) \equiv g(x)h(x)$ (modulo p), the degrees of g and h cannot both be less than n; otherwise we would have $0 = f(\pi) = g(\pi)h(\pi)$, and that cannot happen since $g(\pi) \ne 0$ and $h(\pi) \ne 0$.

Conversely, if x is any root of an nth degree polynomial that is irreducible modulo p, the set $\{1, x, x^2, \ldots, x^{n-1}\}$ forms a basis for the elements of GF(p^n); see [1]. But such a root is not necessarily a primitive element of the field. The polynomial f defined by the coefficients in (5) satisfies a stronger condition than irreducibility, for any root of $f(x)$ will not only generate GF(q) but will also be primitive. Polynomials that satisfy this stronger condition have been called "primitive polynomials"; however, the same term has also been used to describe polynomials whose coefficients are relatively prime. Therefore we will call $f(x)$ an *indexing polynomial*.

In view of these remarks, the construction of tables for GF(q) is reduced to a hunt for an indexing polynomial modulo p, of degree n. No general method for constructing such polynomials is known, and they apparently conform to no pattern.

If $n + 1$ is prime and if the prime number p is a primitive root of $n + 1$, it is known that the polynomial $f(x) = x^n + x^{n-1} + \cdots + x + 1$

is irreducible modulo p. For if $g(x)$ is an irreducible factor of $f(x)$ of degree m, then $g(x)$ will generate the finite field $GF(p^m)$, so $x^{p^m-1} - 1$ will be divisible by $g(x)$, modulo p. But $x^k - 1$ has a factor in common with $f(x)$ if and only if k is a multiple of $n+1$, when $n+1$ is prime; hence $p^m - 1$ is a multiple of $n + 1$, so we must have $m = n$.[1] Swift [10] has utilized this fact in the construction of fields $GF(2^n)$. But this choice of $f(x)$ is not always an indexing polynomial; it is merely irreducible.

A search for indexing polynomials is thus largely a matter of trial and error, as far as anybody knows. In the next section, however, some results will be illustrated to show how the search can be considerably reduced by making educated guesses.

3. Exponent of a Polynomial

In this paper we assume that all polynomials have integer coefficients. We write
$$f(x) \equiv 0 \quad (\text{modd } g(x), h(x))$$
if $f(x) = a(x)g(x) + b(x)h(x)$ for some polynomials $a(x)$ and $b(x)$. With this notation, the field $GF(q)$ with indexing polynomial $f(x)$ can be thought of as a set of residues (modd $f(x), p$).

Suppose $f(x)$ is a polynomial whose constant term is not a multiple of p. We say m is the *exponent* of $f(x)$ modulo p, if m is the least positive integer such that
$$x^m \equiv 1 \quad (\text{modd } f(x), p).$$

Another way of phrasing this condition is to say that $f(x)$ is a factor of $x^m - 1$, modulo p.

Since there are only p^n residues (modd $f(x), p$), we must have $x^r \equiv x^s$ (modd $f(x), p$) for some integers $r > s$. The constant term of $f(x)$ is nonzero, so $f(x)$ and x^s are relatively prime (modulo p), and there are polynomials $g(x), h(x)$ such that
$$g(x)f(x) + h(x)x^s \equiv 1 \quad (\text{modulo } p).$$

Thus $g(x)f(x)x^{r-s} + h(x)x^r \equiv x^{r-s}$ (modulo p), proving that
$$x^{r-s} \equiv h(x)x^r \equiv h(x)x^s \equiv 1 \quad (\text{modd } f(x), p).$$

---

[1] This argument can be generalized to the case $n = \varphi(r)$ and $f(x) = \Psi_r(x)$, if the order of p modulo r is n. See Appendix 1 for definitions of $\varphi(r)$ and $\Psi_r(x)$.

This reasoning shows that every polynomial with a nonzero constant term has a finite exponent m modulo p, and that the polynomials

$$1, \quad x, \quad x^2, \quad \ldots, \quad x^{m-1}$$

are mutually incongruent (modd $f(x), p$).

Clearly $x^r \not\equiv 0$ (modd $f(x), p$), so we have $m \leq p^n - 1$, there being p^n residues possible. It is not true in general that m must divide $p^n - 1$; for example,[2] the polynomial $x^5 + x + 1$ has exponent 21 modulo 2.

If, however, $m = p^n - 1$, we can conclude that $f(x)$ is irreducible, for we can use $f(x)$ to construct a finite field. In this case $f(x)$ is in fact an indexing polynomial. Thus we can state

Lemma 1. *A polynomial of degree n is an indexing polynomial modulo p if and only if its exponent is $p^n - 1$.* (No test for irreducibility need be made.) □

If $x^r \equiv a$ (modd $f(x), p$), where a is an integer, we have $x^{r+s} \equiv ax^s$. From this fact we can easily derive the following law:

Lemma 2. *Let r be the least positive integer such that there is an integer a with $x^r \equiv a$ (modd $f(x), p$), and let t be the least positive integer such that $a^t \equiv 1$ (modulo p). Then the exponent of $f(x)$ modulo p is rt.* □

Let $f(x)$ be an indexing polynomial modulo p. If $f(\pi) = 0$, we have

$$f(\pi^p) \equiv f(\pi)^p \equiv 0 \quad \text{(modulo } p\text{)}$$

by Fermat's theorem. Therefore the roots of $f(x)$ in $\mathrm{GF}(q)$ are

$$\pi, \quad \pi^p, \quad \pi^{p^2}, \quad \ldots, \quad \pi^{p^{n-1}},$$

and we must have

$$f(x) = (x - \pi)(x - \pi^p)(x - \pi^{p^2})\ldots(x - \pi^{p^{n-1}}). \tag{8}$$

Consequently the constant term a_n of $f(x)$ is

$$(-1)^n \pi \pi^p \pi^{p^2} \ldots \pi^{p^{n-1}} = (-1)^n \pi^r, \quad r = (p^n - 1)/(p - 1).$$

Comparing this result with Lemma 2, we obtain the following result:

---

[2] The exponent of a polynomial that is irreducible mod p must divide $p^n - 1$, since an irreducible polynomial generates a field. The converse is false; that is, there are reducible polynomials whose exponent divides $p^n - 1$. If $f(x)$ has no multiple factors (mod p), the exponent of $f(x)$ divides $p^t - 1$ where t is the least common multiple of the degrees of the irreducible factors of $f(x)$. The polynomial $x^5 + x + 1$ is equal to $(x^3 + x^2 + 1)(x^2 + x + 1)$, modulo 2.

Theorem. $f(x)$ is an indexing polynomial modulo p if and only if

1) $(-1)^n a_n$ is a primitive root of p;
2) the smallest positive integer r for which $x^r \equiv$ integer $(\text{modd } f(x), p)$ is $r = (p^n - 1)/(p - 1)$, and in this case $x^r \equiv (-1)^n a_n$. □

We can now devise a fairly efficient algorithm for a machine to determine whether a given polynomial is indexing modulo p or not.

a) $(-1)^n a_n$ must be a primitive root of p.
b) $f(a) \not\equiv 0 \pmod{p}$ for $a = 1, 2, \ldots, p - 1$. (This test rules out linear factors in a hurry.)
c) If the recursion relationship (6)–(7) is used to find the residues of $x^r \ (\text{modd } f(x), p)$ for $r = 1, 2, \ldots$ until $x^r \equiv$ integer, then r must be $(p^n - 1)/(p - 1)$ and the integer must be $(-1)^n a_n$.

(A method that significantly improves the speed of step (c) is given in Appendix 2.)

4. Description and Use of the Tables

For $n = 1$ the indexing polynomials are the polynomials $x - a$, where a is a primitive root of p. Tables of primitive roots are fairly common [9], so we shall restrict our tabulation to the cases where $n > 1$.

Section 5 contains a complete table of the elements and their indices for all such fields with 128 elements or less. When only a few elements are present, a single table suffices to specify both the exp and ind functions. The parentheses and commas of the vector notation may be dropped; that is, we write simply 142 for the vector $(1, 4, 2)$. For example, the table for the field GF(9) is

ind	exp	*L*		ind	exp	*L*		ind	exp	*L*		ind	exp	*L*
0	01	*4*		*2*	21	*3*		*4*	02	*1*		*6*	12	*1*
1	10	*7*		*3*	22	*5*		*5*	20	*2*		*7*	11	*6*

based on the indexing polynomial $x^2 + x + 2$ (modulo 3), and we can immediately read off values such as $\exp 3 = 22$ and $\text{ind } 20 = 5$; notice that indices are given in *italic type* to distinguish them from vectors. A typical calculation like $(11 - 22 \cdot 10)/21 + (1/20)$ goes as follows:

$$\frac{11 - \mathit{3} \cdot \mathit{1}}{21} + \frac{01}{20} = \frac{11 - \mathit{4}}{21} + \frac{\mathit{0}}{5} = \frac{11 - 02}{21} + \mathit{3}$$

$$= \frac{\mathit{12}}{21} + 22 = \frac{\mathit{6}}{\mathit{2}} + 22 = \mathit{4} + 22 = 02 + 22 = 21 = \mathit{2}$$

because the product and quotient of indices (italic numbers) is obtained by addition and subtraction mod $p^n - 1 = 8$, while the sum and difference of vectors (roman digits) is obtained by addition and subtraction mod $p = 3$.

The table for GF(9) also shows values of the "Jacobi logarithm,"[3] defined as

$$L(\pi^k) = \text{ind}(1 + \pi^k). \tag{9}$$

This function (see [7]) allows us to work entirely with index numbers instead of vectors, because we have

$$\pi^j + \pi^k = \pi^{j+L(k-j)} \tag{10}$$

unless one of the terms is zero. More precisely, if we let π^Λ denote zero, we have

$$\pi^j + \pi^k = \begin{cases} k, & \text{if } j = \Lambda; \\ j, & \text{if } k = \Lambda; \\ \Lambda, & \text{if } L(k \ominus j) = \Lambda; \\ j \oplus L(k \ominus j), & \text{otherwise;} \end{cases} \tag{11}$$

here $\oplus$ and $\ominus$ denote addition and subtraction mod $p^n - 1$. For example, $3 + 4 = 3 \oplus L(1) = 3 \oplus 7 = 2$. And the "typical" calculation becomes

$$\frac{7 + 4 \cdot 3 \cdot 1}{2} + \frac{0}{5} = \frac{7 + 0}{2} + 3 = \frac{7 \oplus L(1)}{2} + 3$$

$$= (7 \oplus 7 \ominus 2) + 3 = 4 \oplus L(7) = 4 \oplus 6 = 2.$$

The condition $L(k \ominus j) = \Lambda$ occurs if and only if $k = j$ when $p = 2$; it occurs if and only if $k = j \oplus (p^n - 1)/2$ when p is odd.

Many applications can be made of these tables, only one of which is hand computation. For fields with more than 128 elements, the authors feel that hand computation will not be an important application, but the calculations will rather be done by computer. In this case only a single indexing polynomial is needed, and the computer program can easily make up its own index and exponent tables. So for the cases $p^n > 128$, only polynomials are tabulated.

Although there is essentially only one field with p^n elements, there are many ways of looking at it, and so there are in general many indexing polynomials of the same degree modulo p. In certain applications, a

---

[3] Jacobi logarithms were not included when these tables were originally published in 1964.

complete list of these polynomials is necessary; therefore Section 6 lists all indexing polynomials for $p^n \leq 1024$.

In Section 7, a single indexing polynomial is given for each field with $p^n < 10^9$ and $p < 50$. The polynomials that are listed are "simple" in the sense that, wherever possible, all of the coefficients except the constant term are 0 or 1, with as many 0s as possible. Such polynomials are in some sense the easiest to work with, because the recursion relationship used in calculating index tables involves at most one multiplication and the minimum number of subtractions.[4]

The tables of Section 5 have been constructed using the polynomials listed in Section 7. A number of checks have been made to ensure the validity of the polynomials of Section 7. The tables of Section 6 have been checked using Equation (A.10) of Appendix 1.

Appendix 1 describes an algorithm for extending the tables of Section 6, using the tables of Section 7. From a single indexing polynomial one can therefore find them all.

5. Complete Index Tables for Fields with $p^n \leq 128$

GF(2^2)

ind	exp	L	ind	exp	L	ind	exp	L	ind	exp	L
Λ	00	0	0	01	Λ	1	10	2	2	11	1

GF(2^3)

ind	exp	L	ind	exp	L	ind	exp	L	ind	exp	L
Λ	000	0	1	010	5	3	101	2	5	011	1
0	001	Λ	2	100	3	4	111	6	6	110	4

GF(2^4)

ind	exp	L	ind	exp	L	ind	exp	L	ind	exp	L
Λ	0000	0	3	1000	4	7	0111	13	11	1101	14
0	0001	Λ	4	1001	3	8	1110	6	12	0011	1
1	0010	12	5	1011	10	9	0101	2	13	0110	7
2	0100	9	6	1111	8	10	1010	5	14	1100	11

GF(2^5)

ind	exp	L	ind	exp	L	ind	exp	L	ind	exp	L
Λ	00000	0	7	01101	16	15	00110	22	23	01110	11
0	00001	Λ	8	11010	19	16	01100	7	24	11100	9
1	00010	14	9	11101	24	17	11000	18	25	10001	4
2	00100	28	10	10011	6	18	11001	17	26	01011	29
3	01000	5	11	01111	23	19	11011	8	27	10110	21
4	10000	25	12	11110	20	20	11111	12	28	00101	2
5	01001	3	13	10101	30	21	10111	27	29	01010	26
6	10010	10	14	00011	1	22	00111	15	30	10100	13

[4] Subtractions are preferred to additions here because it is generally more efficient for a computer to evaluate the residue $(a - b) \bmod p$ than it is to evaluate $(a + b) \bmod p$, due to the existence of a built-in machine instruction that tests whether a number is negative. An indexing polynomial with $a_1, \ldots, a_{n-1} \leq 1$ was found for every field GF(q) except in the case $q = 3^3$.

$GF(2^6)$

ind	exp	L	ind	exp	L	ind	exp	L	ind	exp	L
Λ	000000	0	15	010101	55	31	001101	60	47	110111	30
0	000001	Λ	16	101010	46	32	011010	29	48	001111	40
1	000010	58	17	110101	33	33	110100	17	49	011110	11
2	000100	53	18	001011	54	34	001001	3	50	111100	28
3	001000	34	19	010110	26	35	010010	22	51	011001	61
4	010000	43	20	101100	24	36	100100	45	52	110010	38
5	100000	6	21	111001	42	37	101001	56	53	000101	2
6	100001	5	22	010011	35	38	110011	52	54	001010	18
7	100011	44	23	100110	8	39	000111	59	55	010100	15
8	100111	23	24	101101	20	40	001110	48	56	101000	37
9	101111	27	25	111011	14	41	011100	13	57	110001	62
10	111111	12	26	010111	19	42	111000	21	58	000011	1
11	011111	49	27	101110	9	43	010001	4	59	000110	39
12	111110	10	28	111101	50	44	100010	7	60	001100	31
13	011101	41	29	011011	32	45	100101	36	61	011000	51
14	111010	25	30	110110	47	46	101011	16	62	110000	57

$GF(2^7)$

ind	exp	L	ind	exp	L	ind	exp	L	ind	exp	L
Λ	0000000	0	31	0001011	15	63	0001001	3	95	0100101	65
0	0000001	Λ	32	0010110	97	64	0010010	67	96	1001010	111
1	0000010	7	33	0101100	77	65	0100100	95	97	0010111	32
2	0000100	14	34	1011000	86	66	1001000	27	98	0101110	117
3	0001000	63	35	0110011	109	67	0010011	64	99	1011100	103
4	0010000	28	36	1100110	106	68	0100110	45	100	0111011	39
5	0100000	54	37	1001111	46	69	1001100	107	101	1110110	84
6	1000000	126	38	0011101	58	70	0011011	91	102	1101111	80
7	0000011	1	39	0111010	100	71	0110110	79	103	1011101	99
8	0000110	56	40	1110100	51	72	1101100	85	104	0111001	59
9	0001100	90	41	1101011	75	73	1011011	78	105	1110010	25
10	0011000	108	42	1010101	114	74	0110101	92	106	1100111	36
11	0110000	87	43	0101001	17	75	1101010	41	107	1001101	69
12	1100000	125	44	1010010	94	76	1010111	116	108	0011001	10
13	1000011	55	45	0100111	68	77	0101101	33	109	0110010	35
14	0000101	2	46	1001110	37	78	1011010	73	110	1100100	26
15	0001010	31	47	0011111	22	79	0110111	71	111	1001001	96
16	0010100	112	48	0111110	119	80	1101110	102	112	0010101	16
17	0101000	43	49	1111100	122	81	1011111	118	113	0101010	115
18	1010000	53	50	1111011	83	82	0111101	23	114	1010100	42
19	0100011	29	51	1110101	40	83	1111010	50	115	0101011	113
20	1000110	89	52	1101001	93	84	1110111	101	116	1010110	76
21	0001111	57	53	1010001	18	85	1101101	72	117	0101111	98
22	0011110	47	54	0100001	5	86	1011001	34	118	1011110	81
23	0111100	82	55	1000010	13	87	0110001	11	119	0111111	48
24	1111000	123	56	0000111	8	88	1100010	61	120	1111110	121
25	1110011	105	57	0001110	21	89	1000111	20	121	1111111	120
26	1100101	110	58	0011100	38	90	0001101	9	122	1111101	49
27	1001001	66	59	0111000	104	91	0011010	70	123	1111001	24
28	0010001	4	60	1110000	124	92	0110100	74	124	1110001	60
29	0100010	19	61	1100011	88	93	1101000	52	125	1100001	12
30	1000100	62	62	1000101	30	94	1010011	44	126	1000001	6

GF(3^2)

ind	exp	L	ind	exp	L	ind	exp	L	ind	exp	L
0	01	4	2	21	3	4	02	Λ	6	12	1
1	10	7	3	22	5	5	20	2	7	11	6

GF(3^3)

ind	exp	L	ind	exp	L	ind	exp	L	ind	exp	L
0	001	13	7	101	3	13	002	Λ	20	202	15
1	010	18	8	112	19	14	020	24	21	221	9
2	100	7	9	222	6	15	200	16	22	111	8
3	102	2	10	121	4	16	201	20	23	212	25
4	122	12	11	012	1	17	211	23	24	021	5
5	022	14	12	120	10	18	011	11	25	210	17
6	220	21				19	110	22			

GF(3^4)

ind	exp	L	ind	exp	L	ind	exp	L	ind	exp	L
0	0001	40	20	1202	70	40	0002	Λ	60	2101	50
1	0010	77	21	1021	32	41	0020	28	61	2012	26
2	0100	58	22	2211	7	42	0200	25	62	1122	24
3	1000	71	23	0112	78	43	2000	4	63	0221	16
4	2001	31	24	1120	47	44	1002	3	64	2210	22
5	1012	59	25	0201	18	45	2021	51	65	0102	2
6	2121	14	26	2010	72	46	1212	75	66	1020	21
7	2212	64	27	1102	79	47	1121	62	67	2201	36
8	0122	69	28	0021	37	48	0211	34	68	0012	1
9	1220	53	29	0210	48	49	2110	12	69	0120	74
10	1201	20	30	2100	60	50	2102	30	70	1200	10
11	1011	5	31	2002	43	51	2022	19	71	1001	44
12	2111	13	32	1022	66	52	1222	9	72	2011	61
13	2112	49	33	2221	15	53	1221	52	73	1112	57
14	2122	35	34	0212	29	54	1211	46	74	0121	8
15	2222	17	35	2120	6	55	1111	73	75	1210	54
16	0222	38	36	2202	39	56	0111	23	76	1101	27
17	2220	33	37	0022	41	57	1110	55	77	0011	68
18	0202	42	38	0220	63	58	0101	65	78	0110	56
19	2020	45	39	2200	67	59	1010	11	79	1100	76

GF(5^2)

ind	exp	L	ind	exp	L	ind	exp	L	ind	exp	L
0	01	6	6	02	18	12	04	Λ	18	03	12
1	10	17	7	20	21	13	40	22	19	30	8
2	43	5	8	31	4	14	12	15	20	24	7
3	42	2	9	34	19	15	13	10	21	21	23
4	32	11	10	14	1	16	23	20	22	41	3
5	44	13	11	33	9	17	11	14	23	22	16

GF(5³)

ind	exp	L	ind	exp	L	ind	exp	L	ind	exp	L
0	001	93	31	003	62	62	004	Λ	93	002	31
1	010	29	32	030	6	63	040	22	94	020	49
2	100	55	33	300	96	64	400	51	95	200	70
3	403	117	34	204	95	65	102	101	96	301	20
4	132	42	35	341	107	66	423	106	97	214	50
5	223	21	36	114	30	67	332	110	98	441	121
6	031	115	37	043	91	68	024	94	99	012	80
7	310	71	38	430	46	69	240	10	100	120	58
8	304	33	39	402	3	70	201	24	101	103	113
9	244	69	40	122	41	71	311	15	102	433	120
10	241	27	41	123	108	72	314	7	103	432	102
11	211	13	42	133	76	73	344	112	104	422	66
12	411	47	43	233	87	74	144	85	105	322	119
13	212	45	44	131	4	75	343	73	106	424	19
14	421	104	45	213	97	76	134	81	107	342	75
15	312	89	46	431	103	77	243	9	108	124	100
16	324	116	47	412	56	78	231	57	109	143	74
17	444	92	48	222	5	79	111	52	110	333	90
18	042	37	49	021	122	80	013	84	111	034	32
19	420	14	50	210	11	81	130	44	112	340	35
20	302	86	51	401	39	82	203	34	113	104	2
21	224	123	52	112	59	83	331	67	114	443	17
22	041	18	53	023	68	84	014	1	115	032	60
23	410	12	54	230	78	85	140	88	116	320	25
24	202	82	55	101	65	86	303	8	117	404	64
25	321	105	56	413	26	87	234	54	118	142	109
26	414	23	57	232	43	88	141	118	119	323	16
27	242	77	58	121	40	89	313	72	120	434	38
28	221	48	59	113	36	90	334	61	121	442	114
29	011	99	60	033	111	91	044	63	122	022	53
30	110	79	61	330	83	92	440	98	123	220	28

GF(7²)

ind	exp	L	ind	exp	L	ind	exp	L	ind	exp	L
0	01	16	12	14	14	24	06	Λ	36	63	2
1	10	31	13	34	20	25	60	29	37	43	15
2	64	35	14	15	5	26	13	12	38	62	36
3	53	6	15	44	10	27	24	21	39	33	13
4	56	41	16	02	8	28	21	47	40	05	24
5	16	1	17	20	28	29	61	38	41	50	18
6	54	23	18	51	45	30	23	27	42	26	17
7	66	25	19	36	9	31	11	11	43	41	44
8	03	32	20	35	19	32	04	40	44	42	37
9	30	22	21	25	42	33	40	43	45	52	3
10	45	46	22	31	34	34	32	39	46	46	33
11	12	26	23	55	4	35	65	7	47	22	30

GF(11²)

The digit '10' (ten) is denoted by an asterisk (*)

ind	exp	L	ind	exp	L	ind	exp	L	ind	exp	L
0	01	36	30	58	93	60	0*	Λ	90	63	63
1	10	71	31	39	110	61	*0	28	91	82	76
2	*4	66	32	61	33	62	17	99	92	5*	25
3	57	30	33	62	90	63	64	52	93	59	92
4	29	58	34	72	80	64	92	45	94	49	51
5	78	18	35	66	67	65	43	23	95	55	112
6	16	62	36	02	48	66	*5	82	96	09	60
7	54	95	37	20	42	67	67	106	97	90	118
8	*9	11	38	98	47	68	12	103	98	23	104
9	*7	43	39	*3	2	69	14	22	99	18	113
10	87	59	40	47	20	70	34	41	100	74	74
11	**	61	41	35	116	71	11	68	101	86	10
12	07	108	42	21	107	72	04	24	102	9*	97
13	70	111	43	*8	8	73	40	55	103	13	69
14	46	40	44	97	38	74	75	81	104	24	15
15	25	19	45	93	89	75	96	44	105	28	4
16	38	31	46	53	7	76	83	114	106	68	26
17	51	86	47	99	102	77	6*	85	107	22	98
18	79	115	48	03	72	78	42	65	108	08	96
19	26	29	49	30	117	79	95	75	109	80	50
20	48	94	50	81	91	80	73	100	110	3*	49
21	45	14	51	4*	73	81	76	83	111	71	34
22	15	6	52	65	35	82	*6	9	112	56	3
23	44	21	53	*2	39	83	77	5	113	19	88
24	05	84	54	37	16	84	06	12	114	84	56
25	50	17	55	41	78	85	60	32	115	7*	13
26	69	77	56	85	101	86	52	46	116	36	54
27	32	119	57	8*	109	87	89	57	117	31	27
28	*1	53	58	2*	37	88	1*	1	118	91	64
29	27	105	59	88	87	89	94	79	119	33	70

6. Complete Tables of Indexing Polynomials for $p^n \leq 1024$

The following tables list the coefficients of every indexing polynomial $x^n + a_1 x^{n-1} + \cdots + a_{n-1} x + a_n$, in right-to-left lexicographic order.

$p = 2$ $n = 2$ $q = 4$ $3 = $ prime $\varphi(3)/2 = 1$

a_1	a_2
1	1

$p = 2$ $n = 3$ $q = 8$ $7 = $ prime $\varphi(7)/3 = 2$

a_1	a_2	a_3		a_1	a_2	a_3
1	0	1		0	1	1

$p = 2$ $n = 4$ $q = 16$ $15 = 3 \cdot 5$ $\varphi(15)/4 = 2$

a_1	a_2	a_3	a_4		a_1	a_2	a_3	a_4
1	0	0	1		0	0	1	1

$$p = 2 \qquad n = 5 \qquad q = 32 \qquad 31 = \text{prime} \qquad \varphi(31)/5 = 6$$

a_1	a_2	a_3	a_4	a_5
0	1	0	0	1
0	0	1	0	1

a_1	a_2	a_3	a_4	a_5
1	1	1	0	1
1	1	0	1	1

a_1	a_2	a_3	a_4	a_5
1	0	1	1	1
0	1	1	1	1

$$p = 2 \qquad n = 6 \qquad q = 64 \qquad 63 = 3^2 \cdot 7 \qquad \varphi(63)/6 = 6$$

a_1	a_2	a_3	a_4	a_5	a_6
1	0	0	0	0	1
1	0	1	1	0	1

a_1	a_2	a_3	a_4	a_5	a_6
0	0	0	0	1	1
1	1	0	0	1	1

a_1	a_2	a_3	a_4	a_5	a_6
0	1	1	0	1	1
1	0	0	1	1	1

$$p = 2 \qquad n = 7 \qquad q = 128 \qquad 127 = \text{prime} \qquad \varphi(127)/7 = 18$$

a_1	a_2	a_3	a_4	a_5	a_6	a_7
1	0	0	0	0	0	1
0	0	1	0	0	0	1
1	1	1	0	0	0	1
0	0	0	1	0	0	1
0	1	1	1	0	0	1
1	1	0	0	1	0	1

a_1	a_2	a_3	a_4	a_5	a_6	a_7
1	0	1	0	1	0	1
0	0	1	1	1	0	1
1	1	1	1	1	0	1
0	0	0	0	0	1	1
1	0	1	0	0	1	1
1	0	0	1	0	1	1

a_1	a_2	a_3	a_4	a_5	a_6	a_7
0	1	0	1	0	1	1
0	1	0	0	1	1	1
1	1	1	0	1	1	1
0	0	0	1	1	1	1
1	1	0	1	1	1	1
0	1	1	1	1	1	1

$$p = 2 \qquad n = 8 \qquad q = 256 \qquad 255 = 3 \cdot 5 \cdot 17 \qquad \varphi(255)/8 = 16$$

a_1	a_2	a_3	a_4	a_5	a_6	a_7	a_8
0	1	1	1	0	0	0	1
1	0	1	0	1	0	0	1
0	1	1	0	1	0	0	1
0	1	1	0	0	1	0	1

a_1	a_2	a_3	a_4	a_5	a_6	a_7	a_8
1	1	1	1	0	1	0	1
1	0	0	0	1	1	0	1
0	1	0	0	1	1	0	1
0	0	1	0	1	1	0	1

a_1	a_2	a_3	a_4	a_5	a_6	a_7	a_8
0	0	0	1	1	1	0	1
1	1	0	0	0	0	1	1
0	1	1	0	0	0	1	1
0	0	1	0	1	0	1	1

a_1	a_2	a_3	a_4	a_5	a_6	a_7	a_8
1	0	0	0	0	1	1	1
1	1	1	0	0	1	1	1
1	1	0	0	1	1	1	1
0	1	0	1	1	1	1	1

$$p = 2 \qquad n = 9 \qquad q = 512 \qquad 511 = 7 \cdot 73 \qquad \varphi(511)/9 = 48$$

a_1	a_2	a_3	a_4	a_5	a_6	a_7	a_8	a_9
0	0	0	1	0	0	0	0	1
1	0	1	1	0	0	0	0	1
0	0	0	0	1	0	0	0	1
0	1	1	0	1	0	0	0	1
1	0	0	1	1	0	0	0	1
0	0	1	1	0	1	0	0	1
1	1	1	1	1	0	0	0	1
0	0	1	0	1	1	0	0	1
1	1	1	0	1	1	0	0	1
1	1	0	1	1	1	0	0	1
0	1	1	1	1	1	0	0	1
1	1	0	0	0	0	1	0	1
0	1	0	1	0	0	1	0	1
1	0	0	0	1	0	1	0	1
0	1	0	0	1	0	1	0	1
1	1	1	0	1	0	1	0	1

a_1	a_2	a_3	a_4	a_5	a_6	a_7	a_8	a_9
1	1	0	1	1	0	1	0	1
0	1	1	1	1	0	1	0	1
1	1	1	0	0	1	1	0	1
0	0	0	1	0	1	1	0	1
1	0	1	1	0	1	1	0	1
0	1	0	1	1	1	1	0	1
0	0	1	1	1	1	1	0	1
1	0	0	1	0	0	0	1	1
0	1	0	1	0	0	0	1	1
1	1	1	1	0	0	0	1	1
1	0	0	0	1	0	0	1	1
0	0	0	1	1	0	0	1	1
1	0	1	1	1	0	0	1	1
1	1	1	0	0	1	0	1	1
1	0	1	1	0	1	0	1	1
0	0	0	0	1	1	0	1	1

a_1	a_2	a_3	a_4	a_5	a_6	a_7	a_8	a_9
1	0	1	0	1	1	0	1	1
0	1	1	0	1	1	0	1	1
1	0	0	1	1	1	0	1	1
1	1	1	1	1	1	0	1	1
0	1	0	0	0	0	1	1	1
1	1	1	0	0	0	1	1	1
0	1	0	1	1	0	1	1	1
0	0	1	1	1	0	1	1	1
1	1	0	0	0	1	1	1	1
1	0	1	0	0	1	1	1	1
0	1	1	0	0	1	1	1	1
0	1	0	1	0	1	1	1	1
0	0	1	1	0	1	1	1	1
1	0	0	0	1	1	1	1	1
0	0	1	0	1	1	1	1	1
1	0	1	1	1	1	1	1	1

$p = 2$ $n = 10$ $q = 1024$ $1023 = 3 \cdot 11 \cdot 31$ $\varphi(1023)/10 = 60$

a_1	a_2	a_3	a_4	a_5	a_6	a_7	a_8	a_9	a_{10}
0	0	1	0	0	0	0	0	0	1
1	0	1	1	0	0	0	0	0	1
1	1	0	0	1	0	0	0	0	1
0	1	1	0	1	0	0	0	0	1
0	1	0	0	1	1	0	0	0	1
1	1	1	0	1	1	0	0	0	1
0	0	0	0	0	0	1	0	0	1
1	0	1	0	0	0	1	0	0	1
0	1	0	0	0	1	1	0	0	1
1	1	0	1	0	1	1	0	0	1
1	1	0	0	1	1	1	0	0	1
1	0	0	1	1	1	1	0	0	1
1	1	1	1	1	1	1	0	0	1
0	1	1	0	0	0	0	1	0	1
0	0	1	1	0	0	0	1	0	1
1	0	0	0	1	0	0	1	0	1
0	0	0	1	1	0	0	1	0	1
0	1	1	1	1	0	0	1	0	1
1	0	0	0	0	1	0	1	0	1
1	1	0	1	0	1	0	1	0	1

a_1	a_2	a_3	a_4	a_5	a_6	a_7	a_8	a_9	a_{10}
1	0	1	0	1	1	0	1	0	1
0	1	0	0	0	0	1	1	0	1
1	1	1	0	0	0	1	1	0	1
1	1	0	1	0	0	1	1	0	1
0	0	0	0	1	0	1	1	0	1
1	1	0	0	0	1	1	1	0	1
0	1	0	0	1	1	1	1	0	1
1	1	0	1	1	1	1	1	0	1
1	0	1	1	1	1	1	1	0	1
1	0	0	1	0	0	0	0	1	1
0	1	0	1	0	0	0	0	1	1
0	1	0	0	1	0	0	0	1	1
1	1	0	1	1	0	0	0	1	1
1	0	0	0	0	1	0	0	1	1
1	1	1	0	0	1	0	0	1	1
1	0	1	1	0	1	0	0	1	1
0	0	1	1	1	1	0	0	1	1
1	1	1	1	1	1	0	0	1	1
0	0	1	0	0	0	1	0	1	1
0	1	0	1	1	0	1	0	1	1

a_1	a_2	a_3	a_4	a_5	a_6	a_7	a_8	a_9	a_{10}
0	0	0	0	0	1	1	0	1	1
1	0	0	1	0	1	1	0	1	1
1	1	1	1	0	1	1	0	1	1
0	1	1	1	1	1	1	0	1	1
1	1	0	1	0	0	0	1	1	1
0	1	1	1	0	0	0	1	1	1
0	0	0	0	1	0	0	1	1	1
0	0	1	1	1	0	0	1	1	1
1	1	0	0	0	1	0	1	1	1
0	1	1	0	0	1	0	1	1	1
0	1	0	1	0	1	0	1	1	1
0	0	1	1	0	1	0	1	1	1
1	0	0	0	1	1	0	1	1	1
0	1	1	1	1	0	1	1	1	1
0	1	1	0	0	0	1	1	1	1
1	0	0	1	0	0	1	1	1	1
0	0	0	1	1	0	1	1	1	1
1	0	1	1	0	1	1	1	1	1
1	0	0	1	1	1	1	1	1	1
0	0	1	1	1	1	1	1	1	1

$p = 3$ $n = 2$ $q = 9$ $8 = 2^3$ $\varphi(8)/2 = 2$

a_1	a_2
1	2

a_1	a_2
2	2

$p = 3$ $n = 3$ $q = 27$ $26 = 2 \cdot 13$ $\varphi(26)/3 = 4$

a_1	a_2	a_3
2	0	1

a_1	a_2	a_3
2	1	1

a_1	a_2	a_3
0	2	1

a_1	a_2	a_3
1	2	1

$p = 3$ $n = 4$ $q = 81$ $80 = 2^4 \cdot 5$ $\varphi(80)/4 = 8$

a_1	a_2	a_3	a_4
1	0	0	2
2	0	0	2

a_1	a_2	a_3	a_4
0	0	1	2
2	1	1	2

a_1	a_2	a_3	a_4
2	2	1	2
0	0	2	2

a_1	a_2	a_3	a_4
1	1	2	2
1	2	2	2

$p = 3$ $n = 5$ $q = 243$ $242 = 2 \cdot 11^2$ $\varphi(242)/5 = 22$

a_1	a_2	a_3	a_4	a_5
2	0	0	0	1
1	2	0	0	1
1	0	1	0	1
0	2	1	0	1
2	2	1	0	1
0	1	2	0	1
1	2	2	0	1
2	0	0	1	1
0	1	0	1	1
1	1	0	1	1
1	0	1	1	1

a_1	a_2	a_3	a_4	a_5
2	1	1	1	1
1	2	1	1	1
0	0	2	1	1
1	1	2	1	1
0	2	2	1	1
0	0	0	2	1
1	0	0	2	1
2	2	0	2	1
1	1	1	2	1
2	0	2	2	1
0	1	2	2	1

$p = 3 \qquad n = 6 \qquad q = 729 \qquad 728 = 2^3 \cdot 7 \cdot 13 \qquad \varphi(728)/6 = 48$

$a_1\,a_2\,a_3\,a_4\,a_5\,a_6$	$a_1\,a_2\,a_3\,a_4\,a_5\,a_6$	$a_1\,a_2\,a_3\,a_4\,a_5\,a_6$	$a_1\,a_2\,a_3\,a_4\,a_5\,a_6$
1 0 0 0 0 2	1 0 2 2 0 2	0 1 2 1 1 2	1 1 0 1 2 2
2 0 0 0 0 2	1 2 2 2 0 2	0 1 0 2 1 2	0 2 0 1 2 2
1 0 1 0 0 2	0 0 0 0 1 2	2 2 0 2 1 2	0 1 1 1 2 2
2 0 2 0 0 2	0 0 1 0 1 2	1 0 1 2 1 2	2 0 2 1 2 2
1 2 0 1 0 2	1 1 1 0 1 2	2 1 1 2 1 2	2 1 2 1 2 2
2 2 0 1 0 2	1 2 1 0 1 2	1 2 1 2 1 2	0 2 2 1 2 2
1 2 1 1 0 2	0 1 2 0 1 2	2 1 2 2 1 2	0 1 0 2 2 2
2 2 2 1 0 2	2 1 0 1 1 2	0 0 0 0 2 2	1 2 0 2 2 2
1 1 0 2 0 2	0 2 0 1 1 2	0 1 1 0 2 2	1 1 1 2 2 2
2 1 0 2 0 2	1 0 1 1 1 2	0 0 2 0 2 2	2 0 2 2 2 2
2 0 1 2 0 2	1 1 1 1 1 2	2 1 2 0 2 2	1 1 2 2 2 2
2 2 1 2 0 2	0 2 1 1 1 2	2 2 2 0 2 2	2 2 2 2 2 2

$p = 5 \qquad n = 2 \qquad q = 25 \qquad 24 = 2^3 \cdot 3 \qquad \varphi(24)/2 = 4$

a_1	a_2	a_1	a_2	a_1	a_2	a_1	a_2
1	2	4	2	2	3	3	3

$p = 5 \qquad n = 3 \qquad q = 125 \qquad 124 = 2^2 \cdot 31 \qquad \varphi(124)/3 = 20$

a_1	a_2	a_3	a_1	a_2	a_3	a_1	a_2	a_3	a_1	a_2	a_3
1	0	2	3	2	2	2	0	3	3	2	3
3	0	2	0	3	2	4	0	3	0	3	3
3	1	2	0	4	2	1	1	3	0	4	3
4	1	2	2	4	2	2	1	3	1	4	3
2	2	2	4	4	2	2	2	3	3	4	3

$p = 5 \qquad n = 4 \qquad q = 625 \qquad 624 = 2^4 \cdot 3 \cdot 13 \qquad \varphi(624)/4 = 48$

a_1	a_2	a_3	a_4	a_1	a_2	a_3	a_4	a_1	a_2	a_3	a_4	a_1	a_2	a_3	a_4
1	2	0	2	2	2	2	2	2	2	0	3	2	1	2	3
4	2	0	2	3	3	2	2	3	2	0	3	3	4	2	3
2	3	0	2	1	0	3	2	1	3	0	3	2	0	3	3
3	3	0	2	3	0	3	2	4	3	0	3	4	0	3	3
3	0	1	2	0	1	3	2	1	0	1	3	0	1	3	3
4	0	1	2	3	2	3	2	2	0	1	3	3	1	3	3
1	2	1	2	2	3	3	2	1	1	1	3	2	4	3	3
4	3	1	2	1	0	4	2	0	4	1	3	3	0	4	3
0	4	1	2	2	0	4	2	4	4	1	3	4	0	4	3
2	0	2	2	4	2	4	2	1	0	2	3	0	4	4	3
4	0	2	2	1	3	4	2	3	0	2	3	4	1	4	3
0	1	2	2	0	4	4	2	0	1	2	3	1	4	4	3

$p = 7 \qquad n = 2 \qquad q = 49 \qquad 48 = 2^4 \cdot 3 \qquad \varphi(48)/2 = 8$

a_1	a_2	a_1	a_2	a_1	a_2	a_1	a_2
1	3	5	3	2	5	4	5
2	3	6	3	3	5	5	5

$p = 7 \qquad n = 3 \qquad q = 343 \qquad 342 = 2\cdot 3^2 \cdot 19 \qquad \varphi(342)/3 = 36$

a_1	a_2	a_3	a_1	a_2	a_3	a_1	a_2	a_3	a_1	a_2	a_3
1	1	2	5	4	2	3	0	4	5	3	4
6	1	2	0	5	2	5	0	4	4	4	4
3	2	2	1	5	2	6	0	4	6	4	4
4	2	2	3	5	2	2	1	4	1	5	4
0	3	2	5	5	2	3	1	4	3	5	4
4	3	2	0	6	2	1	2	4	6	5	4
5	3	2	2	6	2	5	2	4	2	6	4
6	3	2	3	6	2	3	3	4	5	6	4
2	4	2	6	6	2	4	3	4	6	6	4

$p = 11 \qquad n = 2 \qquad q = 121 \qquad 120 = 2^3\cdot 3\cdot 5 \qquad \varphi(120)/2 = 16$

a_1	a_2	a_1	a_2	a_1	a_2	a_1	a_2
4	2	2	6	1	7	1	8
5	2	3	6	4	7	3	8
6	2	8	6	7	7	8	8
7	2	9	6	10	7	10	8

$p = 13 \qquad n = 2 \qquad q = 169 \qquad 168 = 2^3\cdot 3\cdot 7 \qquad \varphi(168)/2 = 24$

a_1	a_2	a_1	a_2	a_1	a_2	a_1	a_2	a_1	a_2	a_1	a_2
1	2	9	2	4	6	2	7	10	7	6	11
4	2	12	2	9	6	3	7	11	7	7	11
6	2	2	6	10	6	6	7	4	11	8	11
7	2	3	6	11	6	7	7	5	11	9	11

$p = 17 \qquad n = 2 \qquad q = 289 \qquad 288 = 2^5\cdot 3^2 \qquad \varphi(288)/2 = 48$

a_1	a_2	a_1	a_2	a_1	a_2	a_1	a_2	a_1	a_2	a_1	a_2
1	3	8	5	11	6	1	10	8	11	14	12
6	3	9	5	15	6	3	10	9	11	15	12
7	3	12	5	1	7	4	10	10	11	4	14
10	3	14	5	4	7	13	10	15	11	6	14
11	3	2	6	5	7	14	10	2	12	7	14
16	3	6	6	12	7	16	10	3	12	10	14
3	5	8	6	13	7	2	11	5	12	11	14
5	5	9	6	16	7	7	11	12	12	13	14

$p = 19 \qquad n = 2 \qquad q = 361 \qquad 360 = 2^3\cdot 3^2 \cdot 5 \qquad \varphi(360)/2 = 48$

a_1	a_2	a_1	a_2	a_1	a_2	a_1	a_2	a_1	a_2	a_1	a_2
1	2	1	3	2	10	3	13	1	14	4	15
4	2	7	3	4	10	4	13	6	14	5	15
7	2	8	3	6	10	6	13	7	14	6	15
8	2	9	3	9	10	9	13	8	14	9	15
11	2	10	3	10	10	10	13	11	14	10	15
12	2	11	3	13	10	13	13	12	14	13	15
15	2	12	3	15	10	15	13	13	14	14	15
18	2	18	3	17	10	16	13	18	14	15	15

$p = 23$ $n = 2$ $q = 529$ $528 = 2^4 \cdot 3 \cdot 11$ $\varphi(528)/2 = 80$

a_1 a_2	a_1 a_2	a_1 a_2	a_1 a_2	a_1 a_2	a_1 a_2	a_1 a_2	a_1 a_2
2 5	4 7	13 10	16 11	5 15	6 17	12 19	16 20
4 5	9 7	17 10	20 11	9 15	11 17	16 19	19 20
5 5	14 7	20 10	1 14	10 15	12 17	21 19	5 21
8 5	19 7	21 10	3 14	11 15	17 17	22 19	6 21
15 5	21 7	3 11	5 14	12 15	19 17	4 20	7 21
18 5	22 7	7 11	10 14	13 15	20 17	7 20	9 21
19 5	2 10	8 11	13 14	14 15	1 19	8 20	14 21
21 5	3 10	9 11	18 14	18 15	2 19	10 20	16 21
1 7	6 10	14 11	20 14	3 17	7 19	13 20	17 21
2 7	10 10	15 11	22 14	4 17	11 19	15 20	18 21

$p = 29$ $n = 2$ $q = 841$ $840 = 2^3 \cdot 3 \cdot 5 \cdot 7$ $\varphi(840)/2 = 96$

a_1 a_2	a_1 a_2	a_1 a_2	a_1 a_2	a_1 a_2	a_1 a_2
5 2	1 8	6 11	7 15	2 19	5 26
7 2	7 8	9 11	9 15	4 19	6 26
11 2	10 8	10 11	11 15	7 19	8 26
14 2	14 8	11 11	12 15	8 19	12 26
15 2	15 8	18 11	17 15	21 19	17 26
18 2	19 8	19 11	18 15	22 19	21 26
22 2	22 8	20 11	20 15	25 19	23 26
24 2	28 8	23 11	22 15	27 19	24 26
1 3	3 10	1 14	4 18	3 21	2 27
2 3	5 10	3 14	8 18	4 21	3 27
9 3	9 10	8 14	13 18	6 21	6 27
14 3	10 10	13 14	14 18	12 21	13 27
15 3	19 10	16 14	15 18	17 21	16 27
20 3	20 10	21 14	16 18	23 21	23 27
27 3	24 10	26 14	21 18	25 21	26 27
28 3	26 10	28 14	25 18	26 21	27 27

$p = 31$ $n = 2$ $q = 961$ $960 = 2^6 \cdot 3 \cdot 5$ $\varphi(960)/2 = 128$

a_1 a_2	a_1 a_2	a_1 a_2	a_1 a_2	a_1 a_2	a_1 a_2	a_1 a_2	a_1 a_2
2 3	2 11	1 12	1 13	1 17	2 21	1 22	1 24
5 3	3 11	3 12	4 13	2 17	5 21	4 22	3 24
6 3	4 11	4 12	6 13	3 17	7 21	5 22	4 24
7 3	5 11	10 12	8 13	6 17	8 21	7 22	5 24
8 3	6 11	11 12	9 13	7 17	11 21	9 22	7 24
10 3	9 11	12 12	10 13	8 17	12 21	10 22	8 24
14 3	11 11	14 12	12 13	9 17	13 21	14 22	12 24
15 3	15 11	15 12	13 13	11 17	15 21	15 22	13 24
16 3	16 11	16 12	18 13	20 17	16 21	16 22	18 24
17 3	20 11	17 12	19 13	22 17	18 21	17 22	19 24
21 3	22 11	19 12	21 13	23 17	19 21	21 22	23 24
23 3	25 11	20 12	22 13	24 17	20 21	22 22	24 24
24 3	26 11	21 12	23 13	25 17	23 21	24 22	26 24
25 3	27 11	27 12	25 13	28 17	24 21	26 22	27 24
26 3	28 11	28 12	27 13	29 17	26 21	27 22	28 24
29 3	29 11	30 12	30 13	30 17	29 21	30 22	30 24

7. Indexing Polynomials for $p^n < 10^9$ and $p < 50$*

This table shows the coefficients $a_1 \ldots a_{n-1}|a_n$ of indexing polynomials $x^n + a_1 x^{n-1} + \cdots + a_{n-1} x + a_n$, where only a_n can exceed 2.

2^2	1|1	5^2	1|2	19^2	1|2
2^3	10|1	5^3	10|2	19^3	10|16
2^4	100|1	5^4	101|3	19^4	100|2
2^5	0100|1	5^5	0010|2	19^5	0001|16
2^6	10000|1	5^6	10000|2	19^6	00001|3
2^7	000001|1	5^7	100000|2	19^7	010000|9
2^8	1100001|1	5^8	0010100|3		
2^9	00010000|1	5^9	01100000|3	23^2	1|7
2^{10}	001000000|1	5^{10}	101000000|3	23^3	10|16
2^{11}	0100000000|1	5^{11}	1000000000|2	23^4	001|11
2^{12}	11000001000|1	5^{12}	00001001000|3	23^5	1000|18
2^{13}	110010000000|1			23^6	10000|7
2^{14}	1100000000010|1	7^2	1|3		
2^{15}	10000000000000|1	7^3	11|2	29^2	1|3
2^{16}	101000000001000|1	7^4	110|3	29^3	01|18
2^{17}	0010000000000000|1	7^5	1000|4	29^4	100|2
2^{18}	00000010000000000|1	7^6	11000|3	29^5	0100|26
2^{19}	110010000000000000|1	7^7	010000|4	29^6	00001|3
2^{20}	0010000000000000000|1	7^8	1000000|3		
2^{21}	01000000000000000000|1	7^9	10000100|2	31^2	1|12
2^{22}	100000000000000000000|1	7^{10}	110000000|3	31^3	01|28
2^{23}	0000100000000000000000|1			31^4	100|13
2^{24}	11000010000000000000000|1	11^2	1|7	31^5	0100|20
2^{25}	001000000000000000000000|1	11^3	10|5	31^6	10000|12
2^{26}	1100010000000000000000000|1	11^4	001|2		
2^{27}	110010000000000000000000000|1	11^5	0110|9	37^2	1|5
2^{28}	0010000000000000000000000000|1	11^6	10001|7	37^3	10|24
2^{29}	01000000000000000000000000000|1	11^7	100000|5	37^4	001|2
		11^8	0001001|2	37^5	0001|32
3^2	1|2				
3^3	20|1	13^2	1|2	41^2	1|12
3^4	100|2	13^3	10|7	41^3	01|35
3^5	1010|1	13^4	101|2	41^4	001|17
3^6	10000|2	13^5	0101|11	41^5	1000|35
3^7	101000|1	13^6	10100|6		
3^8	0010000|2	13^7	001000|6	43^2	1|3
3^9	01010000|1	13^8	0110000|2	43^3	01|40
3^{10}	101000000|2			43^4	001|20
3^{11}	1000001000|1	17^2	1|3	43^5	1000|40
3^{12}	10001000000|2	17^3	01|14		
3^{13}	100000100000|1	17^4	100|5	47^2	1|13
3^{14}	1000000000000|2	17^5	1000|14	47^3	10|42
3^{15}	10000000001000|1	17^6	10000|3	47^4	100|5
3^{16}	000000100000000|2	17^7	000100|14	47^5	0001|42
3^{17}	1000000010000000|1				
3^{18}	10000000000010000|2				

*Indexing polynomials for 2^n, $n \le 100$, have been computed by Watson [11].

Appendix 1

In a cyclic group $\{1, \pi, \ldots, \pi^{m-1}\}$ of order m, the primitive elements are those π^s where s is relatively prime to m. The number of such elements is $\varphi(m)$, where φ is Euler's "totient function"; if $m = p_1^{e_1} \ldots p_r^{e_r}$ is the decomposition of m into prime factors, then

$$\varphi(m) = \left(1 - \frac{1}{p_1}\right) \ldots \left(1 - \frac{1}{p_r}\right) m. \tag{A.1}$$

In the case of fields $\mathrm{GF}(q)$, $m = q-1 = p^n - 1$. Several primitive elements of the field will have the same indexing polynomial; in fact, exactly n distinct elements correspond to each indexing polynomial, namely

$$\pi^s, \ \pi^{sp}, \ \pi^{sp^2}, \ \ldots, \ \pi^{sp^{n-1}},$$

because of (8). Note that $x^{sp^n} = x^s$. From these facts we see that the field $\mathrm{GF}(q)$ has exactly $\varphi(q-1)/n$ indexing polynomials.

Let $f(x)$ be an indexing polynomial for $\mathrm{GF}(q)$. We will consider an algorithm that calculates the "field polynomial" $f_s(x)$, a polynomial of degree n that has π^s as a root. If s is relatively prime to $q-1$, $f_s(x)$ will be an indexing polynomial. If s is not prime to $q-1$, and if $m(x)$ is the monic polynomial of least degree such that $m(\pi^s) = 0$, then $f_s(x)$ will be congruent to $\big(m(x)\big)^r$ (modulo p) for some integer r.

The algorithm actually deals with any polynomials that have integer coefficients, and the operations performed during the algorithm are operations on integers, not in the field. It is a general method that takes as input any monic polynomial $f(x)$ and produces as output the monic polynomial $f_s(x)$ whose roots are the sth powers of the roots of $f(x)$.

For example,[5] if $f(x) = (x-1)(x+1)(x+3) = x^3 + 3x^2 - x - 3$, we would obtain $f_3(x) = (x-1)(x+1)(x+27) = x^3 + 27x^2 - x - 27$.

---

[5] In general if $f(x) = x^3 + ax^2 + bx + c$, then $f_3(x) = x^3 + (a^3 - 3(ab-c))x^2 + (b^3 - 3c(ab-c))x + c^3$. If $f(x) = x^n + a_1 x^{n-1} + \cdots + a_n$, we will have the general formula

$$f_3(x) = \sum_{\substack{b_1 + 2b_2 + \cdots + nb_n = 3r \\ b_0 + b_1 + \cdots + b_n = 3}} \frac{c(+)}{b_0! \, b_1! \ldots b_n!} a_1^{b_1} a_2^{b_2} \ldots a_n^{b_n} x^{n-r}$$

where $b_i \geq 0$, and $c(+) = -3$ if $b_0 + b_3 + b_6 + \cdots = 1$, $c(+) = 6$ otherwise. The sum is taken over all partitions of $3r$ into at most 3 parts none of which exceeds n (see [8]).

First the algorithm will be given, then we will sketch a proof to indicate why it works. Let $f(x) = x^n + a_1 x^{n-1} + \cdots + a_n$. We construct n-place vectors

$$V_i = (C_{i,1}, C_{i,2}, \ldots, C_{i,n}) \qquad (A.2)$$

by putting $V_0 = (1, 0, \ldots, 0)$ and

$$V_{i+1} = (0, C_{i,1}, \ldots, C_{i,n-1}) - C_{i,n}(a_n, a_{n-1}, \ldots, a_1). \qquad (A.3)$$

Then we define numbers t_i:

$$t_i = C_{i,1} + C_{i+1,2} + \cdots + C_{i+n-1,n}. \qquad (A.4)$$

Finally, we compute numbers $(b_1, b_2, \ldots, b_n)$ such that

$$\begin{aligned}
-b_1 &= t_s \\
-2b_2 &= b_1 t_s + t_{2s} \\
-3b_3 &= b_2 t_s + b_1 t_{2s} + t_{3s} \\
&\;\;\vdots \\
-nb_n &= b_{n-1} t_s + b_{n-2} t_{2s} + \cdots + t_{ns}.
\end{aligned} \qquad (A.5)$$

Then $f_s(x) = x^n + b_1 x^{n-1} + \cdots + b_n$.

Proof. The matrix

$$F = \begin{pmatrix} V_1 \\ V_2 \\ \vdots \\ V_n \end{pmatrix}$$

is the companion matrix for $f(x)$, so $\det(xI - F) = f(x)$. Therefore we have $\det(xI - F^s) = f_s(x)$. But

$$F^r = \begin{pmatrix} V_r \\ V_{r+1} \\ \vdots \\ V_{r+n-1} \end{pmatrix} \qquad (A.6)$$

and therefore $t_r = \operatorname{trace} F^r$. The final result follows from the identity

$$\det(xI - A) = x^n + b_1 x^{n-1} + \cdots + b_n \qquad (A.7)$$

where $-b_1 = \operatorname{trace} A$, $-2b_2 = b_1 \operatorname{trace} A + \operatorname{trace} A^2$, etc. (See [6].) □

When using this method it is necessary to *divide* by 2, 3, ..., n, so the arithmetic should not[6] in general be done modulo p. Since we cannot guarantee that $t_s = t_{s+q-1}$ it becomes advisable to keep s as small as possible when looking for indexing polynomials. The largest value of s we will need, if p is an odd prime, will be approximately $\frac{1}{2}p^n - p^{n-1}$.

For certain special values of s the calculation is simpler. We observe that

$$a_n \pi^{-n} + a_{n-1} \pi^{-(n-1)} + \cdots + a_1 \pi^{-1} + 1 = 0;$$

therefore

$$f_{-1}(x) = x^n + a_{n-1}a_n^{-1}x^{n-1} + \cdots + a_1 a_n^{-1}x + a_n^{-1}. \tag{A.8}$$

This identity effectively cuts the work of finding indexing polynomials in half, because $f_{-1}(x) \neq f(x)$ when $p^n > 4$. (Indeed, if $f_{-1}(x) = f(x)$ there is an exponent u with $p^u \equiv -1$ (modulo $p^n - 1$) and $0 \leq u < n$; this condition can hold only when $p^n = 2^1$ or 2^2 or 3^1.)

Another simplification often applies when p is odd. If $q \equiv 1$ (modulo 4), it is not hard to show that $-\pi$ is a primitive element of the field if and only if π is primitive; thus, $f(-x)$ will be an indexing polynomial. More generally, we can take a given indexing polynomial $f(x)$ with its root π, and ask which multiples $a\pi$ will be primitive, for $a = 1, 2, \ldots, p - 1$; if $a\pi$ is also primitive, then $f(a^{-1}x)$ will be an indexing polynomial. These constants a will be powers of $\pi^r = (-1)^n a_n$, where $r = (p^n - 1)/(p - 1)$. Using such ideas, together with the fact that $(kr + 1)$ is prime to $(p^n - 1)$ if and only if $(kn + 1)$ is prime to $(p - 1)$, we easily derive the following rule: Let A be the set of all integers k, $0 \leq k < p - 1$, for which $(kn + 1)$ is relatively prime to $(p - 1)$. For any such k, the polynomial

$$(-a_n)^{kn} f((-1)^{kn} a_n^{-k} x) = f_{kr+1}(x) \tag{A.9}$$

---

[6] If $p > n$ one *can* work modulo p. In that case division by 2, 3, ..., n is done in the field GF(p), and we have always $t_s = t_{s+q-1}$, because V_i represents $\pi^i = C_{i,n}\pi^{n-1} + \cdots + C_{i,1}$. If $p \leq n$, however, one cannot work modulo p and it will not be true that $t_s = t_{s+q-1}$. If necessary it will always be safe to work modulo p^r where p^r does not divide $n!$. Then great care must be taken in the division, which can be somewhat troublesome. For example, if $p = 2$ and $n = 6$, we can compute the t_i modulo 2^5. Then $-b_1 \equiv t_1$ (modulo 2^5); $-2b_2 \equiv b_1 t_1 + t_2$ (modulo 2^5), so $b_2 \equiv -\frac{1}{2}(b_1 t_1 + t_2)$ (modulo 2^4); $-3b_3 \equiv b_2 t_1 + b_1 t_2 + t_3$ (modulo 2^4), so $b_3 \equiv 5(b_2 t_1 + b_1 t_2 + t_3)$ (modulo 2^4), since $-3 \cdot 5 \equiv 1$; and so on until we have determined $6b_6$ modulo 2^2. Then $3b_6$ will be determined modulo 2, and we'll know b_6 mod 2.

is an indexing polynomial, if $f(x)$ is an indexing polynomial with constant term a_n. We will have $f_{kr+1}(x) \neq f(x)$ if $0 < k < p - 1$, and also $f_{kr+1}(x) \neq f_{-1}(x)$ if $0 < k < p - 1$ provided that $n > 2$. If $n = 2$, however, the polynomial $f_{(p-2)r+1}(x)$ will be the same as the polynomial $f_{-1}(x)$.

Let m be the number of elements in the set A constructed in the preceding paragraph. Then from a single indexing polynomial $f_s(x)$ we obtain m distinct indexing polynomials $f_{s(kr+1)}(x)$, one for each k in A, using (A.9);[7] and if $n > 2$ we obtain m additional polynomials $f_{-s(kr+1)}(x)$, using (A.8).

For rather large values of q, the most straightforward application of the algorithm (A.2)–(A.5) above would entail the storage of a great many values in the memory of the computer. To conserve space, we may start by determining all values of s for which $f_s(x)$ will actually need to be computed, proceeding as follows:

Step 1. Set $s = 1$. Determine the set A mentioned above and the number of elements in it, m.

Step 2. Add s to the list of values for which $f_s(x)$ will be computed. If the list now contains $\varphi(q - 1)/(2mn)$ elements (or in the case $n = 2$, $\varphi(q - 1)/(2m)$ elements), it is complete.

Step 3. Increase s to the next value that is relatively prime to $p^n - 1$.

Step 4. Calculate the $2mn$ values of the form $\pm s(kr + 1)p^u$, where k is in the set A and $0 \leq u < n$. Reduce these to the range between 0 and $p^n - 2$. If any of these numbers is *less* than s, go back to Step 3; otherwise continue again with Step 2.

Once this list has been obtained, the algorithm (A.2)–(A.5) for calculating $f_s(x)$ is carried out simultaneously for all s in the list, by using the necessary values t_i immediately as they are being computed. This process requires storage for only n of the vectors V_i, and also for the n values b_j corresponding to each s in the list.[8]

A convenient check on the computations in this algorithm can be made by computing $b_{n+1} = 0$ in every case, by analogy with equation (A.5).

---

[7] Note that the number a_n that occurs in Equation (A.9) is to be the constant term of $f_s(x)$, not $f(x)$, in this case.

[8] If the memory capacity is still exceeded, the process can be repeated in several steps, each time processing as many items on the list as possible.

A further check on the overall computations can be made if desired by using the formula

$$g_1(x)g_2(x)\ldots g_t(x) \equiv \Psi_{q-1}(x) \quad (\text{modulo } p), \qquad (A.10)$$

where $t = \varphi(q-1)/n$ and $g_1(x)\ldots g_t(x)$ is the product of all the indexing polynomials. Here $\Psi_{q-1}(x)$ is the cyclotomic polynomial of rank $q - 1$. Cyclotomic polynomials can be readily computed from the following series of identities when p is any prime:

$$
\begin{aligned}
\Psi_p(x) &= x^{p-1} + x^{p-2} + \cdots + x + 1; \\
\Psi_{2n}(x) &= \Psi_n(-x), && \text{if } n \text{ is odd}; \\
\Psi_{pn}(x) &= \Psi_n(x^p)/\Psi_n(x), && \text{if } n \text{ is not divisible by } p; \\
\Psi_{pn}(x) &= \Psi_n(x^p), && \text{if } n \text{ is divisible by } p.
\end{aligned}
\qquad (A.11)
$$

Examples

(A) Find all indexing polynomials for GF(32), given the polynomial $f(x) = x^5 + x^2 + 1$. Since $\varphi(31) = 30$, there are 6 indexing polynomials in all.

In this field, π^1, π^2, π^4, π^8, and π^{16} are the roots of $f(x)$, so π^{30}, π^{29}, π^{27}, π^{23}, and π^{15} are the roots of $f_{-1}(x) = x^5 + x^3 + 1$; we have found two of the six polynomials desired.

The next step is to find $f_3(x)$. We can avoid negative numbers in the algorithm if we change to $f(x) = x^5 - x^2 - 1$, because this polynomial is equivalent to $x^5 + x^2 + 1 \pmod 2$. The calculation proceeds as follows:

i	V_i					t_i
0	(1	0	0	0	0)	5
1	(0	1	0	0	0)	0
2	(0	0	1	0	0)	0
3	(0	0	0	1	0)	3
4	(0	0	0	0	1)	0
5	(1	0	1	0	0)	5
6	(0	1	0	1	0)	3
7	(0	0	1	0	1)	0
8	(1	0	1	1	0)	8
9	(0	1	0	1	1)	3
10	(1	0	2	0	1)	5
11	(1	1	1	2	0)	11
12	(0	1	1	1	2)	3

i	V_i					t_i
13	(2	0	3	1	1)	13
14	(1	2	1	3	1)	14
15	(1	1	3	1	3)	8
16	(3	1	4	3	1)	24
17	(1	3	2	4	3)	17
18	(3	1	6	2	4)	21
19	(4	3	5	6	2)	38
20	(2	4	5	5	6)	25
21	(6	2	10	5	5)	45
22	(5	6	7	10	5)	55
23	(5	5	11	7	10)	46
24	(10	5	15	11	7)	83
25	(7	10	12	15	11)	80

In this case one can show that $t_i = 5C_{i,1} + 3C_{i,4}$.

Now for $s = 3$ we can find $f_3(x)$ by

$$-b_1 = t_3 = 3, \quad b_1 = -3;$$
$$-2b_2 = b_1t_3 + t_6 = -9 + 3, \quad b_2 = 3;$$
$$-3b_3 = b_2t_3 + b_1t_6 + t_9 = 9 - 9 + 3, \quad b_3 = -1;$$
$$-4b_4 = b_3t_3 + b_2t_6 + b_1t_9 + t_{12} = -3 + 9 - 9 + 3 = 0, \quad b_4 = 0;$$
$$-5b_5 = b_4t_3 + b_3t_6 + b_2t_9 + b_1t_{12} + t_{15} = 0 - 3 + 9 - 9 + 8 = 5, \quad b_5 = -1,$$

and $-6b_6 = -3 + 0 - 3 + 9 - 24 + 21 = 0$ as a check. Therefore $f_3(x) = x^5 - 3x^4 + 3x^3 - x^2 - 1$ is the polynomial whose roots are the cubes of the roots of $x^5 - x^2 - 1$. Thus, reducing modulo 2, we have shown that π^3, π^6, π^{12}, π^{24}, and π^{17} are roots of $f_3(x) \equiv x^5 + x^4 + x^3 + x^2 + 1$ in the field; π^{28}, π^{25}, π^{19}, π^7, and π^{14} are roots of $f_{-3}(x) \equiv x^5 + x^3 + x^2 + x + 1$. In a similar fashion we find that π^5, π^{10}, π^{20}, π^9, and π^{18} are roots of $f_5(x) = x^5 - 5x^4 + 10x^3 - 11x^2 + 5x - 1 \equiv x^5 + x^4 + x^2 + x + 1$; π^{26}, π^{21}, π^{11}, π^{22}, and π^{13} are roots of $f_{-5}(x) \equiv x^5 + x^4 + x^3 + x + 1$. As a final check we can verify that

$$f(x)f_{-1}(x)f_3(x)f_{-3}(x)f_5(x)f_{-5}(x) \equiv$$
$$\Psi_{31}(x) = x^{30} + x^{29} + \cdots + x + 1 \quad \text{(modulo 2)}.$$

(B) Find all indexing polynomials for GF(125), given the polynomial $f(x) = x^3 + x^2 + 2$. Here $n = 3$ and $\varphi(124) = 60$, so there are 20 answers in all. The set A in this case is $\{0, 2\}$, hence $m = 2$. The polynomials are obtained in sets of $2m = 4$ at a time. The given polynomial readily yields $f_s(x)$ for the following values of s:
$$1, 5, 25; \quad 63, 67, 87; \quad 123, 119, 99; \quad 61, 57, 37.$$

From $f_3(x)$ we obtain the additional values
$$3, 15, 75; \quad 65, 77, 13; \quad 121, 109, 49; \quad 59, 47, 111.$$

From $f_7(x)$ we obtain
$$7, 35, 51; \quad 69, 97, 113; \quad 117, 89, 73; \quad 55, 27, 11.$$

From $f_9(x)$ we obtain
$$9, 45, 101; \quad 71, 107, 39; \quad 115, 79, 23; \quad 53, 17, 85.$$

From $f_{19}(x)$ we obtain
$$19, 95, 103; \quad 81, 33, 41; \quad 105, 29, 21; \quad 43, 91, 83.$$

Therefore the general algorithm is applied for $s = 3, 7, 9$, and 19, after which the answers will then be

$$f_s(x), \quad f_{-s}(x), \quad f_s(-x), \quad f_{-s}(-x), \quad \text{for } s = 1, 3, 7, 9, 19.$$

In this case $p > n$, so we can do the computation modulo p. The values needed for f_3 are:

i	V_i			t_i
0	(1	0	0)	3
1	(0	1	0)	4
2	(0	0	1)	1
3	(3	0	4)	3
4	(2	3	1)	4
5	(3	2	2)	4

i	V_i			t_i
6	(1	3	0)	0
7	(0	1	3)	2
8	(4	0	3)	0
9	(4	4	2)	0
10	(1	4	2)	
11	(1	1	2)	

$$-b_1 \equiv t_3 \equiv 3, \qquad\qquad b_1 \equiv 2\,;$$

$$-2b_2 \equiv b_1 t_3 + t_6 = 1, \qquad\qquad b_2 \equiv 2\,;$$

$$-3b_3 \equiv b_2 t_3 + b_1 t_6 + t_9 \equiv 1, \qquad b_3 \equiv 3\,.$$

$$f_3(x) = x^3 + 2x^2 + 2x + 3\,;$$
$$f_{-3}(x) = x^3 + 4x^2 + 4x + 2\,;$$
$$f_{65}(x) = -f_3(-x) = x^3 + 3x^2 + 2x + 2\,;$$
$$f_{-65}(x) = -f_{-3}(-x) = x^3 + x^2 + 4x + 3\,.$$

(Compare these results with the tables for GF(125) in Sections 5 and 6.)

Appendix 2

A rapid procedure for computing x^m for large m, using relatively few operations, is well known: Take the representation of m in the binary number system and replace each '0' by the letter 'S', each '1' by the letters 'SX'. Cancel off the 'SX' that appears at the left, and the result is a rule for computing x^m, if 'S' is interpreted as the operation of squaring the preceding result, and if 'X' is interpreted as the operation of multiplying the preceding result by x. For example, the binary representation of 21 is 10101; to calculate x^{21} we get the rule $SSXSSX$, that is, "Square, square, multiply by x, square, square, multiply by x." The sequence of results is x, x^2, x^4, x^5, x^{10}, x^{20}, x^{21}.

In the algorithm described in Section 3, we used only the operation of multiplying a polynomial by x (modd $f(x), p$), as given by the recursion relation (A.6)–(A.7). We can improve the efficiency greatly, for large q, if we introduce the additional operation of *squaring* a polynomial (modd $f(x), p$). Given the polynomial

$$c_1 x^{n-1} + c_2 x^{n-2} + \cdots + c_n,$$

its square is $c_1^2 x^{2n-2} + (c_1 c_2 + c_2 c_1)x^{2n-3} + \cdots + c_n^2$, and the powers x^{2n-2}, x^{2n-3}, ..., x^n can all be expressed as known polynomials of degree less than n (modd $f(x), p$). Therefore the operation of squaring can easily be implemented.

The operation of multiplying by x (modd $f(x), p$) requires approximately n multiplications and n additions, and the operation of squaring requires approximately $n^2 + n$ multiplications and $n^2 + n$ additions.[9] Consequently we need to do only about $(n^2 + \frac{3}{2}n)\log_2 m$ operations with the binary method, compared to mn operations when only multiplication by x is used. For example, if $q = 13^5 = 371293$, we want to calculate $x^{(q-1)/12} = x^{30941}$. Using the binary method, the number of operations is reduced from 156000 to about 480.

We need to test whether the smallest m for which $x^m \equiv$ integer (modd $f(x), p$) is equal to $(q-1)/(p-1)$. Since $x^m \equiv$ integer implies that $x^{mk} \equiv$ integer, for all integers k, we see that this condition can be tested using the binary method, by checking only a few of the most crucial values of m, as follows. Let

$$r = \frac{q-1}{p-1} = p_1^{e_1} p_2^{e_2} \ldots p_k^{e_k},$$

where the p_i are distinct primes.

We first check if $x^r \equiv (-1)^n a_n$ (modd $f(x), p$). This test will eliminate a great deal of the unwanted cases at once. But if $f(x)$ passes this test, we must also verify that $x^m \not\equiv$ integer for $m = r/p_1$, r/p_2, ..., r/p_k. If all k of these additional conditions are satisfied, $f(x)$ is an indexing polynomial, since x^m will not be congruent to any integer for any divisor m of r.

In fact, not all k of these additional conditions need to be checked. If p_i is a divisor of $p-1$, it is not necessary to make the test for $m = r/p_i$; for if $x^m \equiv$ integer it would be impossible for $x^r = (x^m)^{p_i}$ to be a primitive root of p.

Some experimental results may be of interest. In the case $p^n = 3^{11}$, we examined 41 polynomials for which the condition $x^r \equiv (-1)^n a_n$ was satisfied; and 40 of them were indexing polynomials. The only exception was $f(x) = x^{11} + x^9 + x^6 + x + 1$, for which $x^{3851} \equiv 2$.

---

[9] Actually it is often possible to multiply polynomials more rapidly on a computer by means of the multiplication instruction that is used for ordinary numbers: We simply need to separate the coefficients by an appropriate number of zeros, to ensure that carries will not propagate too far.

This method depends on a knowledge of the factors of r. For comparatively small r, those factors are easily found by trial and error on a computer. In the case $p = 2$, $r = 2^n - 1$ is a Mersenne number and the factors of such numbers are reasonably well known (see [2]). Numerous special methods are available for factoring $(p^n - 1)/(p - 1)$, and values are given in [5]; those tables are, however, known to contain a number of errors, as occasionally noted in the errata sections of the journal *Mathematical Tables and Other Aids to Computation* over the last twenty years.

Acknowledgment is made to Case Institute of Technology and to Yale University for the use of digital computers.

References

[1] A. Adrian Albert, *Fundamental Concepts of Higher Algebra* (Chicago: University of Chicago Press, 1956).

[2] John Brillhart, "On the factors of certain Mersenne numbers, II," *Mathematics of Computation* **18** (1964), 87–92.

[3] W. H. Bussey, "Galois field tables for $p^n \leq 169$," *Bulletin of the American Mathematical Society* **12** (1905), 22–38.

[4] W. H. Bussey, "Tables of Galois fields of order less than 1,000," *Bulletin of the American Mathematical Society* **16** (1910), 188–206.

[5] Allan J. C. Cunningham and H. J. Woodall, *Factorization of $y^n \pm 1$: $y = 2, 3, 5, 6, 7, 10, 11, 12$ up to high powers (n)* (London: Francis Hodgson, 1925).

[6] Paul S. Dwyer, *Linear Computations* (New York: Wiley, 1951), xi+ 344 pages.

[7] C. G. J. Jacobi, "Über die Kreistheilung und ihre Anwendung auf die Zahlentheorie," *Journal für die reine und angewandte Mathematik* **30** (1846), 166–182. Reprinted in his *Gesammelte Werke* **6** (1891), 254–274.

[8] D. E. Knuth, "m-th powers of algebraic roots," unpublished notes (Pasadena: California Institute of Technology, 1961), 7 pages.

[9] D. H. Lehmer, *Guide to Tables in the Theory of Numbers* (Washington: National Research Council, 1941), xiv + 177 pages.

[10] J. D. Swift, "Construction of Galois fields of characteristic two and irreducible polynomials," *Mathematics of Computation* **14** (1960), 99–103.

[11] E. J. Watson, "Primitive polynomials (mod 2)," *Mathematics of Computation* **16** (1962), 368–369.

[12] Oscar Zariski and Pierre Samuel, *Commutative Algebra* **1** (Princeton, New Jersey: Van Nostrand, 1958).

Addendum

Further techniques of interest can be found in John Conway's paper, "A tabulation of some information concerning finite fields," *Computers in Mathematical Research*, edited by R. F. Churchhouse and J.-C. Herz (Amsterdam: North-Holland, 1968), 37–50.

See also the definitive book *Finite Fields* by Rudolf Lidl and Harald Niederreiter (Reading, Massachusetts: Addison–Wesley, 1983); second edition (Cambridge: Cambridge University Press, 1997), xiv+755 pages.

The authors' suggested name "indexing polynomial" did not catch on, so *The Art of Computer Programming* now refers instead to a "primitive polynomial modulo p."

Chapter 20

Finite Semifields and Projective Planes

*[Originally published in Journal of Algebra **2** (1965), 182–217.]*

1. Introduction

In this paper the term *semifield* is used to describe an algebraic system that satisfies all properties of a field except for the commutativity and associativity of multiplication. Semifields are of special interest today because the projective planes constructed from them have rather remarkable properties. This paper gives a unified presentation of the theory of finite semifields. By incorporating some new approaches in the investigation, it has been possible to obtain simple proofs of the known results and an insight into their interrelationships, as well as a number of new properties.

A brief review of the fundamental definitions and properties of semifields appears in Section 2, along with illustrations of some interesting semifields of order 16.

In Section 3, we introduce a "homogeneous" notation for representing points and lines of an arbitrary projective plane in terms of its ternary ring. The concept of isotopy is generalized to apply to arbitrary ternary rings, and a simple method for mechanically constructing all ternary rings isotopic to a given one is presented. Some known theorems about collineations of planes, and of semifield planes in particular, are proved concisely using the homogeneous notation. Finally, the question of whether a nonlinear "isotopy" can yield new semifields is considered.

Cubical arrays of numbers, of arbitrary finite dimension, are the subject of Section 4. First, operations of transposition and multiplication are discussed. Then the notion of a *nonsingular* hypercube is introduced. Semifields are shown to be equivalent to a certain type of 3-dimensional cubical array, and the projective planes coordinatized by semifields are in 1–1 correspondence with equivalence classes of nonsingular 3-cubes.

By transposition of a 3-cube, up to five further projective planes can be constructed from a single semifield plane. This construction is the topic of Section 5. By exhibiting all of the semifield planes of order 32, with their interrelations and collineation groups, we give examples of transposed planes.

The finite semifields that have appeared in the literature are surveyed briefly in Section 6, including a discussion of all semifields of order 16; of some commutative semifields due to Dickson; of the twisted fields due to Albert; of seminuclear extensions due to Sandler; and of binary semifields due to the author.

A new class of quadratic extensions is considered in Section 7, in which the semifield is a vector space of dimension 2 over a so-called weak nucleus F. In particular, all quadratic extensions of F, for which F is equal to any two of the nuclei, are constructed, thus generalizing a result due to Hughes and Kleinfeld.

The symbol ' $\square$ ' will be used throughout this paper to mean: "This completes the proof of the theorem," or "This is all the proof of the theorem that will be given here."

The principal theorems of this paper are those numbered 3.3.1, 3.3.2, 3.3.4, 4.4.2, 4.5.2, 5.1.1, 5.2.1, 7.2.1, 7.4.1. Several new results about algebraic characterizations and automorphisms of the semifields of orders 16 and 32 appear as examples in Sections 2, 5, and 6.

2. Semifields and Pre-Semifields

We are concerned with a certain type of algebraic system, called a *semifield*. Such a system has several names in the literature, where it is also known, for example, as a "nonassociative division ring" or a "distributive quasifield." Since those terms are somewhat lengthy, and since we make frequent references to such systems in this paper, the more convenient name semifield will be used.

2.1 Definition of semifield. A finite semifield S is a finite algebraic system containing at least two elements, 0 and 1, and possessing two binary operations, addition and multiplication, which satisfy the following axioms:

A1. Addition is a group, with identity element 0.

A2. If $ab = 0$, then either $a = 0$ or $b = 0$.

A3. $a(b + c) = ab + ac$; $(a + b)c = ac + bc$.

A4. $1a = a1 = a$.

Throughout this paper, the term semifield will always be used to denote a *finite* semifield. The definition given here would actually be insufficient to define infinite semifields, because a stronger condition (namely that the equations $ax = b$ and $ya = b$ are uniquely solvable for x and y when $a \neq 0$) would necessarily replace axiom A2. In the finite case we can prove the stronger condition by noting, for example, that $ax = ay$ implies $a(x - y) = 0$, hence $a = 0$ or $x = y$. Notice that a semifield is much like a field, except that multiplication of nonzero elements is required to be merely a loop instead of a group.

Every field is a semifield; the term *proper semifield* denotes a semi-field that is not a field. In other words, every (finite) proper semifield contains elements a, b, and c such that $(ab)c \neq a(bc)$. (When the system is finite, axioms A1–A4 together with associativity of multiplication imply that multiplication is commutative, according to a well-known theorem of Wedderburn [13].)

2.2 Examples. The following remarkable system V is a proper semifield with 16 elements: Let F be the field $GF(4)$, so that F has the four elements 0, 1, ω, and $\omega^2 = 1 + \omega$. The elements of V all have the form $u + \lambda v$, where $u, v \in F$. Addition is defined in an obvious way,

$$(u + \lambda v) + (x + \lambda y) = (u + x) + \lambda(v + y), \tag{2.1}$$

using the addition of F. Multiplication is also defined in terms of the multiplication and addition of F, using the following rule:

$$(u + \lambda v)(x + \lambda y) = (ux + v^2 y) + \lambda(vx + u^2 y + v^2 y^2). \tag{2.2}$$

It is clear that F is embedded in V, and also that A1, A3, and A4 hold. To demonstrate A2, suppose that $(u + \lambda v)(x + \lambda y) = 0$. Then in particular $ux + v^2 y = 0$, so that, if neither of the original factors is zero, there is a nonzero element $z \in F$ such that

$$x = v^2 z, \qquad y = uz.$$

Then

$$vx + u^2 y + v^2 y^2 = v^3 z + u^3 z + u^2 v^2 z^2 = 0.$$

But this is impossible in F, unless $u = v = 0$.

The system V just described is certainly a proper semifield, since V is not commutative. But a good deal of associativity is still present; in fact $(ab)c = a(bc)$ if *any two* of $\{a, b, c\}$ are in F.

The remarkable property of V is that it possesses 6 automorphisms, while the field of 16 elements possesses only 4. The automorphisms σ_{ij} are given by

$$(u + \lambda v)\sigma_{ij} = u^j + \lambda \omega^i v^j \qquad \text{for } i = 0, 1, 2, \text{ and } j = 1, 2.$$

(No other semifield of order 16 has as many automorphisms.) Moreover, V is anti-isomorphic to itself, under the mapping $(u + \lambda v)\tau = u + \lambda v^2$, because we can easily check that $(ab)\tau = (b\tau)(a\tau)$.

Other examples of semifields that have properties in common with V will be discussed later. We will remark here, however, that if we had defined multiplication by the rule

$$(u + \lambda v)(x + \lambda y) = (ux + \omega v^2 y) + \lambda(vx + u^2 y) \qquad (2.3)$$

rather than as in (2.2), we would have obtained another proper semifield containing F; and this system has the stronger associativity property that $(ab)c = a(bc)$, whenever *any one* of $\{a, b, c\}$ is in F. Let us call the latter system W.

We will see in Section 6.1 that a proper semifield must contain at least 16 elements.

2.3 Pre-semifields. We say the system S is a *pre-semifield* if it satisfies all the axioms for a semifield, except possibly A4; that is, it need not have a multiplicative identity, although it must have at least two elements.

A simple example of a pre-semifield can be derived from any field F that has more than one automorphism. In fact, let σ be an automorphism, not the identity, and define $x \circ y = (xy)^\sigma$. Then $(F, +, \circ)$ is a pre-semifield, and it has no identity; for $1 \circ 1 = 1$ implies that 1 must be the identity if any exists, yet $1 \circ y = y^\sigma \neq y$ for some y.

We will see in Section 4 that a finite pre-semifield can be thought of as a three-dimensional array of integers, and that a semifield can be constructed from a pre-semifield in several ways.

2.4 The additive group. It is easy to show that the additive group of a pre-semifield S must be commutative: By the distributive laws,

$$(ac + ad) + (bc + bd) = (a + b)(c + d) = (ac + bc) + (ad + bd).$$

Therefore, by A1, $ad + bc = bc + ad$, and any elements that can be written as products commute under addition. But by A2 and finiteness, any element of S can be written as a product; therefore, the additive group is Abelian.

Another simple argument shows that the additive group is *elementary* Abelian. In fact, let $a \neq 0$, and let p be the additive order of a. Then p must be a prime number, since $(ma)(na) = ((mn)a)a$ for integers m and n. The fact that every nonzero element has prime order suffices to show that the group is elementary Abelian, and that all nonzero elements have the same prime order p. This number p is called the *characteristic* of the pre-semifield.

2.5 Vector space representation. Let S be a pre-semifield, and let F be the field $GF(p)$, where p is the characteristic of S. Then we can consider the elements of F to be "scalars," and S is a vector space over F. In particular, S must have p^n elements, where n is the dimension of S over F.

The simple observations just made in the preceding paragraph are surprisingly useful, since many of the concepts of semifields and pre-semifields are fruitfully translated into vector space terminology. This way of thinking will be exploited further in Section 4. We can, for example, rephrase the distributive laws as follows: Let a be a nonzero element of S; we define the functions L_a and R_a by the rules

$$xL_a = ax; \qquad xR_a = xa. \tag{2.4}$$

Then the distributive laws A3 are equivalent to the statement that L_a and R_a are linear transformations of the vector space into itself. Furthermore, the axiom A2 states that these transformations are all nonsingular, if $a \neq 0$. Therefore L_a and R_a can be represented as nonsingular matrices, with elements in $GF(p)$.

2.6 Nuclei.   Various special subsystems are defined for any given semifield S, indicating degrees of associativity. The most important of these are

the left nucleus N_l:	$\{\, x \mid (xa)b = x(ab),\ \text{for all } a, b \in S \,\}$;
the middle nucleus N_m:	$\{\, x \mid (ax)b = a(xb),\ \text{for all } a, b \in S \,\}$;
the right nucleus N_r:	$\{\, x \mid (ab)x = a(bx),\ \text{for all } a, b \in S \,\}$.

The *nucleus* N is the intersection of the left, middle, and right nuclei. The field $GF(p)$, where p is the characteristic of S, is obviously always part of the nucleus.

In many cases, the nuclei are all trivial, that is, equal to $GF(p)$. This is the case for the system V in Section 2.2. But the system W described in that same section has nucleus $F = GF(4)$.

It is easy to verify that each of the nuclei is actually a *field*. Furthermore, S is a vector space over any of its nuclei; it is a left vector space over N_l, N_m, and N; it is a right vector space over N_m, N_r, and N. The operations L_a and R_a defined by equation (2.4) are not necessarily linear transformations over the nuclei; but R_a is a linear transformation over N_l, and L_a is a linear transformation over N_r, when S is regarded as a left or right vector space, respectively.

3. Projective Planes and Isotopy

Perhaps the major application of semifields today is for the construction of combinatorial designs, and of projective planes in particular. Every proper semifield determines a non-Desarguesian projective plane. In this section we discuss projective planes, and the question whether two semifields coordinatize the same plane.

3.1 Homogeneous coordinates. A *projective plane* of order $q \geq 2$ is a system of points and lines for which the following axioms hold:

   P1. Every point belongs to exactly $q + 1$ lines.
   P2. Every line contains exactly $q + 1$ points.
   P3. Every pair of points belongs to exactly one line.
   P4. Every pair of lines intersects in exactly one point.

Consequently there are N points and N lines, where $\binom{N}{2} = \binom{q+1}{2}N$; that is, $N = q^2 + q + 1$.

We let $A \vee B$ denote the line that contains points A and B; similarly, we let $L \wedge M$ denote the point common to lines L and M.

Coordinates can be assigned to all the points in the following way: Let T be a set of q elements, two of which are 0 and 1. Choose three points not on a line, and call them $(1,0,0)$, $(0,1,0)$, $(0,0,1)$. Choose another point on the line $(0,0,1) \vee (0,1,0)$ and call it $(0,1,1)$; then give the names $(1,a,a)$ for all $a \in T$ to the other points of the line $(1,0,0) \vee (0,1,1)$. Finally, for all $a, b \in T$, define

$$(1,a,b) = \big((1,a,a) \vee (0,0,1)\big) \wedge \big((1,b,b) \vee (0,1,0)\big); \qquad (3.1)$$

$$(0,1,a) = \big((1,0,0) \vee (1,1,a)\big) \wedge \big((0,0,1) \vee (0,1,0)\big). \qquad (3.2)$$

It is easy to verify that these definitions are consistent with each other. Each of the $q^2 + q + 1$ points has now received a name (x_1, x_2, x_3), where the first nonzero coordinate is 1.

We can similarly ascribe coordinate names $[y_1, y_2, y_3]$ to the lines:

$$[0,0,1] = (0,0,1) \vee (0,1,0); \qquad (3.3)$$

$$[0,1,a] = (0,0,1) \vee (1,a{\sim},0); \qquad (3.4)$$

$$[1,a,b] = (0,1,a) \vee (1,0,b). \qquad (3.5)$$

Here $a{\sim}$ is a function of a that we will define momentarily.

The line $[1,a,b]$ does not contain $(0,0,1)$, so its points other than $(0,1,a)$ must have the form $(1, x, x \cdot a \circ b)$ for a certain *ternary operation* $x \cdot y \circ z$ defined on the elements of T. We will prove that this ternary operation satisfies the following four laws:

T1. $x \cdot 0 \circ z = 0 \cdot y \circ z = z$.

T2. $x \cdot 1 \circ 0 = x$, $1 \cdot y \circ 0 = y$.

T3. $x \cdot y \circ z = x \cdot y \circ z'$ implies $z = z'$.

T4. $x \cdot y \circ z = x \cdot y' \circ z'$ and $x' \cdot y \circ z = x' \cdot y' \circ z'$ implies $x = x'$ or $y = y'$.

The line $[1, a, b]$ contains $(1, 0, b)$ by definition, hence $x \cdot 0 \circ b = b$ and the first half of T1 is proved. Furthermore the line $[1, 0, b]$ must be $(0, 1, 0) \vee (1, b, b)$, because the latter contains both $(0, 1, 0)$ and $(1, 0, b)$; hence the points of $[1, 0, b]$ other than $(0, 1, 0)$ are $(1, x, b)$ for $x \in T$, and the second half of T1 holds.

The first half of T2 follows from the fact that the line $[1, 1, 0]$ contains all points $(1, x, x)$, according to our construction. To prove the second half, note that $(1, 1, a)$ is on the line $(1, 0, 0) \vee (0, 1, a)$.

The line $(1, x, x \cdot y \circ z) \vee (0, 1, y)$ is $[1, y, z]$; thus T3 is true. Similarly, the line $(1, x, x \cdot y \circ z) \vee (1, x', x' \cdot y \circ z)$ is $[1, y, z]$ when $x \neq x'$; thus T4 holds.

We still need to define the operation $a\sim$ that appears in our formula (3.4) for the line $[0, 1, a]$. Let us define both $a\sim$ and $\sim a$ by the equations

$$(a\sim) \cdot 1 \circ a = 0, \qquad a \cdot 1 \circ (\sim a) = 0. \tag{3.6}$$

Rule T3 implies that the equation $a \cdot 1 \circ x = 0$ has a unique solution x, since T is finite. Furthermore $x \cdot 1 \circ a = 0$ also has a unique solution, for if $x \cdot 1 \circ a = x' \cdot 1 \circ a = b$ we have also $x \cdot 0 \circ b = x' \cdot 0 \circ b = b$ by T1, hence $x = x'$ by T4. Therefore $a\sim$ and $\sim a$ are well defined, and $0\sim = \sim 0 = 0$. The following formulas can readily be verified:

$$\sim(a\sim) = (\sim a)\sim = a;$$

$$[0, 0, 1] = (0, 0, 1) \vee (0, 1, 0); \qquad (0, 0, 1) = [0, 0, 1] \wedge [0, 1, 0];$$

$$[0, 1, \sim a] = (0, 0, 1) \vee (1, a, b); \qquad (0, 1, a) = [0, 0, 1] \wedge [1, a, b];$$

$$[1, a, b] = (0, 1, a) \vee (1, 0, b); \qquad (1, a, b) = [0, 1, \sim a] \wedge [1, 0, b]. \tag{3.7}$$

This method of assigning coordinates parallels Hall's classical development in [8], except that it has been cast in new notation:

	Homogeneous coordinates		Notation in [8]
Points:	$(0, 0, 1)$		(∞)
	$(0, 1, a)$,	$a \in T$	(a)
	$(1, a, b)$,	$a, b \in T$	(a, b)
Lines:	$[0, 0, 1]$		$L(\infty)$
	$[0, 1, a]$,	$a \in T$	$x = a\sim$
	$[1, a, b]$,	$a, b \in T$	$y = x \cdot a \circ b$

The principal feature of homogeneous coordinates is that the point (x_1, x_2, x_3) is incident with the line $[y_1, y_2, y_3]$ if and only if

$$y_1 x_3 = x_2 \cdot y_2 \circ x_1 y_3. \tag{3.8}$$

Since y_1 and x_1 must equal 0 or 1, the "multiplications" $y_1 x_3$ and $x_1 y_3$ in this relation are easily understood. The fact that incidence can be expressed in a single formula means that many special cases can often be eliminated when carrying out proofs.

Theorem 3.1.1. *A ternary relation that satisfies the laws T1–T4 defines a projective plane, when incidence between points and lines is defined by* (3.8).

Proof. If T has q elements it is easy to verify that P1 and P2 hold.

Suppose (x_1, x_2, x_3) and (x_1', x_2', x_3') are distinct points; we want to prove that they lie on one and only one line. Thus if $x_1 = x_1' = 1$, we want to show that the simultaneous equations

$$y_1 x_3 = x_2 \cdot y_2 \circ y_3, \qquad y_1 x_3' = x_2' \cdot y_2 \circ y_3$$

have a unique solution $[y_1, y_2, y_3]$ when $(x_2, x_3) \neq (x_2', x_3')$. If $x_2 \neq x_2'$, rules T3 and T4 guarantee that the pairs $(x_2 \cdot y_2 \circ y_3, \; x_2' \cdot y_2 \circ y_3)$ run through all q^2 possibilities as y_2 and y_3 vary, so there is exactly one solution with $y_1 = 1$ (and no solution with $y_1 = 0$). On the other hand if $x_2 = x_2'$, the only solution is $[y_1, y_2, y_3] = [0, 1, \sim x_2]$.

If $x_1 = x_1' = 0$, the unique solution to $y_1 x_3 = x_2 \cdot y_2 \circ 0$ and $y_1 x_3' = x_2' \cdot y_2 \circ 0$ is $[y_1, y_2, y_3] = [0, 0, 1]$. And if $(x_1, x_1') = (1, 0)$, the solution is $[0, 1, \sim x_2]$ when $x_2' = 0$; otherwise $x_2' = 1$, $y_2 = x_3'$, and $x_3 = x_2 \cdot x_3' \circ y_3$ has a unique solution y_3 by T3. We have verified P3.

Let us define a "dual" ternary operation $a \star b \, \square \, c$ by the rules

$$1 \cdot (a\neg) \circ a = 0, \qquad 1 \cdot a \circ (\neg a) = 0;$$
$$d = a \cdot b \circ c \quad \Longleftrightarrow \quad c\neg = b \star ((\sim a)\neg) \, \square \, (d\neg). \tag{3.9}$$

(Here the operations $\neg a$ and $a\neg$ are analogous to $\sim a$ and $a\sim$, and we have $\neg 1 = \sim 1$.) Axioms T1–T4 are readily verified, and (3.8) becomes

$$x_1 y_3 \neg \; = y_2 \star ((\sim x_2)\neg) \, \square \, (y_1 x_3 \neg)$$

with respect to the dual operation. We can therefore dualize our proof of P3 to establish P4, by showing that the simultaneous equations

$$y_1 x_3 \neg = x_2 \star ((\sim y_2)\neg) \, \square \, (x_1 y_3 \neg), \qquad y_1 x_3' \neg = x_2' \star ((\sim y_2)\neg) \, \square \, (x_1' y_3 \neg)$$

have a unique solution (x_1, x_2, x_3) when $[y_1, y_2, y_3] \neq [y_1', y_2', y_3']$. $\square$

The dual operation in (3.9) corresponds to interchanging the roles of points and lines, under the correspondence

$$(0,0,1)' = [0,0,1]; \qquad [0,0,1]' = (0,0,1);$$
$$(0,1,a)' = [0,1,\neg a]; \qquad [0,1,a]' = (0,1,\neg a);$$
$$(1,a,b)' = [1,a,\neg b]; \qquad [1,a,b]' = (1,(\neg a)\sim,\neg b). \qquad (3.10)$$

3.2 Isotopes. A *ternary ring* is a system T in which axioms T1–T4 hold. Suppose T is a ternary ring with operation $a \cdot b \circ c$ and T' is a ternary ring with operation $a \star b \square c$. An *isotopism* from T' onto T is a set of three functions (F, G, H), each being 1–1 correspondences from T' to T, such that

$$(0)H = 0,$$
$$(a \star b \square c)H = (aF) \cdot (bG) \circ (cH), \qquad \text{for all } a, b, c \in T'. \qquad (3.11)$$

Lemma 3.2.1. *Let (F, G, H) be an isotopism from T' to T. Then*

$$H = FR_{1G} = GL_{1F},$$

where L and R denote left and right multiplication in the ring T:

$$a \cdot b \circ 0 = aR_b = bL_a \qquad \text{for all } a, b \in T. \qquad (3.12)$$

Proof. $aH = (a \star 1 \square 0)H = (aF) \cdot (1G) \circ 0$ and $bH = (1 \star b \square 0)H = (1F) \cdot (bG) \circ 0$. $\square$

Theorem 3.2.2. *Let T be a ternary ring with q elements. The number of nonisomorphic ternary rings isotopic to T is at most $(q-1)^2$.*

Proof. Let T_1 and T_2 be ternary rings isotopic to T, under the functions $(F_1, G_1, H_1) : T_1 \to T$ and $(F_2, G_2, H_2) : T_2 \to T$. We will show that if $1F_1 = 1F_2 = y$ and if $1G_1 = 1G_2 = z$, then T_1 and T_2 are isomorphic. Since there are $(q-1)$ choices for $1F_1$ and $(q-1)$ independent choices for $1G_1$, there will be at most $(q-1)^2$ nonisomorphic ternary rings, as claimed.

By Lemma 3.2.1, we have

$$F_1\psi = G_1\varphi = H_1,$$
$$F_2\psi = G_2\varphi = H_2,$$

where $\varphi = L_y$ and $\psi = R_z$ are permutations of T. Hence,

$$F_1F_2^{-1} = G_1G_2^{-1} = H_1H_2^{-1} = \alpha.$$

Now α is the required isomorphism from T_1 to T_2, since

$$a\alpha \star_1 b\alpha \,\square_1\, c\alpha = (a\alpha F_2 \cdot b\alpha G_2 \circ c\alpha H_2)H_2^{-1}$$
$$= (aF_1 \cdot bG_1 \circ cH_1)H_1^{-1}\alpha = (a \star_2 b \,\square_2\, c)\alpha,$$

for all $a, b, c \in T_1$. $\quad\square$

We note that the upper bound $(q-1)^2$ is best possible; there is a ternary ring with 32 elements that has 31^2 distinct isotopic ternary rings. (See Section 5.) But if T is a field, we have the other extreme where all isotopic rings are isomorphic to T.

Theorem 3.2.3. *Let y and z be elements of a given ternary ring T. Let $\varphi = L_y$, $\psi = R_z$, $F = \psi^{-1}$, and $G = \varphi^{-1}$, and let T' be the system consisting of the elements of T with a new ternary operation defined as follows:*

$$a \star b \,\square\, c = aF \cdot bG \circ c. \tag{3.13}$$

Then T' is also a ternary ring, having the identity element yz. This ring T' is isotopic to T, and all ternary rings isotopic to T can be constructed in this way (up to isomorphism).

Proof. Note that F and G are well-defined since y and z are nonzero. The latter part of this theorem follows from the preceding proof; we must show only that $a \star b \,\square\, c$ satisfies the requirements for a ternary ring.

    T1. $0 \star b \,\square\, c = 0 \cdot bG \circ c = c = aF \cdot 0 \circ c = a \star 0 \,\square\, c$.

    T2. $yz \star b \,\square\, 0 = y \cdot bG \circ 0 = bG\varphi = b$; similarly $a \star yz \,\square\, 0 = a$.

    T3. $x \star y \,\square\, z = x \star y \,\square\, z'$ implies $xF \cdot yG \circ z = xF \cdot yG \circ z'$
        implies $z = z'$.

    T4. $x \star y \,\square\, z = x \star y' \,\square\, z'$ and $x' \star y \,\square\, z = x' \star y' \,\square\, z'$ implies
        $xF \cdot yG \circ z = xF \cdot y'G \circ z'$ and $x'F \cdot yG \circ z = x'F \cdot y'G \circ z'$
        implies $xF = x'F$ or $yG = y'G$ implies $x = x'$ or $y = y'$. $\quad\square$

Theorem 3.2.3 is essentially a converse to Lemma 3.2.1, for it says that the relations $(0)H = 0$ and $H = FR_{1G} = GL_{1F}$ are sufficient to construct a new ternary operation; no stronger condition can be derived from general isotopy. Theorem 3.2.3 also provides a convenient way to calculate all ternary rings isotopic to a given one. An isotopism where H is the identity, as in (3.13), is called a "principal isotope."

3.3 The significance of isotopy in geometry. An isomorphism α between projective planes is a 1–1 correspondence between points and lines that preserves incidence; in other words, point P is on line L if and only if point $P\alpha$ is on line $L\alpha$. An automorphism of a plane is commonly called a *collineation*.

Theorem 3.3.1. *Let π and π' be projective planes and let α be an isomorphism from π' onto π such that*

$$(0,0,1)'\alpha = (0,0,1), \quad (0,1,0)'\alpha = (0,1,0), \quad (1,0,0)'\alpha = (1,0,0).$$

Then the ternary rings of π and π' are isotopic.

Proof. $[0,0,1]'\alpha = (0,0,1)'\alpha\vee(0,1,0)'\alpha = [0,0,1]$. Therefore $(0,1,a)'\alpha$ lies on $[0,0,1]$, and there must be a 1–1 correspondence G such that

$$(0,1,a)'\alpha = (0,1,aG). \tag{3.14}$$

Similarly we find $[0,1,0]'\alpha = [0,1,0]$ and $[1,0,0]'\alpha = [1,0,0]$; hence there are 1–1 correspondences H and F from T' to T such that

$$(1,a,0)'\alpha = (1,aF,0), \qquad (1,0,b)'\alpha = (1,0,bH). \tag{3.15}$$

We can now calculate the images of all lines:

$$[0,1,\sim a]'\alpha = (0,0,1)'\alpha \vee (1,a,0)'\alpha = [0,1,\sim(aF)];$$
$$[1,a,b]'\alpha = (0,1,a)'\alpha \vee (1,0,b)'\alpha = [1,aG,bH]. \tag{3.16}$$

Finally we find the image of every point:

$$(1,a,b)'\alpha = [0,1,\sim a]'\alpha \wedge [1,0,b]'\alpha = (1,aF,bH). \tag{3.17}$$

Now we can derive the desired law: We have

$$(1,x_2,x_3)' \in [1,y_2,y_3]' \iff (1,x_2,x_3)'\alpha \in [1,y_2,y_3]'\alpha \tag{3.18}$$

for all $x_2,x_3,y_2,y_3 \in T'$; that is,

$$x_3 = x_2 \star y_2 \, {}^\square\, y_3 \iff x_3H = x_2F \cdot y_2G \circ y_3H,$$

and this is precisely the relation used for isotopy. □

Theorem 3.3.2 (Converse of Theorem 3.3.1). *Let (F,G,H) be an isotopism from a ternary ring T' to T, and let π' and π be the corresponding planes. Define α by Equations (3.14)–(3.17); then α is an isomorphism between π' and π.*

Proof. Since α is a 1–1 correspondence, and since (3.18) holds directly from the law of isotopy, there are only a few cases to consider:

I. $(1, x_2, x_3)' \in [0, y_2, y_3]' \iff y_2 = 1$ and $y_3 \sim = x_2$;
$(1, x_2, x_3)' \alpha \in [0, y_2, y_3]' \alpha \iff y_2 = 1$ and $(y_3 \sim) F = x_2 F$.

II. $(0, x_2, x_3)' \in [1, y_2, y_3]' \iff x_2 = 1$ and $x_3 = y_2$;
$(0, x_2, x_3)' \alpha \in [1, y_2, y_3]' \alpha \iff x_2 = 1$ and $x_3 G = y_2 G$.

III. $(0, x_2, x_3)' \in [0, y_2, y_3]' \iff x_2 = 0$ or $y_2 = 0$;
$(0, x_2, x_3)' \alpha \in [0, y_2, y_3]' \alpha \iff x_2 = 0$ or $y_2 = 0$. $\square$

We define *autotopism* in an obvious way, as an isotopism of a ternary ring onto itself. If (F, G, H) and (F', G', H') are autotopisms,

$$(a \cdot b \circ c) H H' = (aF \cdot bG \circ cH) H' = aFF' \cdot bGG' \circ cHH',$$

so we define the product of two autotopisms as

$$(F, G, H)(F', G', H') = (FF', GG', HH'). \tag{3.19}$$

An automorphism is the special case (F, F, F) of an autotopism where all three permutations are equal.

Corollary 3.3.3. *All isotopic ternary rings coordinatize the same projective plane. The collineations of a projective plane that fix the points $(0, 0, 1)$, $(0, 1, 0)$, and $(1, 0, 0)$ form a group of autotopisms of the ternary ring.* $\square$

Theorem 3.3.4. *Let T be a ternary ring with q elements, and let h be the number of autotopisms of T. Then*

$$(q - 1)^2 = \sum_{T'} \frac{h}{k(T')} \tag{3.20}$$

where T' ranges over all nonisomorphic ternary rings isotopic to T, and where $k(T')$ is the number of automorphisms of T'.

Proof. Let y and z range over the nonzero elements of T, and consider the $(q - 1)^2$ ternary rings constructed in Theorem 3.2.3. If T' is any of these ternary rings, we will show that exactly $h/k(T')$ ternary rings of the set are isomorphic to T'; this will prove formula (3.20).

We need only show that $h/k(T)$ of the ternary rings are isomorphic to T, because the autotopism group of T' is conjugate to the autotopism group of T and therefore has the same order, and because the $(q - 1)^2$ ternary rings formed from T' are isomorphic in some order to the $(q-1)^2$ ternary rings formed from T (using Theorem 3.2.2, since the rings are determined by $1F$ and $1G$).

Let α be an isomorphism from T to the ring $T'(R_z^{-1}, L_y^{-1}, 1)$. There are $k(T)$ such isomorphisms. Then

$$(a \cdot b \circ c)\alpha = a\alpha \star b\alpha \mathbin{\square} c\alpha = a\alpha R_z^{-1} \cdot b\alpha L_y^{-1} \circ c\alpha;$$

that is, $(\alpha R_z^{-1}, \alpha L_y^{-1}, \alpha)$ is an autotopism. By Lemma 3.2.1, every autotopism is of this form, and defines y and z; therefore, if r of the pairs (y, z) yield isomorphic rings, there are $h = rk(T)$ autotopisms. □

3.4 Isotopy of semifields. A semifield is a particular type of ternary ring, where $a \cdot b \circ c = ab + c$ and axioms A1–A4 hold. We now specialize the material of the preceding sections to the case of semifields.

Theorem 3.4.1. *Let S be a semifield of characteristic p. All ternary rings isotopic to S are semifields. Moreover, (F, G, H) is an isotopism from T to S if and only if F, G, and H are nonsingular linear transformations from T to S over $\mathrm{GF}(p)$, satisfying the condition*

$$(ab)H = (aF)(bG). \tag{3.21}$$

Proof. Suppose T is isotopic to S, and $(a \cdot b \circ c)H = (aF)(bG) + cH$. Then $(a \cdot 1 \circ c)H = (aF)(1G) + cH = (a \cdot 1 \circ 0)H + cH = aH + cH$, so H is an isomorphism between the "addition" of T and the addition of S; that is, H is a nonsingular linear transformation over $\mathrm{GF}(p)$. Now $H = FR_{1G} = GL_{1F}$ by Lemma 3.2.1, where R_{1G} and L_{1F} are nonsingular linear transformations over $\mathrm{GF}(p)$ of S into itself; hence F and G are also nonsingular linear transformations. Properties A1 through A4 are now verified immediately, as is the converse portion of the theorem. □

Theorem 3.4.2 (Albert). *Every collineation of a plane coordinatized by a proper semifield fixes $(0, 0, 1)$ and $[0, 0, 1]$.*

Proof. This theorem is well known, but its proof requires the development of more geometrical tools than are appropriate here. Proofs may be found in [4] and [14]. □

The previous theorem holds primarily because any semifield already has a great number of collineations. If it were permitted to move the point $(0, 0, 1)$ or the line $[0, 0, 1]$, it would have so many more collineations that it would satisfy the classical theorem of Desargues in projective geometry; but then it would be coordinatized by a field.

The standard collineations, holding in any semifield plane, are the translations $\tau(h, k)$ and the generalized shears $\sigma(h, k)$, defined for all h

and k in the semifield as follows:

$$(x_1, x_2, x_3)\tau(h, k) = (x_1, x_2 + x_1 h, x_3 + x_1 k),$$
$$[y_1, y_2, y_3]\tau(h, k) = [y_1, y_2, y_3 - hy_2 + y_1 k];$$
$$(x_1, x_2, x_3)\sigma(h, k) = (x_1, x_2, x_3 + x_2 h + x_1 k),$$
$$[y_1, y_2, y_3]\sigma(h, k) = [y_1, y_2 + y_1 h, y_3 + y_1 k]. \tag{3.22}$$

The proof that these are collineations is very simple with homogeneous coordinates:

$$y_1 x_3 = x_2 y_2 + x_1 y_3$$

if and only if

$$y_1(x_3 + x_1 k) = (x_2 + x_1 h)y_2 + x_1(y_3 - hy_2 + y_1 k)$$

if and only if

$$y_1(x_3 + x_2 h + x_1 k) = x_2(y_2 + y_1 h) + x_1(y_3 + y_1 k),$$

remembering that x_1 and y_1 are restricted to be 0 or 1.

The following relations are easily computed, using the formulas already derived; here $\alpha(F, G, H)$ represents a collineation corresponding to an autotopism:

$$\tau(0, k) = \sigma(0, k);$$
$$\tau(h, k)\tau(h', k') = \tau(h + h', k + k');$$
$$\sigma(h, k)\sigma(h', k') = \sigma(h + h', k + k');$$
$$\alpha(F, G, H)\alpha(F', G', H') = \alpha(FF', GG', HH');$$
$$\tau(h, k)^{-1}\sigma(l, m)\tau(h, k) = \sigma(l, m - hl);$$
$$\sigma(h, k)^{-1}\tau(l, m)\sigma(h, k) = \tau(l, m + lh);$$
$$\alpha(F, G, H)^{-1}\tau(h, k)\alpha(F, G, H) = \tau(hF, kH);$$
$$\alpha(F, G, H)^{-1}\sigma(h, k)\alpha(F, G, H) = \sigma(hG, kH). \tag{3.23}$$

Theorem 3.4.3 (Albert). *Two semifields coordinatize the same plane if and only if they are isotopic.*

Proof. If β is an isomorphism between two semifield planes, then $(0, 0, 1)\beta = (0, 0, 1)$ and $[0, 0, 1]\beta = [0, 0, 1]$, because this point and line are characterized by Theorem 3.4.2 and the standard collineations.

Hence $(0,1,0)\beta = (0,1,a)$ and $(1,0,0)\beta = (1,b,c)$, for some a, b, and c. Let

$$\alpha = \beta\sigma(-a,0)\tau(-b, ba - c).$$

Then

$$(0,0,1)\alpha = (0,0,1); \quad (1,0,0)\alpha = (1,0,0); \quad (0,1,0)\alpha = (0,1,0);$$

and Theorem 3.3.1 applies. The converse is part of Corollary 3.3.3.   □

Theorem 3.4.4. *Let G be the collineation group of a semifield plane; let T be the subgroup of translations, S the subgroup of generalized shears, H the subgroup corresponding to autotopisms, and A the elementary Abelian additive group of the semifield. Then we have the following normal series for G:*

$$I \lhd T \wedge S \lhd T \lhd T \cup S \lhd T \cup S \cup H = G.$$

The quotient groups are equal respectively to A, A, A, and H.

Proof. This follows from formulas (3.23), and the fact that

$$T \cup S \cup H = G$$

as shown in the preceding proof.   □

Theorems 3.4.3 and 3.4.4 are well known; they indicate how important isotopy is when considering semifield planes. The use of homogeneous coordinates simplifies and clarifies previous proofs of those theorems.

3.5 Nonlinear isotopes. One might pose an interesting problem here, concerning whether linearity of F, G, H is really a necessary condition for constructing semifields. Suppose we have a semifield S and we have defined a new multiplication $*$ on S by the formula

$$(a * b)H = (aF)(bG).$$

We require that F, G, H be permutations of S, and that

$$(a + b) * c = a * c + b * c,$$
$$c * (a + b) = c * a + c * b,$$
$$1 * a = a * 1 = a. \tag{3.24}$$

Question: Does this imply that F, G, H are linear?

The following theorem does not settle this question by any means, but it does provide some insight into the matter. We say an element $x \in S$ is *noncentral* if $xy = yx$ implies that

$$y = r + sx, \qquad \text{for some } r, s \in \mathrm{GF}(p).$$

Theorem 3.5.1. *If the semifield S has characteristic 2, and if every element of $S \setminus \mathrm{GF}(2)$ is noncentral, then the functions F, G, H of (3.24) must be linear.*

Proof. The conditions imply that $aH^{-1} = aF^{-1}*1G^{-1} = 1F^{-1}*aG^{-1}$, so we can write $H = PF = QG$ where P and Q are linear. Thus

$$(aH)(bH) = (aP * bQ)H. \tag{3.25}$$

Let x be such that $xH = 1$. Then since $(aH)(xH) = (xH)(aH)$, Eq. (3.25) shows us that $aP * xQ = xP * aQ$, for all $a \in S$. Adding $aP * aQ$ to each side, we obtain $aP * (a + x)Q = (a + x)P * aQ$; hence

$$(aH)((a + x)H) = ((a + x)H)(aH),$$

for all a. This is the relation we will use in order to show that H is linear.

We can at least prove that $0H = 0$, since $(0H)(aH) = (0P * aQ)H = 0H$ for all a. Therefore $x \neq 0$. We will now show that

$$(a + x)H = aH + xH = aH + 1, \qquad \text{for all } a. \tag{3.26}$$

If $a = 0$ or if $a = x$, this is obviously true (because $x + x = 0$). Otherwise aH is noncentral, by hypothesis. By the definition, this implies that $(a + x)H = 0$ or 1 or aH or $aH + 1$. The first three possibilities are clearly impossible, so (3.26) is established.

Since $H = G\varphi$, where φ is linear, we have

$$(a + x)G = (a + x)H\varphi^{-1} = aH\varphi^{-1} + xH\varphi^{-1} = aG + xG.$$

Finally, then, let a and b be arbitrary nonzero elements of S; define c and d such that $a = c * x$, $b = c * d$. Then

$$\begin{aligned}
(a + b)H &= (c * (x + d))H \\
&= (cF)((x + d)G) \\
&= (cF)(xG + dG) \\
&= (cF)(xG) + (cF)(dG) = aH + bH.
\end{aligned}$$

Thus H is linear and therefore so are F and G.   □

Remark. Theorem 3.5.1 is not trivial, because there exist systems in which the hypothesis is satisfied (for example, V in Section 2.2). Furthermore, some hypothesis is necessary for the theorem, because there are examples in which F, G, H are *not* linear. Such an example is the field $S = \mathrm{GF}(8)$. Let x be a primitive element with minimum polynomial $x^3 + x^2 + 1$; then define H by

$$(x^k \bmod (x^3 + x^2 + 1))H = x^k \bmod (x^3 + x + 1).$$

If we write $(a * b)H = (aH)(bH)$, the equations (3.24) are satisfied, but

$$(1 + x^2)H = 1 + x \neq 1 + x^2 = 1H + x^2 H.$$

Such nonlinear isotopisms do not have geometric significance, however, according to Theorem 3.4.1, and they will not be considered further in this paper.

4. Nonsingular Hypercubes

In this section we turn to another way of looking at semifields and their isotopisms, where we think of 3-dimensional matrices. The 3D representation allows us to see several symmetries in the situation that are not otherwise apparent; and it also allows us to work with pre-semifields, in cases when pre-semifields are more convenient.

4.1 Elementary operations on hypercubes. An m-dimensional hypercube A, for $m \geq 1$, is an array of n^m elements belonging to a field; the elements are denoted by $A_{ij\ldots r}$ where there are m subscripts, and each subscript varies from 1 to n. We consider n to be a fixed integer throughout the entire discussion.

It is easy to devise extremely cumbersome notations for such systems, so an attempt will be made to keep the notation as simple as possible.

Let σ be a permutation of the elements 1, 2, ..., m. Then A^σ will represent the m-cube A with subscripts permuted by σ; that is, the $k\sigma$th subscript of A^σ is the kth subscript of A. This is a generalization of the concept of transposition of matrices; if A is a matrix ($m = 2$), $A^T = A^{(12)}$. In the 3-dimensional case, if $A = (A_{ijk})$, then if $B = A^{(123)} = (B_{ijk})$ we have $B_{ijk} = A_{jki}$ for all i, j, k. We also have the general law $(A^\sigma)^\tau = A^{\sigma\tau}$.

If $m > 1$, we form $(m-1)$-dimensional subcubes of an m-cube A, denoted by $A^{[t]}$, as follows:

$$B = A^{[t]} = (B_{ij...r}) \iff B_{ij...r} = A_{tij...r}. \tag{4.1}$$

Here $1 \leq t \leq n$. Additional subcubes that hold other positions fixed can be formed by combining the operations A^{σ} and $A^{[t]}$; actually all conceivable subcubes can be obtained in this manner.

4.2 Sums and products of hypercubes. The sum of two m-cubes A and B is simply defined as the m-cube consisting of the sums of the components:

$$C = A + B = (C_{ij...r}) \iff C_{ij...r} = A_{ij...r} + B_{ij...r}. \tag{4.2}$$

The products of a p-cube B times an m-cube A, yielding an $(m+p-2)$-cube, can be defined in the same way as the tensor dot product of tensors, as follows:

$$C = B \overset{1}{\times} A = (C_{i...jk...r}) \iff C_{i...jk...r} = \sum_t B_{i...jt} A_{tk...r}.$$

Actually, m of such products can be defined, and we write in general

$$C = B \overset{l}{\times} A = (C_{i...jk...qr...s})$$
$$\iff C_{i...jk...qr...s} = \sum_t B_{k...qt} A_{i...jtr...s} \tag{4.3}$$

where '$i \ldots j$' represents $l-1$ subscripts, '$r \ldots s$' represents $m-l$.

Notice that $A^{[t]} = B \overset{1}{\times} A$, where B is the 1-cube $(B_i) = (\delta_{it})$. We also see that if A, B, C are matrices, $B \overset{1}{\times} A = BA$ and $C \overset{2}{\times} A = AC^T$, expressing the familiar fact that premultiplication corresponds to operating on rows of a matrix, while postmultiplication corresponds to operating on columns.

The associative law for matrix multiplication, $C(AB)T = (CA)B^T$, can be written in the form

$$C \overset{1}{\times} (B \overset{2}{\times} A) = B \overset{2}{\times} (C \overset{1}{\times} A).$$

This relation is a special case of a general rule for products of hypercubes.

Theorem 4.2.1 (Generalized associative law). *If $u < v$, and if the dimension of C is $f + 2$, then*

$$C \overset{u}{\times} (B \overset{v}{\times} A) = B \overset{v+f}{\times} (C \overset{u}{\times} A). \tag{4.4}$$

Proof. Let the dimensions of A and B be m and p, respectively. The general element $D_{i...jk...qr...sw...xy...z}$ of the product $C \overset{u}{\times} (B \overset{v}{\times} A)$, where '$i...j$' represents $u - 1$ subscripts, '$k...q$' represents $f + 1$ subscripts, '$r...s$' represents $v - u - 1$ subscripts, '$w...x$' represents $p - 1$ subscripts, and '$y...z$' represents $m - v$ subscripts, is

$$\sum_t C_{k...qt} \sum_h B_{w...xh} A_{i...jtr...shy...z}$$

by definition. The same general element of the right-hand product is

$$\sum_t B_{w...xt} \sum_h C_{k...qh} A_{i...jhr...sty...z},$$

and these are clearly equal.   □

Corollary 4.2.2. *Let m be the dimension of A, and let B_1, B_2, ..., B_m be matrices. Then the m multiplications $B_l \overset{l}{\times} A$ can be performed on A in any order; that is, if β is a permutation of $\{1, 2, ..., m\}$,*

$$B_1 \overset{1}{\times} (B_2 \overset{2}{\times} (\cdots (B_m \overset{m}{\times} A) \cdots)) = B_{1\beta} \overset{1\beta}{\times} (B_{2\beta} \overset{2\beta}{\times} (\cdots (B_{m\beta} \overset{m\beta}{\times} A) \cdots)).$$

Proof. If the dimension is 2, then f in Theorem 4.2.1 is always zero, so this result follows by repeated application of that theorem.   □

We will write

$$[B_1, B_2, ..., B_m] \times A \tag{4.5}$$

to denote the multiplication operations expressed in Corollary 4.2.2.

The following law generalizes the matrix equation $(CA)^T = A^T C^T$:

Theorem 4.2.3. *If C is a matrix, and if σ is a permutation,*

$$(C \overset{u}{\times} A)^\sigma = C \overset{u\sigma}{\times} A^\sigma. \tag{4.6}$$

Proof. Let the subscripts be i_1, i_2, ..., i_m. The desired result follows from the formulas

$$(A^\sigma)_{i_1 i_2...i_m} = A_{i_{1\sigma} i_{2\sigma}...i_{m\sigma}};$$

$$(C \overset{u}{\times} A)_{i_1 i_2...i_m} = \sum_t C_{i_u t} A_{i_1...i_{u-1} t i_{u+1}...i_m};$$

$$(C \overset{u\sigma}{\times} A^\sigma)_{i_1 i_2...i_m} = \sum_t C_{i_{u\sigma} t} A_{i_{1\sigma}...i_{(u-1)\sigma} t i_{(u+1)\sigma}...i_{m\sigma}}.   □$$

Several other associative laws can be formulated to include the cases not meeting the hypothesis of Theorem 4.2.1; that is, when we are given a product $C \overset{u}{\times} (B \overset{v}{\times} A)$ such that $u \geq v$ and $u + 2 - \dim C \leq v$. But only one of these is of concern to us here:

Theorem 4.2.4. *Using the notation* (4.5),

$$[C_1, C_2, \ldots, C_m] \times ([B_1, B_2, \ldots, B_m] \times A) = [C_1 B_1, C_2 B_2, \ldots, C_m B_m] \times A.$$

Proof. In view of Theorems 4.2.1 and 4.2.3, it suffices to show that

$$(C \overset{1}{\times} (B \overset{1}{\times} A)) = (CB) \overset{1}{\times} A \tag{4.7}$$

when C and B are matrices. The general element $D_{ij\ldots r}$ of the left hand side is

$$\sum_t C_{it} \sum_h B_{th} A_{hj\ldots r};$$

on the right hand side it is

$$\sum_t \left(\sum_h C_{ih} B_{ht} \right) A_{tj\ldots r},$$

and these are equal.   □

4.3 Nonsingular hypercubes. The concept of a nonsingular matrix can be generalized to m-cubes in a meaningful way. The definition proceeds inductively: We say that a vector, or 1-cube, A, is *singular* if and only if $A = 0$. For $m > 1$, we say that an m-cube A is *singular* if and only if there exists a nonsingular vector B such that $B \overset{1}{\times} A$ is singular.

Another way to state this definition is that an m-cube A is *nonsingular* if the following condition is satisfied:

$$x_1 A^{[1]} + x_2 A^{[2]} + \cdots + x_n A^{[n]} \text{ is singular implies that}$$
$$x_1 = x_2 = \cdots = x_n = 0.$$

In other words, any nonzero linear combination of the subcubes $A^{[t]}$ must be nonsingular. This definition certainly includes the ordinary definition of a nonsingular matrix, for in the special case $m = 2$ it says that the rows of A are linearly independent.

Theorem 4.3.1. *Let A be an m-cube, and let σ be a permutation of $\{1, 2, \ldots, m\}$. Then A is nonsingular if and only if A^σ is nonsingular.*

Proof. We use induction on m. For $m = 1$ it is trivial, and for $m = 2$ it is a special case of the theorem that "row rank equals column rank of a matrix." We assume, therefore, that $m > 2$.

Without loss of generality we may assume that σ is either (12) or $(23 \ldots m)$, since these two permutations generate all of the others.

If $\sigma = (12)$, we argue as follows, where B and C denote vectors:

A is nonsingular

$\Longleftrightarrow (C \neq 0 \Rightarrow C \overset{1}{\times} A$ is nonsingular$)$

$\Longleftrightarrow (C \neq 0 \Rightarrow (B \neq 0 \Rightarrow B \overset{1}{\times}(C \overset{1}{\times} A)$ is nonsingular$))$

$$\text{since } m > 2$$

$\Longleftrightarrow (B \neq 0 \Rightarrow (C \neq 0 \Rightarrow B \overset{1}{\times}(C \overset{1}{\times} A)$ is nonsingular$))$

$\Longleftrightarrow (B \neq 0 \Rightarrow (C \neq 0 \Rightarrow C \overset{1}{\times}(B \overset{2}{\times} A)$ is nonsingular$))$

$$\text{by Theorem 4.2.1}$$

$\Longleftrightarrow (B \neq 0 \Rightarrow B \overset{2}{\times} A$ is nonsingular$)$

$\Longleftrightarrow (B \neq 0 \Rightarrow B \overset{1}{\times} A^\sigma$ is nonsingular$)$

$$\text{since } B \overset{1}{\times} A^{(12)} = B \overset{2}{\times} A$$

$\Longleftrightarrow A^\sigma$ is nonsingular.

On the other hand if $\sigma = (23 \ldots m)$, let $\tau = (12 \ldots (m-1))$. Then

$$(B \overset{1}{\times} A)^\tau = B \overset{1}{\times} A^\sigma$$

and we can argue by induction on m as follows:

A is nonsingular $\Longleftrightarrow (B \neq 0 \Rightarrow B \overset{1}{\times} A$ is nonsingular$)$

$\Longleftrightarrow (B \neq 0 \Rightarrow (B \overset{1}{\times} A)^\tau$ is nonsingular$)$

$\Longleftrightarrow (B \neq 0 \Rightarrow B \overset{1}{\times} A^\sigma$ is nonsingular$)$

$\Longleftrightarrow A^\sigma$ is nonsingular. $\quad\square$

Theorem 4.3.2. *Let A be an m-cube and let C_1, C_2, $\ldots$, C_m be nonsingular matrices. Then A is nonsingular if and only if*

$$[C_1, C_2, \ldots, C_m] \times A$$

is nonsingular.

Proof. For $m = 1$ this is merely the definition of a nonsingular matrix C_1. For $m > 1$, it suffices to show that A is singular if, and only if,

$C \overset{2}{\times} A$ is singular, because of Theorem 4.2.3 and Theorem 4.3.1. Using Theorem 4.2.1 with $f = -1$, and induction on m,

$$C \overset{2}{\times} A \text{ is nonsingular} \iff (B \neq 0 \Rightarrow B \overset{1}{\times} (C \overset{2}{\times} A) \text{ is nonsingular})$$

$$\iff (B \neq 0 \Rightarrow C \overset{1}{\times} (B \overset{1}{\times} A) \text{ is nonsingular})$$

$$\iff (B \neq 0 \Rightarrow B \overset{1}{\times} A \text{ is nonsingular})$$

$$\iff A \text{ is nonsingular.} \quad \square$$

Finally, we define an equivalence relation between m-cubes as follows:

$$A \equiv B \qquad \text{if and only if} \qquad A = [C_1, C_2, \dots, C_m] \times B \qquad (4.8)$$

for nonsingular matrices C_1, C_2, ..., C_m. By Theorem 4.2.4 this is an equivalence relation, and by the preceding theorem it preserves singularity.

4.4 Pre-semifields represented as 3-cubes. We now specialize to the case $m = 3$. Let S be a pre-semifield of characteristic p. According to Section 2.5, S is a vector space over $F = \mathrm{GF}(p)$. Let $\{x_1, x_2, \dots, x_n\}$ be a basis of S over F. We can write the multiplication in terms of the basis elements:

$$x_i x_j = \sum_k A_{ijk} x_k. \qquad (4.9)$$

This gives us a 3-cube, A, with entries in F; we will say that A is a *cube corresponding to* S.

The multiplication in S is completely determined by the products of the basis elements, according to the distributive laws, since

$$\left(\sum_i b_i x_i\right)\left(\sum_j c_j x_j\right) = \sum_i \sum_j b_i c_j x_i x_j = \sum_i \sum_j \sum_k b_i c_j A_{ijk} x_k. \qquad (4.10)$$

Theorem 4.4.1. *A cube corresponding to a pre-semifield is nonsingular. Conversely, if A is any nonsingular 3-cube, Eq. (4.10) defines a pre-semifield S, with the x_i being formal basis elements.*

Proof. Because of our other observations, we need show only that multiplication defined by Eq. (4.10) satisfies axiom A2 if and only if A is nonsingular. In the proof of Theorem 4.3.1, we have shown in the case $m = 3$ that A is nonsingular if and only if

$$B \neq 0 \text{ and } C \neq 0 \quad \Rightarrow \quad C \overset{1}{\times} (B \overset{1}{\times} A) \neq 0.$$

But $C \overset{1}{\times} (B \overset{1}{\times} A)$ is the vector $D = (D_k)$ where

$$D_k = \sum_r B_r \sum_s C_s A_{rsk};$$

therefore, this is precisely the condition that

$$a \neq 0 \text{ and } b \neq 0 \quad \Rightarrow \quad ab \neq 0. \qquad \square$$

We observe that a nonsingular 4-cube would correspond to an algebraic system with a ternary multiplication abc satisfying three distributive laws, and with no "zero-divisors." In general, a nonsingular m-cube will lead to an $(m-1)$-ary operation.

If we change to a different basis $\{y_1, \ldots, y_n\}$ of S over F, we have

$$y_i = \sum_j C_{ij} x_j,$$

where $C = (C_{ij})$ is a nonsingular matrix. This introduces a corresponding change in A; let the new cube be B. Then

$$y_i y_j = \left(\sum_r C_{ir} x_r \right) \left(\sum_s C_{js} x_s \right) = \sum_r \sum_s \sum_k C_{ir} C_{js} A_{rsk} x_k$$

$$= \sum_r \sum_s \sum_k \sum_t C_{ir} C_{js} A_{rsk} C_{kt}^{-1} y_k,$$

and B_{ijk} is the coefficient of y_k. Hence $B \equiv A$:

$$B = [C, C, C^{-T}] \times A, \tag{4.11}$$

where $-T$ denotes inverse transpose.

Let us now consider what would happen if we were to define a new multiplication on the elements of S:

$$(a * b)H = (aF)(bG) \tag{4.12}$$

where F, G, H are arbitrary nonsingular linear transformations of S into itself. This gives us another pre-semifield S', which is said to be *isotopic* to S.

Let B be a cube corresponding to the derived pre-semifield S'. We may assume that S' has the same basis $\{x_1, x_2, \ldots, x_n\}$ as S, and that A is the cube for S with this basis. Consider F, G, H as matrices:

$$x_i F = \sum_r F_{ir} x_r, \quad x_j G = \sum_r G_{jr} x_r, \quad x_k H = \sum_r H_{kr} x_r. \tag{4.13}$$

Then

$$x_i * x_j = ((x_iF)(x_jG))H^{-1} = \left(\left(\sum_r F_{ir}x_r\right)\left(\sum_s G_{js}x_s\right)\right)H^{-1}$$

$$= \left(\sum_r\sum_s\sum_t F_{ir}G_{js}A_{rst}x_t\right)H^{-1}$$

$$= \sum_r\sum_s\sum_t\sum_k F_{ir}G_{js}H_{tk}^{-1}A_{rst}x_k. \tag{4.14}$$

Therefore

$$B = [F, G, H^{-T}] \times A. \tag{4.15}$$

We have proved the following fundamental result.

Theorem 4.4.2. *Let S and S' be pre-semifields, and let A and A' be any cubes corresponding to S and S'. Then S is isotopic to S' if and only if $A \equiv A'$, where equivalence is defined in Eq. (4.8).* □

4.5 Semifields and equivalent pre-semifields. The preceding discussion applies to semifields as special cases of pre-semifields. If S is a semifield, we can assume that its basis has the form $\{1, x_2, \ldots, x_n\}$. The corresponding cube A then has a special property.

Let us write

$$A^{r**} = A^{[r]};$$
$$A^{*r*} = (A^{(132)})^{[r]};$$
$$A^{**r} = (A^{(123)})^{[r]}. \tag{4.16}$$

In other words, A^{r**}, A^{*r*}, A^{**r} are the matrices (A_{rij}), (A_{jri}), (A_{ijr}) respectively; A^{r**} is the matrix L_{x_r} of left multiplication by x_r, and A^{*r*} is the transpose of the matrix for right multiplication, R_{x_r}. The following formulas are readily verified:

$$([F, G, H] \times A)^{r**} = F_{r1}(GA^{1**}H^T) + \cdots + F_{rn}(GA^{n**}H^T);$$
$$([F, G, H] \times A)^{*r*} = G_{r1}(HA^{*1*}F^T) + \cdots + G_{rn}(HA^{*n*}F^T);$$
$$([F, G, H] \times A)^{**r} = H_{r1}(FA^{**1}G^T) + \cdots + H_{rn}(FA^{**n}G^T). \tag{4.17}$$

We say A is in *standard form* if $A^{1**} = A^{*1*} = I$.

Theorem 4.5.1. *Let S be a semifield with basis $\{1, x_2, \ldots, x_n\}$; then the cube corresponding to S is nonsingular and in standard form. Conversely, every nonsingular 3-cube in standard form yields a semifield, if multiplication is defined by (4.10).*

Proof. Because of Theorem 4.4.1, we must merely observe that standard form is equivalent to axiom A4. But this is obvious, since standard form is nothing but the statement that $L_1 = R_1 = I$. □

Theorem 4.5.2. *Let S and S' be semifields with corresponding cubes A and A'. Then S and S' coordinatize the same projective plane if and only if $A \equiv A'$. Semifield planes are in 1–1 correspondence with equivalence classes of nonsingular 3-cubes.*

Proof.   Apply Theorems 4.4.2 and 3.4.3. $\square$

Theorem 4.5.3. *If A is a cube corresponding to a proper semifield S, the autotopism group of S is isomorphic to the group of all triples of matrices (F, G, H) such that*

$$[F, G, H] \times A = A. \tag{4.18}$$

Proof.   This follows from Eq. (4.15). $\square$

We note that the same result holds for any 3-cube, if $B \equiv A$, even if B is not in standard form, since the corresponding groups are conjugate.

Now we turn to the question of constructing a semifield from a pre-semifield. This is merely a question of finding three nonsingular matrices such that $[F, G, H] \times A$ is in standard form. Such a construction can be done in several ways; for example:

1) Set $G = (A^{1**})^{-1}$, $B = G \overset{2}{\times} A$, $F = (B^{*1*})^{-T}$, $H = I$.
2) Set $F = (A^{*1*})^{-T}$, $B = F \overset{1}{\times} A$, $G = (B^{1**})^{-1}$, $H = I$.
3) Set $H = (A^{1**})^{-T}$, $F = A^{1**}(A^{*1*})^{-T}$, $G = I$.
4) Set $H = (A^{*1*})^{-1}$, $G = A^{*1*}(A^{1**})^{-1}$, $F = I$.

The proofs that these do the job are similar, using (4.17). We consider, for example, method (3):

$$
F = \begin{pmatrix} a_{111} & a_{112} & \cdots & a_{11n} \\ a_{121} & a_{122} & \cdots & a_{12n} \\ \vdots & \vdots & & \vdots \end{pmatrix} \begin{pmatrix} a_{111} & a_{112} & \cdots & a_{11n} \\ a_{211} & a_{212} & \cdots & a_{21n} \\ \vdots & \vdots & & \vdots \end{pmatrix}^{-1}
$$

$$
= \begin{pmatrix} 1 & 0 & \cdots & 0 \\ * & * & \cdots & * \\ \vdots & \vdots & & \vdots \end{pmatrix}.
$$

Therefore

$$([F, G, H] \times A)^{1**} = GA^{1**}H^T = I;$$

and

$$([F, G, H] \times A)^{*1*} = HA^{*1*}F^T = I.$$

The four methods above can also be translated into algebraic terms. For example, here is method (3) in this form:

Theorem 4.5.4. *Let $(S, +, \circ)$ be a pre-semifield, and let $u \in S$. Then if we define a new multiplication $*$ by the rule*

$$(a \circ u) * (u \circ b) = a \circ b \qquad (4.19)$$

we obtain a semifield $(S, +, *)$ with unit element $u \circ u$.   □

Notice that if $(S, +, \circ)$ is commutative, so is $(S, +, *)$.

Formula (4.19) is actually an instance of Eq. (3.13), which is more general. A less symmetric way to write this formula can be read directly from method (3), namely

$$u \circ (a * b) = (u \circ a) R_u^{-1} \circ b,$$

which yields an isomorphic system with unit element u. Method (4) gives another isomorphic system, "dual" to this one.

Thus, we can obtain semifields from pre-semifields in several ways; in each method we were able to leave F, G, or H equal to the identity. But no matter which way we choose, the result is equivalent, in the sense that the same projective plane will result.

The extra degree of freedom we seem to have in this standardization process indicates that we should seek a "more standard" standard form, so that it might actually be a canonical form. One idea suggests itself immediately: We could require that $A^{**1} = I$ also.

If S is a field with p^2 elements, it can always be put into the form

$$A^{1**} = \begin{pmatrix} 1 & 0 \\ 0 & 1 \end{pmatrix}, \qquad A^{2**} = \begin{pmatrix} 0 & 1 \\ 1 & a \end{pmatrix},$$

if a is chosen such that $x^2 - ax - 1$ is irreducible (modulo p).

The field GF(8) can be put into the forms

$$
\begin{array}{ccc}
100 & 010 & 001 \\
010 & 101 & 011 \\
001 & 011 & 111
\end{array}
\quad \text{or} \quad
\begin{array}{ccc}
100 & 010 & 001 \\
010 & 111 & 011 \\
001 & 011 & 110
\end{array}
$$

where we adopt the convention of writing A^{1**}, A^{2**}, ..., A^{n**} in this order. It is unclear whether the condition $A^{1**} = A^{*1*} = A^{**1} = I$ is possible in general; these examples show that it is not a canonical form, in any event. But if all nonsingular cubes can indeed be standardized in this way, we could eliminate many cases when constructing the systems.

One way to utilize the extra freedom, which has been very fruitful, is to adjust A^{2**}. We can essentially perform any desired similarity transformation on A^{2**} while the standardization is taking place. Therefore, in particular, if the characteristic equation of A^{2**} after standardization is irreducible, we can transform it into a companion matrix and restandardize; this means we are taking the basis of the vector space to be of the form $\{1, x, x^2, x(x^2), x(x(x^2)), \dots \}$. Operations of this kind can be exploited when all possible semifields of a given order are to be constructed (see [4]).

5. Transpose of a Plane

Now we are ready to discuss an interesting relationship between some planes coordinatized by semifields. The relationship is somewhat peculiar, since it has algebraic significance but it does not seem to have any obvious geometric significance.

5.1 Transposition. Let π be a projective plane, let S be a semifield of coordinates for π, and let A be a cube corresponding to S. Let S_1 be the pre-semifield described by $A^{(23)}$, and let S_2 be a semifield constructed from S_1 by isotopy. Then we define π^T, the transpose of π, to be the plane coordinatized by S_2.

Theorem 5.1.1. π^T is uniquely defined.

Proof. No matter which choice of S is made, the resulting A's will be equivalent, by Theorem 4.5.2. The 3-cube $A^{(23)}$ is nonsingular, by Theorem 4.3.1. If $A \equiv B$, then $A^{(23)} \equiv B^{(23)}$, because of Theorem 4.2.3. Therefore $A^{(23)}$ is uniquely determined up to equivalence, and the plane π^T is uniquely determined.   □

Corollary 5.1.2. $(\pi^T)^T = \pi$.   □

5.2 Dualization. The dual π^D of a projective plane coordinatized by a semifield, namely the plane obtained by interchanging the roles of points and lines, is well known to be determined by constructing the anti-isomorphic semifield, that is, by using ab in place of ba when multiplying. (With homogeneous coordinates we actually get the dual plane by mapping

$$(0,0,1) \longleftrightarrow [0,0,1], \quad (0,1,a) \longleftrightarrow [0,1,-a], \quad (1,a,b) \longleftrightarrow [1,a,-b],$$

according to (3.10). Our homogeneous coordinates are somewhat unsymmetrical because they have been designed to handle a general ternary ring.) This definition can be phrased in the same way as the definition for π^T, by using $A^{(12)}$ rather than $A^{(23)}$. We may write $\pi^D = \pi^{(12)}$, $\pi^T = \pi^{(23)}$.

Theorem 5.2.1. *The operations of dual and transpose generate a series of at most six planes, according to the following scheme:*

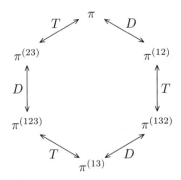

Proof. This theorem is clear from the manner in which dual and transpose have been defined. □

Corollary 5.2.2. *If $\pi = \pi^D \neq \pi^T$ there exists a third plane $\pi^{TD} \neq \pi$, $\pi^{TD} \neq \pi^T$.*

Proof. If $\pi^{TD} = \pi$ then $\pi^T = \pi^D = \pi$. If $\pi^{TD} = \pi^T$, then $\pi^{TD} = (\pi^D)^{TD} = \pi^{(13)} = (\pi^{TD})^T = \pi^{TT} = \pi$. □

5.3 Collineations. Now let us consider the automorphisms of a transposed plane.

Theorem 5.3.1. *The collineation group of π^T has the same order as the collineation group of π.*

Proof. Refer to Theorem 3.4.4. The groups of translations and shears are isomorphic. The autotopism groups are also isomorphic, by Theorem 4.2.3, since

$$[F, G, H] \times A = A \iff [F, H, G] \times A^{(23)} = A^{(23)}. \qquad (5.1)$$

Therefore the collineation groups have the same order. (But Eqs. (3.23) show that the interrelations of autotopisms, translations, and shears are different.) □

The concept of transpose can be generalized to apply to Veblen–Wedderburn systems, but that theory will be reported elsewhere.

5.4 Examples. The transpose of a Desarguesian plane is Desarguesian, because all matrices of left multiplication are powers of the same matrix, and this property remains under transposition.

The semifield planes of order 32 have been completely tabulated by R. J. Walker [16]. Besides the Desarguesian plane, there are five others. Walker's calculations were independently checked by the present author and some interesting structural relationships were found.

Plane $P(1)$ has the 3-cube representation

$$
\begin{array}{ccccc}
10000 & 01000 & 00100 & 00010 & 00001 \\
01000 & 00100 & 00010 & 00001 & 10100 \\
00100 & 00010 & 01001 & 10100 & 00101 \\
00010 & 00001 & 11010 & 11110 & 10111 \\
00001 & 10010 & 11011 & 10000 & 01110
\end{array}
\qquad (5.2)
$$

No way to construct this plane, except by trial and error, is known. The 3-cube in (5.2) has in some sense a relatively high degree of symmetry, when compared to all other representations of $P(1)$. This plane was first discovered in December, 1961, by the author with a program written for a Burroughs 220 digital computer.

There are no autotopisms except the identity; hence the collineation group of $P(1)$ consists entirely of translations and shears. This example is the only known autotopism-free semifield plane (except for its dual). A further consequence is that there are 31^2 distinct semifields of order 32, mutually isotopic but not isomorphic (see Eq. (3.20)).

Plane $P(2)$ is the dual and also the transpose of $P(1)$, and it has the same properties.

Plane $P(3)$ is constructed from the binary semifield of order 32 (see [12]). It has the following 3-cube representation:

$$
\begin{array}{ccccc}
10000 & 01000 & 00100 & 00010 & 00001 \\
01000 & 00100 & 00010 & 00001 & 11000 \\
00100 & 00010 & 11001 & 01010 & 00111 \\
00010 & 00001 & 01010 & 11011 & 11100 \\
00001 & 11000 & 00111 & 11100 & 10111
\end{array}
\qquad (5.3)
$$

This particular representative has five automorphisms and no other autotopisms; there are 192 other distinct systems isotopic to this one, and none of them have any automorphisms. Thus we have

$$
31^2 = 5\left(\frac{1}{5} + \frac{192}{1}\right)
$$

in agreement with formula (3.20).

Plane $P(3)$ is self-dual since its semifield is commutative, but it is not self-transpose. Therefore by Corollary 5.2.2 there are two other planes, $P(4) = P(3)^T$ and $P(5) = P(4)^D$. Those planes likewise contain one system with automorphisms and 192 other systems without.

Exhaustive enumeration on a computer shows that no other proper semifield planes of order 32 exist.

6. Some Known Semifields

In this section, we discuss many of the known semifields and consider what the possible orders of semifields are. Throughout this section the letter p will denote a prime number.

6.1 Orders that are excluded. We have seen that semifields must have p^n elements. For each p and n there is a unique field with p^n elements. What can be said about proper semifields?

If the order is p, a semifield must be the field $\mathrm{GF}(p)$. Furthermore, if the order is p^2, it is well known that a semifield must be the field $\mathrm{GF}(p^2)$: Let $\{1, x\}$ be a basis for the semifield; the multiplication is determined by the definition of $x^2 = ax + b$. But the polynomial $x^2 - ax - b$ has no roots in $\mathrm{GF}(p)$, else we would have $(x-r)(x-s) = x^2 - ax - b = 0$ contradicting axiom A2. Thus, $x^2 - ax - b$ is irreducible, and the multiplication is that of $\mathrm{GF}(p^2)$.

If the order is 8, one easily verifies that we have only the field $\mathrm{GF}(8)$. This must be true since there is a unique projective plane of order 8, but we can verify this particular case directly: Let $\{1, x, y\}$ be a basis, and let L be the matrix A^{2**} for left multiplication by x. If the characteristic equation of L is $\lambda^3 + a\lambda^2 + b\lambda + c = 0$, L satisfies this equation, so we find that $xx^2 + ax^2 + bx + c = 0$. This polynomial must have no linear factors; hence, it is irreducible and takes one of two forms $xx^2 + x + 1 = 0$ or $xx^2 + x^2 + 1 = 0$. Replacing x by $x + 1$, if necessary, we can assume $xx^2 + x + 1 = 0$. In particular, $\{1, x, x^2\}$ is a basis.

The remainder of the proof considers various nontrivial possibilities and will only be sketched:

> If $x^2x = x^2 + 1$ and $x^2x^2 = x$, then $(x^2 + x)^2 = 0$.
> If $x^2x = x^2 + 1$ and $x^2x^2 = x^2 + x$, then $(x^2 + 1)(x^2 + x + 1) = 0$.
> If $x^2x = x + 1$ and $x^2x^2 = 1$, then $(x^2 + 1)^2 = 0$.
> If $x^2x = x + 1$ and $x^2x^2 = x$, then $(x^2 + x)(x^2 + x + 1) = 0$.
> If $x^2x = x + 1$ and $x^2x^2 = x^2 + 1$, then $(x^2 + 1)(x^2 + x) = 0$.
> If $x^2x = x + 1$ and $x^2x^2 = x^2 + x$, then we have the field $\mathrm{GF}(8)$.

We have proved the following theorem:

Theorem 6.1. *Every proper semifield has p^n elements, where $n \geq 3$ and $p^n \geq 16$.* □

We will show by constructions in this section that these easily obtained, *necessary* conditions on the order are actually *sufficient*.

6.2 Semifields of order 16. The semifields of order 16 have been tabulated by Kleinfeld [11], and his results have been independently checked by the author. There are 23 nonisomorphic, proper semifields of order 16. They are all isotopic to either system V or to system W of Section 2.2; consequently two projective planes are formed.

The first plane, consisting of those semifields isotopic to V, contains 18 distinct semifields. There is one (namely V) with 6 automorphisms, another with 3 automorphisms, 8 with 2 automorphisms, and 8 with only the identity automorphism. Hence there must be 18 autotopisms, in agreement with formula (3.20):

$$15^2 = 18\left(\frac{1}{6} + \frac{1}{3} + \frac{8}{2} + \frac{8}{1}\right). \tag{6.1}$$

The second plane has only five distinct semifields. One of these has 4 automorphisms, one (W) has 3, and the other three have 2 automorphisms. Thus, by formula (3.20), there are 108 autotopisms, and

$$15^2 = 108\left(\frac{1}{4} + \frac{1}{3} + \frac{3}{2}\right). \tag{6.2}$$

Since the number of autotopisms is different, we can conclude that each plane is self-dual and self-transpose.

6.3 Early work of Dickson. The study of semifields was originated by L. E. Dickson in 1906 [5]. In two early papers on the subject, he considered the construction of all possible semifields of order p^3, and of all possible commutative semifields of order p^4, where p is odd.

Perhaps the simplest way to construct proper semifields is analogous to our construction of V and W in Section 2; we start with a field $\mathrm{GF}(p^m)$ and construct a semifield of order p^{2m}, having elements

$$a + \lambda b, \qquad \text{for } a, b \in \mathrm{GF}(p^m).$$

One must merely define

$$(a + \lambda b)(c + \lambda d) = f(a, b, c, d) + \lambda g(a, b, c, d) \tag{6.3}$$

where f and g are linear in all four variables, and where the equations

$$f(a, b, c, d) = g(a, b, c, d) = 0$$

imply that $a = b = 0$ or $c = d = 0$. There are many ways to do this, seemingly unrelated; a quite general class of suitable constructions is investigated in Section 7 below.

Dickson [6] gave a particularly simple construction of a *commutative* semifield of this type, for p odd:

$$(a + \lambda b)(c + \lambda d) = (ac + b^\sigma d^\sigma f) + \lambda(ad + bc) \qquad (6.4)$$

where σ is an automorphism and where f is a nonsquare of the field $\mathrm{GF}(p^m)$. The condition that f must be a nonsquare element is clear, for if $f = a^2$ we could choose b such that $b^\sigma = a^{-1}$, obtaining

$$(1 + \lambda b)(1 - \lambda b) = 0.$$

It is easy to verify that the condition is also sufficient (see Section 7).

The complex numbers are a particular case of the system (6.4), although f being a nonsquare is not always sufficient in the infinite case. The system (6.4) is associative if and only if σ is the identity.

6.4 Twisted fields. The following construction is due to A. A. Albert [1, 2, 4]. Define a new multiplication on the elements of $\mathrm{GF}(p^n)$ by

$$x \circ y = xy^q - cx^q y, \qquad (6.5)$$

where $q = p^m$, $1 \le m < n$, and where $c \ne a^{q-1}$ for $a \in \mathrm{GF}(p^n)$. Then we obtain a pre-semifield, since $x^q + y^q = (x + y)^q$, and since

$$x \circ y = 0 \qquad \text{implies} \qquad x = 0 \quad \text{or} \quad y = 0 \quad \text{or} \quad c = (y/x)^{q-1}.$$

Now pass to a semifield, using (4.19); the result is called a *twisted field*.

The construction can be carried out only if c exists subject to the required conditions. This will be the case when $q - 1$ and $p^n - 1$ have a common factor, since multiplication is cyclic of order $p^n - 1$. But if p is odd, there is always the common factor $p - 1$; if $p = 2$ we have $\gcd(2^m - 1, 2^n - 1) = 2^{\gcd(m,n)} - 1$, so we need $\gcd(m, n) > 1$. If $n = mk$, where $k > 2$, it has been shown that the semifield constructed is nonassociative.

Twisted fields exist for nearly all orders not excluded by Theorem 6.1.1. The missing orders are 2^4 and 2^p, where p is a prime greater than 3.

6.5 Sandler's construction. An interesting class of semifields has been constructed by R. Sandler [15]. His semifields have p^{nm^2} elements, where m is greater than 1.

Let $q = p^n$; the elements of S are

$$a_0 + \lambda a_1 + \cdots + \lambda^{m-1} a_{m-1}, \qquad a_k \in \mathrm{GF}(q^m). \tag{6.6}$$

Multiplication is defined as follows: $\lambda^m = \delta$, and

$$(\lambda^i x)(\lambda^j y) = \lambda^{i+j} x^{q^j} y, \qquad \text{for } 0 \leq i < m, \ 0 \leq j < \infty, \tag{6.7}$$

with the convention that λ^k denotes left powers of λ, that is, $\lambda^{k+1} = \lambda \lambda^k$. If δ is chosen such that it satisfies no polynomial of degree less than m over $\mathrm{GF}(q)$, this construction yields a semifield.

For example, let us construct such a system S of order 2^9. The elements are

$$a + \lambda b + \lambda^2 c, \qquad a, b, c \in \mathrm{GF}(8).$$

Multiplication is defined by the rule

$$(a + \lambda b + \lambda^2 c)(d + \lambda e + \lambda^2 f)$$
$$= ad + \lambda a^2 e + \lambda^2 a^4 f + \delta b^4 f + \lambda bd + \lambda^2 b^2 e + \lambda c^2 e + \lambda \delta c^4 f + \lambda^2 cd. \tag{6.8}$$

Note the similarity between this formula and the definition of system W in Section 2.2. Sandler's construction can be written in matrix form, since S is a right vector space over $\mathrm{GF}(8)$; the matrix of left multiplication by $(a + \lambda b + \lambda^2 c)$ is then

$$L = \begin{pmatrix} a & \delta c^2 & \delta b^4 \\ b & a^2 & \delta c^4 \\ c & b^2 & a^4 \end{pmatrix}. \tag{6.9}$$

The determinant of L is

$$a^7 + \delta b^7 + \delta^2 c^7 - \delta(a^2 b^4 c + ab^2 c^4 + a^4 bc^2) = r + s\delta + t\delta^2.$$

Since $r^2 = r$, $s^2 = s$, and $t^2 = t$, this quantity is a polynomial of degree 2 or less over $\mathrm{GF}(2)$, and it cannot be zero by hypothesis unless all of the coefficients vanish; but then $a = b = c = 0$.

6.6 Binary semifields. If n is odd and $mn > 3$, a proper semifield of order 2^{mn} can be constructed, thus filling in all orders not excluded by Theorem 6.1. Let $f : \mathrm{GF}(2^{mn}) \to \mathrm{GF}(2^m)$ be the unique linear functional such that $f(1) = 1$ and $f(x) = f(x^{2^m})$. Define

$$a \circ b = ab + (f(a)b + f(b)a)^2;$$
$$a \circ b = (1 \circ a) * (1 \circ b). \tag{6.10}$$

The resulting system $(\mathrm{GF}(2^{mn}), +, *)$ is a proper semifield with mn automorphisms generated by $x \to xx$, and with exactly $mn(2^m - 1)$ autotopisms. Proofs of these statements may be found in [12].

7. Weak Nuclei and Quadratic Extensions

The concept of nucleus (Section 2.6) is generalized here, and this leads to an important class of semifields of the form (6.3).

7.1 Definition of weak nucleus. Let F be a field contained in a semifield S. We say F is a *weak nucleus* for S if $(ab)c = a(bc)$ whenever *any two* of $\{a, b, c\}$ are in F. In general, S will be a vector space over F, but left and right multiplication need not be linear transformations over F.

   An example of such a system is the semifield V of Section 2.2. Here F is contained in neither the left, middle, or right nucleus of V. There are, on the other hand, semifields of order 16 whose left nucleus is not a weak nucleus. In other words, the statement "F is a left nucleus" does not imply and is not implied by the statement "F is a weak nucleus." But if F is contained in any two of the nuclei N_l, N_m, or N_r, F must necessarily be a weak nucleus.

7.2 Quadratic extensions. Consider the simplest case:

Theorem 7.2.1. *Let F be a weak nucleus for S, and let S have dimension 2 over F. Then the elements of S have the form*

$$a + \lambda b, \qquad a, b \in F; \tag{7.1}$$

and $\lambda \in S$ can be chosen such that

$$a\lambda = \lambda a^\sigma \tag{7.2}$$

for all $a \in F$, where σ is an automorphism of F.

Proof. Let x be a primitive element of the field F, and let

$$x\lambda = a + \lambda b. \tag{7.3}$$

   Case 1, $b = x$. Then we can prove by induction that

$$x^n \lambda = nax^{n-1} + \lambda x^n; \tag{7.4}$$

for

$$x^{n+1}\lambda = (xx^n)\lambda = x(x^n\lambda) = x(nax^{n-1} + \lambda x^n) = nax^n + x(\lambda x^n)$$
$$= nax^n + (x\lambda)x^n = nax^n + (a + \lambda x)x^n = (n+1)ax^n + \lambda x^{n+1}.$$

Notice that the associative law was applied here only when permissible. If F has p^m elements, we can put $n = p^m$; then by (7.4),

$$x\lambda = \lambda x, \tag{7.5}$$

since $p^m a = 0$ and $x^{p^m} = x$.

Case 2, $b \neq x$. Then we may replace λ by $\lambda' = \lambda + a(b - x)^{-1}$, and we find $x\lambda' = \lambda'b$. We easily prove by induction that $x^n\lambda' = \lambda'b^n$. Since x is a primitive element, b is a power of x, say x^σ. This argument shows that $y\lambda' = \lambda'y^\sigma$ for all nonzero y, and in fact

$$(yz)\lambda' = \lambda'(y^\sigma z^\sigma), \tag{7.6}$$

for all nonzero y and z, since y and z are powers of x. Furthermore,

$$(y + z)\lambda' = \lambda'(y^\sigma + z^\sigma), \tag{7.7}$$

by the distributive laws. These equations hold trivially if y or z is zero; therefore, we have shown that $(yz)^\sigma = y^\sigma z^\sigma$, $(y + z)^\sigma = y^\sigma + z^\sigma$. In other words, σ is an automorphism.   □

7.3 Construction of semifields.   The rule for multiplication in a quadratic extension of a weak nucleus must be

$$(a + \lambda b)(c + \lambda d) = ac + \lambda(a^\sigma d + bc) + (\lambda b)(\lambda d), \tag{7.8}$$

according to the preceding theorem. But the fact that F is a weak nucleus tells us nothing about what the product $(\lambda b)(\lambda d)$ must be; such products never occur unless two elements *not* in F are multiplied together. This product can be defined as $f(b, d) + \lambda g(b, d)$ where f and g are bilinear functions, as long as no zero divisors are introduced.

Although many choices might be made for f and g, we will make the assumption here that they have a certain simple form. Let

$$(\lambda b)(\lambda d) = b^\alpha d^\beta f + \lambda(b^\gamma d^\delta g) \tag{7.9}$$

where f and g are now elements of F, and where α, β, γ, and δ are *automorphisms* of F. Under what conditions can we conclude that no zero divisors have been introduced?

Suppose $(a + \lambda b) \neq 0$ and $(c + \lambda d) \neq 0$, but $(a + \lambda b)(c + \lambda d) = 0$. Then, in the first place, we have

$$ac + b^\alpha d^\beta f = 0;$$

this implies there is an element $x \neq 0$ in F such that

$$a = xd^\beta, \qquad c = -x^{-1}b^\alpha f. \tag{7.10}$$

The other condition is that

$$a^\sigma d + bc + b^\gamma d^\delta g = 0,$$

that is,

$$x^\sigma d^{\beta\sigma+1} - b^{\alpha+1} x^{-1} f + b^\gamma d^\delta g = 0. \tag{7.11}$$

Case 1, g = 0. If $g = 0$, we have $x^\alpha d^{\beta\sigma+1} = b^{\alpha+1} x^{-1} f$. Here b and d must be nonzero, and $f = x^{\sigma+1} d^{\beta\sigma+1} b^{-1-\alpha}$. Therefore, if p is an odd prime, the exponents $\sigma+1$, $\beta\sigma+1$, and $-1-\alpha$ are even and we could conclude that f is a *square*; so we choose f to be a nonsquare.

Other *a priori* choices of α, β, and f can also be made in which f is a square, as we will see in our discussion of Case 2.

Case 2, g ≠ 0. In this case, we will try to choose α, β, γ, δ so that (7.11) is a nonzero constant times a polynomial in a single variable. This can be done in essentially one way by writing it in the form

$$x^{-1} b^{\alpha+1} (x^{\sigma+1} b^{-1-\alpha} d^{\beta\sigma+1} + x b^{\gamma-1-\alpha} d^\delta g - f) = 0,$$

and we require that

$$x^{\sigma+1} b^{-1-\alpha} d^{\beta\sigma+1} = (x b^{\gamma-1-\alpha} d^\delta)^{\sigma+1}. \tag{7.12}$$

Lemma 7.3.1. *If p is prime we have*

$$p^a + p^b \equiv p^c + p^d \ (\text{modulo } p^m - 1)$$

if and only if $a \equiv c$ and $b \equiv d$ (modulo m) or $a \equiv d$ and $b \equiv c$ (modulo m).

Proof. Since $p^m \equiv 1$ (modulo $p^m - 1$), we may assume that $0 \le a, b, c, d < m$. Then we have $2 \le p^a + p^b \le 2p^{m-1} \le p^m$, and $2 \le p^c + p^d \le 2p^{m-1} \le p^m$, so we may conclude that

$$p^a + p^b = p^c + p^d.$$

We may now divide by p if necessary, until one of a, b, c, d is zero, say a. Then it is clear that either c or d must be zero.   □

Now suppose $F = \mathrm{GF}(p^m)$. Then (7.12) can be written

$$\beta\sigma + 1 \equiv \delta\sigma + \delta, \qquad \gamma\sigma + \gamma \equiv \alpha\sigma + \sigma \qquad (\text{modulo } p^m - 1). \tag{7.13}$$

Each of α, β, γ, δ, σ is an automorphism, hence a power of p. Thus by Lemma 7.3.1, these two congruences have four solutions:

$$(\beta, \delta) = (\tau^2, \tau) \text{ or } (1,1), \qquad (\alpha, \gamma) = (\tau, 1) \text{ or } (\sigma, \sigma), \qquad (7.14)$$

where $\tau = \sigma^{-1} = p^m/\sigma$. Using any of these four solutions, we can conclude that there are no zero divisors if and only if the polynomial

$$y^{\sigma+1} + gy - f = 0 \qquad (7.15)$$

has no solutions in F.

There are in general many such polynomials, as long as F is not a prime field. For example, if p is odd, we may take $g = 0$, $\sigma = p$, and f can be any element not of the form y^{p+1}. If $p = 2$, we can take $g = 1$ and $\sigma = 2$; then f can be any element not of the form $y^3 + y$. (Such an element must exist, since $0^3 + 0 = 1^3 + 1$.)

7.4 The four types. We obtain many proper semifields, all of which are quadratic over a weak nucleus, by the construction in the previous section. Those of Case 1 have the multiplication rule

$$(a + \lambda b)(c + \lambda d) = (ac + b^\alpha d^\beta f) + \lambda(a^\sigma d + bc) \qquad (7.16)$$

where p is odd and f is a nonsquare; α, β, and σ are arbitrary automorphisms, but not all the identity. Dickson's construction (see Section 6.3) is a special case, indeed the only commutative semifield of this type.

Other semifields, those of Case 2, are constructed by finding an automorphism $\sigma \neq 1$ and elements $f, g \in F$ such that

$$y^{\sigma+1} + gy - f \neq 0, \qquad \text{for } y \in F.$$

Then four types of semifields are produced, all for the same σ, f, g, and for $\tau = \sigma^{-1}$:

$$(a + \lambda b)(c + \lambda d) = \begin{cases} (ac + b^\sigma d^{\tau^2} f) + \lambda(bc + a^\sigma d + b^\sigma d^\tau g); & \text{(I)} \\ (ac + b^\sigma d f) + \lambda(bc + a^\sigma d + b^\sigma dg); & \text{(II)} \\ (ac + b^\tau d^{\tau^2} f) + \lambda(bc + a^\sigma d + bd^\tau g); & \text{(III)} \\ (ac + b^\tau d f) + \lambda(bc + a^\sigma d + bdg). & \text{(IV)} \end{cases} \qquad (7.17)$$

For example, the system V of Section 2.2 is of type I. The system W of that section is actually of all four types. The systems of type II were discovered by Hughes and Kleinfeld [9]. Autotopisms of systems of type II are discussed in [3, 10, 15], and the autotopisms of type III systems are obtained by duality. The autotopisms of types I and IV are not yet known.

Theorem 7.4.1. *Let S be a proper semifield, quadratic over a finite field F. Let N_l, N_m, N_r be the nuclei of S. Then*

(a) $F = N_r = N_m$ *if and only if S is of type II.*
(b) $F = N_l = N_m$ *if and only if S is of type III.*
(c) $F = N_l = N_r$ *if and only if S is of type IV.*

Proof. The "if" parts of this theorem can be obtained by direct calculation, and will be omitted. The "only if" parts follow since, first, F is a weak nucleus if it equals any two of the nuclei, so Theorem 7.2.1 holds. We need then merely calculate $(\lambda b)(\lambda d)$ under the given assumptions (a), (b), or (c).

Let $\lambda^2 = f + \lambda g$. Then

(a) $(\lambda b)(\lambda d) = ((\lambda b)\lambda)d = (\lambda(b\lambda))d = (\lambda(\lambda b^\sigma))d = \lambda((\lambda b^\sigma)d)$
$\qquad = \lambda(\lambda(b^\sigma d)) = \lambda^2(b^\sigma d) = b^\sigma d f + \lambda b^\sigma d g.$

(b) $(\lambda b)(\lambda d) = (b^\tau \lambda)(d^\tau \lambda) = b^\tau(\lambda(d^\tau \lambda)) = b^\tau((\lambda d^\tau)\lambda)$
$\qquad = b^\tau((d^{\tau^2}\lambda)\lambda) = b^\tau(d^{\tau^2}\lambda^2) = (b^\tau d^{\tau^2})\lambda^2$
$\qquad = b^\tau d^{\tau^2}(f + \lambda g) = b^\tau d^{\tau^2}f + \lambda b d^\tau g.$

(c) $(\lambda b)(\lambda d) = (b^\tau \lambda)(\lambda d) = b^\tau(\lambda(\lambda d)) = b^\tau(\lambda^2 d)$
$\qquad = b^\tau((f + \lambda g)d) = b^\tau d f + \lambda b d g.$

The proof is completed by observing that either of the three types implies the existence of the polynomial (7.15), which has no solutions in F. □

No characterization of type I is known.

Corollary 7.4.2. *Under the hypotheses of Theorem 7.4.1,*

$$F = N_l = N_m = N_r$$

if and only if S is of all four types I, II, III, IV, where $\sigma^2 = 1$ and $g = 0$.

Proof. By the theorem, we must have S of types II, III, and IV at least. These types are the same if and only if $\sigma^2 = 1$ and $g = 0$, since $\sigma \neq 1$. □

This corollary and part (b) of Theorem 7.4.1 were proved by Hughes and Kleinfeld in their original paper [9].

References

[1] A. A. Albert, "On nonassociative division algebras," *Transactions of the American Mathematical Society* **72** (1952), 296–309.

[2] A. A. Albert, "Finite noncommutative division algebras," *Proceedings of the American Mathematical Society* **9** (1958), 928–932.

[3] A. A. Albert, "On the collineation groups of certain non-Desarguesian planes," *Portugaliae Mathematica* **18** (1959), 207–224.

[4] A. A. Albert, "Finite division algebras and finite planes," *Proceedings of Symposia in Applied Mathematics* **10** (Providence, Rhode Island: American Mathematical Society, 1960), 53–70.

[5] Leonard Eugene Dickson, "Linear algebras in which division is always uniquely possible," *Transactions of the American Mathematical Society* **7** (1906), 370–390, 514–527.

[6] L. E. Dickson, "Linear algebras with associativity not assumed," *Duke Mathematical Journal* **1** (1935), 113–125.

[7] René Gouarné and Isaac Samuel, "Introduction à l'étude des matrices multidimensionelles," *Cahiers de Physique* **16** (1962), 133–142; Isaac Samuel and René Gouarné, "Déterminants des matrices à N dimensions d'ordre n," *Cahiers de Physique* **16** (1962), 143–152.

[8] Marshall Hall, "Projective planes," *Transactions of the American Mathematical Society* **54** (1943), 229–277; **65** (1949), 473–474. See also Marshall Hall, Jr., *The Theory of Groups* (New York: Macmillan, 1959), 346–420.

[9] D. R. Hughes and Erwin Kleinfeld, "Semi-nuclear extensions of Galois fields," *American Journal of Mathematics* **82** (1960), 389–392.

[10] D. R. Hughes, "Collineation groups of non-Desarguesian planes II," *American Journal of Mathematics* **82** (1960), 113–119.

[11] Erwin Kleinfeld, "Techniques for enumerating Veblen–Wedderburn systems," *Journal of the Association for Computing Machinery* **7** (1960), 330–337.

[12] Donald E. Knuth, "A class of projective planes," *Transactions of the American Mathematical Society* **115** (1965), 541–549. [Reprinted as Chapter 21 of the present volume.]

[13] J. H. Maclagan-Wedderburn, "A theorem on finite algebras," *Transactions of the American Mathematical Society* **6** (1905), 349–352.

[14] Günter Pickert, *Projektive Ebenen* (Berlin: Springer, 1955).

[15] Reuben Sandler, "Autotopism groups of some finite non-associative algebras," *American Journal of Mathematics* **84** (1962), 239–264.

[16] R. J. Walker, "Determination of division algebras with 32 elements," *Proceedings of Symposia in Applied Mathematics* **15** (Providence, Rhode Island: American Mathematical Society, 1963), 83–85.

Addendum

This chapter was extracted from my Ph.D. thesis, written for my advisor Marshall Hall in 1963. I have expanded Section 3 to make it self-contained, since the original paper implicitly assumed that the reader was quite familiar with Hall's publications. I have also decided to use the notations $A \vee B$ and $L \wedge M$ here, in place of what I had called $A : B$ and $L \cap M$ in my thesis.

Although Theorem 3.1.1 is essentially present in Marshall Hall's fundamental paper [8], the proof I have given in the newly expanded Section 3 may well be new. In particular, the dual ternary operation defined in (3.9) has not been discussed explicitly before, in the literature known to me. If we dualize twice, we do not get the original ternary operation back again; but Theorem 3.3.1 tells us that the double dual is isotopic to the original. I have not been able to think of a system of homogeneous coordinates that is truly symmetrical between points and lines in an arbitrary projective plane.

I never published the promised remarks about transposes of Veblen–Wedderburn systems. The basic idea is simply that such a system can be viewed as a set of nonsingular $n \times n$ matrices $\{A_1, \ldots, A_{p^n-1}\}$ over $\mathrm{GF}(p)$ whose differences $A_x - A_y$ are also nonsingular when $x \neq y$. Therefore we get another Veblen–Wedderburn system by transposing all the matrices.

See M. Cordero and G. P. Wene, "A survey of finite semifields," *Discrete Mathematics* **208/209** (1999), 125–137, for the next 30 or so years of the story.

Chapter 21

A Class of Projective Planes

[Originally published in Transactions of the American Mathematical Society **115** *(1965), 541–549.]*

1. Introduction

Finite non-Desarguesian projective planes have been known for all orders p^n, where p is prime, $n \geq 2$, and $p^n \geq 9$, except when $p = 2$ and n is a prime ≥ 5. In this paper a new class of projective planes is defined, having the orders 2^n where $n \geq 5$ is not a power of two, thus establishing in particular the existence of non-Desarguesian planes of the missing orders. The new planes are coordinatized by *semifields* (sometimes called division algebras, nonassociative division rings, or distributive quasifields), which are algebraic systems that satisfy the axioms for a field except with a loop replacing the multiplicative group. It is easy to show that finite *proper* semifields (the finite semifields that are not fields) must have the orders p^n where p is prime, $n \geq 3$, and $p^n \geq 16$. Proper semifields of those orders have been known to exist except as above, when $p = 2$ and n is prime; therefore the new systems show that proper semifields do exist for any order not excluded by simple arguments. A detailed treatment of the general theory of semifields and their relation to projective planes may be found in [3].

The manner in which these planes were discovered is perhaps as interesting as the planes themselves, since computers played a key role in the discovery. The author had received copies of two tables prepared by R. J. Walker (see [4]), which listed all commutative semifields of order 32 for which, if x is a generating element, $x(x(x(x^2))) = x + 1$ or $x^2 + 1$, respectively. There were 24 solutions in each table, and so it seemed plausible that a rule could be found yielding a correspondence between one set of solutions and the other. A few hours of "cryptanalysis" did, in fact, result in the discovery of such a rule; and, since one of the solutions for the table $x(x(x(x^2))) = x^2 + 1$ was the field GF(32), the corresponding system for the other table could be written in a simple algebraic form [2]. After this, there was little difficulty showing that the same construction could be generalized to the construction of proper semifields of all orders 2^{2k+1} with $k > 1$, and subsequent generalizations yielded the systems described in this paper. Therefore, we have an example in which a rather well-known conjecture in combinatorial analysis has been completely resolved because the smallest unknown case was analyzed by computer. Presumably many more such examples will be known in future years.

The new semifields are defined in §2 of the present paper, and certain automorphisms are exhibited in §3. The collineation group of the corresponding projective planes is completely determined in §4. Since the semifields constructed are commutative, the new planes are self-dual.

2. Construction of the semifields

Let $K = \mathrm{GF}(2^{mn})$ where n is odd, $n > 1$; let K_0 be the subfield $\mathrm{GF}(2^m)$. Considering K as a vector space over K_0, let f be any nonzero linear functional from K to K_0; that is,

$$f(\lambda a + \mu b) = \lambda f(a) + \mu f(b), \tag{2.1}$$

for all $a, b \in K$ and all $\lambda, \mu \in K_0$.

Define a new multiplication in K as follows:

$$a \circ b = ab + \big(f(a)b + f(b)a\big)^2. \tag{2.2}$$

Theorem 1. *The algebraic system $(K, +, \circ)$ is a pre-semifield; that is, it satisfies all properties of a semifield except that it may lack a multiplicative identity.*

Proof. Since the mapping $a \to a^2$ is an automorphism of K, the product $a \circ b$ is clearly linear in both variables, so both distributive laws hold. Therefore, we need only show that there are no zero divisors.

Suppose $a \circ b = 0$, and $a, b \neq 0$; then let $x = ab^{-1}$. This implies

$$x + f(a)^2 + f(b)^2 x^2 = 0.$$

We have a quadratic equation with coefficients in K_0; but since the degree of K/K_0 is odd, this equation must be reducible. Therefore, $x \in K_0$. But then, $a = xb$ implies

$$a \circ b = ab + \big(f(xb)b + f(b)xb\big)^2 = ab \neq 0,$$

and this contradiction completes the proof. $\square$

There are, in general, many ways to convert the pre-semifield $(K, +, \circ)$ into a semifield. Perhaps the simplest way is to define a new product $a * b$ by the equation

$$(1 \circ a) * (1 \circ b) = a \circ b. \tag{2.3}$$

Then $(K, +, *)$ is a commutative semifield, since it is easily verified that the distributive laws hold, there are no zero divisors, and $1 = 1 \circ 1$ is a multiplicative identity.

Notice that we have now defined three different "multiplications" on the elements of K: ab, $a \circ b$, and $a * b$. It is important to keep this distinction in mind, since all three multiplications are used simultaneously in several proofs of this paper. The powers of an element, a^2, a^3, etc., will always refer to the multiplication of the *field*.

Theorem 2. *If $mn > 3$, it is possible to choose the function f of (2.1) in such a way that the semifield $(K, +, *)$ is not a field.*

Proof. Let $\{1, x, x^2, \ldots, x^{n-1}\}$ be a basis of K over K_0; set

$$f(1) = f(x) = \cdots = f(x^{n-2}) = 0, \quad f(x^{n-1}) = 1. \tag{2.4}$$

With this definition we find that, for $\lambda \in K_0$,

$$1 \circ \lambda = \lambda, \ 1 \circ \lambda x = \lambda x, \ \ldots, \ 1 \circ \lambda x^{n-2} = \lambda x^{n-2}, \ 1 \circ \lambda x^{n-1} = \lambda x^{n-1} + \lambda^2;$$

and therefore, for all $a, b \in K$ we have in particular

$$1 \circ (1 \circ a) = a, \qquad a * b = (1 \circ a) \circ (1 \circ b). \tag{2.5}$$

Now if $n > 3$, let $k = (n-1)/2$; then $1 < k < n - 2$, and

$$x * (x^k * x^k) = x * x^{n-1} = x^n + x^2 + x \neq x^n = x^{k+1} * x^k = (x * x^k) * x^k.$$

Thus, multiplication is not associative in this case.

If $n = 3$, let λ be an element of K_0; we have

$$(x * x) * \lambda x = x^2 * \lambda x = (x^2 + 1) \circ \lambda x = \lambda x(x^2 + 1) + \lambda^2 x^2,$$
$$x * (x * \lambda x) = x * \lambda x^2 = x \circ (\lambda x^2 + \lambda^2) = (\lambda x^2 + \lambda^2)x + \lambda^2 x^2.$$

Thus, multiplication is not associative unless $\lambda^2 = \lambda$. We can always choose $\lambda \neq \lambda^2$ unless $K = \mathrm{GF}(8)$, which is excluded by hypothesis. This choice of λ completes the proof. $\square$

Let us now consider the effect of choosing different functions f in equation (2.1).

Lemma 1. *If f and g are nonzero linear functionals from K to K_0, there exists an element $z \in K$ such that $f(az) = g(a)$ for all a in K.*

Proof. A simple counting argument will prove this lemma. If

$$\{x_1, x_2, \ldots, x_n\}$$

is a basis of K over K_0, a linear functional f is completely determined by the n choices of $f(x_i) \in K_0$, $1 \le i \le n$, and these choices are independent provided that they are not all zero. Hence, there are $2^{mn} - 1$ nonzero linear functionals.

Suppose f is a nonzero linear functional; then if we define $g_z(a) = f(az)$, for $z \ne 0 \in K$, g_z is also a nonzero linear functional. There are $2^{mn} - 1$ such elements z, so we need only show that no two of them give the same function. But if $f(az_1) = f(az_2)$ for all a, we have $f(a(z_1 - z_2)) = 0$ for all a, hence $z_1 - z_2 = 0$ as desired.   □

An *isotopism* between algebraic systems $(S, +, *)$ and $(S', +', *')$ is a triple (F, G, H) of one-to-one functions from S onto S', such that

$$(a + b)F = aF +' bF, \qquad (a + b)G = aG +' bG,$$
$$(a + b)H = aH +' bH, \qquad (a * b)H = aF *' bG.$$

Theorem 3. *For a given K and K_0, any two semifields $(K, +, *)$ determined by different functionals f in (2.1) are isotopic.*

Proof. Each semifield is isotopic to its corresponding pre-semifield, so we need only show that any two of the pre-semifields are isotopic. Suppose we have

$$a \circ b = ab + \big(f(a)b + f(b)a\big)^2,$$
$$a \cdot b = ab + \big(g(a)b + g(b)a\big)^2.$$

Apply Lemma 1 to find $z \in K$ with $f(az) = g(a)$ for all a. Then

$$az \circ bz = abz^2 + \big(g(a)bz + g(b)az\big)^2 = (a \cdot b)z^2.   □$$

Corollary. *If $mn > 3$, all systems $(K, +, *)$ defined in this section are proper semifields.*

Proof. This corollary follows from Theorem 2 and the well-known fact that a field is never isotopic to a proper semifield. Moreover, the content of Theorem 3 is that all semifields constructed for K and K_0 coordinatize the same projective plane, by the well-known theorem of Albert [1] that two finite semifields coordinatize the same plane if and only if they are isotopic.   □

3. The binary semifield of K/K_0

In this section we will show that if the functional f is chosen appropriately we obtain a semifield possessing at least mn automorphisms. This particular semifield, with f defined by Theorem 4, will be called the binary semifield of K/K_0. We assume henceforth that $mn > 3$.

Theorem 4. Let $q = 2^m$, and let f be such that $f(a) = \lambda$ whenever

$$a = \lambda + b + b^q, \qquad \lambda \in K_0, \qquad b \in K. \tag{3.1}$$

Then f is a linear functional from K to K_0, and $f(a^2) = f(a)^2$.

Proof. First we show that f is well defined. Suppose $\lambda + b + b^q = \mu + c + c^q$ for $\lambda, \mu \in K_0$. Then $(b + c)^q = (b + c) + \lambda + \mu$; that is, $a^q = a + \nu$ for some $a \in K$ and $\nu \in K_0$. Applying the rule again, we find

$$a^{q^2} = a^q + \nu^q = a^q + \nu = a.$$

Since n is odd, we have $a^{q^{n+1}} = a$. But $a^{q^{n+1}} = a^q$, hence $\nu = 0$. Thus $\lambda = \mu$, and $f(\lambda + b + b^q)$ is well defined.

Furthermore, every element of K can be represented in the form $\lambda + b + b^q$, since there are precisely q elements $c \in K$ for which $b + b^q = c + c^q$. The reason is that $(b+c) = (b+c)^q$ holds if and only if $b+c \in K_0$. Therefore, f is *uniquely* defined. Finally,

$$f(\lambda + b + b^q + \mu + c + c^q) = f(\lambda + \mu + (b+c) + (b+c)^q)$$
$$= \lambda + \mu = f(\lambda + b + b^q) + f(\mu + c + c^q);$$
$$f(\mu(\lambda + b + b^q)) = f(\mu\lambda + \mu b + (\mu b)^q)$$
$$= \mu\lambda = \mu f(\lambda + b + b^q);$$
$$f((\lambda + b + b^q)^2) = f(\lambda^2 + b^2 + (b^2)^q)$$
$$= \lambda^2 = f(\lambda + b + b^q)^2. \quad \square$$

Theorem 5. The binary semifield of K/K_0 has at least mn automorphisms, generated by the mapping $a \to a^2$.

Proof. First we can observe that $a \to a^2$ is an automorphism of the pre-semifield:

$$(a \circ b)^2 = \left(ab + \left(f(a)b + f(b)a\right)^2\right)^2$$
$$= a^2 b^2 + \left(f(a)^2 b^2 + f(b)^2 a^2\right)^2$$
$$= a^2 b^2 + \left(f(a^2)b^2 + f(b^2)a^2\right)^2 = a^2 \circ b^2.$$

This automorphism carries over to the semifield, since

$$((1 \circ a) * (1 \circ b))^2 = (a \circ b)^2 = a^2 \circ b^2$$
$$= (1 \circ a^2) * (1 \circ b^2)$$
$$= (1^2 \circ a^2) * (1^2 \circ b^2) = (1 \circ a)^2 * (1 \circ b)^2. \quad \square$$

Theorem 9 below shows, conversely, that all automorphisms are given by Theorem 5.

4. Collineations

If G is the collineation group of a projective plane coordinatized by a finite proper semifield, it is well known [1] that G has subgroups G_1 and G_2, where G_1 is normal and G/G_1 is isomorphic to G_2. Here G_1, the group of "translations and shears," is essentially the same for all semifields; so the collineation group is known once G_2, the group of all collineations fixing three special points, has been determined. Furthermore, G_2 is isomorphic to the group of all *autotopisms* of the semifield, that is, the isotopisms of the semifield onto itself. Therefore we will investigate the autotopisms of $(K, +, *)$ in this section.

The autotopisms of the semifields can easily be derived from autotopisms of their pre-semifields, so we will first consider the latter. Thus, if $aF \circ bG = (a \circ b)H$ for all $a, b \in K$, we need to find F, G, and H. The following theorem reduces these two degrees of freedom to essentially only one degree.

Theorem 6. *If* (F, G, H) *is an autotopism of the pre-semifield* $(K, +, \circ)$, *we have* $F = Gz$ *for some* $z \neq 0$ *in* K_0; *that is,*

$$aF = (aG)z \quad \text{for all } a \in K. \tag{4.1}$$

Proof. We have, for all $a, b \in K$,

$$aF \circ bG = (a \circ b)H = (b \circ a)H = bF \circ aG = aG \circ bF. \tag{4.2}$$

(Our proof will rest solely on the fact that $aF \circ bG = aG \circ bF$.) Thus,

$$(aF)(bG) + \left(f(aF)bG + f(bG)aF \right)^2$$
$$= (aG)(bF) + \left(f(aG)bF + f(bF)aG \right)^2. \tag{4.3}$$

Let $V_1 = \{\, a \mid f(aF) = 0 \,\}$, $V_2 = \{\, a \mid f(aG) = 0 \,\}$, $V_3 = V_1 \cap V_2$. Here V_1 and V_2 are vector spaces of dimension $m(n-1)$ over $\mathrm{GF}(2)$, since the kernel of $f : K \to K_0$ must have dimension $n - 1$ over K_0. Also, $\dim V_3 = \dim V_1 + \dim V_2 - \dim V_1 \cup V_2 \geq 2m(n-1) - mn = m(n-2) \geq 2$. Therefore V_3 contains at least two nonzero elements.

We apply formula (4.3) to find

$$\frac{aF}{aG} = \frac{bF}{bG} = z \quad \text{for all } a, b \in V_3 \setminus \{0\},$$

for some $z \in K$. This proves (4.1) for all elements $a \in V_3$, except that z might not lie in K_0.

Now let $a \in V_3 \setminus \{0\}$, and let b be arbitrary. Then

$$z(aG)(bG) + \big(zf(bG)(aG)\big)^2 = (aG)(bF) + \big(f(bF)(aG)\big)^2;$$

that is,

$$z(bG) + bF = (aG)\big(zf(bG) + f(bF)\big)^2. \tag{4.4}$$

Replace a by another element a' of $V_3 \setminus \{0\}$; since the left-hand side of (4.4) remains constant, but $aG \neq a'G$, we have

$$zf(bG) + f(bF) = 0, \quad \text{for all } b \in K.$$

In particular, if we take $b \notin V_2$, we conclude that $z \in K_0$.

Finally, equation (4.4) becomes

$$z(bG) + bF = 0, \quad \text{for all } b \in K,$$

and this is precisely equation (4.1). $\square$

Corollary. *Let A_0 be the set of all autotopisms of $(K, +, \circ)$ for which $F = G$. Then the complete set of all autotopisms is the set*

$$A = \{ (F\lambda, F\lambda^{-1}, H) \mid 0 \neq \lambda \in K_0 \text{ and } (F, F, H) \in A_0 \}.$$

Proof. It is immediate that if (F, G, H) is an autotopism, so is $(F\lambda, G\lambda^{-1}, H)$, for every nonzero $\lambda \in K_0$, because $a\lambda \circ b\lambda^{-1} = a \circ b$. Therefore every element of A is an autotopism. Conversely, if (F, G, H) is an autotopism we must have $F = Gz$, by Theorem 6. Let $\lambda = \sqrt{z} \in K_0$; then $(F\lambda^{-1}, G\lambda, H)$ is an autotopism belonging to A_0. $\square$

Theorem 7. *If (F, F, H) is an automorphism of $(K, +, \circ)$, there is an automorphism τ of K_0 such that*

$$(\lambda a)F = \lambda^\tau (aF), \quad (\lambda a)H = \lambda^\tau (aH) \quad \text{for all } \lambda \in K_0, \, a \in K. \quad (4.5)$$

Proof. Let λ be a fixed nonzero element of K_0, and define the one-to-one mapping G by the rule

$$aG = (\lambda a)F. \quad (4.6)$$

Then, since $a \circ \lambda b = \lambda a \circ b$ for all $\lambda \in K_0$, we have

$$aF \circ bG = (a \circ \lambda b)H = (\lambda a \circ b)H = aG \circ bF. \quad (4.7)$$

Notice that this result is precisely the same as equation (4.2), which was the basis for the proof of Theorem 6. By the same proof, therefore, we establish the existence of a nonzero element $1/z = \lambda' \in K_0$ such that

$$(\lambda a)F = \lambda'(aF) \quad \text{for all } a \in K.$$

In particular, $\lambda' = (\lambda F)/(1F)$.

Let $0' = 0$, and let $\lambda_1, \lambda_2 \in K_0$. Then

$$(\lambda_1 + \lambda_2)' = (\lambda_1 + \lambda_2)F/1F = \lambda_1' + \lambda_2',$$
$$(\lambda_1 \lambda_2)' = (\lambda_1 \lambda_2)F/1F = \lambda_1'(\lambda_2 F)/1F = \lambda_1' \lambda_2';$$

so $\lambda' = \lambda^\tau$ for some automorphism τ of K_0. The remaining part of the theorem follows immediately from the fact that $a^2 H = (aF)^2$ for all $a \in K$. □

Theorem 7 can now be strengthened, and indeed, the entire collineation group can be determined, as follows:

Theorem 8. *If (F, F, H) is an autotopism of the pre-semifield $(K, +, \circ)$, there is an automorphism σ of K and an element $x \in K$ such that*

$$f(a)^\sigma = f(aF), \quad aF = a^\sigma x, \quad aH = a^\sigma x^2 \quad \text{for all } a \in K. \quad (4.8)$$

Proof. The equation satisfied by H and F is

$$(ab)H + (f(a)^2 b^2)H + (f(b)^2 a^2)H$$
$$= (aF)(bF) + f(aF)^2(bF)^2 + f(bF)^2(aF)^2. \quad (4.9)$$

Let τ be the automorphism of K_0 given by Theorem 7. Since $a^2 H = (aF)^2$, equation (4.9) takes the following form:

$$(ab)H = (aF)(bF) + g(a)^2(b^2 H) + g(b)^2(a^2 H), \qquad (4.10)$$

where

$$g(a) = f(a)^\tau + f(aF). \qquad (4.11)$$

Note that $g\tau^{-1}$ is a linear functional over K_0. Let z be a nonzero element of K such that $g(z) = 0$. We now define new functions g_1, F_1, H_1 as follows:

$$g_1(a) = g(az), \quad aF_1 = (az)F/(zF), \quad aH_1 = (az^2)H/(zF)^2. \quad (4.12)$$

Equation (4.10) transforms into

$$(ab)H_1 = (aF_1)(bF_1) + g_1(a)^2(b^2 H_1) + g_1(b)^2(a^2 H_1) \qquad (4.13)$$

and, in particular, for $b = 1$, we have the important identity

$$aH_1 = aF_1 + g_1(a)^2. \qquad (4.14)$$

We will prove that $g_1(a) = 0$ for all $a \in K$, from which the rest of the theorem follows immediately with $x = 1F$. If $m > 1$ the result is quite easy; for if $\lambda \in K_0$ we have

$$(\lambda a)H_1 = \lambda^\tau(aH_1) = \lambda^\tau(aF_1 + g_1(a)^2) = (\lambda a)F_1 + \lambda^\tau g_1(a)^2,$$

and also $(\lambda a)H_1 = (\lambda a)F_1 + g_1(\lambda a)^2$. But $g_1(\lambda a) = \lambda^\tau g_1(a)$ by (4.12), (4.11), and (4.5); hence $\lambda^\tau = (\lambda^\tau)^2$ or else $g_1(a) = 0$. If $\lambda^\tau = (\lambda^\tau)^2$ for all $\lambda \in K_0$ we have $m = 1$.

In the case $m = 1$ the problem seems to be more difficult. We prove the following lemma:

Lemma 2. Let $V = \{\, a \mid g_1(a) = 0 \,\}$. If $a \in V$ and $a^2 \in V$, then $a^k \in V$ for all $k \geq 0$.

Proof. We show by induction on k that $a^k \in V$ and, simultaneously, that $a^k H_1 = (aF_1)^k$. This is certainly true for $k = 1$. Assume that $k > 1$ and that the hypotheses are true for $k - 1$. Then

$$a^k H_1 = (aF_1)(a^{k-1}F_1) = (aF_1)(a^{k-1}H_1) = (aF_1)^k;$$

$$a^{k+1}H_1 = (a^2 F_1)(a^{k-1}F_1) = (a^2 H_1)(a^{k-1}H_1) = (aF_1)^{k+1};$$

$$a^{k+1}H_1 = (aF_1)(a^k F_1) + g_1(a^k)^2(a^2 H_1)$$
$$= (aF_1)(a^k H_1) + g_1(a^k)^2(aF_1 + (aF_1)^2).$$

Therefore, $g_1(a^k)^2(aF_1 + (aF_1)^2) = 0$. If $g_1(a^k) \neq 0$, we have $aF_1 = (aF_1)^2$ which implies that $a = 0$ or 1. Hence $g_1(a^k) = 0$ and the lemma is proved. $\square$

To complete the proof of Theorem 8, we let $m = 1$, hence $n \geq 5$. The number of elements of K that lie in proper subfields is at most $2^{d_1} + \cdots + 2^{d_r} < 2^{n/3+1}$, where $d_1, \ldots, d_r$ are the divisors of n less than n. Now let $V' = \{ a \mid a^2 \in V \}$, $V'' = V \cap V'$. Since

$$\dim V'' = \dim V + \dim V' - \dim V \cup V'$$
$$\geq (n-1) + (n-1) - n = n - 2,$$

V'' must contain an element a that lies in no proper subfield of K. But then by Lemma 2, V'' contains all polynomials in a, that is, $V'' \supseteq K$. Therefore, $V = K$ and Theorem 8 follows.   □

Theorem 9. *All autotopisms with $F = G$ for the binary semifield of K/K_0 are the automorphisms generated by $a \to a^2$.*

Proof. Let f be the function defined in Theorem 4, and apply Theorem 8 to the resulting pre-semifield. Any autotopism (F, F, H) of $(K, +, \circ)$ must be an automorphism of K, since $f(a^\sigma) = f(a)^\sigma = f(aF) = f(a^\sigma x)$ for all $a \in K$ implies $x = 1$. Therefore, by the Corollary to Theorem 6, all autotopisms (F, G, H) of $(K, +, \circ)$ are given by

$$(F, G, H) = (\sigma\lambda, \sigma\lambda^{-1}, \sigma), \quad \sigma \text{ an automorphism of } K, \ \lambda \in K_0. \tag{4.15}$$

If $aU = 1 \circ a$, the autotopisms of $(K, +, *)$ are simply

$$(U^{-1}FU, \ U^{-1}GU, \ H),$$

where (F, G, H) is an autotopism of $(K, +, \circ)$. The fact that $U^{-1}FU = F$ whenever $F = \sigma$ is an automorphism of K (see the proof of Theorem 5) completes the proof of Theorem 9. Note also that the group of all autotopisms for the binary semifield of K/K_0, being conjugate to (4.15), is isomorphic to the group of permutations of K with elements (σ, λ), where $a(\sigma, \lambda) = a^\sigma\lambda$.   □

Finally, we state the following theorem, which follows immediately from the remarks above.

Theorem 10. *The group of collineations fixing $(0, 0, 1)$, $(0, 1, 0)$, and $(1, 0, 0)$ in the projective plane coordinatized by the binary semifield of K/K_0 is of order $mn(2^m - 1)$. Hence, the projective planes for different choices of K_0 are nonisomorphic. The subgroup of collineations that fix all points of the line $[0, 1, 0]$ is a cyclic, normal subgroup of order $2^m - 1$. The quotient group, isomorphic to the subgroup of all collineations that fix the points $(0, 0, 1)$, $(0, 1, 0)$, $(1, 0, 0)$, and $(1, 1, 1)$, is a cyclic subgroup of order mn.*   □

References

[1] A. A. Albert, "Finite division algebras and finite planes," *Proceedings of Symposia in Applied Mathematics* **10** (Providence, Rhode Island: American Mathematical Society, 1960), 53–70.

[2] Donald E. Knuth, "Non-Desarguesian planes of order 2^{2m+1}," Abstract 62T-137, *Notices of the American Mathematical Society* **9** (1962), 218.

[3] Donald E. Knuth, "Finite semifields and projective planes," *Journal of Algebra* **2** (1965), 182–217. [Reprinted as Chapter 20 of the present volume.]

[4] R. J. Walker, "Determination of division algebras with 32 elements," *Proceedings of Symposia in Applied Mathematics* **15** (Providence, Rhode Island: American Mathematical Society, 1963), 83–85.

Addendum

When I first submitted this paper for publication, I had gone only as far as Theorem 6. I thank the referee for encouraging me strongly to try harder, because those remarks prompted me to discover Theorems 7, 8, 9, and 10. (And I hope the referee is still alive and able to read this note, because I foolishly failed to give any acknowledgment when the paper was originally published. Long live conscientious referees who cajole authors into extending their reach!)

William M. Kantor, "Commutative semifields and symplectic spreads," *Journal of Algebra* (to appear), has recently generalized the construction of binary semifields to show that the number of nonisomorphic semifield planes of order 2^n is not bounded above by any polynomial in 2^n.

Chapter 22

Notes on Central Groupoids

[Originally published in Journal of Combinatorial Theory **8** *(1970), 376–390.]*

A *central groupoid* is an algebraic system with one binary operation, satisfying the identity

$$(x \cdot y) \cdot (y \cdot z) = y \tag{1}$$

for all x, y, and z. Such systems were recently introduced by Evans [1]. The purpose of this paper is to show that central groupoids have some remarkable properties, which can be established in an elementary manner by combining a variety of different mathematical techniques. For example, we will see that central groupoids of order n correspond in a natural manner to the matrices A of zeros and ones that satisfy the relation

$$A^2 = J, \tag{2}$$

where J is the $n \times n$ matrix of all ones. Central groupoids therefore provide a nice classroom example of the manner in which the same mathematical object can be viewed from several different standpoints. The proofs also serve as illustrations of the principle of duality.

1. "Graph-Theoretic" Approach

Let D be a directed graph, and write $x \to y$ if there is an arc in D from vertex x to vertex y. We will say that D has "Property C" if there is a unique oriented path of length two between any two of its vertices. This property implies that no two vertices x and y have two or more arcs from x to y, so we may restrict our attention to directed graphs in which no two arcs have the same initial vertex and the same final vertex. Thus, Property C states that, for any ordered pair of vertices x and y of D, there is exactly one vertex z such that

$$x \to z \to y. \tag{3}$$

As an example of a directed graph with Property C, consider the points of the Euclidean plane as vertices, and let there be an arc from $x = (x_1, x_2)$ to $y = (y_1, y_2)$ if and only if the second component of x equals the first component of y, that is, $x_2 = y_1$. Then (3) holds with $z = (x_2, y_1)$, and for no other z.

It is easy to show that there is a natural correspondence between directed graphs with Property C and central groupoids:

Lemma 1. *Let D be a directed graph with Property C. Define the binary operation $x \cdot y$ on the vertices of D by the rule "$x \cdot y = z$ if and only if z is the vertex satisfying (3)." The vertices of D, with this binary operation, form a central groupoid.*

Proof. Given x, y, and z, we have $x \to x \cdot y \to y \to y \cdot z \to z$ and $x \cdot y \to (x \cdot y) \cdot (y \cdot z) \to y \cdot z$; so, by Property C, $(x \cdot y) \cdot (y \cdot z) = y$. ☐

Lemma 2. *Let C be a central groupoid. Define a directed graph whose vertices are the elements of C by the rule "$x \to z$ if and only if $z = x \cdot y$ for some y." The resulting directed graph satisfies Property C.*

Proof. Let us first prove that, in the directed graph just defined, $z \to y$ if and only if $z = x \cdot y$ for some x. If $z \to y$, then $y = z \cdot w$ for some w, hence $(z \cdot z) \cdot y = (z \cdot z) \cdot (z \cdot w) = z$. Conversely, if $x \cdot y = z$, then $z \to (x \cdot y) \cdot (y \cdot y) = y$.

Now to show that Property C is valid, we have for any vertices x and y the relations
$$x \to x \cdot y \to y.$$
Furthermore if $x \to z \to y$ then $x = w_1 \cdot z$ and $y = z \cdot w_2$ for some w_1 and w_2, so $x \cdot y = (w_1 \cdot z) \cdot (z \cdot w_2) = z$. This establishes Property C. ☐

Theorem 1. *There is a one-to-one correspondence between directed graphs with Property C and central groupoids.*

Proof. This follows immediately from the correspondence defined in Lemmas 1 and 2, since the definitions in those lemmas are inverse to one another. ☐

2. "Constructive" Approach

Let us now consider means of constructing central groupoids. In the preceding section, an example of a directed graph with Property C was given, in which the points of the Euclidean plane were the vertices. In general, if S is any set, the same construction yields a central groupoid on the set $S \times S$ of ordered pairs of elements of S, by the rule

$$(x_1, x_2) \cdot (y_1, y_2) = (x_2, y_1). \tag{4}$$

Let us call this multiplication, which was defined by Evans [1], the *natural* central groupoid on $S \times S$.

A slight modification of this construction gives another class of central groupoids:

Theorem 2. *Let S be a set containing the element 0. Let $\circ$ be a binary operation on S such that $x \circ 0 = 0$, and such that $x \circ z = y$ has a unique solution z whenever y and x are elements of S. Let D be the directed graph on the vertices $S \times S$ having an arc $(x_1, x_2) \to (y_1, y_2)$ if and only if*

$$x_2 = y_1 \quad \text{and} \quad y_2 \neq 0, \quad \text{or} \quad x_1 \circ x_2 = y_1 \quad \text{and} \quad y_2 = 0. \quad (5)$$

Then D has Property C, so it defines a central groupoid.

Proof. Given (x_1, x_2) and (y_1, y_2) in $S \times S$, we must show that there is a unique (z_1, z_2) such that $(x_1, x_2) \to (z_1, z_2) \to (y_1, y_2)$. If $y_1 \neq 0$ and $y_2 \neq 0$ we must have $z_2 = y_1 \neq 0$, hence $z_1 = x_2$. If $y_1 \neq 0$ and $y_2 = 0$, we must have $z_1 \circ z_2 = y_1$, hence $z_2 \neq 0$ and $z_1 = x_2$; and z_2 is the unique solution to the equation $x_2 \circ z_2 = y_1$. Finally, if $y_1 = 0$ then $z_2 = 0$ and $z_1 = x_1 \circ x_2$.

This completes the proof. As a consequence, the corresponding central groupoid has the multiplication rule

$$(x_1, x_2) \cdot (y_1, y_2) = \begin{cases} (x_2, y_1), & \text{if } y_1 \neq 0 \text{ and } y_2 \neq 0; \\ (x_2, z_2), & \text{if } x_2 \circ z_2 = y_1 \neq 0 \text{ and } y_2 = 0; \\ (x_1 \circ x_2, 0), & \text{if } y_1 = 0. \end{cases} \quad (6)$$

We will see later that a system defined in this way is not always isomorphic to a natural central groupoid, so the construction of Theorem 2 is not simply a trivial construction in disguise.   □

3. "Combinatorial" Approach

In any central groupoid C, let

$$R(x) = \{ z \mid x \to z \}, \qquad L(y) = \{ z \mid z \to y \}. \quad (7)$$

Here '$x \to z$' refers to the corresponding directed graph, so $R(x)$ is the "coset" $x \cdot C$ and $L(y)$ is $C \cdot y$.

Lemma 3. *For any x and y in a central groupoid, $R(x)$ and $L(y)$ have the same number of elements; in fact, the function $f(t) = t \cdot y$ is a one-to-one correspondence between $R(x)$ and $L(y)$.*

Proof. If $z \in L(y)$ there is a unique $w \in R(x)$ such that $x \to w \to z \to y$; and $z = w \cdot y = f(w)$.   □

Corollary. *For any x and y in a central groupoid, $R(x)$ and $R(y)$ have the same number of elements.*   □

Theorem 3 (Evans [1]). *A finite central groupoid contains m^2 elements for some integer m.*

First proof. Let C be a central groupoid with n elements, and construct the associated directed graph. If we count how many ordered triples (x, y, z) satisfy $x \to y \to z$, we find there are exactly n^2, one for each choice of x and z. Counting another way, for each fixed y there are m^2 such ordered triples, if m is the number of elements in $R(y)$ and in $L(y)$. Thus $n^2 = nm^2$.

Second proof. Let x be any element of a central groupoid C, and let $R(x)$ contain m elements. The sets $\{\, R(y) \mid y \in R(x) \,\}$ are mutually disjoint and contain all elements of C, by Property C. Since each of these m sets contains m elements, C contains m^2 elements.   □

A third proof of Theorem 3 appears in the next section.

4. "Matrix-Theoretic" Approach

A directed graph D with vertices $V_1, V_2, \ldots, V_n$, in which there is at most one arc going from one vertex to another, can be represented by an $n \times n$ matrix A of zeros and ones, where $A_{ij} = 1$ if and only if there is an arc from V_i to V_j in D. Conversely, any such matrix defines a directed graph, so it is natural to try to reformulate Property C in matrix notation. It is easy to see that a matrix A of zeros and ones corresponds to a directed graph satisfying Property C if and only if

$$A^2 = J, \tag{8}$$

where J is the $n \times n$ matrix whose entries are all equal to one.

This matrix equation can be used to provide alternate proofs of the results considered above. For example, the fact that $R(x)$ and $L(y)$ have the same number of elements is a consequence of the matrix equation $AJ = A^3 = JA$. Moreover, the matrix representation makes it possible to deduce further properties of central groupoids that appear to be quite difficult to derive by any other means:

Theorem 4. *Every finite central groupoid with n elements contains $\sqrt{n}$ idempotent elements (that is, elements x equal to $x \cdot x$).*

Proof. A central groupoid with n elements corresponds to an $n \times n$ matrix A with $A^2 = J$. An element x is idempotent if and only if $x \to x$, so the number of idempotent elements is the *trace* of A.

Let the characteristic equation of A be

$$\det(\lambda I - A) = (\lambda - r_1)(\lambda - r_2) \ldots (\lambda - r_n),$$

where $r_1, \ldots, r_n$ are complex numbers. The characteristic equation of A^2 is then

$$\det(\lambda I - A^2) = \det(\sqrt{\lambda} I - A) \det(\sqrt{\lambda} I + A)$$
$$= (\lambda - r_1^2)(\lambda - r_2^2) \ldots (\lambda - r_n^2).$$

This must be the characteristic equation of J, namely, $(\lambda - n)\lambda^{n-1}$. Hence the characteristic equation of A must be $(\lambda \pm \sqrt{n})\lambda^{n-1}$; and trace($A$) is the negative of the coefficient of λ^{n-1}, namely $\mp\sqrt{n}$. Obviously the trace of A is nonnegative, so it equals $\sqrt{n}$. $\square$

5. "Algebraic" Approach

So far we have done very little, if any, manipulation of algebraic formulas that involve the binary operation and identity (1). Further results can be obtained if we make use of algebraic manipulations.

First let us notice that we always have

$$x \cdot ((x \cdot y) \cdot z) = x \cdot y \tag{9}$$

in any central groupoid, since both of these quantities are equal to

$$((x \cdot x) \cdot (x \cdot y)) \cdot ((x \cdot y) \cdot z)$$

in view of (1). A "dual" argument (namely, the same proof with left and right interchanged) proves the identity

$$(x \cdot (y \cdot z)) \cdot z = y \cdot z. \tag{10}$$

Although (9) and (10) are consequences of (1), we cannot derive (1) from (9) and (10) since the latter are true in any system having a trivial binary operation in which $x \cdot y$ is a constant independent of x and y.

Since central groupoids that are not "natural" have been defined in Section 2 above, it is of interest to find algebraic identities that characterize natural central groupoids. The natural groupoids satisfy a very strong property, namely that

$$(x \cdot y) \cdot z = (w \cdot y) \cdot z \tag{11}$$

for all w, x, y, and z; for, if $x = (x_1, x_2)$, $y = (y_1, y_2)$, and $z = (z_1, z_2)$, the value of $(x \cdot y) \cdot z$ is (y_1, z_1), independent of x.

Identity (11) in a central groupoid implies its dual, since it implies by (9) that

$$x \cdot (y \cdot z) = x \cdot ((x \cdot (y \cdot z)) \cdot y)$$
$$= x \cdot (((x \cdot y) \cdot (y \cdot z)) \cdot y) = x \cdot (y \cdot y);$$

thus we have the identity

$$x \cdot (y \cdot z) = x \cdot (y \cdot w), \tag{12}$$

which is the dual of (11) obtained by interchanging left and right.

It is not difficult to prove in fact that (11) and (1) characterize the natural central groupoids; all identities that hold in a natural central groupoid can be derived from (11) and (1). A slightly stronger result, which starts from a *single* identity, can also be proved:

Theorem 5. *An algebraic system with a single binary operation is a natural central groupoid if and only if it satisfies the identity*

$$(x \cdot ((y \cdot z) \cdot w)) \cdot (z \cdot w) = z. \tag{13}$$

Proof. A natural central groupoid satisfies (13). Conversely, (13) implies that

$$(y \cdot z) \cdot (z \cdot w) = ((x \cdot ((w \cdot (y \cdot z)) \cdot w)) \cdot ((y \cdot z) \cdot w)) \cdot (z \cdot w) = z,$$

so we have a central groupoid. And if we set $y = t \cdot (t \cdot t)$, $z = (t \cdot t) \cdot t$, $w = t \cdot t$ in (13) we find that

$$(x \cdot t) \cdot t = (t \cdot t) \cdot t. \tag{14}$$

Identity (14) is a very special case of (11); it says that $(x \cdot y) \cdot z$ does not depend on x when $y = z$, while (11) says that $(x \cdot y) \cdot z$ does not depend on x when y and z are *arbitrary*. It is a rather surprising fact that the comparatively weak identity (14) can be used to derive (11) in a central groupoid; let us defer the proof of this fact for a moment.

Now suppose that identity (11) is valid in a central groupoid; we want to prove that the groupoid is "natural." If $x = (x_1, x_2)$ in a natural central groupoid, then $x = (x_1, x_1) \cdot (x_2, x_2)$, and we also have

$(x_1, x_1) = (x \cdot x) \cdot x$, $(x_2, x_2) = x \cdot (x \cdot x)$; these formulas suggest that we define

$$x^L = (x \cdot x) \cdot x, \quad x^R = x \cdot (x \cdot x) \tag{15}$$

in any central groupoid. Now identity (11) implies (12) and it therefore implies that

$$(x \cdot (x \cdot x)) \cdot ((y \cdot y) \cdot y) = ((x \cdot x) \cdot (x \cdot x)) \cdot ((y \cdot y) \cdot y)$$
$$= ((x \cdot x) \cdot (x \cdot x)) \cdot ((y \cdot y) \cdot (y \cdot y)) = x \cdot y;$$

in terms of the functions defined in (15), this becomes

$$x^R \cdot y^L = x \cdot y. \tag{16}$$

Multiplying (16) on the right by y and using (10) implies that

$$y^L = (x \cdot y) \cdot y, \tag{17}$$

which is (14) again. Dually, (16) implies that

$$x^R = x \cdot (x \cdot y). \tag{18}$$

We now have

$$((x \cdot y) \cdot (x \cdot y)) \cdot (x \cdot y) = ((y \cdot x) \cdot (x \cdot y)) \cdot (x \cdot y) = x \cdot (x \cdot y),$$

that is,

$$(x \cdot y)^L = x^R. \tag{19}$$

A dual argument shows that (16) also implies the identity

$$(x \cdot y)^R = y^L. \tag{20}$$

[Let us digress for a moment to observe that (19) and (1) imply (20), since $y^L = ((x \cdot y) \cdot (y \cdot z))^L = (x \cdot y)^R$; dually, (20) and (1) imply (19). Furthermore (19) and (1) imply (16), for we can multiply (20) on the left by x to obtain $x \cdot y = x \cdot y^L$, and dually $x \cdot y = x^R \cdot y$.]

Now $(x \cdot x)^L = x^R$ and $(x \cdot x)^R = x^L$, so (16) implies that

$$x^L \cdot x^L = x^L \quad \text{and} \quad x^R \cdot x^R = x^R;$$

that is, x^L and x^R are idempotent. Every element x therefore has a representation of the form

$$x = u \cdot v, \quad u \text{ and } v \text{ idempotent}, \tag{21}$$

since $x = x^L \cdot x^R$. Furthermore this representation is unique, since (21) and (19) imply that $x^L = u^R = u$ and dually $x^R = v$. Finally, if $x = u_1 \cdot v_1$ and $y = u_2 \cdot v_2$ and $x \cdot y = u_3 \cdot v_3$, we have $u_3 = (x \cdot y)^L = x^R = v_1$ and dually $v_3 = u_2$. This rule is isomorphic to multiplication in a natural central groupoid on $S \times S$, where S is the set of idempotent elements, so the proof is complete. [Except that we still must derive (11) from (14).] $\square$

The argument used in our proof of Theorem 5 is certainly longer than necessary, but it demonstrates the interesting fact that all of the identities (11), (12), (16), (17), (18), (19), and (20) are equivalent in a central groupoid; any *one* of these is strong enough to imply that the central groupoid is natural. (Evans [1] showed that (16) and (19) together imply naturalness.)

6. "Automated" Approach

There is still a gap remaining to be closed in the proof of Theorem 5; we must show that the weak identity (14) is strong enough to imply (11) in a central groupoid. Equation (14) is equivalent to

$$x \cdot t = x \cdot t^L, \tag{22}$$

since we can multiply (14) on the left by x to obtain (22), and we can multiply (22) on the right by t to obtain (14).

It appears to be very difficult to derive (11) from (22), and initial attempts by the author to do so were unsuccessful. Indeed, the manipulations of Section 5 seemed to be largely a matter of random trial and error combined with a little bit of luck. This experience provided the motivation to study more general problems of the same nature, and as a result the author was led to develop an algorithm that is able to work systematically with identities in arbitrary universal algebras (see [2]).

An IBM 7094 computer was programmed to perform the algorithm of [2], and this program was presented with the four identities

$$
\begin{aligned}
(x \cdot y) \cdot (y \cdot z) &= y, \\
(x \cdot x) \cdot x &= x^L, \\
x \cdot (x \cdot x) &= x^R, \\
x^R \cdot y &= x \cdot y.
\end{aligned}
\tag{23}
$$

(The last identity is the dual of (22).) In less than nine minutes of calculation, the computer supplied the missing link in our proof of Theorem 5 by demonstrating

Theorem 6. *The four identities* (23) *imply* (22).

(Note that, since (22) is the dual of $x^R \cdot y = x \cdot y$, a dual argument shows that (22) implies (23) in a central groupoid, and so either (22) or (23) implies (16). This is enough to complete the proof of Theorem 5.)

Proof. The following proof was found by the computer; readers may wish to try their hand at discovering a shorter one:

(a) $(x \cdot y) \cdot y^R = (x \cdot y) \cdot (y \cdot (y \cdot y)) = y$, by (23).

(b) $y \cdot (y^R)^R = ((x \cdot y) \cdot y^R) \cdot (y^R)^R = y^R$, by (a) twice.

(c) $(y^R)^R = (y \cdot (y^R)^R) \cdot ((y^R)^R \cdot z) = y^R \cdot ((y^R)^R \cdot z)$
$= y^R \cdot (y^R \cdot z) = y^R \cdot (y \cdot z) = y \cdot (y \cdot z)$, by (b) and (23).

(d) $(x^R)^R = x \cdot (x \cdot x) = x^R$, by (c) and (23).

(e) $y^R = (y^R)^R = y \cdot (y \cdot z)$, by (d) and (c).

(f) $(x \cdot y) \cdot y = (x \cdot y) \cdot ((x \cdot y) \cdot y^R) = (x \cdot y)^R$, by (a) and (e).

(g) $x^L = (x \cdot x) \cdot x = (x \cdot x)^R$, by (23) and (f).

(h) $x^L \cdot y = (x \cdot x)^R \cdot y = (x \cdot x) \cdot y$, by (g) and (23).

(i) $x^L \cdot x^R = (x \cdot x) \cdot x^R = x$, by (h) and (a).

(j) $x^L = (x \cdot x)^R = ((x \cdot x)^R)^R = (x^L)^R$, by (g) and (d).

(k) $y^R = (x \cdot y^R) \cdot (y^R)^R = (x \cdot y^R) \cdot y^R = (x \cdot y^R)^R$,
by (a), (d), and (f).

(l) $(x \cdot y^R) \cdot z = (x \cdot y^R)^R \cdot z = y^R \cdot z = y \cdot z$, by (23) and (k).

(m) $x \cdot y = ((w \cdot x) \cdot (x \cdot y)) \cdot ((x \cdot y) \cdot z) = x \cdot ((x \cdot y) \cdot z)$, by (23).

(n) $x \cdot y^R = x \cdot ((x \cdot y^R) \cdot z) = x \cdot (y \cdot z)$, by (m) and (l).

(o) $x \cdot y^L = x \cdot (y^L)^R = x \cdot (y^L \cdot y^R) = x \cdot y$, by (j), (n), and (i).   □

(Step (e) is the dual of (14), and step (m) is (9). Actually it is not difficult to see that steps (h) and (i) may be eliminated since (i) is an immediate consequence of (23).)

7. "Exhaustive" Approach

Now let us determine all central groupoids of small order n. By Theorem 3, n is a perfect square. By Theorem 1 and the observations of Section 4, we may construct all central groupoids of order n by constructing all matrices A of zeros and ones such that $A^2 = J$. Theorem 4 tells us that such a matrix A has $\sqrt{n}$ ones on the diagonal. Furthermore two central groupoids are isomorphic if and only if the corresponding directed graphs are isomorphic; this means that two matrices A and A' define isomorphic central groupoids if and only if

$$A = P^{-1} A' P$$

for some permutation matrix P.

To find all central groupoids with four elements, we may write

$$
A = \begin{pmatrix} 1 & 1 & 0 & 0 \\ 0 & 0 & 1 & 1 \\ * & * & * & * \\ * & * & * & * \end{pmatrix},
$$

where the $*$'s denote elements yet to be filled in. (The second proof of Theorem 3 shows that we may always write down the first $\sqrt{n}$ rows of A in an analogous manner.) Another 1 and another 0 need to appear on the diagonal, and without loss of generality we may take $A_{33} = 1$, $A_{44} = 0$. Every row and every column must contain $\sqrt{n}$ ones, by Lemma 3; so $A_{34} = 1$, $A_{43} = 0$, $A_{31} = A_{32} = 0$, $A_{41} = A_{42} = 1$. The only solution (up to permutation of rows and columns by P) is therefore

$$
A = \begin{pmatrix} 1 & 1 & 0 & 0 \\ 0 & 0 & 1 & 1 \\ 0 & 0 & 1 & 1 \\ 1 & 1 & 0 & 0 \end{pmatrix};
$$

every central groupoid of order 4 is natural.

A similar but more lengthy analysis shows that there are just six central groupoids of order 9, namely, those defined by the matrices

$$
A_1 = \begin{pmatrix}
1 & 1 & 1 & 0 & 0 & 0 & 0 & 0 & 0 \\
0 & 0 & 0 & 1 & 1 & 1 & 0 & 0 & 0 \\
0 & 0 & 0 & 0 & 0 & 0 & 1 & 1 & 1 \\
1 & 1 & 1 & 0 & 0 & 0 & 0 & 0 & 0 \\
0 & 0 & 0 & 1 & 1 & 1 & 0 & 0 & 0 \\
0 & 0 & 0 & 0 & 0 & 0 & 1 & 1 & 1 \\
1 & 1 & 1 & 0 & 0 & 0 & 0 & 0 & 0 \\
0 & 0 & 0 & 1 & 1 & 1 & 0 & 0 & 0 \\
0 & 0 & 0 & 0 & 0 & 0 & 1 & 1 & 1
\end{pmatrix},
$$

$$
A_2 = \begin{pmatrix}
1 & 1 & 1 & 0 & 0 & 0 & 0 & 0 & 0 \\
0 & 0 & 0 & 1 & 1 & 1 & 0 & 0 & 0 \\
0 & 0 & 0 & 0 & 0 & 0 & 1 & 1 & 1 \\
1 & 1 & 1 & 0 & 0 & 0 & 0 & 0 & 0 \\
0 & 0 & 0 & 0 & 1 & 1 & 1 & 0 & 0 \\
0 & 0 & 0 & 1 & 0 & 0 & 0 & 1 & 1 \\
1 & 1 & 1 & 0 & 0 & 0 & 0 & 0 & 0 \\
0 & 0 & 0 & 0 & 1 & 1 & 1 & 0 & 0 \\
0 & 0 & 0 & 1 & 0 & 0 & 0 & 1 & 1
\end{pmatrix},
$$

$$A_3 = \begin{pmatrix} 1 & 1 & 1 & 0 & 0 & 0 & 0 & 0 & 0 \\ 0 & 0 & 0 & 1 & 1 & 1 & 0 & 0 & 0 \\ 0 & 0 & 0 & 0 & 0 & 0 & 1 & 1 & 1 \\ 1 & 1 & 1 & 0 & 0 & 0 & 0 & 0 & 0 \\ 0 & 0 & 0 & 1 & 1 & 1 & 0 & 0 & 0 \\ 0 & 0 & 0 & 0 & 0 & 0 & 1 & 1 & 1 \\ 1 & 1 & 1 & 0 & 0 & 0 & 0 & 0 & 0 \\ 0 & 0 & 0 & 0 & 1 & 1 & 1 & 0 & 0 \\ 0 & 0 & 0 & 1 & 0 & 0 & 0 & 1 & 1 \end{pmatrix},$$

$$A_4 = \begin{pmatrix} 1 & 1 & 1 & 0 & 0 & 0 & 0 & 0 & 0 \\ 0 & 0 & 0 & 1 & 1 & 1 & 0 & 0 & 0 \\ 0 & 0 & 0 & 0 & 0 & 0 & 1 & 1 & 1 \\ 1 & 1 & 1 & 0 & 0 & 0 & 0 & 0 & 0 \\ 0 & 0 & 0 & 1 & 1 & 1 & 0 & 0 & 0 \\ 0 & 0 & 0 & 0 & 0 & 0 & 1 & 1 & 1 \\ 1 & 0 & 1 & 0 & 1 & 0 & 0 & 0 & 0 \\ 0 & 1 & 0 & 1 & 0 & 1 & 0 & 0 & 0 \\ 0 & 0 & 0 & 0 & 0 & 0 & 1 & 1 & 1 \end{pmatrix},$$

$$A_5 = \begin{pmatrix} 1 & 1 & 1 & 0 & 0 & 0 & 0 & 0 & 0 \\ 0 & 0 & 0 & 1 & 1 & 1 & 0 & 0 & 0 \\ 0 & 0 & 0 & 0 & 0 & 0 & 1 & 1 & 1 \\ 1 & 1 & 1 & 0 & 0 & 0 & 0 & 0 & 0 \\ 0 & 0 & 0 & 1 & 1 & 0 & 0 & 0 & 1 \\ 0 & 0 & 0 & 0 & 0 & 1 & 1 & 1 & 0 \\ 0 & 1 & 1 & 1 & 0 & 0 & 0 & 0 & 0 \\ 1 & 0 & 0 & 0 & 1 & 0 & 0 & 0 & 1 \\ 0 & 0 & 0 & 0 & 0 & 1 & 1 & 1 & 0 \end{pmatrix},$$

$$A_6 = \begin{pmatrix} 1 & 1 & 1 & 0 & 0 & 0 & 0 & 0 & 0 \\ 0 & 0 & 0 & 1 & 1 & 1 & 0 & 0 & 0 \\ 0 & 0 & 0 & 0 & 0 & 0 & 1 & 1 & 1 \\ 1 & 1 & 0 & 0 & 0 & 1 & 0 & 0 & 0 \\ 0 & 0 & 1 & 1 & 1 & 0 & 0 & 0 & 0 \\ 0 & 0 & 0 & 0 & 0 & 0 & 1 & 1 & 1 \\ 1 & 0 & 1 & 0 & 0 & 0 & 0 & 1 & 0 \\ 0 & 0 & 0 & 1 & 1 & 1 & 0 & 0 & 0 \\ 0 & 1 & 0 & 0 & 0 & 0 & 1 & 0 & 1 \end{pmatrix}.$$

Here A_1 represents the natural central groupoid; A_2 is obtained by the construction of Theorem 2 with $0 \circ x = x$, $1 \circ 1 = 2 \circ 1 = 2$, $1 \circ 2 =$

$2 \circ 2 = 1$; and A_3 is obtained by the construction of Theorem 2 with $0 \circ x = 1 \circ x = x$, $2 \circ 1 = 2$, $2 \circ 2 = 1$. The systems A_3 and A_4 are dual to each other or "anti-isomorphic"; this means in matrix terminology that

$$A_3^T = P^{-1} A_4 P$$

for some permutation matrix P. Similarly A_5 and A_6 are dual to each other; neither A_5 nor A_6 arise from Theorem 2. Both A_1 and A_2 are self-dual, that is, anti-isomorphic to themselves.

It is useful to find isomorphic invariants that make it easy to decide which of the six systems above represents a given central groupoid. A useful invariant that was discovered during this enumeration is based on two equivalence relations: Let $x \equiv^L y$ when $x \to z$ is equivalent to $y \to z$, and let $x \equiv^R y$ when $z \to x$ is equivalent to $z \to y$. Thus, $x \equiv^L y$ if and only if row x and row y are equal in the matrix representations, and $x \equiv^R y$ if and only if columns x and y are equal. Let us say *L-classes* and *R-classes* are the equivalence classes defined by these relations, and let us call a class "idempotent" if it contains an idempotent element. Clearly each class contains at most one idempotent element.

The following table lists the number of elements in each of the equivalence classes:

System	idempotent L-classes	other L-classes	idempotent R-classes	other R-classes
1	3, 3, 3		3, 3, 3	
2	3, 2, 2	1, 1	3, 2, 2	1, 1
3	3, 2, 1	2, 1	3, 2, 2	1, 1
4	3, 2, 2	1, 1	3, 2, 1	2, 1
5	2, 2, 1	1, 1, 1, 1	1, 1, 1	2, 2, 1, 1
6	1, 1, 1	2, 2, 1, 1	2, 2, 1	1, 1, 1, 1

For example, if we number the elements from 1 to 9, system 2 has the L-classes $\{1, 4, 7\}$, $\{5, 8\}$, $\{6, 9\}$, $\{2\}$, $\{3\}$, where the first three classes are idempotent. Its R-classes are $\{1, 2, 3\}$, $\{5, 6\}$, $\{8, 9\}$, $\{4\}$, $\{7\}$. In the case of order nine the number of elements in these classes is sufficient to determine the central groupoid, up to isomorphism.

Another isomorphic invariant is the similarity class of the matrix A. Since the Jordan canonical form of J is $\mathrm{diag}(n, 0, \ldots, 0)$ and $A^2 = J$, it follows that the Jordan canonical form of A is $(\sqrt{n}) \oplus C_1 \oplus \cdots \oplus C_k$, where each of the C's is either the 1×1 matrix (0) or the 2×2 matrix $\left(\begin{smallmatrix} 0 & 1 \\ 0 & 0 \end{smallmatrix}\right)$; this is a consequence of the fact that C_j^2 must be zero. Therefore two matrices of zeros and ones satisfying the relation $A^2 = J$ will be similar if and only if they have the same rank; and the rank (which is one plus the number of 2×2 blocks in the Jordan canonical form) can be at most $(n+1)/2$. Since we have observed that the first $\sqrt{n}$ rows of A can be assumed to have a fixed form, the rank of A must be at least $\sqrt{n}$. (These remarks were suggested by N. S. Mendelsohn.)

In the case $n = 9$, the possible ranks are therefore 3, 4, and 5. By inspection A_1 has rank 3; A_2, A_3, and A_4 have rank 4; A_5 and A_6 have rank 5. It is not difficult to see that in general only the matrices of rank $\sqrt{n}$ correspond to natural central groupoids. The construction of Theorem 2 yields matrices of ranks $\sqrt{n}, \sqrt{n} + 1, \ldots, 2\sqrt{n} - 2$, and it is possible to extend Theorem 2 in a straightforward manner to obtain all ranks up to

$$\sqrt{n} + \lfloor \sqrt{n}/3 \rfloor \lfloor (2\sqrt{n} - 1)/3 \rfloor.$$

But it is not clear whether all ranks up to $(n+1)/2$ are always possible.

The examples for the case $n = 9$ show that central groupoids can have a rather complex structure, and that the problem of classifying all finite central groupoids is not trivial (see [3]). The usual technique of studying algebraic systems by investigating homomorphisms and congruence relations does not immediately appear to yield important results in this case; for example, the natural central groupoid with n elements is a homomorphic image of the natural central groupoid with n' elements whenever $n \leq n'$.

As a first step in the study of central groupoids for general n, let us examine the construction of Theorem 2 more closely. If we are given a binary operation $a \circ b$ on a set S satisfying the conditions of Theorem 2, let $f(a) = 0 \circ a$, and define a new binary operation

$$a \mathbin{\square} b = f^{-1}(a \circ b).$$

It follows that $a \mathbin{\square} b$ also satisfies the hypotheses of Theorem 2, and that $0 \mathbin{\square} b = b$ for all $b \in S$. Furthermore the directed graphs (and consequently the central groupoids) constructed by Theorem 2 for $a \circ b$ and for $a \mathbin{\square} b$ are easily shown to be isomorphic, under the mapping

$$\varphi(x, y) = \begin{cases} (f(x), y), & \text{if} \quad y = 0, \\ (x, y), & \text{if} \quad y \neq 0. \end{cases}$$

Therefore we may assume without loss of generality that $0 \circ b = b$ in the construction of Theorem 2.

With this simplification, the question of isomorphism between the central groupoids constructed by Theorem 2 can be resolved as follows:

Theorem 7. *Let S be a finite set on which two binary operations $\circ$ and $\square$ are defined, such that, for all a and b, $0 \circ b = 0 \square b = b$, $a \circ 0 = a \square 0 = 0$, and the equations $a \circ x = b$, $a \square y = b$ have a unique solution (x, y). Let D be the directed graph on $S \times S$ defined by the rule $(x_1, x_2) \to (y_1, y_2)$ if and only if $x_2 = y_1$ and $y_2 \neq 0$ or $x_1 \circ x_2 = y_1$ and $y_2 = 0$; and let D' be the directed graph defined similarly with $\square$ replacing $\circ$. Then D and D' are isomorphic if and only if the binary operations $\circ$ and $\square$ are isomorphic.*

Proof. If $\circ$ and $\square$ are isomorphic there is a permutation f of S such that $f(a) \square f(b) = f(a \circ b)$. If $f(0) = t \neq 0$, let $s = f^{-1}(0)$; then we have

$$a \square t = t, \quad t \square b = b, \quad a \circ s = s, \quad s \circ b = b.$$

Consequently if g is the permutation that transposes 0 and t, the relation $gf(a) \square gf(b) = gf(a \circ b)$ can be verified without difficulty.

We may assume, therefore, that $f(0) = 0$. The function $\varphi(x_1, x_2) = (f(x_1), f(x_2))$ is now a permutation of $S \times S$ such that $(x_1, x_2) \to (y_1, y_2)$ in D if and only if $\varphi(x_1, x_2) \to \varphi(y_1, y_2)$ in D'; hence D and D' are isomorphic.

Conversely let φ be a permutation of $S \times S$ under which D is isomorphic to D'. Since $(x_1, x_2) \to (x_1, x_2)$ if and only if $x_1 = x_2$, the image $\varphi(0, 0)$ must equal (t, t) for some $t \in S$.

Case 1, $t \neq 0$. Let S have m elements; we may assume that $m > 2$. The L-classes under our construction satisfy

$$(x_1, x_2) \stackrel{L}{\equiv} (x_1', x_2') \text{ in } D \iff x_2 = x_2' \text{ and } x_1 \circ x_2 = x_1' \circ x_2.$$

Since the L-class containing $(0, 0)$ in D has m elements, the L-class containing (t, t) in D' must have m elements. Thus

$$(0, t) \stackrel{L}{\equiv} (1, t) \stackrel{L}{\equiv} \cdots \stackrel{L}{\equiv} (m - 1, t) \text{ in } D',$$

and $\varphi(x, 0) = (f(x), t)$ for some permutation f of S. Now for all a, the construction implies

$$(a, 1) \stackrel{R}{\equiv} (a, 2) \stackrel{R}{\equiv} \cdots \stackrel{R}{\equiv} (a, m - 1) \text{ in } D',$$

so the R-class in D' containing (a, t) has at least $m - 1$ elements. Thus the R-class in D containing $(f^{-1}(a), 0)$ has at least $m - 1$ elements. But this R-class can only contain elements of the form $(f^{-1}(a), b)$; hence

$$(f^{-1}(a), 0) \stackrel{R}{\equiv} (f^{-1}(a), g(a)) \text{ in } D$$

for some $g(a) \neq 0$. And this relation implies that

$$(f^{-1}(a), 0) \stackrel{R}{\equiv} (f^{-1}(a), b) \text{ in } D$$

for *all* $b \neq 0$; the R-class has m members. Likewise, the images satisfy

$$(a, 0) \stackrel{R}{\equiv} (a, b) \text{ in } D'$$

for all a and b. We conclude that $a \circ b = b$ and $a \,\square\, b = b$ for all $a, b \in S$; these binary operations are indeed isomorphic.

Case 2, $t = 0$. Since $(0, 0) \stackrel{L}{\equiv} (x, y)$ if and only if $y = 0$, and $(0, 0) \stackrel{R}{\equiv} (x, y)$ if and only if $x = 0$, we have

$$\varphi(x, 0) = (f(x), 0) \quad \text{and} \quad \varphi(0, y) = (0, g(y))$$

for some permutations f and g of S. Now

$$(0, x) \rightarrow (0 \circ x, 0) = (x, 0)$$

in D, and

$$(0, g(x)) \rightarrow (0 \,\square\, g(x), 0) = (g(x), 0)$$

in D'; hence $f = g$. Furthermore $(0, x) \rightarrow (x, x)$ in D and $(0, f(x)) \rightarrow (f(x), f(x))$ in D'; we must have $\varphi(x, x) \rightarrow \varphi(x, x)$ in D', hence $\varphi(x, x) = (f(x), f(x))$.

Finally if $y \neq 0$, the relation $(0, x) \rightarrow (x, y) \rightarrow (y, y)$ in D implies that $(0, f(x)) \rightarrow \varphi(x, y) \rightarrow (f(y), f(y))$ in D', hence

$$\varphi(x, y) = (f(x), f(y)).$$

Now $(a, b) \rightarrow (a \circ b, 0)$ in D and $(f(a), f(b)) \rightarrow (f(a) \,\square\, f(b), 0)$ in D', hence $f(a \circ b) = f(a) \,\square\, f(b)$ and the proof is complete.   □

8. "Free" Approach

Free central groupoids were first constructed by Trevor Evans, who communicated his construction to the author in a letter. The theory behind his construction follows immediately from general principles described in [2], so only the consequences of that theory will be summarized here.

The free central groupoid on any number of generators is the set of all words formed from the generators and the binary operation, having no subwords of the forms $((\alpha \cdot \beta) \cdot (\beta \cdot \gamma))$, $(\alpha \cdot ((\alpha \cdot \beta) \cdot \gamma))$, or $((\alpha \cdot (\beta \cdot \gamma)) \cdot \gamma)$, where α, β, γ are words. For example, the free central groupoid on one generator a has the elements

$$a, \; a \cdot a, \; a \cdot (a \cdot a), \; (a \cdot a) \cdot a, \; a \cdot (a \cdot (a \cdot a)), \; ((a \cdot a) \cdot a) \cdot a,$$
$$a \cdot (a \cdot (a \cdot (a \cdot a))), \; a \cdot (((a \cdot a) \cdot a) \cdot a), \; (a \cdot a) \cdot ((a \cdot a) \cdot a),$$
$$(a \cdot (a \cdot a)) \cdot (a \cdot a), \; (a \cdot (a \cdot (a \cdot a))) \cdot a, \; (((a \cdot a) \cdot a) \cdot a) \cdot a,$$

plus similar words with 6 or more appearances of a.

To multiply two words of the free central groupoids, we write down their product and then replace any subword $((\alpha \cdot \beta) \cdot (\beta \cdot \gamma))$ by β, or any $(\alpha \cdot ((\alpha \cdot \beta) \cdot \gamma))$ by $(\alpha \cdot \beta)$, or any $((\alpha \cdot (\beta \cdot \gamma)) \cdot \gamma)$ by $(\beta \cdot \gamma)$, and repeat these reductions as many times as possible (see equations (1), (9), and (10)). According to [2], the reductions can be done in any order; the final result will be the same.

Theorem 4 states that any central groupoid with a finite number n of elements has $\sqrt{n}$ idempotent elements. It is natural to wonder if any simple formula can be given for the idempotent elements; for example, the formulas $x \cdot (x \cdot x)$ and $(x \cdot x) \cdot x$ describe all the idempotents in a natural central groupoid. But this is not true in general, as the following theorem shows:

Theorem 8. *The free central groupoid on any number of generators contains no idempotent elements.*

Proof (due to Trevor Evans). If α is idempotent, α is not simply a generator; hence α has the form $\beta \cdot \gamma$, which is reduced as far as possible. And since $(\beta \cdot \gamma) \cdot (\beta \cdot \gamma) = \beta \cdot \gamma$, it must be possible to reduce the word $(\beta \cdot \gamma) \cdot (\beta \cdot \gamma)$; hence $\beta = \gamma$. But then $\beta \cdot \beta = (\beta \cdot \beta) \cdot (\beta \cdot \beta) = \beta$, contradicting the fact that $\beta \cdot \gamma$ cannot be reduced. $\square$

Communicated by Marshall Hall, Jr. The work reported in these notes was supported in part by the U.S. Office of Naval Research.

References

[1] Trevor Evans, "Products of points — Some simple algebras and their identities," *American Mathematical Monthly* **74** (1967), 362–372.

[2] Donald E. Knuth and Peter B. Bendix, "Simple word problems in universal algebras," in *Computational Problems in Abstract Algebra*, edited by John Leech (Oxford: Pergamon, 1969), 263–297.

[3] A. J. Hoffman, "Research Problem 2-11," *Journal of Combinatorial Theory* **2** (1967), 393.

Addendum

Upon reading this paper 35 years after it was written, I cannot resist asking another question: How many words in the free central groupoid generated by a have exactly n appearances of a? The sequence begins

$$1, 1, 2, 2, 6, 8, 26, 55, 148, 377, 1066, 2853, 8044, 22298,$$
$$63134, 178399, 509944, 1460728, 4213962, 12194213, \ldots .$$

The rank of a finite central groupoid can be any number between $\sqrt{n}$ and $\lfloor (n+1)/2 \rfloor$; see Leslie E. Shader, "On the existence of finite central groupoids of all possible ranks. I," *Journal of Combinatorial Theory* **A16** (1974), 221–229. His construction for the maximum rank — when the groupoid is "most unnatural" — is particularly interesting. The elements are pairs of integers ab, where $1 \le a, b \le \sqrt{n}$, and where the corresponding directed graph has the following arcs when $1 \le a, b, c, d < m \le \sqrt{n}$:

$$ab \to cm \iff (a = 1 \text{ and } b = c) \text{ or } (a > 1 \text{ and } b = f(a,c) < m-1);$$
$$ab \to mm \iff a > 1 \text{ and } b = m - 1;$$
$$am \to cm \iff a > 1 \text{ and } c = m - 1;$$
$$am \to md;$$
$$am \to mm \iff a = 1;$$
$$mb \to cd \iff (d = 1 \text{ and } b = c) \text{ or } (d > 1 \text{ and } b = f(m,c));$$
$$mb \to cm \iff m = 2 \text{ or } c = g(m,b) + 1 < m - 1;$$
$$mb \to mm \iff b = m - 2;$$
$$mm \to cm \iff c = m - 1 > 1;$$
$$mm \to md;$$
$$mm \to mm \iff m \le 2.$$

Here
$$f(a,c) = \begin{cases} a - 1, & \text{if } a \text{ is odd and } c = 1; \\ 1, & \text{if } a \text{ is odd and } c = a - 1; \\ c, & \text{otherwise;} \end{cases}$$
$$g(m,v) = \begin{cases} m \bmod 2, & \text{if } b = 1; \\ (m + 1) \bmod 2, & \text{if } b = m - 1; \\ b, & \text{otherwise.} \end{cases}$$

Arcs of the forms $ab \to mc$, $am \to cd$, $mb \to md$, and $mm \to cd$ never appear.

For example, when $n = 64$ Shader's matrix A is

```
11111111 00000000 00000000 00000000 00000000 00000000 00000000 00000000
00000000 11111111 00000000 00000000 00000000 00000000 00000000 00000000
00000000 00000000 11111111 00000000 00000000 00000000 00000000 00000000
00000000 00000000 00000000 11111111 00000000 00000000 00000000 00000000
00000000 00000000 00000000 00000000 11111111 00000000 00000000 00000000
00000000 00000000 00000000 00000000 00000000 11111111 00000000 00000000
00000000 00000000 00000000 00000000 00000000 00000000 11111111 00000000
00000000 00000000 00000000 00000000 00000000 00000000 00000000 11111111
11111111 00000000 00000000 00000000 00000000 00000000 00000000 00000000
00000000 11011111 00100000 00000000 00000000 00000000 00000000 00000000
00000000 00100000 11001111 00010000 00000000 00000000 00000000 00000000
00000000 00000000 00010000 11100111 00001000 00000000 00000000 00000000
00000000 00000000 00000000 00001000 11110011 00000100 00000000 00000000
00000000 00000000 00000000 00000000 00000100 11111001 00000010 00000000
00000000 00000000 00000000 00000000 00000000 00000010 11111100 00000001
00000000 00000000 00000000 00000000 00000000 00000000 00000001 11111110
10000000 01011111 00100000 00000000 00000000 00000000 00000000 00000000
01111111 10000000 00000000 00000000 00000000 00000000 00000000 00000000
00000000 00100000 11001111 00010000 00000000 00000000 00000000 00000000
00000000 00000000 00010000 11100111 00001000 00000000 00000000 00000000
00000000 00000000 00000000 00001000 11110011 00000100 00000000 00000000
00000000 00000000 00000000 00000000 00000100 11111001 00000010 00000000
00000000 00000000 00000000 00000000 00000000 00000010 11111100 00000001
00000000 00000000 00000000 00000000 00000000 00000000 00000001 11111110
11111111 00000000 00000000 00000000 00000000 00000000 00000000 00000000
00000000 11101111 00000000 00010000 00000000 00000000 00000000 00000000
00000000 00010000 11101111 00000000 00000000 00000000 00000000 00000000
00000000 00000000 00010000 11100111 00001000 00000000 00000000 00000000
00000000 00000000 00000000 00001000 11110011 00000100 00000000 00000000
00000000 00000000 00000000 00000000 00000100 11111001 00000010 00000000
00000000 00000000 00000000 00000000 00000000 00000010 11111100 00000001
00000000 00000000 00000000 00000000 00000000 00000000 00000001 11111110
10000000 00001000 00000000 01110111 00000000 00000000 00000000 00000000
00000000 11110111 00001000 00000000 00000000 00000000 00000000 00000000
00000000 00000000 11110111 00000000 00001000 00000000 00000000 00000000
01111111 00000000 00000000 10000000 00000000 00000000 00000000 00000000
00000000 00000000 00000000 00001000 11110011 00000100 00000000 00000000
00000000 00000000 00000000 00000000 00000100 11111001 00000010 00000000
00000000 00000000 00000000 00000000 00000000 00000010 11111100 00000001
00000000 00000000 00000000 00000000 00000000 00000000 00000001 11111110
11111111 00000000 00000000 00000000 00000000 00000000 00000000 00000000
00000000 11111011 00000100 00000000 00000000 00000000 00000000 00000000
00000000 00000000 11111011 00000100 00000000 00000000 00000000 00000000
00000000 00000000 00000000 11111011 00000000 00000100 00000000 00000000
00000000 00000100 00000000 00000000 11111011 00000000 00000000 00000000
00000000 00000000 00000000 00000000 00000100 11111001 00000010 00000000
00000000 00000000 00000000 00000000 00000000 00000010 11111100 00000001
00000000 00000000 00000000 00000000 00000000 00000000 00000001 11111110
10000000 00000010 00000000 00000000 00000000 01111101 00000000 00000000
00000000 11111101 00000010 00000000 00000000 00000000 00000000 00000000
00000000 00000000 11111101 00000010 00000000 00000000 00000000 00000000
00000000 00000000 00000000 11111101 00000010 00000000 00000000 00000000
00000000 00000000 00000000 00000000 11111101 00000000 00000010 00000000
01111111 00000000 00000000 00000000 00000000 10000000 00000000 00000000
00000000 00000000 00000000 00000000 00000000 00000010 11111100 00000001
00000000 00000000 00000000 00000000 00000000 00000000 00000001 11111110
11111111 00000000 00000000 00000000 00000000 00000000 00000000 00000000
00000000 11111110 00000001 00000000 00000000 00000000 00000000 00000000
00000000 00000000 11111110 00000001 00000000 00000000 00000000 00000000
00000000 00000000 00000000 11111110 00000001 00000000 00000000 00000000
00000000 00000000 00000000 00000000 11111110 00000001 00000000 00000000
00000000 00000000 00000000 00000000 00000000 11111110 00000000 00000001
00000000 00000001 00000000 00000000 00000000 00000000 11111110 00000000
00000000 00000000 00000000 00000000 00000000 00000000 00000001 11111110
```

In a Ph.D. dissertation supervised by Trevor Evans, Raymond R. Fletcher III defined a "switch" operation on elements of a directed graph D: If $x \to y$, $x \not\to y'$, $x' \not\to y$, and $x' \to y'$ in D, let D' be the same as D but with $x \not\to y$, $x \to y'$, $x' \to y$, $x' \not\to y$. He observed that if D has Property C, then D' has Property C if and only if

$$\{x, x', y, y'\} \text{ are distinct}, \quad x \overset{R}{\equiv} x', \quad \text{and} \quad y \overset{L}{\equiv} y'.$$

[Raymond Russwald Fletcher III, *Unique Path Property Digraphs* (Atlanta, Georgia: Mathematics Department, Emory University, 1991); see also Raymond R. Fletcher III, "Using the theory of groups to construct unique path property digraphs," *Congressus Numerantium* **153** (2001), 193–209.] All of the central groupoids constructed in Theorem 2 can be obtained from the natural central groupoid by repeated switches, and Fletcher has conjectured that in fact *every* central groupoid is obtainable in this way.

Further results have been discovered by Frank Curtis, John Drew, Chi-Kwong Li, and Dan Pragel ["Central groupoids, central digraphs, and zero-one matrices A satisfying $A^2 = J$," preprint (Williamsburg, Virginia: Department of Mathematics, College of William and Mary, February 2003)], who prove among other things that we always have

$$A = P_1 + \cdots + P_{\sqrt{n}}$$

where the P_j are permutation matrices; P_1 is an involution with $\sqrt{n}$ fixed points, while all cycles of P_j for $j > 1$ have length ≥ 3. Because $A^2 = J$, there is exactly one pair (r, s) such that the permutation $P_r P_s$ maps i to j, for each i and j. (In the natural central groupoid we can let P_j map (a, b) to $(b, 1 + (a + j - 2) \bmod \sqrt{n})$ for $1 \leq a, b \leq \sqrt{n}$.) Curtis, Drew, Li, and Pragel show how to use Fletcher's switches to simplify the proof of Shader's theorem; and they conjecture that every central groupoid of size m^2 contains a central subgroupoid of size $(m - 1)^2$.

Chapter 23

Huffman's Algorithm via Algebra

*[Originally published in Journal of Combinatorial Theory **A32** (1982),
216–224. Dedicated to Professor Marshall Hall, Jr., on the occasion of
his retirement.]*

The well-known algorithm of Huffman [6] for finding minimum redundancy codes has found many diverse applications, and in recent years it has been extended in a variety of ways [2–5, 7, 8, 10]. The purpose of this note is to discuss a simple algebraic approach that seems to fit essentially all the applications of Huffman's method that are presently known.

Let us say that a *Huffman algebra* $(A, <, \circ)$ is a linearly ordered set A on which a binary operator has been defined satisfying the following five axioms:

A0 (Increasing property).	$a \leq a \circ b$.
A1 (Commutative law).	$a \circ b = b \circ a$.
A2 (Medial law).	$(a \circ b) \circ (c \circ d) = (a \circ c) \circ (b \circ d)$.
A3 (Preservation of order).	If $a \leq b$ then $a \circ c \leq b \circ c$.
A4 (Associative inequality).	If $a \leq c$ then $(a \circ b) \circ c \leq a \circ (b \circ c)$.

Given elements $a_1, \ldots, a_n$ of A, not necessarily distinct, an *expression* on $\{a_1, \ldots, a_n\}$ is a formula that computes another element of A by applying the binary operation $n - 1$ times and using each a_i exactly once. For example, the commutative law A1 states that both of the possible expressions on $\{a, b\}$ have the same value, and the medial law A2 states that two particular expressions on $\{a, b, c, d\}$ are equal. (The medial law is sometimes also called the law of bisymmetry.) Expressions are essentially binary trees having the multiset $\{a_1, \ldots, a_n\}$ as "leaves."

Huffman's algorithm forms an expression on $\{a_1, \ldots, a_n\}$ in the following way: If $n > 1$, let a_i and a_j be the smallest and second-smallest elements; replace a_i and a_j by $(a_i \circ a_j)$ and repeat the construction on the remaining $n - 1$ elements, until eventually $n = 1$.

For example, suppose A consists of the nonnegative integers, and let $a \circ b = 2(a + b)$. It is easy to check that axioms A0–A4 hold. Huffman's algorithm applied to the elements $\{1, 3, 5, 7, 9\}$ will produce the expression

$$((5 \circ 7) \circ ((1 \circ 3) \circ 9)) = 116.$$

Note that we have

$$5 \cdot 2^2 + 7 \cdot 2^2 + 1 \cdot 2^3 + 3 \cdot 2^3 + 9 \cdot 2^2 = 116;$$

in the case of this particular operation the value is $\sum a_k \cdot 2^{l_k}$, where l_k is the "level" at which a_k appears in the formula, that is, the depth of parenthesis nesting when parentheses surround each use of the binary operation.

When a binary operator satisfies the commutative and medial laws A1 and A2, the value of any expression on $\{a_1, \ldots, a_n\}$ depends only on the a_k and their levels l_k; in other words, any two expressions in which each a_k appears on a given level l_k will be equal. We can prove this by using terminology from family trees: If '$(a \circ b)$' appears in some formula we can say that a and b are siblings and $(a \circ b)$ is their parent. Two elements a_i and a_j on the same level in some expression are either siblings, or they are cousins (their parents are siblings), or they are second cousins (their parents are cousins), etc. We can transform the expression to an equivalent one using A1 and A2 until a_i and a_j are siblings; for if a_i and a_j are kth cousins and $k = 1$, the axioms directly change cousins into siblings, while if $k > 1$ the transformation for order $k-1$ will make their parents into siblings and one more step will complete the job. Now let E and E' be expressions in $\{a_1, \ldots, a_n\}$ for which the levels $l_k = l'_k$ agree for all k. If $n = 1$, clearly $E = E'$. Otherwise E contains some operation $(a_i \circ a_j)$. Since $l_i = l_j$, we can transform E' to an expression $E'' = E'$ in which a_i and a_j are siblings. Replacing $(a_i \circ a_j)$ by a new symbol in E and E'' yields an expression in $n - 1$ elements having corresponding level numbers equal, hence $E = E''$. Incidentally, a similar proof can be used to show that if we use just the medial law A2, the value of any expression on $\{a_1, \ldots, a_n\}$ depends only on the a_k and their levels l_k and their 'right levels' r_k (the number of enclosing operations where a_k appears to the right of the operator).

The main feature of Huffman's algorithm is that it produces *an expression of minimum value*, from among all expressions on the given elements $\{a_1, \ldots, a_n\}$, whenever axioms A0–A4 hold. We can prove this by starting with any expression E and transforming it into expressions of equal or lesser value until we obtain the result of Huffman's construction.

First we let a_i be the smallest of $\{a_1, \ldots, a_n\}$. If $l_i < \max(l_1, \ldots, l_n)$ (that is, if a_i is not at the deepest level of E), let a_k be an element at the deepest level; then some ancestor of a_k is at the same level as a_i, and we can transform E into $E' = E$ where a_i is a sibling of this ancestor. One of the nieces or nephews of a_i in E' is a_k or an ancestor of a_k; call it x. The other niece or nephew, call it y, is not. Since y has been computed from one or more elements greater than or equal to a_i, we have $a_i \leq y$ by axiom A0. Thus $(a_i \circ x) \circ y \leq a_i \circ (x \circ y)$; replacing $(a_i \circ (x \circ y))$ by $((a_i \circ x) \circ y)$ in E' yields an expression $E'' \leq E'$, because of axiom A3. Furthermore a_i has moved to a deeper level in E'', while a_k is still at the same level, which is still maximum among all levels. After repeating this transformation enough times, a_i will appear at the deepest level. The same process can now be repeated with respect to the second-smallest element, a_j, this time using a_i instead of a_k in the argument. Finally, with both a_i and a_j on the same level, we can make them siblings, and E has been reduced to an expression E''' containing $(a_i \circ a_j)$. Replacing a_i and a_j by $(a_i \circ a_j)$, we can repeat the process until the desired Huffman-expression has been reached.

It is not clear that axiom A0 is necessary for the validity of this result; however, Huffman's construction leads to trees of comparatively little interest if axiom A0 is violated, so there seems to be little harm in assuming A0. It can be shown that axioms A0–A4 do *not* imply the law

$$\text{if } x \leq y \text{ then } ((x \circ a) \circ b) \circ y \leq ((y \circ a) \circ b) \circ x,$$

although this seems but a mild extension of A4. Thus, if we are faced with an expression like $((a \circ b) \circ (c \circ d)) \circ e$ where e is the smallest element, we cannot simply exchange e with d, say, in an attempt to move e to the deepest level; the argument in the previous paragraph used A0 to conclude that $c \circ d \geq e$, so that e could be exchanged with $(c \circ d)$ via A4.

The fact that Huffman's algorithm produces the minimum expression on $\{a_1, \ldots, a_n\}$ does not obviously imply Huffman's original theorem that the minimum value of $\sum a_i l_i$ is obtained, when the a_i are nonnegative real numbers and the operation $a \circ b$ is simply $a + b$. For whenever $a \circ b$ is associative, all expressions on $\{a_1, \ldots, a_n\}$ are equal. Previous papers about abstractions of Huffman's method have therefore worked with two separate operations, one for the values that control the construction of the expression and the other for the evaluation function that is to be minimized. However, it is possible to deduce Huffman's theorem without this extra apparatus, by defining a suitable nonassociative operator that works with pairs of numbers instead of single reals.

Let A be the set of ordered pairs (a, a') of nonnegative real numbers, ordered lexicographically so that $(a, a') \leq (b, b')$ if $a < b$ or if $a = b$ and $a' \leq b'$. The operation

$$(a, a') \circ (b, b') = (a + b, a + b + a' + b') \tag{1}$$

is easily seen to satisfy A0–A4. Therefore, the result of Huffman's construction applied to given pairs $\{(a_1, a_1'), \ldots, (a_n, a_n')\}$ is an expression of minimum value. It is not hard to see that the value of any such expression with respect to this operator is the pair

$$(a_1 + \cdots + a_n, \, a_1 l_1 + \cdots + a_n l_n + a_1' + \cdots + a_n'),$$

where each l_k is the level of (a_k, a_k') as before. Since the first component is independent of the l_k, Huffman's construction does indeed minimize $\sum a_k l_k$.

Another interesting operation on pairs is

$$(a, a') \circ (b, b') = (\max(a, b) + 1, \, a' [a \geq b] + b' [b \geq a]), \tag{2}$$

where '$[a \geq b]$' is 1 or 0 according as $a \geq b$ or $a < b$. This operation also satisfies A0–A4; for example, when verifying A2 we have

$$\begin{aligned}((a, a') \circ (b, b')) \circ ((c, c') \circ (d, d')) = (\max(a, b, c, d) + 2, \\
a' [a \geq b, c, d] + b' [b \geq a, c, d] \\
+ c' [c \geq a, b, d] + d' [d \geq a, b, c]),\end{aligned}$$

which is symmetrical in the four arguments. Huffman's construction produces the expression of minimal value, which in this case is the minimum value of

$$\left(\max_{j=1}^{n}(a_j + l_j), \quad \sum \{ a_k' \mid a_k + l_k = \max_{j=1}^{n}(a_j + l_j) \} \right).$$

It is interesting to search for additional operations that satisfy axioms A0–A4, since each of these corresponds to a minimization algorithm. The quadruple operation

$$\begin{aligned}(a, a', a'', a''') \circ (b, b', b'', b''') \\
= (a + b, \, a' + b', \, a + b + a'' + b'', \, a' + b' + a''' + b''') \tag{3}\end{aligned}$$

illustrates another possibility: The optimum in this case is

$$\left(\sum a_k, \, \sum a_k', \, \sum a_k l_k + \sum a_k'', \, \sum a_k' l_k + \sum a_k''' \right),$$

so we minimize the "weighted path length" $\sum a_k l_k$ and — among all trees for which *this* is minimum — we minimize another weighted path length $\sum a_k' l_k$.

A Huffman algebra of a somewhat different sort can be defined on the set A of all finite multisets of real numbers, ordered by the relation $\{\alpha_1, \ldots, \alpha_r\} < \{\beta_1, \ldots, \beta_s\}$ if $(\alpha_1, \ldots, \alpha_r)$ is lexicographically less than $(\beta_1, \ldots, \beta_s)$, when the elements have been sorted into nonincreasing order $\alpha_1 \geq \cdots \geq \alpha_r$ and $\beta_1 \geq \cdots \geq \beta_s$. The operation

$$a \circ b = (a \uplus b) + 1, \tag{4}$$

that is, addition of unity to all elements of the multiset union of the two multisets (counting multiplicities), satisfies axioms A0–A4; and the value of an expression on $\{a_1, \ldots, a_n\}$ in this case is $\biguplus_{k=1}^{n} \{\alpha + l_k \mid \alpha \in a_k\}$. It is interesting also to consider extensions of this operation and this ordering to infinite multisets.

At first glance, operations (1), (2), (3), (4) may seem very tricky or mysterious or both. Actually there is a fairly simple way to account for all of them: The function

$$a \circ b = x(a + b) \tag{5}$$

satisfies A0–A4 for all $x \geq 1$, over the nonnegative reals. Operation (1) corresponds to the multiplier $x = 1 + \varepsilon$, where the pairs (a, a') correspond to polynomials $a + a'\varepsilon$ in ε, modulo ε^2. Operation (3) is similar but with $x = 1 + \varepsilon^2$. Operation (2) corresponds to large values of x; it records the degree and leading coefficient of a polynomial in x so that $(a, a') \leftrightarrow x^a a' + O(x^{a-1})$. Aczél [1, §6.4] has shown that all *continuous* operations $a \circ b$ on the real numbers that satisfy the medial law and are strictly increasing in both arguments have the form $f^{-1}(xf(a) + yf(b) + z)$ for some constants x, y, z and some continuous, strictly increasing function f, where x and y are positive; we can assume that $z = 0$ if $x + y \neq 1$. It follows that the only continuous and strictly increasing operations that make the real numbers into a Huffman algebra are isomorphic to (5) for some $x \geq 1$; we probably will never find Huffman algebras that have a radically different character from this.

When $a \circ b = a + b$, it is well known that the formula $\sum_{i=1}^{n} a_i l_i$ can be rewritten in the form $\sum_{j=1}^{n} s_j$, where $\{s_1, \ldots, s_{n-1}\}$ is the set of subexpressions of a given expression. For example, in the expression $(((a_1 + a_2) + a_3) + (a_4 + a_5))$, we have $3a_1 + 3a_2 + 2a_3 + 2a_4 + 2a_5 = (a_1 + a_2) + ((a_1 + a_2) + a_3) + (a_4 + a_5) + (((a_1 + a_2) + a_3) + (a_4 + a_5))$. Hu and Tucker [5] proved that the sum of the first k subexpressions formed by Huffman's algorithm is less than or equal to the sum of any k subexpressions built up successively starting with $\{a_1, \ldots, a_n\}$ and replacing a_i and a_j by $a_i + a_j$. Glassey and Karp [2] showed that this result has extensive consequences, for it implies that Huffman's algorithm minimizes not only $\sum_{k=1}^{n-1} s_k$ but also $\sum_{k=1}^{n-1} f(s_k)$ for any nondecreasing concave function f. These facts can be put into our algebraic framework in the following manner.

Let $\square$ be an associative, commutative operator over A, satisfying the following operations:

B1. $(a \circ b) \square (c \circ d) = (a \circ c) \square (b \circ d)$.
B2. If $a \le b$ then $a \square c \le b \square c$.
B3. If $a \le c$ then $(a \circ b) \square c \le a \square (b \circ c)$.

We can now show, by mimicking the previous proof in a straightforward way, that Huffman's procedure has the following strong property: Suppose that the first k steps of Huffman's algorithm have reduced the initial elements $\{a_1, \ldots, a_n\}$ to the elements $\{a'_1, \ldots, a'_{n-k}\}$, and consider any other k-step process that obtains $\{a''_1, \ldots, a''_{n-k}\}$ by repeatedly choosing two elements $\{a_i, a_j\}$ and replacing them by $(a_i \circ a_j)$. Then

$$a'_1 \square \cdots \square a'_{n-k} \le a''_1 \square \cdots \square a''_{n-k}.$$

This generalizes our previous result, which was the special case $k = n-1$. The lemma of Hu and Tucker follows by defining $a \circ b$ as in (1) and taking $(a, a') \square (b, b') = (a + b, a' + b')$.

In a "canonical" Huffman algebra defined by (5) we can satisfy B1–B3 by letting $a \square b = a + b$. Suppose that the kth step of Huffman's algorithm forms the subexpression s_k in this case; and let t_k denote the corresponding combination of elements $a'_1 \square \cdots \square a'_{n-k} = a'_1 + \cdots + a'_{n-k}$. Let $t_0 = a_1 + \cdots + a_n$; then we have $t_k = t_{k-1} - s_k/x + s_k = t_{k-1} + y s_k$, where $y = (x - 1)/x$. It follows as in [2] that Huffman's algorithm minimizes $\sum_{k=1}^{n-1} f(s_k)$ for all concave nondecreasing functions f, in the general case of operation (5) for any $x > 1$. One application of this idea, if we let $x \to \infty$, is to find the binary tree having the smallest $\sum_{k=1}^{n-1} h_k$ where h_k is the height of the kth internal node.

It is easy to see that the subexpressions produced by Huffman's algorithm are nondecreasing: If we number the s_j's in the order they are created, we have $s_1 \le \cdots \le s_{n-1}$. Jan van Leeuwen [13] and others have exploited this fact to show that Huffman's procedure can be carried out in linear time, using two queues, if we assume that the inputs are given in order $a_1 \leq \cdots \leq a_n$: After k steps, the remaining $n - k$ elements will be $\{a_{i+1}, \ldots, a_n\}$ and $\{s_{j+1}, \ldots, s_k\}$, for some $i \leq n$ and $j \leq k$; initially $i = j = k = 0$. Then the smallest remaining element is s_{j+1} if $i = n$, or it is a_{i+1} if $j = k$, otherwise it is $\min(a_{i+1}, s_{j+1})$; if it is a_{i+1}, we increase i by 1, otherwise we increase j by 1. The second-smallest element is then found in the same way. Finally s_{k+1} is computed and k is increased by 1.

Consider the behavior of this procedure in the case of operation (3), when $a_1 \leq \cdots \leq a_n$ and $1 \leq a_i' < 2$ and $a_1'' < \cdots < a_n''$; if $a_i = a_{i+1}$, we require $a_i' \leq a_{i+1}'$. Then $s_j' \geq 2$, so the comparison of $(a_{i+1}, a_{i+1}', a_{i+1}'', a_{i+1}''')$ to $(s_j, s_j', s_j'', s_j''')$ can be based entirely on the first components a_{i+1} and s_j, where we regard a_{i+1} as smaller than s_j in case of equality. The particular values of a_i', a_i'', a_i''' have no effect on the algorithm. It follows that the efficient two-queue procedure can be used on the singleton elements $a_1 \leq \cdots \leq a_n$ instead of the quadruples of (3), with the tie-breaking rule that a_{i+1} should be preferred to s_j in case of equality; we obtain a binary tree (that is, an expression) that minimizes $\sum a_k l_k$. Furthermore, among all binary trees that obtain the minimum of $\sum a_k l_k$, this one also minimizes $\sum a_k' l_k$ for all choices $1 \leq a_k' < 2$. (The special case $a_k' = 1$ for all k was proved by Schwartz in [11].) This same binary tree also attains the lexicographic minimum of

$$\left(\sum a_k l_k, \sum a_k l_k^2, \sum a_k l_k^3, \ldots \right),$$

because the two-queue algorithm will produce the identical tree when operation (5) is used with $x = 1 + \varepsilon$, for all sufficiently small ε, and we have $\sum a_k (1+\varepsilon)^{l_k} = \sum a_k + \varepsilon \sum a_k l_k + \varepsilon^2 \sum a_k \binom{l_k}{2} + \cdots$. In particular, the two-queue method finds the tree with minimum $\sum a_k l_k$ having minimum $\sum a_k l_k^2$, so it has minimum variance; this fact was first pointed out by Tamaki [12].

We can use the algebraic ideas sketched here to obtain an interesting "second-order" extension of Huffman's algorithm in certain cases. Let us say that $(A, <, \circ, \equiv)$ is a *strong* Huffman algebra if $(A, <, \circ)$ is a Huffman algebra satisfying the cancellation law

$$a \circ b = a \circ c \text{ implies } b = c, \tag{6}$$

(or equivalently that the operation $a \circ b$ is *strictly* monotonic), and if $\equiv$ is an equivalence relation on A such that

$$a \equiv c \text{ and } a < b < c \text{ implies } a \equiv b \equiv c, \tag{7}$$

$$(a \circ b) \circ c = a \circ (b \circ c) \text{ if and only if } a \equiv c. \tag{8}$$

For example, operation (1) yields a strong Huffman algebra over pairs of nonnegative reals under the equivalence relation

$$(a, a') \equiv (b, b') \text{ if and only if } a = b; \tag{9}$$

we shall call this the *standard* Huffman algebra, since it seems to be the most interesting example of a strong algebra.

Suppose $(A, <, \circ, \equiv)$ is a strong Huffman algebra and $(B, <', \circ')$ is any other Huffman algebra. We shall define an algorithm on $A \times B$ that finds lexicographically minimum expressions; in other words, it constructs an expression of value (a, b) on given pairs of elements $\{(a_1, b_1), \ldots, (a_n, b_n)\}$ such that a is minimum among all expressions on $\{a_1, \ldots, a_n\}$ and such that b is minimum among all expressions on $\{b_1, \ldots, b_n\}$ in B for which the corresponding expression in A yields a. For example, $(A, <, \circ, \equiv)$ might be the standard Huffman algebra and B might be the integers under the operation $b \circ' b' = \max(b, b') + 1$; then we obtain an algorithm that finds a binary tree with minimum $\sum a_k l_k$, having minimum $\max(b_k + l_k)$ among all such trees.

The second-order algorithm simply proceeds as follows: Given $n > 1$ pairs (a_1, b_1), $\ldots$, (a_n, b_n), let (a_i, b_i) be a pair such that $a_i \equiv \min(a_1, \ldots, a_n)$ and b_i is minimum over all b_h such that $a_h \equiv a_i$. In other words, the idea is to find all a's equivalent to the smallest a, and to choose one among these having the smallest b; we call (a_i, b_i) the smallest pair. Similarly, let (a_j, b_j) be the second-smallest pair, that is, the smallest of the remaining $n - 1$ pairs. Replace (a_i, b_i) and (a_j, b_j) by $(a_i \circ a_j, b_i \circ' b_j)$, and repeat this process on the remaining $n - 1$ pairs.

To show that this algorithm produces the lexicographically minimum expression on $A \times B$, consider an expression E of value (a, b) such that a is the minimum value of any expression on $\{a_1, \ldots, a_n\}$ in A. Let (a_i, b_i) and (a_j, b_j) be the smallest and second-smallest elements found by the second-order algorithm; we shall transform $E = (a, b)$ into another expression $E' = (a, b')$ in which $b' \leq' b$ and in which the elements (a_i, b_i) and (a_j, b_j) are siblings. In fact, the transformation used above to justify Huffman's first-order algorithm works in this case as well; let us consider, for example, the procedure that moves (a_i, b_i) to the deepest level. Suppose the subexpression $(a_i \circ (x \circ y), b_i \circ' (x' \circ' y'))$ is to be replaced by $((a_i \circ x) \circ y, (b_i \circ' x') \circ' y')$ as part of the transformation in question. If $y \not\equiv a_i$, we have $y > a_i$, because the alternative, $\min(a_1, \ldots, a_n) \leq y < a_i \equiv \min(a_1, \ldots, a_n)$, would contradict assumption (7). Hence $y \circ (x \circ a_i) < (y \circ x) \circ a_i$ by (8), and this improvement in the subexpression decreases the value of the entire expression, by (6); but

a is minimum by hypothesis. This contradiction proves that $y \equiv a_i$. Now $b_i <' y'$, because of the way (a_i, b_i) was chosen, hence the transformation does not increase b. By this reasoning we can assume that (a_i, b_i) and (a_j, b_j) are siblings in the lexicographically minimum expression, and the algorithm is valid by induction on n.

Let A be the standard Huffman algebra and let B be the Huffman algebra on multisets using operation (4). It can be shown without difficulty that the two-queue algorithm applied to the numbers $a_1 \leq \cdots \leq a_n$ constructs an expression that obeys the second-order algorithm for $A \times B$, acting on the initial data

$$\big((a_1, 0), \{0\}\big), \ \ldots, \ \big((a_n, 0), \{0\}\big).$$

Thus, the binary tree with minimum $\sum a_k l_k$ produced by the two-queue algorithm not only has minimum $\sum l_k$ and minimum $\sum l_k^2$, it also has the minimum multiset levels $\{l_1, \ldots, l_n\}$. Let the levels of the internal nodes of this binary tree be $\{\lambda_1, \ldots, \lambda_{n-1}\}$; since $z^{l_1} + \cdots + z^{l_n} = 1 + (2z - 1)(z^{\lambda_1} + \cdots + z^{\lambda_{n-1}})$, the tree also has minimum $\{\lambda_1, \ldots, \lambda_{n-1}\}$, a result proved in quite a different way by Kou [9]. Thus, the two-queue algorithm is not only efficient, it produces a binary tree that is simultaneously optimum in many different respects.

I wish to thank D. Stott Parker, Jr., whose paper [10] inspired the present closely related work, and who has contributed several important suggestions; in particular, he led me to [12].

References

[1] J. Aczél, *Functional Equations and Their Applications* (New York: Academic Press, 1966).

[2] C. R. Glassey and R. M. Karp, "On the optimality of Huffman trees," *SIAM Journal on Applied Mathematics* **31** (1976), 368–378.

[3] Martin C. Golumbic, "Combinatorial merging," *IEEE Transactions on Computers* **C-25** (1976), 1164–1167.

[4] T. C. Hu, Daniel Kleitman, and Jeanne K. Tamaki, "Binary trees optimum under various criteria," *SIAM Journal on Applied Mathematics* **37** (1979), 246–256.

[5] T. C. Hu and A. C. Tucker, "Optimal computer search trees and variable length alphabetic codes," *SIAM Journal on Applied Mathematics* **21** (1971), 514–532.

[6] D. A. Huffman, "A method for the construction of minimum redundancy codes," *Proceedings of the IRE* **40** (1952), 1098–1101.

[7] F. K. Hwang, "Generalized Huffman trees," *SIAM Journal on Applied Mathematics* **37** (1979), 124–127.

[8] Alon Itai, "Optimal alphabetic trees," *SIAM Journal on Computing* **5** (1976), 9–18.

[9] Lawrence T. Kou, "Minimal variance Huffman codes," *SIAM Journal on Computing* **11** (1982), 138–148.

[10] D. Stott Parker, Jr., "Conditions for optimality of the Huffman algorithm," *SIAM Journal on Computing* **9** (1980), 470–489; **27** (1998), 317.

[11] Eugene S. Schwartz, "An optimum encoding with minimum longest code and total number of digits," *Information and Control* **7** (1964), 37–44.

[12] Jeanne Keiko Tamaki, *Optimal Binary Trees and Sequences Realized by Eulerian Triangulations* (Ph.D. thesis, Massachusetts Institute of Technology, 1978).

[13] Jan van Leeuwen, "On the construction of Huffman trees," in *Third International Colloquium on Automata, Languages, and Programming* (Edinburgh: Edinburgh University Press, 1976), 382–410.

Chapter 24

Wheels Within Wheels

A simple means of representing the structure of all strongly connected directed graphs is developed and applied to a certain problem involving the construction of "toll booths."

[Originally published in Journal of Combinatorial Theory **B16** (1974), 42–46.]

> Their appearance and their work was, as it were,
> a wheel within a wheel.
>
> —Ezekiel 1 : 16

A *strongly connected digraph* is a nonempty directed graph in which an oriented path exists from any vertex to any other. The following lemma shows that all finite strongly connected digraphs can be constructed in a fairly simple way.

Lemma 1. *Every strongly connected digraph $\mathcal{D}$ (possibly infinite) is either a single vertex with no arcs, or it can be represented as in Figure 1 for some $n \geq 1$. Here $\mathcal{D}_1, \ldots, \mathcal{D}_n$ are strongly connected digraphs; x_k and y_k are (possibly equal) vertices of $\mathcal{D}_k$; and e_k is an arc from y_k to x_{k+1}. The original digraph $\mathcal{D}$ consists of the vertices and arcs of $\mathcal{D}_1, \ldots, \mathcal{D}_n$ plus the arcs $e_1, \ldots, e_n$.*

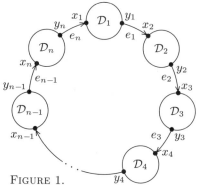

FIGURE 1.

In fact, if σ is any given oriented cycle of $\mathcal{D}$, there exists such a representation in which each of the e_k is contained in σ.

(Each of the $\mathcal{D}_k$ can be decomposed in the same way; thus every strongly connected digraph consists essentially of "wheels within wheels.")

387

Proof. Let e be an arc of $\mathcal{D}$, and let the relation

$$x \longleftrightarrow y \text{ [without } e]$$

mean that oriented paths excluding e exist from x to y and from y to x. This is an equivalence relation that partitions the vertices of $\mathcal{D}$ into components, namely, the so-called strong components of $\mathcal{D} \setminus e$.

Suppose that $e' = (x', y')$ and $e'' = (x'', y'')$ are distinct arcs for which we have $x' \longleftrightarrow x''$, $x' \not\longleftrightarrow y'$, $x'' \not\longleftrightarrow y''$ [without e]. Let $\mathcal{D}^0$, $\mathcal{D}'$, and $\mathcal{D}''$ denote the respective components of x', y', and y''; possibly $\mathcal{D}' = \mathcal{D}''$. The shortest oriented path leading from a vertex of $\mathcal{D}''$ to a vertex of $\mathcal{D}^0$ involves no arcs leading from vertices in $\mathcal{D}^0$, hence $\mathcal{D} \setminus e'$ contains an oriented path from $\mathcal{D}''$ to $\mathcal{D}^0$. Since e'' goes back from $\mathcal{D}^0$ to $\mathcal{D}''$, we have

$$x'' \longleftrightarrow y'' \text{ [without } e'].$$

Furthermore

$$x \longleftrightarrow y \text{ [without } e] \quad \text{implies} \quad x \longleftrightarrow y \text{ [without } e'],$$

for all x and y, since $x \longleftrightarrow y$ [without e] means that x and y belong to the same component, and e' does not lie within any component. Thus $\mathcal{D} \setminus e'$ has fewer strong components than $\mathcal{D} \setminus e$.

Two distinct arcs leading out of the same component, such as e' and e'' in the discussion above, may be called "mates." We shall now prove that there is an arc e in the given cycle σ such that $\mathcal{D} \setminus e$ contains no mates. This will prove the lemma, since the components $\mathcal{D}_1$, $\mathcal{D}_2$, ... must then have the cyclic form shown, with $e = e_1$, say, and with all the other e_k belonging to σ.

If $\mathcal{D}$ is finite we simply let e be an arc in σ such that $\mathcal{D} \setminus e$ has the fewest components. This will imply the nonexistence of mates, since another arc e' of σ that is not included in some component cannot have a mate, lest $\mathcal{D} \setminus e'$ have fewer components.

If $\mathcal{D}$ is infinite the argument is slightly more tricky. We choose e so that the minimum possible number of arcs of σ fail to lie within components of $\mathcal{D} \setminus e$; if two arcs e are equally good by this criterion, we choose one that minimizes the number of arcs of σ that have mates. Now $\mathcal{D} \setminus e$ contains no mates; for if it did, there would be an arc e' of σ that has a mate e''. But that would contradict the choice of e, since e must not lie in a component of $\mathcal{D} \setminus e'$ by the first minimality criterion, and $\mathcal{D} \setminus e'$ contains fewer mated arcs of σ. $\quad \square$

If σ contains an arc e such that $\mathcal{D} \setminus e$ is strongly connected, then $n = 1$ and the lemma holds rather trivially. But if $\mathcal{D}$ contains no such "redundant" arcs, then $n \geq 2$ and the same will be true for the $\mathcal{D}_k$.

The representation is not unique, even if σ is specified to be a simple cycle and $\mathcal{D}$ contains no redundant arcs. For example, Figure 2 shows two representations with $n = 3$ when σ is the outermost cycle.

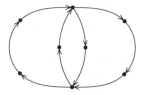

FIGURE 2.

Lemma 1 is sometimes useful when proving properties of strongly connected digraphs by induction, or when finding counterexamples to conjectures. We shall consider one application here, namely the *free route to Las Vegas* problem.

Theorem. *Given a finite, strongly connected network $\mathcal{D}$ of one-way roads between cities, and a designated city called Las Vegas, it is possible to erect toll booths on the roads in such a way that the following three conditions are satisfied:*

i) *There is no toll-free cycle.* (It is impossible to drive indefinitely without paying a toll.)

ii) *Every road is part of a one-toll cycle.* (It is possible to start on any road and return to your starting point, paying only one toll.)

iii) *There is a toll-free route from every city to Las Vegas.*

(Condition (i) calls for comparatively many toll booths, while conditions (ii) and (iii) call for comparatively few. By (iii) and (i), every road leaving Las Vegas must contain a toll booth.)

Proof. We argue by induction on the number of roads (that is, arcs) in $\mathcal{D}$, since the theorem is vacuously true when there are no roads. Using the representation of $\mathcal{D}$ in Lemma 1, we can erect toll booths in each $\mathcal{D}_k$ such that conditions (i) and (ii) hold and such that there are toll-free routes from x_k to y_k, for $1 \leq k \leq n$. Establishing one further toll booth on e_1 makes conditions (i) and (ii) hold in the entire digraph.

Now the proof is completed by applying another lemma.

Lemma 2. *Using the terminology of the theorem, if conditions* (i) *and* (ii) *can be achieved in* $\mathcal{D}$, *it is possible to modify the placement of toll booths so that all three conditions are achieved.*

Proof. The following *move operation* preserves both (i) and (ii), because it neither increases the number of toll booths on any cycle nor decreases that number to zero: "Let C be a set of cities such that each road leading from a city not in C to a city in C contains a toll booth. Destroy all those toll booths, and erect new ones on all roads leading from a city in C to a city not in C, except on those roads that already have toll booths."

The proof will be complete if we can show that a sequence of move operations will produce condition (iii).

Given an arrangement of toll booths in $\mathcal{D}$, satisfying (i) and (ii), let C be the set of all cities from which there exists a toll-free route to Las Vegas (including Las Vegas itself).

If condition (iii) is not satisfied, there will be some city not in C; hence, there will be a road from some city not in C to some city in C, since $\mathcal{D}$ is strongly connected. The move operation removes the toll from that road, and preserves the toll-free property for every city of C. Hence C will grow until eventually every city belongs to it, that is, until condition (iii) holds. □ □

It is plausible to guess that the theorem could be extended, replacing condition (ii) by a similar one:

ii′) *There is a one-toll route from Las Vegas to every other city.*

But the counterexample shown in Figure 3 (obtained by considering Lemma 1) shows that this condition cannot be achieved in general, since there is a unique way to install the toll booths meeting conditions (i) and (iii).

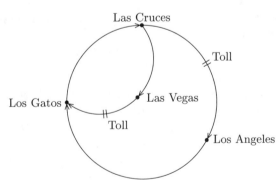

FIGURE 3.

R. Duncan Luce [1] has approached the study of strongly connected digraphs in quite a different fashion. His theorems are related to, but distinct from, the results of the present note, and it is interesting to compare the methods. Instead of building strong digraphs up from smaller ones, he builds them "down" by successively replacing points by cycles. This leads him to a simple inductive proof of a rather curious fact:

A finite strongly connected digraph on two or more points, which fails to be strongly connected when any one of its arcs is removed, always contains at least two points whose in and out degrees are unity.

Acknowledgments

I wish to thank Anatol W. Holt for originally suggesting the toll booth problem to me (without the Las Vegas constraint); this problem originated in his work on Petri nets.

The present proof of Lemma 2 is due to Cláudio Lucchesi, who showed how to simplify my original construction.

I also wish to thank Professor C. St. J. Nash-Williams for pointing out the relevance of reference [1].

This research was supported in part by National Science Foundation grant GJ-992 and in part by Office of Naval Research contract N-00014-67-A-0112-0057 NR 044-402.

References

[1] R. Duncan Luce, "Two decomposition theorems for a class of finite oriented graphs," *American Journal of Mathematics* **74** (1952), 701–722.

Addendum

The concept of "ear decomposition" is also relevant in this connection. See László Lovász, *Combinatorial Problems and Exercises*, 2nd edition (Budapest: Akadémiai Kiadó, 1993), Problem 6.29.

Chapter 25

Complements and Transitive Closures

The complement of the transitive closure of the complement of a transitive relation is transitive. We prove this fact in three ways, analyze the underlying structure, and consider various refinements and applications.

*[Written with R. L. Graham and T. S. Motzkin. Originally published in Discrete Mathematics **2** (1972), 17–29.]*

1. Preliminary Remarks

The purpose of this note is to explore the interaction between two fundamental operations on binary relations. If R is a relation on a set A, the *complement* R^- is defined to be $(A \times A) \setminus R$, and the transitive hull or *transitive closure* R^+ is defined to be the smallest transitive relation containing R. When $(a, b) \in R$ we write aRb. The *composition* $R \circ S$ of two relations R and S is defined to be

$$\{ (a, c) \mid aRb \text{ and } bSc \text{ for some } b \}.$$

It is well known that

$$R^+ = R \cup R \circ R \cup R \circ R \circ R \cup \cdots$$

is the set of all (a, b) such that there exist elements $a_0, a_1, \ldots, a_n$ for some $n \geq 1$ such that

$$a = a_0, \quad a_{i-1} R a_i \text{ for } 1 \leq i \leq n, \quad \text{and } a_n = b.$$

If $R \subseteq S$ it is obvious that $R^- \supseteq S^-$ and $R^+ \subseteq S^+$. In particular we always have

$$R^{-+-} \subseteq R, \tag{1}$$

since $R^- \subseteq R^{-+}$. Another immediate consequence of the definitions is

$$R^+ \setminus R \subseteq R \circ R^+ = R^+ \circ R. \tag{2}$$

By putting these facts together we can derive a less obvious property:

393

Lemma 1. $R^+ \circ R^{+-+-} \subseteq R^{+-+-}$.

Proof. The stated relation is false if and only if there exist a, b, c such that aR^+b, $bR^{+-+-}c$, and $aR^{+-+-}c$. By (1), bR^+c; hence aR^+c, that is, $(a,c) \in R^{+-+} \setminus R^{+-}$. By (2), there exists an element d such that $aR^{+-}d$ and $dR^{+-+}c$. But now if bR^+d, we have aR^+d, contradicting $aR^{+-}d$; and if $bR^{+-}d$, we have $bR^{+-+}c$, contradicting $bR^{+-+-}c$.   □

Theorem 1. *If R is any binary relation, $R^{+-+-+} = R^{+-+-}$. Therefore at most ten relations can be generated from R by taking complements and transitive closures, namely*

$$
\begin{array}{ccccc}
R, & R^+, & R^{+-}, & R^{+-+}, & R^{+-+-}, \\
R^-, & R^{-+}, & R^{-+-}, & R^{-+-+}, & R^{-+-+-}.
\end{array}
\tag{3}
$$

Proof. By the lemma and (1), $R^{+-+-} \circ R^{+-+-} \subseteq R^+ \circ R^{+-+-} \subseteq R^{+-+-}$; that is, R^{+-+-} is transitive. The 10 relations in (3) are now the only possibilities, since $R^{--} = R$ and $R^{++} = R^+$.   □

Theorem 1 is analogous to the well-known "Kuratowski closure and complement problem" [4, 6]; Kuratowski proved in his Ph.D. dissertation that a subset of a topological space generates at most 14 sets under the operations of complementation and (topological) closure.

The following relation on five elements generates all 10 of the distinct possibilities in Theorem 1, hence the result is "best possible":

$$
\begin{array}{ccccc}
 & 00011 & 11000 & 11000 & 00111 \\
 & 01011 & 10000 & 11000 & 00111 \\
R= & R^+= 00111, & R^{+-}=11000, & R^{+-+}=11000, & R^{+-+-}=00111, \\
 & 00111 & 11000 & 11000 & 00111 \\
 & 00111 & 11000 & 11000 & 00111
\end{array}
$$

where

$$
R=\begin{matrix}00011\\01011\\00011,\\00001\\00110\end{matrix} \quad R^+=\begin{matrix}00111\\01111\\00111,\\00111\\00111\end{matrix} \quad R^{+-}=\begin{matrix}11000\\10000\\11000,\\11000\\11000\end{matrix} \quad R^{+-+}=\begin{matrix}11000\\11000\\11000,\\11000\\11000\end{matrix} \quad R^{+-+-}=\begin{matrix}00111\\00111\\00111,\\00111\\00111\end{matrix}
$$

$$\tag{4}$$

$$
R^-=\begin{matrix}11100\\10100\\11100,\\11110\\11001\end{matrix} \quad R^{-+}=\begin{matrix}11100\\11100\\11100,\\11110\\11101\end{matrix} \quad R^{-+-}=\begin{matrix}00011\\00011\\00011,\\00001\\00010\end{matrix} \quad R^{-+-+}=\begin{matrix}00011\\00011\\00011,\\00011\\00011\end{matrix} \quad R^{-+-+-}=\begin{matrix}11100\\11100\\11100.\\11100\\11100\end{matrix}
$$

According to results derived below, this example is the "simplest" relation R that generates all 10 possibilities. The first example of such a relation on five elements was found by Garey [2]. Note that the operation of transposition (often called the converse or *inverse* relation) commutes with complementation and transitive closure; hence at most 20 relations can be generated from a given one under the operations of complementation, closure, and inverse. The example in (4), together with the transposes of each matrix, shows that 20 is best possible.

2. The Underlying Structure

Let us now look at the 10 relations in (3) more closely, so that we can understand what they represent.

If R is not connected, so that $R \subseteq B \times B \cup (A \setminus B) \times (A \setminus B)$ with both B and $A \setminus B$ nonempty, the situation is degenerate. For in this case $R^- \supseteq B \times (A \setminus B) \cup (A \setminus B) \times B$; hence $R^{-+} = A \times A$ and R^{-+-} is empty. Similarly $R^{+-+} = A \times A$, so (3) contains at most 6 different relations. (In fact there are exactly 6 if and only if R is not transitive, when R is not connected.) Therefore the only interesting cases arise when R and R^- are connected.

In general, we can define two important equivalence relations based on a given relation R. Let us write

$$a \longleftrightarrow b \ (R)$$

if $a = b$, or aR^+b and bR^+a. This relation is obviously reflexive, symmetric, and transitive, so it partitions A into equivalence classes; in fact, if we regard R as a directed graph with an arc from a to b if and only if aRb, these classes are precisely the *strong components*.

Another, somewhat coarser, equivalence

$$a \Longleftrightarrow b \ (R)$$

is defined to mean that either $a \longleftrightarrow b \ (R)$ or $a \longleftrightarrow b \ (R^{+-})$. It is not difficult to verify that $\Longleftrightarrow$ is an equivalence relation, because the conditions $a \longleftrightarrow b \ (R)$ and $b \longleftrightarrow c \ (R^{+-})$ imply that $a \longleftrightarrow c \ (R^{+-})$. (Indeed, $aR^{+-+}c$ by Lemma 1, because $bR + a$ and $aR^{+-+-}c$ implies $bR^{+-+-}c$; and if $dR^{+-}b$ we have $dR^{+-}a$, hence $cR^{+-+}b$ implies $cR^{+-+}a$.) Let us call the associated classes the *weak components*.

The conditions $aR^{+-}b$ and $bR^{+-}c$ and aR^+c imply that $bR^{+-}a$ and $cR^{+-}b$. Hence any *minimum-length* chain

$$a = a_0 R^{+-} a_1 R^{+-} \cdots R^{+-} a_n = b$$

of R^{+-} relations between two elements a and b will also be a chain $b = a_n R^{+-} \cdots R^{+-} a_1 R^{+-} a_0 = a$ in the opposite direction, whenever $n \geq 2$. This makes it easy to prove a slightly "stronger" property of the weak components:

Lemma 2. *Let R^M be the symmetric relation*

$$\{ (a,b) \mid aR^{+-}b \text{ and } bR^{+-}a \}.$$

Then $a \Longleftrightarrow b \ (R)$ if and only if either $a \longleftrightarrow b \ (R)$ or $aR^{M+}b$.   □

The importance of the equivalence relations $\longleftrightarrow$ and $\Longleftrightarrow$ is due to the fact that R^+ defines a *partial order* on the strong components, and a *total order* on the weak components. Indeed, the strong components constitute the *finest* partition of A that is partially ordered by R^+, and the weak components constitute the finest partition that is *totally* ordered by R^+. In order to see this, let π be any partition that is totally ordered by R^+, and suppose that a and b are elements of different blocks of π although $a \Longleftrightarrow b$ (R). We may assume that aR^+b and $bR^{+-}a$; hence by Lemma 2 we must have $aR^{M+}b$. But this implies that a and b must belong to the *same* partition of π, contradicting our assumption. In other words, each block of π must be a union of weak components.

The total ordering property allows us to write

$$a < b \ (R)$$

if $a \not\Longleftrightarrow b$ (R) and aR^+b. Every weak component is made up of one or more strong components; we shall call a weak component *simple* if it consists of just one strong component, and we shall call a component *trivial* if it consists of a single element.

These definitions are illustrated in Figure 1, where a relation R on 15 points is shown as a directed graph. The 9 strong components are enclosed in dotted lines, and the 4 weak components are separated by straight vertical lines. Only one of the weak components is simple, and they are all nontrivial; 5 of the strong components are trivial.

We have defined the strong and weak components in such a way that they are unchanged when R is replaced by R^+. Let us now observe what happens when R is replaced by the relation R^{+-}: All arcs within nontrivial strong components disappear, and Lemma 2 tells us that there are paths between any two strong R-components of a single weak R-component. If $a < b$ (R), we have $bR^{+-}a$ by definition; hence all points belonging to *different* weak components are joined in the graph for R^{+-} by an arc from the larger element to the smaller. It follows that elements of different weak R-components belong to different weak R^{+-}-components. Conversely, if a and b belong to the *same* weak R-component, and if this component is simple and nontrivial, then a and b must be unrelated in R^{+-+}; on the other hand if this component is not simple it is easy to see that $a \longleftrightarrow b$ (R^{+-}).

These observations allow us to characterize R^{+-+} completely:

Theorem 2. *For $a \neq b$, $aR^{+-+}b$ if and only if a and b are in the same nonsimple weak R-component or $b < a$ (R). Also, $aR^{+-+}a$ if and only if a is in a nonsimple weak R-component, or a is in a trivial weak R-component and aR^-a.* $\quad\square$

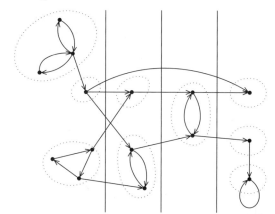

FIGURE 1. Strong and weak components of a relation.

Hence for $a \neq b$, $aR^{+-+-}b$ if and only if a and b both belong to the same simple weak R-component or $a < b$ (R). Also, $aR^{+-+-}a$ if and only if a is in a simple nontrivial weak R-component, or a is in a trivial weak R-component and aRa. This relation is clearly transitive, so we have found the structure underlying Theorem 1.

3. Free Relations

Most relations R seem to generate a complete set of fewer than 10 relations; at least, the authors spent six or seven frustrating hours before finding a single example such as (4), since we had not yet discovered Theorem 2. Let us now determine the structure of "free relations," which generate all 10 distinct possibilities.

So far we have seen connections between R and R^{+-}; there is also an interesting relation between the components of R and R^-:

Lemma 3. *Every nonsimple weak R-component is contained in some simple weak R^--component.*

Proof. Let a and b belong to the same nonsimple weak R-component with $a \neq b$. If $a \nleftrightarrow b$ (R), we have $a \longleftrightarrow b$ (R^{+-}); that is, $aR^{+-+}b$ and $bR^{+-+}a$. But $R^{+-+} \subseteq R^{-+}$; hence $aR^{-+}b$ and $bR^{-+}a$, that is, $a \longleftrightarrow b$ (R^-). On the other hand if $a \longleftrightarrow b$ (R), there is an element c in the same weak R-component but not in the same strong R-component (since the weak component is nonsimple). Again $a \longleftrightarrow b$ (R^-), since $a \longleftrightarrow c$ (R^-) and $b \longleftrightarrow c$ (R^-). Hence a and b belong to the same strong R^--component.

If the weak R^--component containing a and b were not simple, we could show in the same way that its elements all belong to the same strong R-component, since $R^{--} = R$. But that would be absurd. □

Let us say that a weak R-component $\mathcal{W}$ contains an *arc* if there exist elements $a, b \in \mathcal{W}$ such that aRb.

Consider the following four conditions on a relation R:

I) R has a nonsimple weak component containing an arc.

II) R^- has a nonsimple weak component containing an arc.

III) Some simple nontrivial weak R-component intersects some simple nontrivial weak R^--component.

III') The weak R-components are not the same as the weak R^--components.

Theorem 3. *A relation R is free if and only if R satisfies* (I), (II), *and* (III), *or* (I), (II), *and* (III').

Proof. In order for R to be free we must certainly have

i) $R^{+-} \neq R^{+-+}$;

ii) $R^{-+-} \neq R^{-+-+}$;

iii) $R^{+-+-} \neq R^{-+-+}$.

Let us examine these conditions in detail.

Since $aR^{+-}b$ is equivalent to $aR^{+-+}b$ whenever a and b lie in different weak R-components, or if a and b lie in the same strong R-component of a simple weak R-component, condition (i) can hold only if there is a nonsimple weak R-component. Theorem 2 tells us that $aR^{+-+}b$ holds for all a and b within such a component. Thus condition (i) is equivalent to the existence of a nonsimple weak R-component containing elements a and b such that aR^+b, and this is equivalent to (I).

Of course, (ii) is just (i) with R replaced by R^-. Hence, (ii) holds if and only if (II) holds.

Suppose (iii) holds. Since $R^{-+-+} \subseteq R^{+-+-+} = R^{+-+-}$, we must have

$$aR^{+-+-}b \quad \text{and} \quad aR^{-+-+-}b, \quad \text{for some } a \text{ and } b. \tag{5}$$

If $a \neq b$, then Theorem 2 tells us that a and b are in the same simple weak R-component or $a < b$ (R), and a and b are in the same simple weak R^--component or $a < b$ (R^-). If $a = b$, Theorem 2 says that a is in a simple nontrivial weak R-component or a is in a trivial weak R-component and aRa; also a is in a simple nontrivial weak R^--component or a is in a trivial weak R^--component and aR^-a.

We cannot have both $a < b$ (R) and $a < b$ (R^-), because $bR^{+-}a$ implies bR^-a. Therefore property (5) holds if and only if at least one of the following is true:

1) There exists a simple weak R-component not contained in a weak R^--component.

2) There exists a simple weak R^--component not contained in a weak R-component.

3) Some simple nontrivial weak R-component intersects some simple nontrivial weak R^--component.

4) There exists an element a in a trivial weak R-component and a simple nontrivial weak R^--component, with aRa.

5) There exists an element a in a trivial weak R^--component and a simple nontrivial weak R-component, with aR^-a.

Note that by Lemma 3, we may delete the word "simple" in the first two of the conditions. These first two conditions are then exactly equivalent to (III'). Moreover, if they fail to hold, then so do the final two conditions listed. We are left with the third condition, which is, of course, identical to (III).

In summary, we have now shown that (i) $\iff$ (I), (ii) $\iff$ (II), and (iii) $\iff$ (III) or (III'), and the necessity of these conditions has been established.

Assume conversely that (I), (II), and either (III) or (III') hold. To show that R is free, we must prove that all 10 expressions R, R^+, R^{+-}, R^{+-+}, R^{+-+-}, R^-, R^{-+}, R^{-+-}, R^{-+-+}, R^{-+-+-} are distinct. Table 1 indicates the various reasons behind the 45 necessary inequalities.

TABLE 1. Summary of 45 cases.

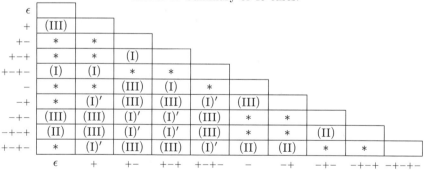

ϵ									
$+$	(III)								
$+-$	*	*							
$+-+$	*	*	(I)						
$+-+-$	(I)	(I)	*	*					
$-$	*	*	(III)	(I)	*				
$-+$	*	(I)$'$	(III)	(III)	(I)$'$	(III)			
$-+-$	(III)	(III)	(I)$'$	(I)$'$	(III)	*	*		
$-+-+$	(II)	(III)	(I)$'$	(I)$'$	(III)	*	*	(II)	
$-+-+-$	*	(I)$'$	(III)	(III)	(I)$'$	(II)	(II)	*	*
	ϵ	$+$	$+-$	$+-+$	$+-+-$	$-$	$-+$	$-+-$	$-+-+$

An entry of (I), (II), or (III) indicates that condition (I), (II), or (III) (or (III')) is used to establish the corresponding inequality. For example, the $(\epsilon, + - + -)$ entry is (I), where ϵ denotes R itself. If $R = R^{+-+-}$,

then
$$R^{+-} = (R^{+-+-})^{+-} = (R^{+-+-+})^{-}$$
$$= (R^{+-+-})^{-} \quad \text{by Theorem 1}$$
$$= R^{+-+},$$

which contradicts (I).

An entry of '$*$' denotes that the corresponding inequality follows from the fact that A is nonempty, so that no relation equals its complement. For example, the $(+, +-+)$ entry is $*$. If $R^{+} = R^{+-+}$ then

$$R^{+-+-} = (R^{+-+})^{-+-} = (R^{+-+-+})^{-} = (R^{+-+-})^{-} \quad \text{by Theorem 1}$$
$$= R^{+-+},$$

which contradicts the nonemptiness of A.

The entry (I)$'$ indicates that the necessary argument uses more than Theorem 1. For example, consider the $(+-+, -+-+)$ entry. By (I), R has a nonsimple weak component. Let a and b belong to this component with $a \neq b$. By Lemma 3, a and b are in a simple R^{-}-component. By Theorem 2, $aR^{+-+}b$ and $aR^{-+-+}b$; that is, $R^{+-+} \neq R^{-+-+}$.

The $(-+, +)$ entry is also (I)$'$, since $R^{-+} = R^{+}$ implies $R^{-+-+} = R^{+-+}$, and the latter is impossible as we have just seen. The reader should have little difficulty in verifying the remaining entries, thus completing the proof of the theorem.   □

With this result, we may now justify the claim made earlier for (4).

Theorem 4. *A relation on fewer than five elements always generates fewer than ten relations under complementation and transitive closure.*

Proof. Suppose R is free and A has ≤ 4 elements. By Theorem 3, (I) and (II) imply that R and R^{-} must each have a nonsimple weak component. By Lemma 3, these components must be disjoint. Hence A must have 4 elements. It is easily seen, though, that in this case (III) and (III$'$) must both fail, contradicting the freeness of R.   □

It can be shown, in fact, that all free relations on 5 elements must have at least 10 ordered pairs. Thus (4) is minimal in a strong sense.

4. Extensions and Applications

Suppose $R \subseteq T$, where T is a total order relation. This relation T may be reflexive, irreflexive, or partly reflexive; the "diagonal" elements are immaterial in the following discussion. We can consider complements with respect to T instead of $A \times A$; thus, let $R^\Delta = T \setminus R$. Then the analog of Lemma 1 does not always hold:

$$R = R^+ = \begin{matrix} 0011 \\ 0000 \\ 0001 \\ 0000 \end{matrix}, \quad T = \begin{matrix} 0111 \\ 0011 \\ 0001 \\ 0000 \end{matrix}, \quad R^{+\Delta+\Delta} = \begin{matrix} 0000 \\ 0000 \\ 0001 \\ 0000 \end{matrix}, \quad R^+ \circ R^{+\Delta+\Delta} = \begin{matrix} 0001 \\ 0000 \\ 0000 \\ 0000 \end{matrix}.$$

On the other hand, the analog of Theorem 1 is true:

Theorem 5. *In terms of the notation above, $R^{+\Delta+\Delta+} = R^{+\Delta+\Delta}$.*

Proof. Assume that $R^{+\Delta+\Delta}$ is not transitive. There must be elements a, b, c such that $aR^{+\Delta+\Delta}b$, $bR^{+\Delta+\Delta}c$, and $aR^{+\Delta+}c$ (since aTc). Hence for some $n \geq 1$ we have elements a_0, a_1, $\ldots$, a_n such that $a = a_0$, $a_0 R^{+\Delta}a_1$, $a_1 R^{+\Delta}a_2$, $\ldots$, $a_{n-1}R^{+\Delta}a_n$, $a_n = c$.

If $b = a_j$ for some j, we would have $aR^{+\Delta+}b$, a contradiction; hence the fact that T is a total order implies that there is some j such that $a_{j-1}Tb$ and bTa_j. Now $a_{j-1}R^+b$ (since $a_{j-1}R^{+\Delta}b$ would imply that $aR^{+\Delta+}b$) and similarly bR^+a_j; hence $a_{j-1}R^+a_j$, a contradiction. ☐

The proof of this theorem makes essential use of the hypothesis that T is a total order. If T were merely assumed to be a partial order containing R, we could not prove Theorem 5, because of the following simple counterexample:

$$R = \begin{matrix} 0010 \\ 0000 \\ 0001 \\ 0000 \end{matrix}, \quad T = \begin{matrix} 0111 \\ 0001 \\ 0001 \\ 0000 \end{matrix}, \quad R^{+\Delta+\Delta} = R \neq R^+.$$

Another common operation of interest is the *reflexive closure* $R^I = R \cup I$ where I is the equality relation. It is not difficult to prove that $R^{I-I} = R^{-I}$ and $R^{+-I-+} = R^{-I-+}$. A somewhat less evident identity, which the reader will find instructive to prove, is

$$R^{+-+I-+-} = R^{-I-+-+}.$$

By using such identities it is possible to establish the analog of Theorem 1 for the three operations $^+$, $^-$, and ^{I}. We state the result without proof.

Theorem 1′. *At most 42 relations can be generated from a relation R by taking complements, transitive closures, and reflexive closures, namely the relations in Figure 2.* ☐

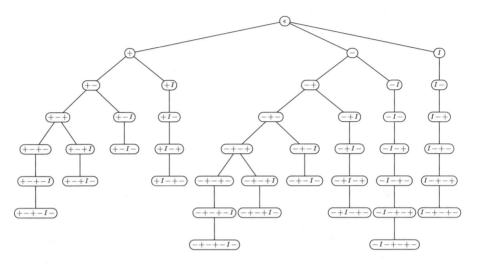

FIGURE 2. Independent relations using $^+$, $^-$, and ^{I}.

The following relation R generates 42 distinct relations under $^+$, $^-$, and ^{I}:

$$R = \begin{array}{l} 000011101 \\ 001011101 \\ 000011101 \\ 000111101 \\ 000111101 \\ 000101101 \\ 000111101 \\ 111111111 \\ 000000000 \end{array}$$

M. R. Garey [2] has recently considered the operation of "transitive reduction," the smallest relation whose transitive closure is the same as R^+. He has shown that any finite relation leads to at most 34 different relations under repeated application of complementation, transitive reduction, and transitive closure, and that this bound actually can be attained.

It is possible to consider other operations on relations and ask similar questions; for example, one might study the difunctional closure $R^D = (R \circ R^T)^+ \circ R$, where R^T is the converse of R (see [5] and [7]). But this will not be done here.

The original application that led to the theorems above was the following: Let R be a transitive relation; find the largest transitive relation contained in R whose complement with respect to $A \times A$ is transitive. By Theorem 1, the answer is simply R^{-+-}. Or, let R be an irreflexive partial ordering contained in the irreflexive total ordering T; find the largest partial ordering contained in R whose complement with respect to T is a partial ordering. By Theorem 4, the answer is $R^{\Delta+\Delta}$.

The latter result applies also to permutations: If $p_1 p_2 \ldots p_n$ is a permutation of $\{1, 2, \ldots, n\}$, an *inversion* is a pair of indices (i, j) such that $i < j$ and $p_i > p_j$. Write iVj if (i, j) is an inversion; then V is transitive, and so is its complement V^Δ with respect to $T = \{(i, j) \mid i < j\}$. Conversely it is not difficult to show [1, pages 114–117] that there is a unique permutation $p_1 p_2 \ldots p_n$ whose inversions correspond in this way to a relation $V \subseteq T$, whenever V and V^Δ are transitive. If R is a transitive subset of T, the relation $V = R^{\Delta+\Delta}$ is the largest subset of R that corresponds to a permutation. The corresponding permutation therefore has the maximum number of inversions, among all permutations whose inversions are contained in R.

If we call a relation *closed* when it is transitive, and *open* when its complement is transitive, then the closure R^+ is the smallest closed relation containing R and the "interior" R^{-+-} is the largest open relation contained in R. In these terms, Theorem 1 asserts that the interior of the closure is closed; dually, the closure of the interior is open.

A result somewhat similar to Theorem 5 has been proved by Guilbaud and Rosenstiehl [3], who discovered that $(R \cup S)^{+\Delta}$ is transitive whenever $R^{+\Delta}$ and $S^{+\Delta}$ are both transitive. The same result holds for $^-$ in place of $^\Delta$. We have been unable to find any other work closely related to the theorems above, in spite of the fact that the operation of transitive closure has been known and applied for so many years. For example, E. Schröder failed to discover any of the theorems of this paper in his "exhaustive" study of identities involving binary relations [8]; he would have dearly loved to know that, in his notation, $a_{00}; (a_{00})_{11} \neq (a_{00})_{11}$ and $((a_{00})_{11})_{00} = (a_{00})_{11}$!

This research was supported in part by the National Science Foundation under grant GJ-992 and in part by the Office of Naval Research under contract N-00014-67-A-0112-0057 NR 044-402.

References

[1] C. Berge, *Principes de Combinatoire* (Paris: Dunod, 1968). English translation, *Principles of Combinatorics* (New York: Academic Press, 1971).

[2] M. R. Garey, personal communication.

[3] G. Guilbaud and P. Rosenstiehl, "Analyse algébrique d'un scrutin," *Mathématiques et Sciences Humaines* **4** (1960), 9–33.

[4] C. Kuratowski, "Sur l'opération $\overline{A}$ de l'Analysis Situs," *Fundamenta Mathematicæ* **3** (1922), 182–199.

[5] Joachim Lambek, "Goursat's theorem and the Zassenhaus lemma," *Canadian Journal of Mathematics* **10** (1958), 45–56.

[6] Eric Langford, "Characterization of Kuratowski 14-sets," *American Mathematical Monthly* **74** (1971), 362–367.

[7] Jacques Riguet, "Quelques propriétés des relations difonctionelles," *Comptes Rendus hebdomadaires des séances de l'Académie des Sciences* **230** (Paris: 1950), 1999–2000.

[8] Ernst Schröder, *Algebra und Logik der Relative*, Volume 3 of *Vorlesungen über die Algebra der Logik* (Leipzig: Teubner, 1895), 346–404.

Addendum

Several further studies of the kind considered here have been made since this paper was published. For example, Peter C. Fishburn ["Operations on binary relations," *Discrete Mathematics* **21** (1978), 7–22] proved that at most 110 relations can be generated using complementation, transitive closure, transposition, asymmetrization (taking $R \setminus R^T$), and symmetrization (taking $R \cap R^T$), not counting the empty and universal relations. David Peleg ["A generalized closure and complement phenomenon," *Discrete Mathematics* **50** (1984), 285–293] showed that similar bounds hold in a large number of similar problems.

The most significant aspect of this work, however, may well be the concept of weak components, which apparently had not been considered before. Efficient algorithms to compute the weak components of a given digraph have been found by Jean François Pacault ["Computing the weak components of a directed graph," *SIAM Journal on Computing* **3** (1974), 56–61] and Robert Endre Tarjan ["A new algorithm for finding weak components," *Information Processing Letters* **3** (1974), 13–15].

Chapter 26

Random Matroids

A simple combinatorial construction capable of producing an arbitrary matroid is introduced, and some of its properties are investigated. The structure of a matroid is defined one rank at a time, and when random choices are made the result might be called a random matroid. Some experimental statistics about such matroids are tabulated. If we specify the subsets of rank $\leq k$, the construction defines a rank function having the richest possible matroid structure on the remaining subsets, in the sense that no new relationships are introduced except those implied by the given subsets of rank $\leq k$. An appendix to this paper presents several computer programs for dealing with matroids over small sets.

*[Originally published in Discrete Mathematics **12** (1975), 341–358.]*

0. Introduction

Mathematical systems called matroids were introduced and named by Whitney in 1935 (see [8]), and the associated theory became extensively developed during the ensuing decades, most notably by Tutte in the late 1950s [6, 7]. The subsequent discovery by Edmonds that most of the known efficient solutions to combinatorial problems can be associated with a matroid structure (see [2, 4]) has led to considerable interest in matroids during recent years.

Matroids are abstract systems, but of course when we deal with them we usually have a more or less concrete model in mind. Much of the theory has been developed from a geometric or algebraic point of view, using the fact that a special type of matroid arises in the study of vector spaces spanned by the rows of a matrix. Other aspects of the theory have been derived using intuition from graph theory, since certain matroids arise naturally in the study of graphs.

The purpose of this paper is to introduce another approach to the study of matroids, a viewpoint that is essentially combinatorial and constructive. The author believes that such an approach may shed new light

405

on the theory, and that many interesting research problems are suggested by this work; but it must be confessed that the present paper contains more definitions than theorems.

The approach we shall discuss rests on a simple technique that constructs all matroids when given a (typically small) number of virtually unconstrained "enlargements" whose consequences fully define the structure. If those "enlargements" are selected at random, we don't obtain truly random matroids, since different matroids will in general be obtained with differing probabilities; but the probability distribution that arises does appear to have interesting properties.

Section 1 of this paper defines matroids and establishes the notational conventions to be used. The main construction appears in Section 2, and an example is given in Section 3. Sections 4, 5, and 6 prove that the algorithm of Section 2 is correct, complete, and well-defined. Section 7 looks at the construction from a more general point of view, and observes that it can be used to define the "free completion" of a matroid above rank k in a meaningful way. Some experimental results are reported in Section 8, and some open problems suggested by this research are listed in Section 9. The Appendix presents detailed computer programs, which implement the construction with reasonable efficiency.

1. Definitions and Notation

Matroids may be defined in many equivalent ways (by their independent sets, their circuits, their bases, their bonds, their rank function, or their closed sets, among others), and for our purposes the definition via closed sets is most convenient. We shall therefore say that a *matroid* $\mathcal{M} = (E, \mathcal{F})$ is a (finite) set E together with a family $\mathcal{F}$ of subsets of E, satisfying the following three axioms:

- i) $E \in \mathcal{F}$;
- ii) if $A \in \mathcal{F}$ and $B \in \mathcal{F}$, then $A \cap B \in \mathcal{F}$;
- iii) if $A \in \mathcal{F}$ and $\{a, b\} \subseteq E \setminus A$, then b is a member of all sets of $\mathcal{F}$ containing $A \cup a$ if and only if a is a member of all sets of $\mathcal{F}$ containing $A \cup b$.

(Here '$A \cup a$' is shorthand for '$A \cup \{a\}$'.) The elements of $\mathcal{F}$ are called the *closed* subsets of E.

If A is any subset of E, we define the "closure" of A,

$$\overline{A} = \bigcap \{ B \in \mathcal{F} \mid B \supseteq A \},$$

to be the intersection of all closed sets containing A. Axiom (i) guarantees that this intersection is nonempty, and Axiom (ii) implies that

$\overline{A}$ is itself closed; consequently $\overline{A}$ is the (unique) smallest closed set containing A, and we have

$$\overline{\overline{A}} = \overline{A}.$$

Axiom (iii) may now be rewritten more simply:

    iii) if $A \in \mathcal{F}$ and $\{a, b\} \subseteq E \setminus A$, then $(b \in \overline{A \cup a} \iff a \in \overline{A \cup b})$.

2. A General Construction

Our goal is to understand the implications of the matroid axioms, and one way to approach them is to try to construct all such families $\mathcal{F}$ over a given set E. We can never hope to look at them all unless E is a rather small set, since there are $2^{2^{n-O(\log n)}}$ possibilities when E has n elements [3]. But it may be helpful to consider algorithms that are at least capable in principle of constructing all matroids.

For this purpose, let us try to construct $\mathcal{F}$ by starting with small closed sets and then defining the larger ones. If A is any closed set, the gist of Axioms (i), (ii), (iii) is that the smallest closed sets properly containing A must partition the elements of $E \setminus A$. In other words, there must exist disjoint sets $\{B_1, \ldots, B_k\}$ such that $B_1 \cup \cdots \cup B_k = E \setminus A$ and such that we have $a \in B_j$ if and only if $\overline{A \cup a} = A \cup B_j$, for $1 \le j \le k$. For if we define a relation on the elements of $E \setminus A$ by saying

$$a \sim b \iff b \in \overline{A \cup a},$$

then Axioms (i) and (ii) imply that

$$a \sim b \iff \overline{A \cup b} \subseteq \overline{A \cup a},$$

and Axiom (iii) tells us that $a \sim b$ if and only if $b \sim a$. Hence $\sim$ is an equivalence relation. The problem is to find a family $\mathcal{F}$ of closed sets that yields such partitions for all $A \in \mathcal{F}$.

The following algorithm, which attempts to find the "finest" partitions consistent with these conditions, now suggests itself.

Step 1. [Initialize.] Set r to 0, and let $\mathcal{F}_0$ be $\{\emptyset\}$, the family of sets consisting of the empty set alone.

Step 2. [Generate covers.] Let $\mathcal{F}_{r+1}$ be the set of all "covers" of the sets in $\mathcal{F}_r$, that is,

$$\mathcal{F}_{r+1} = \{ A \cup a \mid A \in \mathcal{F}_r \quad \text{and} \quad a \in E \setminus A \}.$$

Step 3. [Enlarge.] Add additional sets to $\mathcal{F}_{r+1}$, if desired, where each new set properly contains some element of $\mathcal{F}_r$. (This step is indeterminate. By making different choices we will in general produce different matroids.)

Step 4. [Superpose.] If $\mathcal{F}_{r+1}$ contains any distinct sets A and B whose intersection $A \cap B$ is not contained in C for any $C \in \mathcal{F}_r$, replace A and B in $\mathcal{F}_{r+1}$ by the single set $A \cup B$. Repeat this operation until $A \cap B \subseteq C$ for some $C \in \mathcal{F}_r$ whenever A and B are distinct members of $\mathcal{F}_{r+1}$. (We shall prove later that the replacements can be made in any order without affecting the final result of this step.)

Step 5. [Test for completion.] If $E \in \mathcal{F}_{r+1}$, terminate the construction. Otherwise increase r by 1 and return to Step 2. $\square$

This construction terminates because every member A of $\mathcal{F}_{r+1}$ properly contains some member of $\mathcal{F}_r$, hence A contains at least $r + 1$ elements. We shall prove that the family

$$\mathcal{F} = \mathcal{F}_0 \cup \mathcal{F}_1 \cup \cdots \cup \mathcal{F}_r \cup \mathcal{F}_{r+1}$$

obtained at the conclusion of the construction defines the closed sets of a matroid, no matter what choices are made in Step 3.

3. A "Random" Example

Before going into any details of the proof, let us look at a concrete example in order to fix the ideas. For convenience in notation we shall use the decimal digits $\{0, 1, \ldots, 9\}$ as elements of the set E. Subsets of E will be written without braces or commas, so that '156' stands for the 3-element subset consisting of 1, 5, and 6; and '$\{156, 23\}$' stands for a family of two subsets of E.

The construction of Section 2 begins with $\mathcal{F}_0 = \{\emptyset\}$, then Step 2 tells us that $\mathcal{F}_1$ is the family $\{0, 1, 2, 3, 4, 5, 6, 7, 8, 9\}$ of all singleton subsets. Let us assume that the first execution of Step 3 leaves $\mathcal{F}_1$ unchanged; consequently Step 4 will also leave $\mathcal{F}_1$ unchanged. (It turns out that any changes made to $\mathcal{F}_1$ at this point are equivalent to "shrinking" E into a smaller set. By leaving $\mathcal{F}_1$ unchanged we will be constructing a so-called "combinatorial geometry" [1], namely a matroid in which all one-point sets are closed.) Step 5 sets r to 1 and we return to Step 2.

Now Step 2 causes $\mathcal{F}_2$ to be $\{01, 02, \ldots, 89\}$, the family of all pairs. Let us add further sets to $\mathcal{F}_2$ in a "random" way, using the digits of

$$\pi \approx 3.14159\ 26535\ 89793\ 23846$$

to govern our choices. From '3, 1, 4' we shall add the set 134, from '1, 5, 9' we shall add 159, and similarly we shall add 256 and 358; since '9, 7, 9' involves only two digits, let us include the following digit 3, and then we shall also include 238. Thus, we have added six triples 134, 159, 256, 358, 379, 238 to $\mathcal{F}_2$; and we may as well stop there, since six is a perfect number.

Step 4 is interesting now, since it causes many of the sets in $\mathcal{F}_2$ to be merged together. Since $\mathcal{F}_1$ contains all the one-element sets, any distinct sets A and B in $\mathcal{F}_2$ that have two or more common elements are replaced by $A \cup B$. In particular, sets like 13, which are contained in at least one of the added triples, simply disappear since $13 \cap 134 = 13$ and $13 \cup 134 = 134$. Furthermore, we replace 358 and 238 by their union 2358, which in turn combines with 256 to give 23568. We are ultimately left with 30 subsets, namely

$$\mathcal{F}_2 = \{01, 02, 03, 04, 05, 06, 07, 08, 09, 12, 134,$$
$$159, 16, 17, 18, 23568, 24, 27, 29, 379,$$
$$45, 46, 47, 48, 49, 57, 67, 69, 78, 89\}.$$

Then we set r to 2 and return to Step 2.

Let us leave $\mathcal{F}_3$ untouched when we next reach Step 3; perhaps we don't know any more decimal places of π, or we simply want to see what happens. It turns out that a great deal does happen in Step 4. For example, $235689 \cap 2379 = 239$ is not contained in any member of $\mathcal{F}_2$, so we replace 235689 and 2379 by 2356789. The following 22 subsets are eventually obtained:

$$\mathcal{F}_3 = \{012, 0134, 0159, 016, 017, 018, 023568, 024,$$
$$027, 029, 0379, 045, 046, 047, 048, 049,$$
$$057, 067, 069, 078, 089, 123456789\}.$$

The largest of these subsets, 123456789, has only one proper cover, namely E. So we are bound to have

$$\mathcal{F}_4 = \{0123456789\}$$

regardless of what transpires in the next Step 3, and the construction will terminate with $r = 3$. It is not hard to check that $\mathcal{F} = \mathcal{F}_0 \cup \mathcal{F}_1 \cup \mathcal{F}_2 \cup \mathcal{F}_3 \cup \mathcal{F}_4$ defines the closed sets of a matroid in this particular example.

Incidentally, a set $A \subseteq E$ is called *independent* if $a \notin \overline{A \setminus a}$ for all $a \in A$. It turns out that the *bases*, or maximal independent sets, of the matroid we have constructed are

$$\{0123, 0124, 0125, 0126, 0127, 0128, 0129, 0135, 0136,$$
$$0137, 0138, 0139, 0145, 0146, 0147, 0148, 0149, 0156,$$
$$0157, 0158, 0167, 0168, 0169, 0178, 0179, 0189, 0234,$$
$$0237, 0239, 0245, 0246, 0247, 0248, 0249, 0257, 0259,$$
$$0267, 0269, 0278, 0279, 0289, 0345, 0346, 0347, 0348,$$
$$0349, 0357, 0359, 0367, 0369, 0378, 0389, 0456, 0457,$$
$$0458, 0459, 0467, 0468, 0469, 0478, 0479, 0489, 0567,$$
$$0569, 0578, 0579, 0589, 0678, 0679, 0689, 0789\};$$

and the *circuits*, or minimal dependent sets, are

$$\{134, 159, 235, 236, 238, 256, 258, 268, 356, 358, 368, 379, 568,$$
$$1237, 1239, 1245, 1246, 1247, 1248, 1249, 1257, 1267,$$
$$1269, 1278, 1279, 1289, 1357, 1367, 1369, 1378, 1389,$$
$$1456, 1457, 1458, 1467, 1468, 1469, 1478, 1479, 1489,$$
$$1567, 1578, 1678, 1679, 1689, 1789, 2347, 2349, 2457,$$
$$2459, 2467, 2469, 2478, 2479, 2489, 2579, 2679, 2789,$$
$$3457, 3459, 3467, 3469, 3478, 3489, 4567, 4569, 4578,$$
$$4579, 4589, 4678, 4679, 4689, 4789, 5679, 5789, 6789\}.$$

Notice that our construction needed only six 3-element sets to specify the entire matroid, so this approach has led to economy in specification.

4. Proof of Correctness

Let us now prove that the family $\mathcal{F} = \mathcal{F}_0 \cup \mathcal{F}_1 \cup \cdots \cup \mathcal{F}_r \cup \mathcal{F}_{r+1}$ defined by our construction yields a matroid. We shall prove that $(E, \mathcal{F})$ is a matroid whenever $\mathcal{F} = \mathcal{F}_0 \cup \cdots \cup \mathcal{F}_{r+1}$ is a family of sets with the following properties:

a) $E \in \mathcal{F}_{r+1}$, and $E \notin \mathcal{F}_j$ for $0 \le j \le r$;
b) $\mathcal{F}_0$ consists of a single set;
c) if $A, B \in \mathcal{F}_j$, $A \ne B$, and $j > 0$, then $A \cap B \subseteq C$ for some $C \in \mathcal{F}_{j-1}$;
d) if $A \in \mathcal{F}_j$, $a \in E \setminus A$, and $j \le r$, then $A \cup a \subseteq C$ for some $C \in \mathcal{F}_{j+1}$;
e) if $A, B \in \mathcal{F}_j$ and $A \subseteq B$, then $A = B$.

Properties (a), (b), (c) are immediate from the construction; and so is property (d), since Step 3 and Step 4 do not remove any sets from a family unless a larger set is substituted. Furthermore, Step 2 and Step 3 both ensure that each element of $\mathcal{F}_j$ for $j > 0$ properly contains some element of $\mathcal{F}_{j-1}$. Therefore property (e) follows by induction on j, since $A \subseteq B$ implies that either $A = B$ or $A \subseteq C$ for some $C \in \mathcal{F}_{j-1}$ according to (c), but A properly contains some $D \in \mathcal{F}_{j-1}$.

Axiom (i) holds trivially because of (a).

Given $A \subseteq E$, let q be minimal such that $A \subseteq B$ for some $B \in \mathcal{F}_q$. This B is unique; for if $A \subseteq B$ and $A \subseteq B'$, then $q > 0$ by (b), and $A \subseteq B \cap B' \subseteq C$ for some $C \in \mathcal{F}_{q-1}$ by (c). We shall say that q is the *rank* of A, and we shall define $\overline{A} = B$.

Notice that $\overline{A} = A$ implies that $A \in \mathcal{F}$. Conversely if $A \in \mathcal{F}$ we have $\overline{A} = A$; for if $A \in \mathcal{F}_j$ and the rank of A is $q < j$, then properties (a) and (d) imply that A is properly contained in some $B \in \mathcal{F}_j$, contradicting (e). Hence, in particular,

$$\overline{\overline{A}} = \overline{A}.$$

Let $a \notin A$ and assume that $\overline{A \cup a} \neq \overline{A}$, where A has rank q. Then $A \cup a \not\subseteq C$ for any $C \in \mathcal{F}_j$, $0 \leq j \leq q$, by the uniqueness of $B = \overline{A}$. But we do have $\overline{A} \cup a \subseteq C$ for some $C \in \mathcal{F}_{q+1}$, by property (d); consequently

$$\overline{A \cup a} = \overline{\overline{A} \cup a} \supseteq \overline{A}.$$

We have proved that $\overline{A} \subseteq \overline{A \cup a}$ for all A and a, and by the finiteness of our universe it follows that

$$A \subseteq B \text{ implies } \overline{A} \subseteq \overline{B}$$

for all $A, B \subseteq E$.

If $A, B \in \mathcal{F}$, we now have

$$\overline{A \cap B} \subseteq \overline{A} = A, \qquad \overline{A \cap B} \subseteq \overline{B} = B,$$

hence

$$\overline{A \cap B} \subseteq A \cap B \subseteq \overline{A \cap B};$$

in other words, $A \cap B = \overline{A \cap B}$, and Axiom (ii) is established.

Let $A^\cap$ be the intersection of all closed sets containing a given set A. Clearly $A^\cap \subseteq \overline{A}$, since $\overline{A}$ is such a closed set, and we have proved that $A^\cap$ is closed, hence $A^\cap = \overline{A}$; our definition of $\overline{A}$ in this section agrees with the definition in Section 1.

Now Axiom (iii) follows immediately: We have proved that if $A \in \mathcal{F}_q$ and if the elements of $\mathcal{F}_{q+1}$ that contain A are $B_1, \ldots, B_t$, then the sets $B_1 \setminus A, \ldots, B_t \setminus A$ partition $E \setminus A$.

5. Proof of Completeness

We can also show that *every* matroid is essentially obtainable by the construction in Section 2. Let $(E, \mathcal{F}')$ be a (finite) matroid, and for $A \in \mathcal{F}'$ let rank(A) be the minimal r such that A is the closure of some r-element subset of E. It is well known that $\text{rank}(A \cup a) = \text{rank}(A) + 1$ whenever $a \notin \overline{A}$. Let $\mathcal{F}'_r$ be all the closed sets of rank r, for $r = 0, 1, \ldots$.

If the empty set $\emptyset$ is closed, we can prove that the algorithm in Section 2 is capable of constructing the matroid $(E, \mathcal{F}')$, with $\mathcal{F}_r = \mathcal{F}'_r$ for all r. In fact, this is true for $r = 0$, and it holds for $r + 1$ if we add the elements of $\mathcal{F}'_{r+1}$ to $\mathcal{F}_{r+1}$ in Step 3. The reason is that each cover generated in Step 2 is contained in some unique element of $\mathcal{F}'_{r+1}$, hence Step 4 simply removes everything but $\mathcal{F}'_{r+1}$. (Of course, it is generally possible to obtain the same result with far fewer sets added in Step 3; a study of the minimum number of necessary enlargements should prove to be interesting.)

If the empty set is not closed, then $\overline{\emptyset}$ is contained as "excess baggage" in every closed set, and $(E, \mathcal{F})$ is isomorphic to the matroid $(E \setminus \overline{\emptyset}, \{A \setminus \overline{\emptyset} \mid A \in \mathcal{F}\})$. Our construction would be capable of producing such degenerate matroids if, for example, we would change Step 1 to "Set r to -1, set $\mathcal{F}_{-1}$ empty, set $\mathcal{F}_0$ to $\{\emptyset\}$, and go to Step 3"; but no new cases of interest would be produced.

6. Commutativity

Step 4 of the algorithm in Section 2 is the keystone of our construction, and we should prove that it does not depend on the order in which reductions are made. In general, let $\mathcal{P}$ be any *order ideal* on the subsets of E (that is, if $B \in \mathcal{P}$ and $A \subseteq B$, then $A \in \mathcal{P}$), and let $\mathcal{A}$ be any family of subsets of E; we shall consider the following operation, which generalizes Step 4:

> "If $\mathcal{A}$ contains any distinct sets A and B whose intersection $A \cap B$ is not contained in $\mathcal{P}$, replace A and B in $\mathcal{A}$ by the single set $A \cup B$. Repeat this operation until $A \cap B \in \mathcal{P}$ whenever A and B are distinct members of $\mathcal{A}$."

Let $\mathcal{A} = (A_1, A_2, \ldots, A_m)$ be a sequence of subsets of E, and consider the following operation $\langle i, j \rangle$:

> "If $A_i \neq A_j$ and $A_i \cap A_j \notin \mathcal{P}$, replace both A_i and A_j by $A_i \cup A_j$."

This operation makes two copies of the merged set $A_i \cup A_j$, so that each member of $\mathcal{A}$ retains its original position in the sequence; otherwise it is equivalent to the general operation described above. Suppose we apply such operations repeatedly, obtaining a sequence of sequences $\mathcal{A} = \mathcal{A}^0$, $\mathcal{A}^1$, ..., $\mathcal{A}^k$, where $\mathcal{A}^k = (A_1^k, A_2^k, \ldots, A_m^k)$ is *fixed* in the sense that $A_i^k \neq A_j^k$ implies $A_i^k \cap A_j^k \in \mathcal{P}$ for all i and j. Notice that $A_j^k \supseteq A_j$ for all j. If we apply $\langle i, j \rangle$ operations in another order to the same initial sequence, obtaining $\mathcal{A} = \mathcal{B}^0$, $\mathcal{B}^1$, $\mathcal{B}^2$, ..., it is easy to prove by induction on t that $\mathcal{B}^t = (B_1^t, B_2^t, \ldots, B_m^t)$, where $B_j^t \subseteq A_j^k$ for all j. For if $B_i^{t-1} \neq B_j^{t-1}$ and $B_i^{t-1} \cap B_j^{t-1} \notin \mathcal{P}$, we have $A_i^k \cap A_j^k \supseteq B_i^{t-1} \cap B_j^{t-1}$; hence $A_i^k \cap A_j^k \notin \mathcal{P}$, and $A_i^k = A_j^k \supseteq B_i^{t-1} \cup B_j^{t-1} = B_i^t = B_j^t$.

If $\mathcal{B}^t$ and $\mathcal{A}^k$ are both fixed, we have $B_j^t \subseteq A_j^k \subseteq B_j^t$ by symmetry. The final result is therefore independent of the order in which $\langle i, j \rangle$ operations are applied.

7. Free Completion

It is well known that any matroid can be "truncated to rank k," in the sense that we eliminate all closed sets of rank $\geq k$ except E itself. This truncation operation is equivalent to adding E to $\mathcal{F}_{r+1}$ in Step 3, when $r = k - 1$ in our construction.

Conversely, our construction allows us to add the richest possible structure above rank k to a given matroid, in the sense that we can find the greatest number of closed sets for rank $k + 1$ in any matroid having prescribed closed sets for ranks $\leq k$. If we make no additions in Step 3, let us say that the family $\mathcal{F}_{r+1}$ obtained at the end of Step 4 is the *free completion* of $\mathcal{F}_r$. If $(E, \mathcal{F})$ is a given matroid of rank $> k$, its *free completion above rank k* is the matroid over E whose closed sets are the closed sets $\mathcal{F}_0 \cup \cdots \cup \mathcal{F}_k$ of $\mathcal{F}$ having rank $\leq k$, together with $\mathcal{F}_{k+1}' \cup \mathcal{F}_{k+2}' \cup \cdots \cup \mathcal{F}_{r+1}'$, where $\mathcal{F}_k' = \mathcal{F}_k$ and $\mathcal{F}_{q+1}'$ is the free completion of $\mathcal{F}_q'$ for $k \leq q \leq r$, and $\mathcal{F}_{r+1}' = \{E\}$. In a sense every matroid whose closed sets for ranks $\leq k$ are in $\mathcal{F}_0 \cup \cdots \cup \mathcal{F}_k$ is a "homomorphic image" of this free completion, where the homomorphism corresponds to enlargements made in Step 3.

It should be interesting to explore properties of free completion. Notice that the construction of $\mathcal{F}_{r+1}$ depends only on $\mathcal{F}_r$, so that the matroid is being built up layer by layer. The same construction can be applied in general to any "clutter," that is, to the set of maximal elements of any order ideal, in place of $\mathcal{F}_r$ in Step 2; the procedure then defines the free completion of a clutter, whether or not the clutter can be represented as the sets of rank $\leq k$ in some matroid. However, in every case tried by the author where the order ideal is not that of a matroid, the free completion reduced trivially to $\{E\}$; perhaps such collapsing will always occur in non-matroid situations.

8. Some Experiments

In an attempt to study the behavior of the algorithm when random "coarsening" is applied to the structure in Step 3, several experiments were attempted with small sets E.

The experiments were conducted as follows. Step 4 was performed immediately after Step 2, in order to shorten the list of subsets and to be sure that all consequences of the present structure were taken into account. Then a member A of $\mathcal{F}_{r+1}$ was selected at random, each being equally likely; and when A had been chosen, an element a of $E \setminus A$ was selected at random, each being equally likely. The set A was replaced by $A \cup a$ in $\mathcal{F}_{r+1}$, and Step 4 was performed again. This enlargement process was repeated a specified number of times, e_r, depending on the current rank r; e_0 was always 0, so that the first effects would appear in $\mathcal{F}_2$.

For example, our experiment based on π in Section 3 corresponds roughly to $e_1 = 6$, $e_2 = e_3 = 0$, on a 10-element set E. Thirty random experiments were conducted with these parameters, and in each case the resulting matroid had rank 4. Table 1 shows the number of elements in $\mathcal{F}_2$ and $\mathcal{F}_3$ after reduction, together with the number of bases and circuits in the first ten resulting matroids. (The last of these has, by chance, the same statistics as the "random" matroid in Section 3.) The computation time for these ten experiments, using ALGOL W on an IBM 360/67, was 15.6 seconds.

Table 2 shows the average values obtained for several settings of the parameters. In nearly every case the final rank was reduced by one each time an enlargement was made.

It should be interesting to develop theoretical results that account for this observed behavior.

9. Open Problems

A few research problems have been stated above, and they will be repeated here for emphasis.

1. If an order ideal in the lattice of subsets of E does not correspond to the sets of rank $\leq r$ of any matroid, is the free completion of its maximal elements always trivial, or do we obtain a generalization of matroid behavior?

2. What can be said about the smallest number of enlargements needed to completely specify a given matroid of rank r on n elements? (Computer experiments indicate that $n-r$ suitably chosen enlargements

TABLE 1. Ten random experiments with $(e_1, e_2, e_3) = (6, 0, 0)$.

$|\mathcal{F}_2|$	$|\mathcal{F}_3|$	Bases	Circuits
23	15	48	51
32	24	76	89
23	15	48	51
32	24	76	89
23	15	48	51
31	23	74	82
27	19	62	61
23	10	63	36
23	15	48	51
30	22	71	76

TABLE 2. Observed mean values.

$|E|$	$(e_1, e_2, \dots)$	Trials	Bases	Circuits	$|\mathcal{F}_2|$	$|\mathcal{F}_3|$	$|\mathcal{F}_4|$	$|\mathcal{F}_5|$	$|\mathcal{F}_6|$	$|\mathcal{F}_7|$
10	$(6, 0, 0)$	30	62.1	60.8	25.9	17.8	1.0			
10	$(5, 1, 0)$	10	89.8	86.1	33.0	28.5	1.0			
10	$(5, 2, 0)$	8*	109.9	159.8	32.8	1.0				
10	$(5, 2, 0)$	2†	140.5	91.5	34.5	39.0	1.0			
10	$(6, 1, 0)$	20	102.8	141.5	28.9	1.0				
10	$(4, 2, 0)$	10	114.8	105.9	36.4	37.8	1.0			
10	$(3, 3, 0)$	8†	114.6	112.3	38.8	41.9	1.0			
10	$(3, 3, 0)$	2‡	94.5	55.5	38.5	64.5	36.0	1.0		
10	$(0, 6, 0)$	5†	157.8	159.0	45.0	74.0	1.0			
10	$(0, 6, 0)$	5‡	128.0	92.8	45.0	100.2	68.8	1.0		
10	$(0, 1, 1, 1)$	10	38.3	10.6	43.0	101.8	136.4	96.7	29.1	1.0
13	$(6, 0, 0, 0, 0, 0)$	3	141.7	44.0	63.7	149.7	179.7	107.0	26.3	1.0
13	$(6, 2, 0, 0)$	9	432.8	327.2	64.3	137.3	100.2	1.0		

* Averages for experiments when the final rank was 3.
† Averages for experiments when the final rank was 4.
‡ Averages for experiments when the final rank was 5.

will work in nearly all the small cases, but the construction in [3] shows that considerably more enlargements are needed in general.)

3. Can the stochastic properties of this construction be analyzed carefully enough to narrow the known bounds on the asymptotic number g_n of matroids on n elements? (It is known [3] that $\log_2 \log_2 g_n$ lies between $n - \frac{3}{2} \log_2 n$ and $n - \log_2 n$ plus terms of lower order.)

Appendix. Computer Programs

The computer programs used in this study are presented here for the possible benefit of others who wish to experiment with matroids, and also for the possible interest of language designers, because there is still a relative scarcity of published algorithms dealing with manipulation of sets. The programming has been done in ALGOL W [9], a language chosen by the author primarily because of the excellent debugging facilities available [5]. The programs below can easily be transliterated into other languages if desired.

The running time of the algorithm in the text is governed largely by the speed of Step 4, which would be extremely slow if programmed in a brute force manner based on the definitions. The implementation below reduces this cost substantially by using a routine that maintains a list of subsets satisfying the condition at the end of Step 4 at all times, so that the basic operation is one of inserting into such a list. The inner loop of the insertion process is kept short by using a table that tells whether or not any given subset has rank $\leq r$.

The time and space requirements of these algorithms for manipulating random matroids grow exponentially with $n = |E|$, as one might expect. The program below assumes that $n \leq 13$, but with suitable modifications one could adapt it so that cases as large as $n = 20$ become feasible on contemporary medium-to-large scale computers.

Sets are represented in the program by the so-called **bits** variables of ALGOL W, since **bits** variables are subject to Boolean operations. The program occasionally wants to treat such variables also as binary numbers, so that they can be used as subscripts or in arithmetic operations. If v is of type **bits**, ALGOL W uses the notation *number*(v) for the corresponding integer; if u is of type **integer**, the notation *bitstring*(u) stands for the corresponding bits. Neither *number* nor *bitstring* requires any computation time on a binary computer.

The program deals with linked lists of sets, kept in two arrays S and L; $S[k]$ is the set stored at position k, and $L[k]$ is the position number of the next set in the list. The lists are linked circularly, in most cases; if h is the "head" of a list, then $S[h]$ is irrelevant, the first item of the list is in position $L[h]$, and the last item is in position k, where $L[k] = h$. An empty circular list therefore has $L[h] = h$.

The program is designed to do more than the construction in the text; it prints out the independent sets for each rank as well as the circuits of the matroid. For this purpose it is convenient to have a table that indicates the cardinality of each subset. Hence the *rank* array serves double duty: If v is the bitstring representation of a set A, the table entry $rank[number(v)]$ will be set to $100 + |A|$ at the beginning of the computation and until the true rank of A is computed; then again $100 + |A|$ will be used at the end of the program when the circuits are being tabulated.

With these introductory remarks, it is hoped that the comments on the program below will be sufficiently explanatory. Notice that "long labels" are occasionally used as comments, to help indicate the program structure. The text of procedures has been deferred until after the main program, as a further attempt to make the program readable in one pass.

```
begin comment Exploration of "random" matroids;
    integer n; comment number of elements in universe, must be ≤ 13;
    integer mask; comment 2ⁿ − 1, represents the set E;
    integer i, j, k; comment temporary indices;
    integer r; comment the current rank;
    integer h; comment head of the circular list of closed sets for rank r;
    integer nh; comment head of the circular list of closed sets being
            formed for the closed sets for rank r + 1;
    integer avail; comment beginning of the list of available space;
    bits x; comment a set used to communicate with the insert routine;
    bits array S[0 :: 4999]; integer array L[0 :: 4999];
            comment list memory;
    integer array rank[0 :: 8191]; comment 100 + cardinality,
            or assigned rank of subset;
    procedure ... (see the procedure declarations below);

    read(n); comment the first data entry is the number of elements;
    mask := round(2 ↑ n − 1);
    set initial contents of rank table:
        k := 1;  rank[0] := 100;
        while k ≤ mask do
        begin for i := 0 until k − 1 do rank[k + i] := rank[i] + 1;
            k := k + k;
        end;
    initialize list memory to available:
        for i := 0 until 4998 do L[i] := i + 1;
        L[4999] := −1;  avail := 2;
        L[1] := 0;  S[1] := bitstring(0);
        h := 0; comment a list containing the empty set;
    rank[0] := 0;  r := 0;
```

```
    while rank[mask] > r do
    begin comment pass from rank r to r + 1;
        create empty list:
            nh := avail;  avail := L[nh];  L[nh] := nh;
        generate; comment see procedure below;
        enlarge; comment see procedure below;
        return list h to available storage:
            j := h;  while L[j] ≠ h do j := L[j];
            L[j] := avail;  avail := h;
        r := r + 1;  h := nh;
        printlist; comment see procedure below;
        assign rank to sets and print those that are independent:
            write("Independent sets for rank", r, ":");
            j := L[h];
            while j ≠ h do
            begin mark(number(S[j])); comment see procedure below;
                j := L[j];
            end;
    end;
    printcircuits; comment see procedure below;
end.
```

The procedures mentioned in this program are implemented as follows.

```
procedure generate;
begin comment insert the minimal closed sets for rank r + 1 into
            a circular list headed by nh (see Step 2 in the text);
    bits t, v, y; integer j, k; comment temporary storage;
    j := L[h]; comment prepare to go through the h list;
    while j ≠ h do
    begin y := S[j]; comment closed set of rank r;
        t := bitstring(mask − number(y)); comment set complement;
        find all sets in list nh that already contain y and
                remove excess elements from t:
        k := L[nh];
        while k ≠ nh do
        begin if (S[k] ∧ y) = y then t := t ∧ ¬S[k];
            k := L[k];
        end;
        insert y ∪ a for each a ∈ t:
        while t ≠ bitstring(0) do
        begin x := y ∨ (t ∧ ¬bitstring(number(t) − 1));
```

insert; **comment** insert x into nh, possibly
  enlarging x, see below;
   $t := t \wedge \neg x$;
  end;
  $j := L[j]$;
end;
end;

procedure *insert*;
begin comment insert set x into list nh, but augment x
  if necessary (and delete existing entries of the list)
  so that no two entries have an intersection of rank $> r$;
  integer j, k;
  $j := nh$;
store: $S[nh] := x$;
loop: $k := j$;
continue: $j := L[k]$;
  if $rank[number(S[j] \wedge x)] \leq r$ **then goto** loop;
  if $j \neq nh$ **then**
    begin if $x = (x \vee S[j])$ **then**
      begin remove from list and continue:
        $L[k] := L[j]$; $L[j] := avail$; $avail := j$;
        goto continue;
      end else
      begin augment x and go around again:
        $x := x \vee S[j]$; $nh := j$; **goto** store;
      end;
    end;
  insert new item:
    $j := avail$; $avail := L[j]$;
    $L[j] := L[nh]$; $L[nh] := j$; $S[j] := x$;
end;

procedure *enlarge*;
begin comment insert sets read from data cards until
  encountering an empty set;
  $readon(x)$;
  while $x \neq bitstring(0)$ **do**
  begin if $rank[number(x)] > r$ **then** *insert*;
    $readon(x)$;
  end;
end;

procedure *printlist*;
begin integer j;
  write("Closed sets for rank", r, ":");
  $j := L[h]$;
  while $j \neq h$ **do**
  begin *writeon*($S[j]$);
    $j := L[j]$;
  end;
end;

procedure *mark*(**integer value** m);
begin comment given a binary-coded subset m, this procedure
        sets *rank*$[m'] := r$ for all subsets m' of m whose rank
        is not already $\leq r$, and outputs m' if it is independent
        (that is, if its rank equals its cardinality);
  integer t, v;
  if *rank*$[m] > r$ **then**
    begin if *rank*$[m] = 100 + r$ **then** *writeon*(*bitstring*(m));
    rank$[m] := r$;
    $t := m$;
    while $t \neq 0$ **do**
    begin $v := number(bitstring(t) \wedge bitstring(t-1))$;
      $mark(m - t + v)$;
      $t := v$;
    end;
    end;
end;

procedure *printcircuits*;
begin comment this procedure prints all minimal dependent sets and
        assigns rank ≥ 100 to all dependent sets;
  write("The circuits are:");
  $k := 1$;
  while $k \leq mask$ **do**
  begin for $i := 0$ **until** $k - 1$ **do**
    if *rank*$[k + i] = rank[i]$ **then**
      begin *writeon*(*bitstring*($k + i$));
        $unmark(k + i, rank[i] + 101)$;
      end;
    $k := k + k$;
  end;
end;

```
procedure unmark (integer value m, card );
   comment the parameter card is 100 plus the cardinality of m;
begin integer t, v;
   if rank [m] < 100 then
      begin rank [m] := card ;
        t := mask − m;
        while t ≠ 0 do
        begin v := number (bitstring (t) ∧ bitstring (t − 1));
           unmark (m + t − v, card + 1);
           t := v;
        end;
      end;
end;
```

Further efficiency was gained in practice by sorting the closed sets so that they appear on list h in order of decreasing cardinality when the *generate* procedure is called. Thus, the statement '*sort*;' was inserted just before '*printlist*'. A simple radix list sort was used as follows:

```
procedure sort;
begin integer array hd [100 :: 113], tl [100 :: 113];
   for i := 100 until 100 + n do hd [i] := −1;
   j := L[h];  L[h] := h;
   while j ≠ h do
   begin i := rank [number (S[j])];
      k := L[j];  L[j] := hd [i];
      if L[j] < 0 then tl [i] := j;
      hd [i] := j;  j := k;
   end;
   for i := 100 until 100 + n do
      if hd [i] ≥ 0 then begin L[tl [i]] := L[h];  L[h] := hd [i] end;
end;
```

The effect of sorting was to reduce the number of tests on *rank* in the main *insert* loop from about 7500 to about 1700, when $n = 10$ and $e_1 = 6$, $e_2 = e_3 = 0$. For larger n the gain in efficiency was even more significant, since the lists are never very long when $n = 10$ and $e_1 = 6$.

Acknowledgments

I wish to thank Stein Krogdahl and Daniel Kleitman for their helpful comments on the first draft of this paper.

This research was supported in part by National Science Foundation grant GJ-36473X and in part by Office of Naval Research contract NR 044-402.

References

[1] Henry H. Crapo and Gian-Carlo Rota, *On the Foundations of Combinatorial Theory: Combinatorial Geometries*, preliminary edition (Cambridge, Massachusetts: MIT Press, 1970).

[2] J. Edmonds, "Submodular functions, matroids, and certain polyhedra," *Combinatorial Structures and their Applications* (New York: Gordon and Breach, 1970), 69–87.

[3] Donald E. Knuth, "The asymptotic number of geometries," *Journal of Combinatorial Theory* **A16** (1974), 398–400. [Reprinted as Chapter 27 of the present volume.]

[4] Eugene L. Lawler, *Combinatorial Optimization: Networks and Matroids* (New York: Holt, Rinehart and Winston, 1976).

[5] E. Satterthwaite, "Debugging tools for high level languages," *Software — Practice & Experience* **2** (1972), 197–217.

[6] W. T. Tutte, "A homotopy theorem for matroids. I, II," *Transactions of the American Mathematical Society* **88** (1958), 144–174.

[7] W. T. Tutte, "Matroids and graphs," *Transactions of the American Mathematical Society* **90** (1959), 527–552.

[8] Hassler Whitney, "On the abstract properties of linear dependence," *American Journal of Mathematics* **57** (1935), 509–533.

[9] Niklaus Wirth and C. A. R. Hoare, "A contribution to the development of ALGOL," *Communications of the ACM* **9** (1966), 413–434. (The program above makes use of simplified input/output procedures found in the implemented version of this language. See Henry R. Bauer, Sheldon Becker, Susan L. Graham, and Edwin Satterthwaite, "ALGOL W language description," in *ALGOL W (Revised)*, report STAN-CS-69-110 (Stanford, California: Computer Science Department, Stanford University, September 1969), 1–65.)

Addendum

It is amusing in retrospect to see hints of what I now call "literate programming" in the verbose-label convention used in the ALGOL W program of the appendix.

In most cases, every enlargement in Step 3 of the algorithm tends to hasten the end of the process. But Thomas Brylawski has constructed an interesting example where an additional enlargement actually increases the rank of the final matroid: Let $E = \{0, 1, 2, 3, 4, 5, 6, 7\}$. If we enlarge $\mathcal{F}_3$ by adding the five subsets $\{0145, 0246, 0347, 1256, 2367\}$, then $\mathcal{F}_4$

turns out to equal $\{E\}$. On the other hand if we also include a sixth subset, 1357, in $\mathcal{F}_3$, we get

$$\mathcal{F}_4 = \{012456, 013457, 023467, 123567,$$
$$0123, 0127, 0136, 0167, 0235, 0257, 0356, 0567,$$
$$1234, 1247, 1346, 1467, 2345, 2457, 3456, 4567\}$$

and $\mathcal{F}_5 = \{E\}$. [See Thomas Brylawski, "Constructions," in *Theory of Matroids*, edited by Neil White (Cambridge: Cambridge University Press, 1986), 127–223, especially pages 171–173.]

I learned shortly after writing this paper that the problem of extending the rank-r sets of a matroid to higher ranks was first studied by Henry H. Crapo, "Erecting geometries," in *Proceedings of the Second Chapel Hill Conference on Combinatorial Mathematics and Its Applications* (Chapel Hill, North Carolina: 1970), 74–99. His somewhat embarrassing term *free erection* for what I called "free completion" is still standard terminology among matroid theorists. Leigh Roberts, "All erections of a combinatorial geometry and their automorphism groups," *Lecture Notes in Mathematics* **452** (1975), 210–213, gave a procedure to construct all possible matroids, somewhat more complicated than the method presented here. See also Hien Quang Nguyen, "Constructing the free erection of a geometry," *Journal of Combinatorial Theory* **B27** (1979), 216–224.

Robert Bixby resolved Open Problem 1 by finding an elegant proof that the free completion of every non-matroidal clutter is always just $\{E\}$. See R. E. Bixby, "The solution to a matroid problem of Knuth," *Discrete Mathematics* **21** (1978), 87–88.

The other two problems remain tantalizingly open, as far as I know.

Chapter 27

The Asymptotic Number of Geometries

A simple proof is given that $\lim_{n\to\infty}(\log_2\log_2 g_n)/n = 1$, where g_n denotes the number of distinct combinatorial geometries on n points.

[Originally published in Journal of Combinatorial Theory A16 (1974), 398–400.]

Let g_n be the number of combinatorial geometries on n points (that is, matroids with all 1-point sets independent). The following values of g_n for small n have been determined by Blackburn, Crapo, and Higgs [1]:

$$
\begin{array}{rcccccccc}
n & = & 1 & 2 & 3 & 4 & 5 & 6 & 7 & 8 \\
g_n & = & 1 & 1 & 2 & 4 & 9 & 26 & 101 & 950.
\end{array}
$$

Crapo and Rota [2, page 3.3] noted that the law

$$g_{n+1} \approx g_n^{3/2} \tag{1}$$

"seems approximately correct, on the basis of this data alone." If this approximation were valid for all n, we would have

$$\log_2 g_{n+1} \approx (3/2)\log_2 g_n,$$

so that $\log_2 \log_2 g_n \approx n\log_2(3/2)$. On the other hand, every geometry is defined by specifying a certain set of subsets of the n points (for example, the closed sets, the bonds, the bases, or the circuits), hence obviously

$$g_n \le 2^{2^n}. \tag{2}$$

In other words, $\log_2 \log_2 g_n \le n$.

   M. J. Piff and D. J. A. Welsh [4] have shown that g_n eventually grows more rapidly than the first few values would indicate. In fact, they have proved that

$$\log_2 \log_2 g_n \ge n - \frac{5}{2}\log_2 n + O(\log\log n), \tag{3}$$

so g_{n+1} will be roughly g_n^2 when n is large. On the other hand, Piff [3] has recently improved the upper bound (2) to

$$\log_2 \log_2 g_n \leq n - \log_2 n + O(\log \log n). \tag{4}$$

The purpose of this note is to narrow the gap a little further, by showing that

$$\log_2 \log_2 g_n \geq n - \frac{3}{2} \log_2 n + O(\log \log n), \tag{5}$$

The precise result to be proved is that

$$g_n \geq \frac{2^{\binom{n}{\lfloor n/2 \rfloor}/(2n)}}{n!}, \tag{6}$$

from which (5) follows by Stirling's approximation. Indeed, the lower bound in (6) is less than or equal to the number of geometries of a very special kind, namely the so-called "partitions of type $\lfloor n/2 \rfloor - 1$." This confirms a remark of Crapo and Rota [2, page 3.17], who conjectured that partition geometries would probably predominate in any asymptotic enumeration.

The factor $n!$ in (6) accounts for any isomorphisms between the geometries we shall construct, so we shall ignore isomorphisms in what follows.

Let M be a family of subsets of $\{1, 2, \ldots, n\}$, where each subset contains exactly $\lfloor n/2 \rfloor$ elements, and where no two different subsets have more than $\lfloor n/2 \rfloor - 2$ elements in common. This set M, together with the set of all $(\lfloor n/2 \rfloor - 1)$-element subsets that are not contained in any member of M, constitutes a set of blocks such that every subset of size $\lfloor n/2 \rfloor - 1$ is contained in a unique block; therefore it defines a partition geometry.

If M contains m members, each of the 2^m subfamilies of M will define a partition geometry in the same way. Therefore (6) will follow if we can find such a family M of subsets, containing at least $m \geq \binom{n}{\lfloor n/2 \rfloor}/(2n)$ members. This is essentially the approach used in [4], although the authors of [4] did not construct such a large family M.

The problem is solved by realizing that it is the same as finding m binary code words of length n, each containing exactly $\lfloor n/2 \rfloor$ 1 bits, and single-error correcting. This characterization suggests the following "Hamming code" construction: Let $k = \lfloor \log_2 n \rfloor + 1$, and construct the $n \times k$ matrix H of 0s and 1s whose rows are the numbers from 1 to n

expressed in binary notation. For $0 \leq j < 2^k$, consider the set M_j of all row vectors x of 0s and 1s such that x contains exactly $\lfloor n/2 \rfloor$ 1 bits and the vector xH mod 2 is the binary representation of j. Note that if x and y are distinct elements of M_j, they cannot differ in just two places; otherwise we would have $(x + y)H$ mod 2 = $(0 \ldots 0)$, contradicting the fact that no two rows of H are equal. Therefore M_j defines a family of subsets of $\{1, 2, \ldots, n\}$ having the desired property. Furthermore, at least one of these 2^k families M_j will contain at least

$$\binom{n}{\lfloor n/2 \rfloor} / 2^k$$

elements, since the families are disjoint and they exhaust all possible $\lfloor n/2 \rfloor$-element subsets. This completes the proof, since $2^k \leq 2n$. □

This research was supported in part by National Science Foundation grant GJ-992, in part by Office of Naval Research contract N-00014-67-A-0112-0057 NR 044-402, and in part by Norges Almenvitenskapelige Forskningsråd.

References

[1] John E. Blackburn, Henry H. Crapo, and Denis A. Higgs, "A catalogue of combinatorial geometries," *Mathematics of Computation* **27** (1973), 155–166.

[2] Henry H. Crapo and Gian-Carlo Rota, *On the Foundations of Combinatorial Theory: Combinatorial Geometries*, preliminary edition (Cambridge, Massachusetts: MIT Press, 1970).

[3] M. J. Piff, "An upper bound for the number of matroids," *Journal of Combinatorial Theory* **B14** (1973), 241–245.

[4] M. J. Piff and D. J. A. Welsh, "The number of combinatorial geometries," *Bulletin of the London Mathematical Society* **3** (1971), 55–56.

Chapter 28

Permutations with Nonnegative Partial Sums

*[Originally published in Discrete Mathematics **5** (1973), 367–371.]*

If x_1, x_2, ..., x_n are real numbers whose sum is zero, it is well known that we can find some permutation $p(1)p(2)\ldots p(n)$ such that each of the partial sums

$$x_{p(1)} + \cdots + x_{p(j)} \tag{1}$$

is nonnegative, for $1 \leq j \leq n$. In fact, there are at least $(n-1)!$ such permutations; for if $p(1)p(2)\ldots p(n)$ is any permutation, it is not hard to see that some cyclic shift $p(k+1)\ldots p(n)p(1)\ldots p(k)$ will have the stated property.

Daniel Kleitman [2] has conjectured that the number of such permutations is always at most $2n!/(n+2)$, if the x's are nonzero. The object of this note is to prove the following sharpened form of his conjecture:

Theorem. *Let x_1, x_2, ..., x_n be real numbers with the property that $x_1 + x_2 + \cdots + x_n = 0$; and assume that s elements are greater than zero, t elements are less than zero, and $n - s - t$ elements are equal to zero. Let $P(x_1, x_2, \ldots, x_n)$ denote the number of permutations $p(1)p(2)\ldots p(n)$ such that each of the partial sums (1) is nonnegative. Then*

$$P(x_1, x_2, \ldots, x_n) \leq n!/(\max(s,t) + 1). \tag{2}$$

Furthermore, this bound is best possible, in the sense that equality is achieved for some x's whenever s, t, and n are fixed values with $s + t \leq n$.

Proof. We may obviously assume that s and t are positive. Let ϵ be the smallest positive value such that the sum over a subset of the x_i is equal to ϵ. For example, if $n = 4$ and $(x_1, x_2, x_3, x_4) = (\pi, 2 - \pi, 1, -3)$, then the $2^4 - 2$ sums over proper subsets of the x's are

$$\pi,\ 2-\pi,\ 2,\ 1,\ 1+\pi,\ 3-\pi,\ 3,\ -3,\ -3+\pi,\ -1-\pi,\ -1,\ -2,\ -2+\pi,\ -\pi.$$

429

The smallest positive value among these is $\epsilon = -3 + \pi$. By symmetry, the largest negative value will always be $-\epsilon$, since $x_1 + x_2 + \cdots + x_n = 0$.

Now let $x_0 = -\epsilon$, and consider the $n+1$ values $-\epsilon, x_1, x_2, \ldots, x_n$ whose sum is $-\epsilon$. A permutation $q(0)q(1)\ldots q(n)$ of the indices $\{0, 1, \ldots, n\}$ will be called *special* if each of the partial sums

$$x_{q(0)} + x_{q(1)} + \cdots + x_{q(j)}$$

is nonnegative for $0 \le j < n$. (When $j = n$, of course, this sum will be $-\epsilon$.)

The plan of the proof is to show first that there are exactly $n!$ special permutations. Then we shall map each of the $P(x_1, x_2, \ldots, x_n)$ desirable permutations into $t + 1$ distinct special permutations. This will prove that

$$(t+1)P(x_1, x_2, \ldots, x_n) \le n!, \tag{3}$$

and by symmetry the same will be true with s replacing t, hence (2) will follow.

In order to count the special permutations, we can use a cyclic-equivalence argument to show that exactly $1/(n+1)$ of the permutations are special. Let $q(0)q(1)\ldots q(n)$ be a permutation of the indices $\{0, 1, \ldots, n\}$, and consider the quantity

$$f(j) = x_{q(0)} + x_{q(1)} + \cdots + x_{q(j)} + (j+1)\epsilon/(n+1). \tag{4}$$

This function takes on $n+1$ distinct values for $0 \le j \le n$, since the relations $f(j) = f(j')$ and $j < j'$ imply that

$$x_{q(j+1)} + \cdots + x_{q(j')} = (j - j')\epsilon/(n+1),$$

contrary to our choice of ϵ. The permutation $q(k+1)\ldots q(n)q(0)\ldots q(k)$ is special if and only if

$$x_{q(k+1)} + \cdots + x_{q(j)} \ge 0, \qquad\qquad \text{for } k < j \le n; \tag{5}$$
$$x_{q(k+1)} + \cdots + x_{q(n)} + x_{q(0)} + \cdots + x_{q(j)} \ge 0, \quad \text{for } 0 \le j < k. \tag{6}$$

If $j > k$, we have $f(j) > f(k)$ if and only if

$$x_{q(k+1)} + \cdots + x_{q(j)} > (k - j)\epsilon/(n+1)$$

if and only if $x_{q(k+1)} + \cdots + x_{q(j)} \ge 0$. And if $j < k$, we have $f(j) > f(k)$ if and only if

$$x_{q(j+1)} + \cdots + x_{q(k)} < (j - k)\epsilon/(n+1)$$

if and only if

$$x_{q(k+1)} + \cdots + x_{q(n)} + x_{q(0)} + \cdots + x_{q(j)} > (k - j - n - 1)\epsilon/(n + 1)$$

if and only if

$$x_{q(k+1)} + \cdots + x_{q(n)} + x_{q(0)} + \cdots + x_{q(j)} \geq 0.$$

In other words, $q(k + 1) \ldots q(n)q(0) \ldots q(k)$ is special if and only if

$$f(k) = \min_{0 \leq j \leq n} f(j),$$

and this uniquely characterizes the value of k. It follows that exactly $n!$ of the $(n + 1)!$ possible permutations $q(0)q(1) \ldots q(n)$ are special.

Now let $p(1)p(2) \ldots p(n)$ be a permutation of $\{1, 2, \ldots, n\}$ such that all of the partial sums (1) are nonnegative, and let i be one of the t indices such that $x_{p(i)} < 0$. Then

$$p(1) \ldots p(i - 1)\, 0\, p(i + 1) \ldots p(n)p(i) \qquad (7)$$

is a special permutation of $\{0, 1, \ldots, n\}$, since $x_{p(i)} \leq x_0 = -\epsilon$. Furthermore,

$$p(1)p(2) \ldots p(n)\, 0 \qquad (8)$$

is obviously a special permutation. In this way we can construct $(t + 1)P(x_1, x_2, \ldots, x_n)$ special permutations, which clearly are all distinct. Therefore (3) holds, and (2) must be true.

To complete the proof of the theorem, we must show that (2) is best possible. This is equivalent to finding examples in which the permutations constructed in (7) and (8) exhaust all the special permutations. Such examples obviously arise whenever each negative x_i equals $-\epsilon$. Therefore equality holds in (2) when $1 \leq s \leq t$ and

$$x_1 = t - s + 1, \qquad\qquad x_2 = \cdots = x_s = 1,$$
$$x_{s+1} = \cdots = x_{s+t} = -1, \qquad x_{s+t+1} = \cdots = x_n = 0.$$

(For these x's, $\epsilon = 1$. The fact that $P(x_1, \ldots, x_n) = n!/(t + 1)$ in this case is well known, since it is equivalent to other combinatorial problems; see, for example, Erdélyi and Etherington [1].) □

If none of the $2^n - 2$ sums over proper subsets of the x's is zero, it is easy to see by considering cyclic permutations that $P(x_1, x_2, \ldots, x_n) = (n-1)!$. Conversely, if $P(x_1, x_2, \ldots, x_n) = (n-1)!$ those partial sums must all be nonzero.

In the general case, the possible values of $P(x_1, x_2, \ldots, x_n)$ seem to be spread out rather evenly between $(n-1)!$ and $n!/(\max(s,t)+1)$. For example, let a_1, a_2, b_1, b_2, and b_3 be positive numbers with

$$a_1 + a_2 = b_1 + b_2 + b_3;$$

the theorem tells us that $24 \leq P(a_1, a_2, -b_1, -b_2, -b_3) \leq 30$. In fact, it is not difficult to verify in this case that $P(a_1, a_2, -b_1, -b_2, -b_3)$ equals 24 plus twice the number of pairs (i,j) such that $a_i = b_j$. Thus,

$$P(5,1,-2,-2,-2) = 24, \qquad P(4,1,-1,-2,-2) = 26,$$
$$P(3,1,-1,-1,-2) = 28, \qquad P(2,1,-1,-1,-1) = 30.$$

Kleitman has used the theorem above to determine the asymptotic number of different score sequences possible in an n-person round-robin tournament, to within a factor of 2.

Supported in part by the National Science Foundation under grant GJ-992, and in part by the Office of Naval Research under contract N-00014-67-A-0112-0057 NR 044-402.

References

[1] A. Erdélyi and I. M. H. Etherington, "Some problems of non-associative combinations. II," *Edinburgh Mathematical Notes* **32** (1941), 7–12.

[2] D. Kleitman, "The number of tournament score sequences for a large number of players," *Combinatorial Structures and their Applications* (New York: Gordon and Breach, 1970), 209–213.

Chapter 29

Efficient Balanced Codes

[Originally published in IEEE Transactions on Information Theory IT-32 (1986), 51–53.]

A binary word of length m can be called *balanced* if it contains exactly $\lfloor m/2 \rfloor$ ones and $\lceil m/2 \rceil$ zeros. Let us say that a *balanced code* with n information bits and p parity bits is a set of 2^n balanced binary words, each of length $m = n + p$.

Balanced codes have the property that no codeword is "contained" in another; namely, the positions of the 1s in one codeword will never be a subset of the positions of the 1s in a different codeword. This property makes balanced codes attractive for certain applications, such as the encoding of unchangeable data on a laser disk [2]. Conversely, if we wish to form as many binary words of length m as possible with the property that no word is contained in another, Sperner's theorem [3] tells us that we can do no better than to construct the set of all balanced words of length m.

A balanced code is *efficient* if there is a very simple way to encode and decode n-bit numbers. In other words, we want to find a one-to-one correspondence between the set of all n-bit binary words and the set of all $(n + p)$-bit codewords such that, if w corresponds to w', we can rapidly compute w' from w and vice versa. Furthermore we want p to be very small compared to n, so that the code is efficient in its use of space as well as time. For example, it's trivial to construct a balanced code with n information bits and n parity bits by simply letting the binary word w correspond to the codeword $w' = w\overline{w}$, where $\overline{w}$ is the complement of w. Encoding and decoding is certainly efficient in this case, but memory space is being wasted because $p = n$.

Let $M(m) = \binom{m}{\lfloor m/2 \rfloor}$ be the total number of balanced binary words of length m. In order to have a balanced code with n information bits, we clearly need to have enough parity bits p so that $M(n + p) \geq 2^n$.

433

Stirling's approximation tells us that

$$\lg M(m) = m - \frac{1}{2}\lg m - \frac{1}{2}\lg\frac{\pi}{2} - \frac{\epsilon(m)}{m}$$

where $0 \le \epsilon(m) \le 1.25/\ln 2 < 1.81$; the constant $\frac{1}{2}\lg\frac{\pi}{2}$ is approximately 0.325748. Therefore in particular we must have $p > \frac{1}{2}\lg n + 0.3257$ in any balanced code.

The purpose of this note is to describe a balanced code with $n = 2^p$ information bits and p parity bits, for which serial encoding and decoding is especially simple. This means, for example, that 256-bit words can be encoded efficiently with only 8 parity bits, obtaining 264-bit balanced words; the percentage of memory devoted to overhead in order to satisfy the balance constraint is only $8/264 \approx 3.03\%$.

A similar scheme that allows efficient *parallel* decoding and efficient serial encoding is also described. The parallel method for n information bits takes roughly $\lg n + \frac{1}{2}\lg\lg n$ parity bits in its simplest form, and the $\frac{1}{2}\lg\lg n$ term can be reduced to 1 at the expense of additional complexity. For example, a balanced code with 256 information bits and 9 parity bits will be constructed explicitly. This code has the property that the 256-bit word w corresponds to a balanced 265-bit codeword $w' = uw^{(k)}$, where $w^{(k)}$ denotes w with its first k bits complemented, and where the 9-bit prefix u determines k. It is clearly possible to determine w quickly from w' in such a code.

A Simple Parallel Scheme

Let $\nu(w)$ be the total number of 1s in the binary word w; let $\nu_k(w)$ be the number of 1s in the first k bits of w; and let $w^{(k)}$ be the word w with its first k bits complemented. For example, if $w = 0111010110$, we have $\nu(w) = 6$, $\nu_4(w) = 3$, and $w^{(4)} = 1000010110$. Since $k - \nu_k(w)$ of the first k bits of w are 0s, we have

$$\nu(w^{(k)}) = \nu(w) + k - 2\nu_k(w).$$

This simple relation is the key to all the coding schemes that will be described below.

If w has length n and if we let $\sigma_k(w)$ stand for $\nu(w^{(k)})$, the quantity $\sigma_k(w)$ changes by ± 1 when k increases by 1, so it describes a "random walk" from $\sigma_0(w) = \nu(w)$ to $\sigma_n(w) = n - \nu(w)$.

Now comes the point: The value $\lfloor n/2 \rfloor$ lies in the closed interval between ν and $n - \nu$ for all integers ν, hence there always exists a k such that $\sigma_k(w) = \lfloor n/2 \rfloor$. In other words, every word w can be associated with at least one k such that $w^{(k)}$ is balanced. If we encode k in a balanced word u of length p, and if n and p are not both odd, we can let w correspond to the balanced codeword $uw^{(k)}$. If n and p are both odd we can use a similar construction, but the value of k should be chosen so that $\sigma_k(w) = \lceil n/2 \rceil$; then again $uw^{(k)}$ will be balanced.

For example, suppose that we want a balanced code of this sort having 8 information bits. Every 8-bit word w defines at least one value of k such that $w^{(k)}$ is balanced; we never need to use $k = 8$, so we can assume that $0 \le k < 8$. If we arbitrarily choose eight balanced words $(u_0, \ldots, u_7)$ of length 5, we can represent w by the balanced word $u_k w^{(k)}$. (Such a choice of u's is possible since $M(5) = 10 > 8$.) This gives a code with 8 information bits and 5 parity bits. Parallel decoding is easy, because k is determined from u_k by table-lookup; then w is $w^{(k)(k)}$. Serial encoding is also easy because we can determine k by computing $\sigma_k(w)$ for $k = 0, 1, \ldots$ until finding $\sigma_k(w) = 4$.

A similar scheme gives a balanced code with 256 information bits and 11 parity bits, because $M(11) > 256$. In general, this approach works with n information bits and p parity bits whenever $M(p) \ge 2\lceil n/2 \rceil$.

A Simple Serial Scheme

We can decrease the number of parity bits in the previous construction by using all the bits of u. The idea is to encode w as $uw^{(k)}$ for some u and k, as before, but u does not have to be balanced; any imbalance in u will be compensated by a corresponding imbalance in $w^{(k)}$. For example, when $n = 4$ and $p = 2$ we can simply let $k = 0$ when $0 < \nu(w) < 4$, using $u = 11$, 01, and 00 respectively for $\nu(w) = 1$, 2, and 3. The remaining two cases $w = 0000$ and $w = 1111$ are handled by letting $k = 2$ and $u = 10$.

When $n = 8$ and $p = 3$ an exhaustive analysis shows that there is no similar scheme in which k is determined by u; but we can construct a code in which u is determined by $\nu(w)$ as follows:

$\nu(w)$	u	s	$\nu(w)$	u	s	$\nu(w)$	u	s
0	001	4	3	101	3	6	111	2
1	011	3	4	100	4	7	110	3
2	010	4	5	000	5	8	001	4

The word $uw^{(k)}$ will be balanced in this case if and only if $\nu(uw^{(k)}) = \nu(u) + \sigma_k(w) = \lfloor 11/2 \rfloor = 5$; this happens if and only if $\sigma_k(w) = s$, where the values of s have been tabulated above. Since $\sigma_k(w)$ runs from $\nu(n)$ to $n - \nu(n)$, it is easy to verify in each case that some value of k will make $\sigma_k(w) = s$. The code is defined by choosing the smallest such k.

To decode this scheme, that is, to deduce w given $uw^{(k)} = uv$, we first determine $\nu(w)$ from u. Then we find the smallest k such that $\sigma_k(v) = \nu(w)$. This is the value of k for which $v = w^{(k)}$. (Why? Because the value of k used in the encoding clearly has this property. Furthermore if $w = v^{(k)}$ and if $\sigma_{k'}(v) = \sigma_k(v)$ for some $k' < k$, then $\nu(v^{(k')}) = \nu(v^{(k)})$, hence $\nu_k(v) = \nu_{k'}(v) + \frac{1}{2}(k - k')$; hence $\nu_k(w) = \nu_{k'}(w) + \frac{1}{2}(k - k')$ and $\sigma_{k'}(w) = \sigma_k(w)$, contradicting the minimality of k. We are essentially applying the "reflection principle" of [1].)

There's one complication, however: Two different values of $\nu(w)$ correspond to the same value of u. (Namely, $u = 001$ has both $\nu(w) = 0$ and $\nu(w) = 8$.) This is not really a difficulty, because it arises only for the two words $w = 00000000$ and 11111111 (when we know that $k = 4$); but it's an annoying anomaly. The best way to avoid it is to consider only the values of $\sigma_k(v)$ modulo 8 when decoding. We know $\nu(w)$ mod 8, so we choose the smallest k such that $\sigma_k(v) \equiv \nu(w)$ (modulo 8).

Incidentally, there is no balanced code with $n = 8$ and $p = 2$, since $M(10) = 252$ is less than 256. Therefore the balanced code just defined is optimum for $n = 8$.

A similar balanced code can be constructed with p parity bits and $n = 2^p$ information bits, for all $p \geq 3$, as follows: For $0 \leq l < n$, let u_l be a p-bit word such that the number

$$s_l = n/2 + \lfloor p/2 \rfloor \nu(u_l)$$

lies between l and $n - l$, inclusive. This should be a permutation of the p-bit words; that is, $l \neq l'$ should imply that $u_l \neq u_{l'}$. An n-bit word w is then encoded as $u_l w^{(k)}$, where $l = \nu(w) \bmod n$ and where k is minimal such that $\sigma_k(w) \equiv s_l$ (modulo n). An $(n+p)$-bit word $w' = uv$ is decoded as $v^{(k)}$, where k is minimal such that $\sigma_k(v) \equiv l$ (modulo n) and where l is determined by the condition $u = u_l$.

It remains to specify the correspondence between l and u_l. Since p is much smaller than n, the choice is delicate only when l is near $n/2$. It is not difficult to find a mapping that assigns the balanced words to values of l near $n/2$; the rest of the codes are essentially arbitrary.

For example, let $p = 8$ and $n = 256$. We want to permute the 8-bit words u_{128+t} for $-128 \leq t < 128$ in such a way that

$$0 \leq t + \nu(u_{128+t}) - 4 \leq 2t, \quad \text{when } t \geq 0;$$
$$0 \geq t + \nu(u_{128+t}) - 4 \geq 2t, \quad \text{when } t < 0.$$

The inequalities are always valid when $|t| \geq 4$, so the choice of u_l is important only when $124 < l < 132$. A suitable mapping is obtained by letting $u_l = a_l b_l$, where a_l is the 4-bit binary representation of $(l + 8) \bmod 16$ and b_l is the 4-bit binary representation of

$$\left(\frac{1}{16}(l - 120 - a_l) + \bar{a}_l\right) \bmod 16 .$$

If $120 \leq l < 136$, this makes $b_l = \bar{a}_l$, hence u_l is balanced. Conversely, it is easy to deduce l from a given pair of 4-bit words ab:

$$l = \left(120 + a + 16\big((b - \bar{a}) \bmod 16\big)\right) \bmod 256 .$$

An Optimized Parallel Scheme

We have now constructed two balanced codes with $n = 256$; one has $p = 11$ parity bits to allow parallel decoding, and the other has $p = 8$ parity bits to allow serial decoding. The author has been unable to construct a parallel decoder for such schemes when $n = 256$ and $p = 8$, but the following method gives parallel decoding when $p = 9$ and in general whenever $n = 2^{p-1}$.

The idea is to choose l words $(u_1, \ldots, u_l)$ of p bits each and to choose l values $(k_1, \ldots, k_l)$ in the range $0 \leq k_j \leq n$ such that every random walk

$$\big(0, \sigma_0(w)\big), \ \big(1, \sigma_1(w)\big), \ \ldots, \ \big(n, \sigma_n(w)\big) \tag{$*$}$$

is guaranteed to pass through one of the points

$$P_j = \big(k_j, \lfloor (n + p)/2 \rfloor - \nu(u_j)\big)$$

for some j. We can then encode w as the balanced word $u_j w^{(k_j)}$. Parallel decoding is possible since the p-bit parity word u determines the extent of complementation.

We shall choose the u's and k's in such a way that $\nu(u_{j+1}) - \nu(u_j) = 0$ or 1 and $k_{j+1} - k_j = 1 - \big(\nu(u_{j+1}) - \nu(u_j)\big)$. This means that $P_{j+1} - P_j$ is always either $(1, 0)$ or $(0, -1)$. For example, when $p = 3$ and $n = 4$ we can let the pairs (k_j, u_j) be

$$(0, 001) \quad (1, 010) \quad (2, 100) \quad (2, 011) \quad (3, 101) \quad (4, 110)$$

so that the points P_j are

$$(0, 2) \quad (1, 2) \quad (2, 2) \quad (2, 1) \quad (3, 1) \quad (4, 1) .$$

We shall also choose $k_1 = 0$ and $k_l = n$, so that any random walk $(*)$ must lie entirely "above" or "below" the set of P's.

Let $P_1 = (0, M)$ and $P_l = (n, m)$ be the extreme points. If $(*)$ does not intersect the set $\{P_1, \ldots, P_l\}$, we must have either $(\sigma_0(w) > M$ and $\sigma_n(w) > m)$ or $(\sigma_0(w) < M$ and $\sigma_n(w) < m)$. Since $\sigma_0(w) + \sigma_n(w) = n$, this cannot happen unless $n \geq M + m + 2$ or $n \leq M + m - 2$. Therefore it suffices to design the construction so that $|M + m - n| \leq 1$.

A moment's thought now makes it clear what to do: We list all p-bit numbers u in any order such that the weights $\nu(v)$ are nondecreasing, then we choose $l = n + h + 1$ of these near the "middle" of the sequence such that $\nu(u_l) - \nu(u_1) = h$ for some h. For example, the case $p = 3$ worked out above has $h = 1$ and $l = 6$. When $p = 9$ there are 126 u's of weight 4 and 126 of weight 5; we can take $h = 3$, $l = 260$, starting with any four words $(u_1, \ldots, u_4)$ of weight 3, then $(u_5, \ldots, u_{130})$ of weight 4, then $(u_{131}, \ldots, u_{256})$ of weight 5, and $(u_{257}, \ldots, u_{260})$ of weight 6. In this case $n = 256$, $M = \lfloor 265/2 \rfloor - 3 = 129$, $m = \lfloor 265/2 \rfloor - 6 = 126$; hence $M + m = n - 1$ and we have achieved our objective. It is not difficult to verify that the method works for all $p \geq 3$: When p is odd, h will be odd and we will have $M = \frac{1}{2}(n + h - 1)$, $m = \frac{1}{2}(n - h - 1)$, but when p is even h will be even and we will have $M = \frac{1}{2}(n + h)$, $m = \frac{1}{2}(n - h)$.

The method just described does not depend in any essential way on the assumption that n is a power of 2. We can use it, in fact, to transmit as many as $2^p - p - 1$ information bits if we let $l = 2^p$.

This research was supported in part by National Science Foundation grant DCR-83-00984. The author wishes to thank an anonymous referee for several penetrating observations that substantially improved this note.

References

[1] D. André, "Solution directe du problème resolu par M. Bertrand," *Comptes Rendus hebdomadaires des séances de l'Académie des Sciences* **105** (Paris: 1887), 436–437.

[2] Ernst L. Leiss, "Data integrity in digital optical disks," *IEEE Transactions on Computers* **C-33** (1984), 818–827.

[3] Emanuel Sperner, "Ein Satz über·Untermengen einer endlichen Menge," *Mathematische Zeitschrift* **27** (1928), 544–548.

Addendum

The idea of using $w^{(k)}$ is "optimum" in a certain sense; see N. Alon, E. E. Bergmann, D. Coppersmith, and A. M. Odlyzko, "Balancing sets of vectors," *IEEE Transactions on Information Theory* **34** (1988), 128–130. Yet improvements have been found by Bella Bose, "On unordered codes," *IEEE Transactions on Computers* **40** (1991), 125–131.

Chapter 30

The Knowlton–Graham Partition Problem

[Originally published in Journal of Combinatorial Theory **A73** *(1996),
185–189.]*

A long cable contains n indistinguishable wires. Two people, one at each
end, want to label the wires consistently so that both ends of each wire
receive the same label. An interesting way to achieve this objective was
proposed by K. C. Knowlton [3]: Partition $\{1, \ldots, n\}$ into disjoint sets
in two ways $A_1, \ldots, A_p$ and $B_1, \ldots, B_q$, subject to the condition that at
most one element appears both in an A set of cardinality j and in a B set
of cardinality k, for each j and k. We can then use the coordinates (j, k)
to identify each element. R. L. Graham [2] proved that such partitioning
schemes exist if and only if $n \neq 2$, 5, or 9.

By restating the problem in terms of 0–1 matrices, it is possible to
prove Graham's theorem more simply, and to sharpen the results of [2].

Lemma 1. *Knowlton–Graham partitions for n exist if and only if there
is a matrix of 0s and 1s having row sums $(r_1, \ldots, r_m)$ and column sums
$(c_1, \ldots, c_m)$ such that r_j and c_j are multiples of j and $r_1 + \cdots + r_m =
c_1 + \cdots + c_m = n$.*

Proof. If $A_1, \ldots, A_p$ and $B_1, \ldots, B_q$ are partitions of $\{1, \ldots, n\}$ with
the Knowlton–Graham property, let a_{jk} be the number of elements that
appear in an A set of cardinality j and a B set of cardinality k. Then a_{jk}
is 0 or 1; and $r_j = \sum_k a_{jk}$ is j times the number of A sets of cardinality j,
while $c_k = \sum_j a_{jk}$ is k times the number of B sets of cardinality k.

Conversely, given such a matrix, we can use its rows to define a
set partition $A_1, \ldots, A_p$ such that each 1 in row j is in an A set of
cardinality j; similarly its columns define $B_1, \ldots, B_q$ such that each 1 in
column k is in a B set of cardinality k. □

For example, the symmetric matrix

$$\begin{pmatrix} 0 & 1 & 0 & 0 & 0 & 1 \\ 1 & 1 & 1 & 1 & 1 & 1 \\ 0 & 1 & 0 & 0 & 1 & 1 \\ 0 & 1 & 0 & 1 & 1 & 1 \\ 0 & 1 & 1 & 1 & 1 & 1 \\ 1 & 1 & 1 & 1 & 1 & 1 \end{pmatrix}$$

has row and column sums $(2, 6, 3, 4, 5, 6)$ that satisfy the divisibility condition and sum to 26. To identify 26 wires, we can associate the 1s with arbitrary labels $\{a, \ldots, z\}$,

$$\begin{pmatrix} . & a & . & . & . & b \\ c & d & e & f & g & h \\ . & i & . & . & j & k \\ . & l & . & m & n & o \\ . & p & q & r & s & t \\ u & v & w & x & y & z \end{pmatrix}.$$

The person at one end of the cable attaches arbitrary labels $\{a, \ldots, z\}$ to the wires and makes connections so that each element of row j is connected to exactly $j - 1$ other elements of its row; for example, the connected components might be A_1, ..., $A_9 = \{a\}$, $\{b\}$, $\{c, d\}$, $\{e, f\}$, $\{g, h\}$, $\{i, j, k\}$, $\{l, m, n, o\}$, $\{p, q, r, s, t\}$, $\{u, v, w, x, y, z\}$. The person at the other end now uses properties of conductivity to identify the row that each wire belongs to. The wires at that end can then be labeled $\{A, \ldots, Z\}$ in such a way that $\{A, B\} = \{a, b\}$, $\{C, D, E, F, G, H\} = \{c, d, e, f, g, h\}$, etc. Now the wires at the first end are disconnected, while at the other end they are connected so that each element of column k is connected to exactly $k - 1$ other elements of its column. For example, the connected components might now be B_1, ..., $B_9 = \{C\}$, $\{U\}$, $\{A, D\}$, $\{I, L\}$, $\{P, V\}$, $\{E, Q, W\}$, $\{F, M, R, X\}$, $\{G, J, N, S, Y\}$, $\{B, H, K, O, T, Z\}$. Once this has been done, the people at both ends of the cable can give unique coordinates (j, k) to each wire, knowing its row and column.

Knowlton–Graham partitions are said to have order m if there are m elements in the largest of the sets A_1, ..., A_p, B_1, ..., B_q.

Theorem 1 (Graham). *Knowlton–Graham partitions of n having order m are possible only if $\binom{m+1}{2} \le n \le J(m)$, where*

$$J(m) = \sum_{j=1}^{m} j \lfloor m/j \rfloor.$$

Proof. By Lemma 1, Knowlton–Graham partitions of order m imply the existence of an $m \times m$ matrix of 0s and 1s having row sums $(r_1, \ldots, r_m)$ and column sums $(c_1, \ldots, c_m)$, where both r_j and c_j are multiples of j for $1 \leq j \leq m$, and where $r_m + c_m > 0$. Clearly $r_j \leq m$; so r_j is at most $j\lfloor m/j \rfloor$, the largest multiple of j that does not exceed m. This establishes the upper bound $J(m)$.

If $r_m > 0$, we must have $r_m = m$; this implies $c_j > 0$ for all j, hence $c_m = m$. Similarly, $c_m > 0$ implies that $r_m = c_m = m$. So we must have $r_j > 0$ for all j, hence $r_j \geq j$ for all j, hence $n = \sum_{j=1}^{m} r_j \geq \sum_{j=1}^{m} j = \binom{m+1}{2}$. □

When $m = 1, 2, 3$, and 4, Theorem 1 says that $1 \leq n \leq 1$, $3 \leq n \leq 4$, $6 \leq n \leq 8$, and $10 \leq n \leq 15$, respectively; this reasoning explains why the values $n = 2, 5$, and 9 are impossible. For $m \geq 4$ we have $J(m) \geq \binom{m+2}{2}$, so there are no more gaps. In fact, as $m \to \infty$ we have

$$J(m) = m^2 - \sum_{j=1}^{m} (m \bmod j) = \frac{\pi^2}{12} m^2 + O(m \log m)$$

(see [4, Eq. 4.5.3–(21)]); therefore $J(m) / \binom{m+2}{2}$ approaches the limiting value $\pi^2/6 \approx 1.64$.

The main purpose of this note is to prove the converse of Theorem 1, namely that Knowlton–Graham partitions of order m do exist for all n in the range $\binom{m+1}{2} \leq n \leq J(m)$. This question was left open in [2], where Graham observed that it was not sufficient simply to represent n in the form $r_1 + \cdots + r_m$ where each r_j is a positive multiple of j. For example, there is no suitable 0–1 matrix having row sums $(1, 6, 6, 4, 5, 6)$. If there were, we would necessarily have $c_6 = 6$, $c_5 = 5$, $c_4 = 4$, $c_3 = 3$, and we could not make c_2 even.

Gale [1] and Ryser [5] independently found an elegant necessary and sufficient condition for the existence of 0–1 matrices having given row and column sums, but their theorem does not seem to lead easily to the result needed here. Instead, we can use a direct recursive construction.

Lemma 2. *Let $(r_1, \ldots, r_m)$ and $(c_1, \ldots, c_m)$ be integers satisfying the conditions*

$$r_1 + \cdots + r_m = c_1 + \cdots + c_m \,,$$

$$m \geq r_1 \geq \cdots \geq r_m \geq 0, \quad m \geq c_1 \geq \cdots \geq c_m \geq 0 \,,$$

$$r_{j+1} \geq r_j - 1 \quad \text{and} \quad c_{j+1} \geq c_j - 1 \quad \text{for} \quad 1 \leq j < m \,.$$

Then there exists an $m \times m$ matrix of 0s and 1s having row sums $(r_1, \ldots, r_m)$ and column sums $(c_1, \ldots, c_m)$.

Proof. This claim is obvious when $m = 1$, so we may assume inductively that $m > 1$. Let $p = r_1$ and $q = c_1$, and consider the numbers

$$(r'_1, \ldots, r'_{m-1}) = (r_2 - 1, \ldots, r_q - 1, r_{q+1}, \ldots, r_m),$$
$$(c'_1, \ldots, c'_{m-1}) = (c_2 - 1, \ldots, c_p - 1, c_{p+1}, \ldots, c_m).$$

The lemma will be proved if we construct an $(m-1) \times (m-1)$ matrix of 0s and 1s having row sums $(r'_1, \ldots, r'_{m-1})$ and column sums $(c'_1, \ldots, c'_{m-1})$, because we can achieve the desired result by appending a new first row and a new first column. In fact it suffices, by row and column permutations, to construct a 0–1 matrix with row and column sums equal to the numbers $(r''_1, \ldots, r''_{m-1})$ and $(c''_1, \ldots, c''_{m-1})$ obtained by sorting $(r'_1, \ldots, r'_{m-1})$ and $(c'_1, \ldots, c'_{m-1})$ into nonincreasing order.

Since $r''_1 + \cdots + r''_{m-1} = r_1 + \cdots + r_m + 1 - p - q = c_1 + \cdots + c_m + 1 - p - q = c''_1 + \cdots + c''_{m-1}$, we can use the induction hypothesis if we verify that $r''_1 \le m - 1$, $r''_{m-1} \ge 0$, and $r''_{j+1} \ge r''_j - 1$ for $1 \le j < m - 1$; the similar inequalities for $(c''_1, \ldots, c''_{m-1})$ follow by symmetry.

Suppose $r''_1 = m$; this implies $r_{q+1} = m$ and $q < m$. Therefore $r_q = \cdots = r_2 = r_1 = m$, and we have $(q + 1)m \le r_1 + \cdots + r_m = c_1 + \cdots + c_m \le m c_1 = qm$, a contradiction.

Suppose $r''_{m-1} < 0$; this implies $r_q = 0$. Therefore $(q - 1) + \cdots + 1 + 0 \ge r_1 + \cdots + r_m = c_1 + \cdots + c_m \ge q + (q - 1) + \cdots + 1$, another contradiction.

Suppose finally that $r''_{j+1} < r''_j - 1$. This could happen only if $r''_j = r_k$ and $r''_{j+1} = r_l - 1$ for some k and l with $r_k > r_l$. But we would not decrease r_l unless we had also decreased r_k. □

The construction of Lemma 2 produces a symmetric matrix when $(r_1, \ldots, r_m) = (c_1, \ldots, c_m)$. Let's say that Knowlton–Graham partitions are *symmetric* if they correspond to a symmetric matrix. We are now ready to prove the main result.

Theorem 2. *Symmetric Knowlton–Graham partitions of n having order m exist whenever $\binom{m+1}{2} \le n \le J(m)$.*

Proof. When n is in the stated range but not equal to $\binom{m+1}{2}$, there is a number $s \le m/2$ such that we can write $n = t_1 + \cdots + t_m$ where

$$\begin{aligned}
t_j &= j, & &\text{for } s < j \le m; \\
t_s &= ks, & &\text{for some } k, \, 1 < k \le \lfloor m/s \rfloor; \\
t_j &= j \lfloor m/j \rfloor, & &\text{for } 1 < j < s; \\
m - s &< t_1 \le m, & &\text{if } s > 1.
\end{aligned}$$

When $n = J(m)$, this is true with $s = \lfloor m/2 \rfloor$, $k = \lfloor m/s \rfloor$, and $t_1 = m$. Otherwise we can find such a representation by first representing $n + 1$ and subtracting 1 from t_1; then if $s > 1$ and $t_1 = m - s$, we replace t_1 by m and subtract s from t_s; finally, if $t_s = s$, we decrease s by 1.

The remaining case $n = \binom{m+1}{2}$ is simpler because we can write $n = t_1 + \cdots + t_m$ where $t_j = j$ for all j. This is a representation of essentially the same form but with $s = 0$.

Notice that t_j is a multiple of j, for $1 \leq j \leq m$. We can also verify that the set $\{t_1, \ldots, t_m\}$ consists simply of the consecutive elements $\{t_{s+1}, \ldots, t_m\} = \{s+1, \ldots, m\}$. For we have $t_s > s$; and $t_j > m - s \geq s$ for all $j < s$, because $j \lfloor m/j \rfloor = m - (m \bmod j)$.

Let $(r_1, \ldots, r_m)$ and $(c_1, \ldots, c_m)$ be the numbers $(t_1, \ldots, t_m)$ sorted into nonincreasing order. Lemma 2 tells us how to construct a symmetric 0–1 matrix having these row and column sums. After an appropriate permutation of rows and columns, the row and column sums can be made equal to $(t_1, \ldots, t_m)$; and the resulting matrix yields Knowlton–Graham partitions, by Lemma 1. $\square$

References

[1] David Gale, "A theorem on flows in networks," *Pacific Journal of Mathematics* **7** (1957), 1073–1082.

[2] R. L. Graham, "On partitions of a finite set," *Journal of Combinatorial Theory* **1** (1966), 215–223.

[3] Ronald L. Graham and Kenneth C. Knowlton, "Method of identifying conductors in a cable by establishing conductor connection groupings at both ends of the cable," *U.S. Patent 3,369,177* (13 February 1968).

[4] Donald E. Knuth, *Seminumerical Algorithms*, Volume 2 of *The Art of Computer Programming*, second edition (Reading, Massachusetts: Addison–Wesley, 1981).

[5] H. J. Ryser, "Combinatorial properties of matrices of zeros and ones," *Canadian Journal of Mathematics* **9** (1957), 371–377.

Chapter 31

Permutations, Matrices, and Generalized Young Tableaux

A generalized Young tableau of "shape" $(p_1, p_2, \ldots, p_m)$, where $p_1 \geq p_2 \geq \cdots \geq p_m \geq 1$, is an array Y of positive integers y_{ij}, for $1 \leq j \leq p_i$ and $1 \leq i \leq m$, having monotonically nondecreasing rows and strictly increasing columns. By extending a construction due to Robinson and Schensted, it is possible to obtain a one-to-one correspondence between $m \times n$ matrices A of nonnegative integers and ordered pairs (P, Q) of generalized Young tableaux, where P and Q have the same shape, the integer i occurs exactly $a_{i1} + \cdots + a_{in}$ times in Q, and the integer j occurs exactly $a_{1j} + \cdots + a_{mj}$ times in P. A similar correspondence can be given for the case that A is a matrix of zeros and ones, and the shape of Q is the transpose of the shape of P.

*[Originally published in Pacific Journal of Mathematics **34** (1970), 709–727.]*

Figure 1 shows two arrangements of integers that we will call *generalized Young tableaux* of shape $(6, 4, 4, 1)$. A *generalized Young tableau of shape* $(p_1, p_2, \ldots, p_m)$ is an array of $p_1 + p_2 + \cdots + p_m$ positive integers into m left-justified rows, with p_i elements in row i, where $p_1 \geq p_2 \geq \cdots \geq p_m$; the numbers in each row are in nondecreasing order from left to right, and the numbers in each column are in strictly increasing order from top to bottom. (The special case where the elements are the integers $\{1, 2, \ldots, N\}$, each used exactly once, where

$$N = p_1 + p_2 + \cdots + p_m,$$

was introduced by Alfred Young in 1928 as an aid in the study of irreducible representations of the symmetric group on N letters; see [9].)

445

P

1	1	1	2	4	7
2	3	3	5		
3	4	6	6		
6					

Q

1	2	2	3	3	6
3	3	3	4		
4	5	5	5		
5					

FIGURE 1. Generalized Young tableaux with the same shape.

Consider on the other hand the 6×7 array

$$A = \begin{pmatrix} 0 & 0 & 1 & 0 & 0 & 0 & 0 \\ 0 & 0 & 0 & 0 & 0 & 2 & 0 \\ 1 & 1 & 1 & 1 & 0 & 1 & 0 \\ 0 & 0 & 1 & 0 & 1 & 0 & 0 \\ 2 & 1 & 0 & 1 & 0 & 0 & 0 \\ 0 & 0 & 0 & 0 & 0 & 0 & 1 \end{pmatrix} \tag{1.1}$$

having respective column sums $(c_1, \ldots, c_7) = (3, 2, 3, 2, 1, 3, 1)$ and row sums $(r_1, \ldots, r_6) = (1, 2, 5, 2, 4, 1)$. Notice that in Figure 1 the integer i occurs r_i times in Q, and the integer j occurs c_j times in P.

In this paper we shall give a constructive procedure that yields a one-to-one correspondence between matrices A of nonnegative integers and ordered pairs of equal-shape generalized Young tableaux (P, Q) such that the row and column sums of A correspond in the same manner to the number of occurrences of elements in P and Q. In particular, our procedure shows how to construct (1.1) from Figure 1 and conversely.

Figure 2 shows two generalized Young tableaux whose shapes are *transposes* of each other. A modification of the first construction leads to another procedure that gives a similar correspondence between *zero-one matrices* A and such pairs of tableaux. For example, the second construction associates the matrix

$$A = \begin{pmatrix} 0 & 0 & 1 & 0 & 0 \\ 1 & 1 & 0 & 1 & 0 \\ 1 & 0 & 0 & 0 & 0 \\ 0 & 0 & 0 & 0 & 1 \\ 0 & 0 & 1 & 0 & 0 \end{pmatrix} \tag{1.2}$$

with Figure 2. When the column sums of A are all ≤ 1, the two constructions are essentially identical, differing only in that P is transposed.

Matrices A of nonnegative integers correspond in an obvious way to two-line arrays of positive integers

$$\begin{pmatrix} u_1 & u_2 & \cdots & u_N \\ v_1 & v_2 & \cdots & v_N \end{pmatrix} \tag{1.3}$$

$$P \qquad\qquad Q$$

1	1	3
2	4
3
5

1	2	2	4
2	5
3

FIGURE 2. Generalized Young tableaux with transposed shapes.

where the pairs (u_k, v_k) are arranged in nondecreasing lexicographic order from left to right, and where there are exactly a_{ij} occurrences of the pair (i, j). For example, the matrix (1.1) corresponds in this way to

$$\begin{pmatrix} 1 & 2 & 2 & 3 & 3 & 3 & 3 & 3 & 4 & 4 & 5 & 5 & 5 & 5 & 6 \\ 3 & 6 & 6 & 1 & 2 & 3 & 4 & 6 & 3 & 5 & 1 & 1 & 2 & 4 & 7 \end{pmatrix}. \qquad (1.4)$$

Such two-line arrays can be regarded as generalized permutations, for when A is a permutation matrix the corresponding two-line array is the permutation corresponding to A. When A is a zero-one matrix, the pairs (u_k, v_k) in (1.3) are all distinct.

Our construction works with two-line arrays (1.3) instead of the original matrices, although it is of course possible to translate everything we do into the matrix notation. The special case where $u_k = k$ for $1 \le k \le N$ was treated by Craige Schensted in 1961 [7]; in that case A is a zero-one matrix with N rows, each row-sum being equal to unity. Our first construction is identical to Schensted's in such a case. Another procedure that can be shown to be essentially equivalent to Schensted's construction was published already in 1938 by Gilbert de B. Robinson [5, §5], although he described the algorithm rather vaguely and in quite different terms.

Section 2 below presents Schensted's algorithm in detail, and §3 uses that algorithm to achieve the first correspondence. A graph-theoretical interpretation of the correspondence, given in §4, allows us to conclude that transposition of the matrix A corresponds to interchanging P and Q; hence we obtain a useful one-to-one correspondence between *symmetric* matrices A and (single) generalized Young tableaux.

Section 5 shows how to modify the preceding algorithms to obtain the second correspondence. Finally in §6 a combinatorial characterization is given of all matrices that have a given value of P; this characterization leads to an "algebra of tableaux."

As a consequence of the algorithms in this paper it is possible to obtain a constructive proof of MacMahon's classical formulas for the enumeration of plane partitions, as well as new enumeration formulas for certain rather general kinds of plane partitions. These applications will be reported elsewhere [1].

2. The Insertion and Deletion Procedures

It is convenient in the following discussion to regard a generalized Young tableau of shape $(p_1, p_2, \ldots, p_m)$ as a doubly infinite array

$$
Y = \begin{array}{|ccccc}
y_{00} & y_{01} & y_{02} & y_{03} & \cdots \\
y_{10} & y_{11} & y_{12} & y_{13} & \cdots \\
y_{20} & y_{21} & y_{22} & y_{23} & \cdots \\
\vdots & \vdots & \vdots & \vdots &
\end{array}
\tag{2.1}
$$

with $y_{ij} = 0$ if i or j is zero, $y_{ij} = \infty$ if $i > m$ or $j > p_i$, and $0 < y_{ij} \neq \infty$ when $1 \leq i \leq m$, $1 \leq j \leq p_i$. We simply border the tableau with zeros at the top and left, and we put ∞ symbols elsewhere. Using the further convention

$$\infty < \infty, \tag{2.2}$$

this doubly infinite array satisfies the inequalities

$$y_{ij} \leq y_{i(j+1)}, \quad y_{ij} < y_{(i+1)j}, \qquad \text{for all } i, j \geq 0. \tag{2.3}$$

Convention (2.2) may appear somewhat strange, but it does not make the transitive law invalid, and it is a decided convenience in what follows because of the uniform conditions (2.3).

We can now describe Schensted's procedure for inserting a new positive integer x into a generalized Young tableau Y. The following algorithm contains parenthesized "assertions" about the present state of affairs, each of which is easily verified from previously verified assertions; hence we are presenting a proof of the validity of the algorithm at the same time as the algorithm itself is being presented. (See [2, pages 2–3 and 15–16].)

INSERT(x):

I1. Set $i \leftarrow 1$, set $x_1 \leftarrow x$, and set j to some value such that $y_{1j} = \infty$.

I2. (Now $y_{(i-1)j} < x_i < y_{ij}$, and $x_i \neq \infty$.) If $x_i < y_{i(j-1)}$, decrease j by 1 and repeat this step. Otherwise set $x_{i+1} \leftarrow y_{ij}$ and set $r_i \leftarrow j$.

I3. (Now $y_{i(j-1)} \leq x_i < x_{i+1} = y_{ij} \leq y_{i(j+1)}$, $y_{(i-1)j} < x_i < x_{i+1} = y_{ij} < y_{(i+1)j}$, $r_i = j$, and $x_i \neq \infty$.) Set $y_{ij} \leftarrow x_i$.

I4. (Now $y_{i(j-1)} \leq y_{ij} = x_i < x_{i+1} \leq y_{i(j+1)}$, $y_{(i-1)j} < y_{ij} = x_i < x_{i+1} < y_{(i+1)j}$, $r_i = j$, and $x_i \neq \infty$.) If $x_{i+1} \neq \infty$, increase i by 1 and return to step I2.

I5. Set $s \leftarrow i$ and $t \leftarrow j$, then terminate the algorithm. (Now the conditions

$$y_{st} \neq \infty, \quad x_{s+1} = y_{s(t+1)} = y_{(s+1)t} = \infty \qquad (2.4)$$

hold.)   □

The parenthesized assertions in steps I3 and I4 serve to verify that Y remains a generalized Young tableau throughout the algorithm. The algorithm always terminates in finitely many steps, since Y contains only finitely many positive integers. The procedure not only inserts x into the tableau, it also constructs two sequences of positive integers

$$x = x_1 < x_2 < \cdots < x_s$$
$$r_1 \geq r_2 \geq \cdots \geq r_s = t, \qquad (2.5)$$

where s and t are the quantities specified in the last step of the algorithm.

As an example of this insertion process, let us insert $x = 3$ into the tableau

$$Y = \begin{array}{|c|c|c|c|c|}
\hline
1 & 3 & 3 & 5 & 8 \\
\hline
\multicolumn{1}{|c|}{2} & 4 & 6 & 6 \\
\cline{1-4}
\multicolumn{1}{|c|}{3} & 5 & 8 \\
\cline{1-3}
\multicolumn{1}{|c|}{4} \\
\cline{1-1}
\end{array} \qquad (2.6)$$

The algorithm computes $x_1 = 3$, $r_1 = 4$, $x_2 = 5$, $r_2 = 3$, $x_3 = 6$, $r_3 = 3$, $x_4 = 8$, $r_4 = 2$, $x_5 = \infty$, $s = 4$, $t = 2$, as it changes the tableau to

$$Y = \begin{array}{|c|c|c|c|c|}
\hline
1 & 3 & 3 & 3 & 8 \\
\hline
\multicolumn{1}{|c|}{2} & 4 & 5 & 6 \\
\cline{1-4}
\multicolumn{1}{|c|}{3} & 5 & 6 \\
\cline{1-3}
\multicolumn{1}{|c|}{4} & 8 \\
\cline{1-2}
\end{array} \qquad (2.7)$$

The input value, 3, has "bumped" a 5 from the first row into the second row, where the 5 bumped a 6 to row 3, etc.

The most important property of Schensted's insertion algorithm is that it has an inverse; we can restore Y to its original condition again, given the values of s and t, using the following algorithm.

DELETE(s, t):

D1. Set $j \leftarrow t$, $i \leftarrow s$, $x_{s+1} \leftarrow \infty$.

D2. (Now $y_{ij} < x_{i+1} < y_{(i+1)j}$ and $y_{ij} \neq \infty$.) If $y_{i(j+1)} < x_{i+1}$ and $y_{i(j+1)} \neq \infty$, increase j by 1 and repeat this step. Otherwise set $x_i \leftarrow y_{ij}$ and $r_i \leftarrow j$.

D3. (Now $y_{i(j-1)} \leq y_{ij} = x_i < x_{i+1} \leq y_{i(j+1)}$, $y_{(i-1)j} < y_{ij} = x_i < x_{i+1} < y_{(i+1)j}$, $r_i = j$, and $x_i \neq \infty$.) Set $y_{ij} \leftarrow x_{i+1}$.

D4. (Now $y_{i(j-1)} \leq x_i < x_{i+1} = y_{ij} \leq y_{i(j+1)}$, $y_{(i-1)j} < x_i < x_{i+1} = y_{ij} < y_{(i+1)j}$, $r_i = j$, and $x_i \neq \infty$.) If $i \neq 1$, decrease i by 1 and return to step D2.

D5. Set $x \leftarrow x_1$, and terminate the algorithm. (Now $x \neq \infty$.)   □

This algorithm obviously terminates, since Y contains only finitely many positive integers. The parenthesized assertions in steps D3 and D4 show that Y remains a generalized Young tableau; moreover, those assertions uniquely define the value of j, and they are precisely the same as those of steps I4 and I3, respectively. Hence the deletion algorithm recomputes the sequences (2.5) determined by the insertion algorithm, and it restores Y to its original condition. The reader may verify, for example, that DELETE$(4, 2)$ transforms (2.7) into (2.6).

Conversely, if we start with any generalized Young tableau, Y, and if we choose two integers (s, t) such that (2.4) holds, the procedure DELETE(s, t) will specify some positive integer x in step D5, and will remove x from the tableau; the subsequent operation INSERT(x) will put x back again, recomputing s and t and restoring Y to its original form. Thus INSERT and DELETE are inverses of each other.

We will now establish an important property relating the quantities x, s, t in successive insertions (see Schützenberger [8, Remarque 2]).

Theorem 1. *If* INSERT(x), *determining s and t, is immediately followed by* INSERT(x'), *determining s' and t', then*

$$x \leq x' \quad \text{if and only if} \quad s \geq s' \quad \text{if and only if} \quad t < t'. \qquad (2.8)$$

Proof. We prove first that

$$x \leq x' \quad \text{implies} \quad s \geq s' \quad \text{and} \quad t < t'. \qquad (2.9)$$

Let the sequences (2.5) be denoted by x_i, r_i $(1 \leq i \leq s)$ and x'_i, r'_i $(1 \leq i \leq s')$ when x and x' are respectively inserted. Assume by induction that $s \geq i$ and $s' \geq i$ and $x_i \leq x'_i$; this condition holds initially for

$i = 1$. Consider the state of affairs at the beginning of step I3, when x_i' is about to be inserted. We have $x_i = y_{ij}$ for $j = r_i$, hence $j' = r_i' > j$; it follows that $x_{i+1} \leq y_{i(j+1)} \leq y_{ij'} = x_{i+1}'$. If $s' = i$ then $s \geq s'$ and $t' = j' > j \geq t$, so (2.9) holds. On the other hand if $s' > i$ then $x_{i+1}' \neq \infty$; hence $x_{i+1} \neq \infty$, so $s > i$ and the inductive hypothesis is valid for i replaced by $i + 1$.

The theorem now follows if we can prove that

$$x > x' \quad \text{implies} \quad s < s' \quad \text{and} \quad t \geq t'. \tag{2.10}$$

The proof is like the proof of (2.9), but just different enough to require care. Assume by induction that $s \geq i$ and $s' \geq i$ and $x_i' < x_i$; this condition holds initially for $i = 1$. Consider the state of affairs when x_i' is about to be inserted; we have $j \geq r_i' = j'$, hence $x_{i+1}' = y_{ij'} \leq x_i < x_{i+1}$. In particular, $x_{i+1}' \neq \infty$, so $s' > i$. If $s = i$ then $t = j \geq j' \geq t'$, so (2.10) holds. If $s > i$ then the induction hypothesis is valid for i replaced by $i + 1$. □

3. A One-to-One Correspondence

We are now ready to give a fairly direct correspondence between two-line arrays of positive integers

$$\begin{pmatrix} u_1 & u_2 & \dots & u_N \\ v_1 & v_2 & \dots & v_N \end{pmatrix}, \tag{3.1}$$

and ordered pairs (P, Q) of generalized Young tableaux having the same shape, where the pairs (u_k, v_k) are arranged in nondecreasing lexicographic order from left to right, and where the elements of P and Q are respectively $\{v_1, v_2, \ldots, v_N\}$ and $\{u_1, u_2, \ldots, u_N\}$.

The procedure, which we will call construction A, starts with "empty" tableaux:

$$p_{00} = p_{0j} = p_{i0} = q_{00} = q_{0j} = q_{i0} = 0, \ p_{ij} = q_{ij} = \infty, \quad \text{for all } i, j \geq 1. \tag{3.2}$$

Then we do the following steps for $k = 1, 2, \ldots, N$ (in this order):

A1. INSERT(v_k) into tableau P, determining values s_k and t_k in step I5.

A2. Set $q_{s_k t_k} \leftarrow u_k$.

The reader may verify, for example, that this procedure takes the two-line array (1.4) into the tableaux of Figure 1. It is clear from the construction that P and Q have the same shape, since the insertion procedure removes the ∞ from row s and column t of the tableau. Furthermore, since $u_1 \leq u_2 \leq \cdots \leq u_N$, and since step A2 inserts an element on the "periphery" of Q, it is clear that Q will be a generalized Young tableau if we can verify that no equal elements fall into the same column of Q. The latter property follows immediately from Theorem 1, for $u_k = u_{k+1}$ implies that $v_k \leq v_{k+1}$, hence $t_{k+1} > t_k$.

The inverse construction, which we will call construction B, starts with two generalized Young tableaux, P and Q, of shape $(p_1, p_2, \ldots, p_m)$; let $N = p_1 + p_2 + \cdots + p_m$ be the total number of elements. We do the following steps for $k = N, \ldots, 2, 1$ (in this order):

> B1. Find s_k and t_k such that $q_{s_k t_k}$ is the largest positive integer element of Q, where t_k is as large as possible. Set $u_k \leftarrow q_{s_k t_k}$ and then set $q_{s_k t_k} \leftarrow \infty$.
>
> B2. DELETE(s_k, t_k) from tableau P, determining a value x in step D5; set $v_k \leftarrow x$.

This algorithm clearly reverses construction A. Conversely, if we apply construction B to any given pair of generalized Young tableaux having the same shape, we can see by Theorem 1 that the pairs (u_1, v_1), (u_2, v_2), $\ldots$, (u_N, v_N) will be in lexicographic order, namely that $u_1 \leq u_2 \leq \cdots \leq u_N$ and that $u_k = u_{k+1}$ implies $v_k \leq v_{k+1}$. It follows readily that construction A reverses construction B, hence the one-to-one correspondence is established.

Theorem 2. *Constructions A and B, which are inverses of each other, establish a one-to-one correspondence between two-line arrays and generalized Young tableaux having the properties stated above.*   □

By the previously mentioned correspondence between two-line arrays and matrices of nonnegative integers, we have therefore verified the first result advertised in §1.

4. A Graph-Theoretical Viewpoint

The correspondence in the preceding section can be looked at in another way, if we try to build the P and Q tableaux one row at a time instead of using the insertion procedure. The first rows of P and Q can be interpreted in terms of a certain labeled directed graph, which might be called the "inversion digraph" D_1 of the given generalized permutation (3.1); similarly the second rows of P and Q are related to the "second-order inversion digraph" D_2 derived from D_1, and so on. We will now study this graph theoretical interpretation of Schensted's construction, in order to deduce further properties of the correspondence.

Given a two-line array

$$\begin{pmatrix} u_1 & u_2 & \dots & u_N \\ v_1 & v_2 & \dots & v_N \end{pmatrix}, \tag{4.1}$$

we construct its corresponding "inversion digraph" D_1 as follows: There are N vertices, respectively labeled (u_1, v_1), (u_2, v_2), $\dots$, (u_N, v_N); in the discussion below, when we refer to a vertex (u, v) we mean any one of the vertices whose label is (u, v). When $(u, v) \neq (u', v')$, an arc passes from vertex (u, v) to vertex (u', v') if and only if

$$u \le u' \quad \text{and} \quad v \le v'. \tag{4.2}$$

Furthermore we construct arcs between vertices with identical labels by putting all vertices with given label (u, v) into some arbitrary order, say V_1, V_2, $\dots$, V_k, and drawing arcs from V_i to V_j if and only if $i < j$. For example, Figure 3 shows the inversion digraph corresponding to

$$\begin{pmatrix} 1 & 1 & 1 & 1 & 2 & 3 \\ 2 & 3 & 3 & 3 & 1 & 2 \end{pmatrix}.$$

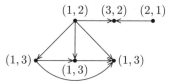

FIGURE 3.

When the two-line array is simply a permutation of the integers $\{1, 2, \dots, N\}$, the number of arcs in D_1 is equal to $\binom{n}{2}$ minus the number of inversions in the permutation according to the classical theory, hence the name "inversion digraph." It is easy to see in the general case that D_1 has no oriented cycles, and in fact it is the digraph of a partial ordering. Notice that our definition of D_1 does not require that the pairs (u_k, v_k) of (4.1) be in lexicographic order from left to right; only the pairs themselves are used. Furthermore (4.2) is symmetric in u and v, hence *the inversion digraph of*

$$\begin{pmatrix} v_1 & v_2 & \dots & v_N \\ u_1 & u_2 & \dots & u_N \end{pmatrix} \tag{4.3}$$

is isomorphic to the inversion digraph of (4.1). This observation will be important to us later.

We now partition the vertices of D_1 into disjoint classes $C_1, C_2, \ldots,$ as follows: Class C_1 contains the "source" vertices, that is, those with no arcs leading in to them; and for $l \geq 1$, class C_{l+1} consists of all vertices that are sources when the vertices of $C_1 \cup \cdots \cup C_l$ (and all arcs touching them) are removed. For example in Figure 3 we have

$$C_1 = \{(1, 2), (2, 1)\},$$
$$C_2 = \{(1, 3), (3, 2)\},$$
$$C_3 = \{(1, 3)\},$$
$$C_4 = \{(1, 3)\}.$$

(The vertices denoted by $(1, 3)$ in C_2, C_3, and C_4 are actually distinct, because of our conventions for dealing with vertices having equal labels.) The reader may easily verify that, in general, C_l consists of all vertices V such that the longest path to V in D_1 has length $l - 1$. This partitioning is closely related to a well-known procedure for "topological sorting" (see [2, §2.2.3]).

If (u, v) and (u', v') are distinct vertices of the same class C_l, there is no arc joining them; it follows from the construction that $u \neq u'$ and $v \neq v'$. Furthermore if $u < u'$ then $v' < v$, and conversely; therefore we can arrange the vertices of C_l into the following order:

$$C_l = (u_{l1}, v_{l1}), (u_{l2}, v_{l2}), \ldots, (u_{ln_l}, v_{ln_l}) \tag{4.4}$$

where

$$u_{l1} < u_{l2} < \cdots < u_{ln_l};$$
$$v_{l1} > v_{l2} > \cdots > v_{ln_l}. \tag{4.5}$$

Lemma 1. *The digraph D_1 constructed above has the following relation to the correspondence defined in §3: If the nonempty vertex classes are $C_1, \ldots, C_d$, the first row of P at the end of the construction is*

$$v_{1n_1}, v_{2n_2}, \ldots, v_{dn_d} \tag{4.6}$$

in the notation of (4.5), and the first row of Q is

$$u_{11}, u_{21}, \ldots, u_{d1}. \tag{4.7}$$

Moreover if we denote the two sequences (2.5) obtained during the operation INSERT(v_k) by

$$v_k = x_{k1} < x_{k2} < \cdots < x_{ks_k},$$
$$r_{k1} \geq r_{k2} \geq \cdots \geq r_{ks_k} = t_k, \tag{4.8}$$

then the pair (u_k, v_k) of (3.1) belongs to class C_l if and only if $l = r_{k1}$.

Proof. We want to show that the vertices of class C_l are those pairs (u_k, v_k) that affect the lth element of the first row of P during the insertion process. The proof is easily carried out by induction on N. For if we add a new vertex (u_{N+1}, v_{N+1}) that is lexicographically greater than all other vertices of D_1, no new arcs lead from this vertex, while there are arcs leading from a vertex of class C_l to this new vertex if and only if $v_{ln_l} \leq v_{N+1}$.   □

Corollary (Schensted). *The number of columns in the generalized Young tableaux (P, Q) corresponding to* (3.1) *is the length of the longest nondecreasing subsequence of the sequence* $v_1, v_2, \ldots, v_N$.

Proof. We have observed that C_d is nonempty if and only if there is a path of length $d - 1$ in D_1; such a path corresponds to a nondecreasing subsequence of length d.   □

We now have characterized the first rows of P and Q in terms of the labeled digraph D_1. Since Schensted's construction behaves on row $i + 1$ in essentially the same way as it does on row i (inserting elements that were bumped down from row i), we can see how to characterize the remaining rows of P and Q. Assuming that the ith-order inversion digraph D_i has been defined, we will construct D_{i+1} by leaving out one vertex of each class and by changing the labels. If class C_l of D_i is given by (4.4) and (4.5), we include $n_l - 1$ vertices labeled

$$(u_{l2}, v_{l1}), (u_{l3}, v_{l2}), \ldots, (u_{ln_l}, v_{l(n_l-1)}) \tag{4.9}$$

in D_{i+1}. After the vertices of D_{i+1} have been determined in this way, from all classes of D_i, the arcs of D_{i+1} are defined in precisely the same manner as we have defined the arcs of D_i.

The vertex labels of D_{i+1} correspond to a two-line array. A few moments' reflection will show that, in view of Lemma 1, construction A in §3 builds rows 2, 3, ... of P and Q in precisely the same way as it would build rows 1, 2, ... if it were given the two-line array corresponding to D_2 instead of the original two-line array. Hence the second rows of P and Q are respectively given by (4.6) and (4.7) corresponding to D_2, and in general *the digraph D_i corresponds to the ith rows of P and Q just as D_1 corresponds to their first rows.* This observation leads to the following result.

Theorem 3. *If the nonnegative integer matrix A corresponds to the generalized Young tableaux (P, Q), then the transposed matrix A^T corresponds to (Q, P).*

Proof. The two-line array corresponding to A^T is obtained from (3.1) by interchanging the two lines and rearranging the columns in lexicographic order. (See (4.3).) We have observed that the resulting graph D_1^T is isomorphic to D_1; this isomorphism associates vertex (v, u) of D_1^T with vertex (u, v) of D_1. The construction of D_{i+1} from D_i shows that the same isomorphism relates D_i^T to D_i, hence the theorem follows from Lemma 1.   □

For the special case of permutation matrices, Robinson [5, page 755] essentially stated Theorem 3 without proof; a proof was subsequently given by Schützenberger [8].

Theorem 4. *The construction of §3 yields a one-to-one correspondence between symmetric matrices A of nonnegative integers, having respective column sums $(c_1, c_2, \ldots, c_n)$, and generalized Young tableaux P, having c_j occurrences of the integer j. In this correspondence the number of columns of P having an odd length is the trace of A.*

Proof. Since A is symmetric, $A = A^T$, hence Theorem 3 tells us that $P = Q$ in the correspondence. It remains only to verify the connection (suggested by Schützenberger) between odd-length columns and the trace of A.

When $A = A^T$, there is a corresponding symmetry in the digraph D_1 since the vertex (u, v) occurs as often as the vertex (v, u). The automorphism $(u, v) \leftrightarrow (v, u)$ shows that each class C_l has the form

$$(u_{l1}, u_{ln_l}), \ (u_{l2}, u_{l(n_l-1)}), \ \ldots, \ (u_{ln_l}, u_{l1}),$$
$$u_{l1} < u_{l2} < \cdots < u_{ln_l}. \tag{4.10}$$

(See (4.5).) Hence C_l contains 0 or 1 elements of the form (u, u) according as n_l is even or odd; and so trace(A) is the number of classes in which n_l is odd. Furthermore the graph D_2 contains as many vertices of the form (u, u) as the number of classes in which n_l is even (see (4.9) and (4.10)); hence it corresponds to a symmetric matrix A_2 such that trace(A)+trace(A_2) = d = number of nonempty classes of D_1 = number of columns of P. Let P_2 be P with its first row removed; by induction on the number of rows of P, we know that P_2 has trace(A_2) odd columns, hence P has $d - $ trace(A_2) = trace(A) odd columns.   □

5. A Dual Correspondence

Let us say a "dual tableau" is an arrangement of positive integers that is like a generalized Young tableau except that the *rows* (instead of the columns) are required to have distinct elements. Thus, every dual tableau is the transpose of a generalized Young tableau and conversely.

If Y is a dual tableau, we can insert a new element x into it using a procedure almost identical to Schensted's construction of §2. The algorithm INSERT*(x) is defined to be the same as INSERT(x), except that the signs $<$ and $\leq$ are interchanged throughout the algorithm. (An element now bumps down another element equal to itself.) Similarly we obtain an algorithm DELETE*(s, t) by changing DELETE(s, t) in the same way. The reader may readily verify that, as before, INSERT*(x) and DELETE*(s, t) are inverse to each other, and that Y remains a dual tableau throughout each algorithm. We also have

Theorem 1*. *If* INSERT*(x), *determining s and t, is immediately followed by* INSERT*(x'), *determining s' and t', then*

$$x < x' \quad \text{if and only if} \quad s \geq s' \quad \text{if and only if} \quad t < t'. \tag{5.1}$$

Proof. In the proof of Theorem 1, change the symbol $<$ to $\leq$ wherever it appears; and change $\leq$ to $<$, except in the two instances

$$\text{`}y_{i(j+1)} \leq y_{ij'}\text{'} \quad \text{and} \quad \text{`}y_{ij'} \leq x_i\text{'}$$

where the $\leq$ is to be retained. (Do not change the symbols $>$ and $\geq$, which have been used consistently for indices instead of elements.) The result is a proof of Theorem 1*. ◻

Now consider a two-line array of positive integers

$$\begin{pmatrix} u_1 & u_2 & \dots & u_N \\ v_1 & v_2 & \dots & v_N \end{pmatrix} \tag{5.2}$$

where the pairs (u_k, v_k) are all *distinct* and arranged in increasing lexicographic order. (Such arrays correspond to matrices A of zeros and ones.) Starting with empty tableaux P and Q as in (3.2), we perform the following steps for $k = 1, 2, \dots, N$ (in this order):

A1*. INSERT*(v_k) into the dual tableau P, determining values s_k and t_k in step I5*.

A2*. Set $q_{s_k t_k} \leftarrow u_k$.

By Theorem 1* and an argument like that of §3, this construction makes Q a generalized Young tableau, while P is a dual tableau. The inverse procedure consists of the following steps for $k = N, \dots, 2, 1$ (in this order):

B1*. Find s_k and t_k such that $q_{s_k t_k}$ is the largest positive integer element of Q, where t_k is as large as possible. Set $u_k \leftarrow q_{s_k t_k}$ and then set $q_{s_k t_k} \leftarrow \infty$.

B2*. DELETE*(s_k, t_k) from tableau P, determining a value x in step D5*; set $v_k \leftarrow x$.

By Theorem 1* this procedure will produce a two-line array (5.2) of distinct pairs in increasing lexicographic order, when given any dual tableau P together with a generalized Young tableau Q of the same shape. Therefore we have a correspondence between such pairs of tableaux and zero-one matrices.

The graph-theoretic equivalent of this construction, corresponding to §4, is obtained by changing (4.2) to

$$u \leq u' \text{ and } v < v'. \tag{5.3}$$

The lack of symmetry in (5.3) makes it impossible to find a simple relation between the tableaux corresponding to a matrix and its transpose; in general the latter two pairs of tableaux can be quite different.

6. Further Properties

Let us now concentrate momentarily on the P tableau, independently of Q. By varying Q, we will in general find many arrays A corresponding to the same P tableau, and it is of interest to look for characteristic properties of such arrays.

Theorem 5. *Let x, x', x'' be positive integers and let Y be a generalized Young tableau. If $x \leq x' < x''$, the sequence of operations*

$$\text{INSERT}(x''), \text{ INSERT}(x), \text{ INSERT}(x') \tag{6.1}$$

has the same effect on Y as the sequence

$$\text{INSERT}(x), \text{ INSERT}(x''), \text{ INSERT}(x'). \tag{6.2}$$

If $x < x' \leq x''$, the sequence

$$\text{INSERT}(x'), \text{ INSERT}(x), \text{ INSERT}(x'') \tag{6.3}$$

has the same effect as

$$\text{INSERT}(x'), \text{ INSERT}(x''), \text{ INSERT}(x). \tag{6.4}$$

Proof. Using notation like that in the proof of Theorem 1, we will show that the sequences in question have the same effect on the first row of Y, and that the corresponding elements x_2, x_2', x_2'' will (by induction) yield the same results on the remainder of Y. By convention let us say that INSERT(∞) is the null operation; when $x'' = \infty$ the theorem holds trivially.

It is not difficult to verify that if x does not displace x'' in (6.1), then the elements x_2, x_2', x_2'' are identical for sequences (6.2) and (6.1), and $x_2 \le x_2' < x_2''$. On the other hand if x does displace x'' in (6.1) then the first row of Y in its original state had the form

$$\ldots\, y\, y'\, y'' \ldots$$

where $y \le x \le x' < x'' < y' \le y''$. After (6.1), the first row of Y becomes

$$\ldots\, y\, x\, x' \ldots$$

and $x_2'' = y'$, $x_2 = x''$, $x_2' = y''$. If instead we use (6.2), the first row becomes

$$\ldots\, y\, x\, x' \ldots$$

and $x_2 = y'$, $x_2'' = y''$, $x_2' = x''$. By induction (using the fact that (6.3) is equivalent to (6.4)),

$$\text{INSERT}(y'),\ \text{INSERT}(x''),\ \text{INSERT}(y'')$$

has the same effect as

$$\text{INSERT}(y'),\ \text{INSERT}(y''),\ \text{INSERT}(x'');$$

hence (6.1) is equivalent to (6.2).

A similar but somewhat simpler proof shows that (6.3) is equivalent to (6.4).   □

Notice that, once the two-line array (3.1) has been put into lexicographic order, the P tableau constructed in §3 is a function of the sequence $v_1, v_2, \ldots, v_N$ only. In general let $(v_1, v_2, \ldots, v_n)$ be any sequence of positive integers, and let $P(v_1, v_2, \ldots, v_n)$ be the generalized Young tableau obtained by starting with an empty tableau and successively performing the operations

$$\text{INSERT}(v_1),\ \text{INSERT}(v_2),\ \ldots,\ \text{INSERT}(v_n).$$

We will write

$$(v_1, \ldots, v_n) \equiv (w_1, \ldots, w_n) \quad \text{if and only if}$$
$$P(v_1, \ldots, v_n) = P(w_1, \ldots, w_n). \tag{6.5}$$

According to Theorem 5, we have

$$(v_1, \ldots, v_n) \equiv (v_1, \ldots, v_{k-1}, v_{k+1}, v_k, v_{k+2}, \ldots, v_n)$$

when

$$v_k \leq v_{k+2} < v_{k+1} \quad \text{or} \quad v_k < v_{k-1} \leq v_{k+1}. \tag{6.6}$$

Symbolically, if $a < b < c$, we have

$$acb \equiv cab, \quad bac \equiv bca, \quad aba \equiv baa, \quad bab \equiv bba. \tag{6.7}$$

Using these elementary transformations we can usually find several sequences equivalent to a given one; for example,

$$(4, 2, 1, 1, 2, 3) \equiv (4, 1, 2, 1, 2, 3) \equiv (1, 4, 2, 1, 2, 3) \equiv (1, 4, 2, 2, 1, 3)$$
$$\equiv (4, 1, 2, 2, 1, 3) \equiv (4, 1, 2, 2, 3, 1) \equiv (1, 4, 2, 2, 3, 1)$$
$$\equiv (1, 2, 4, 2, 3, 1) \equiv (1, 2, 4, 2, 1, 3) \equiv (1, 2, 2, 4, 3, 1).$$

In fact, somewhat surprisingly, the elementary transformations of Theorem 4 precisely characterize the sequences that yield the same P tableau:

Theorem 6. $P(v_1, \ldots, v_n) = P(w_1, \ldots, w_n)$ *if and only if* $(v_1, \ldots, v_n)$ *can be transformed into* $(w_1, \ldots, w_n)$ *by a sequence of the elementary transformations* (6.6).

Proof. Let $P(v_1, \ldots, v_n)$ be a generalized Young tableau Y of shape $(p_1, \ldots, p_m)$; and let

$$(u_1, \ldots, u_n) = (y_{m1}, \ldots, y_{mp_m}, \ldots, y_{21}, \ldots, y_{2p_2}, y_{11}, \ldots, y_{1p_1}).$$

It is easy to show that $P(v_1, \ldots, v_n) = P(u_1, \ldots, u_n)$; hence in view of Theorem 5 it suffices to prove that $(v_1, \ldots, v_n)$ can be transformed into the "canonical" sequence $(u_1, \ldots, u_n)$ for the tableau Y using only the operations (6.6). By induction on n, we need only prove the following statement: *If x is a positive integer, if $(u_1, \ldots, u_n)$ is the canonical sequence for a generalized Young tableau Y, and if $(u'_1, \ldots, u'_n, u'_{n+1})$ is the canonical sequence for Y after the operation* INSERT(x) *has been* *performed, then* $(u_1, \ldots, u_n, x)$ *can be transformed into* $(u'_1, \ldots, u'_{n+1})$ *using the operations* (6.6).

This necessary condition is easily proved, for if we assume that

$$y_1 \leq \cdots \leq y_{j-1} \leq x < y_j \leq y_{j+1} \leq \cdots \leq y_k,$$

where $1 \leq j \leq k$, we have

$$(y_1, \ldots, y_k, x) \equiv \cdots \equiv (y_1, \ldots, y_{j-1}, y_j, x, y_{j+1}, \ldots, y_k)$$
$$\equiv \cdots \equiv (y_j, y_1, \ldots, y_{j-1}, x, y_{j+1}, \ldots, y_k)$$

by rules (6.6); hence we can simulate the insertion algorithm (since $y_j = x_2$, etc.).   □

Corollary. *If* $P(v_1, \ldots, v_m) = P(w_1, \ldots, w_m)$ *and* $P(v_{m+1}, \ldots, v_n) = P(w_{m+1}, \ldots, w_n)$, *then*

$$P(v_1, \ldots, v_m, v_{m+1}, \ldots, v_n) = P(w_1, \ldots, w_m, w_{m+1}, \ldots, w_n).   □$$

This corollary can be expressed in terms of matrices as follows: We denote the P tableau corresponding to matrix A by $P(A)$. Let

$$A = \begin{pmatrix} A_1 \\ A_2 \end{pmatrix}, \qquad B = \begin{pmatrix} B_1 \\ B_2 \end{pmatrix}$$

be two $(m_1 + m_2) \times n$ matrices partitioned into $m_1 \times n$ and $m_2 \times n$ submatrices. If $P(A_1) = P(B_1)$ and $P(A_2) = P(B_2)$, then $P(A) = P(B)$. A similar statement holds for the Q tableau, by taking transposes (see Theorem 2).

The corollary can also be used to define an associative binary operation on generalized Young tableaux if we let

$$P(v_1, \ldots, v_m) P(v_{m+1}, \ldots, v_n) = P(v_1, \ldots, v_n).$$

Some very special cases of this associative operation were discovered by Schensted in his original paper [7]. Perhaps some further properties of the correspondence can be deduced from a deeper study of this "tableau algebra."

Similar properties can be developed for the dual correspondence of §4. If we interchange $<$ and $\leq$ in Theorem 5 and change INSERT to INSERT*, we obtain Theorem 5*, which leads to a corresponding Theorem 6* and its corollary in an analogous way. We can also show that the elementary transformations (6.6) allow us to build the dual tableau P^* of a sequence by working on columns instead of rows during the insertion algorithm, and inserting the elements in the order $v_n, \ldots, v_2, v_1$. Therefore $P(v_n, \ldots, v_1)$ is the transpose of $P^*(v_1, \ldots, v_n)$; this interesting relation implies that *the number of rows of* $P(v_1, \ldots, v_n)$ *is the length of the longest strictly decreasing subsequence of* $(v_1, \ldots, v_n)$. (The latter result is due to Schensted [7].) The details underlying these observations are straightforward and left to the leader.

Let us conclude our remarks by deriving a further consequence of Theorem 6:

Theorem 7. *Let* $\pi(1)\pi(2)\ldots\pi(m)$ *be a permutation of the integers* $\{1, 2, \ldots, m\}$. *There is a constructive one-to-one correspondence between the set of all nonnegative integer* $m \times n$ *matrices with row sums* $(r_1, \ldots, r_m)$ *and those with row sums* $(r_{\pi(1)}, \ldots, r_{\pi(m)})$ *such that, if* A *corresponds to* B, *we have* $P(A) = P(B)$.

Proof. Since any permutation is a product of adjacent transpositions, we may assume without loss of generality that π merely interchanges k and $k+1$. Let us partition

$$A = \begin{pmatrix} A_1 \\ A_2 \\ A_3 \end{pmatrix},$$

where A_1 has $k-1$ rows, A_2 has 2 rows, and A_3 has $m-k-1$ rows. Let the row sums of A_2 be (r, s); it suffices to construct a $2 \times n$ matrix B_2 such that $P(A_2) = P(B_2)$ and such that the row sums of B_2 are (s, r), since the corollary to Theorem 6 tells us that $P(A) = P(B)$ when

$$B = \begin{pmatrix} A_1 \\ B_2 \\ A_3 \end{pmatrix}.$$

The tableau $Q(A_2)$ has r 1's and $(s-t)$ 2's in its first row, and it has t 2's in its second row, for some $t \leq \min(r, s)$. We now define B_2 by saying that $P(B_2) = P(A_2)$ and $Q(B_2)$ has s 1's and $(r-t)$ 2's in its first row, t 2's in its second row. The correspondence of §3 determines a unique B_2 with this property, so the mapping $A_2 \leftrightarrow B_2$ is reversible. $\square$

Corollary. *Let* $p_1 \geq p_2 \geq \cdots \geq p_m \geq 1$ *and* $p_1 + p_2 + \cdots + p_m = N = r_1 + r_2 + \cdots + r_m$, *where* $r_1, r_2, \ldots, r_m$ *are positive integers. Let* $\pi(1)\pi(2)\ldots\pi(m)$ *be a permutation of the integers* $\{1, 2, \ldots, m\}$. *There is a constructive one-to-one correspondence between the set of all generalized Young tableaux of shape* $(p_1, \ldots, p_m)$ *having* r_1 *1's,* r_2 *2's,*

..., r_m m's, and those tableaux of the same shape having $r_{\pi(1)}$ 1's, $r_{\pi(2)}$ 2's, ..., $r_{\pi(m)}$ m's.

Proof. Let P be any fixed generalized Young tableau of shape $(p_1, p_2, \ldots, p_m)$, and consider the correspondence of Theorem 7 as Q varies over all tableaux of the same shape, having r_i occurrences of element i. □

In other words, the number of ways to fill a shape with specified numbers of elements of different kinds is actually independent of the order of those elements.

7. Generating Functions

Let $x = (x_1, \ldots, x_n)$ be a vector of formal variables, and let

$$\{x; p\} = \sum_Y \prod_{i=1}^{m} Y_{i1} \ldots Y_{ip_i}$$

summed over all generalized Young tableaux Y of shape $p = (p_1, p_2, \ldots, p_m)$ that have been filled with the elements $x_1 < x_2 < \cdots < x_n$ each repeated any number of times. For example, we have

$$\{x_1, x_2, x_3; (2, 1)\} = x_1^2 x_2 + x_1^2 x_3 + x_1 x_2^2 + x_1 x_3^2 + x_2^2 x_3 + x_2 x_3^2 + 2 x_1 x_2 x_3$$

corresponding to the tableaux

$$\boxed{\begin{array}{|c|c|}\hline 1 & 1 \\ \hline 2 \\ \hline\end{array}}, \boxed{\begin{array}{|c|c|}\hline 1 & 1 \\ \hline 3 \\ \hline\end{array}}, \boxed{\begin{array}{|c|c|}\hline 1 & 2 \\ \hline 2 \\ \hline\end{array}}, \boxed{\begin{array}{|c|c|}\hline 1 & 3 \\ \hline 3 \\ \hline\end{array}}, \boxed{\begin{array}{|c|c|}\hline 2 & 2 \\ \hline 3 \\ \hline\end{array}}, \boxed{\begin{array}{|c|c|}\hline 2 & 3 \\ \hline 3 \\ \hline\end{array}}, \boxed{\begin{array}{|c|c|}\hline 1 & 2 \\ \hline 3 \\ \hline\end{array}}, \boxed{\begin{array}{|c|c|}\hline 1 & 3 \\ \hline 2 \\ \hline\end{array}}.$$

Dudley E. Littlewood [4, page 191] has shown by group-theoretic means that $\{x; p\}$ is a symmetric function of the x's, which is identical to a function studied by Jacobi, Schur, and others.

The two correspondences we have exhibited therefore provide a constructive proof of Littlewood's identities [3, Theorem V]

$$\prod_{i=1}^{m} \prod_{j=1}^{n} \frac{1}{1 - x_i y_j} = \sum_p \{x_1, \ldots, x_m; p\}\{y_1, \ldots, y_n; p\}$$

$$\prod_{i=1}^{m} \prod_{j=1}^{n} (1 + x_i y_j) = \sum_p \{x_1, \ldots, x_m; p\}\{y_1, \ldots, y_n; p^T\}$$

summed over all shapes p, where p^T denotes the transposed shape. The Jacobi–Trudi identity and the Naegelsbach–Kostka identity [4, pages 88–89] can also be established combinatorially by means of our correspondences, as shown in [1].

Acknowledgments

I wish to thank Edward A. Bender for many stimulating discussions concerning the problems considered in this paper, and I also wish to thank Prof. Schützenberger for calling my attention to Schensted's construction and to reference [5].

References

[1] Edward A. Bender and Donald E. Knuth, "Enumeration of plane partitions," *Journal of Combinatorial Theory* **A13** (1972), 40–54. [Reprinted as Chapter 32 of the present volume.]

[2] Donald E. Knuth, *Fundamental Algorithms*, Volume 1 of *The Art of Computer Programming* (Reading, Massachusetts: Addison–Wesley, 1968).

[3] D. E. Littlewood, "Some properties of S-functions," *Proceedings of the London Mathematical Society* (2) **40** (1936), 49–70.

[4] Dudley E. Littlewood, *The Theory of Group Characters and Matrix Representations of Groups* (Oxford: Clarendon Press, 1940).

[5] G. de B. Robinson, "On the representations of the symmetric group," *American Journal of Mathematics* **60** (1938), 745–760; **69** (1947), 286–298; **70** (1948), 277–294.

[6] Daniel E. Rutherford, *Substitutional Analysis* (New York: Hafner, 1968).

[7] C. Schensted, "Longest increasing and decreasing subsequences," *Canadian Journal of Mathematics* **13** (1961), 179–191.

[8] M. P. Schützenberger, "Quelques remarques sur une construction de Schensted," *Mathematica Scandinavica* **12** (1963), 117–128.

[9] Alfred Young, "On quantitative substitutional analysis (*third paper*)," *Proceedings of the London Mathematical Society* (2) **28** (1928), 255–292.

Addendum

The formula for $\prod_{i,j} 1/(1 - x_i y_j)$ that I found in Littlewood's book is now known as "Cauchy's identity"; Richard Stanley points out in his *Enumerative Combinatorics*, Volume 2 (Cambridge: Cambridge University Press, 1999), 397–398, that it is closely related to two well-known identities found in Cauchy's works. (Indeed, those two identities appear in exercises 1.2.3–38 and 1.2.3–46 of my own book, *Fundamental Algorithms*, but I did not see the connection.)

Chapter 32

Enumeration of Plane Partitions

*[Written with Edward A. Bender. Originally published in Journal of Combinatorial Theory **A13** (1972), 40–54.]*

Using some recent results involving Young tableaux and matrices of non-negative integers [10], it is possible to enumerate various classes of plane partitions by actual construction. One of the results is a simple proof of MacMahon's generating function for plane partitions [12]. Previous results of this type [12, 4, 3, 8, 7] involved complicated algebraic methods that did not reveal any intrinsic "reason" why the corresponding generating functions have such a simple form.

Introduction

A *plane partition* of n is an array of nonnegative integers

$$
\begin{array}{cccc}
n_{11} & n_{12} & n_{13} & \cdots \\
n_{21} & n_{22} & n_{23} & \cdots \\
\vdots & \vdots & \vdots & \ddots
\end{array}
\tag{1}
$$

for which $\sum_{i,j} n_{ij} = n$ and the rows and columns are in nonincreasing order:

$$
n_{ij} \geq n_{(i+1)j}, \quad n_{ij} \geq n_{i(j+1)}, \quad \text{for all } i, j \geq 1.
$$

If $n_{ij} = 0$ for all $i > r$, we call (1) an *r-rowed* partition. If $n_{ij} = 0$ for all $j > c$, it is *c-columned*. If $n_{ij} \leq m$ for all $i, j \geq 1$, we say the parts *do not exceed* m. If $n_{ij} > n_{i(j+1)}$ whenever $n_{ij} \neq 0$, then the nonzero elements of (1) are strictly decreasing in each row; we shall call such a partition *strict*.

If π_n is the number of partitions of n with certain restrictions, then the *generating function* for π_n is

$$
1 + \sum_{n=1}^{\infty} \pi_n x^n.
\tag{2}
$$

It is interesting to find a representation for (2) that does not involve π_n. Since the π_n's are integers, there exist integers b_n such that

$$1 + \sum_{n=1}^{\infty} \pi_n x^n = \prod_{n=1}^{\infty} (1 - x^n)^{b_n}. \tag{3}$$

Constructive methods for studying (3) in the case of plane partitions are almost nonexistent, in contrast to the theory of linear partitions where such techniques were developed very early [14]. In 1916, P. A. MacMahon devoted a considerable part of his second volume of *Combinatory Analysis* to the determination of b_n for r-rowed, c-columned partitions in which the parts do not exceed m. Recent work by Gordon and Houten [8] determined b_n for strict r-rowed partitions. Gordon [7] solved the strict r-rowed case when all parts are odd. All these proofs involve complicated arguments on generating functions although the b_n's are fairly simple.

We will give combinatorial proofs of these results when $r = \infty$. We also solve (3) for strict partitions whose parts lie in an arbitrary set. If the number of rows or columns is also restricted, (2) will be expressed in terms of generating functions for linear partitions; in this case we do not know how to solve (3). The 2-rowed constructions of Cheema and Gordon [5] and Sudler [13] are special cases of our methods.

A generalized *Young tableau* of shape $(p_1, p_2, \ldots, p_r)$, where $p_1 \geq p_2 \geq \cdots \geq p_r \geq 1$, is an arrangement of $p_1 + p_2 + \cdots + p_r$ positive integers into r left-justified rows with p_i elements in row i, where the numbers in each row are in nondecreasing order from left to right, and the numbers in each column are in strictly increasing order from top to bottom. Fortunately the relevant theory still holds if "nondecreasing" is replaced by "nonincreasing" and "strictly increasing" by "strictly decreasing." (The mapping $n \to$ constant $- n$ provides a simple proof of this fact.) We will still call the result a Young tableau. Thus a Young tableau is the transpose of a strict partition. We shall need the following results of Knuth [10].

Theorem A. *There is a one-to-one correspondence between Young tableaux P and symmetric matrices A of nonnegative integers. In this correspondence,*

$$k \text{ appears in } P \text{ exactly } \sum_i a_{ik} \text{ times,}$$

and trace(A) *is the number of columns of P having odd length.* $\square$

Theorem B. *There is a one-to-one correspondence between matrices A of nonnegative integers and ordered pairs (P, Q) of Young tableaux of the same shape. In this correspondence,*

$$k \text{ appears in } P \text{ exactly } \sum_i a_{ik} \text{ times,}$$

and

$$k \text{ appears in } Q \text{ exactly } \sum_j a_{kj} \text{ times.} \quad \square$$

Theorem C. *There is a one-to-one correspondence between matrices A of zeros and ones and ordered pairs (P, Q) of Young tableaux whose shapes are transposes of each other. In this correspondence,*

$$k \text{ appears in } P \text{ exactly } \sum_i a_{ik} \text{ times,}$$

and

$$k \text{ appears in } Q \text{ exactly } \sum_j a_{kj} \text{ times.} \quad \square$$

Strict Partitions

Since a Young tableau is the transpose of a strict partition we can easily use Theorem A to prove the following:

Theorem 1. *The generating function for strict plane partitions whose parts lie in a set S of positive integers is*

$$\prod_{i \in S} \left((1 - x^i)^{-1} \prod_{\substack{j \in S \\ j > i}} (1 - x^{i+j})^{-1} \right). \tag{4}$$

Corollary (Gordon and Houten [8]). *The generating function for strict plane partitions is*

$$\prod_{n=1}^{\infty} (1 - x^n)^{-\lfloor (n+1)/2 \rfloor}.$$

Corollary (Gordon [7]). *The generating function for strict plane partitions with odd parts is*

$$\prod_{n=1}^{\infty} (1 - x^{2n-1})^{-1} (1 - x^{2n})^{-\lfloor n/2 \rfloor}.$$

This function also generates the number of symmetric plane partitions (that is, plane partitions with $n_{ij} = n_{ji}$ in (1)).

Proof. Partitions of the type desired in the theorem are in one-to-one correspondence with Young tableaux whose entries lie in S. By Theorem A, there is a one-to-one correspondence between partitions of n

of the desired type and symmetric matrices with nonnegative integer entries such that, for any such matrix A,

$$a_{ij} = 0 \text{ if } j \notin S; \quad \text{and} \quad \sum_{i,j} j a_{ij} = n.$$

Hence the desired generating function is

$$\sum \prod_{i,j} x^{j a_{ij}},$$

where the sum is over all matrices of the type described above. By rewriting we obtain

$$\sum_{A} \Big(\prod_{i} x^{i a_{ii}} \prod_{j>i} x^{(i+j) a_{ij}} \Big)$$

$$= \prod_{i \in S} \Big(\sum_{a_{ii}=0}^{\infty} x^{i a_{ii}} \prod_{\substack{j \in S \\ j>i}} \sum_{a_{ij}=0}^{\infty} x^{(i+j) a_{ij}} \Big)$$

$$= \prod_{i \in S} \Big((1 - x^i)^{-1} \prod_{\substack{j \in S \\ j>i}} (1 - x^{i+j})^{-1} \Big).$$

The last part of the second corollary requires the well-known correspondence between partitions into distinct odd parts and self-conjugate partitions [9, page 278]. This correspondence is applied to each row.   □

No result like Theorem 1 appears to exist when the strictness condition is removed. Perhaps 3-dimensional partitions that are "strict" in some sense could be enumerated. Practically nothing is known about solid partitions except that MacMahon's conjecture [12, page 175] is false [1].

We now consider extensions of Theorem 1. The restriction that S consist of positive integers is irrelevant; we need only require that S be linearly ordered. (As can be seen from (4), the actual ordering is irrelevant.) Since the product $\prod_{i \in S}(1 - x^i)^{-1}$ in (4) comes from the diagonal of A, we can use Theorem A to show that the generating function for strict partitions with parts in S and exactly k odd rows is the coefficient of y^k in

$$\prod_{i \in S} \Big((1 - x^i y)^{-1} \prod_{\substack{j \in S \\ j>i}} (1 - x^{i+j})^{-1} \Big).$$

We can also keep track of the occurrence of parts. In

$$\prod_{i \in S}\left((1 - x_i y)^{-1} \prod_{\substack{j \in S \\ j > i}}(1 - x_i x_j)^{-1}\right)$$

the coefficient of $y^k x_1^{a_1} x_2^{a_2} \ldots$ is the number of strict partitions of $\sum i a_i$ such that i appears exactly a_i times and there are exactly k rows of odd length.

Unrestricted Plane Partitions

To obtain a generating function for unrestricted plane partitions, we shall make use of a straightforward generalization of a construction due to Frobenius [6, page 523]; see also Sudler [13].

Lemma 1. *There is a one-to-one correspondence between linear partitions $n_1 \geq n_2 \geq \cdots \geq n_k > 0$ and ordered pairs of vectors $(\mathbf{p}, \mathbf{q})$ such that*

$$n_1 = p_1 > p_2 > \cdots > p_t > 0,$$
$$k - 1 = q_1 > q_2 > \cdots > q_t \geq 0,$$
$$\sum (p_i + q_i) = \sum n_j.$$

In this correspondence, another linear partition $n_1' \geq n_2' \geq \cdots$ satisfies

$$n_j' \geq n_j \qquad \text{for all } j \geq 1$$

if and only if $p_i' \geq p_i$ and $q_i' \geq q_i$ for all $i \geq 1$. (We take n, p, and q to be 0 when the subscripts lie outside the range of definition.)

Proof. An illustration of the Frobenius construction is given in Figure 1. Details are left to the reader. $\square$

FIGURE 1.

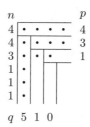

Theorem 2 (MacMahon [12]). *The generating function for r-rowed partitions with parts not exceeding m is*

$$\prod_{i=1}^{r}\prod_{j=1}^{m}(1 - x^{i+j-1})^{-1}.$$

Proof. In Theorem B, consider matrices with r rows and m columns. In the corresponding Young tableaux, $p_{ij} \leq m$ and $q_{ij} \leq r$. Apply the lemma for all $j \geq 1$, with

$$p_i = p_{ij}, \quad q_i = q_{ij} - 1, \quad \text{and} \quad n_i = n_{ij}.$$

This rule gives a one-to-one correspondence with r-rowed partitions with no parts exceeding m. The desired generating function is therefore

$$\sum_{A}\prod_{i,j} x^{(i+j-1)a_{ij}} = \prod_{i=1}^{r}\prod_{j=1}^{m}\sum_{a_{ij}=0}^{\infty} x^{(i+j-1)a_{ij}}. \quad \square$$

For example, the plane partition

$$
\begin{matrix}
n_{11} & n_{12} & n_{13} & n_{14} \\
n_{21} & n_{22} & n_{23} & n_{24} \\
n_{31} & n_{32} & n_{23} & \\
n_{41} & n_{42} & &
\end{matrix}
\quad = \quad
\begin{matrix}
5 & 3 & 3 & 1 \\
4 & 3 & 2 & 1 \\
4 & 2 & 1 & \\
2 & 2 & &
\end{matrix}
$$

corresponds to the Young tableaux

$$
\begin{matrix}
p_{11} & p_{12} & p_{13} & p_{14} \\
p_{21} & p_{22} & p_{23} & \\
p_{31} & & &
\end{matrix}
\; = \;
\begin{matrix}
5 & 3 & 3 & 1 \\
3 & 2 & 1 & \\
2 & & &
\end{matrix}
\qquad
\begin{matrix}
q_{11} & q_{12} & q_{13} & q_{14} \\
q_{21} & q_{22} & q_{23} & \\
q_{31} & & &
\end{matrix}
\; = \;
\begin{matrix}
4 & 4 & 3 & 2 \\
3 & 3 & 1 & \\
1 & & &
\end{matrix} \;,
$$

which correspond to the 4×5 matrix

$$
\begin{pmatrix}
a_{11} & a_{12} & a_{13} & a_{14} & a_{15} \\
a_{21} & a_{22} & a_{23} & a_{24} & a_{25} \\
a_{31} & a_{32} & a_{33} & a_{34} & a_{35} \\
a_{41} & a_{42} & a_{43} & a_{44} & a_{45}
\end{pmatrix}
=
\begin{pmatrix}
0 & 0 & 1 & 0 & 1 \\
1 & 0 & 0 & 0 & 0 \\
1 & 0 & 2 & 0 & 0 \\
0 & 2 & 0 & 0 & 0
\end{pmatrix}.
$$

A closer examination of the correspondence in this proof yields the following mild generalization:

Theorem 2*. *The generating function for r-rowed partitions with parts not exceeding m and with exactly k parts in the rth row is*

$$x^{rk} \prod_{j=1}^{k} \frac{1 - x^{m+j-1}}{1 - x^j} \cdot \prod_{i=1}^{m} \prod_{j=1}^{r-1} (1 - x^{i+j-1})^{-1}.$$

Proof. A column of the plane partition has r entries if and only if the corresponding Q tableau has an r in that column. Thus we want our generating function sum to range over all $r \times m$ matrices with $\sum a_{rj} = k$. The ith row in our matrix corresponds to a partition of $\sum_j (i - 1 + j)a_{ij}$ into $\sum_j a_{ij}$ parts in the range i to $i - 1 + m$. The generating function for $i = r$ and $\sum a_{rj} = k$ is therefore

$$x^{rk} \frac{(1 - x^m) \dots (1 - x^{m+k-1})}{(1 - x) \dots (1 - x^k)}. \quad \square$$

This theorem has a peculiar consequence when $r = 3$, $k = 1$, and $m = \infty$:

Corollary. *The generating function for three-rowed plane partitions with exactly one part in the third row is $x^3 p(x)^2$, where $p(x)$ is the generating function for linear partitions.* $\square$

Strict Partitions of Given Shape

In Theorem 1 we gave a simple generating function for strict partitions whose parts lie in a given set S. In this section we enumerate those that have a given shape. When S is the set of positive integers we obtain a formula of Gordon and Houten for r-rowed partitions [8, Part II, Eq. (3)], which they simplified after considerable work. When S is the set of odd positive integers we obtain Gordon's r-rowed result [7], which also required considerable algebra to put in a simple form. Unfortunately, we do not know how to avoid this algebra.

Let $\mathbf{p} = (p_1, \dots, p_r)$ be the shape of a Young tableau and let $\mathbf{v}$ be an element of $\mathbf{Z}^n$, the n-tuples of integers. Define $N(\mathbf{p}, \mathbf{v})$ to be the number of ways to fill the shape $\mathbf{p}$ to obtain a Young tableau containing exactly v_k copies of k for $1 \le k \le n$. Clearly $N(\mathbf{p}, \mathbf{v}) = 0$ if some $v_k < 0$, or if $\mathbf{p}$ has more than n rows, or if $\sum v_k \ne \sum p_j$. Define $\epsilon(\mathbf{v})$ to be zero if $\mathbf{v}$ contains two equal components; otherwise $\epsilon(\mathbf{v}) = +1$ or -1 according as the permutation that arranges the components of $\mathbf{v}$ into ascending order is even or odd. In particular,

$$\epsilon((1, 2, \dots, n)) = +1 \quad \text{and} \quad \epsilon((n, n - 1, \dots, 1)) = (-1)^{n(n-1)/2}.$$

Lemma 2. *Let* $\mathbf{r} = (n-1,\ldots,1,0)$ *and* $\mathbf{i} = (0,1,\ldots,n-1)$, *let* $\mathbf{p}$ *and* $\mathbf{q}$ *be shapes of n-rowed tableaux, and let* $\Pi(\mathbf{q})$ *denote all permutations of* $\mathbf{q}$. *Then*

$$\sum_{\mathbf{v} \in \Pi(\mathbf{q})} \epsilon(\mathbf{v}) N(\mathbf{p}, \mathbf{v} - \mathbf{i}) = \delta_{\mathbf{p}(\mathbf{q}-\mathbf{r})},$$

where δ is the Kronecker delta.

Proof. Blaha [2] has proved a result equivalent to this lemma by using group-theoretical calculations; we will present a combinatorial proof. Let $N_k(\mathbf{p}, \mathbf{v})$ be the number of Young tableaux of shape $\mathbf{p}$ containing exactly v_i copies of i and such that the jth row consists entirely of copies of $n+1-j$ for $1 \leq j \leq k$. Thus $N_0(\mathbf{p}, \mathbf{v}) = N(\mathbf{p}, \mathbf{v})$. We will prove that

$$\sum_{\mathbf{v} \in \Pi(\mathbf{q})} \epsilon(\mathbf{v}) N(\mathbf{p}, \mathbf{v} - \mathbf{i}) = \sum_{\mathbf{v}+\mathbf{i} \in \Pi(\mathbf{q})} \epsilon(\mathbf{v} + \mathbf{i}) N_k(\mathbf{p}, \mathbf{v})$$

for all k, by induction on k. Since

$$N_n(\mathbf{p}, \mathbf{v}) = \begin{cases} 1, & \text{if} \quad \mathbf{v} = (p_n, p_{n-1}, \ldots, p_1), \\ 0, & \text{otherwise}, \end{cases}$$

and since $p_n < p_{n-1} + 1 < \cdots < p_1 + n - 1$, the lemma will follow.

Let $k \geq 0$ be fixed, and let T be a tableau that contributes to $N_k(\mathbf{p}, \mathbf{v})$ but not to $N_{k+1}(\mathbf{p}, \mathbf{v})$. This means that the $(k+1)$st row of T contains an entry smaller than $n - k$. Let m be the minimum entry in row $k + 1$. We will define an involution that maps T into a tableau S counted in $N_k(\mathbf{p}, \mathbf{u})$ but not in $N_{k+1}(\mathbf{p}, \mathbf{u})$, where

$$u_i = \begin{cases} v_{m+1} + 1, & \text{if } i = m; \\ v_m - 1, & \text{if } i = m+1; \\ v_i, & \text{otherwise}. \end{cases}$$

This will complete the proof, since $\mathbf{u}+\mathbf{i} \in \Pi(\mathbf{q})$ if and only if $\mathbf{v}+\mathbf{i} \in \Pi(\mathbf{q})$, and since $\epsilon(\mathbf{u} + \mathbf{i}) = -\epsilon(\mathbf{v} + \mathbf{i})$.

Replace the leftmost m in the $(k + 1)$st row of T by $m + 1$. The result is still a Young tableau, say R, since the kth row of T contains only the elements $n + 1 - k > m + 1$, and $\mathbf{u} + \mathbf{i} \in \Pi(\mathbf{q})$ if and only if $\mathbf{v} + \mathbf{i} \in \Pi(\mathbf{q})$.

Any three consecutive rows of a Young tableau R have the form shown in Figure 2, where $0 \leq a \leq b \leq c \leq d$, $x > m + 1$, and $m > y, z$. (Obvious conventions are used when $i = 1$ or when $j + c = 0$.)

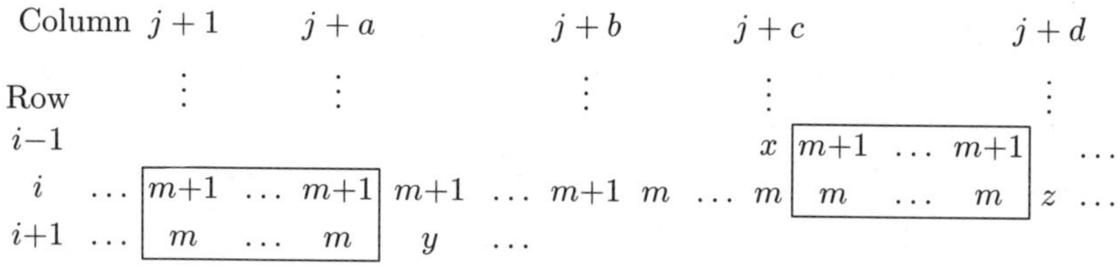

Column $j+1$ $j+a$ $j+b$ $j+c$ $j+d$

Row
$i-1$

i $\ldots$ $\boxed{m+1 \ \ldots \ m+1}$ $m+1$ $\ldots$ $m+1$ m $\ldots$ m $\boxed{m \ \ldots \ m}$ $z \ldots$

Row $i-1$: x $\boxed{m+1 \ \ldots \ m+1}$ $\ldots$

$i+1$ $\ldots$ $\boxed{m \ \ldots \ m}$ y $\ldots$

FIGURE 2. Appearances of m and $m+1$ near row i.

We now define

$$s_{ik} = \begin{cases} m+1, & \text{if } j+a < k \le j+a-b+c; \\ m, & \text{if } j+a-b+c < k \le j+c; \\ r_{ik}, & \text{otherwise.} \end{cases}$$

When carried out for $1 \le i \le n$, this rule defines a map $\varphi_m : R \to S$ on tableaux such that

1) R and S have the same shape;
2) $r_{ij} = s_{ij}$ if $r_{ij} \ne m, m+1$;
3) $r_{ij} = s_{ij}$ if $s_{ij} \ne m, m+1$;
4) the number of m's in R equals the number of $(m+1)$'s in S;
5) the number of $(m+1)$'s in R equals the number of m's in S.

Clearly φ_m is an involution. Since the smallest entry in the $(k+1)$st row of S is m, it is readily verified that the map $T \to S$ is an involution. $\square$

Corollary [10, 11]. *If $\mathbf{p}$ is the shape of a tableau, $\mathbf{v} \in \mathbf{Z}^n$, and π permutes $\{1, 2, \ldots, n\}$, then*

$$N(\mathbf{p}, \mathbf{v}) = N(\mathbf{p}, \pi(\mathbf{v})),$$

where $\pi(\mathbf{v})_i = v_{\pi(i)}$.

Proof. The map φ_m introduced in the proof of the lemma shows that we may interchange v_m and v_{m+1}. $\square$

Corollary. *Let $\mathbf{p}$, $\mathbf{q}$, and $\mathbf{s}$ be tableau shapes. Define*

$$N^{(-1)}(\mathbf{s}, \mathbf{q}) = \sum \epsilon(\mathbf{v} + \mathbf{i}),$$

where the sum ranges over all $\mathbf{v} \in \Pi(\mathbf{s})$ such that $\mathbf{v} + \mathbf{i} \in \Pi(\mathbf{q} + \mathbf{r})$. Then

$$\sum_{\mathbf{s}} N(\mathbf{p}, \mathbf{s}) N^{(-1)}(\mathbf{s}, \mathbf{q}) = \delta_{\mathbf{pq}}.$$

Proof.   Each permutation $\mathbf{w}$ of $\mathbf{q} + \mathbf{r}$ corresponds to a unique $\mathbf{v} = \mathbf{w} - \mathbf{i}$. If $\mathbf{v}$ has no negative coordinates, there's a unique shape $\mathbf{s}$ with $\mathbf{v} \in \Pi(\mathbf{s})$; and $N(\mathbf{p}, \mathbf{s}) = N(\mathbf{p}, \mathbf{v})$. Lemma 2 says that the sum of $\epsilon(\mathbf{v} + \mathbf{i}) N(\mathbf{p}, \mathbf{v})$ for all such $\mathbf{v}$ is $\delta_{\mathbf{pq}}$. $\square$

Theorem 3. *The generating function for strict plane partitions of shape* $\mathbf{p}$, *all of whose parts lie in* S, *is*

$$\det \left| a(p_i - i + j, x) \right|_1^n$$

when $\mathbf{p} \in \mathbf{Z}^n$, *where* $a(k, x)$ *is the generating function for linear partitions into exactly* k *distinct parts all lying in* S.

Proof.   By Lemma 2, the desired generating function $b(\mathbf{p}, x)$ is

$$\sum_{\mathbf{q}} b(\mathbf{q}, x) \sum_{\mathbf{v}} \epsilon(\mathbf{v}) N(\mathbf{q}, \mathbf{v} - \mathbf{i}) = \sum_{\mathbf{v}} \epsilon(\mathbf{v}) \sum_{\mathbf{q}} b(\mathbf{q}, x) N(\mathbf{q}, \mathbf{v} - \mathbf{i}),$$

where $\mathbf{q}$ ranges over all shapes and $\mathbf{v}$ ranges over $\Pi(\mathbf{p} + \mathbf{r})$.

The inner sum can be evaluated by Theorem C: Let $b(\mathbf{q}, x)$ be associated with the tableau P and $N(\mathbf{q}, \mathbf{v} - \mathbf{i})$ with Q; thus $b(\mathbf{q}, x) N(\mathbf{q}, \mathbf{v} - \mathbf{i}) = \sum_{(P,Q)} \prod_{i,j} x^{n_{ij}}$, where P^T is a strict plane partition of shape $\mathbf{q}$ whose parts n_{ij} lie in S and Q is a tableau of shape $\mathbf{q}$ whose elements are enumerated by $\mathbf{v} - \mathbf{i}$. According to Theorem C, the sum over all $\mathbf{q}$ is $\sum_A \prod_{i,j} x^{j a_{ij}}$ where A is a 0–1 integer matrix with $a_{ij} = 0$ if $j \notin S$ and with row sums $\sum_j a_{ij} = v_i - i + 1$. Thus the ith row of the matrix corresponds to a partition into exactly $v_i - i + 1$ distinct parts, all of which lie in S. Hence

$$b(\mathbf{p}, x) = \sum_{\mathbf{v}} \epsilon(\mathbf{v}) \prod_{k=1}^n a(v_k - k + 1, x)$$

$$= \sum_{\pi} \epsilon(\pi(\mathbf{p} + \mathbf{r})) \prod_{k=1}^n a(p_{\pi(k)} + n - \pi(k) - k + 1, x)$$

$$= \sum_{\pi} \epsilon(\pi(\mathbf{r})) \prod_{k=1}^n a(p_k - k + 1 + \pi(\mathbf{r})_k, x)$$

$$= \det \left| a(p_i - i + j, x) \right|_1^n,$$

where $\pi(\mathbf{w})_k = w_{\pi(k)}$. $\square$

Corollary. *The generating function for* r-*rowed strict plane partitions with parts in* S *is*

$$\sum_{v_1 > \cdots > v_r} \det \left| a(v_i + j, x) \right|_1^r. \tag{5}$$

Proof.   Sum the formula of Theorem 3 over all shapes in $\mathbf{Z}^r$. $\square$

Theorem 4. *The generating function for strict plane partitions whose shape is the transpose of* $\mathbf{p} \in \mathbf{Z}^n$ *and all of whose parts lie in* S *is*

$$\det\left|h(p_i - i + j, x)\right|_1^n,$$

where $h(k, x)$ *is the generating function for linear partitions into exactly* k *parts all lying in* S.

Proof. The proof is similar to that for Theorem 3, but Theorem B is used instead of Theorem C. □

Corollary. *The generating function for c-columned strict plane partitions with parts in* S *is*

$$\sum_{v_1 > \cdots > v_c} \det\left|h(v_i + j, x)\right|_1^c. \quad □ \qquad (6)$$

We are now faced with evaluating (5) and (6). This appears to be a difficult problem, although Gordon and Houten have shown how to simplify the equations to some extent.

Lemma 3. *Given any sequence* a_k, *for* $-\infty < k < \infty$, *define*

$$s = \sum_k a_k; \qquad c_v = \sum_k a_k a_{k+v}.$$

Then

$$\sum_{v_1 > \cdots > v_{2n}} \det\left|a_{v_i+j}\right|_1^{2n} = \det\left|c_{i-j} + c_{i+j-1}\right|_1^n;$$

$$\sum_{v_1 > \cdots > v_{2n+1}} \det\left|a_{v_i+j}\right|_1^{2n+1} = s \det\left|c_{i-j} - c_{i+j}\right|_1^n.$$

Proof. Gordon [7] proves this when

$$a_n = x^{n^2} \prod_{i=1}^n (1 - x^{2i})^{-1},$$

but his steps are completely general and prove the lemma. □

Two values of S are of particular interest in Theorem 3:

$$S = \{1, 3, \ldots, 2m - 1\};$$
$$S = \{1, 2, \ldots, m\}.$$

Various special cases have been evaluated:

$$m = \infty \quad \text{(Gordon [7], Gordon and Houten [8])},$$
$$r = \infty \quad \text{(Theorem 1)},$$
$$r = 1 \quad \text{(well known)}.$$

Let $\binom{n}{k}_x$ denote the generalized binomial coefficient:

$$\frac{(1 - x^n)(1 - x^{n-1}) \dots (1 - x^{n-k+1})}{(1 - x)(1 - x^2) \dots (1 - x^k)}.$$

By a straightforward generalization of the methods of Gordon and Houten [8] and Gordon [7], we obtain the following determinants for the generating function $g_r(x)$ of r-rowed strict partitions:

$S = \{1, 3, \dots, 2m - 1\},$ $\qquad$ $S = \{1, 2, \dots, m\},$

$a_k = x^{k^2} \binom{m}{k}_{x^2},$ $\qquad$ $a_k = x^{k(k+1)/2} \binom{m}{k}_x,$

$s = (1+x)(1+x^3) \dots (1+x^{2m-1}),$ $\quad$ $s = (1+x)(1+x^2) \dots (1+x^m),$

$c_k = x^{k^2} \binom{2m}{m+k}_{x^2},$ $\qquad$ $\gamma_k = c_k + c_{k-1} = x^{k(k-1)/2} \binom{2m+1}{m+k}_x,$

$g_{2n}(x) = \det \left| c_{i-j} + c_{i+j-1} \right|_1^n,$ $\quad$ $g_{2n}(x) = \det \left| [i > 1]\gamma_{i-j} + \gamma_{i+j-1} \right|_1^n,$

$g_{2n+1}(x) = s \det \left| c_{i-j} - c_{i+j} \right|_1^n,$ $\quad$ $g_{2n+1}(x) = s \det \left| \gamma_{i-j} - \gamma_{i+j} \right|_1^n.$

The authors have not been able to simplify these determinants any further, even in the limiting case as $x \to 1$ (when only the number of elements in S is relevant). But the known results, and calculations for small r, give overwhelming empirical evidence that the answer has a simple form. MacMahon [12, page 270] conjectured that the generating function for r-rowed strict partitions with odd parts, none exceeding $2m - 1$, is

$$\prod_{i=1}^{m} \left(\frac{1 - x^{r+2i-1}}{1 - x^{2i}} \prod_{j=i+1}^{m} \frac{1 + x^{2(r+i+j-1)}}{1 - x^{2(i+j-1)}} \right). \tag{7}$$

We conjecture that r-rowed strict partitions with no part exceeding m are generated by

$$\prod_{i=1}^{m} \prod_{j=i}^{m} \frac{1 - x^{r+i+j-1}}{1 - x^{i+j-1}}. \tag{8}$$

Symmetric Functions

Most of the preceding can be expressed in the language of symmetric functions. This approach leads to combinatorial proofs of results on Schur functions. Most of the formulas below can be found in Littlewood [11], where they are proved by group theoretic means. We will only sketch the symmetric function viewpoint.

Let $\mathbf{x} = (x_1, \ldots, x_n)$ and $\mathbf{y} = (y_1, \ldots, y_m)$ be vectors of formal variables. We define the following symmetric functions:

$$a_k(\mathbf{x}) = \sum x_{i_1} \ldots x_{i_k} \quad \text{where } 1 \le i_1 < \cdots < i_k \le n$$

$$\text{(combinations without repetition)}$$

$$h_k(\mathbf{x}) = \sum x_{i_1} \ldots x_{i_k} \quad \text{where } 1 \le i_1 \le \cdots \le i_k \le n$$

$$\text{(combinations with repetition)}$$

If $\mathbf{p}$ is a tableau shape, define the symmetric functions

$$a(\mathbf{p}, \mathbf{x}) = a_{p_1}(\mathbf{x}) a_{p_2}(\mathbf{x}) \ldots;$$
$$h(\mathbf{p}, \mathbf{x}) = h_{p_1}(\mathbf{x}) h_{p_2}(\mathbf{x}) \ldots;$$
$$f(\mathbf{p}, \mathbf{x}) = \sum x_{i_1}^{p_1} x_{i_2}^{p_2} \ldots, \text{ summed over all distinct terms}$$
$$\text{with } i_1, i_2, \ldots \text{ distinct;}$$
$$t(\mathbf{p}, \mathbf{x}) = \sum_{\mathbf{q}} N(\mathbf{p}, \mathbf{q}) f(\mathbf{q}, \mathbf{x}), \text{ summed over all tableau shapes } \mathbf{q}.$$

Since $N(\mathbf{p}, \pi(\mathbf{q})) = N(\mathbf{p}, \mathbf{q})$, the function t enumerates tableaux. Define the antisymmetric function

$$\Delta(\mathbf{p}, \mathbf{x}) = \sum_{\pi} \epsilon(\pi) x_{\pi(1)}^{p_1 - 1} x_{\pi(2)}^{p_2 - 2} \ldots = \det \left| x_j^{p_i - i} \right|_1^n,$$

where $\epsilon(\pi)$ is the sign of the permutation π.

Let $n = |S|$; then Theorem 1 is essentially Theorem A rewritten in the form

$$\prod_{i=1}^{n} \left((1 - x_i)^{-1} \prod_{j=i+1}^{n} (1 - x_i x_j)^{-1} \right) = \sum_{\mathbf{p}} t(\mathbf{p}, \mathbf{x}),$$

with $S = \{s_1, \ldots, s_n\}$ and $x_i = x^{s_i}$.

As remarked in [10], Theorems B and C can be written, respectively,

$$\prod_{i=1}^{m}\prod_{j=1}^{n}(1-x_iy_j)^{-1} = \sum_{\mathbf{p}} t(\mathbf{p},\mathbf{x})t(\mathbf{p},\mathbf{y});$$

$$\prod_{i=1}^{m}\prod_{j=1}^{n}(1+x_iy_j) = \sum_{\mathbf{p}} t(\mathbf{p},\mathbf{x})t(\mathbf{p}^T,\mathbf{y}).$$

Here $\mathbf{p}^T$ is the partition conjugate to $\mathbf{p}$; that is, $(\mathbf{p}^T)_i$ is the number of indices j for which $p_j \geq i$. One may also write Theorems B and C in the form

$$\sum_{\mathbf{p}} N(\mathbf{p},\mathbf{q})t(\mathbf{p},\mathbf{x}) = h(\mathbf{q},\mathbf{x})$$

and

$$\sum_{\mathbf{p}} N(\mathbf{p}^T,\mathbf{q})t(\mathbf{p},\mathbf{x}) = a(\mathbf{q},\mathbf{x}),$$

by considering matrices with given row sums. If these equations are inverted, Theorems 4 and 3 are obtained.

We will show that

$$\sum_{\mathbf{p}} N(\mathbf{p}^T,\mathbf{q})\Delta(\mathbf{p},\mathbf{x}) = a(\mathbf{q},\mathbf{x})\Delta(\mathbf{0},\mathbf{x}). \qquad (*)$$

Since the N matrix is nonsingular, this identity together with the interpretation of Theorem C just given provides a combinatorial proof that

$$t(\mathbf{p},\mathbf{x}) = \Delta(\mathbf{p},\mathbf{x})/\Delta(\mathbf{0},\mathbf{x}).$$

This ratio is the "Schur function" for shape $\mathbf{p}$ [11, page 87]. Naegelsbach [11, page 89] proved Theorem 3 for Schur functions, and Jacobi [11, page 87] proved Theorem 4 for Schur functions. The stated relation between t and Δ was discovered by Littlewood [11, page 191], so he was probably aware of Theorems 3 and 4.

Clearly

$$a_k(\mathbf{x})\Delta(\mathbf{p},\mathbf{x}) = \sum_{\mathbf{e}} \Delta(\mathbf{p}+\mathbf{e},\mathbf{x}),$$

where $\mathbf{e}$ ranges over all vectors of zeros and ones with $\sum e_i = k$. Since $\Delta(\mathbf{q},\mathbf{x}) = 0$ if $q_{i+1} = q_i + 1$, we have

$$a_k(\mathbf{x})\Delta(\mathbf{p},\mathbf{x}) = \sum \Delta(\mathbf{s},\mathbf{x}),$$

where **s** ranges over all shapes obtainable from the shape **p** by bordering it with k elements, no two in a row.

We can now prove $(*)$ by induction on the number of nonzero components of **q**. For if $\mathbf{q} = (q_1, \ldots, q_r, 0, 0, \ldots)$ and $\mathbf{q}' = (q_1, \ldots, q_r, k, 0, \ldots)$, we have

$$a(\mathbf{q}', \mathbf{x}) \Delta(\mathbf{0}, \mathbf{x}) = \sum_{\mathbf{p}} N(\mathbf{p}^T, \mathbf{q}) a_k(\mathbf{x}) \Delta(\mathbf{p}, \mathbf{x})$$

$$= \sum_{\mathbf{p}} N(\mathbf{p}^T, \mathbf{q}) \sum_{\mathbf{s}} \Delta(\mathbf{s}, \mathbf{x}) = \sum_{\mathbf{p}} N(\mathbf{s}^T, \mathbf{q}) \Delta(\mathbf{s}, \mathbf{x}). \quad \square$$

Remarks

The r-rowed c-columned plane partitions with no part exceeding m are enumerated by the generating function

$$\prod_{i=1}^{r} \prod_{j=1}^{m} \frac{1 - x^{c+i+j-1}}{1 - x^{i+j-1}} \tag{9}$$

(see [12, page 187]). We do not see how to prove this combinatorially, but it may be possible to do so using an argument like those used in Theorems 1 and 2, because [10] shows that the first row in a tableau associated with a matrix A by Theorem A or B has length

$$\max \sum_{k} a_{i_k j_k}$$

taken over all sequences such that $(i_1, j_1) = (1, 1)$ and

$$(i_{k+1}, j_{k+1}) - (i_k, j_k) = (1, 0) \text{ or } (0, 1).$$

It is plausible that the generating functions (7), (8), and (9) could all be proved by using this fact.

The number of m-dimensional partitions of n with distinct parts all lying in a set S is clearly given by

$$\sum_{k=0}^{\infty} a_m(k) q_k(n), \tag{10}$$

where

$q_k(n)$ is the number of linear partitions of n into exactly k distinct parts all lying in S;

$a_m(k)$ is the number of arrangements of k distinct numbers into an m-dimensional partition.

It is well known that

$$\sum_{n,k} q_k(n) x^n y^k = \prod_{j \in S} (1 + x^j y). \tag{11}$$

Clearly $a_1(k) = 1$. By Theorem A, $a_2(k)$ is the number of symmetric $k \times k$ permutation matrices. This is the number of involutions on $\{1, 2, \ldots, k\}$, which has the well-known generating function

$$\sum a_2(k) \frac{x^k}{k!} = e^{x + x^2/2}. \tag{12}$$

We do not know the value of $a_3(k)$, nor do we know how to use (11) and (12) to simplify (10).

By Theorem C, the generating function for partitions strict in both rows and columns is

$$\sum_A \prod_{i,j} x^{j a_{ij}},$$

where A ranges over all 0–1 matrices such that $P = Q^T$ in the correspondence. Unfortunately no parallel for Theorem A is known.

References

[1] A. O. L. Atkin, P. Bratley, I. G. MacDonald, and J. K. S. McKay, "Some computations for m-dimensional partitions," *Proceedings of the Cambridge Philosophical Society* **63** (1967), 1097–1100.

[2] S. Blaha, "The calculation of the irreducible characters of the symmetric groups in terms of compound characters," preprint (New York: Rockefeller University, 1969).

[3] L. Carlitz, "Rectangular arrays and plane partitions," *Acta Arithmetica* **13** (1967), 29–47.

[4] T. W. Chaundy, "Partition generating functions," *Quarterly Journal of Mathematics* **2** (1931), 234–240.

[5] M. S. Cheema and B. Gordon, "Some remarks on two- and three-line partitions," *Duke Mathematical Journal* **31** (1964), 267–273.

[6] G. Frobenius, "Über die Charaktere der symmetrischen Gruppe," *Sitzungsberichte der Königlich Preußischen Akademie der Wissenschaften zu Berlin* (1900), 417, 516–534. Reprinted in his *Gesammelte Abhandlungen* **3** (Berlin: Springer, 1968), 148–166.

[7] Basil Gordon, "Notes on plane partitions. V," *Journal of Combinatorial Theory* **B11** (1971), 157–168.

[8] Basil Gordon and Lorne Houten, "Notes on plane partitions. I, II," *Journal of Combinatorial Theory* **4** (1968), 72–99.

[9] G. H. Hardy and E. M. Wright, *An Introduction to the Theory of Numbers*, 5th edition (Oxford: Clarendon Press, 1979).

[10] Donald E. Knuth, "Permutations, matrices, and generalized Young tableaux," *Pacific Journal of Mathematics* **34** (1970), 709–727. [Reprinted as Chapter 31 of the present volume.]

[11] Dudley E. Littlewood, *The Theory of Group Characters and Matrix Representations of Groups*, 2nd edition (Oxford: Clarendon Press, 1958).

[12] P. A. MacMahon, *Combinatory Analysis*, Volume 2 (Cambridge: Cambridge University Press, 1916).

[13] Culbreth Sudler, Jr., "A direct proof of two theorems on two-line partitions," *Proceedings of the American Mathematical Society* **16** (1965), 161–168, 558.

[14] J. J. Sylvester, with insertions by Dr. F. Franklin, "A constructive theory of partitions, arranged in three acts, an interact and an exodion," *American Journal of Mathematics* **5** (1882), 251–330; **6** (1883), 334–336. Reprinted in Sylvester's *Collected Mathematical Papers* **4**, 1–83.

Addendum

Shortly after this paper was written in 1969, Basil Gordon proved conjecture (8), and Richard Stanley extended several of the theorems in important ways. [See Richard P. Stanley, "Theory and applications of plane partitions," *Studies in Applied Mathematics* **50** (1971), 167–188, 259–279; Basil Gordon, "A proof of the Bender–Knuth conjecture," *Pacific Journal of Mathematics* **108** (1983), 99–113.]

MacMahon's conjecture (7) was first proved by George Andrews, who also showed that (7) and (8) are essentially equivalent. [George E. Andrews, "Plane partitions (I): The MacMahon conjecture," *Studies in Foundations and Combinatorics*, edited by Gian-Carlo Rota, *Advances in Mathematics Supplementary Studies* **1** (New York: Academic Press, 1978), 131–150; "Plane partitions (II): The equivalence of the Bender–Knuth and MacMahon conjectures," *Pacific Journal of Mathematics* **72** (1977), 283–291.] Another proof was found by Ian Macdonald [I. G. Macdonald, *Symmetric Functions and Hall Polynomials* (Oxford: Clarendon Press, 1979), Section I.5, examples 17 and 19]; and yet another by Robert A. Proctor ["Bruhat lattices, plane partition generating functions, and minuscule representations," *European Journal of Combinatorics* **5** (1984), 331–350]. Generalizations of both (7) and (8) were subsequently discovered by Jacques Désarménien, "Une généralisation des formules de Gordon et de MacMahon," *Comptes Rendus hebdo-madaires des séances de l'Académie des Sciences* (I) **309** (Paris: 1989), 269–272.

An elegant combinatorial proof of MacMahon's generating function (9) for plane partitions with bounded parts has been found by C. Krattenthaler, "Another involution principle-free bijective proof of Stanley's hook-content formula," *Journal of Combinatorial Theory* **A88** (1999), 66–92.

An excellent introduction to the enormous literature that has developed relative to the subjects treated in this paper can be found in Chapter 7 of Richard Stanley's *Enumerative Combinatorics*, Volume 2 (Cambridge: Cambridge University Press, 1999).

In spite of all these advances, there has apparently been no progress on the problem implicitly raised in the final paragraph of the paper, namely to understand the 0–1 matrices A that lead to identical tableaux P^T and Q under the correspondence of Theorem C. The smallest such matrices, omitting cases with all-zero rows, are the following:

$$A = \quad (1) \quad \begin{pmatrix} 1 & 0 \\ 0 & 1 \end{pmatrix} \quad \begin{pmatrix} 0 & 1 \\ 1 & 0 \end{pmatrix} \quad \begin{pmatrix} 1 & 1 \\ 1 & 0 \end{pmatrix}$$

$$P^T = Q = \quad 1 \qquad \begin{matrix} 2 & 1 \end{matrix} \qquad \begin{matrix} 2 \\ 1 \end{matrix} \qquad \begin{matrix} 2 & 1 \\ 1 \end{matrix}$$

$$A = \begin{pmatrix} 1 & 0 & 0 \\ 0 & 1 & 0 \\ 0 & 0 & 1 \end{pmatrix} \begin{pmatrix} 1 & 0 & 0 \\ 0 & 0 & 1 \\ 0 & 1 & 0 \end{pmatrix} \begin{pmatrix} 0 & 1 & 0 \\ 1 & 0 & 0 \\ 0 & 0 & 1 \end{pmatrix} \begin{pmatrix} 0 & 0 & 1 \\ 0 & 1 & 0 \\ 1 & 0 & 0 \end{pmatrix}$$

$$P^T = Q = \quad \begin{matrix} 3 & 2 & 1 \end{matrix} \qquad \begin{matrix} 3 & 1 \\ 2 \end{matrix} \qquad \begin{matrix} 3 & 2 \\ 1 \end{matrix} \qquad \begin{matrix} 3 \\ 2 \\ 1 \end{matrix}$$

$$A = \begin{pmatrix} 1 & 0 & 0 \\ 0 & 1 & 1 \\ 0 & 1 & 0 \end{pmatrix} \begin{pmatrix} 1 & 0 & 1 \\ 0 & 1 & 0 \\ 1 & 0 & 0 \end{pmatrix} \begin{pmatrix} 1 & 1 & 0 \\ 1 & 0 & 0 \\ 0 & 0 & 1 \end{pmatrix} \begin{pmatrix} 0 & 1 & 0 \\ 0 & 1 & 1 \\ 1 & 0 & 0 \end{pmatrix}$$

$$P^T = Q = \quad \begin{matrix} 3 & 2 & 1 \\ 2 \end{matrix} \qquad \begin{matrix} 3 & 1 \\ 2 \\ 1 \end{matrix} \qquad \begin{matrix} 3 & 2 & 1 \\ 1 \end{matrix} \qquad \begin{matrix} 3 & 2 \\ 2 \\ 1 \end{matrix}$$

$$A = \begin{pmatrix} 0 & 1 & 0 \\ 1 & 0 & 1 \\ 0 & 1 & 0 \end{pmatrix} \begin{pmatrix} 0 & 1 & 1 \\ 1 & 1 & 0 \\ 1 & 0 & 0 \end{pmatrix} \begin{pmatrix} 1 & 1 & 0 \\ 1 & 0 & 1 \\ 0 & 1 & 0 \end{pmatrix} \begin{pmatrix} 1 & 1 & 0 \\ 0 & 1 & 1 \\ 1 & 0 & 0 \end{pmatrix} \begin{pmatrix} 1 & 1 & 1 \\ 1 & 1 & 0 \\ 1 & 0 & 0 \end{pmatrix}$$

$$P^T = Q = \quad \begin{matrix} 3 & 2 \\ 2 & 1 \end{matrix} \qquad \begin{matrix} 3 & 2 \\ 2 & 1 \\ 1 \end{matrix} \qquad \begin{matrix} 3 & 2 & 1 \\ 2 & 1 \end{matrix} \qquad \begin{matrix} 3 & 2 & 1 \\ 2 \\ 1 \end{matrix} \qquad \begin{matrix} 3 & 2 & 1 \\ 2 & 1 \\ 1 \end{matrix}$$

Chapter 33

A Note on Solid Partitions

The problem of enumerating partitions that satisfy a given partial order relation can be reduced to the problem of enumerating permutations satisfying that relation. This theorem is applied to the enumeration of solid partitions; existing tables of solid partitions are extended.

[Originally published in Mathematics of Computation **24** (1970), 955–961.]

A *plane partition* of n is an arrangement of nonnegative integers

$$
\begin{array}{cccc}
n_{00} & n_{01} & n_{02} & \cdots \\
n_{10} & n_{11} & n_{12} & \cdots \\
\vdots & \vdots & \vdots & \ddots
\end{array}
$$

that sum to n and are nonincreasing in both rows and columns:

$$ n_{ij} \geq n_{i(j+1)}, \qquad n_{ij} \geq n_{(i+1)j}. $$

For example,

$$
\begin{array}{ccccc}
5 & 4 & 2 & 1 & 1 \\
3 & 2 & & & \\
2 & 2 & & &
\end{array}
\tag{1}
$$

is a plane partition of 22. (Blank entries are zero.)

Let $b(n)$ denote the total number of plane partitions of n. In 1912, Major Percy A. MacMahon triumphantly announced [6] a proof of the remarkable formula

$$ 1 + b(1)z + b(2)z^2 + b(3)z^3 + \cdots = \frac{1}{(1-z)(1-z^2)^2(1-z^3)^3 \ldots}, $$

which he had previously verified by numerous empirical calculations, but which he had been unable to prove six months earlier [5]. In his enthusiasm he concluded his paper by saying, "We have evidently, potentially, the complete solution of the problem of three-dimensional partition, and it remains to work it out and bring it to the same completeness as has been secured in this Part for the problem in two dimensions. This will form the subject of Part VII of this Memoir."

But the problem of enumerating three-dimensional ("solid") partitions has never been resolved, and Part VII of MacMahon's classic Memoir never appeared. No constructive proof of MacMahon's formula for the two-dimensional case was known until 1969 [3].

MacMahon conjectured at one time that, if $c(n)$ is the number of solid partitions on n, the formula

$$1+c(1)z+c(2)z^2+c(3)z^3+\cdots = \frac{1}{(1-z)(1-z^2)^3(1-z^3)^6(1-z^4)^{10}\cdots}$$

might be valid. (See [7, pages 175–176].) But his footnote on page 175 suggests that he doubted it later; and recently computed tables [1] show that this formula gives the wrong answer for $n = 6$. In order to find out the true nature of $c(n)$, the only known approach is to prepare tables, by brute force [2], and to examine those tables with the hope of finding some pattern.

The purpose of this note is to describe a slightly sophisticated computational method for extending the existing tables of $c(n)$, by showing how the coefficients $d(n)$ in the formula

$$1 + c(1)z + c(2)z^2 + \cdots = \frac{1 + d(1)z + d(2)z^2 + \cdots}{(1-z)(1-z^2)(1-z^3)\ldots}$$

can be computed for small n. The computational technique we will discuss was essentially used by MacMahon [5] in his examination of the two-dimensional case, and later also in his preliminary study of solid partitions [6, pages 360–373; 7, pages 247–257]. Since the ideas apply under fairly general circumstances, we will reformulate MacMahon's method for the case of arbitrary partially-ordered sets.

Theory

Let P be a set of elements that are *partially ordered* by the relation $\prec$ and *well-ordered* by the total order relation $<$. We will assume that the partial order is embedded in the total order, in the sense that

$$x \prec y \qquad \text{implies} \qquad x < y. \qquad (2)$$

By a *labeling* of P we mean a function $n(x)$ taking the elements of P into the set N of nonnegative integers, satisfying the two conditions

(L1) $x \prec y$ implies $n(x) \geq n(y)$;
(L2) Only finitely many x have $n(x) > 0$.

We wish to count the number of labelings of P that satisfy certain restrictions.

It is not difficult to show that there is a one-to-one correspondence between labelings of P and pairs of sequences

$$n_1 \geq n_2 \geq \cdots \geq n_m, \qquad (x_1, x_2, \ldots, x_m), \qquad (3)$$

where $m \geq 0$, the n_i are positive integers, and the x_i are distinct elements of P, subject to the following two conditions:

(S1) If $x \prec x_j$ and $x \in P$, then $x = x_i$ for some $i < j$.

(S2) $x_i > x_{i+1}$ implies $n_i > n_{i+1}$, for $1 \leq i < m$.

To construct such a correspondence, we may proceed as follows. Given a labeling, let n_1, n_2, ..., n_m be the nonzero labels in nonincreasing order, and let x_i be such that $n(x_i) = n_i$; the x's are arranged so that we put x before y when $n(x) = n(y)$ and $x < y$. Then (S1) is satisfied, since $x \prec x_j$ implies that $n(x) \geq n(x_j)$ and $x < x_j$; and (S2) is satisfied, since $n_i = n_{i+1}$ implies that $x_i < x_{i+1}$. Conversely, given sequences (3) satisfying (S1) and (S2), we define a labeling by setting $n(x_i) = n_i$ for $1 \leq i \leq m$ and $n(x) = 0$ for all other x. Clearly (L2) is satisfied. And so is (L1); for if $x \prec y$ we have either $n(x) \geq 0 = n(y)$ or $y = x_j$, $x = x_i$, $i < j$, $n(x) = n_i \geq n_j = n(y)$. It is easy to verify that these two constructions are inverses of each other, since (S2) and the relation $n(x_i) = n_i$ uniquely define the sequence of x's.

For example, let P be the set of integer points $\{ ij \mid i, j \geq 0 \}$ of the first quadrant, subject to the partial order

$$ij \preceq i'j' \quad \text{if and only if} \quad i \leq i' \quad \text{and} \quad j \leq j',$$

and to the well-order

$$ij < i'j' \quad \text{if and only if} \quad i < i' \quad \text{or} \quad (i = i' \quad \text{and} \quad j < j').$$

A labeling of this set P is essentially a plane partition; for example, (1) has $n(00) = 5$, $n(01) = 4$, ..., $n(05) = 0$, etc. The sequences (3) corresponding to (1) are

$$\begin{aligned} n_1, \ldots, n_m &= 5, \ 4, \ 3, \ 2, \ 2, \ 2, \ 2, \ 1, \ 1; \\ (x_1, \ldots, x_m) &= 00, \ 01, \ 10, \ 02, \ 11, \ 20, \ 21, \ 03, \ 04. \end{aligned} \qquad (4)$$

We are interested primarily in cases where P is countably infinite, and when the labels sum to a given number n:

$$\sum_{x \in P} n(x) = n. \qquad (5)$$

Let us say that such a labeling is a P-partition of n.

In this situation we can refine the correspondence given above, in order to minimize the dependence of the n's on the x's: Every P-partition of n corresponds uniquely to a pair of sequences

$$n_1 \geq n_2 \geq n_3 \geq \cdots \geq 0, \qquad (x_1, x_2, \ldots, x_m), \qquad (6)$$

where $m \geq 0$, the n_i form an infinite sequence of nonnegative integers with $\sum n_i = n$, the x_i are distinct elements of P, and conditions (S1), (S2), (S3), and (S4) hold whenever $m > 0$, where the two new conditions are

(S3) $n_m > n_{m+1}$.

(S4) There exists $x \in P$ such that $x < x_m$ and $x \neq x_i$, for $1 \leq i \leq m$.

Such pairs of sequences correspond uniquely to pairs of sequences (3) of the former type, when P is infinite, if we extend the n's by adding infinitely many zeros, and if we contract the x's, if necessary, by removing x_m if it is the least element of $P \setminus \{x_1, \ldots, x_{m-1}\}$. For example, (4) corresponds to

$$\begin{aligned} &5, 4, 3, 2, 2, 2, 2, 1, 1, 0, 0, 0, \ldots\,; \\ &00, 01, 10, 02, 11, 20, 21. \end{aligned} \qquad (7)$$

It is not difficult to verify that this process defines a one-to-one correspondence.

Let us say that a sequence of x's satisfying conditions (S1) and (S4) is a *topological sequence*, since such sequences arise in connection with "topological sorting" (see [4, §2.2.3]). The *index* of a topological sequence is defined to be

$$m + \sum \{j \mid 1 \leq j < m \quad \text{and} \quad x_j > x_{j+1}\}. \qquad (8)$$

Given a topological sequence, every sequence $n_1 \geq n_2 \geq \cdots$ of nonnegative integers satisfying (S2) and (S3) corresponds uniquely to a sequence $p_1 \geq p_2 \geq \cdots$ obtained by subtracting unity from $n_1, \ldots, n_j$, in turn, for each j such that $x_j > x_{j+1}$ or $j = m$.

For example, the topological sequence

$$(x_1, \ldots, x_7) = (00, 01, 10, 02, 11, 20, 21) \qquad (9)$$

has index $7 + 3$; and the sequence $p_1 \geq p_2 \geq p_3 \geq \cdots$ corresponding to (7) is obtained by subtracting 1 from n_1, n_2, n_3, then subtracting 1 from each of $n_1, n_2, \ldots, n_7$:

$$p_1, p_2, p_3, \ldots = 3, 2, 1, 1, 1, 1, 1, 1, 1, 0, 0, 0, \ldots. \qquad (10)$$

Notice that the sum of the p_i is the sum of the n_i diminished by the index; this is the only interdependence between the x's, the p's, and n.

In summary, our observations have the following consequence:

Theorem. *Let P be an infinite partially-ordered set. There is a one-to-one correspondence between P-partitions of n and ordered pairs $(\mathbf{x}, \mathbf{p})$ where $\mathbf{x} = (x_1, \ldots, x_m)$ is a topological sequence and $\mathbf{p} = p_1, p_2, \ldots$ is a partition (in the ordinary sense) of $n - k$, where k is the index of $\mathbf{x}$.* $\square$

Since the generating function for ordinary partitions is well known to be $1/\big((1 - z)(1 - z^2)(1 - z^3) \ldots\big)$, we have

Corollary. *Let P be an infinite partially-ordered set; let $s(n)$ be the number of P-partitions of n, and let $t(k)$ be the number of topological sequences of P having index k. Then*

$$1 + s(1)z + s(2)z^2 + \cdots \;=\; \frac{1 + t(1)z + t(2)z^2 + \cdots}{(1 - z)(1 - z^2)(1 - z^3) \ldots}. \quad \square$$

Consequently, we can enumerate P-partitions by enumerating only the topological sequences, and the latter are easier to deal with. Notice that the definition of topological sequence involves an assumed well-ordering $<$; but the corollary shows that *the number of topological sequences of given index is independent of the well-ordering.* Therefore we can choose any convenient well-ordering relation (for example, the lexicographic order in our examples) when doing the enumeration.

An Application

To find the number $c(n)$ of solid partitions for n, let P be the set of three-dimensional lattice points

$$\{\, ijk \mid i, j, k \geq 0 \,\},$$

with the partial ordering

$$ijk \preceq i'j'k' \qquad \text{if and only if} \qquad i \leq i', \quad j \leq j', \quad k \leq k'$$

and with the lexicographic well-ordering $ijk < i'j'k'$ if and only if

$$i < i' \quad \text{or} \quad (i = i' \text{ and } j < j') \quad \text{or} \quad (i = i', \; j = j', \text{ and } k < k').$$

If $d(n)$ is the corresponding number of topological sequences having index n, we have the formula

$$1 + c(1)z + c(2)z^2 + \cdots \;=\; \frac{1 + d(1)z + d(2)z^2 + \cdots}{(1 - z)(1 - z^2)(1 - z^3) \ldots}.$$

The following ALGOL program enumerates $d(0)$, $d(1)$, $\ldots$, $d(n)$, using a fairly standard backtracking method:

procedure *count* $(n, d, pmax)$; **integer** $n, pmax$; **integer array** d;
begin comment $d[i]$ is set to the number of topological sequences of
  index i, for $0 \le i \le n$, in solid partitions with at most *pmax* planes;
  integer array $cc[0:n, 0:n]$;   **comment** columns in given plane and row;
  integer array $rr[0:n]$;   **comment** rows in given plane;
  integer pp;   **comment** planes;
  integer k;   **comment** recursion depth;
  integer array $row[0:n]$;   **comment** row of given move;
  integer array $plane[0:n]$;   **comment** plane of given move;
  integer array $index[0:n]$;   **comment** partial sum for index;
  integer p, r, c;   **comment** current plane, row, column;
  integer t;   **comment** temporary storage;
  for $p := 0$ **step** 1 **until** n **do**
    begin $rr[p] := 0$; **for** $r := 0$ **step** 1 **until** n **do** $cc[p, r] := 0$ **end**;
  $pp := k := index[0] := plane[0] := row[0] := 0$; $d[0] := 1$;
up: $k := k + 1$; $p := pp$; $r := 0$;
  if $p \ge pmax$ **then go to** *again*;
try: $c := cc[p, r]$;
  if $p > 0$ **then if** $cc[p - 1, r] \le c$ **then go to** *again*;
  if $r > 0$ **then if** $cc[p, r - 1] \le c$ **then go to** *again*;
  if $p < plane[k - 1]$ **then go to** *less*;
  if $p = plane[k - 1]$ **then if** $r < row[k - 1]$ **then go to** *less*;
  $t := index[k - 1]$; **go to** *move*;
less: $t := index[k - 1] + k - 1$;
  if $k + t > n$ **then go to** *nope*;
move: **comment** We have now decided to choose the point (p, r, c)
    as the kth element of the topological sequence;
  $index[k] := t$;
  if $p + r > 0$ **then** $d[k + t] := d[k + t] + 1$;
  if $t + k \ge n$ **then go to** *again*;
  if $r + c = 0$ **then** $pp := pp + 1$;
  if $c = 0$ **then** $rr[pp] := rr[pp] + 1$;
  $cc[p, r] := c + 1$;
  $plane[k] := p$; $row[k] := r$; **go to** *up*;
again: **if** $r > 0$ **then begin** $r := r - 1$; **go to** *try* **end**;
  if $p > 0$ **then begin** $p := p - 1$; $r := rr[p]$; **go to** *try* **end**;
nope: $k := k - 1$;
  if $k > 0$ **then begin**
    $p := plane[k]$; $r := row[k]$;
    $c := cc[p, r] - 1$; $cc[p, r] := c$;
    if $c = 0$ **then** $rr[p] := rr[p] - 1$;
    if $r + c = 0$ **then** $pp := pp - 1$;
    go to *again* **end**;
end *count*.

Setting *pmax* to 2 in this program makes it possible to count $c_2(n)$, the number of solid partitions restricted to at most two planes. Table 1 shows the values of $c(n)$, $c_2(n)$, $d(n)$, $d_2(n)$, $e(n)$, and $e_2(n)$, for $n \leq 40$, where the exponents $e(n)$ are defined by the relation

$$1 + c(1)z + c(2)z^2 + \cdots = \frac{1}{(1 - z)^{e(1)}(1 - z^2)^{e(2)}(1 - z^3)^{e(3)} \cdots}$$

and the values of $d_2(n)$ and $e_2(n)$ are defined similarly. Unfortunately no pattern is evident in these numbers, so this table should suffice to disprove most simple conjectures about solid partitions.

References

[1] A. O. L. Atkin, P. Bratley, I. G. Macdonald, and J. K. S. McKay, "Some computations for m-dimensional partitions," *Proceedings of the Cambridge Philosophical Society* **63** (1967), 1097–1100.

[2] P. Bratley and J. K. S. McKay, "Algorithm 313: Multi-dimensional partition generator," *Communications of the ACM* **10** (1967), 666.

[3] Edward A. Bender and Donald E. Knuth, "Enumeration of plane partitions," *Journal of Combinatorial Theory* **A13** (1972), 40–54. [Reprinted as Chapter 32 of the present volume.]

[4] Donald E. Knuth, *Fundamental Algorithms*, Volume 1 of *The Art of Computer Programming* (Reading, Massachusetts: Addison–Wesley, 1968).

[5] P. A. MacMahon, "Memoir on the theory of the partitions of numbers. V: Partitions in two-dimensional space," *Philosophical Transactions of the Royal Society of London* **A211** (1912), 75–110.

[6] P. A. MacMahon, "Memoir on the theory of the partitions of numbers. VI: Partitions in two-dimensional space, to which is added an adumbration of the theory of the partitions in three-dimensional space," *Philosophical Transactions of the Royal Society of London* **A211** (1912), 345–373.

[7] P. A. MacMahon, *Combinatory Analysis*, Volume 2 (Cambridge: Cambridge University Press, 1916).

[8] E. M. Wright, "The generating function of solid partitions," *Proceedings of the Royal Society of Edinburgh* **A67** (1967), 185–195.

TABLE 1

n	$c(n)$	$c_2(n)$	$d(n)$	$d_2(n)$	$e(n)$	$e_2(n)$
0	1	1	1	1	0	0
1	1	1	0	0	1	1
2	4	4	2	2	3	3
3	10	9	5	4	6	5
4	26	22	12	9	10	7
5	59	46	24	16	15	9
6	140	102	56	35	20	10
7	307	206	113	63	26	12
8	684	427	248	129	34	16
9	1464	841	503	234	46	21
10	3122	1658	1043	445	68	29
11	6500	3173	2080	798	97	32
12	13426	6038	4169	1458	120	22
13	27248	11251	8145	2568	112	-2
14	54804	20807	15897	4561	23	-39
15	108802	37907	30545	7924	-186	-67
16	214071	68493	58402	13770	-496	-48
17	416849	122338	110461	23584	-735	64
18	805124	216819	207802	40301	-531	277
19	1541637	380637	387561	68097	779	576
20	2930329	663417	718875	114646	3894	848
21	5528733	1147033	1324038	191336	9323	981
22	10362312	1969961	2425473	317893	16472	771
23	19295226	3359677	4416193	524396	23056	40
24	35713454	5694592	7999516	861054	23850	-1498
25	65715094	9592063	14411507	1405130	10116	-4276
26	120256653	16065593	25837198	2282651	-31613	-8745
27	218893580	26756430	46092306	3688254	-120720	-15062
28	396418699	44328414	81851250	5933463	-283202	-21702
29	714399381	73063013	144691532	9499792	-548924	-24254
30	1281403841	119841187	254682865	15147223	-932162	-13524
31	2287986987	195639752	446399687	24047494	-1380125	23375
32	4067428375	317946756	779302305	38030315	-1655072	99649
33	7200210523	514459448	1355143463	59906474	-1144651	219267
34	12693890803	828956243	2347655027	94025349	1385629	364518
35	22290727268	1330297711	4052251638	147040618	7943203	481760
36	38993410516	2126539101	6969994551	229168751	21083967	471166
37	67959010130	3386536313	11947742454	355966496	42787785	187450
38	118016656268	5373487443	20413108740	551160481	71816191	-545986
39	204233654229	8496146713	34765144046	850708501	98995196	-1891411
40	352245710866	13387678269	59025064235	1309118374	100392874	-3910600

Addendum

Table 1 contains data for $n > 28$ that was not present in the original paper. One conjecture from [1] that is *not* ruled out by these numbers is the hypothesis that the coefficient of z^n in MacMahon's formula

$$\frac{1}{(1-z)(1-z^2)^3(1-z^3)^6(1-z^4)^{10}\ldots}$$

might be an *upper bound* for the number of solid partitions of n. For example, the coefficient of z^{40} is approximately $1.029\,c(40)$.

Chapter 34

Identities from Partition Involutions

*[To George Pólya on the 2^{15}th day after his birth: 31 August 1977. Written with Michael S. Paterson. Originally published in Fibonacci Quarterly **16** (1978), 198–212.]*

Subbarao and Andrews have observed that the combinatorial technique used by F. Franklin to prove Euler's famous partition identity

$$(1-x)(1-x^2)(1-x^3)(1-x^4)\ldots = 1 - x - x^2 + x^5 + x^7 - x^{12} - x^{15} + \cdots$$

can be applied to prove the more general formula

$$1 - x - x^2 y(1 - xy) - x^3 y^2(1 - xy)(1 - x^2 y)$$
$$- x^4 y^3(1 - xy)(1 - x^2 y)(1 - x^3 y) - \cdots$$
$$= 1 - x - x^2 y + x^5 y^3 + x^7 y^4 - x^{12} y^6 - x^{15} y^7 + \cdots,$$

which reduces to Euler's when $y = 1$. This note shows that several finite versions of Euler's identity can also be demonstrated using the same elementary technique. For example,

$$1 - x - x^2 + x^5 + x^7 - x^{12} - x^{15}$$
$$= (1 - x)(1 - x^2)(1 - x^3)(1 - x^4)(1 - x^5)(1 - x^6)$$
$$- x^7(1 - x^2)(1 - x^3)(1 - x^4)(1 - x^5)$$
$$+ x^{7+6}(1 - x^3)(1 - x^4)$$
$$- x^{7+6+5}$$
$$= (1 - x)(1 - x^2)(1 - x^3)$$
$$- x^4(1 - x^2)(1 - x^3)$$
$$+ x^{4+5}(1 - x^3)$$
$$- x^{4+5+6}.$$

By using Sylvester's modification of Franklin's construction, it is also possible to generalize Jacobi's triple product identity.

0. Introduction

Nearly a century ago [7] [16, §12], a young man named Fabian Franklin published what was to become one of the first noteworthy American contributions to mathematics, an elementary combinatorial proof of Euler's well-known identity

$$\prod_{j=1}^{\infty}(1-x^j) = 1 - x - x^2 + x^5 + x^7 - \cdots = \sum_{k=-\infty}^{\infty}(-1)^k x^{(3k^2+k)/2}. \quad (0.1)$$

His approach was to find a nearly one-to-one correspondence between partitions with an even number of distinct parts and those with an odd number of distinct parts, thereby showing that most of the terms on the left-hand side of (0.1) cancel in pairs. Such combinatorial proofs of identities often yield further information, and in the first part of this note we shall demonstrate that Franklin's construction can be used to prove somewhat more than (0.1).

In the second part of this note, we show that Sylvester's modification of Franklin's construction can be applied in a similar way to obtain generalizations of Jacobi's triple product identity

$$\prod_{j=1}^{\infty}(1-q^{2j-1}z)(1-q^{2j-1}z^{-1})(1-q^{2j}) = 1 - q(z+z^{-1}) + q^4(z^2+z^{-2}) - \cdots$$

$$= \sum_{k=-\infty}^{\infty}(-1)^k q^{k^2}z^k. \quad (0.2)$$

1. The Basic Involution

First let us recall the details of Franklin's construction. Let λ be a partition of n into t distinct parts, so that $\lambda = \{a_1, \ldots, a_t\}$ for some integers $a_1 > \cdots > a_t > 0$, where $a_1 + \cdots + a_t = n$. We shall write

$$\Sigma(\lambda) = n, \quad \nu(\lambda) = t, \quad \mu(\lambda) = a_1, \quad (1.1)$$

for the sum, number of parts, and maximum part of λ, respectively; if λ is the empty set, we let $\Sigma(\lambda) = \nu(\lambda) = \mu(\lambda) = 0$. Following Hardy and Wright [9, §19.11], we also define the "base" $\beta(\lambda)$ and "slope" $\sigma(\lambda)$ as follows:

$$\beta(\lambda) = \min\{j \mid j \in \lambda\}, \quad \sigma(\lambda) = \min\{j \mid \mu(\lambda) - j \notin \lambda\}. \quad (1.2)$$

Note that if λ is nonempty we have

$$\mu(\lambda) \geq \beta(\lambda) + \nu(\lambda) - 1 \qquad \text{and} \qquad \nu(\lambda) \geq \sigma(\lambda). \tag{1.3}$$

The partition $F(\lambda)$ corresponding to λ under Franklin's transformation is obtained as follows:

i) If $\beta(\lambda) \leq \sigma(\lambda)$ and $\beta(\lambda) < \nu(\lambda)$, remove the smallest part, $\beta(\lambda)$, and increase each of the largest $\beta(\lambda)$ parts by one.

ii) If $\beta(\lambda) > \sigma(\lambda)$ and either $\sigma(\lambda) < \nu(\lambda)$ or $\sigma(\lambda) \neq \beta(\lambda) - 1$, decrease each of the largest $\sigma(\lambda)$ parts by one and append a new smallest part, $\sigma(\lambda)$.

iii) Otherwise $F(\lambda) = \lambda$. [This case holds if and only if λ is empty or $\sigma(\lambda) = \nu(\lambda) \leq \beta(\lambda) \leq \sigma(\lambda) + 1$.]

These definitions are easily understood in terms of the "Ferrers graph" [16, page 253] for the partition λ, as shown in Figure 1. It is not difficult to verify that F is an involution, namely, that

$$F(F(\lambda)) = \lambda \tag{1.4}$$

for all λ.

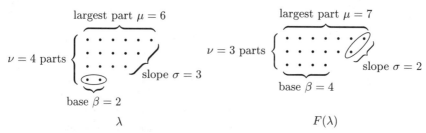

largest part $\mu = 6$ largest part $\mu = 7$

$\nu = 4$ parts $\nu = 3$ parts

slope $\sigma = 3$ slope $\sigma = 2$

base $\beta = 2$ base $\beta = 4$

λ $F(\lambda)$

FIGURE 1. Two partitions of 17 into distinct parts, obtained from each other by moving the two circled elements.

For each $m \geq 0$ there is exactly one partition λ such that $\mu(\lambda) = m$ and $F(\lambda) = \lambda$. (See Figure 2.) We shall denote this fixed point of the mapping by φ_m; it has $\lceil m/2 \rceil$ consecutive parts,

$$\varphi_m = \{m, m - 1, \ldots, \lfloor m/2 \rfloor + 1\}. \tag{1.5}$$

ϵ

$\varphi_0 \qquad \varphi_1 \qquad \varphi_2 \qquad \varphi_3 \qquad \varphi_4 \qquad \varphi_5 \qquad \cdots$

FIGURE 2. The partitions that remain fixed under F.

Let

$$\Phi = \{\varphi_0, \varphi_1, \varphi_2, \ldots\} \tag{1.6}$$

be the set of all such partitions. Notice that the somewhat similar partitions $\{2k+1, 2k, \ldots, k+2\}$ and $\{2k, 2k-1, \ldots, k\}$ are *not* fixed under F, although their bases and slopes do intersect.

2. Extended Generating Functions

If S is any set of partitions, we define the *generating function* of S by the formula

$$G_S(x, y, z) = \sum_{\lambda \in S} x^{\Sigma(\lambda)} y^{\mu(\lambda)} z^{\nu(\lambda)}. \tag{2.1}$$

The identities we shall derive from Franklin's construction are special cases of the following elementary result:

Theorem 1. *If S is any set of partitions,*

$$G_S(x, y, -y) = G_{S \cap \Phi}(x, y, -y) + G_{S \setminus F(S)}(x, y, -y). \tag{2.2}$$

Proof. Let λ be a partition with $\lambda' = F(\lambda) \neq \lambda$. Then $\Sigma(\lambda') = \Sigma(\lambda)$, $\mu(\lambda') = \mu(\lambda) \pm 1$, and $\nu(\lambda') = \nu(\lambda) \mp 1$, hence

$$x^{\Sigma(\lambda)} y^{\mu(\lambda)} (-y)^{\nu(\lambda)} + x^{\Sigma(\lambda')} y^{\mu(\lambda')} (-y)^{\nu(\lambda')} = 0. \tag{2.3}$$

This equation means that λ and λ' do not contribute to $G_S(x, y, -y)$ if they are both members of S. The only terms that fail to cancel out are from partitions $\lambda \in S$ with $F(\lambda) = \lambda$, namely the elements of $S \cap \Phi$, and those from partitions $\lambda \in S$ with $F(\lambda) \notin S$, namely the elements of $S \setminus F(S)$. $\square$

3. Three Identities

In order to get interesting corollaries of Theorem 1, we must find sets S for which the corresponding generating functions are reasonably simple.

First, let S be the set P of all partitions. Theorem 1 implies that

$$G_P(x, y, -y) = G_\Phi(x, y, -y). \tag{3.1}$$

Now

$$G_P(x, y, z) = 1 + \sum_{m=1}^{\infty} x^m y^m z \prod_{j=1}^{m-1} (1 + x^j z) \tag{3.2}$$

and

$$G_\Phi(x, y, z) = 1 + \sum_{m=1}^{\infty} x^{m(m+1)/2 - \lfloor m/2 \rfloor \lfloor m/2+1 \rfloor / 2} y^m z^{\lceil m/2 \rceil} \qquad (3.3)$$

$$= 1 + \sum_{k=1}^{\infty} \left(x^{(3k^2-k)/2} y^{2k-1} z^k + x^{(3k^2+k)/2} y^{2k} z^k \right). \qquad (3.4)$$

Thus we have

Corollary 1.1.

$$\sum_{m=1}^{\infty} x^m y^{m+1} \prod_{j=1}^{m-1} (1 - x^j y)$$

$$= \sum_{k=1}^{\infty} (-1)^{k-1} \left(x^{(3k^2-k)/2} y^{3k-1} + x^{(3k^2+k)/2} y^{3k} \right). \quad \square \qquad (3.5)$$

Franklin essentially considered the special case $y = 1$ of this identity, when the left-hand side reduces to $1 - \prod_{j=1}^{\infty}(1 - x^j)$. Equation (3.5) was originally discovered by L. J. Rogers [12, §10(4)], who gave an analytic proof. The fact that Franklin's correspondence could be used to obtain (3.5) was first noticed by M. V. Subbarao [14] and G. E. Andrews [2].

Although the power series identity for Corollary 1.1 is formally true, it does not converge for all x and y; for example, if we set $y = x^{-1}$ we get the anomalous formula $x^{-1} = x^{-1} + x^{-1} - 1 - x + x^4 + x^6 - \cdots$. To better understand the rate of convergence, we can obtain an exact truncated version of the sum by restricting S to the set

$$P_n = \{ \lambda \mid \mu(\lambda) \le n \}. \qquad (3.6)$$

Since

$$P_n \setminus F(P_n) = \{ \lambda \mid \mu(\lambda) = n \text{ and } \beta(\lambda) \le \sigma(\lambda) \text{ and } \beta(\lambda) < \nu(\lambda) \}$$
$$= \{ \lambda \mid \mu(\lambda) = n \text{ and } \beta(\lambda) \le \sigma(\lambda) \text{ and } \beta(\lambda) \le \tfrac{1}{2}n \}, \qquad (3.7)$$

we have

$$G_{P_n \setminus F(P_n)}(x, y, z) = \sum_{b=1}^{\lfloor n/2 \rfloor} (x^b y^n z) \left(\prod_{j=b+1}^{n-b} (1 + x^j z) \right) \left(\prod_{j=0}^{b-1} x^{n-j} z \right). \qquad (3.8)$$

Thus Theorem 1 yields

Corollary 1.2.

$$\sum_{m=1}^{n} x^m y^{m+1} \prod_{j=1}^{m-1} (1 - x^j y)$$

$$= \sum_{k=1}^{\lceil n/2 \rceil} (-1)^{k-1} x^{(3k^2-k)/2} y^{3k-1} + \sum_{k=1}^{\lfloor n/2 \rfloor} (-1)^{k-1} x^{(3k^2+k)/2} y^{3k}$$

$$+ \sum_{b=1}^{\lfloor n/2 \rfloor} (-1)^b y^{n+b+1} \Big(\prod_{j=b+1}^{n-b} (1 - x^j y) \Big) \Big(\prod_{j=0}^{b-1} x^{n-j+1} \Big). \quad (3.9)$$

For example, the cases $n = 4$ and $n = 5$ of this identity are

$$xy^2 + x^2 y^3 (1 - xy) + x^3 y^4 (1 - xy)(1 - x^2 y)$$
$$+ x^4 y^5 (1 - xy)(1 - x^2 y)(1 - x^3 y)$$
$$= xy^2 + x^2 y^3 - x^5 y^5 - x^7 y^6$$
$$- x^5 y^6 (1 - x^2 y)(1 - x^3 y)$$
$$+ x^{5+4} y^7;$$

$$xy^2 + x^2 y^3 (1 - xy) + x^3 y^4 (1 - xy)(1 - x^2 y)$$
$$+ x^4 y^5 (1 - xy)(1 - x^2 y)(1 - x^3 y)$$
$$+ x^5 y^6 (1 - xy)(1 - x^2 y)(1 - x^3 y)(1 - x^4 y)$$
$$= xy^2 + x^2 y^3 - x^5 y^5 - x^7 y^6 + x^{12} y^8$$
$$- x^6 y^7 (1 - x^2 y)(1 - x^3 y)(1 - x^4 y)$$
$$+ x^{6+5} y^8 (1 - x^3 y).$$

Setting $y = 1$ and subtracting both sides from 1 yields truncated versions of Euler's formula that appear to be new. For example,

$$1 - x - x^2 + x^5 + x^7$$
$$= (1 - x)(1 - x^2)(1 - x^3)(1 - x^4)$$
$$- x^5 (1 - x^2)(1 - x^3)$$
$$+ x^{5+4}; \quad (3.10)$$

$$1 - x - x^2 + x^5 + x^7 - x^{12}$$
$$= (1 - x)(1 - x^2)(1 - x^3)(1 - x^4)(1 - x^5)$$
$$- x^6 (1 - x^2)(1 - x^3)(1 - x^4)$$
$$+ x^{6+5} (1 - x^3); \quad (3.11)$$

$$1 - x - x^2 + x^5 + x^7 - x^{12} - x^{15}$$
$$= (1 - x)(1 - x^2)(1 - x^3)(1 - x^4)(1 - x^5)(1 - x^6)$$
$$- x^7(1 - x^2)(1 - x^3)(1 - x^4)(1 - x^5)$$
$$+ x^{7+6}(1 - x^3)(1 - x^4)$$
$$- x^{7+6+5}. \tag{3.12}$$

Essentially the same formulas, but with n decreased by 2, would have been obtained if we had set $y = x^{-1}$ in the identity of Corollary 1.2.

Let us also consider another family of partition sets with a reasonably simple generating function,

$$S_n = \{\lambda \mid \beta(\lambda) > \mu(\lambda) - n \text{ and } \sigma(\lambda) \geq \mu(\lambda) - n\}. \tag{3.13}$$

These sets are closed under F. For if $\lambda' = F(\lambda) \neq \lambda$ we have either
$$\mu(\lambda') = \mu(\lambda) + 1, \ \beta(\lambda') \geq \beta(\lambda) + 1, \text{ and } \sigma(\lambda') = \beta(\lambda); \text{ or}$$
$$\mu(\lambda') = \mu(\lambda) - 1, \ \beta(\lambda') = \sigma(\lambda), \text{ and } \sigma(\lambda') \geq \sigma(\lambda).$$
Note that S_n is finite, since $\lambda \in S_n$ implies that

$$2\mu(\lambda) - 2n \leq \beta(\lambda) + \sigma(\lambda) - 1 \leq \mu(\lambda),$$

hence $\mu(\lambda) \leq 2n$. The set of fixed points $S_n \cap \Phi$ is $\{\varphi_0, \varphi_1, \ldots, \varphi_{2n}\}$, and

$$G_{S_n}(x, y, z) = G_{P_n}(x, y, z)$$
$$+ \sum_{m=n+1}^{2n} x^m y^m z \left(\prod_{j=m-n+1}^{n} (1 + x^j z) \right) \left(\prod_{j=n+1}^{m-1} x^j z \right), \tag{3.14}$$

so Theorem 1 yields a companion to Corollary 1.2:

Corollary 1.3.

$$\sum_{m=1}^{n} x^m y^{m+1} \prod_{j=1}^{m-1} (1 - x^j y)$$

$$= \sum_{k=1}^{n} (-1)^{k-1} \left(x^{(3k^2-k)/2} y^{3k-1} + x^{(3k^2+k)/2} y^{3k} \right)$$

$$+ \sum_{b=1}^{n} (-1)^b y^{2b+n} \left(\prod_{j=b+1}^{n} (1 - x^j y) \right) \left(\prod_{j=n+1}^{n+b} x^j \right). \tag{3.15}$$

For example, the cases $n = 2$ and $n = 3$ of this identity are

$$xy^2 + x^2 y^3 (1 - xy)$$
$$= xy^2 + x^2 y^3 - x^5 y^5 - x^7 y^6 - x^3 y^4 (1 - x^2 y) + x^{3+4} y^6;$$

$$xy^2 + x^2 y^3 (1 - xy) + x^3 y^4 (1 - xy)(1 - x^2 y)$$
$$= xy^2 + x^2 y^3 - x^5 y^5 - x^7 y^6 + x^{12} y^8 + x^{15} y^9$$
$$- x^4 y^5 (1 - x^2 y)(1 - x^3 y) + x^{4+5} y^7 (1 - x^3 y) - x^{4+5+6} y^9.$$

Setting $y = 1$ and subtracting from 1 leads to formulas somewhat analogous to (3.10) and (3.12):

$$1 - x - x^2 + x^5 + x^7$$
$$= (1 - x)(1 - x^2) - x^3(1 - x^2) + x^{3+4}; \tag{3.16}$$

$$1 - x - x^2 + x^5 + x^7 - x^{12} - x^{15}$$
$$= (1 - x)(1 - x^2)(1 - x^3) - x^4(1 - x^2)(1 - x^3)$$
$$+ x^{4+5}(1 - x^3) - x^{4+5+6}. \tag{3.17}$$

Let us restate the identities arising from Corollaries 1.2 and 1.3 when $y = 1$, where n is even in Corollary 1.2:

$$1 + \sum_{k=1}^{n} (-1)^k \left(x^{(3k^2 - k)/2} + x^{(3k^2 + k)/2} \right)$$

$$= \sum_{k=0}^{n} (-1)^k x^{(2n+2)k - k(k+1)/2} \prod_{j=k+1}^{2n-k} (1 - x^j) \tag{3.18}$$

$$= \sum_{k=0}^{n} (-1)^k x^{nk + k(k+1)/2} \prod_{j=k+1}^{n} (1 - x^j). \tag{3.19}$$

Formula (3.19) was discovered by D. Shanks [13] in the course of some experiments on nonlinear transformations of series; he observed that it can be proved by induction on n without great difficulty. There is also a short proof of (3.18): Let

$$A(k, n) = (1 - x^k) + x^k(1 - x^k)(1 - x^{k+1}) + \cdots$$
$$+ x^{kn}(1 - x^k) \ldots (1 - x^{k+n}), \tag{3.20}$$
$$R(k, n) = x^{(n+1)k}(1 - x^{k+1}) \ldots (1 - x^{k+n}). \tag{3.21}$$

Then $A(0, n) = 0$, $A(k, 0) = 1 - x^k$, $A(k, -1) = 0$, and it is not difficult to show that

$$A(k, n) = 1 - x^{2k+1} - R(k, n) - x^{3k+2} A(k+1, n-2) \text{ if } n > 0. \quad (3.22)$$

Iteration of this recurrence yields identity (3.18). The use of this recurrence is actually only a slight extension of Euler's original technique [6] for proving (0.1).

It is interesting to compare (3.18) and (3.19) to "classical" formulas on terminating basic hypergeometric series, as suggested in a note to the authors by G. E. Andrews. If we set $a = 1$, $b = c = d = \infty$, and $q = x$, in a highly general identity given by R. P. Agarwal [1, Eq. (4.2)], we obtain

$$1 + \sum_{k=1}^{n} (-1)^k \left(x^{(3k^2 - k)/2} + x^{(3k^2 + k)/2} \right)$$

$$= \sum_{k=0}^{n} (-1)^k x^{k(k+1)/2} \left(\prod_{j=k+1}^{2n-k} (1 - x^j) \right) \Big/ \prod_{j=1}^{n-k} (1 - x^j). \quad (3.23)$$

In particular, when $n = 3$ this formula gives the following analog of (3.12) and (3.17):

$$1 - x - x^2 + x^5 + x^7 - x^{12} - x^{15}$$

$$= \frac{(1 - x)(1 - x^2)(1 - x^3)(1 - x^4)(1 - x^5)(1 - x^6)}{(1 - x)(1 - x^2)(1 - x^3)}$$

$$- x^1 \frac{(1 - x^2)(1 - x^3)(1 - x^4)(1 - x^5)}{(1 - x)(1 - x^2)}$$

$$+ x^{1+2} \frac{(1 - x^3)(1 - x^4)}{1 - x} - x^{1+2+3}. \quad (3.24)$$

4. Sylvester's Involution

Let us now turn to Jacobi's identity (0.2), which is formally equivalent under the substitution $q^2 = uv$ and $z^2 = uv^{-1}$ to

$$\prod_{j=1}^{\infty} (1 - u^j v^{j-1})(1 - u^j v^j)(1 - u^{j-1} v^j)$$

$$= 1 + \sum_{k=1}^{\infty} (-1)^k \left(u^{(k^2+k)/2} v^{(k^2-k)/2} + u^{(k^2-k)/2} v^{(k^2+k)/2} \right). \quad (4.1)$$

The left-hand side of this equation can be interpreted as involving partitions of Gaussian integers $m + ni$ into distinct parts of the form $p + qi$, where $\max(p, q) > 0$ and $|p - q| \leq 1$; the coefficient of $u^m v^n$ will be the excess of the number of such partitions with an even number of parts over those with an odd number of parts. The right-hand side says that there exists a nearly one-to-one correspondence between such even and odd partitions, the only unmatched partitions being of the form

$$\{1, 2 + i, \ldots, k + (k - 1)i\} \quad \text{or} \quad \{i, 1 + 2i, \ldots, k - 1 + ki\}. \tag{4.2}$$

An explicit correspondence of this sort was discovered by J. J. Sylvester [16, §§57–61, 64–68] shortly after he had learned of Franklin's construction; at that time Sylvester was a professor at Johns Hopkins University in Baltimore.

We shall represent complex partitions λ by three real partitions, λ_+, λ_0, λ_-, containing the numbers $\max(p, q)$ for those parts $p + qi$ in which $p - q = +1$, 0, or -1, respectively. For example, the complex partition

$$\lambda = \{3 + 2i,\ 2 + i,\ 1,\ 3 + 3i,\ 1 + i,\ 3 + 4i\}$$

of $13 + 11i$ will be represented by

$$\lambda_+ = \{3, 2, 1\}, \quad \lambda_0 = \{3, 1\}, \quad \lambda_- = \{4\}.$$

Sylvester noted that if i is artificially set equal to 2, we obtain a one-to-one correspondence between the relevant complex partitions of $m + ni$ and the real partitions of $m + 2n$ into distinct parts; λ_+, λ_0, λ_- map into the parts congruent respectively to $+1$, 0, and -1 modulo 3, hence Jacobi's identity implies Euler's.

In order to present Sylvester's construction, we recall the definitions of $\Sigma(\lambda)$, $\nu(\lambda)$, $\mu(\lambda)$, $\beta(\lambda)$, and $\sigma(\lambda)$ for real partitions in Section 1 above; we also add two more attributes,

$$\tau(\lambda) = \min\{\, k \mid k + 1 \notin \lambda \,\} \tag{4.3}$$

and

$$\alpha(\lambda) = \min\{\, k \mid k \in \lambda \text{ and } k > \tau(\lambda) \,\}. \tag{4.4}$$

By convention, the minimum over an empty set is ∞; thus $\beta(\lambda) = \infty$ if and only if λ is empty, and $\alpha(\lambda) = \infty$ if and only if λ has the form

$\{1, 2, \ldots, k\}$ for some $k \geq 0$. Sylvester defined an involution $F(\lambda)$ on complex partitions λ by what amounts to the following seven rules:[1]

i) If $\beta(\lambda_0) \leq \sigma(\lambda_+)$, increase each of the largest $\beta(\lambda_0)$ parts of λ_+ by one; then remove the smallest part, $\beta(\lambda_0)$, from λ_0.

ii) If $\beta(\lambda_0) > \sigma(\lambda_+) > 0$ and $\sigma(\lambda_+) \neq \mu(\lambda_+)$, append a new smallest part, $\sigma(\lambda_+)$, to λ_0; then decrease each of the largest $\sigma(\lambda_+)$ parts of λ_+ by one.

iii) If $\beta(\lambda_0) > \sigma(\lambda_+) = \mu(\lambda_+)$ and $\beta(\lambda_0) < \sigma(\lambda_+) + \beta(\lambda_-)$, append a new largest part, $\mu(\lambda_+) + 1$, to λ_+ and a new smallest part, $\beta(\lambda_0) - \mu(\lambda_+)$, to λ_-; then remove the smallest part, $\beta(\lambda_0)$, from λ_0.

iv) If $\beta(\lambda_0) > \sigma(\lambda_+) = \mu(\lambda_+) > 0$ and $\beta(\lambda_0) + 1 > \mu(\lambda_+) + \beta(\lambda_-)$, append a new smallest part, $\mu(\lambda_+) + \beta(\lambda_-) - 1$ to λ_0; then remove the largest part, $\mu(\lambda_+)$, from λ_+ and the smallest part, $\beta(\lambda_-)$, from λ_-.

v) If $\mu(\lambda_+) = 0$ and $\alpha(\lambda_-) > \beta(\lambda_0) + \tau(\lambda_-)$ and $\tau(\lambda_-) > 0$, replace the part $\tau(\lambda_-)$ in λ_- by $\tau(\lambda_-) + \beta(\lambda_0)$; then remove the smallest part, $\beta(\lambda_0)$, from λ_0.

vi) If $\mu(\lambda_+) = 0$ and $\alpha(\lambda_-) < \beta(\lambda_0) + \tau(\lambda_-) + 1$, append a new smallest part, $\alpha(\lambda_-) - \tau(\lambda_-) - 1$, to λ_0; then replace the part $\alpha(\lambda_-)$ in λ_- by $\tau(\lambda_-) + 1$.

vii) Otherwise $F(\lambda) = \lambda$. [This case occurs if and only if λ is one of the special partitions in (4.2).]

In these rules $\infty + x = \infty$; therefore the condition '$\beta(\lambda_0) + 1 > \mu(\lambda_+) + \beta(\lambda_-)$' in (iv) is *not* equivalent to '$\beta(\lambda_0) \geq \mu(\lambda_+) + \beta(\lambda_-)$', and a similar remark applies to rule (vi).

It can be shown that $F(F(\lambda)) = \lambda$, and that in fact rules (i)–(ii), (iii)–(iv), (v)–(vi) undo each other.[2]

---

[1] Zolnowsky [18] independently rediscovered Sylvester's rules (i)–(iv), and observed that these four were sufficient to prove Jacobi's identity since they will handle all cases $m + ni$ with $m \geq n$.

Sylvester's construction seems to have been overlooked by later researchers, possibly because it comes near the end of a very long paper. Furthermore his notation was rather obscure, and he made numerous careless errors that a puzzled reader must rectify. Indeed, the present authors may never have been able to understand what Sylvester was talking about if Zolnowsky's clear presentation had not been available.

[2] At this point one cannot resist quoting Sylvester, who stated that these rules posses what he called Catholicity, Homœogenesis, Mutuality, Inertia, and Enantiotropy: "I need hardly say that so highly organized a scheme ... has not issued from the mind of its composer in a single gush, but is

For example, Sylvester's correspondence pairs up complex partitions in the following way, if we denote partitions by listing the respective elements of λ_+, λ_0, λ_- separated by vertical bars:[3]

$$
\begin{array}{lll}
3|1| & \longleftrightarrow\ 4||, & \text{rules (i) and (ii);} \\
21|1|1 & \longleftrightarrow\ 31||1, & \text{rules (i) and (ii);} \\
1|21| & \longleftrightarrow\ 2|2|, & \text{rules (i) and (ii);} \\
1|3| & \longleftrightarrow\ 21||2, & \text{rules (iii) and (iv);} \\
|2|21 & \longleftrightarrow\ ||41, & \text{rules (v) and (vi);} \\
|1|31 & \longleftrightarrow\ ||32, & \text{rules (v) and (vi).}
\end{array}
$$

5. Generating Functions Revisited

If S is a set of complex partitions, we let

$$
G_S(u,v,y,z) = \sum_{\lambda \in S} u^{\Re\Sigma(\lambda)} v^{\Im\Sigma(\lambda)} y^{\mu(\lambda)} z^{\nu(\lambda_0)}, \tag{5.1}
$$

where

$$
\Re\Sigma(\lambda) = \Sigma(\lambda_+) + \Sigma(\lambda_0) + \Sigma(\lambda_-) - \nu(\lambda_-);
$$
$$
\Im\Sigma(\lambda) = \Sigma(\lambda_+) - \nu(\lambda_+) + \Sigma(\lambda_0) + \Sigma(\lambda_-);
$$
$$
\mu(\lambda) = \begin{cases} \mu(\lambda_+), & \text{if } \mu(\lambda_+) > 0; \\ -\tau(\lambda_-), & \text{if } \mu(\lambda_+) = 0. \end{cases}
$$

These definitions have the property we want, as shown in the following theorem.

Theorem 2. *Let S be any set of complex partitions, and let Φ be the set of all complex partitions of the form* (4.2). *Then*

$$
G_S(u,v,y,-y) = G_{S \cap \Phi}(u,v,y,-y) + G_{S \setminus F(S)}(u,v,y,-y). \tag{5.3}
$$

---

the result of an analytical process of continued residuation of successive heaping of exception upon exception in a manner dictated at each point in its development by the nature of the process and the resistance, so to say, of its subject-matter" [16, page 314].

[3] These are the complex partitions whose sums have the form $k + (11 - 2k)i$. Sylvester gave an incorrect table corresponding to these 12 partitions at the bottom of [16, page 315]; in his notation, he should have written "1st Species. 11 3.8; 6.3.2 6.5; 8.2.1 3.5.2.1. 2d Species. 9.2 5.2.4. 3d Species. 10.1 6.4.1; 7.4 3.7.1."

Proof. As in Theorem 1, we need only verify that if $\lambda' = F(\lambda) \neq \lambda$ we have $\Sigma(\lambda') = \Sigma(\lambda)$, $\mu(\lambda') = \mu(\lambda) \pm 1$, and $\nu(\lambda'_0) = \nu(\lambda_0) \mp 1$. Rules (i), (iii), (v) all leave Σ unchanged, decrease $\nu(\lambda_0)$, and increase $\mu(\lambda)$; rules (ii), (iv), (vi) are the inverses. There is one slightly subtle case worth discussing: Rule (iii) applies when $\mu(\lambda_+) = 0$ and it changes $\mu(\lambda_+)$ to 1; in that case the hypothesis $0 < \beta(\lambda_0) < \beta(\lambda_-)$ implies that $\tau(\lambda_-) = 0$, hence $\mu(\lambda) = 0$. $\quad\square$

6. Jacobi-like Identities

We shall apply Theorem 2 only to two infinite sets of complex partitions, leaving it to the reader to discover interesting finite versions of Jacobi's identity analogous to Corollaries 1.2 and 1.3.

If P is the set of all complex partitions, we have

$$
G_P(u,v,y,z) = \left(\sum_{m=1}^{\infty} u^m v^{m-1} y^m \left(\prod_{j=1}^{m-1}(1+u^j v^{j-1}) \right) \left(\prod_{j=1}^{\infty}(1+u^{j-1}v^j) \right) \right.
$$
$$
\left. + \sum_{m=0}^{\infty} y^{-m} \left(\prod_{j=1}^{m} u^{j-1}v^j \right) \left(\prod_{j=m+2}^{\infty}(1+u^{j-1}v^j) \right) \right) \prod_{j=1}^{\infty}(1+u^j v^j z);
\tag{6.1}
$$

furthermore, (4.2) and (5.1) yield

$$
G_\Phi(u,v,y,z) = 1 + \sum_{k=1}^{\infty} \left(u^{(k^2+k)/2} v^{(k^2-k)/2} y^k + u^{(k^2-k)/2} v^{(k^2+k)/2} y^{-k} \right),
\tag{6.2}
$$

independent of z. According to Theorem 2, setting $z = -y$ in (6.1) gives the identity $G_P(u,v,y,-y) = G_\Phi(u,v,y,-y)$, which can be rewritten as

Corollary 2.1.

$$
\sum_{m=-\infty}^{\infty} \frac{u^m v^{m-1} y^m}{\prod_{j=1}^{\infty}(1 + u^{j+m}v^{j+m-1})} \left(\prod_{j=1}^{\infty}(1+u^j v^{j-1})(1-u^j v^j y)(1+u^{j-1}v^j) \right)
$$
$$
= \sum_{k=-\infty}^{\infty} u^{(k^2+k)/2} v^{(k^2-k)/2} y^k. \quad\square
$$

Our derivation makes it clear that this formula reduces to (4.1) if we set $y = 1$ and replace (u, v) by $(-u, -v)$; it is therefore a three-parameter generalization of Jacobi's two-parameter identity.

The right-hand side of Corollary 2.1 can be expressed as

$$\sum_{k=-\infty}^{\infty} (uy)^{(k^2+k)/2} (vy^{-1})^{(k^2-k)/2}$$

$$= \prod_{j=1}^{\infty} (1 + u^j v^{j-1} y)(1 - u^j v^j)(1 + u^{j-1} v^j y^{-1})$$

by Jacobi's identity (4.1); hence Corollary 2.1 implies that

$$\sum_{m=-\infty}^{\infty} \frac{u^m v^{m-1} y^m}{\prod_{j=0}^{\infty} (1 + u^{j+m} v^{j+m-1})}$$

$$= \prod_{j=1}^{\infty} \frac{(1 + u^j v^{j-1} y)}{(1 + u^j v^{j-1})} \frac{(1 - u^j v^j)}{(1 - u^j v^j y)} \frac{(1 + u^{j-1} v^j y^{-1})}{(1 + u^{j-1} v^j)}.$$

Let us set $a = -v^{-1}$, $q = uv$, and $x = uvy$, to make the structure of this formula slightly more clear; rearranging the factors on the right, we obtain

$$\sum_{n=-\infty}^{\infty} \frac{x^n}{\prod_{j=0}^{\infty} (1 - aq^{j+n})}$$

$$= \prod_{j=0}^{\infty} \frac{(1 - a^{-1} x^{-1} q^{j+1})(1 - axq^j)(1 - q^{j+1})}{(1 - a^{-1} q^{j+1})(1 - aq^j)(1 - xq^j)}. \qquad (6.3)$$

This three-parameter identity turns out to be merely the special case $b = 0$ of a "remarkable formula with many parameters" discovered by S. Ramanujan (see [8, Eq. (12.12.2)]); Ramanujan's formula, for which a surprisingly simple analytic proof has been recently found [5], can be written

$$\sum_{n=-\infty}^{\infty} x^n \prod_{j=0}^{\infty} \frac{1 - bq^{j+n}}{1 - aq^{j+n}}$$

$$= \prod_{j=0}^{\infty} \frac{(1 - ba^{-1} q^j)(1 - a^{-1} x^{-1} q^{j+1})(1 - axq^j)(1 - q^{j+1})}{(1 - ba^{-1} x^{-1} q^j)(1 - a^{-1} q^{j+1})(1 - aq^j)(1 - xq^j)}. \qquad (6.4)$$

The series converges when $|b/a| < |x| < 1$ and $|q| < 1$.

If we let S be the set of all complex partitions with λ_+ nonempty, $G_S(u, v, y, z)$ and $G_{S \cap \Phi}(u, v, y, z)$ are given by the terms in (6.1) and (6.2) involving y^m for $m \geq 1$. The set $S \backslash F(S)$ consists of those partitions with $\lambda_+ = \{1\}$ and $\beta(\lambda_-) < \beta(\lambda_0)$, hence

$$G_{S \backslash F(S)}(u, v, y, z) = uy \sum_{b=1}^{\infty} u^{b-1} v^b \prod_{j=b+1}^{\infty} (1 + u^j v^j z)(1 + u^{j-1} v^j).$$

By Theorem 2, we obtain

Corollary 2.2.

$$\left(\sum_{m=1}^{\infty} u^m v^{m-1} y^m \prod_{j=1}^{m-1} (1 + u^j v^{j-1}) \right) \left(\prod_{j=1}^{\infty} (1 - u^j v^j y)(1 + u^{j-1} v^j) \right)$$

$$= \sum_{k=1}^{\infty} u^{(k^2+k)/2} v^{(k^2-k)/2} y^k + y \sum_{b=1}^{\infty} u^b v^b \prod_{j=b+1}^{\infty} (1 - u^j v^j y)(1 + u^{j-1} v^j). \quad \square$$

If we subtract this identity from that of Corollary 2.1, we get the formula for the complement of S, namely

$$\sum_{m=0}^{\infty} y^{-m} \left(\prod_{j=1}^{m} u^{j-1} v^j \right) \left(\prod_{j=m+2}^{\infty} (1 + u^{j-1} v^j) \right) \left(\prod_{j=1}^{\infty} (1 - u^j v^j y) \right)$$

$$= \sum_{k=0}^{\infty} u^{(k^2-k)/2} v^{(k^2+k)/2} y^{-k} - y \sum_{b=1}^{\infty} u^b v^b \prod_{j=b+1}^{\infty} (1 - u^j v^j y)(1 + u^{j-1} v^j).$$

$$(6.5)$$

Putting $y = 1$ reduces the left-hand side to $\prod_{j=1}^{\infty} (1 - u^j v^j)(1 + u^{j-1} v^j)$; hence we obtain

$$\sum_{b=0}^{\infty} u^b v^b \prod_{j=b+1}^{\infty} (1 - u^j v^j)(1 + u^{j-1} v^j) = \sum_{k=0}^{\infty} u^{(k^2-k)/2} v^{(k^2+k)/2}. \quad (6.6)$$

Let $q = uv$ and $x = -u^{-1}$; this formula is equivalent to the identity

$$\sum_{b=0}^{\infty} q^b \prod_{j=b+1}^{\infty} (1 - q^j)(1 - q^j x) = \sum_{k=0}^{\infty} (-x)^k q^{(k^2+k)/2}. \quad (6.7)$$

Equation (6.7) can be derived readily from known results about basic hypergeometric functions. Let us first divide both sides of the equation by $\prod_{j=1}^{\infty} (1 - q^j)(1 - q^j x)$, obtaining

$$\sum_{n=0}^{\infty} \frac{q^n}{\prod_{j=1}^{n} (1 - xq^j)(1 - q^j)} = \left(\frac{1}{\prod_{j=1}^{\infty} (1 - xq^j)(1 - q^j)} \right) \sum_{k=0}^{\infty} (-x)^k q^{(k^2+k)/2}.$$

Now we use E. Heine's important transformation of such series, an identity in five parameters [10, Equation 79], which essentially states that

$$f(u, v; a, b; q) = f(v, u; b, a; q)$$

if $f(u, v; a, b; q)$ is the function

$$\left(\sum_{n=0}^{\infty} u^n \prod_{j=0}^{n-1} \frac{(1-aq^j)(1-vq^j)}{(1-bvq^j)(1-q^{j+1})} \right) \prod_{j=0}^{\infty} \frac{1-uq^j}{1-auq^j}. \tag{6.8}$$

In our case we let $u = q$, $v = xq/b$, $a = 0$, and $b \to \infty$, obtaining the desired result:

$$\left(\sum_{n=0}^{\infty} \frac{q^n}{\prod_{j=1}^{n}(1 - xq^j)(1 - q^j)} \right) \prod_{j=1}^{\infty} (1 - q^j)$$

$$= \left(\sum_{n=0}^{\infty} x^n \prod_{j=1}^{n}(-q^j) \right) \prod_{j=1}^{\infty} \frac{1}{(1 - xq^j)}.$$

It is not clear whether or not the more general equation (6.5) is related to known formulas in an equally simple way.

An amusing special case of (6.7) can be obtained by setting $q = x^2$ and multiplying both sides by x:

$$\sum_{k \text{ odd}} x^k \prod_{j=k+1}^{\infty} (1 - x^j) = x - x^4 + x^9 - \cdots = \sum_{k=0}^{\infty} (-1)^k x^{(k+1)^2}. \tag{6.9}$$

"The partitions of n into an odd number of distinct parts in which the least part is odd are equinumerous with its partitions into an even number of distinct parts in which the least part is odd, unless n is a perfect square." An equivalent statement was posed as a problem by G. E. Andrews several years ago [3], and he has sketched a combinatorial proof in [4, pages 156–157]. However there must be an involution on partitions that proves this formula! If the reader can find one, it might well lead to a number of interesting new identities.

Acknowledgments

We wish to thank L. Carlitz for calling our attention to reference [13]; R. Askey for pointing out references [5] and [10]; and especially G. E. Andrews, for references [1], [2], [4], [14], and for many detailed comments on an early draft of this paper.

This research was supported in part by National Science Foundation grant MCS 72-03752 A03 and in part by Office of Naval Research contract N00014-76-C-330.

References

[1] R. P. Agarwal, "On the partial sums of series of hypergeometric type," *Proceedings of the Cambridge Philosophical Society* **49** (1953), 441–445.

[2] George E. Andrews, "Two theorems of Gauss and allied identities proved arithmetically," *Pacific Journal of Mathematics* **41** (1972), 563–578.

[3] George E. Andrews, "Problem 5865," *American Mathematical Monthly* **79** (1972), 668; solution by Allen Stenger, **80** (1973), 1148.

[4] George E. Andrews, *The Theory of Partitions* (Reading, Massachusetts: Addison–Wesley, 1977).

[5] George E. Andrews and Richard Askey, "A simple proof of Ramanujan's summation of the $_1\psi_1$," *Æquationes Mathematicæ* **18** (1978), 333–337.

[6] L. Euler, "Demonstratio theorematis circa ordinem in summis divisorum observatum," *Novi commentarii academiæ scientiarum imperialis Petropolitanæ* **5** (1754), 75–83. Reprinted in his *Commentationes arithmeticæ collectæ*, Volume 1 (1849), 234–238. Reprinted in his *Opera Omnia*, Series 1, Volume 2, 390–398.

[7] J. [sic] Franklin, "Sur le développement du produit infini $(1 - x)$ $(1 - x^2)(1 - x^3)(1 - x^4)\ldots$," *Comptes Rendus hebdomadaires des séances de l'Académie des Sciences* **92** (Paris: 1881), 448–450.

[8] G. H. Hardy, *Ramanujan* (Cambridge: Cambridge University Press, 1940).

[9] G. H. Hardy and E. M. Wright, *An Introduction to the Theory of Numbers*, 4th edition (Oxford: Clarendon Press, 1960).

[10] E. Heine, "Untersuchungen über die Reihe $1 + \left((1 - q^\alpha)(1 - q^\beta)/(1 - q)(1 - q^\gamma)\right)x + \left((1 - q^\alpha)(1 - q^{\alpha+1})(1 - q^\beta)(1 - q^{\beta+1})/(1 - q)(1 - q^2)(1 - q^\gamma)(1 - q^{\gamma+1})\right)x^2 + \cdots$," *Journal für die reine und angewandte Mathematik* **34** (1847), 285–328.

[11] C. G. J. Jacobi, *Fundamenta nova theoriæ functionum ellipticarum* (Königsberg: 1829). Reprinted in his *Gesammelte Werke* **1**, 49–239.

[12] L. J. Rogers, "On two theorems of combinatory analysis and some allied identities," *Proceedings of the London Mathematical Society* (2) **16** (1917), 315–336.

[13] Daniel Shanks, "A short proof of an identity of Euler," *Proceedings of the American Mathematical Society* **2** (1951), 747–749.

[14] M. V. Subbarao, "Combinatorial proofs of some identities," *Proceedings of the Washington State University Conference on Number Theory* (Pullman, Washington: Mathematics Department, Washington State University, March 1971), 80–91.

[15] C. Sudler, Jr., "Two enumerative proofs of an identity of Jacobi," *Proceedings of the Edinburgh Mathematical Society* (2) **15** (1966), 67–71.

[16] J. J. Sylvester, with insertions by Dr. F. Franklin, "A constructive theory of partitions, arranged in three acts, an interact and an exodion," *American Journal of Mathematics* **5** (1882), 251–330; **6** (1883), 334–336. Reprinted in Sylvester's *Collected Mathematical Papers* **4**, 1–83.

[17] E. M. Wright, "An enumerative proof of an identity of Jacobi," *Journal of the London Mathematical Society* **40** (1965), 55–57.

[18] J. Zolnowsky, "A direct combinatorial proof of the Jacobi identity," *Discrete Mathematics* **9** (1974), 293–298.

Addendum

Combinatorial proofs of identities usually take one of three forms, either enumerative, bijective, or involutory. We might know two different ways to count certain objects, as in the identities at the end of Chapters 13 and 14; or we might find a one-to-one correspondence between classes of objects that have different generating functions, as in Chapters 31 and 32; or we might find involutions by which positive terms cancel with negative terms, as in the present chapter.

Sylvester gave a bijective proof of Jacobi's triple-product identity in [15, §§37–40], and also in [15, §62] where he credits Arthur S. Hathaway. His bijection has been rediscovered by several other authors; see, for example, [16] and [14].

Ae Ja Yee has recently found a remarkable bijective proof for Ramanujan's remarkable identity (6.4), completely different from the involutory technique that was used to prove the special case $b = 0$ in (6.3). See "Combinatorial proofs of Ramanujan's $_1\psi_1$ summation and the q-Gauss summation," preprint (Urbana, Illinois: Department of Mathematics, University of Illinois, 2003).

Chapter 35

Subspaces, Subsets, and Partitions

To each k-dimensional subspace of an n-dimensional vector space over GF(q) we assign a number p and a partition of p into at most k parts not exceeding n − k, in such a way that exactly q^p of the subspaces map into a given partition. The construction may also be regarded as an order-preserving map from k-dimensional subspaces into k-element subsets.

[Originally published in Journal of Combinatorial Theory A10 (1971), 178–180.]

Let

$$\binom{n}{k}_q = \frac{(q^n - 1) \dots (q^{n-k+1} - 1)}{(q^k - 1) \dots (q - 1)}$$

denote the q-generalization of the binomial coefficient $\binom{n}{k}$, and let V be an n-dimensional vector space over a field of q elements. It is well known [1] that $\binom{n}{k}_q$ is the number of k-dimensional subspaces of V.

It is also well known (although perhaps to a different set of people) that $\binom{n}{k}_q$ is the generating function, in powers of q, for partitions of integers into at most k parts not exceeding n − k; an interesting combinatorial proof of this fact was given by Sylvester [2, page 269].

The purpose of the present note is to point out a fairly natural connection between subspaces and partitions, in order to account for the fact that the same quantity $\binom{n}{k}_q$ occurs in both contexts.

Let us represent the elements of V as n-tuples $(x_1, \dots, x_n)$ of elements of the field. If U is a k-dimensional subspace of V, we can find a "canonical" basis

$$u_1 = (u_{11}, \dots, u_{1n}), \quad \dots, \quad u_k = (u_{k1}, \dots, u_{kn})$$

for U satisfying the conditions

$$u_{in_i} = 1, \quad u_{ij} = 0 \quad \text{for } j > n_i, \quad u_{ln_i} = 0 \quad \text{for } l < i,$$

for $1 \le i \le k$, where $n \ge n_1 > n_2 > \cdots > n_k \ge 1$. For example, if $n = 9$, $k = 4$, $n_1 = 8$, $n_2 = 5$, $n_3 = 3$, and $n_4 = 2$, the basis has the schematic pattern

$$
\begin{aligned}
u_1 &= (u_{11}, \quad 0, \quad 0, \quad u_{14}, \quad 0, \quad u_{16}, \quad u_{17}, \quad 1, \quad 0), \\
u_2 &= (u_{21}, \quad 0, \quad 0, \quad u_{24}, \quad 1, \quad 0, \quad 0, \quad 0, \quad 0), \\
u_3 &= (u_{31}, \quad 0, \quad 1, \quad 0, \quad 0, \quad 0, \quad 0, \quad 0, \quad 0), \\
u_4 &= (u_{41}, \quad 1, \quad 0, \quad 0, \quad 0, \quad 0, \quad 0, \quad 0, \quad 0).
\end{aligned}
$$

The index n_1 is $\max\{\, j \mid \exists (x_1, \ldots, x_n) \in U, \ x_j \ne 0 \,\}$, and in general

$$n_i = \max\{\, j \mid \exists (x_1, \ldots, x_n) \in U, \ x_{n_1} = 0, \ \ldots, \ x_{n_{i-1}} = 0, \ x_j \ne 0 \,\};$$

so it is clear that we can obtain a canonical basis of U by a familiar "triangularization" procedure. There is a unique canonical basis, since u_i is the only vector in U such that $u_{in_j} = \delta_{ij}$, for $1 \le i, j \le k$.

A canonical basis belonging to a certain choice of $n_1, n_2, \ldots, n_k$ has exactly

$$p_i = n_i + i - k - 1$$

unspecified positions in the ith vector u_i. These numbers form a partition

$$p_1 \ge p_2 \ge \cdots \ge p_k \ge 0$$

of the number $p = p_1 + p_2 + \cdots + p_k$ into at most k parts, with no part exceeding $n - k$. Conversely, every partition of a number into at most k parts not exceeding $n - k$ corresponds in this way to a unique choice of indices $n \ge n_1 > n_2 > \cdots > n_k \ge 1$. Since canonical bases are unique, each of the q^p ways to fill in the p unspecified positions in the pattern corresponding to the partition $(p_1, \ldots, p_k)$ leads to a different subspace U.

Consequently the number of k-dimensional subspaces of V is

$$a_0 + a_1 q + a_2 q^2 + \cdots + a_m q^m + \cdots,$$

where a_m is the number of partitions of m into at most k parts not exceeding $n - k$.

It is somewhat more natural to think of this construction in terms of the set $\{n_1, \ldots, n_k\} \subseteq \{1, 2, \ldots, n\}$ instead of the partition $(p_1, \ldots, p_k)$.

For example, if U' is a k'-dimensional subspace of U and if the set of indices determined by U' is $\{n'_1, \ldots, n'_{k'}\}$, then it is easy to prove that $\{n'_1, \ldots, n'_{k'}\} \subseteq \{n_1, \ldots, n_k\}$. (For example, we can count the number of subspaces of U having indices contained in $\{n_1, \ldots, n_k\}$, showing that all subspaces are obtained; or we can extend the basis of U' to a suitable basis of U.) Thus we have a natural order- and rank-preserving map from the lattice of subspaces of V onto the lattice of subsets of $\{1, 2, \ldots, n\}$.

References

[1] Jay Goldman and Gian-Carlo Rota, "On the foundations of combinatorial theory. IV. Finite vector spaces and Eulerian generating functions," *Studies in Applied Mathematics* **49** (1970), 239–258.

[2] J. J. Sylvester, with insertions by Dr. F. Franklin, "A constructive theory of partitions, arranged in three acts, an interact and an exodion," *American Journal of Mathematics* **5** (1882), 251–330; **6** (1883), 334–336. Reprinted in Sylvester's *Collected Mathematical Papers* **4**, 1–83.

Addendum

Richard Stanley informs me that an essentially similar construction arose in Charles Ehresmann's study of Grassman varieties: "Sur la topologie de certains espaces homogènes," *Annals of Mathematics* **35** (1934), 396–443, §10.

Chapter 36

The Power of a Prime That Divides a Generalized Binomial Coefficient

*[Written with Herbert S. Wilf. Originally published in Journal für die reine und angewandte Mathematik **396** (1989), 212–219.]*

The purpose of this note is to generalize the following result of Kummer [8, page 116]:

Theorem. *The highest power of a prime p that divides the binomial coefficient $\binom{m+n}{m}$ is equal to the number of "carries" that occur when the integers m and n are added in p-ary notation.*

For example, $\binom{88}{50}$ is divisible by exactly the 3rd power of 3, because exactly 3 carries occur during the ternary addition

$$(38)_{10} + (50)_{10} = (1102)_3 + (1212)_3 = (10021)_3 = (88)_{10}.$$

The main idea is to consider generalized binomial coefficients that are formed from an arbitrary sequence $\mathcal{C}$, as shown in (3) below. We will isolate a property of the sequence $\mathcal{C}$ that guarantees the existence of a theorem like Kummer's, relating divisibility by prime powers to carries in addition.

A special case of the theorem we shall prove describes the prime power divisibility of Gauss's generalized binomial coefficients [5, §5],

$$\binom{m+n}{m}_q = \frac{(1-q^{m+n})(1-q^{m+n-1})\dots(1-q^{m+1})}{(1-q^n)(1-q^{n-1})\dots(1-q)}, \tag{1}$$

a result that was found first by Fray [4].

Another special case gives a characterization of the highest power to which a given prime divides the "Fibonomial coefficients" of Lucas [9, §9],

$$\binom{m+n}{m}_{\mathcal{F}} = \frac{F_{m+n}F_{m+n-1}\dots F_{m+1}}{F_n F_{n-1}\dots F_1}, \tag{2}$$

515

where $\langle F_1, F_2, \dots \rangle = \langle 1, 1, 2, 3, 5, 8, \dots \rangle$ is the Fibonacci sequence. These coefficients are integers that satisfy the recurrence

$$\binom{m+n}{m}_{\mathcal{F}} = F_{m+1}\binom{m+n-1}{m}_{\mathcal{F}} + F_{n-1}\binom{m+n-1}{n}_{\mathcal{F}}.$$

Generalized Binomial Coefficients

Let $\mathcal{C} = \langle C_1, C_2, \dots \rangle$ be a sequence of positive integers. We define $\mathcal{C}$-nomial coefficients by the rule

$$\binom{m+n}{m}_{\mathcal{C}} = \frac{C_{m+n}C_{m+n-1}\dots C_{m+1}}{C_n C_{n-1}\dots C_1} \tag{3}$$

for all nonnegative integers m and n.

Generalized coefficients of this kind have been studied by several authors. Bachmann [1, page 81], Carmichael [2, page 40], and Jarden and Motzkin [6] have given proofs that if the sequence $\mathcal{C}$ is formed from a three term recurrence

$$C_{j+1} = aC_j + bC_{j-1},$$

with starting values $C_1 = C_2 = 1$, and with integer a, b, then the $\mathcal{C}$-nomial coefficients are integers.

We are interested in the following questions: For a fixed prime p, what is the highest power of p that divides $\binom{m+n}{m}_{\mathcal{C}}$? And under what conditions on the sequence $\mathcal{C}$ is there an analog of Kummer's theorem?

Given integers m and n, let $d_m(n)$ be the number of positive indices $j \le n$ such that C_j is divisible by m. If p is prime and $x \ne 0$ is rational, let $\varepsilon_p(x)$ be the power by which p enters x, that is, the highest power by which p divides the numerator of x minus the highest power by which p divides the denominator. (Thus, x is an integer if and only if $\varepsilon_p(x) \ge 0$ for all p.)

Proposition 1. *The maximum power of a prime p that divides the $\mathcal{C}$-nomial coefficient $\binom{m+n}{m}_{\mathcal{C}}$ is*

$$\varepsilon_p\left(\binom{m+n}{m}_{\mathcal{C}}\right) = \sum_{k \ge 1}\left(d_{p^k}(m+n) - d_{p^k}(m) - d_{p^k}(n)\right). \tag{4}$$

Proof. We can write $\binom{m+n}{m}_C = \Pi(m+n)/(\Pi(m)\Pi(n))$, where $\Pi(n) = C_1 C_2 \ldots C_n$. Now

$$\varepsilon_p\big(\Pi(n)\big) = \sum_{j=1}^{n} \varepsilon_p(C_j)$$

$$= \sum_{j=1}^{n} \sum_{k=1}^{\infty} [p^k \backslash C_j]$$

$$= \sum_{k=1}^{\infty} \sum_{j=1}^{n} [p^k \backslash C_j] = \sum_{k \geq 1} d_{p^k}(n),$$

where $[p^k \backslash C_j]$ denotes 1 if p^k divides C_j, otherwise 0. The result follows since

$$\varepsilon_p\left(\binom{m+n}{m}_C\right) = \varepsilon_p\big(\Pi(m+n)\big) - \varepsilon_p\big(\Pi(m)\big) - \varepsilon_p\big(\Pi(n)\big). \qquad \square$$

Corollary 1. *If $d_k(m+n) \geq d_k(m) + d_k(n)$ for all positive k, m, and n, the C-nomial coefficients are all integers.* $\square$

Regularly Divisible Sequences

We say that the sequence C is *regularly divisible* if it has the following property for each integer $m > 0$: Either there exists an integer $r(m)$ such that C_j is divisible by m if and only if j is divisible by $r(m)$, or C_j is never divisible by m for any $j > 0$. In the latter case we let $r(m) = \infty$. Notice that the d functions for a regularly divisible sequence have the simple form

$$d_m(n) = \left\lfloor \frac{n}{r(m)} \right\rfloor, \tag{5}$$

which satisfies the condition of Corollary 1. Therefore,

Corollary 2. *The C-nomial coefficients corresponding to a regularly divisible sequence are all integers.* $\square$

Regularly divisible sequences can be characterized in another interesting way:

Proposition 2. *The sequence $\langle C_1, C_2, C_3, \ldots \rangle$ is regularly divisible if and only if*

$$\gcd(C_m, C_n) = C_{\gcd(m,n)}, \quad \text{for all } m, n > 0. \tag{6}$$

Proof. Assume first that C is regularly divisible, and let m and n be positive integers. If $g = \gcd(C_m, C_n)$, we know that m and n are divisible by $r(g)$, hence $\gcd(m,n)$ is divisible by $r(g)$, hence $C_{\gcd(m,n)}$ is divisible by g. Also $\gcd(m,n)$ is divisible by $r(C_{\gcd(m,n)})$, hence m and n are divisible by $r(C_{\gcd(m,n)})$, hence C_m and C_n are divisible by $C_{\gcd(m,n)}$, hence g is divisible by $C_{\gcd(m,n)}$. Therefore (6) holds.

Conversely, assume that (6) holds and that m is a positive integer. If some C_j is divisible by m, let $r(m)$ be the smallest such j. Then $\gcd(C_j, C_{r(m)})$ is divisible by m, hence $C_{\gcd(j,r(m))}$ is divisible by m, hence $\gcd(j, r(m)) = r(m)$ by minimality; we have shown that C_j is a multiple of m only if j is a multiple of $r(m)$. And if j is a multiple of $r(m)$ we have $\gcd(C_j, C_{r(m)}) = C_{r(m)}$, hence C_j is a multiple of m. Therefore C is regularly divisible. $\square$

The number $r(m)$ is traditionally called the *rank of apparition* of m in the sequence C. If m' is a multiple of m, the rank $r(m')$ must be a multiple of $r(m)$ in any regularly divisible sequence. Thus, in particular, every prime p defines a sequence of positive integers

$$a_1(p) = r(p), \quad a_2(p) = r(p^2)/r(p), \quad a_3(p) = r(p^3)/r(p^2), \quad \ldots,$$

which either terminates with $a_k(p) = \infty$ for some k or continues indefinitely with $a_k(p) > 1$ for infinitely many k. Conversely, every collection of such sequences, defined for each prime p, defines a regularly divisible sequence C.

Ideal Primes

We say that the prime p is *ideal* for a sequence C if C is regularly divisible and there is a number $s(p)$ such that the multipliers $a_2(p), a_3(p), \ldots$ defined in the previous paragraph are

$$a_k(p) = \begin{cases} 1, & \text{if } 2 \le k \le s(p); \\ p, & \text{if } k > s(p). \end{cases} \tag{7}$$

Thus

$$r(p^k) = \begin{cases} r(p), & \text{if } 1 \le k \le s(p); \\ p^{k-s(p)} r(p), & \text{if } k \ge s(p). \end{cases} \tag{8}$$

Such primes lead to a Kummer-like theorem for generalized binomial coefficients:

Proposition 3. *Let* p *be an ideal prime for a sequence* $\mathcal{C}$. *Then the exponent of the highest power of* p *that divides the* $\mathcal{C}$-*nomial coefficient* $\binom{m+n}{m}_{\mathcal{C}}$ *is equal to the number of carries that occur to the left of the radix point when the rational numbers* $m/r(p)$ *and* $n/r(p)$ *are added in* p-*ary notation, plus an extra* $s(p)$ *if a carry occurs across the radix point itself.*

Proof. We use Proposition 1 and formula (5). If $1 \leq k \leq s(p)$ we have

$$
d_{p^k}(m+n) - d_{p^k}(m) - d_{p^k}(n) = \left\lfloor \frac{m+n}{r(p)} \right\rfloor - \left\lfloor \frac{m}{r(p)} \right\rfloor - \left\lfloor \frac{n}{r(p)} \right\rfloor,
$$

and this is 1 if and only if a carry occurs across the radix point when $m/r(p)$ is added to $n/r(p)$; otherwise it is 0. Similarly if $k > s(p)$,

$$
\begin{aligned}
d_{p^k}(m+n) &- d_{p^k}(m) - d_{p^k}(n) \\
&= \left\lfloor \frac{m+n}{p^{k-s(p)} r(p)} \right\rfloor - \left\lfloor \frac{m}{p^{k-s(p)} r(p)} \right\rfloor - \left\lfloor \frac{n}{p^{k-s(p)} r(p)} \right\rfloor,
\end{aligned}
$$

which is 1 if and only if a carry occurs $k - s(p)$ positions to the left of the radix point.   □

Proposition 3 can be generalized in a straightforward way to multinomial coefficients (see Dickson [3]), in which case we count the carries that occur when more than two numbers are added.

If p is not ideal, a similar result holds, but we must use a mixed-radix number system with radices $a_2(p)$, $a_3(p)$, $a_4(p)$, ...

Gaussian Coefficients

Fix an integer $q > 1$, and let $\mathcal{C}$ be the sequence

$$
\langle q - 1, q^2 - 1, q^3 - 1, \ldots \rangle.
$$

Then the $\mathcal{C}$-nomial coefficients (3) are the Gaussian coefficients (1). It is well known that this sequence $\mathcal{C}$ is regularly divisible; the integer $r(m)$ is called the order of q modulo m, namely the smallest power j such that $q^j \equiv 1$ (modulo m). We denote this quantity $r(m)$ by $r_q(m)$.

If p is a prime that divides q, we have $r_q(p) = \infty$. On the other hand, every odd prime p that does not divide q is ideal for the sequence $\mathcal{C}$. (A proof of this well-known fact can be found, for example, in [7, Lemma 3.2.1.2P].) Therefore Proposition 3 leads to

Theorem 1. *Let $q > 1$ be an integer, and let p be an odd prime. If p divides q, it does not divide the Gaussian coefficient $\binom{m+n}{m}_q$ for any nonnegative m and n. Otherwise $\varepsilon_p\left(\binom{m+n}{m}_q\right)$ is equal to the number of carries that occur to the left of the radix point when $m/r_q(p)$ is added to $n/r_q(p)$ in p-ary notation, plus an additional $s_q(p) = \varepsilon_p(q^{r(p)} - 1)$ if there is a carry across the radix point itself.* □

For example, if $q = 2$ and $p = 7$ we have $r_2(7) = 3$ and $s_2(7) = 1$. If $m = 2$ and $n = 5$ we have $m/3 = (0.444\ldots)_7$ and $n/3 = (1.444\ldots)_7$. The sum is $(m + n)/3 = (2.222\ldots)_7$; a single carry has occurred at the radix point, and we ignore the (infinitely many) carries that occur to the right of the point. Sure enough, $\binom{7}{2}_2 = 2667$ is divisible by 7 but not by 7^2.

The fractions $m/r_q(p)$ and $n/r_q(p)$ are always of the repeating form $(\alpha.ddd\ldots)_p$, where $0 \le d < p - 1$, because $r_q(p)$ is a divisor of $p - 1$.

The case $p = 2$ is slightly special, but it can be handled by almost the same methods. Suppose $q > 1$ is odd. Then there is a unique $f > 1$ such that

$$q \equiv 2^f \pm 1 \pmod{2^{f+1}}.$$

If $q \equiv 2^f + 1$ we have $r_q(2) = r_q(2^2) = \cdots = r_q(2^f) = 1$, and $r_q(2^k) = 2^{k-f}$ for $k \ge f$; but if $q \equiv 2^f - 1$ we have $r_q(2) = 1$, $r_q(2^2) = \cdots = r_q(2^{f+1}) = 2$, and $r_q(2^k) = 2^{k-f}$ for $k > f$.

It follows that the highest power of 2 dividing $\binom{m+n}{m}_q$ is the number of carries when m is added to n in binary notation, plus $f - 1$ if m and n are both odd and if $q \equiv 2^f - 1 \pmod{2^{f+1}}$.

For example, if $q = 23$ we have $f = 3$, so we add $m + n$ in binary and count the carries, throwing in an extra $f - 1 = 2$ if there's a carry out of the rightmost bit position. If $q = 25$ again $f = 3$; but in this case $q \equiv 2^f + 1 \pmod{2^{f+1}}$, so the highest power of 2 dividing $\binom{m+n}{m}_{25}$ is the same as for the ordinary binomial coefficient $\binom{m+n}{m}$.

Fibonacci Coefficients

Now let's turn to the case where the generating sequence $\mathcal{C}$ is the sequence of Fibonacci numbers. This sequence satisfies (6), by a well-known theorem of Lucas [9, page 206]; so it is regularly divisible.

Let $r(p)$ be the least positive integer such that $p\backslash F_{r(p)}$. Then F_j is divisible by p if and only if j is divisible by $r(p)$; indeed it is well known [10] that the period of the Fibonacci sequence modulo p is either $r(p)$, $2r(p)$, or $4r(p)$. It is also well known (see, for example, exercise 3.2.2–11 in [7]) that every odd prime is ideal for the Fibonacci sequence. Special consideration of the prime 2 leads to our second main result:

Theorem 2. *The highest power of the odd prime p that divides the Fibonomial coefficient $\binom{m+n}{m}_{\mathcal{F}}$ is the number of carries that occur to the left of the radix point when $m/r(p)$ is added to $n/r(p)$ in p-ary notation, plus $\varepsilon_p(F_{r(p)})$ if a carry occurs across the radix point. The highest power of 2 that divides $\binom{m+n}{m}_{\mathcal{F}}$ is the number of carries that occur when $m/3$ is added to $n/3$ in binary notation, not counting carries to the right of the binary point, plus 1 if there is a carry from the 1's to the 2's position.*   □

A Cyclotomic Approach

Let us sketch one more proof of Kummer's theorem. This one uses a more powerful apparatus than necessary, but it also sheds additional light on the problem.

If we write $q^n - 1$ in factored form as a product of cyclotomic polynomials,

$$q^n - 1 = \prod_{d \backslash n} \Psi_d(q), \tag{9}$$

we obtain a factorization of Gaussian coefficients by substituting into the right side of (1) and cancelling common factors:

$$\binom{m+n}{m}_q = \prod_{h \in H(m,n)} \Psi_h(q), \tag{10}$$

where

$$H(m,n) = \{\, h \geq 1 \mid m \bmod h + n \bmod h \geq h \,\}.$$

If we now let $q \to 1$, the left side becomes the ordinary binomial coefficient. The right side becomes a product of well-known cyclotomic values,

$$\Psi_h(1) = \begin{cases} p, & \text{if } h = p^k \text{ is a prime power;} \\ 1, & \text{if } h \text{ is not a prime power.} \end{cases} \tag{11}$$

Thus each factor is either 1 or a single prime, and p occurs as often as there are powers of p in the set $H(m,n)$; this is easily seen to be the number of carries in the p-ary addition $m + n$.

A corollary of (10), obtained by matching the degrees, is an identity for Euler's function that we can state as follows: *Fix integers $m, n \geq 0$. The product mn is the sum of $\varphi(h)$, over all integers h for which a carry occurs out of the units position when adding $m + n$ in radix h.*

Some Determinants

The special properties of regularly divisible sequences allow us to evaluate some striking determinants. The genesis of these ideas was in the well-known result that

$$\det\big(\gcd(i,j)\big)_{i,j=1}^{n} = \varphi(1)\varphi(2)\ldots\varphi(n).$$

This identity was generalized in [12] to a theorem about determinants in semi-lattices, which we will quote here in just enough generality to cover the situation at hand. If f is any function of the positive integers, we have

$$\det\big(f\big(\gcd(i,j)\big)\big)_{i,j=1}^{n} = \prod_{m=1}^{n}\left(\sum_{d\backslash m}\mu\Big(\frac{m}{d}\Big)f(d)\right).$$

In view of (6), we find the following evaluation.

Proposition 4. Let $\langle C_1, C_2, \ldots\rangle$ be a regularly divisible sequence. Then

$$\det\big(\gcd(C_i, C_j)\big)_{i,j=1}^{n} = \prod_{m=1}^{n}\left(\sum_{d\backslash m}\mu\Big(\frac{m}{d}\Big)C_d\right). \quad \square \qquad (12)$$

If apply this result to the sequence $\langle q^j - 1\rangle_{j=1}^{\infty}$, we encounter, on the right side for $m > 1$, the quantity

$$M(m,q) = \sum_{d\backslash m}\mu\Big(\frac{m}{d}\Big)q^d, \qquad (13)$$

which is well known to be the number of nonperiodic words of m letters, over an alphabet of q letters. Thus we have the remarkable identity

$$\det\big(\gcd(q^i - 1, q^j - 1)\big)_{i,j=1}^{n} = (q-1)\prod_{m=2}^{n} M(m,q). \qquad (14)$$

Similarly we can apply (12) to the Fibonacci sequence, to find that

$$\det\big(\gcd(F_{qi}, F_{qj})\big)_{i,j=1}^{n} = \prod_{m=1}^{n}\left(\sum_{d\backslash m}\mu\Big(\frac{m}{d}\Big)F_{qd}\right). \qquad (15)$$

Is there a "natural" interpretation of the factors of this product?

Additional Remarks

We have derived our theorems for $\mathcal{C}$-nomial coefficients belonging to regularly divisible sequences, but similar theorems apply in more general situations. For example, we obtain a sequence satisfying the condition of Corollary 1 if we let

$$d_{p^k}(n) = \lfloor \alpha_p n / p^k \rfloor \tag{16}$$

for all primes p and all $k \geq 1$, where α_p is any real number such that $0 \leq \alpha_p \leq p$. Such sequences $\mathcal{C}$ are not regularly divisible, unless each α_p is either zero or p times the reciprocal of an integer. The highest power of p that divides $\binom{m+n}{m}_{\mathcal{C}}$ in such cases is the number of carries that occur to the left of the radix point when $\alpha_p m$ is added to $\alpha_p n$ in p-ary notation.

One special case of this construction occurs when $\alpha_p = 2$ for all p; then it turns out that $C_j = 2j(2j-1)$, and $\binom{m+n}{m}_{\mathcal{C}} = \binom{2m+2n}{2m}$.

Another interesting (and remarkable) case occurs when $\alpha_p = \phi^{-1} = (\sqrt{5}-1)/2$ for all p; then it turns out that $C_{\lceil \phi n \rceil} = n$ and $C_{\lceil \phi^2 n \rceil} = 1$ for all $n \geq 1$.

An ideal prime p is called *simple* if $s(p) = 1$; in such cases Proposition 3 reduces to counting the number of carries to the left of and at the radix point. Nonsimple primes exist for sequences of the form $q^j - 1$; for example $r_3(11) = 10$, and $3^{10} - 1 = 2^3 \cdot 11^2 \cdot 61$. Another example [11] is $q = 2$, $p = 1093$, $r_q(p) = 364$, $s_q(p) = 2$. But in the case of the Fibonacci sequence, calculations by Wall [11] have shown that all primes < 10000 are simple. Does the Fibonacci sequence have any nonsimple primes? Can one prove that it has infinitely many simple primes?

This research was supported in part by the National Science Foundation under grant CCR-86-10181 and in part by Office of Naval Research contracts N00014-87-K-0502 and N00014-85-K-0320.

References

[1] Paul Bachmann, *Niedere Zahlentheorie*, Volume 2 (Leipzig: Teubner, 1910).

[2] R. D. Carmichael, "On the numerical factors of the arithmetic forms $\alpha^n \pm \beta^n$," *Annals of Mathematics* **15** (1913–1914), 30–70.

[3] L. E. Dickson, "Theorems on the residues of multinomial coefficients with respect to a prime modulus," *Quarterly Journal of Pure and Applied Mathematics* **33** (1902), 378–384.

[4] Robert D. Fray, "Congruence properties of ordinary and q-binomial coefficients," *Duke Mathematical Journal* **34** (1967), 467–480.

[5] Carolo Friderico Gauß, "Summatio quarumdam serierum singularium," *Commentationes societatis regiæ scientiarum Gottingensis recentiores* **1** (1808), 147–186. [The page numbers are taken from the Kraus reprint of 1970; in the original publication, each paper had its own page numbers beginning with page 1.] Reprinted in Gauss's *Werke* **2** (1863), 9–45.

[6] Dov Jarden and Theodor Motzkin, "The product of sequences with a common linear recursion formula of order 2," *Riveon Lematematika* **3** (1949), 25–27, 38. Reprinted in Dov Jarden, *Recurring Sequences* (Jerusalem: Riveon Lematematika, 1958).

[7] Donald E. Knuth, *Seminumerical Algorithms*, Volume 2 of *The Art of Computer Programming* (Reading, Massachusetts: Addison–Wesley, 1969).

[8] E. E. Kummer, "Über die Ergänzungssätze zu den allgemeinen Reciprocitätsgesetzen," *Journal für die reine und angewandte Mathematik* **44** (1852), 93–146.

[9] Edouard Lucas, "Théorie des fonctions numériques simplement périodiques," *American Journal of Mathematics* **1** (1878), 184–240.

[10] D. W. Robinson, "The Fibonacci matrix modulo m," *Fibonacci Quarterly* **1**, 2 (April 1963), 29–36.

[11] D. D. Wall, "Fibonacci series modulo m," *American Mathematical Monthly* **67** (1960), 525–532.

[12] Herbert S. Wilf, "Hadamard determinants, Möbius functions, and the chromatic number of a graph," *Bulletin of the American Mathematical Society* **74** (1968), 960–964.

Chapter 37

An Almost Linear Recurrence

*[Originally published in Fibonacci Quarterly **4** (1966), 117–128.]*

A general linear recurrence with constant coefficients has the form

$$u_0 = a_0, \quad u_1 = a_1, \quad \ldots, \quad u_{r-1} = a_{r-1};$$
$$u_n = b_1 u_{n-1} + b_2 u_{n-2} + \cdots + b_r u_{n-r}, \quad \text{for } n \geq r.$$

The Fibonacci sequence is the simplest nontrivial case. Consider, however, the following sequence:

$$\varphi_0 = 1; \quad \varphi_n = \varphi_{n-1} + \varphi_{\lfloor n/2 \rfloor}, \quad \text{for } n > 0. \tag{1}$$

In this case, successive terms are formed from the previous one by adding the term "halfway back." This recurrence, which may be considered as a new kind of generalization of the Fibonacci sequence, has a number of interesting properties that we will examine here.

The sequence begins 1, 2, 4, 6, 10, 14, 20, 26, 36, ... We easily see that all terms except the first are even, and furthermore φ_n is divisible by 4 if and only if $n \equiv 2^{2k-1}$ (modulo 2^{2k}) for some $k \geq 1$. The reader may enjoy discovering further arithmetical properties of φ_n.

Sequence (1) also has an interesting combinatorial interpretation: The number φ_n is precisely the number of partitions of the number $2n$ into powers of 2. For example, $6 = 4 + 2 = 4 + 1 + 1 = 2 + 2 + 2 = 2 + 2 + 1 + 1 = 2 + 1 + 1 + 1 + 1 = 1 + 1 + 1 + 1 + 1 + 1$, and $\varphi_3 = 6$. To verify this interpretation, let $P(m)$ be the number of partitions of m into powers of 2. If $2n = a_1 + a_2 + \cdots + a_k$, where $a_1 \geq a_2 \geq \cdots \geq a_k$ and each a_j is a power of 2, there are two cases:

i) $a_k = 1$; then $a_1 + \cdots + a_{k-1}$ is a partition of $2n - 1$.

ii) $a_k > 1$; then $a_1/2 + \cdots + a_k/2$ is a partition of n.

Conversely, all partitions of $2n$ are obtained from partitions of $2n - 1$ and n in this way, so $P(2n) = P(2n - 1) + P(n)$. We also find that $P(2n + 1) = P(2n)$ by a similar argument; here only case (i) can arise, because $2n + 1$ is an odd number. These recurrence relations for P, together with the initial conditions $P(1) = 1$ and $P(2) = 2$, establish the fact that $\varphi_n = P(2n)$.

The same sequence also arises in other ways. The author first noticed it in connection with the solution of the recurrence relation

$$M(0) = 0; \qquad M(n) = n + \min_{0 \le k < n} \big(2M(k) + M(n - 1 - k)\big), \qquad (*)$$

for which it can be shown that $M(n) - M(n-1) = m$ if $\varphi_m \le 2n < \varphi_{m+1}$, and

$$M\left(\frac{1}{2}\varphi_n - 1\right) = \frac{n - 1}{2}\varphi_n - \left\lfloor \frac{1}{4}\varphi_{2n-1} \right\rfloor, \qquad \text{for } n > 0.$$

Recurrences such as $(*)$ occur in the study of dynamic programming problems, and they will be the subject of another paper.

Let us begin our analysis of φ_n by noticing some of its most elementary properties. By applying the rule (1) repeatedly, we find

$$\varphi_{2n+1} = 2(\varphi_0 + \cdots + \varphi_n). \qquad (2)$$

Another immediate consequence of (1) is

$$\varphi_{2n}^2 - \varphi_{2n+1}\varphi_{2n-1} = \varphi_n^2. \qquad (3)$$

The sequence φ_n grows fairly rapidly (but not, of course, as rapidly as the Fibonacci sequence). For example,

$$\varphi_{500} = 1981471878;$$
$$\varphi_{1000} = 264830889564;$$
$$\varphi_{5000} = 222057486573449444;$$
$$\varphi_{10000} = 214454008193526428202.$$

In fact, we can show that φ_n grows faster than any power of n:

Theorem 1. *Given any power k, there is an integer N_k such that we have $\varphi_n > n^k$ for all $n > N_k$.*

Proof. Let N be such that $2^{k+1} + 1 \ge (2 + 1/N)^{k+1}$, and let

$$\alpha = \min_{N \le n \le 2N} (\varphi_n/n^{k+1}).$$

Then by induction $\varphi_n \geq \alpha n^{k+1}$ for all $n \geq N$, since this is true for $N \leq n \leq 2N$, and if $n > 2N$ we have

$$\begin{aligned}
\varphi_n &= \varphi_{n-1} + \varphi_{\lfloor n/2 \rfloor} \\
&\geq \alpha\big((n-1)^{k+1} + ((n-1)/2)^{k+1}\big) \\
&= \alpha(1 + 2^{-k-1})(n-1)^{k+1} \\
&\geq \alpha(1 + 1/(2N))^{k+1}(n-1)^{k+1} \\
&\geq \alpha(1 + 1/(n-1))^{k+1}(n-1)^{k+1} = \alpha n^{k+1}.
\end{aligned}$$

If we choose $N_k > 1/\alpha$ and $N_k \geq N$, the proof is complete. □

We now consider the generating function for φ_n: Let

$$F(z) = \varphi_0 + \varphi_1 z + \varphi_2 z^2 + \varphi_3 z^3 + \cdots. \tag{4}$$

Notice that

$$\begin{aligned}
(1+z)F(z^2) &= \varphi_0 + \varphi_0 z + \varphi_1 z^2 + \varphi_1 z^3 + \varphi_2 z^4 + \varphi_2 z^5 + \cdots \\
&= \varphi_0 + (\varphi_1 - \varphi_0)z + (\varphi_2 - \varphi_1)z^2 + (\varphi_3 - \varphi_2)z^3 + \cdots \\
&= (1-z)F(z);
\end{aligned}$$

thus

$$F(z) = \frac{1+z}{1-z}F(z^2) = \frac{(1+z)}{(1-z)}\frac{(1+z^2)}{(1-z^2)}F(z^4) = \cdots.$$

We have therefore

$$\begin{aligned}
F(z) &= \frac{(1+z)(1+z^2)(1+z^4)(1+z^8)\ldots}{(1-z)(1-z^2)(1-z^4)(1-z^8)\ldots} \\
&= 1/(1-z)^2(1-z^2)(1-z^4)(1-z^8)\ldots. \tag{5}
\end{aligned}$$

From this form of the generating function, we see that $F(z)$ converges for $|z| < 1$. (As a function of the complex variable z, the infinite product $F(z)$ has the unit circle as a natural boundary, because the series diverges when $z = e^{\pi i m/2^n}$ for integers m and n.) It follows that

$$\limsup_{n\to\infty} \sqrt[n]{\varphi_n} = 1;$$

that is, the sequence φ_n grows more slowly than α^n for any constant $\alpha > 1$. This growth rate stands in marked contrast to that of linear recurrences such as the Fibonacci numbers.

In the remainder of this paper we will determine the true rate of growth of the sequence φ_n; we will prove by elementary methods that

$$\ln \varphi_n \sim \frac{1}{\ln 4} (\ln n)^2,$$

that is,

$$\varphi_n = e^{(\ln n)^2/\ln 4 + o(\log n)^2}. \tag{6}$$

The techniques to be used are similar to others that have been used to determine the order of magnitude of the ordinary partition function (see [2]).

We start by observing that

$$\ln F(z) = -\ln(1-z) + \sum_{k=0}^{\infty} (-\ln(1-z^{2^k}))$$

$$= \sum_{r=1}^{\infty} \frac{z^r}{r} + \sum_{k=0}^{\infty} \sum_{r=1}^{\infty} \frac{z^{2^k r}}{r}$$

and hence by differentiation

$$\frac{F'(z)}{F(z)} = \sum_{r=1}^{\infty} z^{r-1} + \sum_{k=0}^{\infty} \sum_{r=1}^{\infty} 2^k z^{2^k r}$$

$$= 2 + 4z + 2z^2 + 8z^3 + 2z^4 + 4z^5 + \cdots + 2^{\rho(k)+1} z^{k-1} + \cdots,$$

where $2^{\rho(k)}$ is the highest power of 2 dividing k. Therefore

$$\frac{F'(z)}{F(z)} = (1-z)(2 + 6z + 8z^2 + 16z^3 + 18z^4 + 22z^5 + \cdots + \psi_k z^{k-1} + \cdots)$$

where if $k = 2^{a_1} + 2^{a_2} + \cdots + 2^{a_r}$ for $a_1 > a_2 > \cdots > a_r \geq 0$ the coefficient of z^{k-1} in the power series on the right-hand side is

$$\psi_k = 2^{\rho(1)+1} + 2^{\rho(2)+1} + \cdots + 2^{\rho(k)+1} = a_1 2^{a_1} + a_2 2^{a_2} + \cdots + a_r 2^{a_r} + 2k.$$

(The reader will find that the verification of this formula is an interesting exercise in the use of the binary number system.) We can estimate the magnitude of ψ_k as follows:

$$\psi_k \geq a_1 k + 2k - (1 \cdot 2^{a_1 - 1} + 2 \cdot 2^{a_1 - 2} + \cdots + a_1 \cdot 2^0)$$
$$= (a_1 + 2)k - 2^{a_1 + 1} + a_1 + 2 \geq (1 + \log_2 k)k - 2k;$$

hence
$$k \log_2 k - k \le \psi_k \le k \log_2 k + 2k. \tag{7}$$

This estimate and the monotonicity of φ_n are the only facts about $F(z)$ that are used in the derivation below.

Let
$$G(z) = e^{(\ln(1-z))^2/\ln 4}.$$

Then we have
$$\frac{G'(z)}{G(z)} = \frac{-\ln(1-z)}{(1-z)\ln 2}$$
$$= (1-z)\left(\frac{1}{\ln 2}z + \frac{5}{2\ln 2}z^2 + \frac{13}{3\ln 2}z^3 + \frac{77}{12\ln 2}z^4 + \cdots\right).$$

The derivative of $-\ln(1-z)/(1-z)$ is $(1-\ln(1-z))/(1-z)^2$, so we find that the coefficient of z^{k-1} in the power series on the right is
$$\chi_k = \frac{k}{\ln 2}(H_k - 1), \tag{8}$$

where H_k is the harmonic number
$$H_k = 1 + \frac{1}{2} + \cdots + \frac{1}{k}. \tag{9}$$

Since $H_k = \ln k + O(1)$, we have therefore established the equations
$$\frac{F'(z)}{F(z)} = (1-z)\sum_{k=1}^{\infty}\psi_k z^{k-1}, \qquad \frac{G'(z)}{G(z)} = (1-z)\sum_{k=1}^{\infty}\chi_k z^{k-1}, \tag{10}$$

and
$$\psi_k = \chi_k + O(k). \tag{11}$$

These formulas suggest a possible relation between the coefficients of $F(z)$ and those of $G(z)$. Notice that if
$$\frac{F'(z)}{F(z)} = (1-z)f(z),$$

then
$$F(z) = \exp\int_0^z (1-t)f(t)\,dt.$$

Therefore the following lemma shows how relations (10) and (11) might be applied to our problem:

Lemma 1. *Let*

$$A(z) = \exp \int_0^z (1-t)a(t)\,dt, \qquad B(z) = \exp \int_0^z (1-t)b(t)\,dt,$$

where

$$A(z) = \sum_{k=0}^{\infty} A_k z^k, \qquad a(z) = \sum_{k=0}^{\infty} a_k z^{k-1},$$

$$B(z) = \sum_{k=0}^{\infty} B_k z^k, \qquad b(z) = \sum_{k=0}^{\infty} b_k z^{k-1}.$$

Assume that the coefficients of $A(z)$ and of $b(z)$ are nonnegative and nondecreasing. Then if $a_k \le b_k$ for all k, we have $A_k \le B_k$; if $a_k \ge b_k$ for all k, we have $A_k \ge B_k$.

Proof. Clearly $A_0 = B_0 = 1$. Assume that $a_k \le b_k$ for all $k \ge 0$, and that $A_k \le B_k$ for $0 \le k < n$. Then since $A'(z) = (1-z)a(z)A(z)$, we have

$$
\begin{aligned}
nA_n &= a_n A_0 + a_{n-1}(A_1 - A_0) + \cdots + a_1(A_{n-1} - A_{n-2}) \\
&\le b_n A_0 + b_{n-1}(A_1 - A_0) + \cdots + b_1(A_{n-1} - A_{n-2}) \\
&= A_0(b_n - b_{n-1}) + A_1(b_{n-1} - b_{n-2}) + \cdots + A_{n-1}b_1 \\
&\le B_0(b_n - b_{n-1}) + B_1(b_{n-1} - b_{n-2}) + \cdots + B_{n-1}b_1 = nB_n.
\end{aligned}
$$

Essentially the same argument works if $a_k \ge b_k$ for all k.   □

The problem is now one of estimating the coefficients of $G(z)$.

Theorem 2. *If*

$$e^{\alpha(\ln(1-z))^2} = \sum_{n=0}^{\infty} c_n z^n, \tag{12}$$

we have

$$\ln c_n = \alpha(\ln n)^2 + O\big((\log n)(\log\log n)\big). \tag{13}$$

Proof. First we show that

$$\left(\ln \frac{1}{1-z}\right)^m = \sum_{n=m}^{\infty} \frac{m}{n} H_{m,n} z^n, \tag{14}$$

where

$$H_{m,n} = \sum \frac{1}{a_1 \ldots a_{m-1}}$$

summed over all integers $a_1, \ldots, a_{m-1}$ such that $1 \le a_j < n$ and the a_j are *distinct*. This follows inductively, since the derivative of (14) yields

$$\frac{1}{1-z}\left(\ln\frac{1}{1-z}\right)^{m-1} = \sum_{n=m}^{\infty} H_{m,n} z^{n-1},$$

and we have

$$H_{m,n} = H_{m,n-1} + \frac{m-1}{n-1} H_{m-1,n-1}. \tag{15}$$

Turning to equation (12), we have

$$\sum_{n=0}^{\infty} c_n z^n = \sum_{m=0}^{\infty} \frac{\alpha^m (\ln(1-z))^{2m}}{m!}$$

$$= 1 + \sum_{n=1}^{\infty}\left(\sum_{m=1}^{\infty} \frac{\alpha^m}{m!}\left(\frac{2m}{n}\right) H_{2m,n}\right) z^n. \tag{16}$$

(We define $H_{m,n} = 0$ if $m > n$, so the parenthesized summation on the right is actually a finite sum for any fixed value of n.)

Our theorem relies on the estimates

$$(H_{n-1} - H_{m-1})^{m-1} \le H_{m,n} \le H_{n-1}^{m-1}, \qquad \text{if } m \le n. \tag{17}$$

The right-hand inequality is obvious, since H_{n-1}^{m-1} is just the sum $\sum 1/(a_1 \ldots a_{m-1})$ without the restriction that the a's are distinct. On the other hand, given any term of

$$(H_{n-1} - H_{m-1})^{m-1} = \sum_{m \le a_1, \ldots, a_{m-1} < n} \frac{1}{a_1 \ldots a_{m-1}},$$

we form a term $1/(b_1 \ldots b_{m-1})$ belonging to $H_{m,n}$, where $b_k = a_k - r$ if a_k is the rth largest of $\{a_1, \ldots, a_{m-1}\}$. Thus, we decrease the largest element by 1, the second-largest by 2, and so on; in case of ties, an arbitrary order is taken. For example, if $a_1 a_2 a_3 a_4 = 6 \cdot 5 \cdot 6 \cdot 7$, we have $b_1 b_2 b_3 b_4 = 4 \cdot 1 \cdot 3 \cdot 6$. No two terms $1/(a_1 \ldots a_{m-1})$ map into the same $1/(b_1 \ldots b_{m-1})$, and clearly $1/(a_1 \ldots a_{m-1}) \le 1/(b_1 \ldots b_{m-1})$, so the left-hand side of (17) is established.

Putting the right-hand side of (17) into (16), we obtain

$$c_n = \frac{2}{n} \sum_{m=1}^{\infty} \frac{\alpha^m}{(m-1)!} H_{2m,n}$$

$$\le \frac{2\alpha H_{n-1}}{n} \sum_{m=1}^{\infty} \frac{\alpha^{m-1} H_{n-1}^{2m-2}}{(m-1)!} = \frac{2\alpha}{n} e^{\alpha H_{n-1}^2}. \tag{18}$$

On the other hand,

$$c_n > \frac{2}{n} \frac{\alpha^m}{(m-1)!} H_{2m,n} \tag{19}$$

for any particular value of m. We choose m to be approximately $\alpha H_{n-1}^2 + 1$, assuming that n is large. Then we evaluate the logarithm of the term on the right, using Stirling's approximation and the left-hand side of (17):

$$
\begin{aligned}
\ln c_n &> \ln\left(\frac{2\alpha}{n} \frac{\alpha^{m-1}}{(m-1)!} (H_{n-1} - H_{2m-1})^{2m-1}\right) \\
&= \alpha H_{n-1}^2 \ln \alpha + 2\alpha H_{n-1}^2 \ln(H_{n-1} - H_{2m-1}) \\
&\quad - \alpha H_{n-1}^2 \big(\ln(\alpha H_{n-1}^2) - 1\big) + O(\log n) \\
&= \alpha H_{n-1}^2 + 2\alpha H_{n-1}^2 \ln\left(1 - \frac{H_{2m-1}}{H_{n-1}}\right) + O(\log n) \\
&= \alpha H_{n-1}^2 - 2\alpha H_{n-1} H_{2m-1} + O(\log n).
\end{aligned}
$$

This bound together with (18) establishes Theorem 2, since $H_{2m-1} = O(\log \log n)$.   □

Theorem 3. *Let c_n be as in Theorem 2. Then*

$$\lim_{n\to\infty} \frac{c_{n+1}}{c_n} = 1.$$

Proof. Since $H_{m,n+1} \geq H_{m,n}$, we have

$$\frac{c_{n+1}}{c_n} \geq \frac{n}{n+1}$$

by (16). We also observe that $H_{m,n} \leq H_{n-1} H_{m-1,n}$. Hence, by (15),

$$H_{m,n+1} \leq H_{m,n} + \frac{m-1}{n} H_{n-1} H_{m-2,n}, \qquad \text{if } m > 2;$$

and since $H_{2,n+1} = H_n$, we have

$$
\begin{aligned}
c_{n+1} &\leq \sum_{m=1}^{\infty} \frac{\alpha^m}{m!} \left(\frac{2m}{n+1}\right) H_{2m,n} + \frac{2\alpha}{n(n+1)} \\
&\quad + \frac{2\alpha}{n+1} H_{n-1} \sum_{m=2}^{\infty} \frac{\alpha^{m-1}}{(m-1)!} \left(\frac{2m-1}{2m-2}\right) \left(\frac{2(m-1)}{n}\right) H_{2(m-1),n} \\
&\leq \frac{n}{n+1} c_n + \frac{3\alpha H_{n-1}}{n+1} c_n + \frac{2\alpha}{n(n+1)}. \qquad \square
\end{aligned}
$$

Corollary 3. *If $P(z)$ is any polynomial, and if*

$$e^{\alpha(\ln(1-z))^2+P(z)} = \sum_{n=0}^{\infty} C_n z^n,$$

then

$$\ln C_n = \ln c_n + O(1).$$

Proof. If $e^{P(z)} = a_0 + a_1 z + a_2 z^2 + \cdots$, we have

$$\frac{C_n}{c_n} = \frac{a_0 c_n + a_1 c_{n-1} + \cdots + a_n c_0}{c_n} \rightarrow e^{P(1)}. \quad \square$$

Theorem 4. $\ln \varphi_n \sim (\ln n)^2 / \ln 4$.

Proof. Let $\epsilon > 0$ be given. By (11), we can find N so that when $k > N$ we have $(1-\epsilon)\chi_k < \psi_k < (1+\epsilon)\chi_k$. Apply Lemma 1 with $A(z) = F(z)$ and with

$$b(z) = \sum_{k=N+1}^{\infty} (1-\epsilon)\chi_k z^{k-1}.$$

We find $\varphi_n \geq C_n$ where, by Corollary 3,

$$\ln C_n \sim \left(\frac{1-\epsilon}{\ln 4}\right)(\ln n)^2.$$

Then apply Lemma 1 with $A(z) = F(z)$ and with

$$b(z) = \sum_{k=1}^{N} \max(\psi_1, \ldots, \psi_k) z^{k-1} + \sum_{k=N+1}^{\infty} (1+\epsilon)\chi_k z^{k-1}.$$

This gives us $\varphi_n \leq C'_n$, where

$$\ln C'_n \sim \left(\frac{1+\epsilon}{\ln 4}\right)(\ln n)^2.$$

Therefore

$$\left| \frac{\ln \varphi_n}{(\ln n)^2} - \frac{1}{\ln 4} \right|$$

is arbitrarily small when n is large enough. $\quad \square$

Of course, the estimate we have derived in this theorem is very crude as far as the actual value of φ_n is concerned. Empirical tests based on the exact values of φ_n for $n \leq 10000$ reveal excellent agreement with the following formula:

$$\ln \varphi_n \approx \frac{\ln n}{\ln 4}(\ln n - 2 \ln \ln n + 1) + \ln n - .843. \tag{20}$$

The error is less than .05 for $n > 10$; it reaches a low of about $-.05$ when n is near 50, then increases to approximately .032 when n is near 5000, and it decreases slowly after that. Thus we can use (20) to calculate

$$\varphi_n \approx .472 n^{1.721} \left(\frac{\sqrt{n}}{\ln n}\right)^{\log_2 n} \tag{21}$$

with an error of at most 5% when $10 < n \leq 10000$.

Although formula (20) gives reasonably good accuracy, it should be remembered that only the first term of the expansion has been verified, and the comparatively small values of $\ln \ln n$ in the range of n considered makes it quite possible that (20) is not the true asymptotic result. On the assumption that the true formula is a relatively "simple" one, however, (20) does give striking agreement.

A similar situation exists in the study of the ordinary partition function. The methods used above can be applied with ease to that problem, giving

$$\ln p(n) \ \sim \ \pi \sqrt{\frac{2}{3}n};$$

the actual asymptotic formula for $p(n)$ itself is

$$p(n) \ \sim \ \frac{1}{4\sqrt{3}n} e^{\pi\sqrt{2n/3}},$$

or, more precisely,

$$p(n) = \frac{e^{\pi\sqrt{2n'/3}}}{4n'\sqrt{3}}\left(1 - \frac{1}{\pi}\sqrt{\frac{3}{2n'}}\right)\left(1 + O\left(e^{-\pi\sqrt{n/6}}\right)\right), \quad n' = n - \frac{1}{24}.$$

It is doubtful that it could have been guessed empirically in either of these two forms. For an account of these estimates and a bibliography, see [1].

The preparation of this paper was supported, in part, by NSF grant GP-212. Acknowledgment is also made to the Burroughs Corporation for the use of a B5000 computer.

References

[1] G. H. Hardy, *Ramanujan* (Cambridge: Cambridge University Press, 1940), Chapter 8.

[2] K. Knopp and I. Schur, "Elementarer Beweis einiger asymptotischer Formeln der additiven Zahlentheorie," *Mathematische Zeitschrift* **24** (1925–1926), 559–574.

Addendum

When I wrote this paper I was blissfully unaware of the extensive history of related work. For example, I didn't know that Kurt Mahler had already deduced the true asymptotic growth of φ_n, which is not (20) but

$$\ln \varphi_n = \frac{(\ln n - \ln \ln n)^2}{\ln 4} + A \ln n - B \ln \ln n + O(1),$$

where $A = 1/2 + 1/\ln 2 + (\ln \ln 2)/\ln 2$ and $B = 1 + (\ln \ln 2)/\ln 2$. ["On a special functional equation," *Journal of the London Mathematical Society* **15** (1940), 115–123.] Section 3 of the following chapter discusses what I learned about the subject during the next ten years.

Chapter 38

Recurrence Relations Based on Minimization

This paper investigates solutions of the general recurrence

$$M(0) = g(0), \quad M(n{+}1) = g(n{+}1) + \min_{0 \le k \le n} \big(\alpha M(k) + \beta M(n{-}k)\big)$$

for various choices of α, β, and $g(n)$. In a large number of cases the solution $M(n)$ is shown to be a convex function whose values can be computed much more efficiently than would be suggested by the defining recurrence. The asymptotic behavior of $M(n)$ can be deduced using combinatorial methods in conjunction with analytic techniques. In some cases there are strong connections between $M(n)$ and the function $H(x)$ defined by

$$H(x) = 1 \text{ for } x < 1, \quad H(x) = H((x{-}1)/\alpha) + H((x{-}1)/\beta) \text{ for } x \ge 1.$$

Special cases of these recurrences lead to a surprising number of interesting problems involving both discrete and continuous mathematics.

[Written with Michael L. Fredman. Originally published in *Journal of Mathematical Analysis and Applications* **48** (1974), 534–559.]

0. Introduction

Let α and β be positive real constants, and let $g(n)$ be a real-valued function over the nonnegative integers. Consider the new function $M_{g\alpha\beta}(n)$ over the nonnegative integers, defined as follows.

$$M_{g\alpha\beta}(0) = g(0),$$
$$M_{g\alpha\beta}(n + 1) = g(n + 1) + \min_{0 \le k \le n} (\alpha M_{g\alpha\beta}(k) + \beta M_{g\alpha\beta}(n - k)). \quad (0.1)$$

We shall occasionally write $M(n)$ instead of $M_{g\alpha\beta}(n)$. Functions of this type occur in discrete dynamic programming situations, where it is important to study the behavior of $M_{g\alpha\beta}(n)$ for large n.

The purpose of this paper is to introduce some techniques that are useful in the investigation of $M_{g\alpha\beta}(n)$, and in many cases to obtain ways of computing $M_{g\alpha\beta}(n)$ with much less work than definition (0.1) implies. Particular attention is paid to the cases $g(n) = \delta_{n0}$, $g(n) = 1$, $g(n) = n$, and $g(n) = n^2$, where asymptotic formulas are derived.

1. A Convexity Theorem

A real-valued function $g(n)$ over the nonnegative integers is called convex if its second difference is nonnegative, that is, if

$$g(n+2) - g(n+1) \geq g(n+1) - g(n) \quad \text{for all } n \geq 0.$$

The following result due to N. G. de Bruijn [4], generalizing earlier work of the authors, shows that "convolution minimization" of convex functions is particularly simple.

Lemma. *Let $a(n)$ and $b(n)$ be convex functions over the nonnegative integers, and define the new function*

$$c(n) = \min_{0 \leq k \leq n} \big(a(k) + b(n-k) \big), \quad \text{for } n \geq 0. \tag{1.1}$$

Then $c(n)$ is convex. Moreover, if $c(n) = a(k) + b(n-k)$, then

$$c(n+1) = \min \big(a(k) + b(n+1-k), \, a(k+1) + b(n-k) \big). \tag{1.2}$$

Proof. Let $\Delta f(n)$ stand for $f(n+1) - f(n)$. The two sequences

$$\begin{aligned}
\Delta a(0), \, \Delta a(1), \, \Delta a(2), \, \Delta a(3), \, \ldots, \\
\Delta b(0), \, \Delta b(1), \, \Delta b(2), \, \Delta b(3), \, \ldots,
\end{aligned} \tag{1.3}$$

are nondecreasing by hypothesis.

Suppose that the smallest n elements of (1.3) are $\Delta a(0)$, $\Delta a(1)$, ..., $\Delta a(i-1)$ and $\Delta b(0)$, $\Delta b(1)$, ..., $\Delta b(j-1)$, where $i + j = n$. If $k < i$, then $n - k - 1 \geq j$; hence $\Delta b(n - k - 1) \geq \Delta a(k)$, that is,

$$a(k) + b(n-k) \geq a(k+1) + b(n-k-1).$$

But if $k \geq i$, then $n - k - 1 < j$, and we have $\Delta b(n - k - 1) \leq \Delta a(k)$; that is,

$$a(k) + b(n-k) \leq a(k+1) + b(n-k-1).$$

It follows that $c(n) = a(i) + b(j)$.

As we increase n to $n+1$, we increase i or j by one, depending on which of $\{\Delta a(i), \Delta b(j)\}$ is larger. In other words, the sequence

$$\Delta c(0),\ \Delta c(1),\ \Delta c(2),\ \Delta c(3),\ \ldots, \tag{1.4}$$

is precisely the result obtained when *merging* the two sequences (1.3) into nondecreasing order. The lemma now follows immediately.   □

(Notice that (1.4) need not contain all the elements of the sequences being merged. If some $\Delta a(i)$ is larger than all the $\Delta b(j)$, or if some $\Delta b(j)$ is larger than all the $\Delta a(i)$, sequence (1.4) will omit much of (1.3).)

The lemma allows us to show that many of the most important functions $M_{g\alpha\beta}$ are convex.

Theorem 1. *The function $M_{g\alpha\beta}$ defined in (0.1) is convex if the function g satisfies the following two conditions:*

i) $g(n+2) - g(n+1) \geq g(n+1) - g(n)$ for all $n \geq 1$;

ii) $g(2) - g(1) + \min(\alpha d, \beta d) \geq d$, where

$$d = g(1) + (\alpha + \beta - 1)g(0). \tag{1.5}$$

Proof.   The value of d is just $M(1) - M(0)$, and condition (ii) says that

$$M(2) - M(1) \geq M(1) - M(0).$$

Assume that the sequence $\langle M(0), \ldots, M(n) \rangle$ has been proved convex, for some $n \geq 2$, in the sense that all $n - 1$ second-differences are nonnegative. The lemma shows that the sequence $\langle c(0), \ldots, c(n) \rangle$ is also convex, where $c(n) = \min_{0 \leq k \leq n}(\alpha M(k) + \beta M(n - k))$. Therefore the sequence

$$\langle M(1), \ldots, M(n+1) \rangle = \langle g(1) + c(0), \ldots, g(n+1) + c(n) \rangle,$$

being the sum of two convex sequences, is convex. Since $n \geq 2$, the sequence $\langle M(0), \ldots, M(n+1) \rangle$ must be convex for all n.   □

Let

$$D(n) = \Delta M(n) = M(n+1) - M(n). \tag{1.6}$$

Then the proofs of the lemma and the theorem show that $D(0) = d$ and $D(n) = g(n+1) - g(n) + F(n)$ for $n \geq 1$, where the infinite sequence

$$F(1),\ F(2),\ F(3),\ \ldots,$$

is the result of merging the two sequences

$$\alpha D(0),\ \alpha D(1),\ \alpha D(2),\ \ldots,$$
$$\beta D(0),\ \beta D(1),\ \beta D(2),\ \ldots,$$

into increasing order. Therefore we can compute the M function using the following simple algorithm, whenever $g(n)$ satisfies the hypotheses of Theorem 1.

```
begin integer i, j, n; real M, F; array D[0:N];
    i := j := 0;  D[0] := g(1) + (α + β − 1) × g(0);
    for n := 1 step 1 until N do
        begin if α × D[i] ≤ β × D[j] then
            begin F := α × D[i];  i := i + 1 end
        else begin F := β × D[j];  j := j + 1 end;
        D[n] := g(n + 1) − g(n) + F;
    end computation of D;
    M := g(0);
    for n := 0 step 1 until N do
        begin print("n=", n, ";  D[n]=", D[n], ";  M[n]=", M);
        M := M + D[n];
        end printing the table of D and M;
end.
```

This algorithm takes only $O(N)$ steps to compute $M[0], M[1], \ldots, M[N]$, instead of the order N^2 steps that are implied by the original definition of $M_{g\alpha\beta}(n)$ in (0.1).

Theorem 1 also has a useful corollary when α and β are equal.

Corollary. *Let* $\alpha = \beta$ *and let* $g(n)$ *be as in Theorem 1. Then*

$$M_{g\alpha\alpha}(n) = g(n) + \alpha\big(M_{g\alpha\alpha}(\lfloor (n-1)/2 \rfloor) + M_{g\alpha\alpha}(\lceil (n-1)/2 \rceil)\big) \quad (1.7)$$

for all $n \geq 1$. *(Here* $\lfloor x \rfloor$ *and* $\lceil x \rceil$ *denote respectively the greatest integer* $\leq x$ *and the least integer* $\geq x$.)

Proof. By Theorem 1 with $\beta = \alpha$, M is convex. It is easy to prove for any convex function M that the minimum value of $M(k) + M(n - k)$ occurs for $k = \lfloor n/2 \rfloor$. (Notice further that we have

$$\langle F(1), F(2), F(3), F(4), \ldots \rangle = \langle \alpha D(0), \alpha D(0), \alpha D(1), \alpha D(1), \ldots \rangle$$

in this case.)   □

2. The Case $g(n) = n$: "Optimal Trees"

When $g(n) = n$, so that $D(0) = 1$ and $D(n) = 1 + F(n)$ in the discussion above, we are soon led to an interpretation of $M_{g\alpha\beta}(n)$ in terms of binary trees. In this section we shall develop the tree relationship in an independent manner, without explicitly using the result of Theorem 1. Our general plan is to define a weighting function for the nodes of a binary tree; $M_{g\alpha\beta}(n)$ will turn out to be the minimum total weight of any binary tree with n nodes. (See [12, §2.3] for an introduction to the well-known properties of binary trees.)

A binary tree T is, by definition, either empty or it consists of a left subtree $l(T)$, a right subtree $r(T)$, and an apex or root node $a(T)$; the subtrees $l(T)$ and $r(T)$ are themselves binary trees. Let Λ denote the empty binary tree, and let $|T|$ be the number of nodes of T. Thus,

$$|T| = \begin{cases} 0, & \text{if } T = \Lambda; \\ 1 + |l(T)| + |r(T)|, & \text{if } T \neq \Lambda. \end{cases} \tag{2.1}$$

Now consider the function

$$\mathcal{M}(T) = \begin{cases} 0, & \text{if } T = \Lambda; \\ |T| + \alpha\mathcal{M}(l(T)) + \beta\mathcal{M}(r(T)), & \text{if } T \neq \Lambda; \end{cases} \tag{2.2}$$

and let

$$M(n) = \min_{|T|=n} \mathcal{M}(T). \tag{2.3}$$

We shall say T is "optimal" if $\mathcal{M}(T) = M(|T|)$. It is easy to see that the "principle of optimality" of dynamic programming is satisfied, in the sense that all subtrees of an optimal tree must be optimal. Consequently, for $n > 0$ we have

$$M(n) = n + \min_{0 \leq k < n} (\alpha M(k) + \beta M(n - 1 - k));$$

that is, $M(n) = M_{g\alpha\beta}(n)$.

Another way to view the situation is to consider finite strings, namely sequences of the letters L and R. If σ is such a string, define $w(\sigma)$ by the following rules.

$$w(\epsilon) = 1; \quad w(L\sigma) = 1 + \alpha w(\sigma); \quad w(R\sigma) = 1 + \beta w(\sigma). \tag{2.4}$$

Here ϵ denotes the empty string. As an example of this definition,

$$w(LRRLL) = 1 + \alpha + \alpha\beta + \alpha\beta^2 + \alpha^2\beta^2 + \alpha^3\beta^2.$$

Any node in a binary tree may be uniquely identified by a sequence of L's and R's [8, page 249]. We denote $a(T)$ by ϵ, and denote the nodes of $l(T)$ and $r(T)$ by placing an L or R, respectively, before the denotations in $l(T)$ and $r(T)$. Thus if $\mathcal{S}(T)$ is the set of all such strings, we have

$$\mathcal{S}(T) = \begin{cases} \emptyset, & \text{if } T = \Lambda; \\ \{\epsilon\} \cup L\mathcal{S}(l(T)) \cup R\mathcal{S}(r(T)), & \text{if } T \neq \Lambda. \end{cases}$$

It is easy to see that a set of strings S is equal to $\mathcal{S}(T)$ for some T if and only if

$$\sigma L \in S \quad \text{or} \quad \sigma R \in S \quad \text{implies} \quad \sigma \in S. \tag{2.5}$$

Furthermore, $\mathcal{M}$ is a "total weight" function, in the sense that

$$\mathcal{M}(T) = \sum_{\sigma \in \mathcal{S}(T)} w(\sigma). \tag{2.6}$$

This is the basic relation we shall use; it is easily verified by induction.

Now consider a sequence of strings $\langle \sigma_1, \sigma_2, \sigma_3, \dots \rangle$ such that, for each n, the weight $w(\sigma_n)$ is minimum among all strings that are not in $\{\sigma_1, \dots, \sigma_{n-1}\}$. Thus, $\sigma_1 = \epsilon$; $\sigma_2 = L$ if $\alpha < \beta$; $\sigma_2 = R$ if $\alpha > \beta$. (For some choices of α and β, such as $\alpha = \frac{1}{2}$ and $\beta = \frac{2}{3}$, there are infinitely many strings that will never appear in the sequence.) For each n, the set $S_n = \{\sigma_1, \dots, \sigma_n\}$ defines an optimal binary tree; this follows from (2.5), because $w(\sigma L)$ and $w(\sigma R)$ are always greater than $w(\sigma)$. Consequently,

$$M_{g\alpha\beta}(n) = \sum_{1 \leq k \leq n} w(\sigma_k). \tag{2.7}$$

This explicit interpretation of $M_{g\alpha\beta}$ is essentially that of the remarks following Theorem 1, since $\langle D(0), D(1), \dots \rangle$ is precisely the sequence $\langle w(\sigma_1), w(\sigma_2), \dots \rangle$.

As a simple application of these ideas, we can derive an asymptotic formula.

Theorem 2. *Let* $g(n) = n$, $0 < \alpha < \beta$, *and* $\alpha < 1$. *Then*

$$M_{g\alpha\beta}(n) \sim \frac{n}{1 - \alpha}. \tag{2.8}$$

Proof. If σ is a string of length $\geq m$, we have

$$w(\sigma) \geq w(L^m) = 1 + \alpha + \cdots + \alpha^m = (1 - \alpha^{m+1})/(1 - \alpha).$$

There are only finitely many strings of length $< m$; hence

$$\liminf_{n \to \infty} M(n)/n \geq (1 - \alpha^{m+1})/(1 - \alpha) \quad \text{for all } m.$$

On the other hand, $\limsup_{n \to \infty} M(n)/n \leq 1/(1 - \alpha)$, since the sequence of strings $\epsilon, L, L^2, L^3, \dots$, gives an upper bound. $\quad\square$

3. The Case $g(n) = n$: Asymptotics when $\min(\alpha, \beta) = 1$

Theorem 2 shows how M grows when $\min(\alpha, \beta) < 1$. When $\alpha = \beta = 1$, we have $w(\sigma) = m + 1$ for all strings σ of length m; hence we can obtain the well-known explicit formula

$$M_{g11}(n) = \sum_{1 \le k \le n} \lceil \log_2(k+1) \rceil = (n+1)\lceil \log_2(n+1) \rceil - 2^{\lceil \log_2(n+1) \rceil} + 1$$

$$= n \log_2 n + O(n). \tag{3.1}$$

When $\alpha = 1$ and $\beta > 1$, the problem of estimating $M_{g\alpha\beta}(n)$ is considerably more difficult. In this case, the weight function $w(\sigma)$ is related to partitions into powers of β; for example,

$$w(LRRLL) = 1 + 1 + \beta + \beta^2 + \beta^2 + \beta^2.$$

The weights take the form of polynomials with nonnegative coefficients,

$$a_0 + a_1 \beta + a_2 \beta^2 + \cdots + a_k \beta^k, \tag{3.2}$$

such that there are no "gaps":

$$a_j > 0 \Rightarrow a_{j-1} > 0. \tag{3.3}$$

An expression of the form (3.2) may be called a *partition* into powers of β; if condition (3.3) is also satisfied, we shall call it a *gapless partition*. It is convenient to regard the case $a_0 = a_1 = \cdots = 0$ as a gapless partition, even though it is not the weight of any string σ; the nonzero gapless partitions are in one-to-one correspondence with strings of L's and R's, since (3.2) is the weight of $L^{a_0-1}RL^{a_1-1}R \ldots RL^{a_k-1}$.

Let $P(x)$ denote the number of partitions into powers of β whose value is $\le x$, and let $H(x)$ be the corresponding number of gapless partitions. Thus, $H(x)$ is the number of strings of weight $\le x$, plus one. We have $P(x) = H(x) = 1$ for $0 \le x < 1$, and it is not difficult to deduce the following recurrence relations for $x \ge 1$.

$$P(x) = P(x - 1) + P(x/\beta); \tag{3.4}$$

$$H(x) = H(x - 1) + H((x - 1)/\beta). \tag{3.5}$$

As a consequence, we have the following relation between partitions and gapless partitions.

Lemma 3.1.

$$c_1^{-1}P(x+1/(\beta{-}1)) \leq H(x) \leq c_2^{-1}P(x+1/(\beta{-}1)),$$

where

$$c_1 = P(1+1/(\beta{-}1)-0) \quad \text{and} \quad c_2 = P(1/(\beta{-}1)).$$

(The notation $P(x-0)$ means $\lim_{\epsilon \to 0} P(x-\epsilon^2)$.)

Proof. Let $H_1(x) = P(x+1/(\beta{-}1))$. For $x \geq 1$, we have

$$\begin{aligned}
H_1(x) &= P(x{-}1+1/(\beta{-}1)) + P((x{-}1)/\beta+1/(\beta{-}1)) \\
&= H_1(x{-}1) + H_1((x{-}1)/\beta);
\end{aligned}$$

and for $0 \leq x < 1$, we have $c_2 \leq H_1(x) \leq c_1$. Thus,

$$c_1^{-1}H_1(x) \leq H(x) \leq c_2^{-1}H_1(x) \quad \text{for all } x,$$

by induction on $\lfloor x \rfloor$. $\square$

When $\beta = 2$, we have $c_1 = c_2 = 2$, so Lemma 3.1 shows that the number of gapless partitions of n into powers of 2 is exactly half the number of ordinary partitions of $n+1$ into powers of 2, for all positive integers n. A combinatorial proof of this result is also possible. The number of ordinary partitions (3.2) of n in which $a_k = 1$ is the same as the number with $a_k > 1$, under the correspondence

$$(a_0, a_1, \ldots, a_{k-1}, 1) \longleftrightarrow (a_0, a_1, \ldots, a_{k-1}+2).$$

The number of ordinary partitions of n in which $a_k = 1$ is the same as the number of gapless partitions of $n-1$, under the correspondence

$$(a_0, a_1, \ldots, a_{k-1}, 1) \longleftrightarrow (a_0+1, a_1+1, \ldots, a_{k-1}+1).$$

The H function has a comparatively simple relation to M, namely,

$$M(H(x)-1) = \int_0^x t\, dH(t) = xH(x) - \int_0^x H(t)\, dt, \qquad (3.6)$$

since $M(H(x)-1)$ is the sum of all gapless partitions whose value is $\leq x$ (see (2.7)). Therefore, we can use known results about partitions into powers of β in order to deduce the asymptotic behavior of M.

Theorem 3. *When $\beta > 1$ and $g(n) = n$, we have*

$$M_{g1\beta} \sim \frac{1}{e}\left(\frac{2\ln n}{\beta \ln \beta}\right)^{1/2} n^{1+(2\ln \beta/\ln n)^{1/2}}. \qquad (3.7)$$

Proof. N. G. de Bruijn [3] has proved that

$$\ln P(x) = \ln\left(\frac{y^2}{2}+y\right) + \left(\frac{1}{\ln \beta}-\frac{1}{2}\right)\ln x + p(y) + O\left(\frac{(\log\log x)^2}{\log x}\right), \quad (3.8)$$

where $y = \log_\beta x - \log_\beta \log_\beta x$, and where p is a rather horrible looking function of period 1, namely,

$$p(y) = \left(\frac{\pi^2}{12} - \gamma_1 - \frac{\gamma^2}{2}\right) \Big/ \ln \beta + \frac{\ln \beta}{12} - \frac{\ln 2\pi}{2}$$
$$+ \sum_{k \neq 0} \Gamma\left(\frac{2\pi i k}{\ln \beta}\right)\zeta\left(1 + \frac{2\pi i k}{\ln \beta}\right)\frac{e^{2\pi i k y}}{\ln \beta}, \qquad (3.9)$$

where $\zeta(z+1) = z^{-1} + \gamma - \gamma_1 z/1! + \gamma_2 z^2/2! - \cdots$.

Now we wish to show that the integral $\int_0^x H(t)\,dt$ in (3.6) is small with respect to the other term $xH(x)$. We have

$$\int_0^x H(t)\,dt = \int_0^x (H(\beta t + 1) - H(\beta t))\,dt$$
$$= \beta^{-1}\left(\int_{\beta x}^{\beta x+1} H(u)\,du - \int_0^1 H(u)\,du\right) = O(H(\beta x)).$$

By (3.8) and Lemma 3.1, $\ln(H(\beta x)/H(x)) = y\ln \beta + O(1)$; hence,

$$H(\beta x) = O(xH(x)/\log x). \qquad (3.10)$$

If we set $n = H(x) - 1$ and $M(n) = ne^{f(n)}$, we now have

$$M(n) = xn + O(xn/\log x), \qquad (3.11)$$
$$f(n) = \ln x + O(1/\log x), \qquad (3.12)$$

and it remains to express $\ln x$ in terms of n.

We have

$$\ln x = y\ln \beta + \ln y + O(\log\log x/\log x);$$

hence by (3.8) and Lemma 3.1,

$$\ln n = \frac{\ln \beta}{2} y^2 + \left(1 + \frac{\ln \beta}{2}\right) y + \left(\frac{1}{\ln \beta} - \frac{1}{2}\right) \ln y + O(1).$$

Consequently,

$$y = (2 \log_\beta n)^{1/2} - \frac{1}{2} - \frac{1}{\ln \beta} + O\left(\frac{\log \log n}{(\log n)^{1/2}}\right),$$

and (3.7) follows immediately for those values of n having the special form $H(x) - 1$. In general, suppose that

$$H(x - 0) - 1 = n_0 < n \leq n_1 = H(x) - 1.$$

Then,

$$n_1 - n_0 \leq H(x) - H(x - 1) = H((x - 1)/\beta) = O(H(x) \log x/x) = o(n_1);$$

hence, $n_0/n_1 \to 1$ as $n \to \infty$. By (3.11), $M(n_0)/M(n_1) \to 1$. $\quad\square$

This proof can be extended to obtain slightly more information than is stated in Theorem 3; we could evaluate $f(n)$ with a relative error of $O(1/(\log n)^{1/2})$. But the complicated form of (3.9) shows that it is inherently very difficult to go any further than this.

Before moving to the next topic, let us digress for a moment to summarize the interesting history of the present case. Euler gave the generating function for partitions into powers of 2 in his famous paper on partitions [6]. A. Cayley [1] proved that the number of sequences $\langle a_1, a_2, \ldots, a_k \rangle$ such that $a_1 = 1$ and $1 \leq a_{i+1} \leq 2a_i$ is equal to the number of partitions of $2^k - 1$ into powers of 2; he proved this by using the corresponding generating function. Binary partitions were independently studied by Tanturri [17]. The behavior of the generating function in the neighborhood of unity was investigated about 1923 by C. L. Siegel, in an unpublished work. P. Erdős [5] found the leading term of (3.8), and K. Mahler [14] found the other terms, except with $O(1)$ instead of the periodic function $p(y)$, when β is an integer. N. G. de Bruijn [3] obtained (3.8) for all $\beta > 1$, and his work was generalized further by W. B. Pennington [15]. The connection between binary partitions and the M_{g12} function was pointed out by Knuth [11], who gave an elementary derivation of the leading term in (3.8) when $\beta = 2$. Heller [9] found the leading terms of (3.8) using a different approach. Arithmetic properties of β-ary partitions have been studied by Churchhouse [2] and Rödseth [16].

4. The Case $g(n) = n$: Asymptotics when $\min(\alpha, \beta) > 1$

When $\alpha = \beta$, the weight of any string σ is simply $1 + \alpha + \cdots + \alpha^{|\sigma|}$; so it is easy to obtain an "explicit" formula for $M_{g\alpha\beta}$ when $g(n) = n$ and $\alpha = \beta$:

$$
\begin{aligned}
M_{g\alpha\alpha}(2^m + k - 1) &= 2^m(1 + \alpha + \cdots + \alpha^{m-1}) \\
&\quad - (1 + 2\alpha + \cdots + 2^{m-1}\alpha^{m-1}) + k(1 + \alpha + \cdots + \alpha^m)
\end{aligned}
\tag{4.1}
$$

for $0 \le k \le 2^m$. It follows that for $0 \le \theta \le 1$ and $\alpha > 1$ we have

$$
\lim_{m \to \infty} \frac{M_{g\alpha\alpha}((1+\theta)2^m)}{2^m \alpha^m} = \frac{1}{\alpha - 1} - \frac{1}{2\alpha - 1} + \frac{\theta\alpha}{\alpha - 1}.
\tag{4.2}
$$

Replacing $(1 + \theta)2^m$ by n, it follows that

$$
M_{g\alpha\alpha}(n) \sim c(\theta)n^{1 + \log_2 \alpha},
$$

where

$$
\theta = 2^{(\log_2 n) \bmod 1} - 1
$$

and

$$
c(\theta) = \left(\frac{1 + \theta\alpha}{\alpha - 1} - \frac{1}{2\alpha - 1} \right)(1 + \theta)^{-(1 + \log_2 \alpha)}
$$

is a periodic function of $\log_2 n$. For example, when $\alpha = \beta = 2$, the asymptotic form of $M_{g22}(n)$ varies between $\frac{2}{3}n^2$ (when $n \approx 2^m$) and $\frac{3}{4}n^2$ (when $n \approx \frac{4}{3}2^m$).

We shall see that such behavior is typical of the case $\min(\alpha, \beta) > 1$. If we define the constant γ by the relation

$$
\alpha^{-\gamma} + \beta^{-\gamma} = 1,
\tag{4.3}
$$

we will find that $M_{g\alpha\beta}(n)$ grows approximately as $n^{1+1/\gamma}$. When $\log\alpha/\log\beta$ is irrational, it turns out that $M_{g\alpha\beta}(n)/n^{1+1/\gamma}$ actually approaches a limit as $n \to \infty$. On the other hand, in many cases when $\log\alpha/\log\beta$ is rational, $M_{g\alpha\beta}(n)/n^{1+1/\gamma}$ oscillates between two different limits, as in the case $\alpha = \beta$.

We shall begin our analysis of the general case $g(n) = n$ when $\min(\alpha, \beta) > 1$ by generalizing the H function used in Section 3. Let $h(x)$ be the number of strings σ whose weight $w(\sigma)$ is $\le x$, and let

$$
H(x) = h(x) + 1.
\tag{4.4}
$$

We have $H(x) = 1$ for $0 \leq x < 1$, and for $x \geq 1$ the rule for defining weights implies that

$$H(x) = H((x-1)/\alpha) + H((x-1)/\beta). \qquad (4.5)$$

The basic relation (3.6) between H and M, namely,

$$M(h(x)) = xH(x) - \int_0^x H(t)\,dt = xh(x) - \int_0^x h(t)\,dt, \qquad (4.6)$$

is still valid for this generalized H function. Indeed, by separating the strings σ that begin with L from those that begin with R (see (2.4)), we obtain the formula

$$M(h(x)) = h(x) + \alpha M(h((x-1)/\alpha)) + \beta M(h((x-1)/\beta)). \qquad (4.7)$$

Therefore if we can determine the asymptotic behavior of h (or H), we will be able to see how M grows, and to see how the value of k for which the minimum occurs in (0.1) depends on n.

Now that the problem has been set up in this way, it is comparatively easy to deduce the order of growth of M.

Lemma 4.1. *Let γ be the positive constant defined by (4.3). There exist positive constants c_1, c_2, C_1, C_2, such that*

$$c_1 x^\gamma \leq H(x) \leq c_2 x^\gamma, \qquad (4.8)$$
$$C_1 x^{1+1/\gamma} \leq M(x) \leq C_2 x^{1+1/\gamma}, \qquad (4.9)$$

for all sufficiently large x.

Proof. Choose c_2 so that $H(x) \leq c_2 x^\gamma$ for $1 \leq x < 2$. Then we can prove by induction on n that $H(x) \leq c_2 x^\gamma$ for $1 \leq x < n$, since

$$H(x) = H((x-1)/\alpha) + H((x-1)/\beta),$$

which (by induction) is at most

$$c_2((x-1)/\alpha)^\gamma + c_2((x-1)/\beta)^\gamma = c_2(x-1)^\gamma < c_2 x^\gamma.$$

The lower bound is a little trickier. If we assume that there is a positive constant a such that $H(x) \geq ax^{\gamma-\epsilon}$ for $x < x_0$, then we have

$$H(x_0) \geq a(x_0 - 1)^{\gamma-\epsilon} K,$$

where

$$K = \frac{\alpha^\epsilon}{\alpha^\gamma} + \frac{\beta^\epsilon}{\beta^\gamma} > \min(\alpha^\epsilon, \beta^\epsilon) > 1. \tag{4.10}$$

For sufficiently large x_0, we will have $a(x_0 - 1)^{\gamma - \epsilon} K \geq a x_0^{\gamma - \epsilon}$ for $x_0 \leq x < x_0 + 1$. Indeed, we can clearly extend this to *all* $x \geq x_0$. Since such an a exists for arbitrarily small ϵ, we must have $H(x)/x^{\gamma - \epsilon} \to \infty$ as $x \to \infty$.

Let c be a constant such that $x^\gamma - (x-1)^\gamma \leq c x^{\gamma - 1}$ for all large x; and let R be a constant such that $RK > R + c$, where $K = \alpha^{1-\gamma} + \beta^{1-\gamma} > 1$, as in (4.10). For sufficiently large x_0, we will have $(x_0 - 1)^{\gamma - 1} RK \geq x_0^{\gamma - 1}(R + c)$ and $H(x) > R x^{\gamma - 1}$ for all $x \geq x_0$. Thus, there will be a positive constant $c_1 \leq 1$ such that

$$H(x) \geq c_1 x^\gamma + R x^{\gamma - 1} \quad \text{for } x_0 \leq x \leq \max(\alpha, \beta) x_0 + 1. \tag{4.11}$$

We will show that this relation holds for all $x \geq x_0$. Let $x_n = \max(\alpha, \beta) x_0 + n$; we will show by induction on n that (4.11) holds for $x_n \leq x \leq x_{n+1}$, and this estimate will establish (4.8). The calculation is not difficult, and it reveals why we have been foresighted enough to choose c and R in such a mysterious way:

$$\begin{aligned}
H(x) &= H((x - 1)/\alpha) + H((x - 1)/\beta) \\
&\geq c_1((x - 1)^\gamma/\alpha^\gamma + (x - 1)^\gamma/\beta^\gamma) \\
&\quad + R((x - 1)^{\gamma - 1}/\alpha^{\gamma - 1} + (x - 1)^{\gamma - 1}/\beta^{\gamma - 1}) \\
&= c_1(x - 1)^\gamma + RK(x - 1)^{\gamma - 1} \\
&\geq c_1 x^\gamma - c_1 c x^{\gamma - 1} + (R + c) x^{\gamma - 1} \\
&\geq c_1 x^\gamma + R x^{\gamma - 1}.
\end{aligned}$$

Now to obtain bounds on $M(x)$, we may use (2.7). By the definition of H, we have

$$w(\sigma_n) \leq x \quad \text{if and only if} \quad H(x) > n; \tag{4.12}$$

hence by (4.8),

$$c_2^{-1/\gamma} n^{1/\gamma} < w(\sigma_n) \leq c_1^{-1/\gamma}(n + 1)^{1/\gamma} \tag{4.13}$$

for all large n. It follows that $M(n)$, the sum of the first n weights, satisfies

$$\liminf_{n \to \infty} \frac{M(n)}{n^{1+1/\gamma}} \geq \frac{\gamma}{\gamma + 1} c_2^{-1/\gamma}, \quad \limsup_{n \to \infty} \frac{M(n)}{n^{1+1/\gamma}} \leq \frac{\gamma}{\gamma + 1} c_1^{-1/\gamma}. \tag{4.14}$$

The desired relation (4.9) is an immediate consequence. $\quad\square$

The latter part of this proof suggests the following result.

Lemma 4.2. *Let γ be as in Lemma 4.1. Then $\lim_{x\to\infty} H(x)/x^\gamma$ exists if and only if $\lim_{n\to\infty} M(n)/n^{1+1/\gamma}$ exists.*

Proof. If $\lim_{x\to\infty} H(x)/x^\gamma = c$, then by (4.14),

$$\lim_{n\to\infty} \frac{M(n)}{n^{1+1/\gamma}} = \left(\frac{\gamma}{\gamma+1}\right)c^{-1/\gamma}.$$

Conversely, if $M(n) \sim Cn^{1+1/\gamma}$, we must have $w(\sigma_n) \sim (1+1/\gamma)Cn^{1/\gamma}$ since $w(\sigma_n)$ is a nondecreasing function of n. (This follows from a straightforward "Tauberian" argument; we have

$$M\big(\lfloor(1+\epsilon)n\rfloor\big) - M(n) \geq \big(\lfloor(1+\epsilon)n\rfloor - n\big)\sigma(w_n);$$

hence, $\limsup_{n\to\infty} w(\sigma_n)/n^{1/\gamma} \leq C\epsilon^{-1}((1+\epsilon)^{1+1/\gamma} - 1)$ for all $\epsilon > 0$. Similarly, $\liminf_{n\to\infty} w(\sigma_n)/n^{1/\gamma} \geq C\epsilon^{-1}(1-(1-\epsilon)^{1+1/\gamma})$.) Relation (4.12) completes the proof. $\square$

Now let us investigate whether or not the limits do exist, for various α and β. We have seen that the limit does not exist when $\alpha = \beta$; similarly we can construct a large number of further examples, including all cases where α and β are integers and $\log\alpha/\log\beta$ is rational.

Theorem 4.1. *Let $\min(\alpha,\beta) > 1$, and let γ be defined by (4.3). If $\log\alpha/\log\beta$ is rational, and if $\gamma < 1$ (that is, if $\alpha^{-1} + \beta^{-1} < 1$), then $\lim_{n\to\infty} M(n)/n^{1+1/\gamma}$ does not exist.*

Proof. We have $\alpha = \theta^p$ and $\beta = \theta^q$, where p and q are relatively prime positive integers and $\theta > 1$. Without loss of generality, we may assume that $p < q$. We will show that large "gaps" exist between weights, in the sense that there are positive real numbers $x < y$ such that no string weights lie between $\theta^m x$ and $\theta^m y - 1$ for any m. This is enough to prove the theorem, since existence of the limit would imply that

$$H(\theta^n x)/H(\theta^n y - 1) \to x^\gamma/y^\gamma \neq 1 = H(\theta^m x)/H(\theta^m y - 1).$$

The weight of every string is a polynomial in θ, namely,

$$\theta^{a_t} + \theta^{a_{t-1}} + \cdots + \theta^{a_1} + \theta^{a_0} \tag{4.15}$$

where $a_0 = 0$ and $a_{i+1} - a_i = p$ or q for $0 \leq i < t$. We may think of (4.15) as a number written with radix θ and digits 0 or 1, subject to the requirement that exactly $p - 1$ or $q - 1$ 0s occur between adjacent 1s. For convenience, we shall call (4.15) a *weight of order* a_t.

Let S be the set of all infinite expansions $\theta^{-b_0} + \theta^{-b_1} + \theta^{-b_2} + \cdots$, where $b_0 = 0$ and $b_{i+1} - b_i = p$ or q for all $i \geq 0$. Thus, S is a set of real numbers that satisfies

$$S = (1 + \theta^{-p}S) \cup (1 + \theta^{-q}S). \tag{4.16}$$

The largest element of S is $1/(1 - \theta^{-p})$. This set contains large gaps, since the largest element of $1 + \theta^{-q}S$ is $1 + \theta^{-q}/(1 - \theta^{-p})$, which is smaller than the smallest element $1 + \theta^{-p}/(1 - \theta^{-q})$ of $1 + \theta^{-p}S$. (We have

$$\theta^{-q}/(1 - \theta^{-p}) < \theta^{-p}/(1 - \theta^{-q})$$

since this relation is equivalent to $\theta^{-q} - \theta^{-2q} < \theta^{-p} - \theta^{-2p}$, that is, $(\theta^{-q} - \theta^{-p})(1 - \theta^{-q} - \theta^{-p}) < 0$.) Equation (4.16) now shows that there are many further gaps. For example,

$$1 + \theta^{-q} + \theta^{-q}S < 1 + \theta^{-q} + \theta^{-p}S < 1 + \theta^{-p} + \theta^{-q}S < 1 + \theta^{-p} + \theta^{-p}S,$$

and we see that S is contained in something like a "Cantor ternary set": Every point not in S lies in an interval that is not in S, and S has measure zero.

Since every element of $\theta^{q-p}S$ is greater than every element of S, we can find positive numbers $x < y$ such that the interval $(x \mathbin{.\,.} y)$ contains no points of

$$S_1 = \cdots \cup \theta^{-2}S \cup \theta^{-1}S \cup S \cup \theta S \cup \theta^2 S \cup \cdots. \tag{4.17}$$

If w is the weight of a string such that $\theta^m x < w < \theta^m y - 1$, then $w_1 = w + \theta^{-q}/(1 - \theta^{-q}) \in S_1$; hence, $\theta^{-m}w_1$ is an element of $S_1 \cap (x \mathbin{.\,.} y)$. This contradicts the choice of x and y; so there are no string weights between $\theta^m x$ and $\theta^m y - 1$. □

The next theorem shows why the hypothesis $\gamma < 1$ is necessary in Theorem 4.1, since there are infinitely many examples when $M(n)/n^{1+1/\gamma}$ approaches a limit even though $\log \alpha / \log \beta$ is rational.

Theorem 4.2. *Let* $\alpha^{-1} + \beta^{-1} = 1$. *If* $\log \alpha / \log \beta$ *is rational and* $\alpha \neq \beta$, *then*

$$M(n) \sim \frac{\beta - \alpha}{2(\log \beta - \log \alpha)}(\alpha^{-1} \log \alpha + \beta^{-1} \log \beta)n^2. \tag{4.18}$$

Proof.   We have $\alpha = \theta^p$ and $\beta = \theta^q$ where p and q are relatively prime positive integers and where θ is the unique real root > 1 of the equation $1 - \theta^{-p} - \theta^{-q} = 0$. Since $\alpha \neq \beta$, we may assume that $p < q$. To prove this theorem, we shall refine the observations made in the proof of Theorem 4.1 by studying the weights of order m more closely. Since p and q are relatively prime, there will be weights of order m for all large m.

The weights of order m have the form $\theta^m + w$, where w is a weight of order $m - p$ or $m - q$. For large m, the largest weight of order $m - q$ is

$$\theta^{m-q} + \theta^{m-q-p} + \theta^{m-q-2p} + \cdots + \theta^{uq} + \theta^{uq-q} + \cdots + 1$$

$$= \frac{\theta^{m-q+p} - \theta^{uq}}{\theta^p - 1} + \frac{\theta^{uq} - 1}{\theta^q - 1}$$

$$= \theta^m + a_u,$$

where

$$(u+1)q \equiv m \quad (\text{modulo } p), \qquad 0 \le u < p,$$

and

$$a_u = \theta^{uq}(\theta^{p-q} - \theta^{q-p}) - \theta^{p-q}. \tag{4.19}$$

Similarly the smallest weight of order $m - p$ is $\theta^m + b_v$, where

$$(v+1)p \equiv m \quad (\text{modulo } q), \qquad 0 \le v < q,$$

and

$$b_v = \theta^{vp}(\theta^{q-p} - \theta^{p-q}) - \theta^{q-p}. \tag{4.20}$$

We have

$$a_u \le a_0 = -\theta^{q-p} < -\theta^{p-q} = b_0 \le b_v; \tag{4.21}$$

hence, the weights of order m appear in increasing order if we read their radix θ representations in lexicographic order.

Let $r = q - p$ and let $0 \le s < r$. We shall divide the weights into r disjoint classes, where the weights of class s consist of all weights of order s, $s+r$, $s+2r$, $\ldots$, $s+kr$, $\ldots$. From the argument in the preceding paragraph, we see that the weights of class s appear in increasing order if we treat their radix θ representations as binary numbers; furthermore, the difference between consecutive weights of class s is bounded. (The set of all such differences contains pq elements $\{ b_v - a_u \mid 0 \le u < p, 0 \le v < q \}$, plus perhaps a finite number of other differences that might appear for small m.)

Let f_m be the number of weights of order m, so that $f_0 = 1$, $f_m = 0$ for $m < 0$, and $f_m = f_{m-p} + f_{m-q}$ for $m > 0$. (In the special case $p = 1$

and $q = 2$, for example, f_m is a Fibonacci number and $\theta = (1 + \sqrt{5})/2$.)
Let

$$g_m = f_m + f_{m-r} + f_{m-2r} + \cdots$$

be the number of weights of order $\leq m$ belonging to class m mod r; and
finally let $h_0 = 1$ and $h_m = g_{m-r} - g_{m-r-q}$ for $m > 0$. We shall prove
that if

$$w = \theta^{a_t} + \theta^{a_{t-1}} + \cdots + \theta^{a_1} + \theta^{a_0} \tag{4.22}$$

is the nth smallest weight of class s, we have

$$n = h_{a_t} + h_{a_{t-1}} + \cdots + h_{a_1} + h_{a_0}. \tag{4.23}$$

The proof is by induction on $m = a_t$; since $n = 1 = h_0$ when $m = 0$,
we may assume that $m > 0$. Let (4.22) be the kth smallest weight of
order m. Then $n = g_m - f_m + k$, where $1 \leq k \leq f_m$. If $a_{t-1} = a_t - q$,
then $w - \theta^m$ is the kth smallest weight of order $m - q$; hence by induction,
(4.23) holds if and only if $n - h_m = g_{m-q} - f_{m-q} + k$. The latter is true
by the definition of h_m and k. Similarly, if $a_{t-1} = a_t - p$, the quantity
$w - \theta^m$ is the $(k - f_{m-q})$th smallest weight of order $m - p$. Hence by
induction, (4.23) holds if and only if

$$n - h_m = g_{m-p} - f_{m-p} + k - f_{m-q};$$

and $g_{m-p} - f_{m-p} = g_{m-q}$, so the proof of (4.23) is complete.

Notice that the generating function for the h's is

$$\sum_m h_m z^m = 1 + \frac{z^r - z^{r+q}}{(1 - z^r)(1 - z^p - z^q)} = \frac{1 - z^p}{(1 - z^r)(1 - z^p - z^q)}. \tag{4.24}$$

This formula can be written

$$\sum_m h_m z^m = \frac{c}{1 - \theta z} + R(z), \quad c = \frac{1 - \theta^{-p}}{(1 - \theta^{-r})(p\theta^{-p} + q\theta^{-q})}, \tag{4.25}$$

where $R(z)$ has no singularities in $|z| \leq \theta^{-1} + \epsilon$, since θ^{-1} is the smallest
root of $1 - z^p - z^q = 0$. (If $1 = z^p + z^q$, then $1 \leq |z|^p + |z|^q$; hence
$|z| \geq \theta^{-1}$, with equality if and only if $z = \theta^{-1}$.) Consequently,

$$h_m = c\theta^m + o(\theta^m). \tag{4.26}$$

Let $H_s(x)$ be the number of weights of class s that are $\leq x$, so that we
have

$$H(x) = H_0(x) + H_1(x) + \cdots + H_{r-1}(x) + 1. \tag{4.27}$$

For fixed s, we will show that $\lim_{x\to\infty} H_s(x)/x = c$. Let w be the largest weight of class s that is $\le x$; we have observed that $x - w$ is bounded. If w is given by (4.22), the value of $H_s(x)$ is given by (4.23), which equals $cw + o(w) = cx + o(x)$. It follows that

$$H(x) \sim rcx, \tag{4.28}$$

and the theorem is obtained by applying Lemma 4.2, since we have

$$rc = \frac{\log\beta - \log\alpha}{(\beta - \alpha)(\alpha^{-1}\log\alpha + \beta^{-1}\log\beta)}. \quad \square$$

A more detailed examination of the simplest case of Theorem 4.2, when $\alpha = \phi = (1+\sqrt{5})/2$ and $\beta = \phi^2$, actually yields an explicit formula for the nth weight:

$$w(\sigma_n) = \phi^{-1}\lfloor n\phi^{-1}\rfloor + n. \tag{4.29}$$

(See [12, exercise 1.2.8–36].) We also have

$$M(F_n) = \frac{1}{2\sqrt{5}}(\phi^{2n-1} - 2\phi^{n-1} - (-1)^n(2 - \phi^{-n+2})) + F_n \tag{4.30}$$

in this case, when $F_n = (\phi^n - (-\phi)^{-n})/\sqrt{5}$ is a Fibonacci number.

A completely different approach seems to be necessary when the ratio $\log\alpha/\log\beta$ is irrational. The following discussion is based on Dirichlet integrals.

Theorem 4.3. *Assume that $\min(\alpha, \beta) > 1$ and $\log\alpha/\log\beta$ is irrational, and let γ be defined by (4.3). Then $\lim_{n\to\infty} M(n)/n^{1+1/\gamma}$ exists.*

Proof. We shall make use of the following result from the analytic theory of numbers.

Lemma 4.3. *Let $f(t)$ be a nondecreasing function of the real variable t, with $f(t) \ge 0$. Assume that $G(s) = \int_1^\infty f(t)\,dt/t^{s+1}$ is an analytic function of the complex variable s when $\Re(s) \ge \gamma > 0$, except for a first-order pole at $s = \gamma$ with positive residue C. Then $f(t) \sim Ct^\gamma$.* $\square$

A proof of this lemma appears in the appendix below. Let us apply Lemma 4.3 to the function $f(t) = h(t) = H(t) - 1$. By Lemma 4.1, the integral

$$G(s) = \int_1^\infty h(t)\,dt/t^{s+1} \tag{4.31}$$

diverges when $s = \gamma$, but it converges absolutely and uniformly in any bounded region such that $\Re(s) \geq \gamma + \epsilon$, for all fixed $\epsilon > 0$. It follows that $G(s)$ is analytic in the half-plane $\Re(s) > \gamma$.

We will now show that $G(s)$ has a simple pole at $s = \gamma$, by analytically continuing G to the left of the line $\Re(s) = \gamma$. Consider the function $G_1(s) = (1 - \alpha^{-s} - \beta^{-s})G(s)$; when $\Re(s) > \gamma$, we have

$$
\begin{aligned}
G_1(s) &= \int_0^\infty h(t+1)\frac{dt}{(t+1)^{s+1}} - \int_\alpha^\infty h\left(\frac{t}{\alpha}\right)\frac{dt}{t^{s+1}} - \int_\beta^\infty h\left(\frac{t}{\beta}\right)\frac{dt}{t^{s+1}} \\
&= \int_0^\infty \left(1 + h\left(\frac{t}{\alpha}\right) + h\left(\frac{t}{\beta}\right)\right)\frac{dt}{(t+1)^{s+1}} - \int_0^\infty \left(h\left(\frac{t}{\alpha}\right) + h\left(\frac{t}{\beta}\right)\right)\frac{dt}{t^{s+1}} \\
&= \frac{1}{s} + \int_{\min(\alpha,\beta)}^\infty \left(h\left(\frac{t}{\alpha}\right) + h\left(\frac{t}{\beta}\right)\right)dt((t+1)^{-s-1} - t^{-s-1}). \quad (4.32)
\end{aligned}
$$

(This derivation uses (4.5) together with the fact that $h(t) = 0$ for $t < 1$.) Since $(t+1)^{-s-1} - t^{-s-1} = O((s+1)t^{-s-2})$, the latter integral converges whenever $\Re(s) > \gamma - 1$. Therefore we can analytically continue $G(s)$ into this region, by using the formula

$$
G(s) = G_1(s)/(1 - \alpha^{-s} - \beta^{-s}), \qquad (4.33)
$$

and letting $G_1(s)$ be defined by (4.32). The only singularities of $G(s)$ in this region are the poles at $s = 0$ (if $\gamma < 1$) and possibly at the zeros of $1 - \alpha^{-s} - \beta^{-s}$. For $s = \gamma$, we have a simple pole since we know this is a singularity of $G(s)$; the corresponding residue

$$
G_1(\gamma)/(\ln\alpha \cdot \alpha^{-\gamma} + \ln\beta \cdot \beta^{-\gamma})
$$

must be positive, since $(s - \gamma)G(s)$ is positive when s approaches γ from the right. Furthermore, γ is the only singularity of $G(s)$ when $\Re(s) \geq \gamma$, for if we write $s = \sigma + i\tau$ we have

$$
|\alpha^{-s} + \beta^{-s}| \leq \alpha^{-\sigma} + \beta^{-\sigma} \leq \alpha^{-\gamma} + \beta^{-\gamma} = 1
$$

where equality holds if and only if $\alpha^{-s} = \alpha^{-\gamma}$ and $\beta^{-s} = \beta^{-\gamma}$. This condition implies that $\tau = 2\pi p/\ln\alpha$ and $\tau = 2\pi q/\ln\beta$ for some integers p and q; hence $\tau = 0$, because $\log\alpha/\log\beta$ is irrational.

We have now shown that $G(s)$ satisfies the hypotheses of Lemma 4.3, so $h(t) \sim Ct^\gamma$. This completes the proof (see Lemma 4.2). □

Incidentally, if we attempt to apply this same method of proof when $\log \alpha / \log \beta$ is rational, we find that $1 - \alpha^{-s} - \beta^{-s}$ has infinitely many zeros on the line $\Re(s) = \gamma$. But by an amazing coincidence, when $\gamma = 1$ and $\alpha \neq \beta$, $G_1(s)$ happens to be zero at all but one of these points.

It is possible to evaluate the residue C, when $\gamma = 1$; in fact, (4.18) holds also when $\log \alpha / \log \beta$ is irrational, since the residue is a continuous function of α.

The reader will notice that Theorems 4.1–4.3 do not cover all cases. If $\gamma > 1$ and $\log \alpha / \log \beta$ is rational, the authors conjecture that $\lim_{n \to \infty} M(n)/n^{1+1/\gamma}$ does not exist. It can be shown that this conjecture holds "almost always," with at most countably many counterexamples (see Fredman [7]).

5. The Case $g(n) = 1$

Another interesting case of the general problem we are considering occurs when $g(n) = 1$ for all n. The problem breaks into two subcases.

Theorem 5.1. *Assume that $g(n) = 1$ for all n, and that $\min(\alpha, \beta) > 1$. Let γ be defined by (4.3). Then there exist positive constants C_1 and C_2 such that $C_1 n^{1+1/\gamma} < M(n) < C_2 n^{1+1/\gamma}$ for all n. Furthermore, $\lim_{n \to \infty} M(n)/n^{1+1/\gamma}$ exists if and only if $\log \alpha / \log \beta$ is irrational.*

Proof. Theorem 1 applies to $M_{g\alpha\beta}$; in this case, $D(n) = F(n)$ for $n \geq 1$, and it is easy to see that again we obtain a tree interpretation as in Section 2 above. This time we have $D(n) = (\alpha + \beta)w(\sigma_{n+1})$, where the weights are defined by the rules

$$w(\epsilon) = 1, \quad w(L\sigma) = \alpha w(\sigma), \quad w(R\sigma) = \beta w(\sigma). \tag{5.1}$$

The new rule is simpler than (2.4); the new weights are given by the final term of the previous weights. For example, $w(LRRLL) = \alpha^3 \beta^2$. The weight $\alpha^i \beta^j$ occurs $\binom{i+j}{i}$ times, since this is the number of strings containing i L's and j R's.

To deduce the asymptotic behavior of $M_{g\alpha\beta}(n)$, we proceed as above, letting $h(x)$ be the number of weights $\leq x$, and $H(x) = h(x) + 1$. This time $H(x) = 1$ for $0 \leq x < 1$, and

$$H(x) = H(x/\alpha) + H(x/\beta) \quad \text{for } x \geq 1, \tag{5.2}$$

a relation simpler than (4.5). It is now easy to prove Lemma 4.1 for the new H and M functions, and Lemma 4.2 follows as before. Once again we can use the idea in the proof of Theorem 4.3. Let $G(s)$ be defined by (4.31), and $G_1(s) = (1 - \alpha^{-s} - \beta^{-s})G(s)$. When $\Re(s) > \gamma$, we have

$$G_1(s) = \int_1^\infty \left(h(t) - h(t/\alpha) - h(t/\beta)\right)\frac{dt}{t^{s+1}} = \int_1^\infty \frac{dt}{t^{s+1}} = \frac{1}{s}. \qquad (5.3)$$

Thus, $G(s)$ can be analytically extended to the entire plane by using the formula $G(s) = 1/\big(s(1 - \alpha^{-s} - \beta^{-s})\big)$. When $\log\alpha/\log\beta$ is irrational, we argue as in Theorem 4.3 that Lemma 4.3 applies; hence,

$$h(x) \sim Cx^\gamma \qquad \text{and} \qquad M(n) \sim (\alpha + \beta)C^{-1/\gamma}\gamma n^{1+1/\gamma}/(\gamma + 1),$$

where $C = 1/(\alpha^{-\gamma}\ln\alpha^\gamma + \beta^{-\gamma}\ln\beta^\gamma)$. When $\log\alpha/\log\beta$ is rational, on the other hand, there is no analog to Theorem 4.2; the limit $h(x)/x^\gamma$ never exists. The reason is that $\alpha = \theta^p$ and $\beta = \theta^q$ for some θ, and all weights are powers of θ. Thus, $h(\theta^n) = h(\theta^{n+1} - 0)$; that is, there are large gaps between the weights, as in Theorem 4.1. $\square$

Theorem 5.2. *If $g(n) = 1$ for all n and if $\alpha \leq \min(1, \beta)$, then*

$$M_{g\alpha\beta}(n) = 1 + (\alpha + \beta)(1 + \alpha + \cdots + \alpha^{n-1}). \qquad (5.4)$$

Proof. In this case Theorem 1 does *not* apply, and in fact the function $M_{g\alpha\beta}$ turns out to be *concave*! We will prove (5.4) by induction. For $k < n/2$, we have $M(k) \leq M(n - k)$; hence,

$$\alpha M(k) + \beta M(n - k) \geq \alpha M(n - k) + \beta M(k).$$

For $k > n/2$, we have

$$\alpha M(k) + \beta M(n - k) - (\alpha M(k - 1) + \beta M(n - k + 1))$$
$$= (\alpha + \beta)(\alpha^k - \beta\alpha^{n-k}) \leq (\alpha + \beta)(\alpha^k - \alpha^{n+1-k}) \leq 0.$$

Hence, $M(n + 1) = 1 + \alpha M(n) + \beta M(0)$, and the proof by induction is complete. $\square$

6. The Case $g(n) = \delta_{n0}$

When $g(0) = 1$ and $g(n) = 0$ for all $n > 0$, we obtain a case strongly related to the previous one. Let $M^*(n) = M_{g\alpha\beta}(n) - 1/(\alpha + \beta)$; then,

$$M^*(0) = 1 - 1/(\alpha + \beta),$$
$$M^*(n + 1) = \min_{0 \leq k \leq n}(\alpha M^*(k) + \alpha/(\alpha + \beta) + \beta M^*(n - k) + \beta/(\alpha + \beta))$$
$$- 1/(\alpha + \beta)$$
$$= 1 - 1/(\alpha + \beta) + \min_{0 \leq k \leq n}(\alpha M^*(k) + \beta M^*(n - k)).$$

In other words $M^*(n) = M_{g^*\alpha\beta}(n)$, where $g^*(n) = 1 - 1/(\alpha + \beta)$. If $\alpha + \beta > 1$, the new function $M^*(n)$ is therefore just $1 - 1/(\alpha + \beta)$ times the function in Theorem 5.1 or 5.2.

If $\alpha + \beta = 1$, the function $M(n)$ is trivially equal to 1 for all n. The remaining case has a new twist.

Theorem 6. *Let* $g(n) = \delta_{n0}$ *and* $\alpha + \beta < 1$, *and let* γ *be defined by* (4.3). *If* $\log \alpha / \log \beta$ *is irrational,*

$$M_{g\alpha\beta}(n) \sim (\alpha + \beta - 1)C^{-1/\gamma}\gamma n^{1+1/\gamma}/(\gamma + 1),$$

where $C = 1/(\alpha^{-\gamma}\ln\alpha^\gamma + \beta^{-\gamma}\ln\beta^\gamma)$. *On the other hand, if* $\log\alpha/\log\beta$ *is rational,*

$$\infty > \limsup_{n\to\infty} M_{g\alpha\beta}(n)/n^{1+1/\gamma} > \liminf_{n\to\infty} M_{g\alpha\beta}(n)/n^{1+1/\gamma} > 0.$$

(Notice that γ is negative, between -1 and 0.)

Proof. Theorem 1 applies to this case, since $D(0) = \alpha+\beta-1$ is *negative*. It follows that the D's are all negative (we have $M(n) > M(n+1)$, but M is still convex); in fact, $D(n) = (\alpha + \beta - 1)/w(\sigma_{n+1})$, where the weights $w(\sigma)$ are defined now by the inverse rules

$$w(\epsilon) = 1, \quad w(L\sigma) = \alpha^{-1}w(\sigma), \quad w(R\sigma) = \beta^{-1}w(\sigma). \tag{6.1}$$

The function $H(x)$ of Theorem 5.1 applies, but with α, β, γ replaced, respectively, by $\alpha^{-1}, \beta^{-1}, -\gamma$; we have $c_1 x^{-\gamma} \leq H(x) \leq c_2 x^{-\gamma}$. Therefore (compare with (4.13)), we have

$$(\alpha + \beta - 1)c_2^{-1/\gamma}n^{1/\gamma} < D(n) \leq (\alpha + \beta - 1)c_1^{-1/\gamma}(n + 1)^{1/\gamma}.$$

The theorem now follows as in Lemma 4.2 and Theorem 5.1, provided we can prove that $M(n) \to 0$ as $n \to \infty$, since

$$M(n) = 1 + D(0) + \cdots + D(n - 1).$$

By definition, $M(n + 1) \leq \alpha M(k) + \beta M(n - k)$ for all k; and since the D's are negative, $M(n) < M(n - 1)$. Hence,

$$M(n + 1) \leq \alpha M(\lfloor n/2 \rfloor) + \beta M(\lceil n/2 \rceil)$$
$$\leq \alpha M(\lfloor n/2 \rfloor) + \beta M(\lfloor n/2 \rfloor).$$

But $\alpha + \beta < 1$; hence $M(n) \to 0$. $\square$

7. The Case $g(n) = n^2$

We shall conclude our study of the M functions by considering a driver function $g(n)$ that grows more rapidly than those considered so far.

Theorem 7. *If $g(n) = n^2$ for all n and if $\alpha^{-1} + \beta^{-1} > 1$, then*

$$M_{g\alpha\beta}(n) \sim (\alpha + \beta)n^2/(\alpha + \beta - \alpha\beta). \tag{7.1}$$

Proof. We may apply Theorem 1, and we note that $i + j = n - 1$, whenever the **if** test occurs in the algorithm following that theorem. Therefore $F(n+1) = \min(\alpha D(k), \beta D(n-k))$ for some k. If $F(n+1) = \alpha D(k)$, then

$$\alpha D(j) \leq F(n+1) \leq \beta D(n-j) \quad \text{for all } j \leq k,$$

and

$$\beta D(n-j) \leq F(n+1) \leq \alpha D(j) \quad \text{for all } j > k;$$

similarly if $F(n+1) = \beta D(n-k)$, we have

$$\alpha D(j) \leq F(n+1) \leq \beta D(n-j) \quad \text{for all } j < k,$$

and

$$\beta D(n-j) \leq F(n+1) \leq \alpha D(j) \quad \text{for all } j \geq k.$$

Thus in all cases, when $0 \leq j \leq n$ we have

$$\min(\alpha D(j), \beta D(n-j)) \leq F(n+1) \leq \max(\alpha D(j), \beta D(n-j)). \tag{7.2}$$

In particular, (7.2) holds when $j = \lfloor \beta n/(\alpha + \beta) \rfloor$. If we now write $D(n+1) = E(n) + 2Cn$, for $C = (\alpha + \beta)/(\alpha + \beta - \alpha\beta)$, we have $D(n+1) = g(n+2) - g(n+1) + F(n+1)$; hence,

$$\min\left(\alpha D(\lfloor \beta n/(\alpha + \beta) \rfloor), \beta D(\lfloor \alpha n/(\alpha + \beta) \rfloor)\right)$$
$$\leq E(n) + 2Cn - (2n + 3) \tag{7.3}$$
$$\leq \max\left(\alpha D(\lfloor \beta n/(\alpha + \beta) \rfloor), \beta D(\lceil \alpha n/(\alpha + \beta) \rceil)\right).$$

Now $2C\alpha\lfloor \beta n/(\alpha+\beta) \rfloor = 2Cn - 2n + O(1)$, so there is a constant A such that

$$|E(n)| \leq \max\left(\alpha|E(\lfloor \beta n/(\alpha+\beta) \rfloor)|, \beta|E(\lceil \alpha n/(\alpha+\beta) \rceil)|\right) + A. \tag{7.4}$$

From these relations we can prove that $|E(n)|$ does not grow too rapidly. Since $\alpha\beta/(\alpha + \beta) < 1$, there is a constant $\lambda < 1$ such that

$$\max\big(\alpha(\beta/(\alpha + \beta))^\lambda, \beta(\alpha/(\alpha + \beta))^\lambda\big) < 1;$$

let this maximum be ρ. There is a constant n_0 such that

$$\rho(n + \alpha + \beta)^\lambda + A \leq n^\lambda \quad \text{for all } n \geq n_0,$$

and we can find $C_1 \geq 1$ with $|E(n)| \leq C_1 n^\lambda$ for all $n < n_0$. By induction, (7.4) shows that $|E(n)| \leq C_1 n^\lambda$ for *all* n.

We have proved that $D(n) = 2Cn + O(n^\lambda)$; consequently,

$$M(n) = Cn^2 + O(n^{\lambda+1}). \quad \square$$

When $\alpha + \beta = \alpha\beta$, that is, $\alpha^{-1} + \beta^{-1} = 1$, we can use the same technique to show that $M_{g\alpha\beta}$ grows as $n^2 \log n$. Assume that $\alpha \leq \beta$. If we write

$$D(n+1) = E_\alpha(n) + 2n \log n/\log \alpha,$$

we find

$$E_\alpha(n) \leq \max\big(\alpha E_\alpha(\lfloor n/\alpha \rfloor), \beta E_\alpha(\lceil n/\beta \rceil)\big) + O(\log n);$$

this bound implies that $E_\alpha(n) \leq C_\alpha n$ for some C_α. Similarly if we write

$$D(n+1) = E_\beta(n) + 2n \log n/\log \beta,$$

we find

$$E_\beta(n) \geq \min\big(\alpha E_\beta(\lfloor n/\alpha \rfloor), \beta E_\beta(\lceil n/\beta \rceil)\big) + O(\log n),$$

so $E_\beta(n) \geq -C_\beta n$. Therefore, $M_{g\alpha\beta}(n)$ lies between $n^2 \log n/\log \beta + O(n^2)$ and $n^2 \log n/\log \alpha + O(n^2)$. It would be interesting to discover if $\lim_{n\to\infty} M_{g\alpha\beta}(n)/(n^2 \log n)$ exists. In the case $\alpha = \beta = 2$, our derivation proves that

$$M_{g22}(n) = n^2 \log_2 n + O(n^2), \tag{7.5}$$

a formula analogous to (3.1).

Incidentally when $\alpha = \beta = 2$, it is possible to give "explicit" formulas for $M(n)$, in terms of the binary representation of n. Let

$$n + 1 = 2^{a_1} + 2^{a_2} + \cdots + 2^{a_r},$$

where $a_1 > a_2 > \cdots > a_r \geq 0$. Then,

$$D(n) = 1 + 2(a_1 \cdot 2^{a_1} + (a_2 + 1) \cdot 2^{a_2} + \cdots + (a_r + 1) \cdot 2^{a_r})$$

and

$$M(n) = \sum_{1 \leq i,j \leq r} 2^{a_i + a_j} \left(\max(a_i, a_j) + 1 - 2\delta_{ij} \right)$$

$$- 2^{a_1 + 1}(n - 2^{a_1}) - \tfrac{2}{3}(2^{2a_1} - 1) + 2n - 1.$$

In particular,

$$M(2^a - 1) = (a - \tfrac{5}{3}) \cdot 2^{2a} + 2^{a+1} - \tfrac{1}{3}.$$

When $\alpha + \beta < \alpha\beta$, we have $(\alpha - 1)(\beta - 1) > 1$, so $\min(\alpha, \beta) > 1$. Now $g(n) = n^2 \geq n$, so we know from the results of Section 4 that $M_{g\alpha\beta}(n) \geq C_1 n^{1+1/\gamma}$ for some C_1, where $\alpha^{-\gamma} + \beta^{-\gamma} = 1$ (hence $\gamma < 1$ and $1 + 1/\gamma > 2$). It can be shown that $M_{g\alpha\beta}(n)$ is also $\leq C_2 n^{1+1/\gamma}$ in this case; in fact, whenever $\min(\alpha, \beta) > 1$, the general upper bound $M_{g\alpha\beta}(n) = O(n^{1+1/\gamma})$ holds for *all* functions $g(n)$ that are $O(n^{1+1/\gamma-\epsilon})$. This result appears in [7].

Appendix: A Tauberian Theorem

Now let us return to Lemma 4.3, on which we based our proofs of Theorems 4.3, 5.1, and 6. Results of this type were originally given by N. Wiener [18, 19] and S. Ikehara [10], in a rather complicated form somewhat more general than we need. Landau [13] simplified the ideas and used them to give a new proof of the prime number theorem, but he gave a slightly less general result than Lemma 4.3. The following proof is based on that of Landau, with minor modifications in order to prove what we need. (At this point, the reader should refer back to the statement of Lemma 4.3.)

Let $g(s) = G(s) - C/(s - \gamma)$, a function that is analytic for $\Re(s) \geq \gamma$. We now introduce two parameters, y and λ, which will eventually approach infinity. By the Riemann–Lebesgue lemma and the fact that g is analytic,

$$\left| \int_{-2}^{2} \left(1 - \frac{|t|}{2} \right) e^{i\lambda yt} g(\gamma + \epsilon + i\lambda t) \, dt \right| \leq \frac{2K(\lambda)}{\lambda y} \quad \text{for } 0 \leq \epsilon \leq 1, \quad (A.1)$$

where $K(\lambda)$ depends only on λ. Let

$$\phi(x) = f(e^{y + x/\lambda}) e^{-\gamma(y + x/\lambda)}; \quad (A.2)$$

then for fixed $0 < \epsilon \le 1$ and for $n \to \infty$, we have

$$\int_{-\lambda y}^{\lambda(n-y)} \phi(x) e^{-\epsilon(y+x/\lambda)} \left(\frac{\sin x}{x}\right)^2 dx$$

$$= \frac{1}{2} \int_{-\lambda y}^{\lambda(n-y)} \phi(x) e^{-\epsilon(y+x/\lambda)} \int_{-2}^{2} \left(1 - \frac{|t|}{2}\right) e^{-ixt} \, dt \, dx$$

$$= \frac{1}{2} \int_{-2}^{2} \left(1 - \frac{|t|}{2}\right) e^{i\lambda yt} \int_{-\lambda y}^{\lambda(n-y)} f(e^{y+x/\lambda}) e^{-(\gamma+\epsilon+i\lambda t)(y+x/\lambda)} \, dx \, dt$$

$$= \frac{\lambda}{2} \int_{-2}^{2} \left(1 - \frac{|t|}{2}\right) e^{i\lambda yt} \int_{0}^{n} f(u) u^{-(\gamma+\epsilon+i\lambda t+1)} \, du \, dt$$

$$= \frac{\lambda}{2} \int_{-2}^{2} \left(1 - \frac{|t|}{2}\right) e^{i\lambda yt} G(\gamma + \epsilon + i\lambda t) \, dt + o(1),$$

as $n \to \infty$. (The parameter n was introduced in order to justify the change in order of integration.) Note that in the special "ideal" case $f(u) = Cu^\gamma$ we have $\phi(x) = C$ and $G(s) = C/(s - \gamma)$; subtracting this particular case from the general case and combining the result with (A.1) yields

$$\left| \int_{-\lambda y}^{\infty} \phi(x) e^{-\epsilon(y+x/\lambda)} \left(\frac{\sin x}{x}\right)^2 dx \right.$$

$$\left. - C \int_{-\lambda y}^{\infty} e^{-\epsilon(y+x/\lambda)} \left(\frac{\sin x}{x}\right)^2 dx \right| \le \frac{K(\lambda)}{y}. \quad \text{(A.3)}$$

Now we can let $\epsilon \to 0$, because if ϵ is extremely small the integrals clearly approach their value when $\epsilon = 0$. Therefore we have proved

$$\left| \int_{-\lambda y}^{\infty} \phi(x) \left(\frac{\sin x}{x}\right)^2 dx - C \int_{-\lambda y}^{\infty} \left(\frac{\sin x}{x}\right)^2 dx \right| \le \frac{K(\lambda)}{y}. \quad \text{(A.4)}$$

(This is the key inequality that gives us a handle on the problem. When y is very large and $|x|$ is bounded, $\phi(x)$ must be very nearly equal to C.)

From the monotonicity of f, we now have the following inequalities when $-\lambda^{1/2} \le x \le \lambda^{1/2}$.

$$e^{-\gamma(y+1/\lambda^{1/2})} f(e^{y-1/\lambda^{1/2}}) \le \phi(x) \le e^{-\gamma(y-1/\lambda^{1/2})} f(e^{y+1/\lambda^{1/2}}). \quad \text{(A.5)}$$

Hence,

$$\frac{f(e^{y-1/\lambda^{1/2}})}{e^{\gamma(y-1/\lambda^{1/2})}} \int_{-\lambda^{1/2}}^{\lambda^{1/2}} \left(\frac{\sin x}{x}\right)^2 dx \le e^{2\gamma/\lambda^{1/2}} \int_{-\lambda^{1/2}}^{\lambda^{1/2}} \phi(x) \left(\frac{\sin x}{x}\right)^2 dx$$

$$\le e^{2\gamma/\lambda^{1/2}} \left(C \int_{-\lambda y}^{\infty} \left(\frac{\sin x}{x}\right)^2 dx + \frac{K(\lambda)}{y} \right)$$

for all fixed λ and $y > 1$. If we let $y \to \infty$, we obtain

$$\left(\limsup_{u \to \infty} \frac{f(u)}{u^\gamma} \right) \int_{-\lambda^{1/2}}^{\lambda^{1/2}} \left(\frac{\sin x}{x} \right)^2 dx \le e^{2\gamma/\lambda^{1/2}} C \int_{-\infty}^{\infty} \left(\frac{\sin x}{x} \right)^2 dx;$$

and if we now let $\lambda \to \infty$, we have $\limsup f(u)/u^\gamma = C$. A similar argument, using the other half of relation (A.5), proves that

$$\liminf_{u \to \infty} \frac{f(u)}{u^\gamma} = C. \quad \Box$$

This research was supported in part by the National Science Foundation under grant GJ-992 and in part by Office of Naval Research contract N-00014-67-A-0112-0057 NR 044-402.

References

[1] A. Cayley, "On a problem in the partition of numbers," *Philosophical Magazine* **13** (1857), 245–248. Reprinted in his *Collected Mathematical Papers* **3**, 247–249.

[2] R. F. Churchhouse, "Binary partitions," in *Computers in Number Theory*, edited by A. O. L. Atkin and B. J. Birch (New York: Academic Press, 1971), 397–400.

[3] N. G. de Bruijn, "On Mahler's partition problem," *Indagationes Mathematicæ* **10** (1948), 210–220.

[4] N. G. de Bruijn, personal communication (January 1973).

[5] P. Erdős, "On an elementary proof of some asymptotic formulas in the theory of partitions," *Annals of Mathematics* (2) **43** (1942), 437–450, especially pages 447–448.

[6] L. Euler, "De partitione numerorum," *Novi commentarii academiæ scientiarum imperialis Petropolitanæ* **3** (1750), 125–169, especially pages 162–164. Reprinted in his *Commentationes arithmeticæ collectæ*, Volume 1 (1849), 73–101. Reprinted in his *Opera Omnia*, Series 1, Volume 2, 254–294.

[7] Michael L. Fredman, *Growth Properties of a Class of Recursively Defined Functions*, report STAN-CS-72-296 (Ph.D. thesis, Stanford University, June 1972).

[8] Francis Galton, *Natural Inheritance* (New York: Macmillan, 1889).

[9] Sidney Heller, "An asymptotic solution of the difference equation $a_{n+1} - a_n = a_{[n/2]}$," *Journal of Mathematical Analysis and Applications* **34** (1971), 464–469.

[10] S. Ikehara, "An extension of Landau's theorem in the analytical theory of numbers," *Journal of Mathematics and Physics* **10** (1931), 1–12.

[11] Donald E. Knuth, "An almost linear recurrence," *Fibonacci Quarterly* **4** (1966), 117–128. [Reprinted as Chapter 37 of the present volume.]

[12] Donald E. Knuth, *Fundamental Algorithms*, Volume 1 of *The Art of Computer Programming* (Reading, Massachusetts: Addison–Wesley, 1968).

[13] E. Landau, "Ueber den Wienerschen neuen Weg zum Primzahlsatz," *Sitzungsberichte der Preußischen Akademie der Wissenschaften, Physikalisch–Mathematische Klasse* **32** (1932), 514–521.

[14] Kurt Mahler, "On a special functional equation," *Journal of the London Mathematical Society* **15** (1940), 115–123.

[15] W. B. Pennington, "On Mahler's partition problem," *Annals of Mathematics* (2) **57** (1953), 531–546.

[16] Öystein Rödseth, "Some arithmetical properties of m-ary partitions," *Proceedings of the Cambridge Philosophical Society* **68** (1970), 447–453.

[17] Alberto Tanturri, "Sul numero delle partizioni d'un numero in potenze di 2," *Atti Accademia delle Scienze di Torino*, Classe di Scienze Fisiche, Matematiche e Naturali **54** (1918), 69–82; *Rendiconti Accademia Nazionale dei Lincei*, Classe di Scienze Fisiche, Matematiche e Naturali **27** (1918), 399–403.

[18] N. Wiener, "A new method in Tauberian theorems," *Journal of Mathematics and Physics* **7** (1927–1928), 161–184.

[19] N. Wiener, "Tauberian theorems," *Annals of Mathematics* (2) **33** (1932), 1–100.

Addendum

A more direct way to obtain Theorem 5.1, without relying on complex integration, has been found by Nicholas Pippenger, "An elementary approach to some analytic asymptotics," *SIAM Journal on Mathematical Analysis* **24** (1993), 1361–1377. His interesting approach provides additional information and shows that the smoothness of the limiting behavior depends on the size of the partial quotients of $\log \alpha / \log \beta$. (See exercise 9.61 of *Concrete Mathematics* for related ideas.)

Chapter 39

A Recurrence Related to Trees

The asymptotic behavior of the solutions to an interesting class of recurrence relations, which arise in the study of trees and random graphs, is derived here by making uniform estimates on the elements of a basis of the solution space. We also investigate a family of polynomials with integer coefficients, which may be called the "tree polynomials."

[Written with Boris Pittel. Originally published in Proceedings of the American Mathematical Society **105** (1989), 335–349.]

There are $n^{n-2}(n-1)!$ sequences of edges between vertices

$$u_1 \longrightarrow v_1, \quad \ldots, \quad u_{n-1} \longrightarrow v_{n-1}, \qquad 1 \le u_k < v_k \le n, \qquad (0.1)$$

that define a free tree on $\{1, \ldots, n\}$, because there are n^{n-2} free trees on n labeled vertices and every such tree has $n-1$ edges. If we consider each of these $n^{n-2}(n-1)!$ sequences to be equally likely, the probability that u_{n-1} and v_{n-1} belong respectively to components of sizes k and $n-k$ based on the first $n-2$ edges is

$$p_{nk} = \frac{1}{2(n-1)} \binom{n}{k} \left(\frac{k}{n}\right)^{k-1} \left(\frac{n-k}{n}\right)^{n-k-1}, \qquad 0 < k < n. \qquad (0.2)$$

Knuth and Schönhage [9, §§9–12] considered tree-construction algorithms whose analysis depended on the solution of the recurrence

$$x_n = c_n + \sum_{0 < k < n} p_{nk}(x_k + x_{n-k}) \qquad (0.3)$$

for various sequences $\langle c_n \rangle$. The purpose of the present note is to extend the results of [9] and to consider related sequences of functions whose exact and asymptotic values arise in a variety of algorithms.

565

Much of the analysis below, as in [9], depends on properties of the formal power series

$$T(z) = \sum_{n \geq 1} \frac{n^{n-1} z^n}{n!} = z + z^2 + \frac{3}{2} z^3 + \frac{8}{3} z^4 + \frac{125}{24} z^5 + \cdots, \quad (0.4)$$

which is the exponential generating function for labeled, rooted trees. Section 1 discusses the connection between $T(z)$ and functions that generate the sequences $\langle c_n \rangle$ and $\langle x_n \rangle$ of (0.3). Section 2 derives fundamental properties of the *tree polynomials* $t_n(y)$, defined by

$$\frac{1}{(1 - T(z))^y} = \sum_{n \geq 0} t_n(y) \frac{z^n}{n!}, \quad (0.5)$$

for fixed y as $n \to \infty$. Asymptotic properties of the tree polynomials and related quantities are derived in Section 3. These results are applied in Section 4 to derive asymptotic properties of $\langle x_n \rangle$ when the asymptotic behavior of $\langle c_n \rangle$ is given.

We prove in particular that if $c_n = n^\beta$ and $\beta > 1/2$ then

$$x_n = \frac{\Gamma(\beta - 1/2)}{\sqrt{2} \Gamma(\beta)} n^{\beta+1/2} + O(n^\beta) + O(n \log n). \quad (0.6)$$

On the other hand, if $c_n = O(n^\beta)$ and $\beta < 1/2$ then

$$x_n = \kappa n + O(n^{\beta+1/2}) + O(1), \quad (0.7)$$

where

$$\kappa = \sum_{m \geq 1} \frac{c_m}{m^2} \left(1 + \frac{m^m e^{-m}}{m!} (m - Q(m)) \right) \quad (0.8)$$

and $Q(m)$ is Ramanujan's function. Thus x_n is superlinear in n or linear in n dependent upon whether $\beta > 1/2$ or $\beta < 1/2$.

1. Basic Solutions

If $\langle x_n^{(m)} \rangle$ is the solution of (0.3) in the special case $c_n = \delta_{mn}$, the general solution is

$$x_n = x_n^{(1)} c_1 + x_n^{(2)} c_2 + \cdots + x_n^{(n)} c_n. \quad (1.1)$$

It is easy to verify that $x_n^{(1)} = n$ and that $x_n^{(1)} + x_n^{(2)} + \cdots = 2n - 1$. Knuth and Schönhage [9] proved that

$$x_n^{(2)} = \frac{n}{4} \left(1 + \left(1 - \frac{2}{n} \right)^{n-2} \right) \quad (1.2)$$

and, for general m, that

$$x_n^{(m)} = \frac{n}{m^2}\left(m\binom{n}{m}\left(1-\frac{m}{n}\right)^{n-m}\left(\frac{m}{n}\right)^m + 1\right.$$
$$\left. - \sum_{j=0}^{m-1}\binom{n-1}{j}\left(1-\frac{m}{n}\right)^{n-1-j}\left(\frac{m}{n}\right)^j\right). \qquad (1.3)$$

Notice that $x_n^{(m)} = 0$ when $m > n$, and $x_n^{(n)} = 1$.

A new family of special solutions to the recurrence also proves to be important, namely

$$x_n(y) = t_n(y)/n^{n-1} \qquad (1.4)$$

where $t_n(y)$ is the tree polynomial of (0.5). We can find the corresponding driver functions $c_n(y)$ by plugging into (0.3):

$$\frac{t_n(y)}{n^{n-1}} = c_n(y) + \frac{1}{(n-1)n^{n-2}}\sum_{k=1}^{n-1}\binom{n}{k}k^{k-1}t_{n-k}(y). \qquad (1.5)$$

(We have used the fact that $p_{nk} = p_{n(n-k)}$.) This sum is n^{n-1} less than the coefficient of $z^n/n!$ in $T(z)/(1-T(z))^y = 1/(1-T(z))^y - 1/(1-T(z))^{y-1}$, namely $t_n(y) - t_n(y-1)$. Therefore we have

$$c_n(y) = \frac{nt_n(y-1) - t_n(y) + n^n}{(n-1)n^{n-1}}, \qquad \text{for } n \geq 2; \qquad (1.6)$$

and $c_1(y) = t_1(y) = y$.

The value of $c_n(y)$ can also be expressed in another form, because of an interesting identity satisfied by the tree polynomials. The well-known equation

$$T(z) = ze^{T(z)}, \qquad (1.7)$$

due to Eisenstein [1], implies that

$$T'(z) = \frac{T(z)}{z(1-T(z))}. \qquad (1.8)$$

Therefore

$$\sum_{n\geq 0}\frac{nt_n(y-1)z^n}{n!} = z\frac{d}{dz}\frac{1}{(1-T(z))^{y-1}} = \frac{(y-1)T(z)}{(1-T(z))^{y+1}}$$

and we have the difference equation

$$t_n(y+1) = t_n(y) + n\frac{t_n(y-1)}{y-1}. \tag{1.9}$$

It follows that, in addition to (1.6), we have the alternative formula

$$c_n(y) = \frac{(y-1)t_n(y+1) - yt_n(y) + n^n}{(n-1)n^{n-1}}, \qquad \text{for } n \geq 2. \tag{1.10}$$

2. Tree Polynomials

The first few cases of the polynomials $t_n(y)$ defined in (0.5) are

$$\begin{aligned}
t_0(y) &= 1; \\
t_1(y) &= y; \\
t_2(y) &= y^2 + 3y; \\
t_3(y) &= y^3 + 9y^2 + 17y; \\
t_4(y) &= y^4 + 18y^3 + 95y^2 + 142y.
\end{aligned} \tag{2.1}$$

Our derivation of (1.6) from (1.5) was, in essence, a proof of the difference equation

$$t_n(y) - t_n(y-1) = \sum_{k=1}^{n} \binom{n}{k} k^{k-1} t_{n-k}(y). \tag{2.2}$$

Therefore, by (1.9), we have

$$t_n(y-2) = \frac{y-2}{n} \sum_{k=1}^{n} \binom{n}{k} k^{k-1} t_{n-k}(y), \tag{2.3}$$

a recurrence by which $t_n(y)$ can be obtained when $t_{n-1}(y), \ldots, t_0(y)$ are known. Two analogous formulas can also be proved:

$$t_n(y+1) = \sum_{k=0}^{n} \binom{n}{k} k^k t_{n-k}(y); \tag{2.4}$$

$$t_n'(y) = \sum_{k=1}^{n} \binom{n}{k} k^{k-1} Q(k) t_{n-k}(y). \tag{2.5}$$

Here $Q(n)$ is Ramanujan's function

$$Q(n) = 1 + \frac{n-1}{n} + \frac{n-1}{n}\frac{n-2}{n} + \frac{n-1}{n}\frac{n-2}{n}\frac{n-3}{n} + \cdots$$
$$= \sqrt{\frac{\pi n}{2}} - \frac{1}{3} + \frac{1}{12}\sqrt{\frac{\pi}{2n}} - \frac{4}{135n} + O(n^{-3/2}),$$
(2.6)

which arises in the analysis of many algorithms (see [5]). Equation (2.4) follows from the well-known identity (Riordan [12])

$$\frac{1}{1-T(z)} = \sum_{n\geq 0} \frac{n^n z^n}{n!},$$
(2.7)

which gives the exponential generating function for n^n, the number of all mappings of the set $\{1,\dots,n\}$ into itself. Equation (2.5) follows from

$$\ln\frac{1}{1-T(z)} = \sum_{n\geq 1} \frac{C(n) z^n}{n!}$$
(2.8)

where $C(n)$ is the total number of all one-cycle mappings [12], and from a formula

$$C(n) = n^{n-1}Q(n),$$

which is due to Rényi [11]. In general, we have the formal relation

$$a_1 T(z) + a_2 T(z)^2 + a_3 T(z)^3 + \cdots$$
$$= \sum_{n\geq 1} \frac{n^{n-1} z^n}{n!}\left(a_1 + 2a_2\frac{n-1}{n} + 3a_3\frac{n-1}{n}\frac{n-2}{n} + \cdots\right),$$
(2.9)

by Lagrange's inversion formula; equations (2.7) and (2.8) are special cases of this identity.

Equation (2.4) is also a special case of the convolution formula

$$t_n(y_1 + y_2) = \sum_{k=0}^{n}\binom{n}{k} t_k(y_1) t_{n-k}(y_2).$$
(2.10)

If $F(z)$ is any formal power series such that $\ln F(z) = \sum_{n\geq 1} b_n z^n/n!$, the coefficient of $y^m z^n/n!$ in $F(z)^y = e^{y\ln F(z)}$ is a multinomial convolution

$$\frac{1}{m!} \sum_{\substack{k_1+\cdots+k_m=n \\ k_1,\dots,k_m > 0}} \binom{n}{k_1,\dots,k_m} b_{k_1}\cdots b_{k_m}$$
$$= \sum_{\substack{r_1+r_2+\cdots=m \\ r_1+2r_2+\cdots=n \\ r_1,r_2,\dots\geq 0}} \frac{n!}{1!^{r_1}r_1!\,2!^{r_2}r_2!\cdots} b_1^{r_1} b_2^{r_2}\cdots.$$
(2.11)

(This is essentially Arbogast's formula [6, exercise 1.2.5–21].) In our case each $b_k = C(k)$ is an integer, so the coefficients of the tree polynomials $t_n(y)$ are integers.

Indeed, the coefficients of $t_n(y)$ have a natural combinatorial interpretation, as noted by Flajolet and Soria [4, Example 4]: The coefficient of y^k is the number of mappings of $\{1, 2, \ldots, n\}$ into itself such that the corresponding functional digraph has exactly k components. (Equivalently, the mapping has exactly k different cycles; see [7, exercise 3.1–14].) The reason is simply that the corresponding exponential generating function, according to (2.8) and well-known techniques for enumerating labeled configurations, is $\exp\bigl(y \ln(1/(1 - T(z)))\bigr)$, and this is just (0.5). Stepanov [13] showed that these coefficients are asymptotically normal with mean and variance $\frac{1}{2} \ln n + O(1)$; Flajolet and Soria [4] extended his theorem to a general result on the number of components in random labeled structures.

We wish to study the asymptotic behavior of $t_n(y)$ for fixed y as $n \to \infty$, and for this we need another formula based on (2.9): The binomial expansion $(1 - T(z))^{-y} = \sum_{k \geq 0} \binom{y+k-1}{k} T(z)^k$ and equation (2.9) imply that

$$
t_n(y) = n^{n-1} \left(\frac{y}{0!} + \frac{y(y+1)}{1!} \frac{n-1}{n} + \frac{y(y+1)(y+2)}{2!} \frac{n-1}{n} \frac{n-2}{n} + \cdots \right)
$$
$$
= \frac{n^{n-1}}{\Gamma(y)} \sum_{k \geq 0} \binom{n-1}{k} \frac{\Gamma(y+k+1)}{n^k}. \tag{2.12}
$$

Integral formulas for $t_n(y)$ are also of interest:

$$
t_n(y) = n^n y \int_0^\infty \sum_k \binom{n-1}{k} \binom{y+k}{k} t^k e^{-nt} \, dt
$$
$$
= n^n + n^{n+1} \int_0^\infty \sum_k \binom{n-1}{k} \binom{y+k-1}{k+1} t^k e^{-nt} \, dt. \tag{2.13}
$$

3. Asymptotic Lemmas

The terms of (2.12) increase until k is approximately $\sqrt{yn}$, after which they begin to decrease rather rapidly. More explicitly, when y and $\epsilon > 0$ are fixed and $0 < k < n^{1/2+\epsilon}$, we have

$$
\frac{\Gamma(y+k+1)}{k!} = k^y \left(1 + \frac{y^2 + y}{2k} + O(k^{-2}) \right), \tag{3.1}
$$

$$\frac{(n-1)!}{(n-1-k)!\,n^k} = e^{-k^2/2n}\left(1 - \frac{k}{2n} - \frac{k^3}{6n^2} + O(n^{-1+6\epsilon})\right), \qquad (3.2)$$

by Stirling's approximation. Therefore the terms of (2.12) are exponentially small when $k > n^{1/2+\epsilon}$, and we can replace them with other exponentially small terms to get

$$t_n(y) = \frac{n^{n-1}}{\Gamma(y)}\left(S_{2n}(y) + \frac{y^2+y}{2}S_{2n}(y-1)\right.$$
$$\left. - \frac{1}{2n}S_{2n}(y+1) - \frac{1}{6n^2}S_{2n}(y+3) + R_n(y)\right), \quad (3.3)$$

where

$$S_n(y) = \sum_{k\geq 1} k^y e^{-k^2/n} \qquad (3.4)$$

and where $R_n(y)$ is an error term to be determined below.

When y is a nonnegative integer, the sum $S_n(y)$ is essentially a theta function whose asymptotic behavior is well known. When $y = -1$, the analysis of bubble sort in Knuth [8, exercise 5.2.2–4] shows that $S_n(-1) = \frac{1}{2}\ln n + \frac{1}{2}\gamma + O(n^{-1})$. For our purposes we want to consider $S_n(y)$ when y is a real number larger than -1. It turns out that $S_n(y) = \int_0^\infty x^y e^{-x^2/n}\,dx + O(1)$ in this case, and we can continue the asymptotic series explicitly:

Lemma 1. *If $y > -1$, the series $S_n(y)$ defined in (3.4) satisfies*

$$S_n(y) = \frac{1}{2}\Gamma\left(\frac{y+1}{2}\right)n^{(y+1)/2} + \sum_{j=0}^{m}\frac{(-1)^j\zeta(-y-2j)}{j!\,n^j} + O(n^{-m-1}), \quad (3.5)$$

for all $m \geq 0$, as $n \to \infty$.

Proof. Consider the related sum

$$\overline{S}_{n,N} = \sum_{k=1}^{N\sqrt{n}-1} k^y\left(e^{-k^2/n} - \frac{1}{0!} + \frac{k^2}{1!\,n} - \cdots + \frac{(-1)^{m+1}k^{2m}}{m!\,n^m}\right)$$
$$= n^{y/2}\sum_{k=1}^{N\sqrt{n}-1} g_m\left(\frac{k}{\sqrt{n}}\right), \qquad (3.6)$$

where

$$g_m(x) = x^y\left(e^{-x^2} - \sum_{j=0}^{m}\frac{(-x^2)^j}{j!}\right). \qquad (3.7)$$

Euler's summation formula tells us that

$$\overline{S}_{n,N} = n^{(y+1)/2} \int_0^N g_m(x)\,dx + \sum_{l=1}^p \frac{B_l}{l!} g_m^{(l-1)}(N) n^{(y+1-l)/2} + R_p, \quad (3.8)$$

where

$$R_p = \frac{(-1)^{p+1}}{p!} n^{(y+1-p)/2} \int_0^N B_p(\{x/\sqrt{n}\}) g_m^{(p)}(x)\,dx, \quad (3.9)$$

if p is chosen small enough that $\lim_{x\to 0} g_m^{(p-1)}(x) = 0$. The latter condition is satisfied if $p < y + 2m + 3$, since $g_m(x) = O(x^{y+2m+2})$ as $x \to 0$. If we choose p so that $y + 2m + 1 < p < y + 2m + 3$, we can in fact prove that the integral $\int_0^\infty |g_m^{(p)}(x)|\,dx$ is finite, because $g_m^{(p)}(x)$ is $O(x^{y+2m+2-p})$ as $x \to 0$ and its terms that do not involve e^{-x^2} are $O(x^{y+2m-p})$ as $x \to \infty$. Therefore $R_p = O(n^{(y+1-p)/2}) = O(n^{-m})$ for this choice of p, uniformly in N.

Euler's summation formula also implies the identity

$$\sum_{k=1}^{M-1} k^\alpha = \zeta(-\alpha) + \sum_{l=0}^p \binom{\alpha+1}{l} B_l \frac{M^{\alpha+1-l}}{\alpha+1} + O(M^{\alpha-p}); \quad (3.10)$$

indeed, this is the traditional way to define the zeta function. Thus if

$$h_m(x) = x^y \sum_{j=0}^m \frac{(-x^2)^j}{j!} \quad (3.11)$$

is the "smoothing function" included in $g_m(x)$, we have

$$n^{y/2} \sum_{k=1}^{N\sqrt{n}-1} h_m\left(\frac{k}{\sqrt{n}}\right) = n^{(y+1)/2} \int_0^N h_m(x)\,dx + \sum_{j=0}^m \frac{(-1)^j \zeta(-y-2j)}{j!\,n^j}$$

$$+ \sum_{l=1}^p \frac{B_l}{l!} h_m^{(l-1)}(N) n^{(y+1-l)/2} + O(n^{-m}). \quad (3.12)$$

Adding (3.8) to (3.12) and letting $N \to \infty$ completes the proof of (3.5), except that the error estimate is $O(n^{-m})$ instead of $O(n^{-m-1})$. To obtain the sharper estimate, we simply replace m by $m+1$.   □

Lemma 2. *The tree polynomial $t_n(y)$ defined in (0.5) satisfies*

$$t_n(y) = \frac{\sqrt{2\pi}n^{n-1}}{2^{y/2}}\left(\frac{n^{(y+1)/2}}{\Gamma(y/2)} + \frac{\sqrt{2}y}{3}\frac{n^{y/2}}{\Gamma((y-1)/2)}\right.$$

$$\left. + O(n^{(y-1)/2}) + O(1)\right), \qquad (3.13)$$

for fixed y as $n \to \infty$.

Proof. We may assume that $y > -1$, for if $y = -1$ the stated result is consistent with the fact that $t_n(-1) = -n^{n-1}$ and if $y < -1$ we easily have $t_n(y) = O(n^{n-1})$.

The contribution to the error term $R_n(y)$ in (3.3) that arises from $O(n^{-1+6\epsilon})$ in (3.2) is $O(n^{(y-1)/2+6\epsilon})$, by Lemma 1; this can be improved to $O(n^{(y-1)/2})$ by carrying out the expansion in (3.2) a bit further. The contribution to $R_n(y)$ from $O(k^{-2})$ in (3.1) is $O(n^{(y-1)/2})$ if $y > 1$. If $y = 1$, the $O(k^{-2})$ of (3.1) is actually zero; and if $y < 1$ the contribution to $R_n(y)$ is $O(1)$ because $S_n(y-2) < \sum_{k\geq 1} k^{y-2}$.

To complete the estimation of $t_n(y)$ from (3.3), we need only the weak form $S_n(y) = \frac{1}{2}\Gamma((y+1)/2)n^{(y+1)/2} + O(1)$ of Lemma 1. Application of the duplication formula $\Gamma(y) = 2^{y-1}\Gamma(y/2)\Gamma((y+1)/2)/\sqrt{\pi}$ completes the proof.   □

Lemma 2 can be verified, for example, in the special cases

$$t_n(1) = n^n; \quad t_n(2) = n^n(Q(n) + 1). \qquad (3.14)$$

Computer calculations show that the series in Lemma 2 continues as follows:

$$t_n(y) = n^{n-1}\left(f_n(y) + \frac{y}{3}f_n(y-1) + \frac{(y-1)(y+1)(2y-3)}{36(y-2)}f_n(y-2)\right.$$

$$+ \frac{y(y+2)(10y^2 - 35y + 24)}{1620(y-3)}f_n(y-3)$$

$$\left. + O(n^{(y-3)/2}) + O(1)\right), \qquad (3.15)$$

where $f_n(y) = \sqrt{2\pi}n^{(y+1)/2}/(2^{y/2}\Gamma(y/2))$. Stronger methods than those above are needed to penetrate the $O(1)$ term.

Philippe Flajolet has pointed out, in a personal communication to the authors, that Lemma 2 can also be deduced directly from a singularity analysis as in [3]. Indeed, we have

$$T(z) = 1 - S(\sqrt{2 - 2ez}), \qquad (3.16)$$

where the analytic function

$$S(z) = z - \frac{1}{3}z^2 + \frac{11}{72}z^3 - \frac{43}{540}z^4 + \frac{769}{17280}z^5 - \frac{221}{8505}z^6 + \cdots \quad (3.17)$$

satisfies the functional relation

$$(1 - S(z))e^{S(z)} = 1 - \frac{z^2}{2} \quad (3.18)$$

and has a Taylor series that converges for $|z| < \sqrt{2}$. This leads to an asymptotic expansion of the coefficients of $(1 - T(z))^{-y} = S(\sqrt{2 - 2ez})^{-y}$. And in fact, the resulting expansion holds for all y; we can actually eliminate the term $O(1)$ from (3.13) and (3.15). (We do not, however, require this strengthening of Lemma 2 in the calculations below.)

It is interesting to observe that (1.9) implies

$$t_n(y) = n\left(\frac{t_n(y-2)}{y-2} + \frac{t_n(y-3)}{y-3} + \cdots + \frac{t_n(y-m)}{y-m}\right)$$
$$+ t_n(y - m + 1); \quad (3.19)$$

the terms on the right decrease by factors of order $\sqrt{n}$ until order n^{n-1} is reached.

Lemma 3. *If $y > 0$, we have*

$$\frac{t_n'(y)}{t_n(y)} = \frac{1}{2}\left(\ln\frac{n}{2} + \gamma - H_{y/2-1}\right) + O(n^{-1/2}), \quad (3.20)$$

where H_y denotes the harmonic number

$$H_y = \sum_{k \geq 1}\left(\frac{1}{k} - \frac{1}{k+y}\right). \quad (3.21)$$

Proof. The calculations in the proofs of Lemmas 1 and 2 also support differentiation by y. The right-hand side of (3.20) is $f_n'(y)/f_n(y)$, because of the well-known relation

$$H_y = \frac{\Gamma'(y+1)}{\Gamma(y+1)} + \gamma. \quad (3.22)$$

(See [6, exercise 1.2.7–23].) □

The special case $y = 1$ of Lemma 3 is of particular interest. According to (0.5), (2.7), and (2.8), we have

$$t'_n(1) = n! \, [z^n] \, \frac{1}{1-T(z)} \ln \frac{1}{1-T(z)}$$

$$= \sum_{k=1}^{n} \binom{n}{k} k^{k-1} Q(k)(n-k)^{n-k}. \tag{3.23}$$

By our combinatorial interpretation of $t_n(y)$, this is the total number of cycles among all mappings of $\{1, 2, \ldots, n\}$ into itself. Lemma 3 and equation (3.14) imply that

$$t'_n(1) = \tfrac{1}{2} n^n (\ln n + \gamma + \ln 2 + O(n^{-1/2})), \tag{3.24}$$

a formula first obtained by Martin Kruskal [10]. The $O(n^{-1/2})$ can be refined further to $\frac{1}{6}\sqrt{2\pi/n} + O(n^{-1} \log n)$. In fact, the general formula (3.15) can be differentiated with respect to y; the $O(1)$ term can be eliminated, as remarked earlier.

The derivative of the formula in Lemma 1 with respect to y is sufficiently interesting to be stated explicitly. We have

$$\sum_{k=1}^{\infty} (k^y \ln k) e^{-k^2/n} = \frac{1}{4} \Gamma\left(\frac{y+1}{2}\right) n^{(y+1)/2} (\ln n - \gamma + H_{(y-1)/2})$$

$$+ \sum_{j=0}^{m} \frac{(-1)^j \zeta'(-y-2j)}{j! \, n^j} + O(n^{-m-1}), \quad (3.25)$$

for all $y > -1$.

We will also need asymptotic estimates of the basic functions $x_n^{(m)}$ of (1.3), valid for all m and n.

Lemma 4. *Let*

$$\alpha_m = \frac{1}{m^2}\left(1 + \frac{m^m e^{-m}}{m!}(m - Q(m))\right), \tag{3.26}$$

where $Q(m)$ is Ramanujan's function (2.6). The quantity $x_n^{(m)}$ defined in (1.3) satisfies

$$x_n^{(m)} = n\alpha_m + O(m^{-1/2}) + O((n-m)^{-1/2}), \tag{3.27}$$

uniformly for $1 \le m \le n - 1$.

Proof.   With Stirling's approximation, we can rewrite the main term of (1.3) as follows:

$$\binom{n}{m}\left(1-\frac{m}{n}\right)^{n-m}\left(\frac{m}{n}\right)^m = \frac{n!}{n^n e^{-n}}\,\frac{m^m e^{-m}}{m!}\,\frac{(n-m)^{n-m}e^{-(n-m)}}{(n-m)!}$$

$$= \frac{m^m e^{-m}}{m!}\sqrt{\frac{n}{n-m}}\left(1+O\left(\frac{1}{n-m}\right)\right). \quad (3.28)$$

Since

$$\sqrt{\frac{n}{n-m}} = 1 + \frac{m}{\sqrt{n-m}\,(\sqrt{n}+\sqrt{n-m})},$$

the right-hand side of (3.28) is

$$\frac{m^m e^{-m}}{m!}\left(1+O\left(\frac{1}{n-m}\right)+O\left(\frac{m}{\sqrt{n(n-m)}}\right)\right). \qquad (3.29)$$

The other terms of (1.3) are

$$\binom{n-1}{j}\left(1-\frac{m}{n}\right)^{n-1-j}\left(\frac{m}{n}\right)^j$$

$$= \frac{m^j}{j!}\,\frac{n!}{n^n e^{-n}}\,\frac{(n-1-j)^{n-1-j}}{(n-1-j)!}\,\frac{e^{j+1-n}}{e^{j+1}}\left(\frac{n-m}{n-1-j}\right)^{n-1-j}.$$

If $m \le n^{1/2}$ we have

$$\left(\frac{n-m}{n-1-j}\right)^{n-1-j} = e^{j+1-m}\left(1+O\left(\frac{(m-1-j)^2}{n}\right)\right);$$

so the entire term is

$$\frac{m^j e^{-m}}{j!}\left(1+O\left(\frac{(m-1-j)^2}{n}\right)\right),$$

uniformly in j and m. Summing over j gives

$$\sum_{j=0}^{m-1}\binom{n-1}{j}\left(1-\frac{m}{n}\right)^{n-1-j}\left(\frac{m}{n}\right)^j$$

$$= \frac{m^m e^{-m}}{m!}Q(m)\left(1+O\left(\frac{m}{n}\right)\right), \qquad (3.30)$$

because $\sum_{j=0}^{m-1} m^j/j! = m^m Q(m)/m!$ and

$$\sum_{j=0}^{m-1} \frac{m^j}{j!} (m-j-1)^2 = \sum_{j=0}^{m-1} \frac{m^j}{j!} \left((m-1)^2 - (2m-3)j + j(j-1)\right)$$

$$= \frac{m^m}{m!} \left((m+1)Q(m) - 2m\right).$$

Hence, for $m \leq n^{1/2}$ we have (see (3.28), (3.29))

$$x_n^{(m)} = \frac{n}{m^2} \left(m \frac{m^m e^{-m}}{m!} \left(1 + O\left(\frac{m}{n}\right)\right) + 1 - \frac{m^m e^{-m}}{m!} Q(m) \left(1 + O\left(\frac{m}{n}\right)\right)\right)$$

$$= n\alpha_m + O(m^{-1/2}).$$

We now consider the case $m > n^{1/2}$. The sum in (3.30) is the probability P_{nm} that the total number of successes in $n-1$ Bernoulli trials does not exceed $m-1$, where the probability of success in each trial equals $p = m/n$. By the Berry–Esseen theorem (Feller [2]), we can write then

$$P_{nm} = \Phi\left(-\left(\frac{n-m}{m(n-1)}\right)^{1/2}\right) + O\left(\frac{\rho}{\sigma^3 \sqrt{n}}\right); \qquad (3.31)$$

here $\Phi(\cdot)$ is the standard Gaussian distribution, and

$$\sigma^2 = pq = \frac{m}{n}\left(1 - \frac{m}{n}\right),$$

$$\rho = pq(p^2 + q^2) \leq pq.$$

Since $\Phi(0) = 1/2$, the relation (3.31) leads to

$$P_{nm} = 1/2 + O((m(1 - m/n))^{-1/2}). \qquad (3.32)$$

Applying the same theorem to the Poisson distribution, or using (2.6), we also obtain

$$\sum_{j=0}^{m-1} \frac{m^j e^{-m}}{j!} = \frac{1}{2} + O(m^{-1/2}). \qquad (3.33)$$

According to (1.3), (3.29), (3.32), and (3.33), we can write then for $m > n^{1/2}$,

$$x_n^{(m)} = \frac{n}{m^2} \left(m \frac{m^m e^{-m}}{m!} \left(1 + O\left(\frac{m}{n^{1/2}(n-m)^{1/2}}\right)\right) + 1 \right.$$

$$\left. - \sum_{j=0}^{m-1} \frac{m^j e^{-m}}{j!} + O\left(\frac{n^{1/2}}{m^{1/2}(n-m)^{1/2}}\right)\right)$$

$$= n\alpha_m + O\left(\frac{n^{1/2}}{m^{1/2}(n-m)^{1/2}}\right).$$

By considering two subcases $m < n/2$ and $n/2 \leq m < n$ separately, we immediately get (3.27).

The proof of Lemma 4 is now complete.   □

4. The Recurrence with Polynomial Growth

We are now ready to return to the recurrence (0.3), and to derive the asymptotic behavior of x_n when c_n is a power of n.

Theorem 1. *Let $c_n = n^\beta$ for all $n > 0$, and let*

$$l(\beta) = \frac{\Gamma(\beta - 1/2)}{\sqrt{2}\Gamma(\beta)}. \tag{4.1}$$

Then the solution to (0.3) satisfies

$$x_n = \begin{cases} l(\beta)n^{\beta+1/2} + O(n^\beta), & \text{if } \beta > 1; \\ l(\beta)n^{3/2} + O(n\log n), & \text{if } \beta = 1; \\ l(\beta)n^{\beta+1/2} + O(n), & \text{if } \frac{1}{2} < \beta < 1; \\ (2\pi)^{-1/2}n\ln n + O(n), & \text{if } \beta = \frac{1}{2}; \\ O(n), & \text{if } \beta < \frac{1}{2}. \end{cases} \tag{4.2}$$

Proof.   The special cases $\beta = 1$ and $\beta = \frac{1}{2}$ have already been considered by Knuth and Schönhage [9, (12.7) and (12.9)], who showed in fact that the recurrence with $c_n = n$ has a closed form solution

$$x_n = (n - 1)Q(n) + nK(n). \tag{4.3}$$

Here $K(n)$ is Kruskal's function from (3.24),

$$K(n) = \frac{t'_n(1)}{t_n(1)} = 1 + \frac{1}{2}\frac{n-1}{n} + \frac{1}{3}\frac{n-1}{n}\frac{n-2}{n} + \cdots. \tag{4.4}$$

Another closed form that works when $\beta = \frac{1}{2}$ is the pair of sequences

$$c_n = \frac{n}{n-1}(Q(n) - 1), \quad x_n = nK(n) - Q(n), \quad n > 2; \tag{4.5}$$

in this case we let $c_1 = x_1 = 0$.

For other values of β we use the sequences $c_n(y)$ and $x_n(y)$ of (1.5) and (1.10). Lemma 2 tells us that

$$c_n(y) = \frac{(y-1)t_n(y+1) - yt_n(y) + n^n}{(n-1)n^{n-1}}$$

$$= \frac{\sqrt{4\pi}\, n^{y/2}}{2^{y/2}\Gamma((y-1)/2)} + O(n^{(y-1)/2}) + 1 + O(n^{-1}); \quad (4.6)$$

$$x_n(y) = \frac{t_n(y)}{n^{n-1}} = \frac{\sqrt{2\pi}}{2^{y/2}} \frac{n^{(y+1)/2}}{\Gamma(y/2)} + O(n^{y/2}) + O(1). \quad (4.7)$$

Let $y = 2\beta$. If $\beta \leq 0$, so that $c_n = n^\beta = O(1)$, we clearly have $x_n = O(n)$, hence we may assume that $\beta > 0$. Therefore $y > 0$, and the terms $O(n^{-1})$ and $O(1)$ in (4.6) and (4.7) are irrelevant for our present purposes.

The special sequence $c_n = n^\beta$ is equal to

$$2^{y/2}\Gamma((y-1)/2)c_n(y)/\sqrt{4\pi} + c_n',$$

where $c_n' = 1 + O(n^{\beta-1/2})$. Thus

$$x_n = 2^{y/2}\Gamma((y-1)/2)x_n(y)/\sqrt{4\pi} + x_n'$$
$$= l(\beta)n^{\beta+1/2} + x_n' + O(n^\beta).$$

If $0 < \beta < 1/2$, we have $x_n = O(n)$ because $x_n' = O(n)$. If $\beta > 1/2$, the estimates in (4.2) for $\beta > \frac{1}{2}$ now follow by induction on the integer $\lfloor 2\beta - 1 \rfloor$. $\square$

The asymptotic growth of x_n when $c_n = n^\beta \ln n$ can be obtained by differentiating the results of Theorem 1 with respect to β.

To complete our study, we would like to investigate the case $\beta < 1/2$ in greater detail, by showing that x_n/n often approaches a constant κ (depending on $\langle c_n \rangle$), and by estimating the difference $x_n - \kappa n$ for large n.

If $\langle c_n \rangle$ is any sequence such that $\sum_{m \geq 1} c_m/m^{3/2}$ converges absolutely, let us define

$$\kappa\langle c_n \rangle = \sum_{m \geq 1} c_m \alpha_m, \quad (4.8)$$

where α_m is the quantity defined in Lemma 4, equation (3.26).

Theorem 2. *If $\beta < \frac{1}{2}$ and $c_n = O(n^\beta)$, the solution x_n to (0.3) satisfies*

$$x_n = n\kappa\langle c_n \rangle + O(n^{\beta+1/2}). \tag{4.9}$$

Proof. We have $x_n = \sum_{m \geq 1} x_n^{(m)} c_m$, so we need to apply the estimates of Lemma 4 to the sum $\sum_{m \geq 1} \varepsilon_n^{(m)} c_m$ where

$$\varepsilon_n^{(m)} = x_n^{(m)} - n\alpha_m. \tag{4.10}$$

The sum splits into four parts, of which the first and second are sufficiently small because of (3.27):

$$\sum_{m=1}^{n/2} \varepsilon_n^{(m)} c_m = \sum_{m=1}^{n/2} O(m^{\beta-1/2}) = O(n^{\beta+1/2});$$

$$\sum_{m=n/2}^{n-1} \varepsilon_n^{(m)} c_m = \sum_{m=n/2}^{n-1} O((n-m)^{-1/2} n^\beta)$$

$$= \sum_{m=1}^{n/2} O(m^{1/2} n^\beta) = O(n^{\beta+1/2});$$

$$\varepsilon_n^{(n)} c_n = (1 - n\alpha_n) c_n = O(n^\beta);$$

$$\sum_{m>n} \varepsilon_n^{(m)} c_m = \sum_{m>n} (-n) \alpha_m c_m = O(n^{\beta+1/2}). \qquad \square$$

The quantity $\kappa\langle c_n \rangle$ can, incidentally, be written as an integral instead of as a sum:

$$\kappa\langle c_n \rangle = c_1 + \int_0^{e^{-1}} \frac{c(w)\,dw}{wT(w)} = c_1 + \int_0^1 u^{-2} C(ue^{-u})(1-u)\,du, \tag{4.11}$$

where

$$C(z) = \sum_{n \geq 2} \frac{(n-1)n^{n-1} c_n}{n!} z^n. \tag{4.12}$$

Some sort of smoothness condition is necessary for the validity of (4.9); we cannot conclude that x_n/n approaches a limit if we know only that $\sum_{m \geq 1} c_m/m^{3/2}$ exists. For example, we might have

$$c_n = \begin{cases} 0, & \text{if } n \text{ is not a power of 2;} \\ 3^k/k^2, & \text{if } n = 2^k. \end{cases} \tag{4.13}$$

Then $x_n = n^{3/2}/(\lg n)^2 + O(n)$ when $n = 2^k$, but $x_n = O(n)$ when $n = 2^k - 1$.

It is, however, possible to prove that

$$\liminf_{n \to \infty} \frac{x_n}{n} = \kappa \langle c_n \rangle, \tag{4.14}$$

if $\sum_{m \geq 1} c_m/m^{3/2}$ exists and if all c_n are nonnegative. For in this case we can define $c'_n = c_n$ for $n \leq N$ and $c'_n = 0$ for $n > N$; then $x_n/n \geq x'_n/n \to \sum_{m=1}^{N} \alpha_m c_m$ for all N, and we can let $N \to \infty$ to obtain

$$\liminf_{n \to \infty} \frac{x_n}{n} \geq \kappa \langle c_n \rangle.$$

To prove that also

$$\liminf_{n \to \infty} \frac{x_n}{n} \leq \kappa \langle c_n \rangle,$$

we argue as follows. Set $c'_1 = 0$, and $c'_n = c_n$ for $n > 1$. Then $x_n = nc_1 + x'_n$. Introducing

$$G(z) = \sum_{n \geq 1} \frac{n^{n-1} x'_n}{n!} z^n$$

we have (see (4.12) and [9, equation (11.10)])

$$G(z) = \frac{T(z)}{1 - T(z)} \int_0^z \frac{C(w)\, dw}{w T(w)}.$$

Since $1 - T(z) \sim 2^{1/2}(1 - ez)^{1/2}$ as z increases to e^{-1}, we have then

$$\lim_{z \uparrow e^{-1}} 2^{1/2}(1 - ez)^{1/2} G(z) = \int_0^{e^{-1}} \frac{C(w)\, dw}{w T(w)} = \kappa \langle c_n \rangle - c_1.$$

By the Tauberian theorem for power series (see [2], for instance), we can write subsequently

$$\sum_{j=1}^{n} \frac{j^{j-1} x'_j e^{-j}}{j!} = (1 + o(1))(2\pi)^{-1/2} 2n^{1/2} (\kappa \langle c_n \rangle - c_1).$$

According to Stirling's formula for factorials this leads to

$$\liminf_{n \to \infty} \frac{x'_n}{n} \leq \kappa \langle c_n \rangle - c_1.$$

Given a sequence c_n such that $\sum_{m \geq 1} |c_m|/m^{3/2}$ exists, we can define the *residual sequence*

$$r_n = x_n - n\kappa\langle c_n \rangle.$$

Residuals act linearly on $\langle c_n \rangle$, hence we can obtain further information about their asymptotic value by taking linear combinations of special sequences.

The residual of the constant sequence $\langle c_n \rangle$ with $c_n = 1$ is -1, because $x_n = 2n - 1$. It follows from (4.6) that the residual of the sequence $\langle c_n(2\beta) \rangle$ is $l(\beta)n^{1/2}c_n(2\beta) + O(n^\beta) + O(1)$, when $\beta < \frac{1}{2}$. Therefore if $c_n = n^\beta$ we can put the estimates of Theorem 1 and Theorem 2 into a more precise form:

$$x_n = n\kappa\langle n^\beta \rangle + l(\beta)n^{\beta+1/2} + O(n^\beta) + O(1), \quad \text{if } \beta < \tfrac{1}{2}. \tag{4.15}$$

Acknowledgment

We wish to thank Philippe Flajolet for his penetrating remarks on a draft of this paper.

This research was supported in part by the National Science Foundation under grant CCR-86-10181, and by Office of Naval Research contract N00014-87-K-0502.

References

[1] Gotthold Eisenstein, "Entwicklung von $\alpha^{\alpha^{\alpha^{\cdots}}}$," *Journal für die reine und angewandte Mathematik* **28** (1844), 49–52. Reprinted in his *Mathematische Werke* **1**, 122–125.

[2] William Feller, *An Introduction to Probability Theory and Its Applications*, Volume 2 (New York: Wiley, 1971).

[3] Philippe Flajolet and Andrew Odlyzko, "Singularity analysis of generating functions," *SIAM Journal of Discrete Mathematics* **3** (1990), 216–240.

[4] Philippe Flajolet and Michèle Soria, "Gaussian limiting distributions for the number of components in combinatorial structures," *Journal of Combinatorial Theory* **A53** (1990), 165–182.

[5] Donald E. Knuth, "An analysis of optimum caching," *Journal of Algorithms* **6** (1985), 181–199. [Reprinted as Chapter 17 of *Selected Papers on Analysis of Algorithms*, CSLI Lecture Notes 102 (Stanford, California: Center for the Study of Language and Information, 2000), 235–255.]

[6] Donald E. Knuth, *Fundamental Algorithms*, Volume 1 of *The Art of Computer Programming* (Reading, Massachusetts: Addison-Wesley, 1968).

[7] Donald E. Knuth, *Seminumerical Algorithms*, Volume 2 of *The Art of Computer Programming* (Reading, Massachusetts: Addison-Wesley, 1969).

[8] Donald E. Knuth, *Sorting and Searching*, Volume 3 of *The Art of Computer Programming* (Reading, Massachusetts: Addison-Wesley, 1973).

[9] Donald E. Knuth and Arnold Schönhage, "The expected linearity of a simple equivalence algorithm," *Theoretical Computer Science* **6** (1978), 281–315. [Reprinted with additional material as Chapter 21 of *Selected Papers on Analysis of Algorithms*, CSLI Lecture Notes 102 (Stanford, California: Center for the Study of Language and Information, 2000), 341–389.]

[10] Martin Kruskal, "The expected number of components under a random mapping function," *American Mathematical Monthly* **61** (1954), 392–397.

[11] Alfred Rényi, "Some remarks on the theory of trees," *Magyar Tudományos Akadémia Matematikai Kutató Intézetének Közleményei* **4** (1959), 73–85. Reprinted in *Selected Papers of Alfréd Rényi* **2**, 363–374.

[12] John Riordan, *Combinatorial Identities* (New York: John Wiley & Sons, 1968).

[13] V. E. Stepanov, "Предельные распределения некоторых характеристик случайных отображений," *Teoriĩa Veroĩatnosteĩ i ee Primeneniĩa* **14** (1969), 639–653. English translation, "Limit distributions of certain characteristics of random mappings," *Theory of Probability and Its Applications* **14** (1969), 612–626.

Chapter 40

The First Cycles in an Evolving Graph

If successive connections are added at random to an initially disconnected set of n points, the expected length of the first cycle that appears will be proportional to $n^{1/6}$, with a standard deviation proportional to $n^{1/4}$. The size of the component containing this cycle will be of order $n^{1/2}$, on the average, with standard deviation of order $n^{7/12}$. The average length of the kth cycle is proportional to $n^{1/6}(\log n)^{k-1}$. Furthermore, the probability is $\sqrt{2/3} + O(n^{-1/3})$ that the graph has no components with more than one cycle at the moment when the number of edges passes $\frac{1}{2}n$. These results can be proved with analytical methods based on combinatorial enumeration with multivariate generating functions, followed by contour integration to derive asymptotic formulas for the quantities of interest.

[Written with Philippe Flajolet and Boris Pittel. Originally published in Combinatorics 1988, Proceedings of the Cambridge Conference in Honour of Paul Erdős, edited by Béla Bollobás (North-Holland, 1989), 167–215 = Discrete Mathematics **75** (1989), 167–215.]

0. Introduction

A classic paper by Erdős and Rényi [6] inaugurated the study of the random graph process, in which we begin with a totally disconnected graph and enrich it by successively adding edges at random. Algorithms that deal with graphs often mimic such a process, inputting a sequence of edges until some stopping criterion occurs, based on the configuration of edges seen so far. To analyze such algorithms, we wish to estimate relevant characteristics of the resulting graph. For example, we might stop when the graph first contains a particular kind of subgraph, and we might ask how large that subgraph is.

The purpose of this paper is to introduce analytical methods by which such questions can be answered systematically. In particular, we will apply the ideas to an interesting question posed by Paul Erdős and communicated by Edgar Palmer to the 1985 Seminar on Random Graphs in Poznań: "What is the expected length of the first cycle in an evolving graph?" The answer turns out to be rather surprising: The first cycle has length $Kn^{1/6} + O(n^{3/22})$ on the average, where

$$ K = \frac{1}{\sqrt{8\pi}\, i} \int_{-\infty}^{\infty} \int_{\Gamma} e^{(\mu + 2s)(\mu - s)^2/6} \, \frac{ds}{s} \, d\mu \approx 2.0337 $$

for a certain contour Γ. The form of this result suggests that the expected behavior may be quite difficult to derive using techniques that do not use contour integration.

The methods to be described start with comparatively easy techniques of combinatorial analysis based on generating functions, and finish with more difficult (yet standard) techniques of complex analysis. The main novelty in this approach is the use of contour integration to give parametric estimates of a function that appears within an ordinary integral. Such methods may well find application in other studies of random graphs, so they are presented here in an expository fashion and in somewhat greater generality than is needed to solve the special problems used as examples.

Section 1 introduces two basic models of evolving graphs that will be studied in the sequel, corresponding roughly to sampling with and without replacement. Section 2 discusses bivariate generating functions suitable for studying these graphs. Such generating functions can be used to derive probabilities in both of the models, as shown in Section 3. Asymptotic calculations in Section 4, based on the saddle point method, lead to results in Section 5 about the limiting distribution of first cycle lengths. Section 6 proves the main theorem about expected cycle length, and Section 7 derives auxiliary results about the expected waiting time and expected component sizes. The joint distribution of cycle lengths and edges is studied in Section 8, which also demonstrates a connection between waiting times and the parametric functions of Section 3. Section 9 extends the ideas to another problem in which we consider the first "bicyclic" component instead of the first cycle. An alternative approach to waiting times is considered in Section 10, where we also give an affirmative answer to a long-standing conjecture of Erdős and Rényi about the probability that a graph is planar. Finally we consider the first k cycles, in Section 11.

1. Models of Graph Evolution

We shall consider two related ways to enrich an initially empty graph on the vertices $\{1, 2, \ldots, n\}$. The first procedure, called the *uniform model*, is the simplest: At each step we generate an ordered pair $\langle x, y \rangle$, where x and y are uniformly distributed between 1 and n, and all n^2 pairs are equally likely. The (undirected) edge $x - y$ is then added to the graph. In this way we obtain a multigraph, which may have duplicate edges or self-loops $x - x$. Interesting variants of this model can be obtained by imposing other distributions on the pairs $\langle x, y \rangle$, but we shall not pursue such generalizations in the present paper.

Another way to generate a sequence of random edges may be called the *permutation model*; this model corresponds directly to random graphs as studied in the classic papers by Erdős and Rényi [6, 7]. Here we consider all $\binom{n}{2}!$ permutations of the pairs $\langle x, y \rangle$ with $1 \leq x < y \leq n$ to be equally likely, and we generate new edges $x - y$ by considering the pairs as they occur in such a permutation. The resulting graph contains no self-loops and no multiple edges; we are essentially sampling without replacement. The permutation model can be derived from the uniform model if we generate $\langle x, y \rangle$ uniformly but disregard any pairs with $x = y$ or pairs that duplicate a previous edge.

Our goal is to study the generation of random edges in such models until a cycle first appears in the resulting graph. (Equivalently, we are studying the first time that a sequence of random "union" operations specifies a redundant union; see [11].) In the uniform model, the process might stop with a self-loop $\langle x, x \rangle$, which is a cycle of length 1. Or it might stop with a duplicate edge (a pair $\langle x, y \rangle$ such that either $\langle x, y \rangle$ or $\langle y, x \rangle$ has occurred before); this is a cycle of length 2. In the permutation model all cycles have length 3 or more.

For example, Figure 1 illustrates a "random" graph on $n = 100$ vertices, based on the representation of $\pi = 3.1415926\ldots$ in decimal notation. (Here the vertices have been labeled 00 to 99 instead of 1 to n.) A cycle first appears when the 45th random pair, $\langle 05, 55 \rangle$, is added. In this case the uniform and permutation models produce identical graphs, because the first cycle has length > 2; in other words, no duplicate edges or self-loops are generated before there is a cycle. (We will see in Theorem 2 below that both models give the same graph with probability approaching 8/15.)

The permutation model of graph evolution is often called the "random graph process." In these terms we can call the uniform graph model the "random multigraph process."

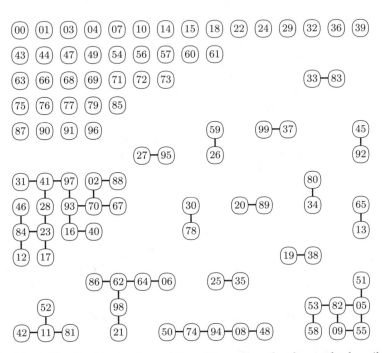

FIGURE 1. The final state of a graph on 100 vertices that has evolved until a cycle first appears. The successive ordered pairs ⟨31, 41⟩ ⟨59, 26⟩ ⟨53, 58⟩ ⟨97, 93⟩ ⟨23, 84⟩ ⟨62, 64⟩ ⟨33, 83⟩ ⟨27, 95⟩ ⟨02, 88⟩ ⟨41, 97⟩ ⟨16, 93⟩ ⟨99, 37⟩ ⟨51, 05⟩ ⟨82, 09⟩ ⟨74, 94⟩ ⟨45, 92⟩ ⟨30, 78⟩ ⟨16, 40⟩ ⟨62, 86⟩ ⟨20, 89⟩ ⟨98, 62⟩ ⟨80, 34⟩ ⟨82, 53⟩ ⟨42, 11⟩ ⟨70, 67⟩ ⟨98, 21⟩ ⟨48, 08⟩ ⟨65, 13⟩ ⟨28, 23⟩ ⟨06, 64⟩ ⟨70, 93⟩ ⟨84, 46⟩ ⟨09, 55⟩ ⟨05, 82⟩ ⟨23, 17⟩ ⟨25, 35⟩ ⟨94, 08⟩ ⟨12, 84⟩ ⟨81, 11⟩ ⟨74, 50⟩ ⟨28, 41⟩ ⟨02, 70⟩ ⟨19, 38⟩ ⟨52, 11⟩ produce nothing but free trees in the initially empty graph, but then ⟨05, 55⟩ yields a cycle of length 4. (This cycle appears in the lower right corner.) At this point there still are 40 isolated vertices (shown at the top and upper left) that have not yet been mentioned.

Let us recall briefly some of the main results of Erdős and Rényi from [7], to establish a context for the facts proved below. (A detailed discussion of the theory appears in [2] and [12].) The following properties hold "almost surely" (that is, with probability tending to 1 as $n \to \infty$) at the time when m random edges have been added to an initially disconnected set of n vertices: Only isolated vertices and edges will be present when $m \ll n^{1/2}$; but trees of order 3 will start to appear at time $m \approx n^{1/2}$, and trees of order 4 at time $m \approx n^{2/3}$, ..., trees of order $k + 1$ at time $m \approx n^{1-1/k}$. There is (almost surely) no cycle while $m \ll n$.

Later, when $m = \lambda n/2$ and $\lambda < 1$, there is at most one cycle in each component, and the largest component almost surely has size $\Theta(\log n)$. A dramatic phase transition occurs near $m = n/2$, when one or several large components of size about $n^{2/3}$ appear. Still later, when $m = \lambda n/2$ and $\lambda > 1$, we find a single "giant" component of size $\Theta(n)$.

We wish to examine the state of the graph when the first cycle appears. According to [7], this almost surely happens at some time $m < n/2$; we will see (Section 7, Corollary 3) that the expected time is $m \approx n/3$ in the uniform model, $m \approx 0.44n$ in the permutation model. There still remain $\Theta(n)$ isolated vertices when the first cycle is formed (Corollary 4). And at this time the expected cycle length is of order $n^{1/6}$ (Section 6, Theorem 3), with standard deviation of order $n^{1/4}$ (Section 7). The expected size of the component containing the first cycle will be $\Theta(n^{1/2})$, with standard deviation of order $n^{7/12}$ (Corollary 1). We can also characterize the limit distribution of the first cycle length (Section 5, Theorem 2), as well as the limit distribution of the first cyclic component size (Section 7, Corollary 2). These distributions have a very slowly decaying tail and an infinite mean; hence their expected values of order $n^{1/6}$ and $n^{1/2}$ do not contradict the fact that the largest component almost surely has size $O(\log n)$ when $m/n \leq \frac{1}{2} - \delta$.

With the same methods we will also gain some insight into events that take place around the time $m \approx n/2$. The first bicyclic component (Section 9) appears at time $n/2 + \Theta(n^{2/3})$, and its size is then of order $n^{2/3}$ (Corollary 5). However, at time $m = n/2$ and a little beyond, there still is a positive probability that the graph will have no bicyclic component (Theorem 5 and Corollary 7); it will therefore still be planar.

2. Generating Functions for Stopping Configurations

Probabilities and expected values in such random models can be obtained from generating functions whose coefficients count the number of graphs with specified characteristics and specified weights.

In our case we wish to count graphs that have a single cycle. Such graphs can conveniently be regarded as an unordered set of unrooted trees (representing the acyclic components) together with an ordered sequence of rooted trees (representing the component that has a cycle). For example, the graph of Figure 1 contains 40 isolated vertices, 11 vertex pairs that are (unrooted) trees of size 2, and additional trees of respective sizes 4, 5, 6, and 16; these are the acyclic components. The cyclic component is represented by a sequence of l rooted trees, where l is the length of the cycle, and the roots are the vertices of the cycle. In

Figure 1, this sequence is

$$
\left\langle \; \overset{\textstyle\fbox{55}}{}, \; \fbox{09}, \; \overset{\textstyle\fbox{82}}{\underset{\fbox{05}}{\rule{0pt}{1em}}}, \; \overset{\displaystyle\fbox{58}}{\overset{\textstyle\fbox{53}}{}} \; \fbox{51} \; \right\rangle .
\tag{2.1}
$$

If the final cycle-completing edge in the random model was $\langle x, y\rangle$, we arrange the sequence of rooted trees so that the first root is y and the last root is x. We shall say that a collection of unrooted and rooted trees as just described is a *stopping configuration*.

The enumeration of such labeled objects with exponential generating functions is a standard exercise in combinatorial analysis (see, for example, [5], [8], or [9]), but it will be helpful to review the basic ideas briefly. If $F(z)$ is a power series, we write $[z^n]\,F(z)$ for the coefficient of z^n. We say that $F(z)$ is the *exponential generating function* (egf) for a collection F of labeled objects if $n!\,[z^n]\,F(z)$ is the number f_n of ways to attach labels to objects in F that have n elements, that is, if

$$
F(z) = \sum_{n\geq 0} f_n \frac{z^n}{n!} = \sum_{A\in F} \frac{z^{|A|}}{|A|!} .
\tag{2.2}
$$

If $F_1(z), \ldots, F_k(z)$ are egfs for $F_1, \ldots, F_k$, respectively, then the product $F_1(z) \ldots F_k(z)$ is the egf for all ordered sequences $\langle A_1, \ldots, A_k\rangle$ where A_j is an element of F_j with an appropriate relabeling. In particular, if $F_1(z) = \cdots = F_k(z) = F(z)$, then the functions

$$
F(z)^k \qquad \text{and} \qquad F(z)^k/k!
$$

exponentially generate *sequences* and *sets*, respectively, of k objects from F. Summing over k, we deduce that the functions

$$
\frac{1}{1 - F(z)} \qquad \text{and} \qquad \exp F(z)
\tag{2.3}
$$

are the respective egfs for sequences and sets of all lengths $k \geq 0$.

We can, for instance, use these ideas to discover the egf for labeled, rooted trees, which we shall call $T(z)$. Every such tree is an ordered pair $\langle A, B\rangle$ where A is the root node and B is a set of rooted trees (the children of the root). The egf for A is simply z, and the egf for B is $\exp T(z)$ according to (2.3); hence we have the well-known relation

$$
T(z) = z\,e^{T(z)} .
\tag{2.4}
$$

Let $U(z)$ be the egf for labeled, unrooted trees. We can represent every rooted tree T on the labels $\{1, \ldots, n\}$ as either an unrooted tree U (if 1 is the root of T) or as a unordered pair $\{A, B\}$ where A and B are rooted trees (if 1 is not the root of T). In the latter case, either A or B contains the node 1, say A does; we add an edge from the root of A to the root of B. This construction is reversible, hence we have another well-known relation:

$$T(z) = U(z) + \tfrac{1}{2} T(z)^2 \,. \tag{2.5}$$

We can now enumerate stopping configurations that contain k unrooted trees in the acyclic components and l rooted trees in the cyclic components: The egf is

$$T(z)^l U(z)^k / k! \,. \tag{2.6}$$

Summing over k and l, the total number of stopping configurations for cycles of length $\geq l$ has the egf

$$\frac{T(z)^l}{1 - T(z)} \, e^{U(z)} \,. \tag{2.7}$$

We get all stopping configurations for the uniform model when $l = 1$, and for the permutation model when $l = 3$.

For our purposes we need additional information provided by a *bivariate generating function* (bgf), which enumerates stopping configurations by the number of edges as well as the number of vertices. A bgf is a power series

$$F(w, z) = \sum_{m, n \geq 0} f_{m,n} \, w^m \, \frac{z^n}{n!} \tag{2.8}$$

in which $f_{m,n}$ is the number of stopping configurations with m edges and n vertices, weighted by some criterion. Notice that the coefficients are "exponential" in n (that is, they include a factor $1/n!$), but not in m; setting $w = 1$ converts the bgf into an egf.

The bgf for unrooted trees is $U(wz)/w$, because every tree with n vertices contains $n - 1$ edges. Similarly, the bgf for rooted trees is $T(wz)/w$. The bgf for stopping configurations with k unrooted trees and l rooted trees, corresponding to (2.6), is

$$w^l \big(T(wz)/w \big)^l \big(U(wz)/w \big)^k / k! = T(wz)^l \big(U(wz)/w \big)^k / k! \,, \tag{2.9}$$

because we implicitly associate l additional edges with the edges of the rooted trees. (These are the edges of the cycle.) Summing over k and l

gives us the bgf analogous to (2.7) for the set of all stopping configurations with cycle length $\geq l$:

$$S_l(w, z) = \frac{T(wz)^l}{1 - T(wz)} e^{U(wz)/w}. \qquad (2.10)$$

When $l = 1$, for example, we obtain the bgf for all configurations in which the uniform model can stop:

$$S_1(w, z) = wz + (w + 2w^2) z^2 + \left(\frac{w + 5w^2 + 9w^3}{2}\right) z^3$$
$$+ \left(\frac{w + 9w^2 + 36w^3 + 64w^4}{6}\right) z^4 + \cdots. \qquad (2.11)$$

In particular, when there are $n = 3$ vertices the coefficient of $z^n/n!$ is

$$3w + 15w^2 + 27w^3; \qquad (2.12)$$

this formula means that there are 3 stopping configurations in which the uniform model stops after $m = 1$ steps, plus 15 in which it stops after $m = 2$ steps, plus 27 in which it stops after $m = 3$ steps. The 27 with $m = 3$ have the following forms:

6 cases	6 cases	6 cases	6 cases	3 cases
	a	b	a	a b
	$\mid$	$\mid$	$\mid$	$\backslash\,/$
$\langle a, b, c\rangle$	$\langle b, c\rangle$	$\langle a, c\rangle$	b	$\langle c\rangle$
			$\mid$	
			$\langle c\rangle$	

The 15 with $m = 2$ include 6 with a 2-cycle and 9 with a 1-cycle.

   Setting $l = 3$ in (2.10) gives the bgf for all stopping configurations in the permutation model:

$$S_3(w, z) = w^3 z^3 + (w^3 + 4w^4) z^4 + \left(\frac{w^3 + 9w^4 + 25w^5}{2}\right) z^5 + \cdots. \qquad (2.13)$$

In both models the process must stop after at most n edges have appeared.

3. Probabilities from Generating Functions

We need to multiply the coefficient of $u^m z^n/n!$ in a bgf by a suitable function of m and n, in order to compute the probability that a given stopping configuration occurs in the dynamic evolution process.

In the uniform model, a given stopping configuration with m edges $\{u_1 - v_1, \ldots, u_m - v_m\}$ can arise from exactly $2^{m-1}(m-1)!$ sequences of ordered pairs $\langle x_1, y_1 \rangle, \ldots, \langle x_m, y_m \rangle$. (The values of x_m and y_m are determined, since they are roots of specific trees in the cycle; the other $m-1$ edges can be permuted in $(m-1)!$ ways, and there is an additional factor of 2^{m-1} because each of these edges can be written as an ordered pair in two ways.) Therefore the probability of obtaining any given m-edge, n-vertex stopping configuration in this model is $2^{m-1}(m-1)!/n^{2m}$.

For example, we can check this calculation when $n = 3$, using (2.12):

$$\frac{3 \cdot 2^0 \cdot 0!}{9^1} + \frac{15 \cdot 2^1 \cdot 1!}{9^2} + \frac{27 \cdot 2^2 \cdot 2!}{9^3} = \frac{1}{3} + \frac{10}{27} + \frac{8}{27}.$$

The probabilities sum to 1, as they should; we note in particular that the process stops in this case after the second step with probability $\frac{10}{27}$.

Given a bgf $F(w, z) = \sum_{m,n} f_{m,n} w^m z^n/n!$ as in (2.8), we want to calculate the corresponding probability

$$\Phi_n F = \sum_{m \geq 1} \frac{2^{m-1}(m-1)!}{n^{2m}} f_{m,n} \tag{3.1}$$

for problems of size n. The linear functional Φ_n can be obtained in two steps. If we know

$$F_n(w) = n! \, [z^n] \, F(w, z) = \sum_{m \geq 1} f_{m,n} w^m, \tag{3.2}$$

then

$$\Phi_n F = \frac{1}{2} \int_0^\infty e^{-n^2 t/2} F_n(t) \, \frac{dt}{t}, \tag{3.3}$$

because the operation $f(w) \mapsto \frac{1}{2} \int_0^\infty e^{-n^2 t/2} f(t) \, dt/t$ maps w^m into

$$\frac{1}{2} \int_0^\infty e^{-n^2 t/2} t^{m-1} \, dt = \frac{2^{m-1}}{n^{2m}} \int_0^\infty e^{-u} u^{m-1} \, du = \frac{2^{m-1}(m-1)!}{n^{2m}}.$$

And we do know $F_n(w)$, because Cauchy's integral formula gives

$$F_n(w) = \frac{n!}{2\pi i} \oint F(w, z) \, \frac{dz}{z^{n+1}}, \tag{3.4}$$

if we integrate around a small circle enclosing the origin. Therefore Φ_n is determined.

A similar method applies to the permutation model. In this case any stopping configuration with m edges arises from $\frac{1}{2}(m-1)!$ sequences of pairs $\langle x, y \rangle$ having $x < y$. (The factor $\frac{1}{2}$ comes from the fact that a cycle can be oriented in two ways. Strictly speaking, our definitions impose an ordering on the nodes in the cycle so that exactly half of all stopping configurations with $l \geq 3$ are forbidden.) A given sequence of m edges occurs with probability $1/(N(N-1)\ldots(N-m+1))$, where $N = \binom{n}{2}$. Hence the weighting function that converts m-edge, n-vertex stopping configurations to probabilities in the permutation model is

$$\frac{(m-1)!}{2N(N-1)\ldots(N-m+1)} = \frac{1}{2m\binom{N}{m}}.$$

For example, we can check this calculation by looking at the case $n = 5$, when there are $N = 10$ possible edges. The coefficient of $z^5/5!$ in (2.13) is $60w^3 + 540w^4 + 1500w^5$, and

$$\frac{60}{2 \cdot 3 \cdot \binom{10}{3}} + \frac{540}{2 \cdot 4 \cdot \binom{10}{4}} + \frac{1500}{2 \cdot 5 \cdot \binom{10}{5}} = \frac{1}{12} + \frac{9}{28} + \frac{25}{42} = 1.$$

The relevant linear functional in the permutation model is

$$\Psi_n F = \sum_{m \geq 1} \frac{f_{m,n}}{2m\binom{N}{m}}, \qquad \text{where } N = \binom{n}{2}, \qquad (3.5)$$

and in this case the integral formula analogous to (3.3) is

$$\Psi_n F = \frac{1}{2} \int_0^\infty \frac{1}{(1+t)^N} F_n(t) \frac{dt}{t(1+t)}. \qquad (3.6)$$

(The substitution $u = t/(1+t)$ converts $\int_0^\infty t^{m-1} dt/(1+t)^{N+1}$ into

$$\int_0^1 u^{m-1}(1-u)^{N-m}\,du = B(N+1-m, m) = \frac{1}{m\binom{N}{m}},$$

by well-known formulas.) Notice that the kernel factor $(1+t)^{-N} = e^{-N(t+O(t^2))}$ in (3.6) is analogous to the $e^{-n^2 t/2}$ in (3.3).

Our formulas for Φ_n and Ψ_n in (3.3) and (3.6) evaluate $F_n(w)$ only at positive real values of the parameter w. However, $F(w, z)$ is evaluated for (small) complex values of z in (3.4). We can think of w as a positive real parameter in that formula.

In our applications the bgf $F(w, z)$ actually has the special form

$$F(w, z) = f\big(w, T(wz)\big), \tag{3.7}$$

where T is the tree function (2.4). For example, the bgf $S_l(w, z)$ of (2.10) is the function $s_l\big(w, T(wz)\big)$ defined by

$$s_l(w, z) = \frac{z^l}{1 - z} e^{(z - z^2/2)/w}, \tag{3.8}$$

because of (2.5). The linear functionals Φ_n and Ψ_n can be simplified in such cases because we can "invert" functions of T.

Namely, the relation $T(z)e^{-T(z)} = z$ of (2.4) implies that T is the inverse of the function ue^{-u}; hence

$$T(ue^{-u}) = u \tag{3.9}$$

when $|u|$ is small. The contour integral (3.4) now becomes

$$
\begin{aligned}
F_n(w) &= \frac{n!}{2\pi i} \oint f\big(w, T(wz)\big) \frac{dz}{z^{n+1}} \\
&= \frac{n! \, w^n}{2\pi i} \oint f(w, u) \, e^{nu}(1 - u) \frac{du}{u^{n+1}} \\
&= n! \, w^n \, [u^n] \, f(w, u)e^{nu}(1 - u)
\end{aligned}
\tag{3.10}
$$

if we make the substitution $wz = ue^{-u}$. (This is a special case of a trick often used to prove Lagrange's inversion theorem.)

It turns out that a rescaling of the parameters is quite helpful: We can replace w by λ/n, thereby introducing a factor n^{-n} that nicely tames the effect of $n!$ in (3.10). An additional factor of e^n will reduce the coefficient to polynomial growth; such transformations lead to the following convenient reformulation of the operations described above:

Theorem 1. Let $f\big(w, T(wz)\big)$ be the bivariate generating function (2.8) for a collection F of stopping configurations. Then the probability that a random graph will lie in F, if the graph is constructed by the process described in Section 1, is $\phi_n f$ for the uniform model and $\psi_n f$ for the *permutation model, where ϕ_n and ψ_n can be computed as follows:*

$$\phi_n f = \int_0^\infty f_n(\lambda)\,d\lambda\,; \tag{3.11}$$

$$\psi_n f = \int_0^\infty \left(\frac{e^\lambda}{(1+\lambda/n)^{n-1}}\right)^{n/2} \frac{f_n(\lambda)}{1+\lambda/n}\,d\lambda\,; \tag{3.12}$$

$$f_n(\lambda) = \frac{n!\,n^{-n}e^{-n\lambda/2}}{4\pi i\lambda} \oint f\left(\frac{\lambda}{n},z\right)(1-z)\left(\frac{\lambda e^z}{z}\right)^n \frac{dz}{z}. \quad \square \tag{3.13}$$

These formulas appear formidable at first glance — there is a contour integral inside a real integral — but we will see that they lead to asymptotic results without great difficulty. The value of $f_n(\lambda)$ is nonnegative, and it decreases rapidly to zero when λ is greater than 1 because of the factor $e^{-n\lambda/2}$. The difference between ϕ_n and ψ_n is a rather horrid-looking fudge factor, but we can replace it by its asymptotic value

$$\left(\frac{e^\lambda}{(1+\lambda/n)^{n-1}}\right)^{n/2} \frac{1}{1+\lambda/n}$$
$$= e^{\lambda/2+\lambda^2/4}\left(1 - \frac{12\lambda+3\lambda^2+2\lambda^3}{12n} + O\left(\frac{\lambda^2+\lambda^6}{n^2}\right)\right) \tag{3.14}$$

uniformly for $0 \le \lambda \le n^{1/3}$. Thus the two models are roughly the same, except that the permutation model calls for an additional factor of $e^{\lambda/2+\lambda^2/4}$ in the integral transform. We can take comfort in the fact that some simplification must be possible, since

$$\phi_n s_1 = 1 \quad \text{for all } n \ge 1; \tag{3.15}$$

$$\psi_n s_3 = 1 \quad \text{for all } n \ge 3. \tag{3.16}$$

(The function s_l is defined in (3.8); it yields the probability that the first cycle has length $\ge l$. Thus, formulas (3.15) and (3.16) simply state that the algorithm stops with probability 1 in both models.)

4. Asymptotic Distributions

Let's try to get a concrete idea of what the abstract formulas in Theorem 1 mean, by working out some of their simplest consequences. Our goal in this section will be to derive asymptotic formulas for the probability that the cycle length is $\ge l$. Once the methods are understood, more difficult applications will not be much of a challenge.

According to Theorem 1 and equation (3.8), the probability $P_{\geq l,n}$ that the uniform model produces a cycle of length $\geq l$ is $\int_0^\infty s_{l,n}(\lambda)\,d\lambda$, where

$$s_{l,n}(\lambda) = \frac{n!\,n^{-n}e^{-n\lambda/2}}{4\pi i\lambda} \oint z^{l-1}e^{nh(z)}\,dz\,, \tag{4.1}$$

$$h(z) = \frac{z - z^2/2}{\lambda} + z - \ln z + \ln \lambda\,. \tag{4.2}$$

The contour integral in (4.1) is a polynomial in λ of degree $n-1$, and this polynomial is a multiple of λ^{l-1}, because

$$\frac{1}{2\pi i} \oint e^{n((z-z^2/2)/\lambda+z)}\lambda^{n-1}\frac{dz}{z^{n-l+1}}$$

$$= [z^{n-l}]\,\lambda^{n-1}\exp\left(n\left(\frac{z - z^2/2}{\lambda} + z\right)\right). \tag{4.3}$$

For example, when $n = 3$ we have

$$s_{1,3}(\lambda) = \tfrac{1}{6}\left(3 + 5\lambda + 3\lambda^2\right)e^{-3\lambda/2}\,,$$

$$s_{2,3}(\lambda) = \tfrac{1}{3}\left(\lambda + \lambda^2\right)e^{-3\lambda/2}\,, \tag{4.4}$$

$$s_{3,3}(\lambda) = \tfrac{1}{9}\,\lambda^2\,e^{-3\lambda/2}\,.$$

Integrating over $0 \leq \lambda < \infty$ gives the respective values 1, $\frac{28}{81}$, $\frac{16}{243}$. Hence when $n = 3$ the probability that the uniform model produces a cycle of length 1 is $1 - \frac{28}{81} = \frac{53}{81}$; the cycle has length 2 with probability $\frac{28}{81} - \frac{16}{243} = \frac{68}{243}$; and it has length 3 with probability $\frac{16}{243}$.

The coefficients of the polynomial part of $s_{l,n}(\lambda)$ always have "mirror symmetry" in the sense that

$$e^{n\lambda/2}s_{l,n}(\lambda) = \lambda^{n+l-2}e^{n\lambda^{-1}/2}s_{l,n}(\lambda^{-1})\,. \tag{4.5}$$

For example, we have

$$s_{1,5}(\lambda) = \tfrac{1}{50}\left(25 + 70\lambda + 93\lambda^2 + 70\lambda^3 + 25\lambda^4\right)e^{-5\lambda/2}\,. \tag{4.6}$$

Relation (4.5) is important because it says that we can deduce the value of $s_{l,n}(\lambda)$ for all $\lambda \geq 0$ if we know its value for $0 \leq \lambda \leq 1$. The proof is immediate:

$$[z^{n-l}]\,\lambda^{n-1}\exp\left(n\left(\frac{z - z^2/2}{\lambda} + z\right)\right)$$

$$= \lambda^{n-1-(n-l)}\,[z^{n-l}]\exp\left(n\left(\frac{\lambda z - (\lambda z)^2/2}{\lambda} + \lambda z\right)\right)$$

$$= \lambda^{n+l-2}\,[z^{n-l}]\,\lambda^{1-n}\exp\left(n(\lambda(z - z^2/2) + z)\right)\,.$$

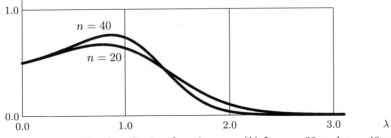

FIGURE 2. The distribution functions $s_{1,n}(\lambda)$ for $n = 20$ and $n = 40$.

Figure 2 shows $s_{1,n}(\lambda)$ when $n = 20$ and $n = 40$. These functions both yield 1 when integrated from 0 to ∞; notice that when n increases, more of the "mass" is concentrated in the range $0 \leq \lambda \leq 1$. In fact, we shall soon prove that $\lim_{n\to\infty} \int_0^1 s_{1,n}(\lambda)\,d\lambda = 1$. (A "physical" interpretation of this fact appears in Section 8.)

Let us first attempt to find a uniform estimate for $s_{1,n}(\lambda)$ when $0 \leq \lambda < 1$. Integrals of the type (4.1) are well suited to the "saddle point method" [4, §5.7]; hence we investigate the roots of $h'(z) = 0$:

$$h'(z) = \frac{1-z}{\lambda} + 1 - \frac{1}{z}\,; \tag{4.7}$$

$$h''(z) = -\frac{1}{\lambda} + \frac{1}{z^2}\,. \tag{4.8}$$

There are two saddle points, at $z = \lambda$ and $z = 1$. We notice that $h''(\lambda) = (1-\lambda)/\lambda^2 > 0$ and $h''(1) = (\lambda-1)/\lambda < 0$; also $h(1) - h(\lambda) > 0$. Hence we want our path of integration to pass vertically through the point $z = \lambda$.

If we integrate around a circle $|z| = r$, where r is any radius between 0 and $(1+\lambda)/2$, we can show that $|e^{h(z)}|$ takes its maximum value at $z = r$ and its minimum value at $z = -r$, with no other local maxima or minima. For if $z = re^{i\theta}$ we have*

$$|e^{h(z)}| = e^{\Re h(z)} = e^{f(r,\theta)} \tag{4.9}$$

where

$$f(r,\theta) = \lambda^{-1}(r\cos\theta - \tfrac{1}{2}r^2\cos 2\theta) + r\cos\theta - \ln r + \ln\lambda\,, \tag{4.10}$$

---

*We use the notation $\Re z$ for the real part of z and $\Im z$ for the imaginary part.

and the first derivative

$$f'(r, \theta) = -\lambda^{-1}(r\sin\theta - 2r^2\sin\theta\cos\theta) - r\sin\theta \qquad (4.11)$$

is zero only when $\sin\theta = 0$. (We cannot have $2r\cos\theta = 1+\lambda$.) Therefore the integrand in (4.1) makes most of its contributions near $\theta = 0$.

Let us now assume that $\lambda < 1$. On the circular path $z = \lambda e^{i\theta}$, equation (4.1) takes the form

$$s_{l,n}(\lambda) = \frac{n!\, n^{-n} e^{-n\lambda/2} \lambda^{l-1}}{4\pi} \int_{-\pi}^{\pi} e^{il\theta + nh(\lambda e^{i\theta})} \, d\theta. \qquad (4.12)$$

And by what we have just proved, we can integrate from $-\theta_0$ to θ_0 instead of from $-\pi$ to π, for any desired $\theta_0 < \pi$, using $O\big(\big|\exp\big(nh(\lambda e^{i\theta_0})\big)\big|\big)$ as an upper bound for the omitted portion of the integral.

The main point of the saddle point method is that $h'(\lambda) = 0$ and $h''(\lambda) > 0$, hence $nh(\lambda e^{i\theta})$ is approximately $nh(\lambda) - n\lambda^2 h''(\lambda)\,\theta^2/2$ in the neighborhood of $\theta = 0$. Therefore we will be able to estimate the integrand with a formula like $ae^{-nb\theta^2/2}$, plus terms that are asymptotically negligible when $|\theta|$ is small. Let's see what that will buy us, saving the justification for later: If the integrand is replaced by $e^{nh(\lambda) - n\lambda^2 h''(\lambda)\theta^2/2}$, the integral reduces to

$$s_{l,n}(\lambda) \approx \frac{n!\, n^{-n} e^n \lambda^{l-1}}{4\pi} \int_{-\infty}^{\infty} e^{-n(1-\lambda)\theta^2/2} \, d\theta. \qquad (4.13)$$

And this is just a multiple of the familiar integral for a normally distributed random variable with mean 0 and variance $1/(n(1-\lambda))$. In general, if k is any nonnegative even integer we have the well-known identity

$$\int_{-\infty}^{\infty} \theta^k e^{-a\theta^2/2} \, d\theta = 2 \int_0^\infty \left(\frac{2t}{a}\right)^{k/2} e^{-t} \frac{dt}{\sqrt{2at}}$$

$$= \frac{2^{(k+1)/2}}{a^{(k+1)/2}} \int_0^\infty t^{(k+1)/2 - 1} e^{-t} \, dt$$

$$= \frac{2^{(k+1)/2}\, \Gamma\big((k+1)/2\big)}{a^{(k+1)/2}}$$

$$= \frac{\sqrt{2\pi}}{a^{(k+1)/2}} \prod_{j=1}^{k/2} (2j-1). \qquad (4.14)$$

(The corresponding integral is zero when k is odd.) Therefore our approximation (4.13) simplifies to

$$s_{l,n}(\lambda) \approx \frac{n!\, n^{-n} e^n \lambda^{l-1}}{2\sqrt{2\pi n(1-\lambda)}} \approx \frac{\lambda^{l-1}}{2\sqrt{1-\lambda}}.$$

In fact, it is possible to prove a stronger result, without handwaving:

Lemma 1. *If $0 \le \lambda \le 1 - (\ln n)^2/n^{1/2}$ and $l \ge 1$, the function $s_{l,n}(\lambda)$ defined in (4.1) satisfies*

$$s_{l,n}(\lambda) = \frac{\lambda^{l-1}}{2\sqrt{1-\lambda}} + O\!\left(\frac{l^2 \lambda^{l-1}}{n(1-\lambda)^{3/2}}\right) + O\!\left(\frac{\lambda^{l-1}}{n(1-\lambda)^{7/2}}\right), \quad (4.15)$$

uniformly in λ and l, as $n \to \infty$.

Proof. On the circle $|z| = \lambda$ the function $h(z)$ is simply

$$h(\lambda e^{i\theta}) = (1+\lambda)e^{i\theta} - \frac{1}{2}\lambda e^{2i\theta} - i\theta$$

$$= 1 + \frac{1}{2}\lambda + \sum_{k \ge 2}\left(1 - (2^{k-1}-1)\lambda\right)\frac{(i\theta)^k}{k!}$$

$$= 1 + \frac{1}{2}\lambda - \frac{1}{2}(1-\lambda)\theta^2 + i\,O(\theta^3) + O(\theta^4), \quad (4.16)$$

where the quantities represented by $O(\theta^3)$ and $O(\theta^4)$ are real. To evaluate (4.12) we want to know the value of

$$\int_{-\pi}^{\pi} \exp\!\left(il\theta + n(h(\lambda e^{i\theta}) - 1 - \tfrac{1}{2}\lambda)\right) d\theta\,;$$

and we have observed that it suffices to integrate from $-\theta_0$ to θ_0, for any convenient value θ_0, using the magnitude of the integrand at θ_0 to bound the resulting error.

Let $\lambda = 1 - \mu n^{-1}$, so that $\sqrt{n}\,(\ln n)^2 \le \mu = n(1-\lambda) \le n$. We will integrate from $-\theta_0$ to θ_0, where $\theta_0 = \mu^{-1/2}\ln n$. The resulting error will then be exponentially small, because

$$\left|\exp\!\left(il\theta_0 + n(h(\lambda e^{i\theta_0}) - 1 - \tfrac{1}{2}\lambda)\right)\right| = \exp\!\left(-\tfrac{1}{2}n(1-\lambda)\theta_0^2 + O(n\theta_0^4)\right)$$

$$= O(e^{-(\ln n)^2/2})\,.$$

The substitution $\theta = t\mu^{-1/2}$ yields

$$\int_{-\theta_0}^{\theta_0} \exp\left(il\theta - n(1-\lambda)\theta^2/2 + iO(n\theta^3) + O(n\theta^4)\right) d\theta$$

$$= \mu^{-1/2} \int_{-\ln n}^{\ln n} \exp\left(-t^2/2 + O(n\mu^{-2}t^4)\right) \cos\left(l\mu^{-1/2}t + O(n\mu^{-3/2}t^3)\right) dt$$

$$= \mu^{-1/2} \int_{-\ln n}^{\ln n} e^{-t^2/2} \left(1 + O(n\mu^{-2}t^4)\right) \cos\left(l\mu^{-1/2}t + O(n\mu^{-3/2}t^3)\right) dt$$

$$= \mu^{-1/2} \int_{-\infty}^{\infty} e^{-t^2/2} \left(1 + O(n\mu^{-2}t^4)\right) \left(1 + O(l^2\mu^{-1}t^2) + O(n^2\mu^{-3}t^6)\right) dt$$

$$\qquad + O(e^{-(\ln n)^2/3})$$

$$= \sqrt{\frac{2\pi}{\mu}} \left(1 + O(n\mu^{-2}) + O(l^2\mu^{-1}) + O(n^2\mu^{-3})\right).$$

(We are allowed to replace $e^{O(x)}$ by $1 + O(x)$ when $x = O(1)$, hence the replacement of $\exp\left(O(n\mu^{-2}t^4)\right)$ by $1 + O(n\mu^{-2}t^4)$ is legitimate when $|t| \le \ln n$. Other estimates made in this derivation, where we replaced $\cos x$ by $1 + O(x^2)$ and $(x+y)^2$ by $O(x^2) + O(y^2)$, are valid without restrictions on x and y.)   □

The procedure used in the proof of Lemma 1 can be used to obtain as many further terms of the asymptotic expansion as desired (using a computer). For example, the O terms of (4.15) can be shown to equal

$$\frac{\lambda^{l-1}}{2\sqrt{1-\lambda}} \left(1 + \frac{1}{12n} - \frac{l^2}{2n(1-\lambda)} - \frac{l(3\lambda-1)+7\lambda-1}{2n(1-\lambda)^2} - \frac{5(3\lambda-1)^2}{24n(1-\lambda)^3}\right) \quad (4.17)$$

plus terms of lesser order. However, we reach a point of diminishing returns in these estimates when λ becomes larger than $1 - n^{-1/3}$ or when l becomes larger than $n^{1/2}$.

Lemma 2. *If $\lambda \ge 1 + (\ln n)^2/n^{1/2}$ and $l \ge 1$, the function $s_{l,n}(\lambda)$ defined in (4.1) satisfies*

$$s_{l,n}(\lambda) = \frac{e^{-n(\lambda-1/\lambda)/2}\lambda^n}{2\sqrt{\lambda(\lambda-1)}} \left(1 + O\left(\frac{l^2\lambda}{n(\lambda-1)}\right) + O\left(\frac{\lambda^3}{n(\lambda-1)^3}\right)\right), \quad (4.18)$$

uniformly in λ and l, as $n \to \infty$.

Proof. We could prove this by contour integration, using an argument almost identical to that in Lemma 1 but choosing the other saddle point and integrating around $|z| = 1$. But (4.18) is actually an immediate consequence of Lemma 1 and the reflection law (4.5).   □

For fixed $\lambda > 1$, equation (4.18) implies that $s_{l,n}(\lambda)$ decreases exponentially to zero as $n \to \infty$, because $\lambda - 1/\lambda > 2 \ln \lambda$. (If $\lambda = e^t$, we always have $e^t - e^{-t} > 2t$.) On the other hand, the difference between $\lambda - 1/\lambda$ and $2 \ln \lambda$ is of order $(\lambda - 1)^3$, so formula (4.18) says that

$$s_{l,n}\bigl(1 + \epsilon n^{-1/3}\bigr)$$
$$= \frac{e^{-\epsilon^3/6} n^{1/6}}{2\sqrt{\epsilon}} \left(1 + O\Bigl(\frac{l^2}{n^{2/3}\epsilon}\Bigr) + O\Bigl(\frac{1}{\epsilon^3}\Bigr) + O\Bigl(\frac{\epsilon + \epsilon^4}{n^{1/3}}\Bigr)\right),$$

when $n^{-1/6}(\ln n)^2 \le \epsilon < \frac{1}{2}n^{1/3}$. Thus $s_{l,n}(1 + n^{-1/3})$ is of order $n^{1/6}$, but the nearby value $s_{l,n}(1 + n^{-1/3}\ln n)$ is already exponentially small. In other words $s_{l,n}(\lambda)$ is unbounded as $n \to \infty$, but it decreases very rapidly when λ passes 1.

Lemma 1 tells us that $s_{l,n}(1 - n^{1/3})$ is of order $n^{1/6}$ and that $s_{l,n}(1 - n^{-1/3}\ln n) \sim \frac{1}{2}n^{1/6}(\ln n)^{-1/2}$. But the error estimates in both Lemmas 1 and 2 blow up when λ is near 1, because the two saddle points at λ and 1 come together; indeed, we have $h'(1) \approx 0$ and $h''(1) \approx 0$ but $h'''(1) \ne 0$ when $\lambda \approx 1$, so the magnitude of $e^{h(z)}$ near $z = 1$ has a graph that looks like a three-legged saddle — as used perhaps by Martian horsemen. A third lemma closes the gap in our knowledge by focusing on the region near $\lambda = 1$:

Lemma 3. *If $|\mu| \le n^{1/12}$ and $l \le n^{1/4}$, the function $s_{l,n}(\lambda)$ defined in (4.1) satisfies*

$$s_{l,n}\bigl(e^{-\mu n^{-1/3}}\bigr) = \frac{n^{1/6}}{\sqrt{8\pi}\,i} \int_{\Gamma^{(\alpha)}} e^{P(\mu,s)}\, ds + O\Bigl(\frac{l + l|\mu|^{1/2} + \mu^4}{n^{1/6}}\Bigr) \quad (4.19)$$

uniformly in μ and l as $n \to \infty$, where

$$P(\mu, s) = \frac{(\mu + 2s)(\mu - s)^2}{6}; \quad (4.20)$$

the contour $\Gamma^{(\alpha)}$ is defined for any positive number α by

$$s(t) = \begin{cases} e^{-\pi i/3}|t|, & \text{for } t \le -2\alpha; \\ \alpha + \frac{i}{2}\sqrt{3}\,t, & \text{for } -2\alpha \le t \le 2\alpha; \\ e^{+\pi i/3}\,t, & \text{for } t \ge 2\alpha. \end{cases} \quad (4.21)$$

Proof. The integral over $\Gamma^{(\alpha)}$ does not depend on α, since $e^{P(\mu,s)}$ has no singularities. We will find it convenient to let α be the positive solution to

$$\mu = \alpha - \alpha^{-1}. \qquad (4.22)$$

Let $\nu = n^{-1/3}$ and $\lambda = e^{-\mu\nu}$. A straightforward calculation proves that, for any s,

$$-\tfrac{1}{2}e^{-\mu\nu} + h(e^{-s\nu}) = 1 + \nu^3 P(\mu, s) + R', \qquad (4.23)$$

where the remainder term is

$$R' = \sum_{k \geq 4}\left((\mu - s)^k - \tfrac{1}{2}(\mu - 2s)^k + (-s)^k + \tfrac{1}{2}\mu^k\right)\frac{\nu^k}{k!}$$

$$= \sum_{k \geq 4}\frac{O\big((|\mu| + |s|)\nu\big)^k}{k!} = O\big((|\mu| + |s|)\nu\big)^4,$$

uniformly in any region where $(|\mu| + |s|)\nu$ is bounded. The terms in ν and ν^2 have cancelled out beautifully from the right-hand side of (4.23), thereby making the asymptotic behavior simple when we multiply by $n = \nu^{-3}$.

Formula (4.1), with $z = e^{-s\nu}$, becomes

$$s_{l,n}(e^{-\mu\nu}) = \frac{n!\,n^{-n}e^n\nu}{4\pi i e^{-\mu\nu}} \oint \exp\big(-sl\nu + P(\mu, s) + O((|\mu|+|s|)^4\nu)\big)\,ds, \qquad (4.24)$$

where s traverses a path from $\beta - i\pi n^{1/3}$ to $\beta + i\pi n^{1/3}$ for some β. We will choose $\beta = 2n^{1/12}$ and let $s = \beta + i\theta$ for $-\pi n^{1/3} \leq \theta \leq -\sqrt{3}\beta$; this brings s to the point $\beta - \sqrt{3}\beta i = s(-2\beta)$ on the contour $\Gamma^{(\alpha)}$ defined by (4.21) and (4.22). (Notice that $\beta > \alpha$, since $\mu \leq n^{1/12}$.) Then we shall continue with $s = s(t)$ on $\Gamma^{(\alpha)}$ for $-2\beta \leq t \leq 2\beta$. Finally we take $s = \beta + i\theta$ again, for $\sqrt{3}\beta \leq \theta \leq \pi n^{1/3}$. This contour keeps $z = e^{-s\nu}$ inside the unit circle.

Let $\Gamma_n^{(\alpha)}$ be the portion of $\Gamma^{(\alpha)}$ that we are traversing, namely the portion for $-2\beta \leq t \leq 2\beta$, and let C_n be the other part of the contour just described. We will show that the integral over C_n is negligibly small. Indeed, the integrand has the nice monotonicity property that we described earlier in connection with (4.11), because C_n corresponds to a circle in the z plane of radius $r = e^{-\beta\nu} < (1+e^{-\mu\nu})/2 = (1+\lambda)/2$. Therefore the integrand is largest at the point where C_n meets $\Gamma_n^{(\alpha)}$, and we will see that it is exponentially small there.

On $\Gamma_n^{(\alpha)}$ we have $s = O(n^{1/12})$, hence the O term in (4.24) is bounded and so is the term $-sl\nu$. We can therefore write this part of the integral as

$$\int_{\Gamma_n^{(\alpha)}} e^{P(\mu,s)} \left(1 + O(sl\nu) + O(s^4\nu) + O(\mu^4\nu)\right) ds. \tag{4.25}$$

Now we have $P(\mu, s) = \frac{1}{6}\mu^3 - \frac{1}{2}\mu s^2 + \frac{1}{3}s^3$, so when $s = e^{\pi i/3}t$ the real part of $P(\mu, s)$ is

$$\tfrac{1}{6}\mu^3 - \tfrac{1}{2}\mu t^2 \cos\tfrac{2\pi}{3} + \tfrac{1}{3}t^3 \cos\tfrac{3\pi}{3} = \tfrac{1}{6}\mu^3 + \tfrac{1}{4}\mu t^2 - \tfrac{1}{3}t^3 \,;$$

its first derivative, $\frac{1}{2}\mu t - t^2$, is negative when $t \geq 2\alpha$. When $t = 2\beta$ at the end of $\Gamma_n^{(\alpha)}$, we have $\mu \leq \frac{1}{4}t$, hence the real part is at most

$$\tfrac{1}{384}t^3 + \tfrac{1}{64}t^3 - \tfrac{1}{3}t^3 \,, \qquad t = 4n^{1/12};$$

the integrand is indeed exponentially small when this point is reached.

Furthermore when $s = \alpha + \frac{i}{2}\sqrt{3}t$ the real part of $P(\mu, s)$ turns out to be

$$\tfrac{1}{2}\alpha^{-1} - \tfrac{1}{6}\alpha^{-3} - \tfrac{3}{8}(\alpha + \alpha^{-1})t^2 \leq \tfrac{1}{3}. \tag{4.26}$$

Therefore $\left|e^{P(\mu,s)}\right|$ is uniformly bounded on $\Gamma_n^{(\alpha)}$, and we have

$$\int_{\Gamma_n^{(\alpha)}} \left|s^p e^{P(\mu,s)}\right| ds = O(1) + O\left(|\alpha|^{p-1/2}\right)$$

for any fixed nonnegative power p. (When α is large, the integrand is $O(\alpha^p)$ when $|t| = O(\alpha^{-1/2})$, and approximately zero for larger $|t|$.) The O terms can now be removed from the integral (4.25). Finally we can extend the domain of integration from $\Gamma_n^{(\alpha)}$ to the full contour $\Gamma^{(\alpha)}$, obtaining (4.19). □

The integral (4.19) is investigated further in the appendix below, where the following result is derived as a special case of a general series expansion:

$$s_{l,n}(1) \sim \frac{\Gamma\left(\tfrac{1}{3}\right)}{3^{1/6}\sqrt{8\pi}}\, n^{1/6}. \tag{4.27}$$

5. Distribution of Cycle Lengths

We can now combine the three lemmas with Theorem 1 and obtain the limiting probability distribution of cycle lengths:

Theorem 2. *For fixed l as $n \to \infty$, the graph evolution procedure of Section 1 generates a first cycle of length l with probability*

$$P_{l,n} = \frac{1}{3} \prod_{k=1}^{l-1} \left(\frac{2k}{2k+3}\right) + O(n^{-1/6}), \qquad l \geq 1, \qquad (5.1)$$

in the uniform model, and with probability

$$\widehat{P}_{l,n} = \frac{1}{2} \int_0^1 \lambda^{l-1} \sqrt{1-\lambda}\, e^{\lambda/2 + \lambda^2/4}\, d\lambda + O(n^{-1/6}), \qquad l \geq 3, \quad (5.2)$$

in the permutation model.

Proof.   In the uniform model we have

$$P_{l,n} = \int_0^\infty \left(s_{l,n}(\lambda) - s_{l+1,n}(\lambda)\right) d\lambda, \qquad (5.3)$$

so it suffices to determine $\int_0^\infty s_{l,n}(\lambda)\, d\lambda$. The integral for $\lambda \geq 1 - n^{-1/3}$ is $O(n^{-1/6})$, by Lemmas 2 and 3. Therefore we may restrict consideration to the interval $0 \leq \lambda \leq 1 - n^{-1/3}$, when we find that the total error in (4.15) is

$$n^{-1} \int_0^{1-n^{-1/3}} O\left((1-\lambda)^{-7/2}\right) d\lambda = n^{-1}O(n^{5/6}) = O(n^{-1/6}).$$

The integral from $1 - n^{-1/3}$ to 1 of $(1 - \lambda)^{-1/2}$ is also $O(n^{-1/6})$; hence

$$P_{\geq l,n} = \int_0^\infty s_{l,n}(\lambda)\, d\lambda = \frac{1}{2} \int_0^1 \frac{\lambda^{l-1}\, d\lambda}{\sqrt{1-\lambda}} + O(n^{-1/6}). \qquad (5.4)$$

And this is a Beta integral,

$$P_{\geq l,n} \sim \frac{1}{2}\, B(l, \tfrac{1}{2}) = \frac{1}{2}\, \frac{\Gamma(l)\Gamma(\tfrac{1}{2})}{\Gamma(l + \tfrac{1}{2})} = \prod_{k=1}^{l-1} \left(\frac{2k}{2k+1}\right). \qquad (5.5)$$

The difference $P_{\geq l,n} - P_{\geq l+1,n}$ is, similarly, $\frac{1}{2} B(l, \tfrac{3}{2}) + O(n^{-1/6})$, and we obtain (5.1). Equation (5.2) follows from (3.12) and (3.14).   □

Thus the cycle lengths approach a stationary distribution, without any normalization. Formula (5.2) was first obtained (without the error bound) by Svante Janson [10], using a general theory of Poisson processes, and independently by Béla Bollobás [3], using the theory of martingales.

Since the extra factor $e^{\lambda/2+\lambda^2/4}$ lies between 1 and $e^{3/4} \approx 2.11700002$ for $0 \le \lambda \le 1$, both probabilities $P_{l,n}$ and $\hat{P}_{l,n}$ have the same order of growth as l increases. Indeed, let

$$p_l = \frac{1}{3} \prod_{k=1}^{l-1} \left(\frac{2k}{2k+3} \right) = \frac{1}{2} \int_0^1 \lambda^{l-1} \sqrt{1-\lambda}\, d\lambda \, ;$$

$$\hat{p}_l = \frac{1}{2} \int_0^1 \lambda^{l-1} \sqrt{1-\lambda}\; e^{\lambda/2+\lambda^2/4}\, d\lambda . \tag{5.6}$$

Then we can write $\lambda/2 + \lambda^2/4 = 3/4 + O(\lambda - 1)$, obtaining

$$p_l = \frac{\sqrt{\pi}}{4\, l^{3/2}} + O(l^{-5/2}) ; \quad \hat{p}_l = \frac{\sqrt{\pi}\, e^{3/4}}{4\, l^{3/2}} + O(l^{-5/2}) . \tag{5.7}$$

In both cases the average value $\sum_l l p_l$ is infinite; therefore the expected cycle length must be unbounded as $n \to \infty$.

The limiting probabilities p_l for the uniform model obey simple recurrence relations:

$$p_{l+1} = \frac{2l}{2l+3}\, p_l \, ; \qquad p_{\ge l+1} = \frac{2l}{2l+1}\, p_{\ge l} = 2l\, p_l \, . \tag{5.8}$$

Hence it is natural to wonder if the corresponding numbers $\hat{p}_l$ and $\hat{p}_{\ge l}$ for the permutation model satisfy similar recurrences, and in fact they do. First we note that

$$\hat{p}_{\ge 3} = \frac{1}{2} \int_0^1 \frac{\lambda^2}{\sqrt{1-\lambda}}\, e^{\lambda/2+\lambda^2/4}\, d\lambda = -\sqrt{1-\lambda}\, e^{\lambda/2+\lambda^2/4} \Big|_0^1 = 1 . \tag{5.9}$$

A similar integration shows that

$$\hat{p}_{\ge l+3} = 2l\, \hat{p}_l \, , \tag{5.10}$$

and it follows that we have the recurrence

$$\hat{p}_{l+2} = 2(l-1)\hat{p}_{l-1} - 2l\, \hat{p}_l \, . \tag{5.11}$$

(Is there a "simple" graph-theoretic explanation of (5.10)?) Setting $l = 3$ in (5.10) yields

$$\hat{p}_5 = 1 - 7\hat{p}_3 - \hat{p}_4 \, ; \tag{5.12}$$

hence the values of $\hat{p}_l$ can all be expressed in the form $a_l + b_l\hat{p}_3 + c_l\hat{p}_4$ where a_l, b_l, and c_l are integers. Recurrence (5.11) is numerically unstable, but we can obtain accurate values

$$\hat{p}_3 \approx 0.12160\,82217\,14483\,58918\,, \qquad (5.13)$$

$$\hat{p}_4 \approx 0.08491\,50995\,26335\,99860\,, \qquad (5.14)$$

by calculating $\hat{p}_l$ and $\hat{p}_{l+1}$ accurately for some large l and then solving backwards.

Do the fundamental quantities $s_{l,n}(\lambda)$ defined in (4.1) obey a recurrence relation? Yes, but it is a bit more complicated: We have

$$s_{l+2,n}(\lambda) = (1 + \lambda)s_{l+1,n}(\lambda) - \lambda\left(1 - \frac{l}{n}\right)s_{l,n}(\lambda)\,. \qquad (5.15)$$

This relation follows since $s_{l+1,n}(\lambda)$ has the form

$$\frac{f(n, \lambda)}{2\pi i} \oint z^l e^{nh(z)}\, dz$$

where $f(n, \lambda)$ does not depend on l. Differentiating the integrand with respect to z yields a function with nothing but zero residues, hence

$$0 = \frac{f(n, \lambda)}{2\pi i} \oint z^l e^{nh(z)} \left(\frac{l}{z} + nh'(z)\right) dz$$

$$= l s_{l,n}(\lambda) + n\left(\frac{s_{l+1,n}(\lambda) - s_{l+2,n}(\lambda)}{\lambda} + s_{l+1,n}(\lambda) - s_{l,n}(\lambda)\right).$$

The recurrence (5.15) can be used to calculate $s_{l,n}(\lambda)$ "backwards," starting with the values

$$s_{n+1,n}(\lambda) = 0\,, \qquad s_{n,n}(\lambda) = \tfrac{1}{2}\,n!\,n^{-n}\lambda^{n-1}e^{-n\lambda/2}\,, \qquad (5.16)$$

and working down to $s_{1,n}(\lambda)$. This observation does not appear to lead to any simple consequences about the asymptotic behavior of $s_{l,n}(\lambda)$. We can, however, use (5.15) to prove by induction that the coefficients of the polynomials $\bigl(s_{l,n}(\lambda) - s_{l+1,n}(\lambda)\bigr)e^{n\lambda/2}$ are nonnegative. Furthermore, (5.15) implies the remarkable identity

$$\sum_{l \geq 1} l s_{l,n}(\lambda) = n\bigl(s_{1,n}(\lambda) - \lambda^{-1}s_{2,n}(\lambda)\bigr)\,, \qquad (5.17)$$

which can be used to study the variance of the cycle lengths.

6. The Average Cycle Length

We have seen in Section 3 how to set up a bivariate generating function $F(w, z)$ for a set of stopping configurations, thereby allowing us to compute the probability $\Phi_n F$ that such configurations occur in a graph of n vertices. But we can, of course, also use $\Phi_n F$ to compute expected values, if $F(w, z)$ is a bgf in which each stopping configuration has been multiplied by a weight representing the random variable in question.

For example, $T(wz)^l e^{U(wz)/w}$ is the bgf for stopping configurations with cycles of length l, hence

$$A(w, z) = \big(T(wz) + 2T(wz)^2 + 3T(wz)^3 + \cdots\big) e^{U(wz)/w}$$

$$= \frac{T(wz)}{\big(1 - T(wz)\big)^2} \, e^{U(wz)/w} \tag{6.1}$$

is a bgf such that $\Phi_n A$ is the expected cycle length in the uniform model. According to Theorem 1, this expected cycle length is

$$\int_0^\infty a_n(\lambda) \, d\lambda, \qquad a_n(\lambda) = \frac{n! \, n^{-n} e^{-n\lambda/2}}{4\pi i \lambda} \oint \frac{1}{1 - z} \, e^{nh(z)} \, dz, \tag{6.2}$$

where $h(z)$ is the familiar function of (4.2). Notice that we have

$$a_n(\lambda) = \sum_{l \geq 1} s_{l,n}(\lambda) = \sum_{l=1}^n s_{l,n}(\lambda). \tag{6.3}$$

Since $s_{l,n}(\lambda)$ is exponentially small for $\lambda \geq 1 + n^{-1/3} \ln n$, we need not consider large values of λ. However, the presence of $1 - z$ in the denominator of (6.2) means that values of $a_n(\lambda)$ near $\lambda = 1$ will be crucial.

A slight modification to the proof of Lemma 1 shows that the asymptotic formula

$$a_n(\lambda) = \frac{1}{2(1 - \lambda)^{3/2}} + O\Big(\frac{1}{n(1 - \lambda)^{9/2}}\Big) \tag{6.4}$$

holds uniformly for $0 \leq \lambda \leq 1 - n^{-1/2}(\ln n)^2$ as $n \to \infty$. If we integrate this quantity as λ varies from 0 to $1 - n^{-1/3} \ln n$, say, we get

$$\frac{n^{1/6}}{\sqrt{\ln n}} + O\Big(\frac{n^{1/6}}{(\ln n)^{7/2}}\Big); \tag{6.5}$$

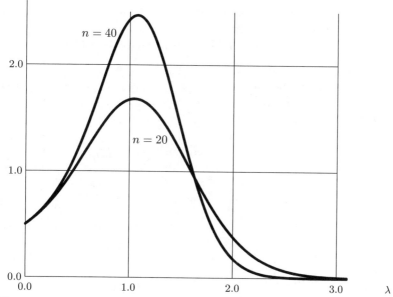

FIGURE 3. The weighted distribution functions $a_n(\lambda)$ for $n = 20$ and $n = 40$.

hence we may conclude that the value of $a_n(\lambda)$ is negligible except when $|\lambda - 1| \leq n^{-1/3} \ln n$, if we can show that the integral of $a_n(\lambda)$ over that range is of order $n^{1/6}$.

Figure 3 shows the behavior of $a_n(\lambda)$ for $n = 20$ and $n = 40$. As n increases, the function has sharper and sharper peaks, apparently reaching a maximum when λ is very slightly greater than 1.

The contour integral that arises when λ is near 1 is just like the integral we studied in Lemma 3, except that there is an additional factor $(1 - z)^{-1}$. If we set $z = e^{-s\nu}$ as in that lemma, we have

$$\frac{1}{1-z} = -\sum_{k \geq 0}(-s\nu)^{k-1}\frac{B_k}{k!} = \frac{n^{1/3}}{s} + O(1), \qquad (6.6)$$

uniformly in s, provided only that $|s| \leq 6n^{1/3}$, since the series converges for $|s| < 2\pi n^{1/3}$. Therefore the calculations of Lemma 3 can be applied almost without change, and we obtain

$$a_n\left(e^{-\mu n^{-1/3}}\right) = \frac{n^{1/2}}{\sqrt{8\pi}\,i}\int_{\Gamma_n^{(\alpha)}} e^{P(\mu,s)}\frac{ds}{s} + O\left((1+\mu^4)n^{1/6}\right), \qquad (6.7)$$

uniformly over all μ such that $|\mu| \leq n^{1/12}$. Once again α can be any positive constant.

Finally we can compute the asymptotic path length, proving the formula claimed in the introduction:

Theorem 3. *The expected length of the first cycle in an evolving graph is $Kn^{1/6} + O(n^{3/22})$ in the uniform model, and $e^{3/4}Kn^{1/6} + O(n^{3/22})$ in the permutation model, where*

$$K = \frac{1}{\sqrt{8\pi}\,i} \int_{-\infty}^{\infty} \int_{\Gamma} e^{(\mu+2s)(\mu-s)^2/6}\,\frac{ds}{s}\,d\mu \qquad (6.8)$$

and $\Gamma = \Gamma^{(1)}$ is the contour defined in (4.21).

Proof. Setting $\lambda = e^{-\mu n^{-1/3}}$, $\lambda_1 = e^{-n^{-3/11}}$, and $\lambda_2 = e^{+n^{-3/11}}$, we have

$$\int_{\lambda_1}^{\lambda_2} a_n(\lambda)\,d\lambda = \frac{n^{1/6}}{\sqrt{8\pi}\,i} \int_{\mu_2}^{\mu_1} e^{-\mu n^{-1/3}} \int_{\Gamma} e^{P(\mu,s)}\,\frac{ds}{s}\,d\mu + O(n^{3/22}), \quad (6.9)$$

where $\mu_1 = n^{2/33}$ and $\mu_2 = -n^{2/33}$. (These magic constants will be explained below.) When μ is between μ_2 and μ_1, the integrand factor $\exp(-\mu n^{-1/3})$ is $1 + O(n^{-3/11})$, so we can ignore it. Thus we obtain an integral whose integrand matches (6.8).

This integrand is exponentially small as $\mu \to -\infty$, and we will prove in the appendix that it is $O(\mu^{-3/2})$ as $\mu \to \infty$. Extending the integral from $-\infty$ to ∞, instead of from μ_2 to μ_1, therefore introduces an error of $n^{1/6}O(\mu_1^{-1/2}) = O(n^{3/22})$. To obtain the total expected length $\int_0^{\infty} a_n(\lambda)\,d\lambda$, we must add $\left(\int_0^{\lambda_1} + \int_{\lambda_2}^{\infty}\right) a_n(\lambda)\,d\lambda$; this gives a further error of $O(n^{3/22})$, by (6.4), so we have established the result claimed for the uniform model. (If we had chosen $\mu_1 = n^x$, these error estimates would have been $O(n^{1/6-x/2})$, while the error in (6.9) would have been $O(n^{5x-1/6})$; the value $x = \frac{2}{33}$ gives the best bounds.)

The permutation model requires an additional factor

$$\exp\left(\tfrac{1}{2}\lambda + \tfrac{1}{4}\lambda^2\right) = \exp\left(\tfrac{3}{4} + O\left(\mu n^{-1/3}\right)\right),$$

which is treated similarly. There also is a (negligible) factor $e^{-2s\nu}$ in the inner integral, because the numerator of the bgf in (6.1) must be changed from $T(wz)$ to $T(wz)^3$ in order to get the expected value of $l - 2$. $\quad \square$

7. Additional Statistics

To find the variance of the cycle length, we can compute

$$\int_0^\infty b_n(\lambda)\, d\lambda, \qquad B(w,z) = \frac{T(wz)}{\left(1 - T(wz)\right)^3}\, e^{U(wz)/w}, \qquad (7.1)$$

which is the expected value of $\frac{1}{2}l(l+1)$. We get $b_n(\lambda)$ from $a_n(\lambda)$ by essentially changing (6.6) to

$$\frac{n^{2/3}}{s^2} + O\!\left(sn^{1/3}\right). \qquad (7.2)$$

The net effect is to multiply the formula for $\int_0^\infty a_n(\lambda)\, d\lambda$ by $n^{1/3}$, and to change the constant of proportionality by replacing s by s^2 in the denominator of (6.8). Thus the expected value of $\frac{1}{2}l(l+1)$ is of order $n^{1/2}$; the standard deviation is therefore asymptotically proportional to $n^{1/4}$, somewhat greater than the mean.

In general, if we have a bgf of the form

$$C_{l,k}(w,z) = \frac{T(wz)^l}{\left(1 - T(wz)\right)^k}\, e^{U(wz)/w}, \qquad k > \frac{3}{2}, \qquad (7.3)$$

where l is fixed as $n \to \infty$, the resulting value of $\int_0^\infty c_{l,k,n}(\lambda)\, d\lambda$ will be of order $n^{(2k-3)/6}$, by the same argument.

Therefore we can grind out more facts by setting up appropriate bgfs. Let us introduce (temporarily) a trivariate generating function

$$D_l(\zeta, w, z) = \frac{T(\zeta wz)^l}{1 - T(\zeta wz)}\, e^{U(wz)/w}, \qquad (7.4)$$

in which the coefficient of $\zeta^j w^m z^n/n!$ is the number of stopping configurations with cycles of length $\geq l$ having m edges and n vertices, with j vertices in the cyclic component. If we take the partial derivative with respect to ζ and then set $\zeta = 1$, we get a bgf for the expected value of j, namely

$$D_l'(1, w, z) = \frac{T(wz)^{l+1}}{\left(1 - T(wz)\right)^2} \left(\frac{l}{T(wz)} + \frac{1}{1 - T(wz)}\right) e^{U(wz)/w}. \qquad (7.5)$$

(This follows from the well-known relation

$$T'(z) = \frac{T(z)}{z\left(1 - T(z)\right)}, \qquad (7.6)$$

a consequence of (2.4).) Another derivative gives the expected value of $j(j-1)$ and introduces another $\left(1 - T(wz)\right)^2$ in the denominator. Therefore (7.3) applies and we can state:

Corollary 1. *The expected size of the first cyclic component in an evolving graph on n vertices is asymptotically proportional to $n^{1/2}$, and the standard deviation is of order $n^{7/12}$.*

Proof. Take $l = 1$ in (7.4) for the uniform model and $l = 3$ for the permutation model; use $k = 3$ in (7.3) for the mean and $k = 5$ for the variance. $\square$

A similar derivation, with $T(\zeta wz)T(wz)^l e^{U(wz)/w} / \big(1 - T(wz)\big)$ in place of (7.4), shows that the expected number of vertices in the tree that leads into the first vertex x_m of the first cycle is the same as the expected length of that cycle. The same holds for any individual tree in the cyclic component. Thus the cyclic component consists of $\Theta(n^{1/6})$ trees, on the average, each of which has $\Theta(n^{1/6})$ vertices, on the average; a dependency between these two statistics causes the overall expected size to be $\Theta(n^{1/2})$.

We can find the limiting distribution of the number of vertices in the first cyclic component by considering the coefficient of ζ^j in (7.4). Indeed, we have

$$D_1(\zeta, w, z) = \sum_{j \geq 1} \zeta^j \frac{j^j}{j!} (wz)^j e^{U(wz)/w} \, ;$$

and we can write this as a function of w and $T(wz)$ by using identity (2.4), which says that $wz = T(wz)e^{-T(wz)}$. Our general method now tells us to evaluate the integral

$$\oint (ze^{-z})^j \frac{1-z}{z} e^{nh(z)} \, dz$$

asymptotically as $n \to \infty$. We find as before that the only relevant contributions occur when $\lambda < 1$, and an argument like that of Theorem 2 shows that a proper probability distribution appears in the limit:

Corollary 2. *For fixed j as $n \to \infty$, the random graph evolution procedure generates a first cyclic component of size j with probability*

$$\theta_{j,n} = \frac{1}{2} \frac{j^j}{j!} \int_0^1 \lambda^{j-1} e^{-j\lambda} \sqrt{1 - \lambda} \, d\lambda + O(n^{-1/6}), \quad j \geq 1, \qquad (7.7)$$

in the uniform model, and with probability

$$\hat{\theta}_{j,n} = \frac{1}{2} \frac{j^{j-2}(j-1)(j-2)}{j!} \int_0^1 \lambda^{j-1} e^{-j\lambda + \lambda/2 + \lambda^2/4} \sqrt{1 - \lambda} \, d\lambda$$

$$+ O(n^{-1/6}), \quad j \geq 3, \qquad (7.8)$$

in the permutation model. $\square$

These limiting probabilities q_j and $\hat{q}_j$ sum to 1. We have, for example,

$$\sum_{j \geq 1} q_j = \frac{1}{2} \sum_{j \geq 1} \frac{j^j}{j!} \int_0^1 \lambda^{j-1} e^{-j\lambda} \sqrt{1-\lambda}\, d\lambda$$

$$= \frac{1}{2} \int_0^1 \lambda^{-1} \left(\sum_{j \geq 1} \frac{(j\lambda e^{-\lambda})^j}{j!} \right) \sqrt{1-\lambda}\, d\lambda$$

$$= \frac{1}{2} \int_0^1 \lambda^{-1} \frac{T(\lambda e^{-\lambda})}{1 - T(\lambda e^{-\lambda})} \sqrt{1-\lambda}\, d\lambda = \frac{1}{2} \int_0^1 (1-\lambda)^{-1/2}\, d\lambda = 1.$$

Both q_j and $\hat{q}_j$ are of order $j^{-5/4}$ as j grows; indeed, the substitution $\lambda = 1 - \sqrt{2t/n}$ shows that $q_j = cj^{-5/4} + O(j^{-7/4})$ and $\hat{q}_j = e^{3/4} cj^{-5/4} + O(j^{-7/4})$, where $c = 2^{-7/4} \pi^{-1/2} \Gamma(\frac{3}{4})$. We have seen in Corollary 1 that the expected value of the component size is unbounded. Here is a table of approximate probabilities when j is small:

$$\begin{array}{ll}
q_1 \approx .23096\,; & \hat{q}_3 \approx .01804\,; \\
q_2 \approx .09501\,; & \hat{q}_4 \approx .02181\,; \\
q_3 \approx .05649\,; & \hat{q}_5 \approx .02153\,; \\
q_4 \approx .03909\,; & \hat{q}_6 \approx .02015\,; \\
q_{10} \approx .01214\,; & \hat{q}_{10} \approx .01436\,; \\
q_{20} \approx .00504\,; & \hat{q}_{20} \approx .00754\,.
\end{array}$$

The value of q_1 is $\frac{1}{2} + \frac{1}{4} e^{-1} \sqrt{\pi}\, i\, \mathrm{erf}(i)$, according to MACSYMA.

To get the expected value of m, the number of edges, we can use the fact that $n - m$ is the number of acyclic components. The relevant trivariate generating function is

$$E_l(\zeta, w, z) = \frac{T(wz)^l}{1 - T(wz)}\, e^{\zeta U(wz)/w}, \tag{7.9}$$

and we have

$$E_l'(1, w, z) = \frac{U(wz)}{w}\, E_l(1, w, z)$$

$$= \frac{1}{w} \frac{T(wz)^{l+1} - T(wz)^{l+2}/2}{1 - T(wz)}\, e^{U(wz)/w}. \tag{7.10}$$

The factor w^{-1} contributes a factor of n/λ to the corresponding function $e_{l,n}(\lambda)$, according to (3.13), hence we have

$$e_{l,n}(\lambda) = \frac{n}{\lambda} \left(s_{l+1,n}(\lambda) - \frac{s_{l+2,n}(\lambda)}{2} \right). \tag{7.11}$$

The integral $\int_1^\infty s_{l,n}(\lambda)\, d\lambda/\lambda$ is of order $n^{-1/6}$, by the results of Section 4, and we have in fact

$$\int_0^\infty s_{l,n}(\lambda)\,\frac{d\lambda}{\lambda} = \frac{1}{2}\int_0^1 \frac{\lambda^{l-2}}{\sqrt{1-\lambda}}\, d\lambda + O(n^{-1/6}). \qquad (7.12)$$

Therefore the waiting time has a simple relation to cycle length probabilities:

Corollary 3. *The expected number of edges when an evolving graph obtains its first cycle is* $\frac{1}{3}n + O(n^{5/6})$, *in the uniform model. It is* $\frac{1}{2}(1 - \hat{p}_3)n + O(n^{5/6})$ *in the permutation model, where* $\hat{p}_3$ *is the constant in* (5.13).

Proof. Take $l = 1$ and $l = 3$ in (7.11), getting $n(P_{\geq 1,n} - \frac{1}{2}P_{\geq 2,n})$ or $n(\widehat{P}_{\geq 3,n} - \frac{1}{2}\widehat{P}_{\geq 4,n}) + O(n^{5/6})$ as the expected values of $n - m$. $\square$

The variance can be shown, similarly, to have the respective values

$$\left(\frac{1}{45}\,n^2\,,\ \frac{\hat{p}_3 - \hat{p}_3^2 - \hat{p}_4}{4}\, n^2\right) + O(n^{11/6}). \qquad (7.13)$$

We will examine another way to compute the expected waiting time in Section 10 below.

Finally, let us investigate the number of vertices that remain isolated when the first cycle appears. The relevant trivariate generating function is

$$F_l(\zeta, w, z) = \frac{T(w,z)^l}{1 - T(w,z)}\, e^{(U(wz) - wz + \zeta wz)/w}, \qquad (7.14)$$

since we put the ζ marker on the unordered components of size 1. In this case we find

$$F_l'(1, w, z) = z S_l(w, z) = w^{-1}\, T(wz)\, e^{-T(wz)} S_l(w, z). \qquad (7.15)$$

Corollary 4. *The expected number of isolated vertices when the first cycle appears in an evolving graph is*

$$\frac{n}{2}\int_0^1 \frac{e^{-\lambda}\, d\lambda}{\sqrt{1-\lambda}} + O(n^{5/6}) \qquad (7.16)$$

in the uniform model, and

$$\frac{n}{2}\int_0^1 \frac{\lambda^2 e^{-\lambda/2 + \lambda^2/4}}{\sqrt{1-\lambda}}\, d\lambda + O(n^{5/6}) \qquad (7.17)$$

in the permutation model. $\square$

MACSYMA finds the integral in (7.16) to be $-e^{-1}\sqrt{\pi}\, i\operatorname{erf}(i)$; the coefficient of n is therefore $1 - 2q_1 \approx 0.53808$. The corresponding coefficient in (7.17) is ≈ 0.42046.

8. Cycle Lengths Versus Edges

Let us now try to study the joint distribution of l and m, the cycle length and the number of edges when the evolution procedure of Section 1 is applied to n initially disconnected vertices. The corresponding probabilities will be called $P_{l,m,n}$ in the uniform model and $\widehat{P}_{l,m,n}$ in the permutation model.

We can express these probabilities directly from univariate generating functions, instead of using the more elaborate machinery of Theorem 1. Let $C_{l,m,n}$ be the number of stopping configurations in which the process can stop with a cycle of length l and with m edges on n vertices. Then there are $n - m$ components in the acyclic part, and we have

$$C_{l,m,n} = \frac{n!}{(n-m)!} \, [z^n] \, T(z)^l \, U(z)^{n-m} . \tag{8.1}$$

These numbers, incidentally, satisfy the recurrences

$$C_{l+1,m,n} = \sum_{k \geq 1} \binom{n}{k} k^{k-1} C_{l,m-k,n-k} ; \tag{8.2}$$

$$(n - m)C_{l,m,n} = \sum_{k \geq 1} \binom{n}{k} k^{k-2} C_{l,m+1-k,n-k} . \tag{8.3}$$

The corresponding probabilities, as we have seen in Section 3, are

$$P_{l,m,n} = \frac{2^{m-1}(m-1)!}{n^{2m}} \, C_{l,m,n} ; \tag{8.4}$$

$$\widehat{P}_{l,m,n} = \frac{1}{2m\binom{(n^2-n)/2}{m}} \, C_{l,m,n} = e^{m/n + m^2/n^2} \, P_{l,m,n} + O(n^{-1}). \tag{8.5}$$

Let us set $\lambda = 2m/n$. Erdős and Rényi [7] observed that an evolving graph on n vertices changes its character dramatically when m grows so that λ passes the critical value $\lambda = 1$. It turns out that, for sub-critical graphs ($\lambda < 1$), the quantity $P_{\geq l,m,n}$ behaves very much like the function $s_{l,n}(\lambda)$ in Lemma 1, except for a factor $2/n$ (which corresponds to $d\lambda$):

Theorem 4. *If $2m/n = \lambda < 1$ as $n \to \infty$, where $\delta \leq \lambda \leq 1 - \delta$, we have*

$$P_{l,m,n} = \frac{\lambda^{l-1}\sqrt{1-\lambda}}{n} \left(1 + O\left(\frac{l^2}{n\delta}\right) + O\left(\frac{1}{n^{1/2}\delta^{3/2}}\right)\right), \tag{8.6}$$

uniformly in $\delta > 0$ and $l \geq 1$.

Proof. We will apply the saddle point method to estimate the coefficient of z^n in $T(z)^l U(z)^{n-m}/(1-T(z))$, thereby obtaining an asymptotic value of $C_{\geq l,m,n}$. Again we replace z by ze^{-z} in order to obtain a simpler integral:

$$\frac{1}{2\pi i} \oint \frac{T(z)^l U(z)^{n-m}\, dz}{(1 - T(z))\, z^{n+1}} = \frac{1}{2\pi i} \oint z^l (z - z^2/2)^{n-m} e^{nz}\, \frac{dz}{z^{n+1}}$$

$$= \frac{1}{2\pi i} \oint z^l e^{nh(z)}\, \frac{dz}{z}, \qquad (8.7)$$

where in this case we have

$$h(z) = z - \frac{\lambda}{2} \ln z + \left(1 - \tfrac{1}{2}\lambda\right) \ln\left(1 - \tfrac{1}{2}z\right), \qquad (8.8)$$

$$h'(z) = 1 - \frac{\lambda}{2z} - \frac{2 - \lambda}{4 - 2z}, \qquad (8.9)$$

$$h''(z) = \frac{1}{2}\left(\frac{\lambda}{z^2} - \frac{2 - \lambda}{(2 - z)^2}\right). \qquad (8.10)$$

There are two saddle points, at $z = \lambda$ and $z = 1$, just as we observed for a different function $h(z)$ in Section 4. (Is there an "obvious" reason why this should be true?) Again we have $h''(\lambda) > 0$ and $h''(1) < 0$, so we want to integrate on the circular path $|z| = \lambda$.

The real part of $h(\lambda e^{i\theta})$ is now

$$\lambda \cos\theta - \tfrac{1}{2}\lambda \ln \lambda + \tfrac{1}{2}\left(1 - \tfrac{1}{2}\lambda\right) \ln L, \quad L = 1 + \tfrac{1}{4}\lambda^2 - \lambda \cos\theta, \quad (8.11)$$

and its second derivative is

$$\frac{\lambda}{4}\left(\cos\theta\left(\frac{2 - \lambda}{L} - 4\right) - \frac{\lambda(2 - \lambda)\sin^2\theta}{L^2}\right).$$

This quantity is negative when $\cos\theta \geq 0$, because

$$\frac{2 - \lambda}{L} - 4 \leq \frac{2 - \lambda}{(1 - \lambda/2)^2} - 4 = -4\left(\frac{1 - \lambda}{2 - \lambda}\right).$$

Furthermore (8.11) is less than

$$\Re h(\lambda) - \lambda - \left(1 - \tfrac{1}{2}\lambda\right)\left(\ln(1 - \tfrac{1}{2}\lambda) - \ln(1 + \tfrac{1}{2}\lambda)\right) < \Re h(\lambda) - \tfrac{1}{3}\lambda^2$$

when $\cos\theta < 0$. Therefore we can restrict attention once again to the neighborhood of $\theta = 0$, and the result is

$$C_{\geq l,m,n} = \frac{n!}{(n-m)!} \frac{1}{2\pi i} \oint z^l e^{nh(z)} \frac{dz}{z}$$

$$= \frac{n!}{(n-m)!} \frac{\lambda^l e^{nh(\lambda)}}{\sqrt{2\pi n(1-\lambda)}} \sqrt{\frac{n-m}{m}}$$

$$\times \left(1 + O\left(\frac{l^2}{n\lambda(1-\lambda)}\right) + O\left(\frac{1}{n^{1/2}\lambda^{3/2}(1-\lambda)^{3/2}}\right)\right)$$

$$= \frac{\lambda^l}{\sqrt{1-\lambda}} \frac{n! \, e^n}{\sqrt{2\pi n} \, n^{n-2m}} \frac{\sqrt{2\pi(n-m)} \, (n-m)^{n-m}}{(n-m)!} \frac{e^m \, 2^{-m}}{e^{n-m}} \frac{e^m \, 2^{-m}}{\sqrt{2\pi m} \, m^m}$$

$$\times \left(1 + O(l^2 n^{-1}\delta^{-1}) + O(n^{-1/2}\delta^{-3/2})\right)$$

$$= \frac{\lambda^l}{\sqrt{1-\lambda}} \frac{n^{2m}}{m! \, 2^m} \left(1 + O(l^2 n^{-1}\delta^{-1}) + O(n^{-1/2}\delta^{-3/2})\right). \quad (8.12)$$

Now we can use (8.4) to conclude that

$$P_{\geq l,m,n} \sim \frac{\lambda^{l-1}}{\sqrt{1-\lambda} \, n}, \quad (8.13)$$

as desired.   □

Theorem 4 gives us the promised "physical" interpretation of the parameter λ in the machinery of Theorem 1: The running time m of the random process is represented by $\frac{1}{2}\lambda n$, at least when $\lambda < 1$. Thus, Figure 2 shows the approximate distribution of running times in the uniform model, when $n = 20$ and $n = 40$. A similar statement holds for the permutation model; but in that case we should consider the graph of $s_{3,n}(\lambda)e^{\lambda/2+\lambda^2/4}$ instead of $s_{1,n}(\lambda)$, because of (3.14) and (8.5).

It is interesting to note that, for fixed ratio $\lambda = 2m/n < 1$ and for varying $l \ll \sqrt{n}$, the distribution of cycle lengths over all graphs whose first cycle occurs at time m is approximately geometric in l, with parameter λ, except for a normalization factor.

If we set $l = 3$ in (8.13) and apply (8.5), we get

$$\Pr(\text{Graph with } n \text{ vertices and } m \text{ edges has no cycle})$$

$$= \sum_{\mu > m} \widehat{P}_{\geq 3,\mu,n} \sim \frac{1}{2} \int_\lambda^1 \frac{\lambda^2}{\sqrt{1-\lambda}} e^{\lambda/2+\lambda^2/4} \, d\lambda$$

$$= \sqrt{1-\lambda} \, e^{\lambda/2+\lambda^2/4}, \quad (8.14)$$

if $\lim_{n\to\infty} 2m/n = \lambda < 1$; this formula is a classic result of Erdős and Rényi [7, Theorem 5b].

The situation changes when $m > \frac{1}{2}n$; in this "supercritical" case the ratio $2m/n$ no longer represents the parameter λ in Theorem 1 and Figure 2. (We might expect the relationship to break down when λ is large, because the evolution process always stops with $m \le n$; the λ of Theorem 1 and Figure 2 is a continuous parameter that defines a positive but exponentially small function as $\lambda \to \infty$.) We can use the method of Theorem 3 when $\lambda > 1$, integrating on the circle $|z| = 1$, to deduce that

$$P_{\ge l,m,n} \sim \frac{e^{-n(\lambda-1)}}{\lambda n\sqrt{\lambda-1}} \frac{\lambda^{m+1/2}}{(2-\lambda)^{n-m+1/2}}, \qquad 1+\delta \le \lambda \le 2-\delta. \quad (8.15)$$

(Compare with (4.18).) The probability $P_{l,m,n}$ is obtained if we insert the factor $(1-z)$ into the contour integrand; this introduces the factor $\frac{1}{2}\theta^2$ at the saddle point $\theta = 0$, and the result is

$$P_{l,m,n} \sim \frac{1}{2n(\lambda-1)} P_{\ge l,m,n} = \frac{1}{4m-2n} P_{\ge l,m,n}, \quad 1+\delta \le \lambda \le 2-\delta.$$
$$(8.16)$$

The method of Theorem 1 seems preferable to working directly with the actual probabilities $P_{l,m,n}$ for $m \ge \frac{1}{2} n$, because $s_{l,n}(\lambda)$ is a "smooth" function of λ by which we can use uniform methods like Lemma 3 to span the critical region near $\lambda = 1$.

9. Bicyclic Components

Let's turn now to a related problem that can be handled with similar techniques. Instead of stopping the random graph or multigraph process when a cycle appears, let's keep it running until the first time there is a *bicyclic component*—a component with more than one cycle. If the first such component contains j vertices it will have $j + 1$ edges. The solution to this problem sheds more light on the generating-function-based techniques we have been discussing.

As before, we begin by defining and enumerating all of the stopping configurations in which our random process might terminate. The first bicyclic component can arise in one of two ways: Either (1) the final edge lies entirely within a component that was already unicyclic (a component that already contained a cycle), or (2) the final edge joins two different unicyclic components.

Our experiences so far suggest that we ought to look first at the uniform model, in which each step selects from n^2 ordered pairs $\langle x, y \rangle$

at random, since the uniform model tends to give formulas that are less frightening than the ones that arise in the permutation model.

The generating function for unicyclic components on n labeled vertices turns out to be

$$V(z) = \frac{1}{2} \ln \frac{1}{1 - T(z)} = \frac{T(z)}{2} + \frac{T(z)^2}{4} + \frac{T(z)^3}{6} + \cdots . \qquad (9.1)$$

Here's why: Every cycle of length $l \geq 3$ corresponds to $2l$ sequences of l rooted trees, because we can list the trees of the cycle by starting at l different places and we can traverse the cycle in two directions. Cycles of length $l < 3$ have the forms $\langle x, x \rangle$ or $\langle x, y \rangle\langle x, y \rangle$; we will want to divide by $2l$ in these cases also, because of the weighting function $2^{m-1}(m-1)!$ that will be applied later. (This weighting function assumes that a given multiset of m edges containing no bicyclic components can arise in $2^{m-1}(m-1)!$ ways as a sequence $\langle x_1, y_1 \rangle \ldots \langle x_{m-1}, y_{m-1} \rangle$ of ordered pairs; but the actual number of ways is $2^{m-1-k}(m-1)!$, where k is the number of 1-cycles and 2-cycles, so we want to introduce a factor of $\frac{1}{2}$ for every such cycle.)

In case (1) the stopping configuration consists of a unicyclic component together with two special vertices $\langle x, y \rangle$ in that component, plus a set of any number of additional acyclic or unicyclic components. In case (2) the stopping configuration consists of an ordered pair of unicyclic components together with a vertex x in the first and a vertex y in the second, plus a set of additional acyclic or unicyclic components as before.

Let $\vartheta = z\frac{d}{dz}$ be the operator that multiplies the coefficient of z^n by n. Then the egf for stopping configurations in case (1) is $(\vartheta^2 V(z))$ $\exp(U(z) + V(z))$, and in case (2) it is $(\vartheta V(z))^2 \exp(U(z) + V(z))$. (The operator ϑ selects a vertex, and $U(z) + V(z)$ enumerates acyclic and unicyclic components.) Using (7.6) we have

$$\vartheta V(z) = \tfrac{1}{2}T(z)/(1 - T(z))^2 ; \qquad (9.2)$$

$$\vartheta^2 V(z) = \tfrac{1}{2}T(z)(1 + T(z))/(1 - T(z))^4 . \qquad (9.3)$$

Therefore the overall egf for stopping configurations comes to

$$\left(\vartheta^2 V(z) + (\vartheta V(z))^2\right) e^{U(z)+V(z)} = \frac{T(z)(2 + 3T(z))}{4(1 - T(z))^{9/2}} e^{U(z)} ; \qquad (9.4)$$

this is only slightly more complex than formula (2.7), the analogous egf for stopping configurations in the first cycle problem.

Once again we need to work with bgf's, so that we have access to the number of edges. The appropriate bivariate generating function for stopping configurations in the uniform model is easily deduced from our derivation of (9.4): We have

$$S(w, z) = \frac{wT(wz)\big(2 + 3T(wz)\big)}{4\big(1 - T(wz)\big)^{9/2}} e^{U(wz)/w} . \tag{9.5}$$

And as in Section 3, we can state that $S(w, z)$ expands to the sum $\sum_{m,n} s_{m,n} w^m z^n/n!$, where $2^{m-1}(m-1)! \, s_{m,n}/n^{2m}$ is the probability that the process stops when the mth edge is introduced.

As a check, let's look at the coefficients for small n:

$$S(w, z) = \frac{w^2}{2} z + \frac{w^2 + 7w^3}{2} z^2 + \frac{4w^2 + 60w^3 + 261w^4}{16} z^3 + \cdots .$$

When $n = 3$, the respective probabilities that we stop at time $m = 2, 3, 4$ are

$$\frac{2^1 \cdot 1! \cdot 3! \cdot 4}{3^4 \cdot 16} = \frac{1}{27} , \qquad \frac{2^2 \cdot 2! \cdot 3! \cdot 60}{3^6 \cdot 16} = \frac{20}{81} , \qquad \frac{2^3 \cdot 3! \cdot 3! \cdot 261}{3^8 \cdot 16} = \frac{58}{81} ,$$

and these sum to 1 as they should. In general, we have

$$\Phi_n S = 1 , \qquad \text{for all } n \geq 1; \tag{9.6}$$

the operator Φ_n of Section 3 applies to the bicyclic problem as well as to the unicyclic problem, and we can use the simplifications of Theorem 1 just as we did before.

Now let's turn to the permutation model, in which cycles of lengths 1 and 2 are forbidden. The appropriate egf for cycles is therefore

$$\widehat{V}(z) = V(z) - \frac{T(z)}{2} - \frac{T(z)^2}{4} = \frac{1}{2} \ln \frac{1}{1-T(z)} - \frac{T(z)}{2} - \frac{T(z)^2}{4} , \tag{9.7}$$

a formula noted by Wright [15]. The egf for stopping configurations in case (2) is $\big(\vartheta \widehat{V}(z)\big)^2 \exp\big(U(z) + \widehat{V}(z)\big)$, because we choose x and y in distinct components as before. But in case (1) the number of ordered pairs $\langle x, y \rangle$ is twice the number of edges not already present in the unicyclic component, so the appropriate egf for case (1) is

$$(\vartheta^2 - 3\vartheta)\widehat{V}(z) \exp\big(U(z) + \widehat{V}(z)\big) .$$

Adding these cases together and introducing w as before gives us the bgf for stopping configurations in the permutation model:

$$\hat{S}(w, z) = w \frac{T(wz)^4 \left(10 - 6T(wz) + T(wz)^2\right)}{4\left(1 - T(wz)\right)^{9/2}}$$

$$\times \exp\left(\frac{U(wz)}{w} - \frac{T(wz)}{2} - \frac{T(wz)^2}{4}\right). \qquad (9.8)$$

If we write $\hat{S}(w, z) = \sum_{m,n} \hat{s}_{m,n} w^m z^n / n!$, then $\hat{s}_{m,n} / \left(2m\binom{n(n-1)/2}{m}\right)$ is the probability that the first bicyclic component arises in the permutation model when the mth edge appears. The coefficients for small n are

$$\hat{S}(w, z) = \frac{5w^5}{2} z^4 + \frac{5w^5 + 37w^6}{2} z^5 + \frac{5w^5 + 79w^6 + 367w^7}{4} z^6 + \cdots;$$

thus when $n = 6$, the process stops at time $m = 5, 6, 7$ with respective probabilities $\frac{30}{1001}, \frac{237}{1001}, \frac{734}{1001}$. We have

$$\Psi_n \hat{S} = 1, \qquad \text{for all } n \geq 4, \qquad (9.9)$$

where Ψ_n is the operator of Section 3. Notice that the coefficient of z^3 in $\hat{S}(w, z)$ is zero; a graph on 3 vertices never has more than one cycle, so we shouldn't look for bicyclic components in the permutation model unless $n \geq 4$. But when $n \geq 4$, we obtain a bicyclic component after at most $n + 1$ edges have been added.

What is the size of the first bicyclic component? In the uniform model, the generating function

$$\frac{w\,T(\zeta wz)\left(2 + 3T(\zeta wz)\right)}{4\left(1 - T(\zeta wz)\right)^4} \frac{e^{U(wz)/w}}{\left(1 - T(wz)\right)^{1/2}} \qquad (9.10)$$

puts ζ^j into each stopping configuration whose bicyclic component contains j vertices. After differentiating with respect to ζ and setting $\zeta = 1$, we obtain an expression for the expected bicyclic component size:

$$\Phi_n \frac{2wT(wz)\left(1 + 6T(wz) + 3T(wz)^2\right)}{\left(1 - T(wz)\right)^{13/2}} e^{U(wz)/w}. \qquad (9.11)$$

A similar formula, with the same denominator $\left(1 - T(wz)\right)^{13/2}$, applies to the permutation model. If the factor w weren't present, we would have a generating function of the form (7.3), with $k = \frac{13}{2}$; the Φ_n operator would then produce a result of order $n^{(2k-3)/6} = n^{5/3}$. The factor w changes the integrand by λ/n, and $\lambda \approx 1$ in the region where the integral becomes unbounded; hence the w essentially divides by n, and we can state the following result:

Corollary 5. *The expected size of the first bicyclic component in an evolving graph is of order $n^{2/3}$. The standard deviation is also of order $n^{2/3}$.* □

Corollary 6. *The expected number of cycles in unicyclic components, at the moment when the first bicyclic component appears, is $\frac{1}{6}\ln n + O(1)$. The expected total length of these cycles is proportional to $n^{1/3}$.*

Proof. For the first result, replace $V(wz)$ in the exponent of $S(w,z)$ by $\zeta V(wz)$; for the second, replace $(1 - T(wz))^{1/2}$ in the denominator by $(1 - \zeta T(wz))^{1/2}$. Then in both cases, differentiate with respect to ζ, and set $\zeta = 1$.   □

We can find the expected waiting time m by using the trick (7.9) that led to Corollary 3. In this case $n + 1 - m$ is the number of acyclic components, so the expected value of $n + 1 - m$ is

$$\Phi_n\left(\frac{U(wz)}{w}S(w,z)\right) \quad \text{or} \quad \Psi_n\left(\frac{U(wz)}{w}\hat{S}(w,z)\right), \qquad (9.12)$$

depending on the model. In both cases the multiplication by $U(wz) = T(wz) - T(wz)^2/2$ yields a numerator polynomial in $T(wz)$ whose value $\mod(1 - T(wz))$ is half what it was before. Since $\Phi_n S(w,z) = \Psi_n \hat{S}(w,z) = 1$, and since division by w contributes a factor of n, the waiting time must be asymptotically $\frac{1}{2}n$.

10. Waiting Times Revisited

When our goal is to find the average value of m, we can use another method based on generating functions for the "going configurations" instead of the stopping configurations. Namely, if $f_{m,n}$ is the number of graphs with n vertices and m edges such that the random process is *not* stopped, we can use this information to calculate the probability that the process is still going after m steps. The sum of these probabilities, over all m, is the expected waiting time.

In the first cycle problem, a going configuration is simply a forest (a collection of edges with no cycles); hence the bgf for going configurations is just

$$F(w,z) = \sum_{m,n} f_{m,n}w^m\frac{z^n}{n!} = e^{U(wz)/w}. \qquad (10.1)$$

Each going configuration occurs with probability $2^m m!/n^{2m}$ in the uniform model, so the expected waiting time for a graph with n vertices is $\sum_m 2^m m! \, f_{m,n}/n^{2m}$. The operator Φ_n of Section 3 computes

$\sum_m 2^{m-1}(m-1)! f_{m,n}/n^{2m}$, so it's almost what we want. We can obtain the desired operator for going configurations by first multiplying by w, getting the bgf $\sum_n f_{m,n}w^{m+1}z^n/n!$; then we apply Φ_n to get $\sum_m 2^m m! f_{m,n}/n^{2m+2}$; and finally we multiply by n^2. In other words, the expected waiting time in the uniform model is

$$n^2\Phi_n w F(w,z). \tag{10.2}$$

Alternatively, we can obtain the desired operator by first differentiating with respect to w (getting $\sum_m mf_{m,n}w^{m-1}z^n/n!$), then multiplying by $2w$ and applying Φ_n. In other words,

$$2\Phi_n w \frac{\partial}{\partial w}F(w,z) + f_{0,n} \tag{10.3}$$

gives the same result as (10.2). (We must add in the term $f_{0,n}$, which is annihilated by differentiation.) Indeed, we have the operator identity

$$n^2\Phi_n w = 2\Phi_n w \frac{\partial}{\partial w}, \tag{10.4}$$

valid when applied to any bgf with $F(0,z) = 0$. Since $w\frac{\partial}{\partial w} = \frac{\partial}{\partial w}w - 1$, we can rewrite (10.4) as follows:

$$\frac{n^2}{2}\Phi_n = \Phi_n\left(\frac{\partial}{\partial w} - w^{-1}\right). \tag{10.5}$$

Comparing (10.3) with our formula (7.10) for the average of $n - m$ yields the interesting identity

$$\Phi_n\left(\left(\frac{1}{1-T} - 1 - T + 3T^2\right)\frac{e^{U(wz)/w}}{2w}\right) = n-1, \quad T \equiv T(wz), \tag{10.6}$$

which does not obviously follow from (10.5) and any other identities that we know. It may be possible to find a family of formulas such as this, allowing us to deduce nonobvious relations between different statistics on random graphs.

In the permutation model, the relevant formula for expected waiting time is

$$2\Psi_n w \frac{\partial}{\partial w}F(w,z) + f_{0,n} \tag{10.7}$$

as in (10.3). There is apparently no simple analog of (10.2), although we can derive a formula that is somewhat like (10.5):

$$\left(\frac{n(n-1)}{2} + 1\right)\Psi_n = \Psi_n\left((1+w)\frac{\partial}{\partial w} - w^{-1}\right). \tag{10.8}$$

The identity analogous to (10.6) is

$$\Psi_n\left(\left(\frac{1}{1-T} - 1 - T + T^2 - T^3 + T^4\right)\frac{e^{U(wz)/w}}{2w}\right) = n - 1, \qquad (10.9)$$

valid for $n \geq 3$.

The bgf for going configurations in the problem of bicyclic components is

$$e^{U(wz)/w + V(wz)} \qquad \text{or} \qquad e^{U(wz)/w + \widehat{V}(wz)}, \qquad (10.10)$$

depending on the model, because the process keeps going if and only if the graph components are acyclic or unicyclic. The formulas in (9.12) now lead, via (10.3) and (10.7), to further identities like (10.6) and (10.9):

$$\Phi_n\left(\left(\frac{T^2}{w} - \frac{3}{8} + \frac{3 - 4T + 6T^2}{8(1-T)^4}\right)\frac{e^{U(wz)/w}}{\sqrt{1-T}}\right) = n; \qquad (10.11)$$

$$\Psi_n\left(\left(\frac{T^2}{w} + \frac{1+4T^3+T^4}{8} + \frac{-1+4T-6T^2+8T^3}{8(1-T)^4}\right)\right.$$
$$\left.\times\frac{e^{U(wz)/w - T/2 - T^2/4}}{\sqrt{1-T}}\right) = n. \qquad (10.12)$$

(Again T stands for $T(wz)$, and the identity for Ψ_n holds only when $n \geq 3$.) Is there a simple combinatorial or algebraic principle that accounts for amazing formulas like this?

We have observed in Section 9 that the waiting time for the first bicyclic component is approximately $\frac{1}{2}n$; thus, the graph tends to become bicyclic when m passes the critical value where random graphs rapidly gain a complex structure. It is interesting to look more closely at this transitional phase, by studying the probability that there is not yet a bicyclic component when $m \approx \frac{1}{2}n$. For this purpose we can combine the ideas used to prove Lemma 3 in Section 4 and Theorem 4 in Section 8.

Theorem 5. *Let $\lambda = 2m/n = e^{-\mu\nu}$, where $\nu = n^{-1/3}$. Then the probability that a random graph with n vertices and m edges has no bicyclic component is*

$$\pi_{m,n} = \frac{1}{\sqrt{2\pi i}}\int_\Gamma s^{1/2}e^{(\mu+2s)(\mu-s)^2/6}\,ds + O\left(\frac{1+\mu^4}{n^{1/3}}\right), \qquad (10.13)$$

uniformly for $|\mu| \leq n^{1/12}$, where $\Gamma = \Gamma^{(1)}$ is the contour defined in (4.21).

Proof.   We have

$$\pi_{m,n} = \frac{n!}{(n-m)!\binom{N}{m}}\,[z^n]\,U(z)^{n-m}e^{\widehat{V}(z)}\,, \tag{10.14}$$

where $\widehat{V}(z)$ is defined in (9.7) and $N = \binom{n}{2}$. Let $h(z)$ be the function defined in (8.8); then, as in that derivation,

$$[z^n]\,U(z)^{n-m}e^{\widehat{V}(z)} = \frac{1}{2\pi i}\oint \sqrt{1-z}\,e^{-z/2-z^2/4}\,e^{nh(z)}\,\frac{dz}{z}. \tag{10.15}$$

Let $z = e^{-s\nu}$. A tedious but straightforward calculation shows that (10.15) equals

$$\frac{n^{-1/2}e^{-3/4-\mu^3/6+n}}{2\pi\,2^{n-m}\,i}\left(\int_{\Gamma} s^{1/2}e^{P(\mu,s)}\,ds + O\big((1+\mu^4)\nu\big)\right), \tag{10.16}$$

if we argue as in Lemma 3. Here are the new details: We rewrite (8.8) in the form

$$h(z) = z - \frac{n-m}{n}\ln 2 - \ln z - \left(1-\frac{\lambda}{2}\right)\ln\frac{1}{1-(z-1)^2}$$

and analyze $h(e^{-s\nu})$ by using the uniform estimates $e^{-\mu\nu} = 1 - \mu\nu + O(\mu^2\nu^2)$ and $\ln\big(1-(e^{-s\nu}-1)^2\big) = -s^2\nu^2 + s^3\nu^3 + O(s^4\nu^4)$. To show that the integral over C_n is negligibly small for this new integrand, we note that for $\sqrt{3}\beta \le \theta \le \pi n^{1/3}$ the derivative of the real part of $h(e^{-\beta\nu+i\theta\nu})$ is

$$\nu r\sin\theta\nu\left(-1+\frac{\frac{1}{2}(1-\frac{1}{2}\lambda)}{1-r\cos\theta\nu+\frac{1}{4}r^2}\right) \le \nu r\sin\theta\nu\left(-1+\frac{\frac{1}{2}(1-\frac{1}{2}\lambda)}{(1-\frac{1}{2}r)^2}\right)$$

where $r = e^{-\beta\nu}$. This is negative, because $1-\frac{1}{2}r > \frac{1}{2}$ and $1-\frac{1}{2}r \ge 1-\frac{1}{2}\lambda$ when $|\mu| \le n^{1/12}$. The integral corresponding to (4.25) is

$$\int_{\Gamma_n^{(\alpha)}} s^{1/2}e^{P(\mu,s)-\mu^3/6}\big(1+O(s\nu)+O(s^2\mu^2\nu)+O(s^4\nu)\big)\,ds.$$

Instead of (4.26) we need the real part of $P(\mu,s) - \mu^3/6$, which is

$$-\tfrac{1}{6}\alpha^3 + \tfrac{1}{2}\alpha - \tfrac{3}{8}(\alpha+\alpha^{-1})t^2 \;\le\; \tfrac{1}{3}\,;$$

hence $\left|e^{P(\mu,s)-\mu^3/6}\right|$ is uniformly bounded on $\Gamma_n^{(\alpha)}$.

Furthermore the quantity $n!/\big((n-m)!\binom{N}{m}\big)$ in (10.14) can be shown to equal

$$\sqrt{2\pi n}\,e^{3/4+\mu^3/6-n}\,2^{n-m}\big(1+O\big((1+\mu^4)\nu\big)\big)\,. \tag{10.17}$$

Multiplying (10.16) by (10.17) yields (10.13).   □

(Theorem 5 applies also to multigraphs: If we use

$$\pi_{m,n} = \frac{n!\,2^m\,m!}{(n-m)!\,n^{2m}}\,[z^n]\,U(z)^{n-m}e^{V(z)}$$

in place of (10.14), we obtain the same asymptotic result (10.13). Multigraphs are assumed to be generated as in the uniform model, with each of the n^2 edges $\langle x,y\rangle$ equiprobable. Hence each self-loop $\langle x,x\rangle$ occurs with probability $1/n^2$, while edges $x-y$ with $x \neq y$ occur with probability $2/n^2$ since they arise from either $\langle x,y\rangle$ or $\langle y,x\rangle$.)

When $\mu \to -\infty$, the value of the integral in (10.13) is exponentially small; in fact it is $O(e^{\mu^3/6-\mu/2})$, because

$$\Re P(\mu, 1+iy) = \frac{(\mu+2)(\mu-1)^2 - (6-3\mu)y^2}{6}\,.$$

On the other hand, when $\mu \to +\infty$ we can prove that the integral is $1+O(\mu^{-3})$, by integrating on the path $s = \mu + iy/\sqrt{\mu}$ for $-\infty < y < \infty$. For we have $P(\mu, \mu + iy/\sqrt{\mu}) = -y^2/2 - iy^3/(3\mu^{3/2})$; the integral can be restricted to $|y| \leq \ln n$, in which range the integrand is $e^{-y^2/2}i$ times $1 + \frac{1}{2}iy\mu^{-3/2} - \frac{1}{2}iy^3\mu^{-3/2} + O\big((y^2+y^6)\mu^{-3}\big)$. Therefore the random graph process almost always keeps going without bicyclic components until the number of edges is on the order of $\frac{1}{2}ne^{-\mu\nu} \approx \frac{1}{2}n - \frac{1}{2}\mu n^{2/3}$. If we take M large enough, the probability is $\geq 1-\epsilon$ that the first bicyclic component occurs when $\frac{1}{2}n - Mn^{2/3} \leq m \leq \frac{1}{2}n + Mn^{2/3}$. Informally we can say that the graph almost certainly becomes bicyclic when the number of edges is $\frac{1}{2}n + \Theta(n^{2/3})$.

When λ is strictly less than 1, say $\lambda \leq 1 - \delta$, we can show that $\pi_{m,n} = 1 - O(n^{-1}\delta^{-1}) - O(n^{-1/2}\delta^{-3/2})$ by integrating on the contour $z = \lambda e^{i\theta}$ as in the proof of Theorem 4. (See [7, Theorem 5e].) We can now sharpen the result of Erdős and Rényi stated in (8.14):

Corollary 7. *Let L be a set of positive integers, and say that an L-cycle is a cycle whose length is in L. Then*

Pr(graph or multigraph with n vertices and m edges has no L-cycle)

$$= \sqrt{1-\lambda} \, \exp\!\left(\sum_{\substack{l \geq 1 \\ l \notin L}} \frac{\lambda^l}{2l} \right) + O(n^{-1/2}), \tag{10.18}$$

if $\lim_{n \to \infty} 2m/n = \lambda < 1$.

(This result applies to graphs as well as multigraphs; we assume that $1 \notin L$ and $2 \notin L$ when we are considering graphs. A multigraph can have self loops (1-cycles) and/or repeated edges (2-cycles), but a graph cannot.)

Proof. The multigraph either contains a bicyclic component or it does not. The first case occurs with probability $O(n^{-1/2})$. In the second case we want the probability of a "going configuration" that consists entirely of acyclic components and unicyclic components whose cycle lengths are $\notin L$. The number of such configurations is

$$[z^n] \, U(z)^{n-m} \exp\!\left(\sum_{\substack{l \geq 1 \\ l \notin L}} \frac{T(z)^l}{2l} \right),$$

so we are able to complete the estimates by repeating almost verbatim the argument of Theorem 4.   □

If we set $L = \{1,2\}$ in (10.18), we get the asymptotic probability that a random multigraph is a graph, namely $e^{-\lambda/2-\lambda^2/4}$. If we set $L = \{3,5,7,9,\dots\}$, we get the asymptotic probability that a random graph is 2-colorable (bipartite), namely

$$\sqrt{1-\lambda} \, \exp\!\left(\tfrac{1}{2}\lambda - \tfrac{1}{4}\ln(1-\lambda^2)\right) = \left(\frac{1-\lambda}{1+\lambda}\right)^{1/4} e^{\lambda/2}. \tag{10.19}$$

Otherwise [7, §10], such a graph is almost surely 3-colorable when $\lambda < 1$.

Choosing $L = \{k+1, k+2, \dots\}$ in (10.18) gives the limiting distribution of the longest cycle in a random graph: All cycle lengths are $\leq k$ with probability

$$\sqrt{1-\lambda} \, \exp\!\left(\sum_{l=1}^{k} \frac{\lambda^l}{2l} \right) + O(n^{-1/2}). \tag{10.20}$$

(An analogous result has been derived by Pittel [13, Theorem 1], for random graphs in which each edge occurs independently with probability λ/n.)

Erdős and Rényi [7, §8] stated that, if r is any real number, the probability that a graph with n vertices and $\frac{1}{2}n + rn^{1/2}$ edges is nonplanar "has a positive lower limit, but we cannot calculate its value. It may even be 1, though this seems unlikely." We can now show that this probability is definitely less than 1. Indeed, a graph with n vertices and $\frac{1}{2}n + rn^{1/2}$ edges has $\mu \approx 2rn^{-1/6}$ in the hypothesis of Theorem 5, so the probability that it has no bicyclic component (and is therefore planar) approaches the limiting value stated for $\mu = 0$ and $\alpha = 1$. We can prove, in fact, that this limiting value $\pi_{n/2,n}$ is rather large:

Corollary 8. *The probability that a graph with n vertices and $\frac{1}{2}n$ edges has no bicyclic component is $\sqrt{2/3} + O(n^{-1/3})$.*

Proof. The contour integral in (10.15) is particularly interesting when $\lambda = 1$ because it has a three-legged saddle point. One way to evaluate it is to consider a path of the form $z = 1 + te^{2\pi i/3}n^{-1/3}$ for $t \geq 0$; this accounts for half of (10.15), and the result turns out to be

$$\frac{n^{-1/2}e^{n-3/4}}{2^{n/2}\pi} \int_0^\infty \sqrt{t}\, e^{-t^3/3}\, dt \left(1 + O(n^{-1/3})\right).$$

We will see in formula (A.8) below that this integral is $\frac{1}{3}\sqrt{3\pi}$. The auxiliary coefficient $n!/\left((n-m)!\binom{N}{m}\right)$ is $\sqrt{2\pi n}\, e^{3/4-n}2^{n/2}\left(1 + O(n^{-1})\right)$ when $m = \frac{1}{2}n$. $\square$

Erdős and Rényi [7] also remarked that a graph with $\frac{1}{2}n + \omega_n\sqrt{n}$ edges has a cycle with any given number of diagonals, with probability $\to 1$ when $\omega_n \to +\infty$ and $n \to \infty$. However, we have just proved that this is not true when $\omega_n = n^{1/6}$. Therefore their claim that a graph with exactly $\frac{1}{2}n$ edges has positive probability of nonplanarity might also be false; an explicit proof or disproof would be desirable.

11. The First k Cycles

As a final example of the techniques we have been considering, let us study the distribution of the first k cycles that appear in an evolving graph. We have seen in Section 10 that this problem is well-defined, at least asymptotically, because the first cycles in a sufficiently large graph will almost always occur in distinct components.

For simplicity let us once again consider the uniform model first. We will run the random multigraph process until there is either a bicyclic component or a set of k unicyclic components, whichever occurs first. In the latter case we let $l_1, l_2, \ldots, l_k$ be the lengths of the first k cycles, in order of appearance.

A stopping configuration in the non-bicyclic case will consist of a sequence of cycles of rooted trees, having respective lengths $(l_1, \ldots, l_{k-1})$, together with a sequence of l_k rooted trees, plus a set of any number of unrooted trees. A cycle of l rooted trees has the egf $T(z)^l/(2l)$, as discussed in (9.1). Therefore if we form the multivariate generating function

$$S(\zeta_1, \ldots, \zeta_k, w, z) =$$

$$= \sum_{l_1, \ldots, l_k \geq 1} \frac{T(wz)^{l_1}}{2l_1} \cdots \frac{T(wz)^{l_{k-1}}}{2l_{k-1}} T(wz)^{l_k} \zeta_1^{l_1} \cdots \zeta_k^{l_k} e^{U(wz)/w}$$

$$= \left(\prod_{j=1}^{k-1} \left(\frac{1}{2} \ln \frac{1}{1 - \zeta_j T(wz)} \right) \right) \frac{\zeta_k T(wz)}{1 - \zeta_k T(wz)} e^{U(wz)/w} \qquad (11.1)$$

the coefficient $n! [\zeta_1^{l_1} \cdots \zeta_k^{l_k} w^m z^n] S(\zeta_1, \ldots, \zeta_k, w, z)$ will be the number of stopping configurations with m edges, n vertices, and cycle lengths $(l_1, \ldots, l_k)$.

In order to convert these coefficients to probabilities, we need to consider how many of the n sequences $\langle x_1, y_1 \rangle, \ldots, \langle x_m, y_m \rangle$ of edges will yield a stopping configuration with parameters $m, n, l_1, \ldots, l_k$. For this we need a slight generalization of the argument at the beginning of Section 3; the appropriate factor is now not $2^{m-1}(m-1)!/n^{2m}$ but rather

$$\frac{2^{m-1}(m-1)!}{n^{2m}} \frac{l_1}{L_1} \frac{l_2}{L_2} \cdots \frac{l_{k-1}}{L_{k-1}}, \qquad (11.2)$$

where

$$L_j = l_1 + \cdots + l_j. \qquad (11.3)$$

The reason is that the $(m-1)!$ permutations of the $m-1$ non-final edges are not all admissible. Exactly l_{k-1}/L_{k-1} of them have the final edge of the $(k-1)$st cycle occurring after all the edges of the first $k-2$ cycles; and l_{k-2}/L_{k-2} of these have the final edge of the $(k-2)$nd cycle occurring after all the edges of the first $k-3$; and so on.

The stopping configurations in the bicyclic case can be ignored, because we know that this case occurs with vanishing probability as $n \to \infty$; but we might as well describe the generating function, so that we can see how rapidly the probability approaches zero. We mimic the derivation of (9.4): Either $k \geq 2$ and there is a unicyclic component with two marked vertices, plus an additional set of acyclic and (at most $k-2$) unicyclic components; or $k \geq 3$ and there are two unicyclic components with marked vertices plus an additional set of acyclic and (at most $k-3$) unicyclic components. The egf is therefore

$$\left(\vartheta^2 V(z)\right) e^{U(z)} \sum_{j=0}^{k-2} \frac{V(z)^j}{j!} + \left(\vartheta\, V(z)\right)^2 e^{U(z)} \sum_{j=0}^{k-3} \frac{V(j)^j}{j!} . \qquad (11.4)$$

Converting to a bgf gives a formula like (9.5) except that it has the form

$$\frac{w\, f_k\!\left(T(wz)\right)}{\left(1 - T(wz)\right)^4}\, e^{U(wz)/w} \qquad (11.5)$$

where f_k is a polynomial. By reasoning as we did after (9.11), we conclude that the Φ_n operator produces a result that is $O(n^{-1} n^{(8-3)/6}) = O(n^{-1/6})$, for every fixed k.

We can now determine the asymptotic probability that a given sequence of cycle lengths will appear:

Theorem 6. *The probability that the random multigraph process produces the first k cycles in distinct components with respective lengths $(l_1, l_2, \ldots, l_k)$ is*

$$\frac{2^{1-k}}{L_1 L_2 \ldots L_{k-1}}\, p_{L_k} + O(n^{-1/6}), \qquad (11.6)$$

for all fixed $l_1, \ldots, l_k \geq 1$, where L_j is defined in (11.3) and p_l is defined in (5.6). The same formula holds for the random graph process, if p is replaced by $\hat{p}$ and if we require $l_1, \ldots, l_k \geq 3$.

Proof. The desired probability, according to (11.1), (11.2), and (3.1), is

$$\Phi_n\!\left(\frac{l_1}{L_1} \frac{l_2}{L_2} \cdots \frac{l_{k-1}}{L_{k-1}} \left[\zeta_1^{l_1} \cdots \zeta_k^{l_k}\right] S(\zeta_1, \ldots, \zeta_k, w, z)\right)$$

$$= \frac{2^{1-k}}{L_1 L_2 \ldots L_{k-1}}\, \Phi_n\!\left(T(wz)^{L_k}\, e^{U(wz)/w}\right), \qquad (11.7)$$

plus $O(n^{-1/6})$ for the probability of failure due to the early occurrence of a bicyclic component. And $\Phi_n\!\left(T(wz)^l\, e^{U(wz)/w}\right)$ is the probability that the first cycle has length l, computed in Theorem 2.

This proves (11.6) in the uniform model; the same ideas apply to the permutation model, with minor changes.   □

The probability distribution in Theorem 6 was first derived by Svante Janson [10], without the error term, in the case of random graphs. We can show that the sum of probabilities (11.6) over all $(l_1, \ldots, l_k)$ equals 1, by using the identities

$$\sum_{k=l+1}^{\infty} p_k = 2l_{p_l}, \qquad \sum_{k=l+3}^{\infty} \hat{p}_k = 2l_{\hat{p}_l} \qquad (11.8)$$

already mentioned in (5.8) and (5.10). Notice that the asymptotic probability in the uniform model that the first k cycles will all be loops of length 1 is $p_k/(2^{k-1}(k-1)!) = 1/(3 \cdot 5 \cdot \ldots \cdot (2k+1))$.

On intuitive grounds we expect the second cycle to be larger than the first, and the third should be larger yet, because the trees that yield cycles gradually get bigger. And indeed, this is true:

Theorem 7. *The average length of the kth cycle, for fixed k, is of order $n^{1/6}(\log n)^{k-1}$.*

Proof. It suffices to give the proof for the uniform model, since the other model is similar.

The basic idea is to apply the identity

$$\frac{z}{(1-z)^2} \left(\ln \frac{1}{1-z} \right)^{k-1} = (k-1)! \sum_{l_1, \ldots, l_k \geq 1} \frac{l_k \, z^{L_k}}{L_1 L_2 \ldots L_{k-1}}, \qquad (11.9)$$

which is readily verified by induction. The average value of l_k is

$$\sum_{l_1, \ldots, l_k \geq 1} l_k \, P(l_1, \ldots, l_k, n), \qquad (11.10)$$

where $P(l_1, \ldots, l_k, n)$ is the probability in (11.7); thus we want to apply Φ_n to the bgf

$$\frac{1}{2^{k-1}(k-1)!} \frac{T}{(1-T)^2} \left(\ln \frac{1}{1-T} \right)^{k-1} e^{U(wz)/w}, \qquad (11.11)$$

where $T \equiv T(wz)$. And it should be clear from the calculations in Sections 5 and 6 that the principal effect of each additional factor $\ln 1/(1-T)$ is to multiply the inner integral by

$$\ln 1/(1 - e^{-\alpha\nu}) = \left(\tfrac{1}{3} \ln n + O(\ln(1 + |\mu|)) \right) \left(1 + O(n^{-1/3}) \right).$$

Therefore the result is $\Theta(\log n)^{k-1}$ times the result of Theorem 3.   □

In this proof we have defined the random variable l_k to be zero if the first k cycles are not well separated, that is, if they do not fall in distinct components. This seems reasonable because the concept of kth cycle becomes murky when many cycles are formed simultaneously.

The conditional probability that the kth cycle in the uniform model has length l, given that the first k cycles are in different components, has the limiting value

$$p_l^{(k)} = \sum_{l_1,\ldots,l_{k-1}\geq 1} \frac{2^{1-k}}{L_1 L_2 \ldots L_{k-1}} \, p_{L_{k-1}+l}, \qquad (11.12)$$

and it is interesting to observe that this sum is always rational. Indeed, relation (11.8) implies that

$$p_{>l}^{(k)} = p_{>l}^{(k-1)} + 2l p_l^{(k)}, \qquad \text{for } l > 0. \qquad (11.13)$$

We can now prove by induction that $p_1^{(k)} = 3^{-k}$ and that

$$p_l^{(k)} = p_l \sum_{1\leq j_1 \leq \cdots \leq j_{k-1}\leq l} \frac{1}{(2j_1 + 1)\ldots(2j_{k-1} + 1)}. \qquad (11.14)$$

By (5.7) we have

$$p_l^{(k)} = \frac{\sqrt{\pi}\,(\ln l)^{k-1}}{2^{k+1}(k-1)!\, l^{3/2}} \left(1 + O(l^{-1})\right) \qquad (11.15)$$

for fixed k as $l \to \infty$.

A somewhat paradoxical situation arises if we ask for the *conditional* expected length of the *first* cycle, given that the first k cycles appear in different components. For example, suppose $k = 2$. Let a_n be the unconditional expected length of the first cycle; let b_n be the probability that the first two cycles are well separated; and let c_n/b_n be the conditional expected length of the first cycle given that the first two cycles are well separated. Then we find

$$a_n = \Phi_n\left(\frac{T}{(1-T)^2}\, e^{U(wz)/w}\right) \sim K n^{1/6};$$

$$b_n = 1 - \Phi_n\left(\frac{w\,T\,(1+T)}{2(1-T)^4}\, e^{U(wz)/w}\right) = 1 - O(n^{-1/6});$$

$$c_n = \sum_{l_1,l_2\geq 1} l_1\, P(l_1, l_2, n) = \Phi_n\left(\frac{T^2}{2(1-T)^2}\, e^{U(wz)/w}\right),$$

where $T \equiv T(wz)$. Since $T^2/(1-T)^2 = T/(1-T)^2 - T/(1-T)$, we have $c_n = \frac{1}{2}(a_n - 1)$ exactly. Thus the expected value c_n/b_n is asymptotically only half of a_n, although both quantities represent the expected length of the first cycle, and although we are conditioning on an event that almost surely occurs! The reason is that the distribution of first cycle lengths has a tail that decays very slowly; and cases when the first cycle is extremely long are much more likely to attract the second cycle into the same component.

Similarly, it can be shown that the conditional expected length of the first cycle, given that the first k cycles appear in separate components, is asymptotic to $2^{1-k}a_n$.

12. Concluding Remarks

We have shown that a combination of generating functions and contour integration can resolve problems that apparently could not be treated successfully with the techniques that have previously been applied to random graphs. Many of the previous techniques, like the laws of large numbers, can be based on special cases of contour integration with the saddle point method; the approach in this paper may have succeeded primarily because we were free to use the saddle point method in a more general context.

It would be interesting to push the techniques further, for example to determine the asymptotic value of $L_n - Kn^{1/6}$ when L_n denotes the expected first cycle length.

Appendix. Evaluation of Integrals

Let us complete this discussion by studying the behavior of the integral in Lemma 3, equation (4.19), and by finding a numerical estimate for the constant K in (6.8). This proves to be an interesting exercise in the theory of functions.

First let's warm up by discussing some simplified functions that will help us get to know the territory. If x is a real number, we define

$$f(x) = \int_{-\infty}^{\infty} \exp(-it - xt^2 + it^3/3)\, dt\,, \qquad (A.1)$$

$$g(x) = \int_{-\infty}^{\infty} \exp(ixt + it^3/3)\, dt\,. \qquad (A.2)$$

The motivation for $f(x)$ comes from the integral in (4.19), which reduces to a multiple of $f\big((\alpha + \alpha^{-1})/2\big)$ under the substitutions $s = \alpha - it$

and $\mu = \alpha - \alpha^{-1}$, if we integrate on a path from $\alpha - i\infty$ to $\alpha + i\infty$ instead of on the contour $\Gamma^{(\alpha)}$. Since our main application of $f(x)$ has $x = (\alpha + \alpha^{-1})/2 \geq 1$, we can assume that $x \geq 1$ in $f(x)$. We have

$$f(x) = \int_{-\infty}^{\infty} e^{-xt^2} \cos(-t + t^3/3)\,dt\,, \qquad (A.3)$$

so $f(x)$ clearly converges for all $x > 0$. We will prove that the related function $g(x)$ converges for all real x (even though the integrand in its definition always has magnitude 1).

If a is any positive number, we have

$$\left| \int_R^{R+ia} \exp\!\left(ixt + \frac{it^3}{3}\right) dt \right| \leq \int_0^a \left| \exp\!\left(ix(R+it) + \frac{i(R+it)^3}{3}\right) \right| dt$$

$$= \int_0^a e^{-xt - R^2 t + t^3/3}\,dt$$

$$\leq e^{a^3/3 + a|x|} \int_0^a e^{-R^2 t}\,dt = O(R^{-2})\,;$$

a similar bound applies if we integrate from $-R$ to $-R + ia$. Hence we can shift the path of integration upward, without affecting the value or the convergence of the integral:

$$g(x) = \int_{-\infty}^{\infty} \exp\!\left(ix(t + ia) + i(t + ai)^3/3\right) dt\,, \qquad \text{for } a \geq 0. \qquad (A.4)$$

There is now a term $-at^2$ in the exponent, so $g(x)$ must indeed converge.

In particular, we have

$$g(a^2 - 1) = \int_{-\infty}^{\infty} \exp\!\left(a - \tfrac{2}{3}a^3 - it - at^2 + \tfrac{1}{3}it^3\right) dt$$

$$= e^{a - 2a^3/3} f(a)\,. \qquad (A.5)$$

Thus $f(x) = e^{2x^3/3 - x} g(x^2 - 1)$. When x is large, we have $f(x) = \sqrt{2\pi}\,x^{-1/2} + O(x^{-3/2})$, hence $g(x^2 - 1)$ must be mighty small.

Another formula for $g(x)$ can be obtained by rotating the path of integration:

$$g(x) = 2\Re \int_0^{\infty} \exp(ixt + it^3/3)\,dt$$

$$= 2\Re\!\left(\zeta \int_0^{\infty} \exp(ix\zeta t + i\zeta^3 t^3/3)\,dt \right)$$

$$= 2\Re\!\left(\zeta \int_0^{\infty} \exp(ix\zeta t - t^3/3)\,dt \right)\,, \qquad (A.6)$$

where

$$\zeta = e^{\pi i/6} = \frac{\sqrt{3}}{2} + \frac{1}{2}i \,. \tag{A.7}$$

(The integral on the arc $Re^{i\theta}$ for $0 \le \theta \le \pi/6$ is negligible for large R, because the magnitude of the integrand is $\exp(-\frac{1}{3}R^3 \sin 3\theta - xR\sin\theta)$.)

Equation (A.6) will be our key to evaluating $f(x)$ via $g(x)$, because we can expand $\exp(ix\zeta t)$ into a convergent power series in t. Then we can interchange summation and integration, evaluating the resulting integrals by using an analog of (4.14):

$$\int_0^\infty t^k e^{-t^3/a} \, dt = \int_0^\infty (ua)^{k/3} e^{-u} \frac{a \, du}{3(ua)^{2/3}}$$

$$= \frac{1}{3} a^{(k+1)/3} \int_0^\infty u^{(k-2)/3} e^{-u} \, du$$

$$= \frac{1}{3} a^{(k+1)/3} \Gamma\big((k+1)/3\big) \,. \tag{A.8}$$

It follows that

$$g(x) = \frac{2}{3^{2/3}} \, \Re\left(\sum_{k \ge 0} \zeta^{k+1} i^k (3x)^k \frac{\Gamma\big((k+1)/3\big)}{k!} \right) . \tag{A.9}$$

The real part of $\zeta^{k+1} i^k$ is $\cos(\frac{2}{3}k + \frac{1}{6})\pi$, which is respectively $(\frac{1}{2}\sqrt{3}, -\frac{1}{2}\sqrt{3}, 0)$ when $k = (0, 1, 2) \bmod 3$; hence

$$g(x) = 3^{-1/6} \sum_{k \ge 0} \frac{(3^{1/3}x)^{3k}}{(3k+1)!} \left((3k+1)\Gamma(k+\tfrac{1}{3}) - 3^{1/3}x\Gamma(k+\tfrac{2}{3}) \right) . \tag{A.10}$$

This series converges for all x; hence $g(z)$ and $f(z)$ are actually analytic functions in the entire complex plane.

We can write (A.10) as a difference of two hypergeometric series of type $_0F_1$:

$$g(x) = 3^{-1/6}\Gamma(\tfrac{1}{3})F(\,;\,\tfrac{2}{3}\,;\tfrac{1}{9}x^3) - 3^{1/6}x\Gamma(\tfrac{2}{3})F(\,;\,\tfrac{4}{3}\,;\tfrac{1}{9}x^3) \,. \tag{A.11}$$

This representation allows us to deduce that $g(z)$ can be expressed as an Airy function, hence as a modified Bessel function of fractional order:

$$g(z) = 2\pi \operatorname{Ai}(z) = \frac{2z^{1/2}}{3^{1/2}} K_{1/3}\big(\tfrac{2}{3}z^{3/2}\big) \,. \tag{A.12}$$

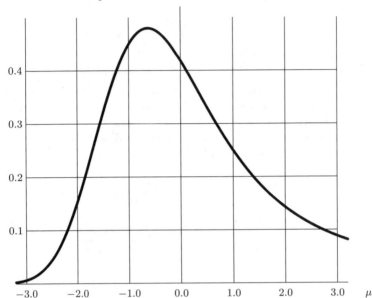

FIGURE 4. The function $K(\mu)$ whose integral yields the first-cycle constant K.

Equation (4.27) follows from the fact that $f(1) = e^{-1/3}g(0)$. In general, our derivation leads from (4.19) to the asymptotic formula

$$s_{l,n}(e^{-\mu\nu}) \sim \frac{e^{\mu^3/12}g(\mu^2/4)}{\sqrt{8\pi}}\, n^{1/6}, \qquad (\text{A}.13)$$

if we assume that the O term in (4.19) is of lesser order.

A somewhat different approach appears to be necessary if we want to evaluate the constant K numerically. Let us consider the value of the inner integral in (6.8),

$$K(\mu) = \frac{1}{\sqrt{8\pi}\,i} \int_\Gamma e^{P(\mu,s)}\, \frac{ds}{s}, \quad P(\mu,s) = \frac{(\mu+2s)(\mu-s)^2}{6}, \qquad (\text{A}.14)$$

for fixed μ; this is the quantity that yields K when integrated over the range $-\infty < \mu < \infty$. It is plotted for $-3 \le \mu \le 3$ in Figure 4.

We can argue, as we did following Theorem 5 in Section 10, that $K(\mu)$ is exponentially small when $\mu \to -\infty$, and that $K(\mu)$ is of order $\mu^{-3/2}$ when $\mu \to +\infty$. Our strategy for evaluating K will be to find a reasonable way to compute $K(\mu)$ when $|\mu|$ is small, together with a precise asymptotic estimate of $K(\mu)$ when μ is large.

First let's assume that μ is near zero. We have, by definition,

$$K(\mu) = \frac{1}{\sqrt{8\pi}\,i}\, e^{\mu^3/6}\, I\!\left(\frac{\mu}{2}\right), \qquad I(x) = \int_\Gamma e^{-xs^2+s^3/3}\, \frac{ds}{s}, \qquad (A.15)$$

where the contour Γ begins at $\infty e^{-i\pi/3}$ and ends at $\infty e^{+i\pi/3}$ after crossing the positive real axis. Then the quantity $u = s^3/3$ describes a contour that starts at $-\infty$ just below the negative real axis, hugs the bottom edge of that axis and circles the origin counterclockwise, then returns (just above the axis) to $-\infty$. This is a contour C for which we have Hankel's well-known formula

$$\frac{1}{\Gamma(s)} = \frac{1}{2\pi i} \int_C \frac{e^u\, du}{u^s}. \qquad (A.16)$$

Hence we can use the substitution $s = 3^{1/3} u^{1/3}$ to write

$$I(x) = \frac{1}{3} \int_C e^u \exp(-3^{2/3}x u^{2/3}) \frac{du}{u} = \frac{1}{3} \int_C e^u \sum_{k \geq 0} \frac{(-3^{2/3}x)^k u^{2k/3-1}}{k!}\, du$$

$$= \frac{2\pi i}{3} \sum_{k \geq 0} \frac{(-3^{2/3}x)^k}{k!\,\Gamma(1 - 2k/3)}. \qquad (A.17)$$

(The series are absolutely convergent.) This is the desired formula by which we can compute $K(\mu)$ when $|\mu|$ isn't too large. It can be expressed hypergeometrically in the form

$$I(x) = \frac{2\pi i}{3}\left(1 - \frac{3^{2/3}x}{\Gamma(\frac{1}{3})} F(\tfrac{5}{6}, \tfrac{1}{3}; \tfrac{4}{3}, \tfrac{2}{3}; -\tfrac{4}{3}x^3)\right.$$

$$\left. - \frac{3^{1/3}x^2}{2\Gamma(\frac{2}{3})} F(\tfrac{7}{6}, \tfrac{2}{3}; \tfrac{4}{3}, \tfrac{5}{3}; -\tfrac{4}{3}x^3)\right). \qquad (A.18)$$

Incidentally, it's interesting to apply this same idea to the integral (4.19). We obtain a formula that looks rather different from (A.11) and (A.13), namely

$$s_{l,n}(e^{-\mu\nu}) \sim e^{\mu^3/6} \sqrt{\frac{\pi}{2}} \left(\frac{3^{-2/3}}{\Gamma(\frac{2}{3})} F(\tfrac{1}{6}; \tfrac{1}{3}; -\tfrac{1}{6}\mu^3)\right.$$

$$\left. - \frac{3^{-1/3}\mu^2}{4\Gamma(\frac{1}{3})} F(\tfrac{5}{6}; \tfrac{5}{3}; -\tfrac{1}{6}\mu^3)\right) n^{1/6}. \qquad (A.19)$$

The quantity in parentheses is a confluent hypergeometric function,

$$\frac{U(\frac{1}{6}, \frac{1}{3}, -\frac{1}{6}\mu^3)}{2^{2/3}\,3^{1/3}\,\pi^{3/2}};$$

equating (A.19) with (A.13) yields a known identity between Airy functions and confluent hypergeometrics [1, Equations 13.1.29 and 13.6.25]. We can also prove equality between individual "halves" of (A.11) and (A.19), using the hypergeometric identity

$$e^{-z/2}\,F(a; 2a; z) = F(\,; a + \tfrac{1}{2}; \tfrac{1}{16}z^2). \qquad (A.20)$$

Now let's consider $K(\mu)$ as $\mu \to \infty$. Our experience with the similar integral (10.13) in the discussion following Theorem 5 suggests that we try integrating along the path $s = \mu + iy/\sqrt{\mu}$. (The contour $\Gamma^{(\mu)}$ can be "straightened out," as we found in (A.6), as long as the tails remain exponentially small.) On this straight line the integral reduces to

$$K(\mu) = \frac{1}{\sqrt{8\pi}\,\mu^{3/2}} \int_{-\infty}^{\infty} \frac{e^{-y^2/2}\exp\!\left(-iy^3/(3\mu^{3/2})\right)dy}{1 + iy/\mu^{3/2}}, \qquad (A.21)$$

and we can obtain an asymptotic formula by expanding the real part of the integrand as $e^{-y^2/2}$ times a power series in y^2 and $1/\mu^3$. Namely, if we set $v = \mu^{-3/2}$ and $x = y^2/3$ for convenience, we have

$$\frac{1}{2}\left(\frac{\exp(-ivx)}{1 + iv} + \frac{\exp(+ivx)}{1 - iv}\right)$$

$$= 1 - \left(1 + x + \frac{x^2}{2!}\right)v^2 + \left(1 + x + \frac{x^2}{2!} + \frac{x^3}{3!} + \frac{x^4}{4!}\right)v^4 - \cdots. \qquad (A.22)$$

Placing this quantity inside (A.21) and applying (4.14) gives

$$K(\mu) \sim \frac{1}{2\mu^{3/2}}\left(1 - \frac{c_1}{\mu^3} + \frac{c_2}{\mu^6} - \frac{c_3}{\mu^9} + \cdots\right), \qquad \text{as } \mu \to \infty, \qquad (A.23)$$

where

$$c_1 = \frac{1}{0!} + \frac{1\cdot 3}{1!\,3^1} + \frac{1\cdot 3\cdot 5}{2!\,3^2} = \frac{17}{6},$$

$$c_2 = \frac{1\cdot 3}{0!} + \frac{1\cdot 3\cdot 5}{1!\,3^1} + \frac{1\cdot 3\cdot 5\cdot 7}{2!\,3^2} + \frac{1\cdot 3\cdot 5\cdot 7\cdot 9}{3!\,3^3} + \frac{1\cdot 3\cdot 5\cdot 7\cdot 9\cdot 11}{4!\,3^4} = \frac{1801}{72},$$

and so on. The general formula is

$$c_n = \sum_{k=0}^{2n} \frac{(2n+2k)!}{2^{n+k}3^k(n+k)!\,k!} \; ; \tag{A.24}$$

and the denominator, incidentally, turns out to be exactly

$$2^{2n-\nu_2(n)} \, 3^{(3n-\nu_3(n))/2} \,, \tag{A.25}$$

where $\nu_r(n)$ denotes the sum of the digits of n in radix-r notation.

We need to justify the expansion (A.23) carefully, because the series $\sum_k (-1)^k c_k/\mu^{3k}$ is divergent. The key idea is to show that it is *strictly enveloping*, in the sense that its partial sums alternately overshoot and undershoot the true value of $K(\mu)$. (See [14, §I.4.1].)

It suffices to prove that (A.22) is strictly enveloping. For this purpose, let

$$f(v) = e^{xv}/(1-v)$$

and consider Taylor's series with remainder:

$$f(v) = f(0) + f'(0)\frac{v}{1!} + \cdots + f^{(n)}(0)\frac{v^n}{n!} + \int_0^v \frac{(v-t)^n}{n!} f^{(n+1)}(t)\,dt.$$

In this case the nth derivative is

$$f_n(v) = n!\,e^{xv} \sum_{k=0}^n \frac{x^k}{(1-v)^{n+1-k}k!}. \tag{A.26}$$

Therefore in particular we have

$$\frac{1}{n!}|f^{(n)}(iv)| \le \sum_{k=0}^n \frac{x^k}{|1-iv|^{n+1-k}k!} \le \sum_{k=0}^n \frac{x^k}{k!} = \frac{f^{(n)}(0)}{n!}$$

for all real v. Setting $n = 2m+1$ now yields

$$\left| \frac{1}{2}\big(f(iv)+f(-iv)\big) - \Big(f(0) - f''(0)\frac{v^2}{2!} + \cdots + (-1)^m f^{(2m)}(0)\frac{v^{2m}}{(2m)!}\Big) \right|$$

$$= \left| \frac{i}{2}\int_0^v \frac{(iv-it)^{2m+1}}{(2m+1)!} f^{(2m+2)}(it)\,dt \right.$$

$$\left. - \frac{i}{2}\int_0^v \frac{(-iv+it)^{2m+1}}{(2m+1)!} f^{(2m+2)}(-it)\,dt \right|$$

$$\le \int_0^v \frac{|v-t|^{2m+1}}{(2m+1)!} |f^{(2m+2)}(it)|\,dt \le \frac{v^{2m+1}}{(2m+2)!} f^{(2m+2)}(0)$$

and this is the inequality we needed to prove.

Since the series (A.23) for $K(\mu)$ is enveloping, we can integrate it term by term to get an enveloping series for the tail of the integral,

$$\int_\mu^\infty K(\mu)\,d\mu \sim \frac{1}{\mu^{1/2}}\left(1 - \frac{c_1}{7\mu^3} + \frac{c_2}{13\mu^6} - \frac{c_3}{19\mu^9} + \cdots\right). \qquad \text{(A.27)}$$

For any fixed μ this series is divergent, but we can find a "best" place to stop it (where the terms begin to increase in magnitude). For example, when $\mu = 5$, the sum of the terms involving c_k for $k \le 21$ on the right of (A.27) is 0.4458165587745; and the partial sum for $k \le 22$ is 0.4458165587784. So we know that

$$0.4458165587745 < \int_5^\infty K(\mu)\,d\mu < 0.4458165587784. \qquad \text{(A.28)}$$

These are the best lower and upper bounds attainable from (A.27), because the next two partial sums are

$$0.4458165587744 \qquad \text{and} \qquad 0.4458165587787.$$

We obtain better accuracy as μ grows, and we get almost no information when μ is too small. For example, when $\mu = 2$, the enveloping series (A.27) tells us only that $.671 < \int_2^\infty K(\mu)\,d\mu < .693$.

The integral of $K(\mu)$ from $-\infty$ to -4 is less than 10^{-5}. A numerical integration over the range $-4 \le \mu \le 5$, using enough terms of the convergent series (A.17) to ensure sufficient accuracy, now suffices to establish the value $K \approx 2.0337$, correct to four decimal places, as claimed in the introduction to this paper. (Such calculations are not quite trivial, because there is great cancellation between terms of (A.17); according to (A.15), the value of $I(\mu/2)$ must be extremely small when μ is 3 or more, because $I(\mu/2)$ must be multiplied by $e^{\mu^3/6}$. The arithmetic leading to the stated result was done as far as possible with rational numbers; then high-precision values of $3^{2/3}/\Gamma(\frac{1}{3})$ and $3^{1/3}/\Gamma(\frac{2}{3})$ were used to combine the results.)

Substantially faster methods would need to be devised if we wanted to calculate K to, say, 100 decimal places.

Acknowledgments

It is a pleasure to acknowledge here the help received in early stages of this work from several participants of the *Random Graphs '85* conference organized by M. Karoński in Poznań: J. W. Moon made useful observations concerning Section 2; S. Janson obtained the limiting distribution (5.2) of cycle lengths, thereby greatly helping to guide some of the initial calculations; discussions with H. Prodinger and P. Kirschenhofer led to important clarifications of several points.

This research was supported in part by the National Science Foundation under grant CCR-86-10181, and by Office of Naval Research contract N00014-87-K-0502.

Bibliography

[1] Milton Abramowitz and Irene A. Stegun, *Handbook of Mathematical Functions* (Washington: National Bureau of Standards, 1964).

[2] Béla Bollobás, *Random Graphs* (London: Academic Press, 1985).

[3] Béla Bollobás, "Sharp concentration of measure phenomena in the theory of random graphs," in *Random Graphs '87*, edited by Michał Karoński, Jerzy Jaworski, and Andrzej Ruciński (Chichester: Wiley, 1990), 1–15.

[4] N. G. de Bruijn, *Asymptotic Methods in Analysis* (Amsterdam: North-Holland, 1958).

[5] Louis Comtet, *Analyse Combinatoire*, Tomes I et II (Paris: Presses Universitaires de France, 1970). English translation, revised and enlarged, *Advanced Combinatorics* (Dordrecht: D. Reidel, 1974).

[6] P. Erdős and A. Rényi, "On random graphs I," *Publicationes Mathematicae (Debrecen)* 6 (1959), 290–297. Reprinted in *Paul Erdős: The Art of Counting* (Cambridge, Massachusetts: MIT Press, 1973), 561–568; and in *Selected Papers of Alfréd Rényi* (Budapest: Akadémiai Kiadó, 1976), 308–315.

[7] P. Erdős and A. Rényi, "On the evolution of random graphs," *Magyar Tudományos Akadémia Matematikai Kutató Intézetének Közleményei* 5 (1960), 17–61. Reprinted in *Paul Erdős: The Art of Counting* (Cambridge, Massachusetts: MIT Press, 1973), 574–618; and in *Selected Papers of Alfréd Rényi* (Budapest: Akadémiai Kiadó, 1976), 482–525.

[8] I. P. Goulden and D. M. Jackson, *Combinatorial Enumeration* (New York: Wiley–Interscience, 1983).

[9] Frank Harary and Edgar M. Palmer, *Graphical Enumeration* (New York: Academic Press, 1973).

[10] Svante Janson, "Poisson convergence and Poisson processes with applications to random graphs," *Stochastic Processes and their Applications* 26 (1987), 1–30.

[11] Donald E. Knuth and Arnold Schönhage, "The expected linearity of a simple equivalence algorithm," *Theoretical Computer Science* **6** (1978), 281–315. [Reprinted with additional material as Chapter 21 of *Selected Papers on Analysis of Algorithms*, CSLI Lecture Notes 102 (Stanford, California: Center for the Study of Language and Information, 2000), 341–389.]

[12] Edgar M. Palmer, *Graphical Evolution* (New York: Wiley, 1985).

[13] Boris Pittel, "On a random graph with a subcritical number of edges," *Transactions of the American Mathematical Society* **309** (1988), 51–75.

[14] George Pólya and Gabor Szegő, *Aufgaben und Lehrsätze aus der Analysis* (Berlin: Springer, 1925). English edition, *Problems and Theorems in Analysis* (New York: Springer–Verlag, 1972).

[15] E. M. Wright, "The number of connected sparsely edged graphs," *Journal of Graph Theory* **1** (1977), 317–330.

Chapter 41

The Birth of the Giant Component

*[To Paul Erdős on his 80th birthday, 26 March 1993. Written with Svante Janson, Tomasz Łuczak, and Boris Pittel. Originally published in Random Structures and Algorithms **4** (1993), 233–358.]*

Abstract

Limiting distributions are derived for the sparse connected components that are present when a random graph on n vertices has approximately $\frac{1}{2}n$ edges. In particular, we show that such a graph consists entirely of trees, unicyclic components, and bicyclic components with probability approaching $\sqrt{2/3} \cosh \sqrt{5/18} \approx 0.9325$ as $n \to \infty$. The limiting probability that it consists of trees, unicyclic components, and at most one other component is approximately 0.9957; the limiting probability that it is planar lies between 0.987 and 0.9998. When a random graph evolves and the number of edges passes $\frac{1}{2}n$, its components grow in cyclic complexity according to an interesting Markov process whose asymptotic structure is derived. The probability that there never is more than a single component with more edges than vertices, throughout the evolution, approaches $5\pi/18 \approx 0.8727$. A "uniform" model of random graphs, which allows self-loops and multiple edges, is shown to lead to formulas that are substantially simpler than the analogous formulas for the classical random graphs of Erdős and Rényi. The notions of "excess" and "deficiency," which are significant characteristics of the generating function as well as of the graphs themselves, lead to a mathematically attractive structural theory for the uniform model. A general approach to the study of stopping configurations makes it possible to sharpen previously obtained estimates in a uniform manner and often to obtain closed forms for the constants of interest. Empirical results are presented to complement the analysis, indicating the typical behavior when n is near 20000. Strong components of random digraphs are also considered.

0. Introduction

When edges are added at random to n initially disconnected points, for large n, a remarkable transition occurs when the number of edges becomes approximately $n/2$. Erdős and Rényi [13] studied random graphs with n vertices and $(1 + \mu)n/2$ edges as $n \to \infty$, and discovered that such graphs almost surely have the following properties: If $\mu < 0$, only small trees and "unicyclic" components are present, where a unicyclic component is a tree with one additional edge; moreover, the size of the largest tree component is

$$(\mu - \ln(1 + \mu))^{-1} \ln n + O(\log \log n).$$

If $\mu = 0$, however, the largest component has size of order $n^{2/3}$. And if $\mu > 0$, there is a unique "giant" component whose size is of order n; in fact, the size of this component is asymptotically αn when $\mu = -\alpha^{-1} \ln(1 - \alpha) - 1$. Thus, for example, a random graph with approximately $n \ln 2$ edges will almost surely have a giant component containing $\sim n/2$ vertices.

The research that led to the present paper began in a rather curious way, as a result of a misunderstanding. In 1988, the students in a class taught by Richard M. Karp performed computational experiments in which graphs with a moderately large number of vertices were generated by adding one edge at a time. A rumor spread that these simulations had turned up a surprising fact: As each of the random graphs evolved, the story went, never once was there more than a single "complex" component; that is, there never were two or more components present simultaneously that were neither trees nor unicyclic. Thus, the first connected component that acquired more edges than vertices was apparently destined to be the giant component. As more edges were added, this component gradually swallowed up all of the others, and none of the others ever became complex before they were swallowed.

Reports of those experiments suggested that a great simplification of the theory of evolving graphs might be possible. Could it be that such behavior occurs almost always, that is, with probability approaching 1 as $n \to \infty$? If so, we could hope for the existence of a much simpler explanation of the fact that a giant component emerges, and we could devise rather simple algorithms for online graph updating that would take advantage of the unique-complex-component phenomenon. At that time the authors who began this investigation (DEK and BP) were unaware of Stepanov's posthumous paper [36]. We were motivated chiefly by the work of Bollobás [5], who had shown that a component of size $n^{2/3}$ or more is almost always unique once the number of edges exceeds

$n/2 + 2(\ln n)^{1/2} n^{2/3}$; moreover, Bollobás proved that such a component gets approximately 4 vertices larger when each new edge is added. His results blended nicely with the unique-complex-component conjecture.

However, we soon found that the conjecture is false: There is nonzero probability that a graph with $n/2$ edges will contain several pretenders to the giant throne, and this probability increases when the number of edges is slightly more than $n/2$. We also learned that Stepanov [36] had already obtained similar results. Thus we could not hope for a theory of random graphs that would be as simple as the conjecture promised. On the other hand, we learned that the graph evolution process does satisfy the conjecture with reasonably high probability; hence algorithms whose efficiency rests on the assumption of a unique complex component will not often be inefficient.

Further analysis revealed, in fact, that we must have misunderstood the initial reports of experimental data. The actual probability that an evolving graph never has two complex components approaches the limiting value $5\pi/18 \approx 0.8727$; therefore the rumor that got us started could not have been true. In fact, the computer experiments by Karp's students had simply reported the state of the graph when exactly $n/2$ edges were present, and at certain other fixed reporting times. A false impression arose because there is high probability that a random graph with $n/2$ edges has at most one complex component; indeed, the probability is $0.9957 + O(n^{-1/3})$. More complicated configurations sometimes arise momentarily just after $n/2$ edges are reached, but they tend to disappear quickly, so they weren't seen. The fallacious rumor of 1988 has, however, turned out to have beneficial effects, because it was a significant catalyst for the discovery of some remarkably beautiful patterns.

Sections 1–10 of this paper provide a basic introduction to the theory of evolving graphs and multigraphs, using generating functions as the principal tool. Two models of graph evolution are presented in Section 1, the "graph process" and the "multigraph process." Their generating functions are introduced in Section 2, and special aspects of those functions related to trees and cycles are discussed in Section 3. Section 4 explains how to derive properties of a graph's more complex features by means of differential equations; the equations are solved for multigraphs in Section 5 and for graphs in Section 6. The resulting decomposition of multigraphs turns out to be surprisingly regular. Section 7 explains the regularities and begins to analyze the algebraic properties of the functions obtained in Section 5. Related results for connected graphs are discussed in Section 8. Section 9 explains the combinatorial significance of the algebraic structure derived earlier. Finally, Section 10 presents a quantitative lemma about the characteristics of random graphs near the critical point $\mu = 0$, making it possible to derive exact values for many relevant statistics.

Readers who cannot wait to get to the "good stuff" should skim Sections 1–10 and move on to Section 11, which begins a sequence of applications of the basic theory. The first step is to analyze the distribution of bicyclic components; then, in Section 12, the same ideas are shown to yield the joint distribution of all kinds of components. The formulas obtained there have a simple structure suggesting that the traditional approach of focusing on connected components is unnecessarily complicated; we obtain a simpler and more symmetrical theory if we first consider the *excess* of edges over vertices, exclusive of tree components, then look at other properties like connectedness after conditioning on the excess. Section 13 motivates this principle, and Section 14 derives the probability distribution of a graph's excess as it passes the critical point. These ideas help to nail down the probability that a graph with $n/2$ edges is planar, as shown in Section 15.

Section 16 begins the discussion of what may well be the most important notion in this paper; readers who have time for nothing else are encouraged to look at Figure 1, which shows the initial stages of the "big bang." The evolution of a graph or multigraph passes through discrete transitions as the excess increases, and important aspects of those changes are illustrated in Figure 1. Section 17 proves that this illustration represents a Markov process, which characterizes almost all graph evolutions. The $\frac{5\pi}{18}$ phenomenon alluded to above is discussed in Section 18, which establishes $\frac{5\pi}{18}$ as an upper bound for the probability in question. Section 19 shows that, for small n, the probability of retaining at most one complex component during the critical stage is in fact greater than $\frac{5\pi}{18}$, decreasing monotonically with n.

The excess of a graph is of principal importance at the critical point, but a secondary concept called *deficiency* becomes important shortly thereafter. A graph with deficiency 0 is called "clean"; such graphs are obtained from 3-regular graphs by splitting edges and/or by attaching trees to vertices of cycles. Section 20 explains how deficiency evolves jointly with increasing excess. Figure 2, at the end of that section, illustrates another Markov process that goes on in parallel with Figure 1. Section 21 shows that most graphs stay clean until they have acquired approximately $n/2 + n^{3/4}$ edges. Section 22 looks more closely at the moment a graph first becomes unclean.

Section 23 tracks the growth of excess and deficiency as a multigraph continues to evolve through $n/2 + n^{4/5}$, $n/2 + n^{5/6}$, ... edges.

The excess and deficiency are shown to be approximately normally distributed about certain well-defined values. Specifically, when the number of edges is $(1 + \mu)n/2$, with $\mu = o(1)$, the excess will be approximately $\frac{2}{3}\mu^3 n$ and the deficiency will be approximately $\frac{2}{3}\mu^4 n$. These statistics complement the well-known fact that the emerging giant component has almost surely grown to encompass approximately $2\mu n$ vertices.

Sections 24 to 26 develop a theory of "stopping configurations," by which we can study the first occurrences of various events during a multigraph's evolution. In particular, an explicit formula is derived for the asymptotic distribution of the time when the excess first reaches a given value r. A closed formula is derived for the "first cycle constant" of [14].

Section 27 completes the discussion initiated in Sections 17 and 18, by proving the $\frac{5\pi}{18}$ phenomenon as a special case of a more general result about the infinite Markov process in Figure 1.

Finally, Section 28 presents empirical data, showing to what extent the theory relates to practice when n is not too large. Section 29 discusses a number of open questions raised by this work.

1. Graph Evolution Models

We shall consider two ways in which a random graph on n vertices might evolve, corresponding to sampling with and without replacement. The first of these, introduced implicitly in [4] and explicitly in [7, proof of Lemma 2.7] and [14], turns out to be simpler to analyze and simpler to simulate by computer, therefore more likely to be of importance in applications to computer science: We generate ordered pairs $\langle x, y \rangle$ repeatedly, where $1 \leq x, y \leq n$, and add the (undirected) edge $x \!-\! y$ to the graph. Each ordered pair $\langle x, y \rangle$ occurs with probability $1/n^2$, so we call this the *uniform model* of random graph generation. It may also be called the *multigraph process*, because it can generate graphs with self-loops $x \!-\! x$, and it can also generate multiple edges. Notice that a self-loop $x \!-\! x$ is generated with probability $1/n^2$, while an edge $x \!-\! y$ with $x \neq y$ is generated with probability $2/n^2$ because it can occur either as $\langle x, y \rangle$ or $\langle y, x \rangle$.

The second evolution procedure, introduced by Erdős and Rényi [12], is called the *permutation model* or the *graph process*. In this case we consider all $N = \binom{n}{2}$ possible edges $x \!-\! y$ with $x < y$ and introduce them in random order, with all $N!$ permutations considered equally likely. In this model there are no self-loops or multiple edges.

A multigraph M on n labeled vertices can be defined by a symmetric $n \times n$ matrix of nonnegative integers m_{xy}, where $m_{xy} = m_{yx}$ is the number of undirected edges $x \!-\! y$ in G. For purposes of analysis, we shall assign a *compensation factor*

$$\kappa(M) = 1 \Big/ \prod_{x=1}^{n} \left(2^{m_{xx}} \prod_{y=x}^{n} m_{xy}! \right) \tag{1.1}$$

to M; if $m = \sum_{x=1}^{n} \sum_{y=x}^{n} m_{xy}$ is the total number of edges, the number of sequences $\langle x_1, y_1 \rangle \langle x_2, y_2 \rangle \dots \langle x_m, y_m \rangle$ that lead to M is then exactly

$$2^m \, m! \, \kappa(M) \,. \tag{1.2}$$

(The factor 2^m accounts for choosing either $\langle x, y \rangle$ or $\langle y, x \rangle$; the $2^{m_{xx}}$ in the denominator of $\kappa(M)$ compensates for the case $x = y$. The other factor $m!$ accounts for permutations of the pairs, with $m_{xy}!$ in $\kappa(M)$ to compensate for permutations between multiple edges.)

Equation (1.2) tells us that $\kappa(M)$ is a natural weighting factor for a multigraph M, because it corresponds to the relative frequency with which M tends to occur in applications. For example, consider multigraphs on three vertices $\{1, 2, 3\}$ having exactly three edges. The edges will form the cycle $M_1 = \{1 - 2, \, 2 - 3, \, 3 - 1\}$ much more often than they will form three identical self-loops $M_2 = \{1 - 1, \, 1 - 1, \, 1 - 1\}$, when the multigraphs are generated in a uniform way. For if we consider the 3^6 possible sequences $\langle x_1, y_1 \rangle \langle x_2, y_2 \rangle \langle x_3, y_3 \rangle$ with $1 \le x, y \le 3$, only one of these generates the latter multigraph, while the cyclic multigraph is obtained in $2^3 \, 3! = 48$ ways. Therefore it makes sense to assign weights so that $\kappa(M_2) = \frac{1}{48}\kappa(M_1)$, and indeed (1.1) gives $\kappa(M_1) = 1$, $\kappa(M_2) = \frac{1}{48}$.

Notice that a given multigraph M is a graph — that is, it has no loops and no multiple edges — if and only if $\kappa(M) = 1$. Notice also that if M consists of several disjoint components $M_1, \dots, M_k$, with no edges between vertices of M_i and M_j for $i \ne j$, we have

$$\kappa(M) = \kappa(M_1) \dots \kappa(M_k) \,. \tag{1.3}$$

2. Generating Functions

We shall use bivariate generating functions (bgf's) to study labeled graphs and multigraphs and their connected components. If $\mathcal{F}$ is a family of multigraphs with labeled vertices, the associated bgf is the formal power series

$$F(w, z) = \sum_{M \in \mathcal{F}} \kappa(M) \, w^{m(M)} \frac{z^{n(M)}}{n(M)!} \,, \tag{2.1}$$

where $m(M)$ and $n(M)$ denote the number of edges and the number of vertices of M. We can do many operations on such power series without regard to convergence. It follows from (1.2) and (2.1) that m steps of the uniform evolution model on n vertices will produce a multigraph in $\mathcal{F}$ with probability

$$\frac{2^m\, m!\, n!}{n^{2m}}\, [w^m z^n]\, F(w,z)\,, \tag{2.2}$$

where the symbol $[w^m z^n]$ denotes the coefficient of $w^m z^n$ in the formal power series that follows it. Similarly, if $\mathcal{F}$ is a family of graphs with labeled vertices, the probability that m steps of the permutation model will produce a graph in $\mathcal{F}$ is

$$\frac{n!}{\binom{N}{m}}\, [w^m z^n]\, F(w,z)\,, \qquad N = \binom{n}{2}. \tag{2.3}$$

Formulas (2.2) and (2.3) are asymptotically related by the formula

$$\binom{N}{m} = \frac{n^{2m}}{2^m\, m!} \exp\left(-\frac{m}{n} - \frac{m^2}{n^2} + O\left(\frac{m}{n^2}\right) + O\left(\frac{m^3}{n^4}\right) \right), \quad 0 \le m \le N, \tag{2.4}$$

which follows from Stirling's approximation.

Incidentally, the exponential factor in (2.4) is the probability that m steps of the multigraph process will produce no self-loops or multiple edges. When $m = n/2$, this probability is $e^{-3/4} + O(n^{-1}) \approx 0.472$.

When we say that the n vertices of a multigraph are "labeled," it is often convenient to think of the labeling as an assignment of the numbers 1 to n. But a strict numeric convention would require us to recompute the labels whenever vertices are removed or when multigraphs are combined. The actual value of a label is, in fact, irrelevant; what really counts is the relative order *between* labels. Labeled multigraphs are multigraphs whose vertices have been totally ordered. In this paper all graphs and multigraphs are assumed to be labeled, that is, totally ordered, even when the adjective "labeled" is not stated.

The bgf (2.1) is an exponential generating function in z, and the factor $\kappa(M)$ is multiplicative according to (1.3). Therefore the product of bgf's

$$F_1(w,z) F_2(w,z) \, \ldots \, F_k(w,z)$$

represents ordered k-tuples of labeled multigraphs $\langle M_1, M_2, \ldots, M_k \rangle$, each M_j being from family $\mathcal{F}_j$. Unordered k-tuples $\{M_1, \ldots, M_k\}$ from a common family $\mathcal{F}$ have the bgf $F(w,z)^k/k!$, if $\mathcal{F}$ does not include the empty multigraph. For example, the bgf for a 3-cycle is $w^3z^3/3!$, and the bgf for two isolated vertices is $z^2/2!$; hence the bgf for a 3-cycle and two isolated vertices is $(w^3z^3/6)(z^2/2) = 10w^3z^5/5!$. (There are 10 such graphs, one for each choice of the isolated points.)

Let $C(w, z)$ be the bgf for all connected multigraphs, and let $G(w, z)$ be the bgf for the set of all multigraphs. Then we have

$$e^{C(w,z)} = \sum_{k \geq 0} \frac{C(w, z)^k}{k!} = G(w, z) \tag{2.5}$$

because the term $C(w, z)^k/k!$ is the bgf for multigraphs having exactly k components. Similarly, if $\widehat{C}(w, z)$ and $\widehat{G}(w, z)$ are the corresponding bgf's for graphs instead of multigraphs, we have

$$e^{\widehat{C}(w,z)} = \widehat{G}(w, z), \tag{2.6}$$

a well-known formula due to Riddell [32]. The bgf for all graphs is obviously

$$\widehat{G}(w, z) = \sum_{n \geq 0} (1 + w)^{n(n-1)/2} \frac{z^n}{n!}. \tag{2.7}$$

Therefore (2.6) gives us the bgf for connected graphs,

$$\widehat{C}(w, z) = \ln\left(1 + z + (1 + w)\frac{z^2}{2} + (1 + w)^3 \frac{z^3}{6} + \cdots\right)$$

$$= z + w\frac{z^2}{2} + (3w^2 + w^3)\frac{z^3}{6} + \cdots. \tag{2.8}$$

The bgf $G(w, z)$ for all multigraphs can be found as follows: The coefficient of $z^n/n!$ is $\sum \kappa(M)w^{m(M)}$, summed over multigraphs M on n vertices. This is

$$\prod_{x=1}^{n}\left(\left(\sum_{m_{xx} \geq 0} \frac{w^{m_{xx}}}{2^{m_{xx}} m_{xx}!}\right) \prod_{y=x+1}^{n}\left(\sum_{m_{xy} \geq 0} \frac{w^{m_{xy}}}{m_{xy}!}\right)\right)$$

$$= \prod_{x=1}^{n} e^{w/2}(e^w)^{n-x} = e^{wn^2/2}.$$

Hence the desired formula is slightly simpler than (2.7):

$$G(w, z) = \sum_{n \geq 0} e^{wn^2/2} \frac{z^n}{n!}. \tag{2.9}$$

The corresponding bgf for connected multigraphs is therefore

$$C(w, z) = \ln G(w, z)$$

$$= \left(1 + \frac{1}{2}w + \frac{1}{8}w^2 + \frac{1}{48}w^3 + \cdots\right)z + \left(w + \frac{3}{2}w^2 + \frac{7}{6}w^3 + \cdots\right)\frac{z^2}{2}$$

$$+ \left(3w^2 + \frac{17}{2}w^3 + \cdots\right)\frac{z^3}{6} + \cdots . \tag{2.10}$$

In this case the coefficient of $w^3 z^3$ is $\frac{17}{2}/3!$, because the connected multigraphs with three edges on three vertices have total weight $\frac{17}{2}$. (The 3-cycle has weight 1; there are 9 multigraphs obtainable by adding a self-loop to a tree, each of weight $\frac{1}{2}$; and there are six multigraphs obtainable by doubling one edge of a tree, again weighted by $\frac{1}{2}$.)

Notice that expression (2.2) is $[w^m z^n] F(w, z) / [w^m z^n] G(w, z)$, the ratio of the weight of multigraphs in $\mathcal{F}$ to the weight of all possible multigraphs. Similarly, expression (2.3) is $[w^m z^n] \widehat{F}(w, z) / [w^m z^n] \widehat{G}(w, z)$.

It is convenient to group the terms of (2.8) and (2.10) according to the excess of edges over vertices in connected components. Let $\mathcal{C}_r$ and $\widehat{\mathcal{C}}_r$ denote the families of connected multigraphs and graphs in which there are exactly r more edges than vertices; let $C_r(w, z)$ and $\widehat{C}_r(w, z)$ be the corresponding bgf's. Then we have

$$C(w, z) = \sum_r C_r(w, z) = \sum_r w^r C_r(wz),$$

$$\widehat{C}(w, z) = \sum_r \widehat{C}_r(w, z) = \sum_r w^r \widehat{C}_r(wz), \tag{2.11}$$

where $C_r(z)$ and $\widehat{C}_r(z)$ are univariate generating functions for $\mathcal{C}_r$ and $\widehat{\mathcal{C}}_r$. A univariate generating function $F(z)$ is $\sum \kappa(M) z^n/n!$, summed over all graphs or multigraphs in a given family $\mathcal{F}$, where $n = n(M)$. We obtain it from a bgf by setting $w = 1$, thereby ignoring the number of edges. Univariate generating functions are easier to deal with than bgf's, so we generally try to avoid the need for two independent variables whenever possible.

3. Trees, Unicycles, and Bicycles

Let us say that a connected component has *excess* r if it belongs to $\mathcal{C}_r$, namely, if it has r more edges than vertices. A connected graph on n vertices must have at least $n - 1$ edges. Hence $C_r = 0$ unless $r \geq -1$.

In the extreme case $r = -1$, we have $C_{-1} = \widehat{C}_{-1}$, the family of all unrooted trees, which are *acyclic components*. In the next case $r = 0$, the generating functions C_0 and $\widehat{C}_0$ represent *unicyclic components*, which are trees with an additional edge. Similarly, C_1 and $\widehat{C}_1$ represent *bicyclic components*. In the present paper we shall deal extensively with sparse components of these three kinds, so it will be convenient to use the special abbreviations

$$U(z) = C_{-1}(z) = \widehat{C}_{-1}(z) \qquad \text{for unrooted trees;}$$
$$V(z) = C_0(z) \text{ and } \widehat{V}(z) = \widehat{C}_0(z) \quad \text{for unicyclic components;}$$
$$W(z) = C_1(z) \text{ and } \widehat{W}(z) = \widehat{C}_1(z) \quad \text{for bicyclic components.}$$

According to a well-known theorem of Sylvester [37] and Borchardt [8], often attributed erroneously to Cayley [10] although Cayley himself credited Borchardt, we have $U(z) = \sum_{n=1}^{\infty} n^{n-2} z^n / n!$. The other four generating functions begin as follows:

$$V(z) = \tfrac{1}{2}z + \tfrac{3}{4}z^2 + \tfrac{17}{12}z^3 + \tfrac{71}{24}z^4 + \tfrac{523}{80}z^5 + \tfrac{899}{60}z^6 + \cdots ;$$

$$\widehat{V}(z) = \tfrac{1}{6}z^3 + \tfrac{5}{8}z^4 + \tfrac{37}{20}z^5 + \tfrac{61}{12}z^6 + \tfrac{4553}{336}z^7 + \tfrac{8551}{240}z^8 + \cdots ;$$

$$W(z) = \tfrac{1}{8}z + \tfrac{7}{12}z^2 + \tfrac{101}{48}z^3 + \tfrac{83}{12}z^4 + \tfrac{12487}{576}z^5 + \tfrac{3961}{60}z^6 + \cdots ;$$

$$\widehat{W}(z) = \tfrac{1}{4}z^4 + \tfrac{41}{24}z^5 + \tfrac{95}{12}z^6 + \tfrac{497}{16}z^7 + \tfrac{4003}{36}z^8 + \tfrac{360763}{960}z^9 + \cdots .$$

All of these generating functions can be expressed succinctly in terms of the tree function

$$T(z) = \sum_{n \geq 1} n^{n-1} \frac{z^n}{n!} = z + z^2 + \tfrac{3}{2}z^3 + \cdots , \tag{3.1}$$

which generates *rooted* labeled trees and satisfies the functional relation

$$T(z) = z e^{T(z)} \tag{3.2}$$

due to Eisenstein [11]. Indeed, the relation

$$U(z) = T(z) - \tfrac{1}{2}T(z)^2 \tag{3.3}$$

is well known, as are the formulas

$$V(z) = \frac{1}{2} \ln \frac{1}{1 - T(z)} , \tag{3.4}$$

$$\widehat{V}(z) = \frac{1}{2} \ln \frac{1}{1 - T(z)} - \frac{1}{2}T(z) - \frac{1}{4}T(z)^2 ; \tag{3.5}$$

see [14]. We can prove (3.4) and (3.5) by noting that the univariate generating function for connected unicyclic multigraphs whose cycle has length k is

$$\frac{T(z)^k}{2k} ;$$

summing over $k \geq 1$ gives (3.4), and summing over $k \geq 3$ gives (3.5). (If $k = 1$, the cycle is a self-loop; hence the multigraph is essentially a rooted tree and the compensation factor is $\frac{1}{2}$. If $k = 2$, the cycle is a duplicate edge; hence the multigraph is essentially an unordered pair of rooted trees, and the compensation factor again is $\frac{1}{2}$. If $k \geq 3$, the unicyclic component is essentially a sequence of k rooted trees, divided by $2k$ to account for cyclic order and optional change of orientation.)

The generating function $\widehat{W}(z)$ was shown by G. N. Bagaev [1] to be

$$\widehat{W}(z) = \frac{T(z)^4 \big(6 - T(z) \big)}{24 \big(1 - T(z) \big)^3} . \tag{3.6}$$

Then E. M. Wright made a careful study of all the generating functions $\widehat{C}_k(z)$, which he called W_k, in a series of significant papers [41, 43–45]. We will show below that the bgf for bicyclic connected multigraphs is

$$W(z) = \frac{T(z) \big(3 + 2T(z) \big)}{24 \big(1 - T(z) \big)^3} . \tag{3.7}$$

The coefficients of powers of $1/\big(1 - T(z)\big)$ arise in numerous applications, so Knuth and Pittel [24] began to catalog some of their interesting properties. For each n the function $t_n(y)$ defined by

$$\frac{1}{\big(1 - T(z)\big)^y} = \sum_{n \geq 0} t_n(y) \frac{z^n}{n!} \tag{3.8}$$

is a polynomial of degree n in y, called the *tree polynomial* of order n. The coefficient of y^k in $t_n(y)$ is the number of mappings from an n-element set into itself having exactly k cycles. For fixed y and $n \to \infty$, we have [24, Lemma 2 and (3.16)]

$$t_n(y) = \frac{\sqrt{2\pi}\, n^{n-1/2+y/2}}{2^{y/2}\Gamma(y/2)} + O(n^{n-1+y/2}) . \tag{3.9}$$

We can, for example, express the number of connected bicyclic graphs on n vertices in terms of the tree polynomial t_n, namely

$$\tfrac{5}{24} t_n(3) - \tfrac{19}{24} t_n(2) + \tfrac{13}{12} t_n(1) - \tfrac{7}{12} t_n(0) + \tfrac{1}{24} t_n(-1) + \tfrac{1}{24} t_n(-2), \tag{3.10}$$

because (3.6) can be rewritten

$$\widehat{W}(z) = \frac{5}{24(1 - T(z))^3} - \frac{19}{24(1 - T(z))^2} + \frac{13}{12(1 - T(z))}$$
$$- \frac{7}{12} + \frac{1 - T(z)}{24} + \frac{(1 - T(z))^2}{24}.$$

Equation (3.9) tells us that only the first term $\frac{5}{24} t_n(3)$ of (3.10) is asymptotically significant. Extensions of (3.9), for cases when y varies with n, appear in equations (19.14) and (19.15) below.

We can also express quantities like (3.10) in terms of Ramanujan's function [30]

$$Q(n) = 1 + \frac{n - 1}{n} + \frac{n - 1}{n} \frac{n - 2}{n} + \frac{n - 1}{n} \frac{n - 2}{n} \frac{n - 3}{n} + \cdots$$
$$= \sqrt{\frac{\pi n}{2}} - \frac{1}{3} + \frac{1}{12} \sqrt{\frac{\pi}{2n}} - \frac{4}{135n} + O(n^{-3/2}), \tag{3.11}$$

which Wright [41] called $1 + h(n)/n^n$. For we have

$$t_n(1) = n^n; \qquad t_n(2) = n^n(1 + Q(n));$$
$$t_n(y + 2) = n \frac{t_n(y)}{y} + t_n(y + 1), \quad y \neq 0. \tag{3.12}$$

(See [24, equations (2.7), (3.14), and (1.9)].) Furthermore, we have

$$[z^n] V(z) = \tfrac{1}{2} n^{n-1} Q(n); \tag{3.13}$$

this follows from a well-known formula of Rényi [31].

4. The Cyclic Components

For theoretical purposes it proves to be important to partition a multigraph into its *acyclic part*, consisting entirely of isolated vertices or trees, and its *cyclic part*, consisting entirely of components that each contain at least one cycle. The cyclic part can in turn be partitioned into the *unicyclic part*, consisting entirely of unicyclic components, and the *complex part*, consisting entirely of components that have more edges than vertices. A multigraph is called cyclic if it equals its cyclic part, complex if it equals its complex part. In this section and the next, we will study the generating functions for cyclic and complex multigraphs. The formulas turn out to be surprisingly simple, and they will be the key to much of what follows.

Let $F(w, z)$ be the bgf for all cyclic multigraphs, that is, for all multigraphs whose acyclic part is empty. Formulas (2.5) and (2.11) tell us that

$$F(w, z) = e^{C_0(w,z) + C_1(w,z) + \cdots} = e^{C(w,z) - C_{-1}(w,z)} = G(w, z) e^{-U(wz)/w};$$

in other words,

$$G(w, z) = e^{U(wz)/w} F(w, z). \tag{4.1}$$

Indeed, this makes sense, because $e^{U(wz)/w}$ is the bgf for all acyclic multigraphs. We will analyze $F = F(w, z)$ by studying a linear differential equation satisfied by $G = G(w, z)$, and seeing that a similar equation is satisfied by F.

Let ϑ_w be the differential operator $w \frac{\partial}{\partial w}$, and let ϑ_z be $z \frac{\partial}{\partial z}$. The operator ϑ_w corresponds to the operation of *marking* an edge in a multigraph, that is, giving some edge a special label, because ϑ_w multiplies the coefficient of $w^m z^n$ by m. Similarly, ϑ_z corresponds to marking a vertex, because it multiplies the coefficient of $w^m z^n$ by n. (For a general discussion of marking, see [16, Sections 2.2.24 and following].) We have

$$\vartheta_w G(w, z) = w \sum_{n \geq 0} \frac{n^2}{2} e^{wn^2/2} \frac{z^n}{n!} = \frac{w}{2} \vartheta_z^2 G(w, z);$$

hence G satisfies the differential equation

$$\frac{2}{w} \vartheta_w G = \vartheta_z^2 G. \tag{4.2}$$

Again, this makes sense: The left side represents all multigraphs having a marked edge and an orientation assigned to that edge, and with the edge count decreased by 1. The right side represents all multigraphs with an ordered pair $\langle x, y \rangle$ of marked vertices. Orienting and discounting an edge is the same as marking two vertices.

We can also write (4.2) in the suggestive form

$$G(w, z) = e^z + \frac{1}{2} \int_0^w \vartheta_z^2 G(w, z) \, dw, \tag{4.3}$$

using the boundary condition $G(0, z) = e^z$. (The generating function for all multigraphs with no edges is, of course, e^z.) The operator ϑ_z^2

corresponds to choosing an ordered pair $\langle x, y \rangle$, and the integral operator $\frac{1}{2} \int_0^w \cdots dw$ corresponds to disorienting that edge and blending it into the existing multigraph. (Notice that the English words "differentiation" and "integration" are remarkably apt synonyms for the combinatorial operations of marking and blending.)

Most of our work will involve ϑ_z instead of ϑ_w, so we shall often write simply ϑ without a subscript when we mean ϑ_z. The marking operator ϑ has a simple effect on the generating functions $U(z)$ for unrooted trees and $T(z)$ for rooted trees. Indeed, we have

$$\vartheta\, U(z) \;=\; T(z)\,, \tag{4.4}$$

because an unrooted tree with a marked vertex is the same as a rooted tree. Furthermore

$$\vartheta\, T(z) \;=\; \sum_{k \ge 1} T(z)^k \;=\; \frac{T(z)}{1 - T(z)}\,, \tag{4.5}$$

because a rooted tree with a marked vertex is combinatorially equivalent to an ordered sequence $\langle T_1, T_2, \ldots, T_k \rangle$ of rooted trees, for some $k \ge 1$. The sequence represents a path of length $k - 1$ from the marked vertex to the root, with rooted subtrees sprouting from each point on that path.

Now let $U = U(wz)/w$ be the function $C_{-1}(w, z)$ that appears in (4.1), and let $T = T(wz) = C_0(w, z)$. We have

$$\vartheta_z U = z \frac{\partial}{\partial z} \frac{U(wz)}{w} = z \frac{w U'(wz)}{w} = z \frac{T(wz)}{wz} = \frac{T}{w}\,;$$

$$\vartheta_w U = w \frac{\partial}{\partial w} \frac{U(wz)}{w} = w \left(\frac{z U'(wz)}{w} - \frac{U(wz)}{w^2} \right) = \frac{T - U}{w} = \frac{T^2}{2w}\,.$$

Thus

$$\frac{2}{w}\, \vartheta_w\, U \;=\; (\vartheta_z U)^2\,. \tag{4.6}$$

In words: "Orienting and discounting an edge of an unrooted tree is equivalent to constructing an ordered pair of rooted trees."

We are now ready to convert (4.2) into a differential equation satisfied by $F = F(w, z)$:

$$\vartheta_w G = \vartheta_w(e^U F) = (\vartheta_w e^U) F + e^U(\vartheta_w F) = e^U\big((\vartheta_w U) F + \vartheta_w F\big)\,;$$

$$\vartheta_z G = \vartheta_z(e^U F) = e^U\big((\vartheta_z U) F + \vartheta_z F\big)\,;$$

$$\vartheta_z^2 G = e^U\big((\vartheta_z^2 U) F + (\vartheta_z U)^2 F + 2(\vartheta_z U)(\vartheta_z F) + \vartheta_z^2 F\big)\,.$$

Therefore, using (4.6), we have

$$\frac{2}{w}\vartheta_w F = (\vartheta_z^2 U)F + 2(\vartheta_z U)(\vartheta_z F) + \vartheta_z^2 F. \tag{4.7}$$

And like our other formulas, this one makes combinatorial sense as well as algebraic sense: The left side tells us that the right side should yield all ways that the cyclic part of a multigraph can grow, since $\frac{2}{w}\vartheta_w F$ is the number of ways it can go backward one step. The first term on the right corresponds to marking two vertices of an unrooted tree (in the acyclic part of the multigraph); joining them will produce a unicyclic component, thereby increasing the number of components in F. The middle term corresponds to marking a vertex in some tree of the acyclic part and another vertex in the cyclic part; joining them will add new vertices to one of F's existing components. The remaining term corresponds to marking two vertices in the cyclic part. If such marked vertices belong to the same component, say a component of excess r, a new edge between them will change the excess of the component to $r+1$. Otherwise, the marked vertices belong to different components, having respective excesses r and s, possibly with $r = s$; joining them will merge the components into a new component of excess $r + s + 1$.

Similarly, we can proceed to study the bgf $E(w, z)$ for the complex part of a multigraph, the part whose components all have positive excess. (The letter E stands for excess.) We have

$$F(w, z) = e^{V(wz)} E(w, z), \tag{4.8}$$

where $V = V(wz)$ generates unicyclic components. It is easy to verify the identity

$$\frac{2}{w}\vartheta_w V = \vartheta_z^2 U + 2(\vartheta_z U)(\vartheta_z V), \tag{4.9}$$

which corresponds to a combinatorially evident fact. Indeed,

$$\vartheta_z^2 U = \frac{1}{w}\frac{T}{1 - T}; \qquad \vartheta_w V = \vartheta_z V = \frac{T}{2(1 - T)^2}. \tag{4.10}$$

Therefore we find

$$\frac{2}{w}\vartheta_w E = (\vartheta_z^2 V)E + (\vartheta_z V)^2 E + 2(\vartheta_z U)(\vartheta_z E) + 2(\vartheta_z V)(\vartheta_z E) + \vartheta_z^2 E. \tag{4.11}$$

5. Enumerating Complex Multigraphs

To solve the differential equation (4.11), we can first write it in the form

$$\frac{1}{w}(\vartheta_w - T\vartheta_z)E \;=\; \frac{1}{2}\,e^{-V}\,\vartheta_z^2\,e^V\,E\,. \tag{5.1}$$

Now we partition $E = E(w, z)$ into terms of equal excess, as we did for $C(w, z)$ in (2.11):

$$E(w, z) = \sum_r E_r(w, z) = \sum_r w^r E_r(wz)\,. \tag{5.2}$$

The univariate generating function $E_r(z)$ represents all complex multigraphs that have exactly r more edges than vertices; in particular, $E_0(z) = 1$, since only the empty multigraph is "complex" and has excess 0. Differentiation yields

$$\vartheta_w E(w, z) = \sum_r \big(rw^r E_r(wz) + w^r(\vartheta E_r)(wz)\big)\,,$$

$$\vartheta_z E(w, z) = \sum_r w^r(\vartheta E_r)(wz)\,,$$

where $(\vartheta E_r)(wz)$ here means $\vartheta_z E_r(z)$ with the argument z subsequently replaced by wz, namely $wz E_r'(wz)$. Therefore, if we equate the coefficients of w^{r-1} on both sides of (5.1) and set $w = 1$, we obtain a differential recurrence for the univariate generating functions $E_r = E_r(z)$:

$$(r + \vartheta - T\vartheta)E_r \;=\; \tfrac{1}{2}\,e^{-V}\,\vartheta^2\,e^V\,E_{r-1}\,. \tag{5.3}$$

It is convenient to introduce a new variable

$$\zeta = \frac{T(z)}{1 - T(z)} \tag{5.4}$$

and to express E_r in terms of ζ instead of z. Note that

$$1 + \zeta = \frac{1}{1 - T(z)}\,; \qquad T(z) = \frac{\zeta}{1 + \zeta}\,; \qquad z = \frac{\zeta}{1 + \zeta}\,\exp\!\Big(\frac{-\zeta}{1 + \zeta}\Big)\,. \tag{5.5}$$

Equation (5.3) now takes the form

$$\big(r + (1 + \zeta)^{-1}\vartheta\big)E_r = \tfrac{1}{2}(1 + \zeta)^{-1/2}\vartheta^2(1 + \zeta)^{1/2}E_{r-1}\,, \tag{5.6}$$

since $e^V = 1/(1 - T(z))^{1/2} = (1 + \zeta)^{1/2}$ by (3.4). We will see later that the variable ζ, which represents an ordered sequence of one or more rooted trees, has important significance in the study of graphs and multigraphs.

In the ζ world, with ϑ still denoting $z \frac{d}{dz}$, we have the operator equation

$$\vartheta \cdot f(\zeta) = f'(\zeta)\zeta(1+\zeta)^2 + f(\zeta)\vartheta, \tag{5.7}$$

because

$$z\frac{d\zeta}{dz} = \frac{T(z)}{1-T(z)} z T'(z) \left(\frac{1}{T(z)} + \frac{1}{1-T(z)}\right) = \frac{T}{(1-T)^3} = \zeta(1+\zeta)^2.$$

Equation (5.7) allows us to commute ϑ with functions of ζ. For example, we find

$$(1+\zeta)^{-1/2}\vartheta(1+\zeta)^{1/2}$$
$$= (1+\zeta)^{-1/2}\left(\tfrac{1}{2}(1+\zeta)^{-1/2}\zeta(1+\zeta)^2 + (1+\zeta)^{1/2}\vartheta\right)$$
$$= \tfrac{1}{2}\zeta(1+\zeta) + \vartheta;$$

hence (5.6) can be rewritten

$$\left(r + (1+\zeta)^{-1}\vartheta\right)E_r = \tfrac{1}{2}\left(\tfrac{1}{2}\zeta(1+\zeta) + \vartheta\right)^2 E_{r-1}. \tag{5.8}$$

To simplify the equation even further, we seek a function $f_r(\zeta)$ such that

$$\vartheta \cdot f_r(\zeta) = (1+\zeta)f_r(\zeta)\left(r + (1+\zeta)^{-1}\vartheta\right);$$

then the differential equation (5.8) will become

$$\vartheta\left(f_r(\zeta)E_r\right) = \tfrac{1}{2}(1+\zeta)f_r(\zeta)\left(\tfrac{1}{2}\zeta(1+\zeta) + \vartheta\right)^2 E_{r-1},$$

which can be solved by integration. According to (5.7), the desired factor $f_r(\zeta)$ is a solution to

$$\frac{f_r'(\zeta)}{f_r(\zeta)} = \frac{r}{\zeta(1+\zeta)} = \frac{r}{\zeta} - \frac{r}{1+\zeta};$$

so we let $f_r(\zeta) = \zeta^r(1+\zeta)^{-r}$, which incidentally equals $T(z)^r$. We have derived the equation

$$\vartheta\left(\frac{\zeta^r E_r}{(1+\zeta)^r}\right) = \frac{1}{2}\frac{\zeta^r}{(1+\zeta)^{r-1}}\left(\tfrac{1}{2}\zeta(1+\zeta) + \vartheta\right)^2 E_{r-1}. \tag{5.9}$$

This differential equation determines E_r uniquely when $r > 0$, given E_{r-1}, since ζ^r vanishes when $z = 0$.

Now all the preliminary groundwork has been laid, and we are ready to calculate E_r. We know that $E_0 = 1$. A bit of experimentation soon reveals a fairly simple pattern: We can prove by induction on r that the solution to (5.9) has the form

$$E_r(z) = \sum_{d=0}^{2r} e_{rd}(1 + \zeta)^r \zeta^{2r-d} = \sum_{d=0}^{2r} \frac{e_{rd} T(z)^{2r-d}}{(1 - T(z))^{3r-d}}, \qquad (5.10)$$

where the coefficients e_{rd} are rational numbers, and where $e_{r(2r)} = 0$ for $r > 0$. Let $e_{rd} = 0$ when $d < 0$ or $d > 2r$. Assuming that (5.10) holds for some r, we use (5.7) and (5.8) to compute

$$A_r = \left(\tfrac{1}{2} \zeta (1 + \zeta) + \vartheta \right) E_r$$

$$= \sum_{d=0}^{2r} e_{rd}(1 + \zeta)^r \zeta^{2r-d} \zeta(1 + \zeta)^2 \left(\frac{\tfrac{1}{2}}{1 + \zeta} + \frac{r}{1 + \zeta} + \frac{2r - d}{\zeta} \right)$$

$$= \sum_{d=0}^{2r+1} a_{rd}(1 + \zeta)^{r+1} \zeta^{2r+1-d},$$

$$a_{rd} = (3r + \tfrac{1}{2} - d) e_{rd} + (2r + 1 - d) e_{r(d-1)}; \qquad (5.11)$$

$$B_r = \left(\tfrac{1}{2} \zeta (1 + \zeta) + \vartheta \right) A_r$$

$$= \sum_{d=0}^{2r+1} a_{rd}(1 + \zeta)^{r+1} \zeta^{2r+1-d} \zeta(1 + \zeta)^2 \left(\frac{\tfrac{1}{2}}{1 + \zeta} + \frac{r + 1}{1 + \zeta} + \frac{2r + 1 - d}{\zeta} \right)$$

$$= \sum_{d=0}^{2r+2} b_{rd}(1 + \zeta)^{r+2} \zeta^{2r+2-d},$$

$$b_{rd} = (3r + \tfrac{5}{2} - d) a_{rd} + (2r + 2 - d) a_{r(d-1)}. \qquad (5.12)$$

Moreover, the left side of Equation (5.9) is a polynomial,

$$\vartheta\left(\zeta^r (1 + \zeta)^{-r} E_r \right) = \vartheta \sum_{d=0}^{2r} e_{rd} \zeta^{3r-d} = \sum_{d=0}^{2r} (3r - d) e_{rd}(1 + \zeta)^2 \zeta^{3r-d}.$$

The corresponding polynomial on the right-hand side is

$$\tfrac{1}{2} \zeta^r (1 + \zeta)^{1-r} \left(\tfrac{1}{2} \zeta (1 + \zeta) + \vartheta \right)^2 E_{r-1} = \tfrac{1}{2} \sum_d b_{(r-1)d}(1 + \zeta)^2 \zeta^{3r-d};$$

therefore we can complete the induction proof by setting

$$e_{rd} = \frac{b_{(r-1)d}}{6r - 2d}, \qquad 0 \le d \le 2r. \tag{5.13}$$

It is easy to check that $a_{r(2r+1)} = 0$ and $b_{r(2r+2)} = 0$, hence $e_{r(2r)} = 0$ when $r > 0$.

In particular, $a_{00} = \frac{1}{2}$, $b_{00} = \frac{5}{4}$, $b_{01} = \frac{1}{2}$, and we obtain

$$E_1(z) = (1 + \zeta)\left(\frac{5}{24}\zeta^2 + \frac{1}{8}\zeta\right)$$
$$= \left(\frac{5}{24}\frac{T(z)^2}{(1 - T(z))^3} + \frac{1}{8}\frac{T(z)}{(1 - T(z))^2}\right). \tag{5.14}$$

A complex multigraph of excess 1 must consist of a single bicyclic component, so $E_1(z)$ is the function we called $W(z)$ in (3.7). If our only goal had been to compute $W(z)$, we could of course have gotten this result easily and directly. The more elaborate machinery above has been developed so that the generating function $E_r(z)$ can readily be computed and analyzed for larger values of r.

6. Enumerating Complex Graphs

For graphs instead of multigraphs, the calculations are more intricate, but it is instructive to look at them and see how they differ. As in (4.1) and (4.8), we separate off the cyclic and complex parts of the bgf by writing

$$\widehat{G}(w, z) = e^{U(wz)/w}\,\widehat{F}(w, z); \qquad \widehat{F}(w, z) = e^{\widehat{V}(wz)}\widehat{E}(w, z). \tag{6.1}$$

Adding a new edge to a graph means that we want to mark an unordered pair of *distinct* vertices, and the operator corresponding to this is $\frac{1}{2}(\vartheta_z^2 - \vartheta_z)$. We must also avoid duplicating an edge that's already present, so we must also subtract ϑ_w. Therefore the differential equation satisfied by $\widehat{G}$ is not (4.2) but

$$\frac{1}{w}\vartheta_w\widehat{G} = \left(\frac{\vartheta_z^2 - \vartheta_z}{2} - \vartheta_w\right)\widehat{G}; \tag{6.2}$$

and the integral equation corresponding to (4.3) is

$$\widehat{G}(w, z) = e^z + \int_0^w \left(\frac{\vartheta_z^2 - \vartheta_z}{2} - \vartheta_w\right)\widehat{G}(w, z)\,dw. \tag{6.3}$$

A computation similar to our derivation of (4.7) now leads to a differential equation defining $\widehat{F}$:

$$\frac{1}{w}\vartheta_w\widehat{F} = \left(\left(\frac{\vartheta_z^2 - \vartheta_z}{2} - \vartheta_w\right)U\right)\widehat{F} + (\vartheta_z U)(\vartheta_z\widehat{F})$$
$$+ \left(\frac{\vartheta_z^2 - \vartheta_z}{2} - \vartheta_w\right)\widehat{F}. \qquad (6.4)$$

The analog of (5.1) turns out to be

$$\frac{1}{w}\left(\vartheta_w\widehat{E} - T\vartheta_z\widehat{E}\right) = e^{-\widehat{V}}\left(\frac{\vartheta_z^2 - \vartheta_z}{2} - \vartheta_w\right)e^{\widehat{V}}\widehat{E}; \qquad (6.5)$$

converting to univariate generating functions $\widehat{E}_r(w,z) = w^r\widehat{E}_r(wz)$ yields

$$\left(r + \vartheta - T\vartheta\right)\widehat{E}_r = e^{-\widehat{V}}\left(1 - r + \frac{\vartheta^2 - 3\vartheta}{2}\right)e^{\widehat{V}}\widehat{E}_{r-1}. \qquad (6.6)$$

Again we multiply by the integration factor $\zeta^r/(1+\zeta)^r$, but the differential equation turns out to be rather messy:

$$\vartheta\left(\left(\frac{\zeta}{1+\zeta}\right)^r\widehat{E}_r\right) = \left(\frac{\zeta^r}{(1+\zeta)^{r-1}}\right)\left(1 - r + \frac{\zeta^4(10+14\zeta+5\zeta^2)}{8(1+\zeta)^2}\right.$$
$$\left. + \frac{\zeta^3-3\zeta-3}{2(1+\zeta)}\vartheta + \frac{\vartheta^2}{2}\right)\widehat{E}_{r-1}. \qquad (6.7)$$

At least it is linear, and it allows us to compute $\widehat{E}_r$ for small r. It turns out that the solution has the form

$$\widehat{E}_r = \sum_{d\geq 0}\hat{e}_{rd}\frac{\zeta^{5r-d}}{(1+\zeta)^{2r}} = \sum_{d\geq 0}\hat{e}_{rd}\frac{T(z)^{5r-d}}{(1-T(z))^{3r-d}}, \qquad (6.8)$$

for appropriate coefficients $\hat{e}_{rd}$. We have, of course, $\hat{e}_{00} = 1$ and $\hat{e}_{0d} = 0$ for $d \neq 0$. When $r > 0$, the values of $\hat{e}_{rd}$ satisfy the following recurrence, equivalent to (6.7):

$$(3r - d)\hat{e}_{rd} + (6r - d + 1)\hat{e}_{r(d-1)} = \sum_{j=0}^{6}c_j(r - 1,d)\hat{e}_{(r-1)(d-j)}, \qquad (6.9)$$

where

$$c_0(r, d) = (6r - 2d + 5)(6r - 2d + 1)/8,$$
$$c_1(r, d) = (132r^2 + (166 - 80d)r + 45 - 50d + 12d^2)/4,$$
$$c_2(r, d) = (398r^2 + (584 - 220d)r + 205 - 160d + 30d^2)/4,$$
$$c_3(r, d) = (316r^2 + (515 - 160d)r + 207 - 129d + 20d^2)/2, \qquad (6.10)$$
$$c_4(r, d) = (279r^2 + (484 - 130d)r + 208 - 112d + 15d^2)/2,$$
$$c_5(r, d) = (13r - 3d + 10)(5r - d + 5),$$
$$c_6(r, d) = (25r^2 + (43 - 10d)r + 18 - 9d + d^2)/2.$$

It is not at all obvious that this recurrence has a solution. We can use it to compute $\hat{e}_{rd}$ for $d = 0, 1, \ldots, 3r - 1$, but then the value of $\hat{e}_{r(3r-1)}$ must satisfy a nontrivial equation when we set $d = 3r$. To get the values of $\hat{e}_{rd}$ when $d \geq 3r$, we can start by assuming that $\hat{e}_{rd} = 0$ for $d \geq 6r$ and work backward. We will prove later that the recurrence always does have a solution, and that the last nonzero coefficient for fixed r can be completely characterized by an almost unbelievable (but true) formula:

$$\hat{e}_{r(5r-s)} = \binom{\binom{s}{2}}{s + r} \frac{1}{s!}, \qquad \text{if } \binom{s-2}{2} \leq r < \binom{s-1}{2}. \qquad (6.11)$$

Moreover, $\hat{e}_{rd} = 0$ for all $d > 5r - s$. Here is a table of values for small r, in case the reader would like to check a computer program that is based on the formulas above:

$d =$	0	1	2	3	4	5	6	7	8	9	10
$\hat{e}_{0d} =$	1										
$\hat{e}_{1d} =$	$\frac{5}{24}$	$\frac{1}{4}$									
$\hat{e}_{2d} =$	$\frac{385}{1152}$	$\frac{175}{96}$	$\frac{133}{32}$	$\frac{79}{16}$	$\frac{49}{16}$	$\frac{5}{6}$	$\frac{1}{24}$				
$\hat{e}_{3d} =$	$\frac{85085}{82944}$	$\frac{5005}{512}$	$\frac{97097}{2304}$	$\frac{7777}{72}$	$\frac{43621}{240}$	$\frac{200561}{960}$	$\frac{950569}{5760}$	$\frac{14001}{160}$	$\frac{7021}{240}$	$\frac{773}{144}$	$\frac{3}{8}$

7. A Surprising Pattern

The numbers $\hat{e}_{rd}$ that characterize cyclic graphs of excess r do not appear to have any nice mathematical properties. But when we calculate the corresponding coefficients e_{rd} for multigraphs, as defined in (5.11)–(5.13), we run into patterns that cry out for explanation. For example, here is a table showing the values for small r:

$d =$	0	1	2	3	4	5	6	7	8	9
$e_{0d} =$	1									
$e_{1d} =$	$\frac{5}{24}$	$\frac{1}{8}$								
$e_{2d} =$	$\frac{385}{1152}$	$\frac{35}{64}$	$\frac{91}{384}$	$\frac{1}{48}$						
$e_{3d} =$	$\frac{85085}{82944}$	$\frac{25025}{9216}$	$\frac{23023}{9216}$	$\frac{2849}{3072}$	$\frac{19}{160}$	$\frac{1}{384}$				
$e_{4d} =$	$\frac{37182145}{7962624}$	$\frac{11316305}{663552}$	$\frac{3556553}{147456}$	$\frac{3658655}{221184}$	$\frac{1656083}{294912}$	$\frac{8723}{10240}$	$\frac{1969}{46080}$	$\frac{1}{3840}$		
$e_{5d} =$	$\frac{5391411025}{191102976}$	$\frac{929553625}{7077888}$	$\frac{7994161175}{31850496}$	$\frac{8068525465}{31850496}$	$\frac{341105765}{2359296}$	$\frac{327803333}{7077888}$	$\frac{1606891}{207360}$	$\frac{140569}{245760}$	$\frac{4043}{322560}$	$\frac{1}{46080}$

Anybody who has played with integers knows that the numerator of e_{32}, 23023, is equal to $7 \cdot 11 \cdot 13 \cdot 23$; moreover, the denominator is $9216 = 2^{10} \cdot 3^2$. Further experiments show that the factorization of, say, e_{55}, is $2^{-18} \cdot 3^{-3} \cdot 11 \cdot 13 \cdot 17 \cdot 19 \cdot 47 \cdot 151$. The occurrence of so many small prime factors cannot be a coincidence!

It is, in fact, easy to see the pattern in the numbers e_{r0}, which satisfy the recurrence

$$e_{r0} = \frac{(6r-1)(6r-5)}{24r} e_{(r-1)0} \qquad (7.1)$$

according to rules (5.11)–(5.13). The numbers $\hat{e}_{r0}$ also satisfy the same recurrence, according to (6.9) and (6.10). Therefore we find

$$e_{r0} = \hat{e}_{r0} = \frac{(6r)!}{2^{5r}3^{2r}(3r)!\,(2r)!}. \qquad (7.2)$$

But the recurrence defining e_{rd} for $d > 0$ is much more complex, and we have no a priori reason to expect these numbers to have any mathematical virtues. The following theorem provides an algebraic explanation of what is going on.

Theorem 1. *The numbers e_{rd} defined in (5.10) can be expressed as*

$$e_{rd} = \frac{(6r-2d)!\,P_d(r)}{2^{5r}\,3^{2r-d}\,(3r-d)!\,(2r-d)!}, \qquad (7.3)$$

where $P_d(r)$ is a polynomial of degree d defined by the formulas

$$P_d(r) = [z^d]\,F(z)^{2r-d}, \qquad (7.4)$$

$$F(z) = 3! \sum_{n \geq 0} \frac{(4z)^n}{(n+3)!} = \frac{6}{(4z)^3}\left(e^{4z} - \frac{(4z)^2}{2} - 4z - 1\right). \qquad (7.5)$$

Proof. By the duplication and triplication formulas for the Gamma function, expression (7.3) can also be written

$$e_{rd} = g_{rd}P_d(r), \qquad g_{rd} = \frac{3^r \, \Gamma(r+\frac{5}{6}-\frac{d}{3})\,\Gamma(r+\frac{1}{2}-\frac{d}{3})\,\Gamma(r+\frac{1}{6}-\frac{d}{3})}{2^{r+d}\,2\pi\,\Gamma(r+1-\frac{d}{2})\,\Gamma(r+\frac{1}{2}-\frac{d}{2})}. \qquad (7.6)$$

Therefore recurrence equation (5.11) becomes

$$a_{rd} = 3(r+\tfrac{1}{6}-\tfrac{d}{3})g_{rd}P_d(r) + 2(r+\tfrac{1}{2}-\tfrac{d}{2})g_{r(d-1)}P_{d-1}(r)$$

$$= 3(r+\tfrac{1}{6}-\tfrac{d}{3})g_{rd}A_d(r),$$

$$A_d(r) = P_d(r) + \tfrac{4}{3}P_{d-1}(r). \qquad (7.7)$$

Similarly, but without as much cancellation, (5.12) becomes

$$b_{rd} = 3(r+\tfrac{5}{6}-\tfrac{d}{3})3(r+\tfrac{1}{6}-\tfrac{d}{3})g_{rd}A_d(r)$$

$$\qquad + 2(r+1-\tfrac{d}{2})3(r+\tfrac{1}{2}-\tfrac{d}{3})g_{r(d-1)}A_{d-1}(r)$$

$$= \tfrac{9}{2}g_{r(d-1)}B_d(r),$$

$$B_d(r) = (r+\tfrac{5}{6}-\tfrac{d}{3})(r+\tfrac{1}{2}-\tfrac{d}{2})A_d(r)$$

$$\qquad + \tfrac{4}{3}(r+1-\tfrac{d}{2})(r+\tfrac{1}{2}-\tfrac{d}{3})A_{d-1}(r). \qquad (7.8)$$

Relation (5.13) becomes

$$(3r + 3 - d)g_{(r+1)d}P_d(r+1)$$

$$= \frac{9}{4}\frac{(r+1-\frac{d}{3})(r+\frac{5}{6}-\frac{d}{3})(r+\frac{1}{2}-\frac{d}{3})}{(r+1-\frac{d}{2})}g_{r(d-1)}P_d(r+1) = \frac{1}{2}b_{rd};$$

hence the original recurrence takes the following form:

$$(r+1-\tfrac{d}{3})(r+\tfrac{5}{6}-\tfrac{d}{3})(r+\tfrac{1}{2}-\tfrac{d}{3})P_d(r+1) = (r+1-\tfrac{d}{2})B_d(r). \qquad (7.9)$$

The boundary conditions are

$$P_d(r) = 0 \text{ for } d < 0; \quad P_0(r) = 1; \quad P_{2d}(d) = 0 \text{ for } d > 0. \qquad (7.10)$$

It is by no means obvious that a polynomial $P_d(r)$ will satisfy (7.7), (7.8), and (7.9). The key observation that makes everything work is that a solution to the simpler recurrence

$$(r + \tfrac{1}{2} - \tfrac{d}{3})P_d(r + \tfrac{1}{2}) = (r + \tfrac{1}{2} - \tfrac{d}{2})A_d(r) \qquad (7.11)$$

suffices to solve the more complex one. This new recurrence is sort of a "half step" between solutions of (7.7), (7.8), and (7.9); it tells us about multigraphs whose excess is an integer plus $\tfrac{1}{2}$, whatever that may mean.

A solution to (7.11) in the extended domain implies a solution to (7.9). For we will then have

$$(r+1-\tfrac{d}{3})(r+\tfrac{5}{6}-\tfrac{d}{3})(r+\tfrac{1}{2}-\tfrac{d}{3})P_d(r+1)$$
$$= (r+\tfrac{5}{6}-\tfrac{d}{3})(r+\tfrac{1}{2}-\tfrac{d}{3})(r+1-\tfrac{d}{2})A_d(r+\tfrac{1}{2})$$

and

$$(r+1-\tfrac{d}{2})B_d(r) = (r+1-\tfrac{d}{2})(r+\tfrac{5}{6}-\tfrac{d}{3})(r+\tfrac{1}{2}-\tfrac{d}{3})A_d(r)$$
$$+ \tfrac{4}{3}(r+1-\tfrac{d}{2})^2(r+\tfrac{1}{2}-\tfrac{d}{3})A_{d-1}(r)$$
$$= (r+1-\tfrac{d}{2})(r+\tfrac{5}{6}-\tfrac{d}{3})(r+\tfrac{1}{2}-\tfrac{d}{3})P_d(r+\tfrac{1}{2})$$
$$+ \tfrac{4}{3}(r+1-\tfrac{d}{2})(r+\tfrac{5}{6}-\tfrac{d}{3})(r+\tfrac{1}{2}-\tfrac{d}{3})P_{d-1}(r+\tfrac{1}{2}).$$

Moreover, $P_d(\tfrac{d}{2}) = 0$ when $d > 0$.

We can solve the simultaneous recurrences (7.7) and (7.11) by constructing solutions to (7.7) that have the desired form (7.4), namely

$$P_d(r) = [z^d]\, F(z)^{2r-d}, \qquad A_d(r) = [z^d]\, F(z)^{2r-d}\big(1+\tfrac{4}{3}zF(z)\big),$$

and noting that the function $F(z)$ of (7.5) satisfies

$$\vartheta F(z) = 4zF(z) + 3 - 3F(z). \tag{7.12}$$

Thus we have

$$dP_d(r+\tfrac{1}{2}) = [z^d]\,\vartheta\big(F(z)^{2r+1-d}\big)$$
$$= [z^d]\,(2r+1-d)F(z)^{2r-d}\big(4zF(z)+3-3F(z)\big)$$
$$= (6r+3-3d)\big(A_d(r) - P_d(r+\tfrac{1}{2})\big),$$

and (7.11) holds. $\square$

Incidentally, the theory of confluent hypergeometric functions provides us with alternative expressions for the function $F(z)$ in (7.5). We have, for example,

$$F(z) = F(1; 4; 4z) = 3\int_0^1 e^{4zt}(1-t)^2\, dt \tag{7.13}$$

$$= \frac{3e^{4z}}{64z^3}\,\gamma(3, 4z) = 3e^{4z}\left(\frac{1}{3\cdot 0!} - \frac{4z}{4\cdot 1!} + \frac{4^2 z^2}{5\cdot 2!} - \frac{4^3 z^3}{6\cdot 3!} + \cdots\right).$$

The general theory of [23] also allows us to write

$$P_d(r) = \frac{2r - d}{2r} [z^d] G(z)^{2r}, \tag{7.14}$$

where $G(z) = 1 + z - \frac{1}{5}z^2 + \frac{2}{15}z^3 - \frac{19}{175}z^4 + \frac{2}{21}z^5 - \frac{2018}{23625}z^6 + \cdots$ is defined implicitly by the relation

$$G\big(zF(z)\big) = F(z). \tag{7.15}$$

Corollary. *For fixed $d \geq 0$ we have*

$$e_{rd} = \frac{3^r}{2^r} \frac{(r + d - 1)!}{2\pi\, d!} \big(1 + O(r^{-1})\big) \tag{7.16}$$

as $r \to \infty$. Moreover, e_{rd} is a rational number whose numerator has at most

$$d + O\big(d(\log d)^2 / \log r\big) \tag{7.17}$$

prime factors greater than $6r$, and whose denominator has no prime factors greater than $3r$.

Proof. The obvious bounds

$$\binom{2r - d}{d} = [z^d] (1 + z)^{2r - d} \leq [z^d] F(z)^{2r - d}$$

$$\leq [z^d] \Big(\frac{1}{1 - z}\Big)^{2r - d} = \binom{2r - 1}{d} \tag{7.18}$$

tell us that $P_d(r) = (2r)^d/d! + O(r^{d-1})$. Formula (7.16) now follows from (7.3) and Stirling's approximation. (We will derive a more precise estimate, suitable when d varies with r, in Section 23 below, Lemma 8.)

All prime factors greater than $6r$ must appear as prime factors of $P_d(r)$. We will prove the upper bound (7.17) by showing that $m_d P_d(r)$ is an integer, where

$$m_d = 5^{\lfloor d/2 \rfloor} 6^{\lfloor d/3 \rfloor} 7^{\lfloor d/4 \rfloor} \cdots = \prod_{k \geq 2} (k + 3)^{\lfloor d/k \rfloor}. \tag{7.19}$$

It will follow that the denominator of $P_d(r)$ contains no prime factors greater than $2r + 1$, and that if the numerator contains k prime factors greater than $6r$, we have $(6r)^k < m_d P_d(r) \leq m_d \binom{2r-1}{d} < m_d (2r)^d$; that is, $k \log 6r < d \log 2r + \log m_d = d \log 2r + O\big(d(\log d)^2\big)$.

The coefficient of z^d in any power of $F(z)$ is a sum of terms $f_1^{k_1} f_2^{k_2} f_3^{k_3} \cdots$, where $f_j = [z^j] F(z) = \frac{4}{5} \frac{4}{6} \cdots \frac{4}{(j+3)}$ and $k_1 + 2k_2 + 3k_3 + \cdots = d$. Thus, for example, the factor 7 occurs in the denominator of $f_1^{k_1} f_2^{k_2} f_3^{k_3} \cdots$ exactly $k_4 + k_5 + \cdots \leq d/4$ times. It follows that the denominator of P_d is a divisor of m_d. □

The estimate (7.17) can be sharpened for small d, because $P_d(r)$ always has $(2r - d)$ as a factor when $d > 0$. For example,

$$P_1(r) = 2r - 1, \qquad P_2(r) = \frac{(r-1)(10r-7)}{5},$$

$$P_3(r) = \frac{(2r-3)(10r^2 - 21r + 10)}{15}.$$

There are no prime factors $> 6r$ when $d \leq 1$, and there is at most one when $d \leq 3$.

Instead of writing

$$E_r(z) = \sum_{d=0}^{2r} e_{rd} \frac{T(z)^{2r-d}}{\left(1 - T(z)\right)^{3r-d}},$$

it is sometimes convenient to use coefficients e'_{rd} such that

$$E_r(z) = \sum_{d=0}^{2r} \frac{e'_{rd}}{\left(1 - T(z)\right)^{3r-d}}. \qquad (7.20)$$

The following table shows that the numbers e'_{rd} tend to alternate in sign:

$d =$	0	1	2	3	4	5	6	7	8
$e'_{0d} =$	1								
$e'_{1d} =$	$\frac{5}{24}$	$-\frac{7}{24}$	$\frac{1}{12}$						
$e'_{2d} =$	$\frac{385}{1152}$	$-\frac{455}{576}$	$\frac{77}{128}$	$-\frac{43}{288}$	$\frac{1}{288}$				
$e'_{3d} =$	$\frac{85085}{82944}$	$-\frac{95095}{27648}$	$\frac{119119}{27648}$	$-\frac{201355}{82944}$	$\frac{38623}{69120}$	$-\frac{803}{34560}$	$-\frac{139}{51840}$		
$e'_{4d} =$	$\frac{37182145}{7962624}$	$-\frac{40415375}{1990656}$	$\frac{141292151}{3981312}$	$-\frac{62775713}{1990656}$	$\frac{116866321}{7962624}$	$-\frac{15867137}{4976640}$	$\frac{850003}{4976640}$	$\frac{25129}{1244160}$	$-\frac{571}{2488320}$

Again, patterns lurk beneath the surface, and there is a prevalence of small prime factors; for example, $-e'_{55} = \frac{7541601353}{63700992} = 2^{-18} \cdot 3^{-5} \cdot 11 \cdot 13 \cdot 17 \cdot 19 \cdot 23 \cdot 31 \cdot 229$. We can in fact prove the existence of a pattern similar to that of the original coefficients e_{rd}:

Corollary. *The numbers e'_{rd} defined in (7.20) can be expressed as*

$$e'_{rd} = \frac{(6r - 2d)! \, Q_d(r)}{2^{5r} \, 3^{2r-d} \, (3r - d)! \, (2r - d)!}, \qquad (7.21)$$

where $Q_d(r)$ is a polynomial of degree d for which $Q_d\left(\frac{d}{3} - \frac{1}{2}\right) = 0$ when $d > 0$.

Proof. By definition, we have

$$e'_{rd} = \sum_{k=0}^{d} \binom{2r-k}{d-k}(-1)^{d-k} e_{rk},\qquad (7.22)$$

because the quantity $T^{2r-k} = (1 - (1 - T))^{2r-k}$ contributes exactly $\binom{2r-k}{d-k}(-1)^{d-k}$ to the coefficient of $(1 - T)^{d-3r}$. Now if we plug in equations (7.3) and (7.21), we find that

$$Q_d(r) = \sum_{k=0}^{d} \frac{(-1)^{d-k}P_k(r)}{3^{d-k}(d-k)!} \frac{(6r-2k)!}{(6r-2d)!} \frac{(3r-d)!}{(3r-k)!}$$

$$= \sum_{k=0}^{d} \left(-\frac{4}{3}\right)^{d-k} \binom{3r-k-\frac{1}{2}}{d-k} P_k(r)$$

$$= \sum_{k=0}^{d} \left(-\frac{4}{3}\right)^{d-k} \binom{3r-k+\frac{1}{2}}{d-k} A_k(r),\qquad (7.23)$$

clearly a polynomial in r of degree $\leq d$. In fact, the leading term is

$$\sum_{k} \left(-\frac{4}{3}\right)^{d-k} \frac{(3r)^{d-k}}{(d-k)!} \frac{(2r)^k}{k!} = \frac{(-2)^d r^d}{d!},$$

so the degree is exactly d. If we set $r = \frac{d}{3} - \frac{1}{2}$, the sum reduces to $A_d\left(\frac{d}{3} - \frac{1}{2}\right)$, which we know is zero for $d > 0$ by (7.11). □

It is interesting to try to compute the coefficients e'_{rd} directly, by proceeding as we did in Section 5 but using the variable

$$\xi = 1 + \zeta = (1 - T(z))^{-1}$$

in place of ζ. The calculations are essentially the same, even slightly simpler, until we get to the analog of equation (5.13); the recurrences that replace (5.11)–(5.13) are

$$a'_{rd} = (3r+\tfrac{1}{2}-d)e'_{rd} - (3r+\tfrac{3}{2}-d)e'_{r(d-1)};\qquad (7.24)$$

$$(3r-d)e'_{rd} - (2r+1-d)e'_{r(d-1)}$$
$$= \tfrac{1}{2}\left((3r-\tfrac{1}{2}-d)a'_{(r-1)d} - (3r+\tfrac{1}{2}-d)a'_{(r-1)(d-1)}\right).\qquad (7.25)$$

It appears to be quite difficult to derive (7.21) directly from these recurrences. The recurrence for $Q_d(r)$, corresponding to equation (7.9) for $P_d(r)$, turns out to be

$$(r - \tfrac{d}{3})(r-\tfrac{1}{2}-\tfrac{d}{3})Q_d(r)$$
$$= (r - \tfrac{d}{2})(r-\tfrac{1}{2}-\tfrac{d}{2})Q_d(r-1)$$
$$+ \tfrac{4}{3}(r+\tfrac{1}{6}-\tfrac{d}{3})(r-\tfrac{1}{2}-\tfrac{d}{3})Q_{d-1}(r)$$
$$- 4(r-\tfrac{d}{2})(r-\tfrac{1}{2}-\tfrac{d}{2})(r-\tfrac{d}{3})(r-\tfrac{1}{6}-\tfrac{d}{3})^{-1}Q_{d-1}(r-1)$$
$$+ 4(r+\tfrac{1}{6}-\tfrac{d}{3})(r-\tfrac{1}{2}-\tfrac{d}{3})Q_{d-2}(r-1)\,, \qquad (7.26)$$

and we can proceed to solve it for $d = 1, 2, \ldots$, if we first multiply both sides by the summation factor

$$\Gamma(r-\tfrac{d}{3})\Gamma(r-\tfrac{1}{2}-\tfrac{d}{3})\Gamma(r+1-\tfrac{d}{2})^{-1}\Gamma(r+\tfrac{1}{2}-\tfrac{d}{2})^{-1}.$$

The equation for $d > 0$ then takes the form

$$S_d(r) = S_d(r-1) + g_d(r) + g_d(r - \tfrac{1}{2})\,,$$

$$S_d(r) = \frac{\Gamma(r+1-\tfrac{d}{3})\,\Gamma(r+\tfrac{1}{2}-\tfrac{d}{3})}{\Gamma(r+1-\tfrac{d}{2})\,\Gamma(r+\tfrac{1}{2}-\tfrac{d}{2})}\,Q_d(r)\,,$$

$$g_d(r) = \frac{\Gamma(r+1-\tfrac{d}{3})\,\Gamma(r+\tfrac{1}{2}-\tfrac{d}{3})}{\Gamma(r+1-\tfrac{d}{2})\,\Gamma(r+\tfrac{1}{2}-\tfrac{d}{2})}\,f_d(r)\,,$$

where $f_d(r) = Q_d(r) - \frac{r-d/2}{r-d/3}Q_d(r - \tfrac{1}{2})$ is a polynomial of degree $d - 1$. For example, $f_1(r) = -\tfrac{4}{3}$ and $f_2(r) = \tfrac{8}{3}r - \tfrac{4}{3}$. There is apparently no analog of the simple relation (7.11) that made everything work nicely in Theorem 1.

A generating function for $Q_d(r)$, analogous to (7.4), can be found by analyzing (7.23) more carefully. Let $H(z)$ satisfy

$$H(z) = F\big(zH(z)^{-1/3}\big) = 1 + z + \tfrac{7}{15}z^2 + \tfrac{1}{15}z^3 + \cdots ; \qquad (7.27)$$

then the elementary theory in [23] proves that

$$\big(x - \tfrac{d}{3}\big)\,[z^d]\,H(z)^x = x\,[z^d]\,F(z)^{x-d/3}\,. \qquad (7.28)$$

Hence, by (7.11) and (7.4),

$$A_d(r) = \frac{r+\tfrac{1}{2}-\tfrac{d}{3}}{r+\tfrac{1}{2}-\tfrac{d}{2}}\,[z^d]\,F(z)^{2r+1-d} = [z^d]\,H(z)^{2r+1-2d/3}\,. \qquad (7.29)$$

And (7.23) can therefore be "summed":

$$Q_d(r) = \sum_{k=0}^{d} \left(-\frac{4}{3}\right)^k \binom{3r - d + k + \frac{1}{2}}{k} A_{d-k}(r)$$

$$= \sum_{k=0}^{d} \left(\frac{4}{3}\right)^k \binom{-3r + d - \frac{3}{2}}{k} [z^{d-k}] H(z)^{(-2/3)(-3r-3/2+d-k)}$$

$$= [z^d] \left(\frac{4}{3} z + H(z)^{-2/3}\right)^{-3r-3/2+d} . \tag{7.30}$$

In particular,

$$Q_0(r) = 1; \qquad Q_1(r) = -2(r + \tfrac{1}{6}); \qquad Q_2(r) = 2(r - \tfrac{1}{6})(r - \tfrac{1}{5}).$$

Although $Q_1(r) = -A_1(r)$ and $Q_2(r) = A_2(r)$, the next case is

$$Q_3(r) = -A_3(r) + \tfrac{16}{135}(r - \tfrac{1}{2}).$$

8. Sparse Components

We can readily compute the univariate generating functions $C_1(z)$, $C_2(z)$, $C_3(z)$, ..., $C_r(z)$ for bicyclic, tricyclic, tetracyclic, ..., $(r + 1)$-cyclic components, now that we know the simple form of $E_1(z)$, $E_2(z)$, $E_3(z)$, ..., $E_r(z)$, because of the fact that

$$\sum_{r \geq 0} w^r E_r = \exp\left(\sum_{r \geq 1} w^r C_r\right). \tag{8.1}$$

Differentiating this formula with respect to w and equating coefficients of w^{r-1} leads to the expression

$$r E_r = \sum_{k=1}^{r} k C_k E_{r-k} , \tag{8.2}$$

from which we may find C_r by calculating

$$C_r = E_r - \frac{1}{r} \sum_{k=1}^{r-1} k C_k E_{r-k} . \tag{8.3}$$

Since we know that $E_r = (1 + \zeta)^r \sum_{d=0}^{2r-1} e_{rd} \zeta^{2r-d}$ for $r > 0$, it follows by induction that C_r can be written in the same form,

$$C_r = (1 + \zeta)^r \sum_{d=0}^{2r-1} c_{rd} \zeta^{2r-d} , \tag{8.4}$$

for appropriate coefficients c_{rd}. (Here, as in Section 5, the variable ζ stands for $T(z)/(1 - T(z))$.) Indeed, relation (8.3) tells us that we can compute c_{rd} by evaluating a double sum

$$c_{rd} = e_{rd} - \frac{1}{r} \sum_{k=1}^{r-1} k \sum_j c_{kj} e_{(r-k)(d-j)} ; \qquad (8.5)$$

the inner sum here is over the range $\max(0, d + 1 - 2r + 2k) \leq j \leq \min(d, 2k - 1)$, which is always nonempty for $0 < k < r$ except when $d = 2r - 1$. We always have $c_{r(2r-1)} = e_{r(2r-1)} = 1/(2^{r+1}(r+1)!)$. Here is a table of the coefficients for small r:

$d =$	0	1	2	3	4	5	6	7	8	9
$c_{1d} =$	$\frac{5}{24}$	$\frac{1}{8}$								
$c_{2d} =$	$\frac{5}{16}$	$\frac{25}{48}$	$\frac{11}{48}$	$\frac{1}{48}$						
$c_{3d} =$	$\frac{1105}{1152}$	$\frac{985}{384}$	$\frac{1373}{576}$	$\frac{515}{576}$	$\frac{223}{1920}$	$\frac{1}{384}$				
$c_{4d} =$	$\frac{565}{128}$	$\frac{12455}{768}$	$\frac{26581}{1152}$	$\frac{12227}{768}$	$\frac{2089}{384}$	$\frac{9583}{11520}$	$\frac{27}{640}$	$\frac{1}{3840}$		
$c_{5d} =$	$\frac{82825}{3072}$	$\frac{387005}{3072}$	$\frac{371195}{1536}$	$\frac{10154003}{41472}$	$\frac{121207}{864}$	$\frac{519883}{11520}$	$\frac{1573507}{207360}$	$\frac{2597}{4608}$	$\frac{803}{64512}$	$\frac{1}{46080}$

In applications, the leading coefficients c_{r0} of C_r are the most important, as are the leading coefficients e_{r0} of E_r, because these govern the dominant asymptotic behavior of $[z^n]\,C_r(z)$ and $[z^n]\,E_r(z)$. Therefore it is convenient to write

$$c_r = c_{r0}, \qquad e_r = e_{r0} . \qquad (8.6)$$

We have seen in (7.2) that there is a simple way to express the numbers e_r in terms of factorials. The values c_r are then easily computed by using relation (8.3), but with c_r and e_r substituted respectively for C_r and E_r.

Asymptotically speaking, the values of c_{rd} and e_{rd} are equivalent when r is large.

Theorem 2. *For fixed $d \geq 0$ we have*

$$c_{rd} = e_{rd}\bigl(1 + O(r^{-1})\bigr) = \frac{3^r}{2^r}\,\frac{(r + d - 1)!}{2\pi\,d!}\bigl(1 + O(r^{-1})\bigr) \qquad (8.7)$$

as $r \to \infty$.

Proof. We know the asymptotic value of e_{rd} from (7.16). To complete the proof, we need only show that the double sum in (8.5) is $O_d(e_{rd}/r)$, where O_d implies a bound for fixed d as $r \to \infty$.

Since $c_{rd} \leq e_{rd}$, each term in the double sum is bounded above by an absolute constant (depending on d) times

$$\frac{3^r}{2^r} \frac{k}{r} \frac{(k+j-1)!}{j!} \frac{(r-k+d-j-1)!}{(d-j)!} = \frac{3^r}{2^r} \frac{k}{r} \frac{(r+d-2)!}{d!} \binom{d}{j} \bigg/ \binom{r+d-2}{k+j-1}.$$

We have $\binom{r+d-2}{k+j-1} \geq r+d-2$ except when $k = 1$ and $j = 0$ or $k = r-1$ and $j = d$. Therefore all but one term is $O_d(e_{rd}/r^2)$, and the exceptional term is $O_d(e_{rd}/r)$. There are $O(rd)$ terms altogether, so the overall double sum is $O_d(e_{rd}/r)$. $\qed$

The simple form (8.4) of $C_r(z)$, the generating function for $(r+1)$-cyclic multigraphs, makes it possible for us to deduce a formula for the corresponding graph-based function $\widehat{C}_r(z)$, which turns out to be only about 50% more complicated. In fact, we will prove a result that applies to the generating functions for infinitely many models of random graphs, including both $G(w, z)$ and $\widehat{G}(w, z)$ as special cases.

Our starting point for this calculation is the formal power series relation

$$\widehat{G}(w, z) = G(\ln(1 + w), z/\sqrt{1 + w}). \tag{8.8}$$

which is an immediate consequence of (2.7) and (2.9). It follows that

$$\widehat{C}(w, z) = C(\ln(1 + w), z/\sqrt{1 + w}). \tag{8.9}$$

We can therefore obtain a near-polynomial formula for $\widehat{C}_r(z)$ as a special case of the following result.

Theorem 3. If $f(w) = 1 + f_1 w + f_2 w^2 + \cdots$ and $g(w) = 1 + g_1 w + g_2 w^2 + \cdots$ are arbitrary formal power series with $f(0) = g(0) = 1$, and if

$$\widetilde{C}(w, z) = C\left(wf(w), z\frac{g(w)}{f(w)}\right) = \sum_r w^r \widetilde{C}_r(wz), \tag{8.10}$$

where C is the bgf (2.10) for connected multigraphs, then there exist coefficients $\tilde{c}_{rd}$ such that

$$\widetilde{C}_r(z) = \sum_{d=0}^{3r+2} \tilde{c}_{rd} \zeta^{3r+2-d} (1 + \zeta)^{-2} = \sum_{d=0}^{3r+2} \tilde{c}_{rd} \frac{T(z)^{3r+2-d}}{(1 - T(z))^{3r-d}} \tag{8.11}$$

for all $r > 0$.

Proof. Consider Ramanujan's function $Q(n)$ of (3.11), which has the asymptotic value $\sqrt{\pi n/2} + O(1)$ as $n \to \infty$. Following Knuth [22], we shall say that a function $s(n)$ of the form $p(n) + q(n)Q(n)$ is a *semipolynomial* when p and q are polynomials. The *degree* of a semipolynomial is computed by assuming that $Q(n)$ is of degree $\frac{1}{2}$. For example, $3 + 2n + (1 + n)Q(n)$ is a semipolynomial of degree $\frac{3}{2}$. More formally, if d is any nonnegative integer, the semipolynomial $p(n) + q(n)Q(n)$ has degree $\leq \frac{1}{2}d$ if and only if p has degree $\leq \frac{1}{2}d$ and q has degree $< \frac{1}{2}d$.

The formulas (3.12) of Section 3, taken from [24], show that generating functions of the form $F(z) = \sum_{k=1}^{d} a_k / (1 - T(z))^k$ are precisely those whose coefficients satisfy

$$[z^n]\, F(z) = \frac{n^n s(n)}{n!}$$

where $s(n)$ is a semipolynomial of degree $\leq \frac{1}{2}(d - 1)$.

Consider now the expansion

$$\sum_r w^r f(w)^r C_r\big(zwg(w)\big) = \sum_r w^r \, \widetilde{C}_r(wz)$$

which follows from (8.10) and (2.11). We will study how each term on the left contributes to terms on the right. First, when $r = -1$ we have

$$\frac{U\big(zwg(w)\big)}{wf(w)} = \sum_{n\geq 1} \frac{n^{n-2}z^n w^{n-1}(1 + g_1 w + \cdots)^n}{n!\,(1 + f_1 w + \cdots)}$$

$$= \sum_{n\geq 1} \frac{n^{n-2}z^n w^{n-1}\big(1 + np_0(n)w + np_1(n)w^2 + \cdots\big)}{n!\,(1 + f_1 w + \cdots)}$$

where each $p_l(n)$ is a polynomial of degree $\leq l$. The effect is to make $\widetilde{C}_{-1}(z) = U(z)$, and to contribute a linear combination of $U(z)$, $T(z)$, and $(1 - T(z))^{-1}, \ldots, (1 - T(z))^{-2l+1}$ to $\widetilde{C}_l(z)$ for each $l \geq 0$. Next, when $r = 0$ we have

$$V\big(zwg(w)\big) = \frac{1}{2}\sum_{n\geq 1} n^{n-1}Q(n)z^n w^n \big(1 + np_0(n)w + np_1(n)w^2 + \cdots\big);$$

this contributes $V(z)$ to $\widetilde{C}_0(z)$ and a linear combination of $(1 - T(z))^{-1}$, $\ldots$, $(1 - T(z))^{-2l}$ to $\widetilde{C}_l(z)$ for each $l > 0$. Finally, when $r > 0$ we have, by (5.11),

$$w^r f(w)^r C_r\big(zwg(w)\big)$$
$$= \sum_{n\geq 0} \frac{n^n s(n)}{n!} z^n w^{n+r}\big(1 + np_0(n)w + np_1(n)w^2 + \cdots\big)f(w)^r,$$

where $s(n)$ is a semipolynomial of degree $\leq (3r-1)/2$. This contributes a linear combination of $(1-T(z))^{-1}, \ldots, (1-T(z))^{-2l-r}$ to $\widetilde{C}_l(z)$ for each $l \geq r$. The proof of (8.11) is complete, because $U(z) = \frac{1}{2}\zeta(2+\zeta)/(1+\zeta)^2$ and $T(z) = \zeta/(1+\zeta)$. $\quad\square$

Incidentally, our proof shows that the only contribution to the coefficient of the "leading term" $T(z)^{3l+2}/(1-T(z))^{3l}$ of $\widetilde{C}_l(z)$ comes from $C_l(z)$ itself. Therefore $\widetilde{C}_r(z)$ and $C_r(z)$ have identical leading coefficients. In particular, $\hat{c}_{r0} = c_{r0} = c_r$. We will see below that this gives the same asymptotic characteristics to the limiting distribution of component types in the uniform and permutation models when $m \approx \frac{1}{2}n$.

Theorem 3 justifies our earlier assertion that the recurrence (6.9)–(6.10) for $\hat{e}_{rd}$ has a solution. The coefficients $\hat{c}_{rd}$ can be computed from those coefficients $\hat{e}_{rd}$ using the relation

$$\widehat{C}_r = \widehat{E}_r - \frac{1}{r}\sum_{k=1}^{r-1} k\widehat{C}_k\widehat{E}_{r-k};$$

but that makes $\widehat{C}_r$ a polynomial of degree $5r$ with denominator $(1+\zeta)^{2r}$, so the numerator and denominator must be divided by $(1+\zeta)^{2r-2}$. A simpler recurrence for $\widehat{C}_r$ was found by Wright [41], who proved Theorem 3 in the special case that $\widetilde{C}_r = \widehat{C}_r$ by a different method. Translated into the notation of the present paper, Wright's recurrence is

$$\vartheta\left(\frac{\zeta}{1+\zeta}\right)^r \widehat{C}_r$$
$$= \frac{\zeta^r}{2(1+\zeta)^{r-1}}\left(\sum_{j=0}^{r-1}(\vartheta\widehat{C}_j)(\vartheta\widehat{C}_{r-1-j}) + \left(\vartheta^2 - 3\vartheta - 2(r-1)\right)\widehat{C}_{r-1}\right),$$
$$\text{for } r > 0, \qquad (8.12)$$

with $\vartheta\widehat{C}_0 = \frac{1}{2}\zeta^3(1+\zeta)^{-1}$. As we saw for the related sequence $\widehat{E}_r$ in Section 6, it isn't obvious that this recurrence has a solution of the desired form

$$\widehat{C}_r(z) = \sum_{d=0}^{3r+2} \hat{c}_{rd}\zeta^{3r+2-d}(1+\zeta)^{-2} = \sum_{d=0}^{3r+2} \hat{c}_{rd}\frac{T(z)^{3r+2-d}}{(1-T(z))^{3r-d}}, \quad (8.13)$$

when $r > 0$. Theorem 3 provides an algebraic proof, while Wright proved the existence by a combination of algebraic and combinatorial methods that we will consider in the next section. Here is a table of the first few values of the coefficients:

$d =$	0	1	2	3	4	5	6	7	8	9	10	11	12
$\hat{c}_{1d} =$	$\frac{5}{24}$	$\frac{1}{4}$											
$\hat{c}_{2d} =$	$\frac{5}{16}$	$\frac{55}{48}$	$\frac{73}{48}$	$\frac{3}{4}$	$\frac{1}{24}$								
$\hat{c}_{3d} =$	$\frac{1105}{1152}$	$\frac{395}{72}$	$\frac{15131}{1152}$	$\frac{2399}{144}$	$\frac{8303}{720}$	$\frac{557}{144}$	$\frac{3}{8}$						
$\hat{c}_{4d} =$	$\frac{565}{128}$	$\frac{26165}{768}$	$\frac{133651}{1152}$	$\frac{523789}{2304}$	$\frac{80573}{288}$	$\frac{317611}{1440}$	$\frac{77773}{720}$	$\frac{89}{3}$	$\frac{839}{240}$	$\frac{1}{12}$			
$\hat{c}_{5d} =$	$\frac{82825}{3072}$	$\frac{67005}{256}$	$\frac{1770535}{1536}$	$\frac{31448897}{10368}$	$\frac{438258631}{82944}$	$\frac{1146749}{180}$	$\frac{86265}{16}$	$\frac{304411}{96}$	$\frac{25180997}{20160}$	$\frac{109627}{360}$	$\frac{781}{20}$	$\frac{439}{240}$	$\frac{1}{120}$

Notice that $\hat{c}_{rd} = 0$ for sufficiently large values of d; we do not have to go all the way up to $d = 3r+2$. In fact, we will see in the next section that the final nonzero coefficient is $\hat{c}_{r(3r+2-s)}$ when $\binom{s-2}{2} \leq r < \binom{s-1}{2}$, and it has the value exhibited in (6.11).

The asymptotic value of the leading coefficients $\hat{c}_{r0} = c_{r0} = c_r$ has an interesting history. Wright [44] gave a complicated argument establishing that $\hat{c}_{r0}$ is asymptotically $(3/2)^r (r-1)!$ times a certain constant, for which he obtained the numerical value 0.159155. Stepanov [35] independently computed the numerical value 0.46 for three times that same constant; the approximation 0.48 would have been more accurate, but Stepanov was perhaps conjecturing that the true value would be

$$\frac{1}{3} + \frac{1}{\pi}\left(\sqrt{3} + \ln(2 - \sqrt{3}\,)\right) \approx 0.46546,$$

which he had announced at the same time in connection with another problem concerning the size of the largest component when the centroid is removed from a random tree. Wright's constant was identified as $1/(2\pi)$ by G. N. Bagaev and E. F. Dmitriev [2], who presented without proof a list of asymptotic expressions for the solution of several related enumeration problems. Lambert Meertens independently found a proof in 1986, but did not publish it; his approach was reported later in [3]. A detailed analysis was also carried out by V. A. Voblyĭ [38], who obtained a number of interesting auxiliary formulas. In particular, if we write $c(z) = c_1 z + c_2 z^2 + c_3 z^3 + \cdots$, Voblyĭ proved the formal power series relation

$$\vartheta c(z) = -\frac{1}{6} + \frac{1}{3z}\left(1 - \frac{I_{-2/3}(1/3z)}{I_{1/3}(1/3z)}\right). \tag{8.14}$$

In other words, he proved that the coefficients c_r show up in the asymptotic series

$$\frac{I_{-2/3}(1/3z)}{I_{1/3}(1/3z)} \sim 1 - \frac{z}{2} - 3c_1 z^2 - 6c_2 z^3 - 9c_3 z^3 - \cdots, \tag{8.15}$$

as $z \to 0$. This is interesting because the left-hand side can also be expressed as a continued fraction

$$2z + \cfrac{1}{8z + \cfrac{1}{14z + \cfrac{1}{20z + \cfrac{1}{26z + \cdots}}}}, \tag{8.16}$$

using the standard recurrence $zI_{\nu+1}(z) = zI_{\nu-1}(z) - 2\nu I_\nu(z)$ for the modified Bessel functions $I_\nu(z)$. In the course of his investigation, Voblyĭ noticed that the coefficients of $e^{c(z)}$ have a simple form, although he did not mention their combinatorial significance; these are the numbers we have called e_r. He gave the formulas

$$\frac{2^r}{3^r} e_r = (-1)^r (1/3, r) = \frac{\Gamma(r + 5/6)\, \Gamma(r + 1/6)}{2\pi r!}, \tag{8.17}$$

which are equivalent to (7.2). Here (ν, r) denotes Hankel's symbol,

$$(\nu, r) = \frac{1}{r!} \prod_{k=1}^{r} (\nu + k - \tfrac{1}{2})(\nu - k + \tfrac{1}{2}).$$

9. Structure of Complex Multigraphs

The generating functions E_r, C_r, $(1 + \zeta)^{2r}\widehat{E}_r$, and $(1 + \zeta)^2\widehat{C}_r$ are polynomials in ζ, and these polynomials have a combinatorial interpretation that provides considerable insight into what is happening as a graph or multigraph evolves. The inner structure in the case of $\widehat{C}_r$ was studied by Wright in his original paper [41]; we will see that his results for graphs become simpler when we consider the analogous results for multigraphs.

Let M be a cyclic multigraph of excess r, that is, any multigraph with no acyclic components, having r more edges than vertices. We can "prune" M by repeatedly cutting off any vertex of degree 1 and the edge leading to that vertex; this eliminates as many edges as vertices, so the pruned multigraph $\overline{M}$ still has excess r. Each vertex of $\overline{M}$ has degree at least 2. Such multigraphs are called *smooth*.

Conversely, given any smooth multigraph $\overline{M}$, we obtain all multigraphs M that prune down to it by simply sprouting a tree from each vertex of $\overline{M}$ (namely, identifying that vertex with the root of a rooted tree). Since $T(z)$ is the generating function for rooted trees, it follows that

$$F_r(z) = \overline{F}_r\big(T(z)\big),\tag{9.1}$$

where $F_r(z)$ is the generating function for all cyclic multigraphs of excess r and $\overline{F}_r$ is the generating function for all smooth multigraphs of excess r. Thus, for example, we must have

$$\overline{F}_1(z) = \frac{1}{24}\,z(3+2z)/(1-z)^{7/2},\tag{9.2}$$

because we know from (3.4), (4.8), (5.2), and (5.14) that

$$F_1(z) = e^{V(z)}E_1(z) = \frac{1}{24}\,T(z)\big(3+2T(z)\big)/\big(1-T(z)\big)^{7/2}.$$

The coefficient of z^n in $\overline{F}_1(z)$ is the sum of $\kappa(\overline{M})$ over all multigraphs $\overline{M}$ on n labeled vertices having $n+1$ edges and all vertices of degree 2 or more, divided by $n!$. For example, the coefficient of z is $1/8$; this is the compensation factor of the multigraph with a single vertex x and two loops from x to itself. The coefficient of z^2 is $\frac{25}{48} = \frac{25}{24}/2!$; the smooth labeled multigraphs

have compensation factors $\frac14, \frac14, \frac14, \frac16, \frac{1}{16}$, and $\frac{1}{16}$, respectively, summing to $\frac{25}{24}$.

The smooth multigraph $\overline{M}$ obtained by repeatedly pruning M is called the *core* of M (see [26]). Let $\overline{F}$ be any family of smooth multigraphs, and let F be the set of all cyclic multigraphs whose core is a member of $\overline{F}$. The argument by which we proved (9.1) also proves that the univariate and bivariate generating functions for F and $\overline{F}$ are related by the equations

$$F(z) = \overline{F}\big(T(z)\big)\,;\qquad F(w,z) = \overline{F}\big(w,\,T(wz)/w\big).\tag{9.3}$$

In particular we have $\widehat{E}_r(z) = \widehat{\overline{E}}_r\big(T(z)\big)$, where $\widehat{\overline{E}}_r$ counts all smooth graphs of excess r having no unicyclic components. This relationship accounts for the curious formula (6.11) about the last nonvanishing coefficient $\hat{e}_{rd}$; we can reason as follows: The minimum number of vertices among all graphs of excess r, when $\binom{s-2}{2} \le r < \binom{s-1}{2}$, is s, because a graph on $s-1$ vertices has at most $\binom{s-1}{2}$ edges and $\binom{s-1}{2} < s-1+r$. The coefficient of the minimum power of ζ in $\widehat{E}_r = \overline{\overline{E}}_r\bigl(\zeta/(1+\zeta)\bigr)$ therefore comes entirely from the $\binom{s(s-1)/2}{s+r}$ graphs on s labeled vertices having exactly $s + r$ edges. All such graphs are smooth.

When M has no unicyclic components we can go beyond pruning to another kind of vertexectomy that we will call *cancelling*: If any vertex has degree 2, we can remove it and splice together the two edges that it formerly touched. Repeated application of this process on any smooth multigraph $\overline{M}$ of excess r will lead to a multigraph $\overline{\overline{M}}$ of excess r in which every vertex has degree 3 or more. (A self-loop $\langle x, x \rangle$ is assumed to contribute 2 to the degree of x. A vertex with a self-loop will be connected to at least one other vertex, because there are no unicycles, so we will never cancel it.) The multigraph $\overline{\overline{M}}$ can be called *reduced*. Only the middle two multigraphs of the six pictured above are reduced.

There are only finitely many reduced multigraphs of excess r. For if such a multigraph has n vertices of degrees $d_1, d_2, \ldots, d_n$, it has $n + r = \frac{1}{2}(d_1 + d_2 + \cdots + d_n) \geq \frac{3}{2}n$ edges, hence $n \leq 2r$. The extreme case $n = 2r$ occurs if and only if the multigraph is 3-regular, that is, if and only if every vertex has degree exactly 3. We will see later that such regularity is, in fact, normal: The complex components of a random graph or multigraph with $\frac{1}{2}n + o(n^{3/4})$ edges almost always reduce to components that are 3-regular.

The reduced multigraph $\overline{\overline{M}}$ obtained by pruning and cancellation from a given complex multigraph M is called the *kernel* of M (see [26]). Our immediate goal is to find the generating function for all smooth multigraphs $\overline{M}$ without unicyclic components that have a given reduced multigraph $\overline{\overline{M}}$ as their kernel. For this it will be convenient to introduce another representation of a multigraph M: We label both the vertices and the edges, and we assign an arbitrary orientation to each edge, thereby obtaining a *directed edge-labeled* multigraph. Let $V = V(M)$ be the set of vertex labels and $E = E(M)$ the set of edge labels. Each edge $e \in E$ has a *dual edge* $\bar{e}$, and $\bar{E}$ is the set of all dual edges. The multigraph M is then represented as a mapping M from $E \cup \bar{E}$ to V, with the interpretation that each directed edge e runs from $M(e)$ to $M(\bar{e})$. The dual of $\bar{e}$, namely $\bar{\bar{e}}$, is e; thus $\bar{e}$ runs from $M(\bar{e})$ to $M(e)$.

If the vertex labels are $1, \ldots, n$ and if the edge labels are $1, \ldots, m$, the multigraph mapping M takes the set $\{1, \ldots, m, \bar{1}, \ldots, \bar{m}\}$ into the set $\{1, \ldots, n\}$. Any such mapping is equivalent to a sequence $\langle x_1, y_1 \rangle \cdots \langle x_m, y_m \rangle$ of ordered pairs generated by the multigraph process of Section 1, where $x_k = M(k)$ and $y_k = M(\bar{k})$.

The number of different mappings M that correspond to a given multigraph M is $2^m m! \, \kappa(M)$, where κ is the compensation factor defined in (1.1). This holds because $2^m m!$ is the number of ways to orient the edges and to assign edge labels, and κ accounts for duplicate assignments that leave us with the same mapping M.

Duplicate assignments can be treated more formally as follows. A *signed permutation* σ of a set E and its dual $\overline{E}$ is a permutation of $E \cup \overline{E}$ with the property that $\sigma \overline{e} = \overline{\sigma e}$ for all e. (The group of all signed permutations on a set of m elements is conventionally called the hyperoctahedral group $\mathbf{B}_m$; it is the group of all $2^m m!$ symmetries of an m-cube.) Given a multigraph represented as a mapping M from $E \cup \overline{E}$ to V, an *edge automorphism* is a signed permutation σ of $E \cup \overline{E}$ with the property that $M(\sigma e) = M(e)$.

It is easy to see that the number of edge automorphisms of M is $1/\kappa(M)$. Such a mapping σ must be the product of one of the $2^{m_{xx}} m_{xx}!$ signed permutations of the m_{xx} self-loops from x to x, for each x, times one of the $m_{xy}!$ signed permutations of the m_{xy} edges from x to y, for each $x < y$. Edge automorphisms are the automorphisms of multigraphs with labeled vertices and unlabeled edges; this explains why $\kappa(M)$ is used as a weighting function for each M in the generating functions we have been discussing.

We are now ready to prove a basic lemma about multigraphs, motivated by but noticeably simpler than the corresponding result for graphs obtained by Wright [41]:

Lemma 1. *If $\overline{\overline{M}}$ is a reduced multigraph having ν vertices, μ edges, and compensation factor κ, the generating function for all smooth, complex multigraphs $\overline{M}$ that reduce to $\overline{\overline{M}}$ under cancellation is*

$$\frac{\kappa \, z^\nu}{(1-z)^\mu \, \nu!}. \tag{9.4}$$

Proof. This result is "intuitively obvious," but it requires a formal proof to ensure that everything is counted properly in the presence of compensation factors. We assume that $\overline{\overline{M}}$ is represented by a fixed mapping from edges and dual edges to vertices, where the set of edge labels is $\{[1], \dots, [\mu]\}$ and the set of vertex labels is $\{(1), \dots, (\nu)\}$. The dual of edge $[j]$ will be denoted by $\overline{[j]} = [-j]$. The given multigraph mapping can be represented as a function M from $\{-\mu, \dots, -1, 1, \dots, \mu\}$ to $\{1, \dots, \nu\}$, such that edge $[j]$ runs from $\big(M(j)\big)$ to $\big(M(-j)\big)$ and edge $[-j]$ runs from $\big(M(-j)\big)$ to $\big(M(j)\big)$. Square brackets and round parentheses are used notationally here in order to distinguish edge labels from vertex labels, although M is a function from integers to integers.

Let s_n be the coefficient of z^n in $z^\nu/(1-z)^\mu$. This quantity s_n is the number of solutions $\langle n_1, \ldots, n_\mu \rangle$ to the equation

$$n_1 + \cdots + n_\mu = n - \nu \qquad (9.5)$$

in nonnegative integers. Let $m - \mu = n - \nu$; then m is the number of edges in an n-vertex multigraph that cancels to $\overline{\overline{M}}$.

We will construct $2^m m!\, n!\, s_n/\nu!$ sequences of ordered pairs $\langle x_1, y_1 \rangle$ $\ldots \langle x_m, y_m \rangle$ of integers $1 \le x_j, y_j \le n$ such that (a) every constructed sequence defines a smooth multigraph that cancels to $\overline{\overline{M}}$; (b) every sequence that defines such a smooth multigraph is constructed exactly $1/\kappa$ times. This will prove the lemma, because of (2.2). As noted earlier, constructing a sequence $\langle x_1, y_1 \rangle \ldots \langle x_m, y_m \rangle$ is equivalent to constructing a map $\overline{M}$ from $\{-m, \ldots, -1, 1, \ldots, m\}$ into $\{1, \ldots, n\}$, if we let $x_j = \overline{M}(j)$ and $y_j = \overline{M}(-j)$.

The construction is as follows. For each ordered solution $\langle n_1, \ldots, n_\mu \rangle$ to (9.5), we effectively insert n_j new vertices into edge $[j]$, thereby undoing the effect of cancellation. Formally, we construct a set of m edge labels

$$E = \{\, [j, k] \mid 1 \le j \le \mu,\ 0 \le k \le n_j \,\} \qquad (9.6)$$

and a set of n vertex labels

$$V = \{\, (i) \mid 1 \le i \le \nu \,\} \ \cup \ \{\, (j, k) \mid 1 \le j \le \mu,\ 1 \le k \le n_j \,\}. \qquad (9.7)$$

Edge $[j, k]$ runs from vertex (j, k) to vertex $(j, k+1)$, where we define for convenience

$$(j, 0) = \big(M(j)\big), \qquad (j, n_j + 1) = \big(M(-j)\big). \qquad (9.8)$$

Thus the original edge $[j]$ from $\big(M(j)\big)$ to $\big(M(-j)\big)$ has become a sequence of $n_j + 1$ edges $[j, 0] \ldots [j, n_j]$ between the same two vertices, with intermediate vertices $(j, 1), \ldots, (j, n_j)$.

The dual of edge $[j, k]$ will be denoted by $-[j, k]$. We also define

$$[-j, k] = -[j, n_{|j|} - k], \qquad (-j, k) = (j, n_{|j|} + 1 - k); \qquad (9.9)$$

this means that the original edge $[-j]$ has become the edge sequence $[-j, 0] \ldots [-j, n_j]$, which is the reverse of $[j, 0] \ldots [j, n_j]$. Edge $[-j, k]$ runs from $(-j, k)$ to $(-j, k+1)$.

To complete the construction, let f be any one-to-one mapping from V to $\{1, \ldots, n\}$ that preserves the order of the original labels (1), $\ldots, (\nu)$; and let g be any *signed bijection* from $\overline{E} \cup E$ to $\{-m, \ldots, -1, 1, \ldots, m\}$. (A signed bijection is a one-to-one correspondence such that $g(\overline{e}) = -g(e)$.) Then we define

$$\overline{M}\big(g([j, k])\big) = f\big((j, k)\big), \qquad (9.10)$$

for all $[j, k]$ in $\overline{E} \cup E$. This mapping $\overline{M}$ corresponds to a sequence $\langle x_1, y_1 \rangle \ldots \langle x_m, y_m \rangle$ that defines a multigraph $\overline{M}$ on $\{1, \ldots, n\}$, as stated above. We have constructed $2^m m! \, n! \, s_n / \nu!$ such sequences, since there are $2^m m!$ choices for g and $n!/\nu!$ for f, given any solution $\langle n_1, \ldots, n_\mu \rangle$ to (9.5).

It is clear that $\overline{M}$ is a smooth multigraph on n vertices that cancels to the given reduced multigraph $\overline{\overline{M}}$, and that every such $\overline{M}$ is constructed at least once. We need to verify that every mapping $\overline{M}$ is obtained exactly $1/\kappa$ times among the $2^m m! \, n! \, s_n / \nu!$ constructed mappings.

Suppose $\overline{M}$ has been constructed from $(\langle n_1, \ldots, n_\mu \rangle, f, g)$, and suppose σ is one of the $1/\kappa$ edge automorphisms of $\overline{\overline{M}}$. We will define a new construction $(\langle n_1', \ldots, n_\mu' \rangle, f', g')$ that produces the same mapping $\overline{M}$. Our notational conventions allow us to regard σ as a permutation of $\{-\mu, \ldots, -1, 1, \ldots, \mu\}$, where

$$\sigma(-j) = -\sigma j \qquad \text{and} \qquad M(\sigma j) = M(j). \qquad (9.11)$$

The new construction is defined by

$$
\begin{aligned}
n_j' &= n_{|\sigma(j)|}, & 1 &\leq j \leq \mu; \\
f'\big((i)\big) &= f\big((i)\big), & 1 &\leq i \leq \nu; \\
f'\big((j, k)'\big) &= f\big((\sigma j, k)\big), & 1 &\leq j \leq \mu, & 1 &\leq k \leq n_j'; \qquad (9.12) \\
g'\big([j, k]'\big) &= g\big([\sigma j, k]\big), & 1 &\leq j \leq \mu, & 0 &\leq k \leq n_j'; \\
\overline{M}'\big(g'([j, k]')\big) &= f'\big((j, k)'\big), & 1 &\leq |j| \leq \mu, & 0 &\leq k \leq n_{|j|}'.
\end{aligned}
$$

Here $(j, k)'$ and $[j, k]'$ are the new vertex and edge labels corresponding to $\langle n_1', \ldots, n_\mu' \rangle$; they are defined in (9.6)–(9.9).

It is easy to verify that the definitions in (9.12) imply validity of the same formulas for the whole range of j and k values:

$$
\begin{aligned}
f'\big((j, k)'\big) &= f\big((\sigma j, k)\big), & 1 &\leq |j| \leq \mu, & 0 &\leq k \leq n_{|j|}' + 1; \\
g'\big([j, k]'\big) &= g\big([\sigma j, k]\big), & 1 &\leq |j| \leq \mu, & 0 &\leq k \leq n_{|j|}'.
\end{aligned} \qquad (9.13)
$$

For example, if $j > 0$ we have

$$f'\big((j,0)'\big) = f'\big((M(j))\big) = f\big((M(j))\big) = f\big((M(\sigma j))\big) = f\big((\sigma j,0)\big)\,;$$

$$f'\big((j,n'_j+1)'\big) = f'\big((M(-j))\big) = f\big((M(-j))\big)$$
$$= f\big((M(\sigma(-j)))\big) = f\big((M(-\sigma j))\big) = f\big((\sigma_j, n'_j + 1)\big)\,;$$

$$f'\big((-j,k)'\big) = f'\big((j,n'_j+1-k)'\big) = f\big((\sigma j, n'_j+1-k)\big)$$
$$= f\big((\sigma j, n_{|\sigma j|}+1-k)\big) = f\big((-\sigma j,k)\big) = f\big((\sigma(-j),k)\big)\,.$$

Therefore if l is any value in $\{-m,\dots,-1,1,\dots,m\}$, we can verify that $\overline{M}'(l) = \overline{M}(l)$, as follows: There are unique j and k such that $l = g([\sigma j, k])$. Hence $l = g'([j,k]')$, and

$$\overline{M}'(l) = f'\big((j,k)'\big) = f\big((\sigma j,k)\big) = \overline{M}(l)\,.$$

Conversely, if $(\langle n'_1,\dots,n'_\mu\rangle, f', g')$ is another construction that makes $\overline{M}'(l) = \overline{M}(l)$ for all l, we can reverse this process and find a unique edge automorphism σ satisfying all the conditions of (9.12). Exactly ν of the vertices of $\overline{M} = \overline{M}'$ have degree ≥ 3, since $\overline{\overline{M}}$ is reduced; these are the images under f and f' of $(1),\dots,(\nu)$, and they have the same order in $\overline{M}$. Therefore $f'((i)) = f((i))$ for $1 \leq i \leq \nu$.

Let $l = g'([j,0])$. Since $\overline{M}'(l) = f'((j,0)) = f'((M(j))) = f\big((M(j))\big)$, we know that $\overline{M}(l)$ must be a vertex of degree ≥ 3, so there must be a value j' (either positive or negative) such that $l = g([j',0])$. This rule defines $\sigma j = j'$. We have $\overline{M}(l) = f\big((j',0)\big) = f\big((M(j'))\big)$, hence $M(\sigma j) = M(j)$.

Let us say that the edge $[j,k]'$ of $\overline{M}'$ corresponds to the edge $[j',k']$ of $\overline{M}$ if $g'([j,k]') = g[j',k']$. We have defined σj for $1 \leq j \leq \mu$ in such a way that $[j,0]'$ corresponds to $[\sigma j,0]$. Suppose we know that $[j,k]'$ corresponds to $[\sigma j,k]$ for some $k < n'_j$; then $-[j,k]'$ also corresponds to $-[\sigma j,k]$. Also $\overline{M}'\big(g'(-[j,k]')\big) = \overline{M}'\big(g'([-j,n'_j-k]')\big) = f'\big((-j,n'_j-k)'\big) = f'\big((j,k+1)'\big) = \overline{M}'\big(g'([j,k+1]')\big)$ is a vertex v of degree 2 in $\overline{M}$, which therefore is equal to $\overline{M}\big(g(-[\sigma j,k])\big) = f\big((-\sigma j, n_{|\sigma j|}-k)\big)$. Consequently we have $k < n_{|\sigma j|}$, $f'\big((j,k+1)'\big) = f\big((\sigma j,k+1)\big)$, and $v = \overline{M}'\big(g'([j,k+1]')\big) = \overline{M}\big(g([\sigma j,k+1])\big)$. Now $[j,k+1]'$ must correspond to $[\sigma j,k+1]$, since there is only one value $l \neq -g'([j,k]')$ such that $\overline{M}(l) = v$. In this way we prove inductively that $[j,k]'$ corresponds to $[\sigma j,k]$ for $0 \leq k \leq n'_j$, and that $n'_j = n'_{|\sigma j|}$. Hence (9.12) holds.   □

Let $\overline{\overline{F}}$ be a family of reduced multigraphs, and let $\overline{F}$ be the family of all smooth complex multigraphs that reduce under cancellation to a member of $\overline{\overline{F}}$. The bivariate generating functions of $\overline{F}$ and $\overline{\overline{F}}$ are then related by the equation

$$\overline{F}(w, z) = \overline{\overline{F}}(w/(1 - wz), z), \qquad (9.14)$$

because Lemma 1 establishes this relation in the case that $\overline{\overline{F}}$ has only one member. Equation (9.14) says simply that every edge in $\overline{\overline{F}}$, represented by w, is to be replaced by a sequence of one or more edges, represented by $w/(1 - wz) = w + w^2z + w^3z^2 + \cdots$; perhaps this means that Lemma 1 is indeed obvious and that the lengthy proof was unnecessary. It is, however, comforting to know that a formal verification is possible, when one is beginning to learn the power of generating function techniques. And somehow, examples of multigraphs with numerous selfloops and repeated edges do seem to mandate a formal proof, because compensation factors change when edges are manipulated.

As an example of Lemma 1, let us derive explicitly the generating function $\overline{E}_1(z) = \overline{C}_1(z)$ for all smooth bicyclic multigraphs. All such multigraphs cancel to a reduced multigraph of excess 1, which can have at most 2 vertices and 3 edges. There are only three possibilities,

$$(9.15)$$

having $\kappa = \frac{1}{8}, \frac{1}{4}$, and $\frac{1}{6}$, respectively. Therefore

$$\begin{aligned}
\overline{E}_1(z) = \overline{C}_1(z) &= \frac{z}{8(1 - z)^2} + \frac{z^2}{8(1 - z)^3} + \frac{z^2}{12(1 - z)^3} \\
&= \frac{z(3 + 2z)}{24(1 - z)^3}, \qquad (9.16)
\end{aligned}$$

in agreement with (9.2). Wright [41] states that there are 15 connected, unlabeled, reduced multigraphs of excess 2, and 107 of excess 3.

If a reduced multigraph of excess r has exactly $2r - d$ vertices, we will say that it has *deficiency* d. A reduced multigraph of deficiency 0 is 3-regular; we will call such multigraphs *clean*.

Corollary. *The coefficient e_{rd} in (5.10) and (7.3) is $(2r-d)!^{-1} \sum \kappa(\overline{\overline{M}})$, summed over all reduced, labeled multigraphs $\overline{\overline{M}}$ of excess r and deficiency d. The coefficient c_{rd} in (8.4) can be obtained in the same way, but restricting the sum to connected multigraphs.* □

This corollary leads to a completely different proof of Theorem 1, because it allows us to obtain formula (7.3) for e_{rd} by a combinatorial counting argument. Consider a reduced multigraph that has exactly d_k vertices of degree k, for each $k \geq 3$; then $\underline{d_3 + d_4 + \cdots = n}$ and $3d_3 + 4d_4 + \cdots = 2m$. We can calculate $\sum \kappa(\overline{\overline{M}})$ over all such $\overline{\overline{M}}$ by counting the number of relevant sequences $\langle x_1, y_1 \rangle \ldots \langle x_m, y_m \rangle$ and dividing by $2^m m!$; and the number of ways to choose $\langle x_1, y_1 \rangle \ldots \langle x_m, y_m \rangle$ is clearly a product of multinomial coefficients,

$$\frac{(2m)!}{3!^{d_3}\, 4!^{d_4} \ldots} \qquad \frac{n!}{d_3!\, d_4! \ldots} \, ,$$

since the first factor is the number of ways to partition $2m$ slots into d_k labeled classes of size k for each k, and the second factor counts the assignments of vertex labels to those classes. To obtain all reduced multigraphs of excess r and deficiency d, we sum over all sequences of nonnegative integers $\langle d_3, d_4, \ldots \rangle$ such that $\sum_{k \geq 3} d_k = 2r - d$ and $\sum_{k \geq 3} k d_k = 6r - 2d$, or equivalently

$$\sum_{k \geq 3} (k-3) d_k = d \qquad \text{and} \qquad \sum_{k \geq 3} (k-2) d_k = 2r \, .$$

Let

$$f_{cd} = \sum \Big\{ \prod_{k \geq 3} \frac{1}{k!^{d_k}\, d_k!} \,\Big|\, \sum_{k \geq 3} (k-3) d_k = d \text{ and } \sum_{k \geq 3} (k-2) d_k = c \Big\}. \tag{9.17}$$

We have just proved that

$$e_{rd} = \frac{(6r - 2d)!}{2^{3r-d}(3r - d)!} f_{(2r)d} \, . \tag{9.18}$$

And we can readily calculate a bivariate generating function for the coefficients f_{rd}:

$$\sum_{r,d \geq 0} f_{rd} w^d z^r = \sum_{d_3, d_4, \ldots \geq 0} \prod_{k \geq 3} \frac{w^{(k-3)d_k} z^{(k-2)d_k}}{k!^{d_k}\, d_k!}$$

$$= \prod_{k \geq 3} \sum_{d_k \geq 0} \Big(\frac{w^{k-3} z^{k-2}}{k!} \Big)^{d_k} \frac{1}{d_k!}$$

$$= \prod_{k \geq 3} \exp(w^{k-3} z^{k-2}/k!)$$

$$= \exp\Big(w^{-3} z^{-2} \sum_{k \geq 3} \frac{(wz)^k}{k!} \Big) = \exp\Big(\frac{z}{6} F\Big(\frac{wz}{4}\Big)\Big) \, ,$$

where F is the function defined in (7.5). Comparing (9.18) to (7.3) now yields the promised proof of (7.4):

$$P_d(r) = 2^{2r+d} 3^{2r-d} (2r - d)!\, f_{(2r)d}$$

$$= 2^{2r+d} 3^{2r-d} (2r - d)!\, [w^d z^{2r}]\, \exp\!\big(z F(wz/4)/6\big)$$

$$= 2^{2r+d} 3^{2r-d} (2r - d)!\, [w^d z^{2r-d}]\, \exp\!\big(z F(w/4)/6\big)$$

$$= 2^{2d} (2r - d)!\, [w^d z^{2r-d}]\, \exp\!\big(z F(w/4)\big) = [w^d]\, F(w)^{2r-d}.$$

These observations also allow us to express e_{rd} in the suggestive form

$$e_{rd} = \frac{1}{2^{3r-d}(3r - d)!} \left\{ \begin{matrix} 6r - 2d \\ 2r - d \end{matrix} \right\}_{\geq 3}, \tag{9.19}$$

where $\left\{ {m \atop n} \right\}_{\geq 3}$ denotes the number of ways to partition an m-element set into n subsets, each containing at least 3 elements. The asymptotic behavior of the integers $2^{3r-d}(3r - d)!\, e_{rd}$ will therefore be analogous to the asymptotic behavior of Stirling numbers.

Lemma 1 captures the combinatorial essence of the generating functions for all complex multigraphs. We can obtain a similar generating function for graphs instead of multigraphs, but we must work a bit harder, and the formulas are not as attractive. The following improvement over Wright's original treatment [41] is based on an approach suggested by V. E. Stepanov [36].

Lemma 2. *Let $\overline{\overline{M}}$ be a reduced multigraph having ν vertices, μ edges, compensation factor κ, and μ_{xy} edges between x and y for $1 \leq x \leq y \leq \nu$. The generating function for all smooth, complex graphs $\overline{G}$ that lead to $\overline{\overline{M}}$ under cancellation is*

$$\frac{\kappa\, z^\nu}{(1 - z)^\mu\, \nu!}\, P(\overline{\overline{M}}, z), \tag{9.20}$$

where

$$P(\overline{\overline{M}}, z) = \prod_{x=1}^{\nu} \left(z^{2\mu_{xx}} \prod_{y=x+1}^{\nu} z^{\mu_{xy}-1} \big(\mu_{xy} - (\mu_{xy} - 1)z\big) \right) \tag{9.21}$$

is a polynomial in z such that $P(\overline{\overline{M}}, 1) = 1$.

Proof. We argue as in Lemma 1, but we must restrict the solutions $\langle n_1, \ldots, n_\mu \rangle$ of (9.5) to cases that produce a graph instead of a multigraph. Thus, each n_j that corresponds to a self-loop must be ≥ 2, so we use $z^2/(1-z)$ instead of $1/(1-z)$ in the contribution that n_j makes to the overall generating function. A subsequence $\langle n_j, \ldots, n_{j+k-1} \rangle$ that corresponds to $k = \mu_{xy}$ edges between distinct vertices $x < y$ must have the property that at most one of $\langle n_j, \ldots, n_{j+k-1} \rangle$ is zero; hence we use

$$\frac{z^k}{(1-z)^k} + \frac{kz^{k-1}}{(1-z)^{k-1}} = \frac{z^{k-1}\big(k - (k-1)z\big)}{(1-z)^k}$$

instead of $1/(1-z)^k$ in its contribution. The net effect is to multiply the previous generating function by $P(\overline{\overline{M}}, z)$.   □

Replacing z by $T(z)$ yields the generating function for all graphs that prune and cancel to $\overline{\overline{M}}$. For example, the generating function $\widehat{E}_1(z) = \widehat{C}_1(z) = \widehat{W}(z)$ of (3.6) can be read off from (9.15): It is

$$\frac{T(z)^5}{8\big(1 - T(z)\big)^2} + \frac{T(z)^6}{8\big(1 - T(z)\big)^3} + \frac{T(z)^4\big(3 - 2T(z)\big)}{12\big(1 - T(z)\big)^3}. \tag{9.22}$$

The degree of the polynomial $P(\overline{\overline{M}}, z)$ is the total number of "penalty points" of $\overline{\overline{M}}$, where each self-loop costs two penalty points, and where each cluster of $\mu_{xy} > 1$ multiple edges between distinct vertices costs $\mu_{xy} - 1$. If $\overline{\overline{M}}$ is a graph, the degree is zero and $P(\overline{\overline{M}}, z) = 1$. At the other extreme, if all edges of $\overline{\overline{M}}$ are self-loops, the degree is 2μ.

The quantity $T(z)^\nu / \big(1 - T(z)\big)^\mu$ becomes $\zeta^\nu (1 + \zeta)^{\mu-\nu}$, when we express it in terms of the variable $\zeta = T(z)/\big(1 - T(z)\big)$ introduced in Section 5; the quantity $P\big(\overline{\overline{M}}, T(z)\big)$ becomes $P\big(\overline{\overline{M}}, \zeta/(1 + \zeta)\big)$. If we restrict consideration to connected multigraphs of excess r, we get rational functions of ζ with denominator $(1 + \zeta)^{r+2}$; this denominator occurs when there are $(r + 1)$ self-loops in $\overline{\overline{M}}$. However, we have seen in Theorem 3 that the denominator of $\widehat{C}_r$ is always a divisor of $(1 + \zeta)^2$. There seems to be no easy combinatorial explanation for the cancellation that occurs when the contributions of different $\overline{\overline{M}}$ are added together. Some of the properties of connected graphs are easier to derive by combinatorics, others are easier to derive by algebra.

The actual coefficients of $P\big(\overline{\overline{M}}, \zeta/(1 + \zeta)\big)$ do not make any significant difference asymptotically, when graphs are sparse; we will see later that the asymptotic behavior as $\zeta \to \infty$ is what counts, hence we only need to know that $P(\overline{\overline{M}}, 1) = 1$. We observed earlier that the leading coefficients $\hat{e}_{r0}$ and e_{r0} of $\widehat{E}$ and E are equal, as are the leading coefficients $\hat{c}_{r0}$ and c_{r0}. Now Lemma 2 shows in fact that each reduced multigraph $\overline{\overline{M}}$ makes the same contribution to the leading coefficient for graphs as it does for multigraphs.

10. A Lemma from Contour Integration

Studies of random graphs that have $m \approx n/2$ edges are traditionally broken into two cases, the "subcritical" case where $m < n/2$ and the "supercritical" case where $m > n/2$. It is desirable, however, to have estimates of probabilities that hold uniformly for all m in the vicinity of $n/2$, passing smoothly from one side to the other. The following lemma, based on techniques introduced in [14], will be our key tool for the computation of probabilities.

Lemma 3. *If* $m = \frac{1}{2}n(1 + \mu n^{-1/3})$ *and if* y *is any real constant, we have*

$$
\frac{2^m\, m!\, n!}{(n-m)!\, n^{2m}}\, [z^n]\, \frac{U(z)^{n-m}}{\left(1 - T(z)\right)^y}
$$
$$
= \sqrt{2\pi}\, A(y,\mu)\, n^{y/3-1/6} + O\!\left((1 + |\mu|^B) n^{y/3-1/2}\right) \quad (10.1)
$$

uniformly for $|\mu| \le n^{1/12}$, *where* $B = \max(4, \frac{9}{2} - y)$ *and*

$$
A(y,\mu) = \frac{e^{-\mu^3/6}}{3^{(y+1)/3}} \sum_{k\ge 0} \frac{\left(\frac{1}{2}3^{2/3}\mu\right)^k}{k!\, \Gamma\!\left((y+1-2k)/3\right)}. \quad (10.2)
$$

As $\mu \to -\infty$, *we have*

$$
A(y,\mu) = \frac{1}{\sqrt{2\pi}\, |\mu|^{y-1/2}} \left(1 - \frac{3y^2 + 3y - 1}{6|\mu|^3} + O(\mu^{-6})\right); \quad (10.3)
$$

as $\mu \to +\infty$, *we have*

$$
A(y,\mu) = \frac{e^{-\mu^3/6}}{2^{y/2}\mu^{1-y/2}} \left(\frac{1}{\Gamma(y/2)} + \frac{4\mu^{-3/2}}{3\sqrt{2}\,\Gamma(y/2 - 3/2)} + O(\mu^{-2})\right). \quad (10.4)
$$

Moreover, (10.1) *can be improved to*

$$
\frac{2^m\, m!\, n!}{(n-m)!\, n^{2m}}\, [z^n]\, \frac{U(z)^{n-m}}{\left(1 - T(z)\right)^y}
$$
$$
= \sqrt{2\pi}\, A(y,\mu)\, n^{y/3-1/6}\left(1 + O(\mu^4 n^{-1/3})\right) \quad (10.5)
$$

if $|\mu|$ *goes to infinity with* n *while remaining* $\le n^{1/12}$.

Proof. First we need to derive some auxiliary results about the function A. If α is any positive number, we define a path $\Pi(\alpha)$ in the complex plane that consists of the following three straight line segments:

$$s(t) = \begin{cases} -e^{-\pi i/3}\, t, & \text{for } -\infty < t \le -2\alpha; \\ \alpha + it\sin\pi/3, & \text{for } -2\alpha \le t \le +2\alpha; \\ e^{+\pi i/3}\, t, & \text{for } +2\alpha \le t < +\infty. \end{cases} \qquad (10.6)$$

Now we define

$$A(y,\mu) = \frac{1}{2\pi i} \int_{\Pi(1)} s^{1-y} e^{K(\mu,s)}\, ds\,, \qquad (10.7)$$

where $K(\mu, s)$ is the polynomial

$$K(\mu, s) = \frac{(s+\mu)^2(2s-\mu)}{6} = \frac{s^3}{3} + \frac{\mu s^2}{2} - \frac{\mu^3}{6}. \qquad (10.8)$$

Our first goal is to show that $A(y,\mu)$ satisfies (10.2), (10.3), and (10.4).

To get (10.2), we make the substitution $u = s^3/3$. As s traverses $\Pi(1)$, the variable u traverses an interesting contour Γ that begins at $-\infty$ and hugs the lower edge of the negative axis, then circles the origin counterclockwise and returns to $-\infty$ along the upper edge of the axis. On this contour Γ we have Hankel's well-known formula for the reciprocal Gamma function,

$$\frac{1}{\Gamma(z)} = \frac{1}{2\pi i} \int_{\Gamma} \frac{e^u\, du}{u^z}\,.$$

(See, for example, [18, Theorem 8.4b].) So we can expand (10.7) into an absolutely convergent series, after substituting $3^{1/3} u^{1/3}$ for s:

$$\int_{\Pi(1)} s^{1-y} e^{K(\mu,s)}\, ds = \frac{e^{-\mu^3/6}}{3^{(y+1)/3}} \int_{\Gamma} \frac{e^u \exp\!\left(\frac{1}{2} 3^{2/3}\mu u^{2/3}\right) du}{u^{(y+1)/3}}$$

$$= \frac{e^{-\mu^3/6}}{3^{(y+1)/3}} \int_{\Gamma} \sum_{k\ge 0} \frac{\left(\frac{1}{2} 3^{2/3}\mu\right)^k e^u\, du}{k!\, u^{(y+1-2k)/3}}\,.$$

Interchanging summation and integration, and applying Hankel's formula, gives (10.2).

To get (10.3) and (10.4), we note first that the integral (10.7) can be taken over any path $\Pi(\alpha)$, not just $\Pi(1)$, because $e^{K(\mu,s)}$ has no singularities. Moreover, we can "straighten out" the path $\Pi(\alpha)$, changing it to a single straight line from $\alpha - i\infty$ to $\alpha + i\infty$, if α is sufficiently large. For we can readily verify that the integrand is exponentially small on any large circular arc $s = Re^{i\theta}$, as $|\theta|$ increases from $\pi/3$ to the angle where $R\cos\theta = \alpha$: The real part of s^3 is $R^3\cos 3\theta$, which increases from $-R^3$ to $4\alpha^3 - 3R^2\alpha$; and the real part of s^2 lies between $-R^2$ and $-R^2/2$. Hence the real part of $K(\mu, s)$ will be at most $-cR^2$ for some positive $c = c(\alpha)$ on the entire arc, whenever $\alpha > 0$ and $\alpha > -\frac{1}{2}\mu$; this will make $s^{1-y}e^{K(\mu,s)}$ exponentially small.

If μ is negative, let $\alpha = -\mu$; then

$$A(y, -\alpha) = \frac{1}{2\pi}\int_{-\infty}^{\infty}(\alpha + it)^{1-y}e^{K(-\alpha,\alpha+it)}\,dt \qquad (10.9)$$

$$= \frac{1}{2\pi\sqrt{\alpha}}\int_{-\infty}^{\infty}(\alpha + it/\sqrt{\alpha})^{1-y}e^{K(-\alpha,\alpha+it/\sqrt{\alpha})}\,dt$$

$$= \frac{1}{2\pi\alpha^{y-1/2}}\int_{-\infty}^{\infty}\left(1 + \frac{it}{\alpha^{3/2}}\right)^{1-y}e^{-t^2/2-it^3/(3\alpha^{3/2})}\,dt\,,$$

and we can find the asymptotic value of the remaining integral by using Laplace's standard technique of "tail-exchange" (see [17, §9.4]):

$$\int_{-\infty}^{\infty}\left(1 + \frac{it}{\alpha^{3/2}}\right)^{1-y}e^{-t^2/2-it^3/(3\alpha^{3/2})}\,dt$$

$$= \int_{-\alpha^\epsilon}^{\alpha^\epsilon}\left(1 + \frac{it}{\alpha^{3/2}}\right)^{1-y}e^{-t^2/2-it^3/(3\alpha^{3/2})}\,dt + O\left(e^{-\alpha^{2\epsilon}/3}\right)$$

$$= \int_{-\alpha^\epsilon}^{\alpha^\epsilon}e^{-t^2/2}\left(1 + \frac{(1-y)it}{\alpha^{3/2}} - \frac{it^3}{3\alpha^{3/2}} + O(\alpha^{6\epsilon-3})\right)dt + O\left(e^{-\alpha^{2\epsilon}/3}\right)$$

$$= \sqrt{2\pi} + O\left(\alpha^{6\epsilon-3}\right).$$

If we expand the integrand further, to terms that are $O(\alpha^{12\epsilon-6})$, we obtain

$$A(y, -\alpha) = \frac{1}{\sqrt{2\pi}\,\alpha^{y-1/2}}\left(1 - \frac{3y^2 + 3y - 1}{6\alpha^3} + O\left(\alpha^{12\epsilon-6}\right)\right).$$

The method can clearly be extended, in principle, to give a complete asymptotic series in powers of α^{-3}, beginning as shown in (10.3).

We also want to know the asymptotic value of $A(y, \mu)$ as $\mu \to +\infty$, and for this we need to work a bit harder. A combination of the methods we have used to prove (10.2) and (10.3) will establish (10.4). The idea now is to integrate on the path $\mu^{-1} + it/\sqrt{\mu}$:

$$A(y, \mu) = \frac{e^{K(\mu, \mu^{-1})}}{2\pi\sqrt{\mu}} \int_{-\infty}^{\infty} \left(\mu^{-1} + \frac{it}{\sqrt{\mu}}\right)^{1-y} \exp\left(it(\mu^{-1/2} + \mu^{-5/2})\right.$$

$$\left. - t^2\left(\tfrac{1}{2} + \mu^{-2}\right) - \tfrac{1}{3}it^3\mu^{-3/2}\right) dt$$

$$= \frac{e^{K(\mu, \mu^{-1})}}{2\pi\mu^{1-y/2}} \int_{-\infty}^{\infty} (\mu^{-1/2} + it)^{1-y} e^{-t^2/2} g(it, \mu)\, dt$$

$$= \frac{e^{K(\mu, \mu^{-1})}}{2\pi i\mu^{1-y/2}} \int_{-\infty i}^{\infty i} (v + \mu^{-1/2})^{1-y} e^{v^2/2} g(v, \mu)\, dv\,,$$

where the last step replaces it by v. We can distort the path of v so that it crosses the positive real axis, and then replace $v^2/2$ by u to get Hankel's contour Γ again:

$$A(y, \mu) = \frac{e^{K(\mu, \mu^{-1})}}{2\pi i\mu^{1-y/2}} \int_{\Gamma} (\sqrt{2u} + \mu^{-1/2})^{1-y} e^u g(\sqrt{2u}, \mu) \frac{du}{\sqrt{2u}}$$

$$= \frac{e^{K(\mu, \mu^{-1})}}{2^{1+y/2}\pi i\mu^{1-y/2}} \int_{\Gamma} (1 + (2\mu u)^{-1/2})^{1-y} u^{-y/2} e^u g(\sqrt{2u}, \mu)\, du\,.$$

For definiteness we can stipulate that the contour Γ lies entirely on the negative axis, except for a circular loop about 0 with a radius of 1. When u is on the negative axis, say $u = -t$, the quantity $\sqrt{2u}$ will be $-i\sqrt{2t}$ on the first part of Γ and $+i\sqrt{2t}$ on the last, so we will have

$$g(\sqrt{2u}, \mu) = \exp\left(\mp i\sqrt{2t}(\mu^{-1/2} + \mu^{-5/2}) - 2t\mu^{-2} \pm \tfrac{1}{3}i(2t)^{3/2}\mu^{-3/2}\right).$$

On the portions of Γ for which $|u| \geq \mu^\epsilon$, the integrand is superpolynomially small; in other words, it approaches zero faster than any negative power of the argument. Hence

$$\int_{\Gamma} = \int_{\Gamma[\mu^\epsilon]} + O\left(e^{-\mu^\epsilon/2}\right),$$

where $\Gamma[\mu^\epsilon]$ is the subcontour that runs along the lower edge of the negative axis from $-\mu^\epsilon$ to the circle $u = e^{i\theta}$ and back to $-\mu^\epsilon$ on the top edge of the axis. On $\Gamma[\mu^\epsilon]$ we have

$$(2\mu u)^{-1/2} = O(\mu^{-1/2}),$$
$$g(\sqrt{2u}, \mu) = 1 + \sqrt{2u}\,\mu^{-1/2} + O(\mu^{\epsilon-1});$$

and $\int_\Gamma |u^{-y/2} e^u|\, du$ exists. Hence

$$\int_{\Gamma[\mu^\epsilon]} \left(1 + (2\mu u)^{-1/2}\right)^{1-y} u^{-y/2} e^u g(\sqrt{2u}, \mu)\, du$$

$$= \int_{\Gamma[\mu^\epsilon]} \left(u^{-y/2} + \frac{1}{\sqrt{2\mu}}\left((1-y)u^{-(y+1)/2} + 2u^{-(y-1)/2}\right)\right) e^u\, du + O(\mu^{\epsilon-1})$$

$$= 2\pi i \left(\frac{1}{\Gamma(y/2)} + \frac{1}{\sqrt{2\mu}}\left(\frac{1-y}{\Gamma((y+1)/2)} + \frac{2}{\Gamma((y-1)/2)}\right)\right) + O(\mu^{\epsilon-1}).$$

The coefficient of $\mu^{-1/2}$ vanishes, because $\Gamma((y+1)/2)$ is equal to $\frac{1}{2}(y-1)\Gamma((y-1)/2)$. We can use the same method to expand the integrand further, obtaining (10.4).

Notice that $1/\Gamma(y/2)$ or $1/\Gamma(y/2 - 3/2)$ may be zero, but not both. Therefore (10.4) gives the asymptotically leading term of $A(y, \mu)$ in all cases.

Whew — we have worked pretty hard to establish (10.2)–(10.4), and we still haven't begun to tackle the main assertion of the lemma. Fortunately, the work we have done so far will help streamline the rest of the proof. The next step is to analyze the factor at the left of (10.1); a routine application of Stirling's approximation shows that

$$\frac{2^m\, m!\, n!}{(n-m)!\, n^{2m}} = \sqrt{2\pi n}\, 2^{n-m} e^{-\mu^3/6-n}\left(1 + O\!\left(\frac{1+\mu^4}{n^{1/3}}\right)\right), \quad (10.10)$$

uniformly for $|\mu| \le n^{1/12}$ as $n \to \infty$, when $m = \frac{n}{2}(1 + \mu n^{-1/3})$.

Now we turn to the other parts of (10.1). Equation (3.2) implies that T has an analytic continuation in which $T(ze^{-z}) = z$ for $|z| < 1$. Hence, by (3.3) and Cauchy's formula for $[z^n]\, f(z)$, we can substitute $\tau = ze^{-z}$ and get

$$[z^n] \frac{U(z)^{n-m}}{(1-T(z))^y} = \frac{1}{2\pi i} \oint \frac{U(\tau)^{n-m}\, d\tau}{(1-T(\tau))^y\, \tau^{n+1}}$$

$$= \frac{e^n\, 2^{m-n}}{2\pi i} \oint (1-z)^{1-y} e^{nh(z)}\, \frac{dz}{z}, \quad (10.11)$$

where

$$h(z) = z - 1 - \frac{m}{n} \ln z + \left(1 - \frac{m}{n}\right) \ln (2 - z)$$

$$= z - 1 - \ln z - \left(1 - \frac{m}{n}\right) \ln \frac{1}{1 - (z - 1)^2} . \qquad (10.12)$$

The contour in (10.11) should keep $|z| < 1$. Notice that $h(1) = h'(1) = 0$; if $m = \frac{1}{2}n$ we also have $h''(1) = 0$. This triple zero accounts for the procedure we shall use to investigate the value of (10.11) for large n.

Let $\nu = n^{-1/3}$, and let α be the positive solution to

$$\mu = \alpha^{-1} - \alpha . \qquad (10.13)$$

We will evaluate (10.11) on the path $z = e^{-(\alpha + it)\nu}$, where t runs from $-\pi n^{1/3}$ to $\pi n^{1/3}$:

$$\oint f(z) \frac{dz}{z} = i\nu \int_{-\pi n^{1/3}}^{\pi n^{1/3}} f(e^{-(\alpha + it)\nu}) \, dt . \qquad (10.14)$$

It will turn out that the main contribution to the value of this integral comes from the vicinity of $t = 0$.

The magnitude of $e^{h(z)}$ depends on $\Re h(z)$.† If $z = \rho e^{i\theta}$, we have

$$\Re h(\rho e^{i\theta}) = \rho \cos \theta - 1 - \frac{m}{n} \ln \rho + \frac{1}{2}\left(1 - \frac{m}{n}\right) \ln (4 - 4\rho \cos \theta + \rho^2) .$$
$$(10.15)$$

The derivative with respect to θ is $-\rho g(\theta) \sin \theta$, where

$$g(\theta) = 1 - \frac{2(1 - m/n)}{4 - 4\rho \cos \theta + \rho^2} \geq \frac{(2 - \rho)^2 - 2(1 - m/n)}{4 - 4\rho \cos \theta + \rho^2} ; \qquad (10.16)$$

and $g(\theta)$ is positive when $\rho = e^{-\alpha\nu}$, because $2(1 - m/n) = 1 - \mu\nu < 1 + \alpha\nu < (2 - e^{-\alpha\nu})^2$. (We always have $0 < \alpha\nu < 2$ when $|\mu| \leq n^{1/12}$, and it is not difficult to verify that $(2 - e^{-x})^2 > 1 + x$ when $0 < x < 2$.) Hence $\Re h(e^{-(\alpha + it)\nu})$ decreases as $|t|$ increases, and $|e^{nh(z)}|$ has its maximum on the circle $z = e^{-(\alpha + it)\nu}$ when $t = 0$.

Looking further at $nh(e^{-s\nu})$, we have the asymptotic estimate

$$n h(e^{-s\nu}) = \tfrac{1}{3} s^3 + \tfrac{1}{2}\mu s^2 + O((\mu^2 s^2 + s^4)\nu) , \qquad (10.17)$$

uniformly in any region such that $|s\nu| \leq c$ where $c < \ln 2$. This bound follows from (10.12), using the expansion

$$\ln \frac{1}{1 - (e^u - 1)^2} = u^2 + u^3 + O(u^4) , \qquad |u| \leq c .$$

---

† $\Re(x + iy) = x$ denotes the real part of the complex number $x + iy$.

We also have

$$(1 - e^{-s\nu})^{1-y} = s^{1-y}\nu^{1-y}(1 + O(s\nu)).\tag{10.18}$$

Therefore if $f(z) = (1 - z)^{1-y}e^{nh(z)}$ is the integrand of (10.11) and (10.14), we have

$$e^{-\mu^3/6}f(e^{-s\nu}) = \nu^{1-y}s^{1-y}e^{K(\mu,s)}\big(1 + O(s\nu)$$
$$+ O(\mu^2 s^2 \nu) + O(s^4\nu)\big),\tag{10.19}$$

when $s = O(n^{1/12})$. (This restriction on s ensures that $\mu^2 s^2 \nu$ and $s^4 \nu$ are bounded, hence the O terms of (10.17) can be moved out of the exponent.)

The exponent $K(\mu, s)$ in (10.19), when $s = \alpha + it$, is

$$K(\alpha^{-1} - \alpha, \alpha + it) = \left(\tfrac{1}{2}\alpha^{-1} - \tfrac{1}{6}\alpha^{-3}\right) + it - \tfrac{1}{2}(\alpha + \alpha^{-1})t^2 - \tfrac{1}{3}it^3.$$

The real part is bounded above by $\tfrac{1}{3} - t^2$, for all $\alpha > 0$, since we have $3\alpha^{-1} - \alpha^{-3} \le 2 \le \alpha + \alpha^{-1}$, with equality if and only if $\alpha = 1$. Hence the integrand $f(e^{-s\nu})$ becomes superpolynomially small when $|t|$ grows, and we have

$$\frac{e^{-\mu^3/6}}{2\pi i}\oint f(z)\frac{dz}{z}$$

$$= \frac{\nu e^{-\mu^3/6}}{2\pi}\int_{-n^{1/12}}^{n^{1/12}} f\big(e^{-(\alpha+it)\nu}\big)\,dt + O\big(e^{-(\alpha+\alpha^{-1})n^{1/6}/3}\big)$$

$$= \frac{\nu^{2-y}}{2\pi i}\int_{\alpha-n^{1/12}i}^{\alpha+n^{1/12}i} s^{1-y}e^{K(\mu,s)}\,ds + O(\nu^{3-y}R) + O\big(e^{-(\alpha+\alpha^{-1})n^{1/6}/3}\big)$$

$$= \nu^{2-y}A(y,\mu) + O(\nu^{3-y}R) + O\big(e^{-\max(2,|\mu|)n^{1/6}/3}\big),$$

where $s = \alpha + it$ and

$$R = \int_{-\infty}^{\infty}\big(|s^{2-y}| + \mu^2|s^{3-y}| + |s^{5-y}|\big)\big|e^{K(\mu,s)}\big|\,dt = R_1 + R_2 + R_3.$$

The lemma will be proved if we can show that $R = O(1 + \mu^B)$ and that $R/A(y,\mu) = O(\mu^4)$ as $|\mu| \to \infty$.

To show that each remainder integral R_1, R_2, R_3 is small, we will let $s = \alpha + iu/\beta$, where $u = \beta t$ and

$$\beta = \sqrt{\alpha + \alpha^{-1}} \,. \tag{10.20}$$

Notice that when $\mu \le 0$ we have $\alpha \ge 1$ and $\alpha = |\mu| + O(|\mu|^{-1})$; when $\mu \ge 0$ we have $0 < \alpha \le 1$ and $\alpha^{-1} = \mu + O(\mu^{-1})$. Therefore in both cases

$$\beta = |\mu|^{1/2} + O(|\mu|^{-1/2}) \qquad \text{as } |\mu| \to \infty. \tag{10.21}$$

The first remainder, R_1, is

$$\int_{-\infty}^{\infty} |\alpha + it|^{2-y} \left| e^{K(\mu,\alpha+it)} \right| dt = \frac{e^{\alpha^{-1}/2 - \alpha^{-3}/6}}{\beta} \int_{-\infty}^{\infty} \left| \alpha + \frac{iu}{\beta} \right|^{2-y} e^{-u^2/2} \, du \,.$$

If $\mu < 0$, we have $\alpha\beta \ge \sqrt{2}$, hence

$$R_1 \le \frac{O(1)\,\alpha^{2-y}}{\beta} \int_{-\infty}^{\infty} \max\left(1, \left|1 + \frac{iu}{\sqrt{2}}\right|^{2-y}\right) e^{-u^2/2} \, du \,;$$

and the integral exists, so this is $O(|\mu|^{3/2-y})$ by (10.21). Similarly, $R_2 = O(|\mu|^{2+5/2-y})$ when $\mu < 0$, and $R_3 = O(|\mu|^{9/2-y})$.

On the other hand, when $\mu > 0$ we have $\alpha\beta \le \sqrt{2}$, and we need to be more cautious. Instead of letting t run from $-\infty$ to $+\infty$ through real values in the derivation above, we will distort the path slightly near the origin, so that t passes through the point $-i/\beta$ and so that the quantity $\beta s = \alpha\beta + iu$ never has magnitude less than 1. (We used essentially the same sort of contour when deriving (10.4).) Then u passes through the point $-i$, and we have

$$R_1 \le \frac{O(1)\,e^{-\mu^3/6}}{\beta^{3-y}} \int_{-\infty}^{\infty} \max\left(1, \left|\sqrt{2} + iu\right|^{2-y}\right) e^{-u^2/2} \, du \,.$$

We therefore have $R_1 = O(e^{-\mu^3/6} \mu^{y/2-3/2})$; similarly,

$$R_2 = O(e^{-\mu^3/6} \mu^{y/2-4/2+2}) \quad \text{and} \quad R_3 = O(e^{-\mu^3/6} \mu^{y/2-6/2}).$$

We know that $A(y,\mu)$ grows at least as fast as $e^{-\mu^3/6} \mu^{y/2-5/2}$, because of (10.4). So in this case the remainders behave even better than we have claimed in (10.5), although the error term $O(\mu^4/n^{1/3})$ is still necessary because of (10.10). ☐

If we differentiate the integral (10.7) with respect to s and with respect to μ, we obtain a recurrence relation for $A(y,\mu)$ and a formula for its derivative:

$$(y-2)A(y,\mu) = \mu A(y-2,\mu) + A(y-3,\mu)\,; \qquad (10.22)$$
$$A'(y,\mu) = \tfrac{1}{2}A(y-2,\mu) - \tfrac{1}{2}\mu^2 A(y,\mu)\,. \qquad (10.23)$$

(The prime here denotes differentiation with respect to the second argument, μ. The derivative with respect to y could also be worked out; but it depends on the derivative of the Gamma function in a rather complicated way, and it is not expressible directly in terms of A itself.)

The derivative is more easily investigated if we define

$$B(y,\mu) = e^{\mu^3/6}\, A(y,\mu)\,. \qquad (10.24)$$

Then

$$(y-2)B(y,\mu) = \mu B(y-2,\mu) + B(y-3,\mu)\,; \qquad (10.25)$$
$$B'(y,\mu) = \tfrac{1}{2}B(y-2,\mu)\,. \qquad (10.26)$$

It is easy to verify that the infinite series of (10.2) satisfies these relations. Repeated application of (10.25) and (10.26) leads to a third-order differential equation for $B = B(y,\mu)$:

$$8B''' - 4\mu^2 B'' + 2\mu(2y-9)B' - (y-2)(y-5)B = 0\,. \qquad (10.27)$$

We can see from (10.22) that, for any fixed $\mu \geq 0$, there are infinitely many negative values of y such that $A(y,\mu) = 0$. For if $y < 0$ and there is no root between $y-1$ and y, then $A(y-1,\mu)$ and $A(y,\mu)$ have the same sign; hence $A(y+2,\mu)$ has the opposite sign, and there's a root between y and $y+2$. Therefore we cannot use equation (10.5) until $|\mu|$ is sufficiently large, at least not when $y < 0$ and $\mu \geq 0$.

Lemma 3 implies the nonobvious inequality $A(y,\mu) \geq 0$ for all $y \geq 0$, since $A(y,\mu)$ is proportional to the limiting value of the coefficients of $U(z)^{n-m}/\bigl(1-T(z)\bigr)^y$, and those coefficients are nonnegative. Moreover, $A(y,\mu)$ is strictly positive for $y \geq 2$ and all μ. For if we had $y \geq 2$ and $A(y,\mu_0) = 0$, we would have $B(y,\mu_0) = 0$; but $B'(y,\mu) \geq 0$ by (10.26), hence we must have $B(y,\mu) = 0$ for all $\mu \leq \mu_0$. But that is impossible because $B(y,\mu)$ is a nonconstant analytic function of μ by (10.2).

When $y = 1$ there is a "closed form" in terms of the Airy function:

$$A(1,\mu) = e^{-\mu^3/12}\mathrm{Ai}(\mu^2/4)\,; \qquad (10.28)$$

this identity is proved in [14, (A.12) and (A.19)]. If we differentiate (10.28) with respect to μ, taking note of the fact that (10.22) gives

$$A(-1, \mu) = -\mu A(0, \mu), \tag{10.29}$$

we find

$$A(0, \mu) = -\tfrac{1}{2}\mu e^{-\mu^3/12} \mathrm{Ai}(\mu^2/4) - e^{-\mu^3/12} \mathrm{Ai}'(\mu^2/4). \tag{10.30}$$

Therefore in particular,

$$e^{\mu^3/12} A(1, \mu) \qquad \text{and} \qquad e^{\mu^3/12}\big(A(0,\mu) + \tfrac{1}{2}\mu A(1,\mu)\big)$$

are even functions of μ. The well-known relations between Airy functions and Bessel functions,

$$\mathrm{Ai}(z) = \frac{1}{\pi}\sqrt{\frac{z}{3}}\, K_{1/3}\Big(\tfrac{2}{3}z^{3/2}\Big), \qquad \mathrm{Ai}'(z) = \frac{-z}{\pi\sqrt{3}}\, K_{2/3}\Big(\tfrac{2}{3}z^{3/2}\Big),$$

yield the additional formulas

$$A(1, \mu) = \frac{e^{-\mu^3/12}\mu}{2\pi\sqrt{3}}\, K_{1/3}\Big(\frac{\mu^3}{12}\Big), \tag{10.31}$$

$$A(0, \mu) + \frac{\mu}{2}A(1, \mu) = A(3, \mu) - \frac{\mu}{2}A(1, \mu)$$

$$= \frac{e^{-\mu^3/12}\mu^2}{4\pi\sqrt{3}}\, K_{2/3}\Big(\frac{\mu^3}{12}\Big). \tag{10.32}$$

Since we know $A(y, \mu)$ for $y = -1$, 0, and 1, we can use (10.22) to determine $A(y, \mu)$ for all negative integers y, and for $y = 3$ as indicated in (10.32). But a new idea is needed if we hope to have a closed form when $y = 2$. It is possible to express $A(2, \mu)$ as an infinite sum of Bessel functions,

$$A(2, \mu) = \frac{1}{3}\bigg(e^{-\mu^3/6}$$

$$+ e^{-\mu^3/12}\bigg(\sum_{k \geq 0}(-1)^k \Big(I_{k+1/3}\Big(\frac{\mu^3}{12}\Big) - I_{k+2/3}\Big(\frac{\mu^3}{12}\Big)\Big)\bigg)\bigg), \tag{10.33}$$

but this may be as close to a closed form as possible unless we use general hypergeometric functions. Equation (10.33) follows from (10.2) and the hypergeometric identity

$$F(\tfrac{1}{2}+a,\, 1+2a-b-c;\, 1+2a-b,\, 1+2a-c;\, 2z)$$

$$= \frac{e^z\,\Gamma(a)}{(z/2)^a} \sum_{k \geq 0} \frac{(-1)^k (2a)^{\overline{k}} b^{\overline{k}} c^{\overline{k}}(k+a)I_{k+a}(z)}{(1+2a-b)^{\overline{k}}(1+2a-c)^{\overline{k}}k!} \tag{10.34}$$

[34, equation (2.8)]; here $x^{\overline{k}}$ denotes $\Gamma(x+k)/\Gamma(x)$, and we obtain (10.33) by setting $z = \mu^3/12$ together with $(a, b, c) = (\tfrac{1}{3}, \tfrac{1}{3}, 1)$ and $(\tfrac{2}{3}, \tfrac{2}{3}, 1)$.

The facts that $K_{1/3}(z) = 3^{-1/2}\pi\big(I_{-1/3}(z) - I_{1/3}(z)\big)$, $K_{2/3}(z) = 3^{-1/2}\pi\big(I_{-2/3}(z) - I_{2/3}(z)\big)$, and $e^{-z} = I_0(z) + 2\sum_{k\geq 1}(-1)^k I_k(z)$ suggest that we look for an identity of the form

$$A(y,\mu) = \left(\frac{\mu}{2}\right)^{2-y} e^{-\mu^3/12} \sum_{k\geq 0} a_k(y) I_{(k+y-2)/3}\left(\frac{\mu^3}{12}\right)$$

$$= \frac{e^{-\mu^3/6}}{3^{(y-2)/3}} \sum_{k\geq 0} \frac{a_k(y)}{\Gamma\left(\frac{k+y+1}{3}\right)} \left(\frac{\mu}{2\cdot 3^{1/3}}\right)^k F\left(\frac{2k+2y-1}{6}; \frac{2k+2y-1}{3}; \frac{\mu^3}{6}\right).$$

$$(10.35)$$

Any formal power series in μ has such an expansion, for all $y > -1$. But the coefficients $a_k(y)$ do not appear to have a simple form except in the cases already mentioned. We have

$$a_0(y) = \frac{1}{3}, \quad a_1(y) = \frac{y-1}{3}, \quad a_2(y) = \frac{y(y-3)}{6},$$

$$a_3(y) = \frac{(y^2-1)(y-6)}{18}, \quad a_4(y) = \frac{(y-1)(y+2)(y^2-11y+12)}{72},$$

$$a_5(y) = \frac{(y+3)y(y-1)(y-3)(y-14)}{360}.$$

Splitting (10.2) into three sums according to the value of $k \bmod 3$ yields a closed form for $A(y,\mu)$ in terms of general hypergeometric series:

$$A(y,\mu) = e^{-\mu^3/6}\Bigg(\frac{1}{3^{(y+1)/3}\,\Gamma\big((y+1)/3\big)}\, F\left(\frac{2-y}{6}, \frac{5-y}{6}; \frac{1}{3}, \frac{2}{3}; \frac{\mu^3}{6}\right)$$

$$+ \frac{1}{3^{(y-1)/3}\,\Gamma\big((y-1)/3\big)}\,\frac{\mu}{2}\, F\left(\frac{4-y}{6}, \frac{7-y}{6}; \frac{2}{3}, \frac{4}{3}; \frac{\mu^3}{6}\right)$$

$$+ \frac{1}{3^{(y-3)/3}\Gamma\big((y-3)/3\big)}\,\frac{\mu^2}{8}\, F\left(\frac{6-y}{6}, \frac{9-y}{6}; \frac{4}{3}, \frac{5}{3}; \frac{\mu^3}{6}\right)\Bigg).$$

$$(10.36)$$

11. Application to Bicyclic Components

Now we are ready to begin using the basic theoretical results of the preceding sections. We will start by considering the case when the parameter μ of Lemma 3 is very small, say $\mu = O(n^{-1/3})$. Then there are $m = \frac{1}{2}n + O(n^{1/3})$ edges.

Theorem 4. *The probability that a random graph or multigraph with n vertices and $\frac{1}{2}n + O(n^{1/3})$ edges has exactly r bicyclic components, and no components of higher cyclic order, is*

$$\left(\frac{5}{18}\right)^r \sqrt{\frac{2}{3}} \frac{1}{(2r)!} + O(n^{-1/3}). \tag{11.1}$$

Proof. (The special case $r = 0$ and $m = \frac{1}{2}n$ of this theorem was Corollary 9 of [14].) Consider first the case of random multigraphs, since that case is simpler. If there are n vertices, m edges, r bicyclic components, and no components with higher cyclic order, there must be exactly $n - m + r$ acyclic components. The probability of such a configuration, according to (2.2), is therefore exactly

$$\frac{2^m\, m!\, n!}{n^{2m}} [z^n] \frac{U(z)^{n-m+r}}{(n-m+r)!} e^{V(z)} \frac{W(z)^r}{r!}, \tag{11.2}$$

where $U(z)$, $V(z)$, $W(z)$ are the generating functions (3.3), (3.4), and (3.7). Now

$$W(z) = \frac{5}{24} \frac{1}{(1-T(z))^3} - \frac{7}{24} \frac{1}{(1-T(z))^2} + \frac{1}{12} \frac{1}{(1-T(z))}, \tag{11.3}$$

using the coefficients e'_{1d} of (7.20); so we see that $W(z)^r$ is a polynomial of degree $3r$ in $(1-T(z))^{-1}$, with leading coefficient $\left(\frac{5}{24}\right)^r$. Lemma 3 tells us that the leading term of $W(z)^r$ is the only significant one, asymptotically speaking, because the other terms contribute at most $n^{-1/3}$ times as much as the leading term. We can also write

$$U(z)^r = 2^{-r}\left(1 - \left(1 - T(z)\right)^2\right)^r; \tag{11.4}$$

this identity allows us to replace $U(z)^r$ by 2^{-r} in (11.2). Since $e^{V(z)} = \left(1 - T(z)\right)^{-1/2}$, the value of (11.2) is

$$\frac{(n-m)!}{(n-m+r)!\, r!\, 2^r} \frac{\sqrt{2\pi}\, n^r}{3^{r+1/2}\Gamma(r+1/2)} \left(\frac{5}{24}\right)^r \left(1 + O(n^{-1/3})\right).$$

And this expression simplifies to (11.1), using the fact that

$$\frac{(n-m)!}{(n-m+r)!} = \frac{2^r}{n^r}\left(1 + O(rn^{-2/3} + r^2n^{-1})\right),$$

together with a special case of the duplication formula for the Gamma function,

$$\Gamma(r+1/2) = \frac{(2r)!\,\sqrt{\pi}}{4^r r!}. \tag{11.5}$$

On the other hand if we are dealing with random graphs instead of multigraphs we must replace (11.2) by

$$\frac{n!}{\binom{n(n-1)/2}{m}} \, [z^n] \, \frac{U(z)^{n-m+r}}{(n-m+r)!} \, e^{\widehat{V}(z)} \, \frac{\widehat{W}(z)^r}{r!} \, , \tag{11.6}$$

where $\widehat{V}(z)$ and $\widehat{W}(z)$ appear in (3.5) and (3.6). Again we have $\widehat{W}(z) = \frac{5}{24}(1 - T(z))^{-3}$ plus less significant terms, so $\widehat{W}(z)$ produces an effect similar to $W(z)$. But

$$\widehat{V}(z) = V(z) - \tfrac{1}{2}T(z) - \tfrac{1}{4}T(z)^2;$$

so we now want the coefficient of $[z^n]$ in an expression proportional to

$$\frac{U(z)^{n-m}}{\bigl(1 - T(z)\bigr)^{3r+1/2}} \, e^{-T(z)/2 - T(z)^2/4} \, ,$$

which has an exponential factor not covered by Lemma 3. The proof of Lemma 3 shows, however, that this exponential factor simply changes the result by a factor of $e^{-3/4} + O(n^{-1/3})$: We multiply (10.18) by $\exp(-e^{-s\nu}/2 - e^{-2s\nu}/4) = e^{-3/4} + O(s\nu)$.

Furthermore, (11.6) contains a factor $e^{+3/4}$ to cancel the $e^{-3/4}$, because of (2.4). Therefore the leading term of the asymptotic probability for graphs is the same as it was for multigraphs.   □

Corollary. *The probability that a random graph or multigraph with n vertices and $\frac{1}{2}n$ edges has only acyclic, unicyclic, and bicyclic components is*

$$\sqrt{\frac{2}{3}} \cosh \sqrt{\frac{5}{18}} + O(n^{-1/3}) \approx 0.9325 \, . \tag{11.7}$$

Proof. The sum over r of the estimate made in Theorem 4 clearly gives a lower bound, so we must prove that it is also an upper bound. That sum can be written

$$\frac{2^m \, m! \, n!}{(n-m)! \, n^{2m}} \, [z^n] \, U(z)^{n-m} f_{n-m}(z) \, ,$$

where

$$f_l(z) = \sum_{r \geq 0} \frac{l!}{(l+r)!} \, \frac{\bigl(U(z)W(z)\bigr)^r}{r!} \, e^{V(z)} \, . \tag{11.8}$$

If we look at the proof of Theorem 4, and the proof of Lemma 3 on which it is based, we see that the calculations all depend on $f_l(ze^{-z})$, where $|z| \le e^{-\nu}$ and $\nu = n^{-1/3}$. In this region,

$$|T(ze^{-z})| \le e^{-\nu}, \qquad |1 - T(ze^{-z})| \ge \nu + O(\nu^2). \qquad (11.9)$$

Thus the sum $f_{n-m}(ze^{-z})$ converges *uniformly* for all n and all $|z| \le e^{-\nu}$. Uniform convergence allows us to interchange summation and integration. (Notice that the function $h(z)$ in the proof of Lemma 3, which influences the behavior of the integrand most strongly as $n \to \infty$, is independent of r.)   □

Another proof of (11.7) will be given below.

12. Components of Higher Cyclic Order

Now let's consider components that are tricyclic, tetracyclic, etc. (Notice that tricyclic components correspond to $C_2(z)$, not $C_3(z)$, in the notation of Section 2; our notation has mathematical advantages, but it is slightly out of phase with the traditional terminology.)

Theorem 5. *The probability that a random graph or multigraph with n vertices and $\frac{1}{2}n + O(n^{1/3})$ edges has exactly r_1 bicyclic components, r_2 tricyclic components, $\ldots$, r_q $(q+1)$-cyclic components, and no components of higher cyclic order, is*

$$\left(\frac{4}{3}\right)^r \sqrt{\frac{2}{3}} \frac{c_1^{r_1}}{r_1!} \frac{c_2^{r_2}}{r_2!} \cdots \frac{c_q^{r_q}}{r_q!} \frac{r!}{(2r)!} + O(n^{-1/3}), \qquad (12.1)$$

where $r = r_1 + 2r_2 + \cdots + qr_q$ and the constants c_j are defined in (8.6).

Proof. If there are n vertices and m edges, there must be exactly $n - m + r$ acyclic components. So we can argue as in Theorem 4 to find

$$\frac{2^m\, m!\, n!}{n^{2m}}\, [z^n]\, \frac{U(z)^{n-m+r}}{(n-m+r)!}\, e^{V(z)}\, \frac{C_1(z)^{r_1}}{r_1!}\, \frac{C_2(z)^{r_2}}{r_2!} \cdots \frac{C_q(z)^{r_q}}{r_q!}$$

$$= \frac{c_1^{r_1}}{r_1!} \frac{c_2^{r_2}}{r_2!} \cdots \frac{c_q^{r_q}}{r_q!} \frac{\sqrt{2\pi}}{3^{r+1/2}\Gamma(r + 1/2)} + O(n^{-1/3}).$$

Formula (12.1) now follows from (11.5) as before.   □

Let's illustrate the consequences of Theorem 5 by computing the limiting probabilities for small values of the parameters $(r_1, r_2, \ldots, r_q)$.

Here is a list of all configurations with $r_1+r_2+\cdots+r_q > 1$ that occur with limiting probability .000005 or more, showing the probabilities rounded to five decimal places:

$$[2] \approx .00263\,; \qquad\qquad [0,1,1] \approx .00003\,;$$
$$[1,1] \approx .00105\,; \qquad\qquad [3] \approx .00002\,;$$
$$[1,0,1] \approx .00031\,; \qquad\quad [1,0,0,0,0,1] \approx .00002\,;$$
$$[1,0,0,1] \approx .00010\,; \qquad\qquad [2,1] \approx .00001\,;$$
$$[0,2] \approx .00008\,; \qquad\qquad [0,1,0,1] \approx .00001\,;$$
$$[1,0,0,0,1] \approx .00004\,; \qquad [1,0,0,0,0,0,1] \approx .00001\,.$$

(The notation $[2]$ stands for the case $r_1 = 2$, $r_2 = r_3 = \cdots = 0$; similarly, $[r_1,\ldots,r_q]$ implies that there are no complex components of cyclic order greater than $q + 1$.)

The sum of these probabilities, .00431, is nicely balanced by $\sqrt{2/3}$ plus the sum of probabilities when there is only one complex component, that is, when $r_q = 1$ and all other r's are zero:

$$.81650 + .11340 + .03780 + .01547 + .00678 + .00307 + .00141$$
$$+ .00066 + .00031 + .00015 + .00007 + .00003 + .00002 + .00001\,;$$

this comes to $.99568 = .99999 - .00431$.

Suppose $\mathcal{R}$ is any countably infinite set of configurations $[r_1, r_2, \ldots, r_q]$, where q might be unbounded. We would like to prove that a random graph or multigraph with approximately $\frac{1}{2}n$ edges lies in $\mathcal{R}$ with limiting probability

$$\sum \bigl\{\, P[r_1, r_2, \ldots, r_q] \,\bigm|\, [r_1, r_2, \ldots, r_q] \in \mathcal{R} \,\bigr\}\,, \qquad (12.2)$$

where $P[r_1, r_2, \ldots, r_q]$ is the limiting value stated in Theorem 5. The technique we used to prove (11.7) does not apply, because the infinite sums over which integration takes place might not converge uniformly when q is unbounded.

However, we are obviously justified in claiming that (12.2) is a *lower* bound for the stated probability, because the sum over any finite subset of $\mathcal{R}$ yields a lower bound.

We will prove below that the sum of $P[r_1, r_2, \ldots, r_q]$ over all possible configurations $[r_1, r_2, \ldots, r_q]$ is 1. Consequently, the sum (12.2) must in fact be the limiting probability of a random graph or multigraph being in $\mathcal{R}$, not just a lower bound. If (12.2) were too low, we would not obtain 1 by adding the complementary probabilities $P[r_1, r_2, \ldots, r_q]$ for $[r_1, r_2, \ldots, r_q] \notin \mathcal{R}$. This observation will lead to the promised "second proof" of (11.7), if we also sum less significant terms to obtain the error bound $O(n^{-1/3})$.

13. Excess Edges

The notion of "excess" was used somewhat informally in the introductory sections of this paper. Let us now define it formally, saying that the *excess* of a graph or multigraph is the number of edges plus the number of acyclic components, minus the number of vertices. Thus a $(q + 1)$-cyclic component has excess q, when $q \geq 0$. If a graph or multigraph has r_1 bicyclic components, r_2 tricyclic components, etc., then it has excess $r = r_1 + 2r_2 + 3r_3 + \cdots$.

If G and G' are graphs on the same vertices, and if $G \cup G'$ and $G \cap G'$ denote the graphs obtained by taking the union and intersection of their sets of edges, the excesses satisfy

$$r(G) + r(G') \leq r(G \cup G') + r(G \cap G').$$

For we can start with empty graphs and insert the edges of $G \cap G'$, preserving equality. Then if we insert an edge of $G \setminus G'$ or of $G' \setminus G$, each side of the inequality increases by either 0 or 1; and the left side cannot increase by 1 unless the right side does also. For example, if the left side increases by 1 when we add an edge of $G \setminus G'$, the endpoints of that edge are in non-trees of G, so they surely are in non-trees of $G \cup G'$.

We have seen in Theorem 5 that the limiting joint probability distribution of the random variables $(r_1, r_2, \ldots)$ in a large random graph or multigraph with approximately $\frac{1}{2}n$ edges has the form

$$\frac{c_1^{r_1}}{r_1!} \frac{c_2^{r_2}}{r_2!} \cdots \frac{c_q^{r_q}}{r_q!} f(r), \tag{13.1}$$

where $r = r_1 + 2r_2 + \cdots + qr_q$ is the excess of the graph and $r_l = 0$ for $l > q$. Indeed, this is not surprising, if we look at the problem in another way.

Let $\mathcal{S}$ be the set of all multigraphs of configuration $[r_1, r_2, \ldots, r_q]$, and let $S(w, z)$ be its bgf. The probability that a given multigraph with m edges and n vertices lies in $\mathcal{S}$ is then

$$\mathrm{Pr}_{mn}(\mathcal{S}) = \frac{[w^m z^n] S(w, z)}{[w^m z^n] G(w, z)}. \tag{13.2}$$

We can also express this quantity as

$$\mathrm{Pr}_{mn}(\mathcal{S}) = \mathrm{Pr}_{mn}(\mathcal{S} \mid r)\,\mathrm{Pr}_{mn}(\mathcal{E}_r)\,, \tag{13.3}$$

where $\mathrm{Pr}_{mn}(\mathcal{S} \mid r)$ means the probability of obtaining an element of $\mathcal{S}$ given that the excess is r, and $\mathrm{Pr}_{mn}(\mathcal{E}_r)$ is the probability that a random multigraph has excess r:

$$\mathrm{Pr}_{mn}(\mathcal{S} \mid r) = \frac{[w^m z^n]\,S(w,z)}{[w^m z^n]\,e^{U(w,z)+V(w,z)}\,E_r(w,z)}\,,$$

$$\mathrm{Pr}_{mn}(\mathcal{E}_r) = \frac{[w^m z^n]\,e^{U(w,z)+V(w,z)}\,E_r(w,z)}{[w^m z^n]\,G(w,z)}\,. \tag{13.4}$$

Since all elements of $\mathcal{S}$ have excess r, we can compute $[w^m z^n]\,S(w,z)$ with univariate generating functions:

$$
\begin{aligned}
S(w,z) &= e^{U(w,z)+V(w,z)}\frac{C_1(w,z)^{r_1}}{r_1!}\frac{C_2(w,z)^{r_2}}{r_2!}\cdots\frac{C_q(w,z)^{r_q}}{r_q!}\\[2mm]
&= e^{U(wz)/w+V(wz)}\frac{\bigl(wC_1(wz)\bigr)^{r_1}}{r_1!}\frac{\bigl(w^2C_2(wz)\bigr)^{r_2}}{r_2!}\cdots\frac{\bigl(w^qC_q(wz)\bigr)^{r_q}}{r_q!}\\[2mm]
&= e^{U(wz)/w+V(wz)}w^r\frac{C_1(wz)^{r_1}}{r_1!}\frac{C_2(wz)^{r_2}}{r_2!}\cdots\frac{C_q(wz)^{r_q}}{r_q!}\,;
\end{aligned}
$$

hence

$$[w^m z^n]\,S(w,z) = [z^n]\,\frac{U(z)^{n+r-m}}{(n+r-m)!}\,e^{V(z)}S(z)\,, \tag{13.5}$$

if we let

$$S(z) = \frac{C_1(z)^{r_1}}{r_1!}\frac{C_2(z)^{r_2}}{r_2!}\cdots\frac{C_q(z)^{r_q}}{r_q!}\,.$$

Similarly

$$[w^m z^n]\,e^{U(w,z)+V(w,z)}E_r(w,z) = [z^n]\,\frac{U(z)^{n+r-m}}{(n+r-m)!}\,e^{V(z)}E_r(z)\,.$$

A multigraph with m edges, n vertices, and excess $r > 0$ has $t = n+r-m$ components that are trees (including isolated vertices). Suppose it has n_1 vertices in complex components and n_0 vertices in trees and unicyclic components. Then

$$\Pr_{mn}(\mathcal{S} \mid r) = \frac{[z^n] \dfrac{U(z)^t}{t!} e^{V(t)} S(z)}{[z^n] \dfrac{U(z)^t}{t!} e^{V(t)} E_r(z)}$$

$$= \frac{\displaystyle\sum_{n_0+n_1=n} \left([z^{n_0}] U(z)^t e^{V(z)}\right)\left([z^{n_1}] S(z)\right)}{\displaystyle\sum_{n_0+n_1=n} \left([z^{n_0}] U(z)^t e^{V(z)}\right)\left([z^{n_1}] E_r(z)\right)}$$

$$= \sum_{n_1} \Pr(\mathcal{S} \mid r, n_1) \Pr_{mn}(n_1 \mid \mathcal{E}_r),$$

where

$$\Pr(\mathcal{S} \mid r, n_1) = \frac{[z^{n_1}] S(z)}{[z^{n_1}] E_r(z)}; \tag{13.6}$$

$$\Pr_{mn}(n_1 \mid \mathcal{E}_r) = \frac{\left([z^{n-n_1}] U(z)^t e^{V(z)}\right)\left([z^{n_1}] E_r(z)\right)}{[z^n] U(z)^t e^{V(z)} E_r(z)}. \tag{13.7}$$

Thus, $\Pr(\mathcal{S})$ has been expressed in terms of a simple ratio (13.6), the number of multigraphs consisting of precisely r_j components of excess j for $1 \le j \le q$, divided by the number of complex multigraphs of excess r. We know from Section 9 that there are coefficients s_d such that

$$S(z) = \frac{s_0 T(z)^{2r}}{\left(1 - T(z)\right)^{3r}} + \frac{s_1 T(z)^{2r-1}}{\left(1 - T(z)\right)^{3r-1}} + \cdots + \frac{s_{2r-1} T(z)}{\left(1 - T(z)\right)^{r+1}}.$$

Indeed, Section 9 tells us that s_d is $\sum \kappa(\overline{\overline{M}})/(2r - d)!$, summed over all reduced multigraphs of configuration $[r_1, r_2, \ldots, r_q]$ having exactly $2r-d$ vertices. We can also write

$$S(z) = \frac{s'_0}{\left(1 - T(z)\right)^{3r}} + \frac{s'_1}{\left(1 - T(z)\right)^{3r-1}} + \cdots + \frac{s'_{2r}}{\left(1 - T(z)\right)^r}, \tag{13.8}$$

letting $s'_d = \sum_k \binom{2r-k}{d-k}(-1)^{d-k} s_k$ as in (7.22). Therefore,

$$n! \, [z^n] S(z) = s'_0 t_n(3r) + s'_1 t_n(3r - 1) + \cdots + s'_{2r} t_n(r),$$

expressing the relevant number of multigraphs in terms of the tree polynomials (3.8); and (3.9) tells us that

$$n!\,[z^n]\,S(z) = s_0' \frac{\sqrt{2\pi}\,n^{n-1/2+3r/2}}{2^{3r/2}\Gamma(3r/2)}\big(1 + O(n^{-1/2})\big).$$

Similarly, we have

$$n!\,[z^n]\,E_r(z) = e_{r0}' \frac{\sqrt{2\pi}\,n^{n-1/2+3r/2}}{2^{3r/2}\Gamma(3r/2)}\big(1 + O(n^{-1/2})\big).$$

Therefore the ratio (13.6) is

$$\Pr(S \mid r, n_1) = \frac{s_0'}{e_{r0}'}\big(1 + O(n_1^{-1/2})\big);$$

and we can sum over n_1 to get

$$\Pr_{mn}(S) = \left(\frac{s_0'}{e_{r0}'} + O(\epsilon)\right)\Pr_{mn}(\mathcal{E}_r), \tag{13.9}$$

where ϵ is the expected value of $n_1^{-1/2}$ in the probability distribution (13.7).

Moreover, the leading coefficient is

$$s_0' = s_0 = \frac{c_1^{r_1}\,c_2^{r_2}}{r_1!\,r_2!}\cdots\frac{c_q^{r_q}}{r_q!}; \tag{13.10}$$

and e_{r0}' is just e_r, the sum of (13.10) over all configurations $[r_1, r_2, \ldots, r_q]$ with $r_1 + 2r_2 + \cdots + qr_q = r$. This derivation explains why we obtained a formula of the form (13.1) in Theorem 5.

With graphs instead of multigraphs, the same considerations apply, but we must add more terms to the formulas. For example, (13.8) becomes

$$\widehat{S}(z) = \frac{\hat{s}_0'}{\big(1 - T(z)\big)^{3r}} + \frac{\hat{s}_1'}{\big(1 - T(z)\big)^{3r-1}} + \cdots$$
$$+ \hat{s}_{3r}' + \hat{s}_{3r+1}'\big(1 - T(z)\big) + \hat{s}_{3r+2}'\big(1 - T(z)\big)^2. \tag{13.11}$$

The leading coefficient $\hat{s}_0'$ is the same as s_0, so the asymptotic behavior is the same as before, if we assume that m is large enough to make the expected value of $n_1^{-1/2}$ approach zero.

We can estimate the expected value of $n_1^{-1/2}$ by finding the expected value of

$$\frac{[z^{n_1}]\, S(z) - (s_0/e_r)E_r(z)}{[z^{n_1}]\, E_r(z)}; \tag{13.12}$$

indeed, this expected value is the true error in the approximation (13.9), so it is even more relevant than the expected value of $n_1^{-1/2}$. Since $S(z) - (s_0/e_r)E_r(z)$ can be expressed as $(s_1' - e_{r1}'s_0/e_r)/(1 - T(z))^{3r-1}$ plus less significant terms, the desired expected value times $\Pr_{mn}(\mathcal{E}_r)$ is obtained by applying Lemma 3 as we did in the proof of Theorem 4, but with $3r$ replaced by $3r - 1$. The result, when m is near $n/2$, is proportional to $n^{-1/3}$.

The expected value of n_1^k can be computed if we replace $S(z)$ by $\vartheta^k E_r(z)$ in these formulas, because $[z^n]\vartheta^k E_r(z) = n^k[z^n]\, E_r(z)$. This has the effect of changing the leading term from $e_r/(1 - T(z))^{3r}$ to $(3r)(3r + 2)\ldots(3r + 2k - 2)e_r/(1 - T(z))^{3r+2k}$, so the result when m is near $\frac{1}{2}n$ is proportional to $n^{2k/3}$. We have proved

Corollary. If $m = \frac{1}{2}n(1 + \mu n^{-1/3})$ and $|\mu| \leq n^{1/12}$, the kth moment $\mathrm{E}_{mn}(n_1^k \,|\, r)$ of the number of vertices in complex components, given that the total excess is r, is

$$\alpha_{kr}\frac{\Gamma(r + \frac{1}{2})}{3^{2k/3}\Gamma(r + \frac{1}{2} + \frac{2}{3}k)}\, n^{2k/3}\big(1 + O(\mu) + O(n^{-1/3})\big), \qquad \text{if } \mu = O(1); \tag{13.13}$$

$$\alpha_{kr}\frac{n^{2k/3}}{\mu^{2k}}\big(1 + O(|\mu|^{-3}) + O(\mu^4 n^{-1/3})\big), \qquad\qquad \text{if } \mu \to -\infty; \tag{13.14}$$

$$\alpha_{kr}\frac{n^{2k/3}\mu^k}{2^k}\frac{\Gamma(\frac{3}{2}r + \frac{1}{4})}{\Gamma(\frac{3}{2}r + \frac{1}{4} + k)}\big(1 + O(\mu^{-1}) + O(\mu^4 n^{-1/3})\big), \quad \text{if } \mu \to +\infty; \tag{13.15}$$

here $\alpha_{kr} = (3r)(3r + 2)\ldots(3r + 2k - 2)$.

Proof. These expressions are α_{kr} times the ratios of formulas (10.2), (10.3), and (10.4) when $y = 3r + \frac{1}{2} + 2k$ to the values that they have when $y = 3r + \frac{1}{2}$. □

Notice that when m is approximately $\frac{1}{2}n - n^{3/4}$, the probable value of n_1 is proportional to $n^{2/3-2/12} = n^{1/2}$; when $m \approx \frac{1}{2}n + n^{3/4}$, it is proportional to $n^{2/3+1/12} = n^{3/4}$. These are the extreme cases $|\mu| = n^{1/12}$ at the limits of Lemma 3's range.

We can use formula (13.3) whenever S is a collection of multigraphs whose complex components have total excess r. We can use formula (13.6) whenever S also places no restriction on its non-complex (acyclic and unicyclic) components. For example, we can determine the conditional probability that a random graph with $\frac{1}{2}n$ edges has a bicyclic component of each of the three types in (9.15), given that it has excess 1. The generating functions $S(z)$ for the three cases are respectively $\frac{1}{8}T/(1-T)^2$, $\frac{1}{8}T^2/(1-T)^3$, $\frac{1}{12}T^2/(1-T)^3$; so the respective conditional probabilities are

$$O(n^{-1/3}), \qquad \frac{3}{5} + O(n^{-1/3}), \qquad \frac{2}{5} + O(n^{-1/3}). \qquad (13.16)$$

All probabilities that are conditional on excess r must, of course, be multiplied by $\Pr_{mn}(\mathcal{E}_r)$, the probability that a random multigraph has excess r. Lemma 3 and the method of Theorem 5 make this easy to compute:

Corollary. *A graph or multigraph with $m = \frac{1}{2}n(1 + \mu n^{-1/3})$ edges and n vertices has excess r with probability*

$$\Pr_{mn}(\mathcal{E}_r) = \sqrt{2\pi}\, e_r\, A(3r + \tfrac{1}{2}, \mu) + O\left(\frac{1 + \mu^4}{n^{1/3}}\right), \qquad (13.17)$$

uniformly for $|\mu| \le n^{1/12}$ as $n \to \infty$, where $e_r = e_{r0}$ is given by (7.2) and $A(y, \mu)$ is given by (10.2). When $\mu \to -\infty$, the probability is $O(|\mu|^{-3r})$; when $\mu \to +\infty$ it is $O(\mu^{3r/2} e^{-\mu^3/6})$. $\square$

(The special case $r = 0$ in (13.17), without the error bound, was found by Britikov [9], who proved that a random graph has excess 0 with probability approaching $\sqrt{2\pi}\, A(\frac{1}{2}, \mu)$, for fixed μ as $n \to \infty$.)

Here is a table that shows how the probabilities of having excess r change as the graph or multigraph evolves past the critical point:

	$r=0$	$r=1$	$r=2$	$r=3$	$r=4$	$r=5$	$r=6$	$r=7$	$r=8$	$r=9$	$r=10$
$\mu = -3$	.994	.006	.000	.000	.000	.000	.000	.000	.000	.000	.000
$\mu = -2$	.983	.015	.001	.000	.000	.000	.000	.000	.000	.000	.000
$\mu = -1$	.947	.043	.008	.002	.000	.000	.000	.000	.000	.000	.000
$\mu = 0$	.816	.113	.040	.017	.007	.003	.001	.001	.000	.000	.000
$\mu = 1$	.475	.179	.115	.077	.052	.035	.023	.015	.010	.007	.004
$\mu = 2$	.100	.082	.085	.086	.084	.079	.073	.066	.058	.051	.043
$\mu = 3$	.003	.004	.007	.010	.013	.017	.020	.024	.028	.031	.034

The mean excess is approximately .308, 1.544, 6.364, and 19.009, for $\mu = 0$, 1, 2, and 3.

In this paper we are interested mainly in graphs or multigraphs with approximately $\frac{1}{2}n$ edges, but it is instructive to consider also the formulas that arise when m is somewhat smaller. The excess is then almost surely zero. In fact, we can obtain a formula that has a much better error bound than (13.17), in the case $r = 0$ and $\mu < -n^{-\epsilon}$: If we set $\lambda = 2m/n$, and if $m < \frac{1}{2}n - n^{2/3+\epsilon}$, the probability of excess 0 can be shown to be exactly

$$\frac{2^m m! \, n!}{(n-m)! \, n^{2m}} \, [z^n] \, \frac{U(z)^{n-m}}{(1-T(z))^{1/2}}$$

$$= \frac{S(m)S(n)}{\sqrt{2\pi} \, S(n-m)} \oint \left(1 - \frac{it\beta}{1-\lambda}\right)^{1/2} \left(1 + \frac{it\beta}{\lambda}\right)^{-1} e^{h(n,\lambda,t) - t^2/2} \, dt, \quad (13.18)$$

where

$$S(n) = \frac{n! \, e^n}{n^n \sqrt{2\pi n}} = 1 + O\!\left(\frac{1}{n}\right), \quad (13.19)$$

$$\beta = \sqrt{\frac{\lambda(2-\lambda)}{(1-\lambda)n}}, \quad (13.20)$$

$$h(n,\lambda,t) = nh(\lambda + it\beta) - nh(\lambda) + t^2/2 \quad (13.21)$$

$$= \frac{n}{2} \sum_{k \geq 3} \frac{(it\beta)^k}{k} \left(\frac{(-1)^k}{\lambda^{k-1}} - \frac{1}{(2-\lambda)^{k-1}}\right), \quad (13.22)$$

and the contour of integration makes $z = \lambda + it\beta$ traverse the circle $|z| = \lambda$ as t varies. The function $h(z)$ in (13.21) is the function defined in (10.12). We are essentially simplifying the proof of Lemma 3 by choosing a path of integration through the saddle point $z = \lambda$, as in the proof of Theorem 4 in [14]. The proof of that theorem justifies restricting t to a neighborhood of zero, so that the tail-exchange method can be applied as in the derivation following (10.9). It follows that the probability of excess 0 is $1 - O\big(n^2/(n-2m)^3\big)$ for m in the stated range. We have in fact the estimate

$$\Pr_{mn}(\mathcal{E}_0) = 1 - \frac{5}{24} \, \alpha^{-3} \big(1 + O(\alpha^{-3}) + O(\alpha n^{-1/3})\big) \quad (13.23)$$

when $m = \frac{1}{2}n(1 - \alpha \, n^{-1/3})$, uniformly for $(\ln n)^2 \leq \alpha \leq \frac{1}{2}n^{1/3}$.

It is interesting to note that the tail-exchange method can be used to extend (13.23) to an asymptotic series in α^{-1} and $\alpha n^{-1/3}$, although the integral (13.18) actually diverges if we let t run through all real values from $-\infty$ to $+\infty$ instead of describing the stated contour. Indeed, the magnitude of the integrand in (13.18) for large real values of $|t|$ is approximately $|t|^{n-2m-1/2}$.

14. Probability Distribution of the Excess

One way to check our calculations is to verify that the probabilities in
(13.17) sum to 1. Thus we want to prove that

$$\sum_{k\geq 0} \sqrt{\frac{2\pi}{3}} \frac{\left(\frac{1}{2} 3^{2/3}\mu\right)^k}{k!} \sum_{r\geq 0} \frac{(6r)!}{2^{5r}3^{3r}(2r)!\,(3r)!\,\Gamma(r+\frac{1}{2}-\frac{2}{3}k)} = e^{\mu^3/6}.$$

$$(14.1)$$

The inner sum is a hypergeometric series whose sum is known;

$$\frac{1}{\Gamma(\frac{1}{2}-\frac{2}{3}k)} F\left(\frac{1}{6}, \frac{5}{6}; \frac{1}{2}-\frac{2k}{3}; \frac{1}{2}\right) = \frac{2^{1/2+2k/3}\sqrt{\pi}}{\Gamma((1-k)/3)\,\Gamma((2-k)/3)}. \qquad (14.2)$$

Indeed, the special hypergeometric

$$f(a, b, z) = F(a, 1-a; b; z),$$

which is related to a Legendre function, satisfies

$$\frac{f(a, b, \frac{1}{2})}{\Gamma(b)} = \frac{2^{1-b}\sqrt{\pi}}{\Gamma(\frac{1}{2}(a+b))\,\Gamma(\frac{1}{2}(1-a+b))}. \qquad (14.3)$$

This well-known relation can be obtained by applying Euler's identity

$$F(a, b; c; z) = (1-z)^{c-a-b} F(c-a, c-b; c; z)$$

and Gauss's identities

$$F(a, b; c; 1) = \big(\Gamma(c-a-b)\Gamma(c)\big)/\big(\Gamma(c-a)\Gamma(c-b)\big),$$
$$F(2a, 2b; a+b+\tfrac{1}{2}; z) = F\big(a, b; a+b+\tfrac{1}{2}; 4z(1-z)\big),$$

which can be found, for example, in [17, (5.92), (5.110), exercise 5.28]:

$$(1-z)^{1-b}F(a, 1-a; b; z) = F(b-a, b+a-1; b; z)$$
$$= F\big(\tfrac{1}{2}b - \tfrac{1}{2}a, \tfrac{1}{2}b + \tfrac{1}{2}a - \tfrac{1}{2}; b; 4z(1-z)\big);$$

we obtain (14.3) by letting $z \to \frac{1}{2}$.

The sum (14.2) vanishes except when $k = 3m$, and in this case the
kth term on the left of (14.1) reduces to simply $(\mu^3/6)^m/m!$ because of
the formula

$$\Gamma(\tfrac{1}{3} - m)\,\Gamma(\tfrac{2}{3} - m) = 3^{3m-1/2}\, 2\pi\, \frac{m!}{(3m)!}. \qquad (14.4)$$

Hence (14.1) is true. It is remarkable that so much of nineteenth century mathematics has turned out to be relevant to the study of random graphs.

When $\mu = 0$, the generating function for the limiting probabilities of excess r turns out to have a closed form: It is

$$\sum_{r \geq 0} \left(\frac{4}{3}\right)^r \sqrt{\frac{2}{3}} \frac{e_r r! \, z^r}{(2r)!} = \sqrt{\frac{2}{3}} F\left(\frac{1}{6}, \frac{5}{6}; \frac{1}{2}; \frac{z}{2}\right)$$

$$= \frac{2 \cos\left(\frac{2}{3} \arcsin \sqrt{z/2}\right)}{\sqrt{6 - 3z}}. \qquad (14.5)$$

From this expression it is easy to calculate the limiting value of the mean excess when $m = \frac{1}{2}n$, namely $\frac{1}{2} - 3^{-3/2} \approx 0.308$. The variance, similarly, is $\frac{23}{27} - 3^{-3/2} \approx 0.6594$.

The limiting mean excess when the number of edges is equal to $\frac{1}{2}n(1 + \mu n^{-1/3})$ does not seem to have a simple closed form, although we can express it as a hypergeometric series and find the asymptotic value. Suppose we insert the factor z^r into the left-hand side of (14.1). Then the left-hand side of (14.2) becomes

$$\frac{1}{\Gamma(1/2 - 2k/3)} F\left(\frac{1}{6}, \frac{5}{6}; \frac{1}{2} - \frac{2k}{3}; \frac{z}{2}\right)$$

$$= \frac{1}{\Gamma(1/2 - 2k/3)} f\left(\frac{1}{6}, \frac{1}{2} - \frac{2k}{3}, \frac{z}{2}\right). \qquad (14.6)$$

To evaluate the derivative of such a function at $\frac{1}{2}$, we can use the identity

$$z(1-z)f'(a, b, z) = \left(az + \frac{1-a-b}{2}\right)f(a, b, z) - \frac{1-a-b}{2} f(-a, b, z), \quad (14.7)$$

which is readily verified by checking that the coefficients of z^n agree on both sides. To get the mean value of r, we want to differentiate (14.6) with respect to z and set $z = 1$; and according to (14.7), this is equivalent to replacing (14.6) by

$$\frac{1}{\Gamma(1/2 - 2k/3)} \left(\left(\frac{1}{2} + \frac{2k}{3}\right)f\left(\frac{1}{6}, \frac{1}{2} - \frac{2k}{3}, \frac{1}{2}\right)\right.$$

$$\left. - \left(\frac{1}{3} + \frac{2k}{3}\right)f\left(-\frac{1}{6}, \frac{1}{2} - \frac{2k}{3}, \frac{1}{2}\right)\right). \qquad (14.8)$$

Again, $f(\frac{1}{6}, \frac{1}{2} - \frac{2k}{3}, \frac{1}{2})$ vanishes unless $k = 3m$. The contribution to the mean from this half of (14.8) is just what we had when we were summing the probabilities, but with an additional factor of $(\frac{1}{2} + \frac{2k}{3})$; so it is

$$e^{-\mu^3/6} \sum_{m \geq 0} (\tfrac{1}{2} + 2m) \frac{(\mu^3/6)^m}{m!} = \frac{1}{2} + \frac{\mu^3}{3}. \tag{14.9}$$

The other half of (14.8) is, however, more complicated, since all values of k make a contribution. According to (14.3), we want to evaluate

$$\sum_{k \geq 0} \sqrt{\frac{2\pi}{3}} \frac{(\frac{1}{2} 3^{2/3} \mu)^k}{k!} \frac{2^{1/2 + 2k/3} \sqrt{\pi} \, (1/3 + 2k/3)}{\Gamma(1/6 - k/3)\, \Gamma(5/6 - k/3)} = \Sigma_0 + \Sigma_1 + \Sigma_2,$$

where Σ_j is a hypergeometric series corresponding to $k = 3m + j$:

$$\Sigma_0 = \frac{1}{3\sqrt{3}} F\left(\frac{5}{6}, \frac{7}{6}; \frac{1}{3}, \frac{2}{3}; \frac{\mu^3}{6}\right);$$

$$\Sigma_1 = -\frac{1}{\sqrt{3}} \frac{\mu \sqrt{\pi}}{6^{1/3}\Gamma(5/6)} F\left(\frac{7}{6}, \frac{3}{2}; \frac{2}{3}, \frac{4}{3}; \frac{\mu^3}{6}\right);$$

$$\Sigma_2 = -\frac{5\sqrt{3}}{2} \frac{\mu^2}{6^{2/3}} \frac{\sqrt{\pi}}{\Gamma(1/6)} F\left(\frac{3}{2}, \frac{11}{6}; \frac{4}{3}, \frac{5}{3}; \frac{\mu^3}{6}\right). \tag{14.10}$$

As $z \to +\infty$, such hypergeometric series satisfy the asymptotic formula

$$F(a, b; c, d; z) = \frac{\Gamma(c)\Gamma(d)}{\Gamma(a)\Gamma(b)} z^{\delta} e^z \Bigg(1 + \frac{\delta(a + b - 1) - ab + cd}{z}$$

$$+ O(z^{-2}) \Bigg) \tag{14.11}$$

where $\delta = a + b - c - d$; this follows by plugging the right-hand side into the differential equation

$$\vartheta(\vartheta + c - 1)(\vartheta + d - 1)F = z(\vartheta + a)(\vartheta + b)F$$

satisfied by the left. We obtain

$$e^{-\mu^3/6} \Sigma_0 = \frac{1}{3\sqrt{3}} \frac{\Gamma(\frac{1}{3})\Gamma(\frac{2}{3})}{\Gamma(\frac{5}{6})\Gamma(\frac{7}{6})} \left(\frac{\mu^3}{6} + \frac{1}{4} + O(\mu^{-3})\right);$$

$$e^{-\mu^3/6} \Sigma_1 = -\frac{1}{\sqrt{3}} \frac{\sqrt{\pi}}{\Gamma(\frac{5}{6})} \frac{\Gamma(\frac{2}{3})\Gamma(\frac{4}{3})}{\Gamma(\frac{7}{6})\Gamma(\frac{3}{2})} \left(\frac{\mu^3}{6} + \frac{1}{4} + O(\mu^{-3})\right);$$

$$e^{-\mu^3/6} \Sigma_2 = -\frac{5\sqrt{3}}{2} \frac{\sqrt{\pi}}{\Gamma(\frac{1}{6})} \frac{\Gamma(\frac{4}{3})\Gamma(\frac{5}{3})}{\Gamma(\frac{3}{2})\Gamma(\frac{11}{6})} \left(\frac{\mu^3}{6} + \frac{1}{4} + O(\mu^{-3})\right); \tag{14.12}$$

therefore $e^{-\mu^3/6}(\Sigma_0+\Sigma_1+\Sigma_2) = (\frac{2}{3}-\frac{4}{3}-\frac{4}{3})(\frac{\mu^3}{6}+\frac{1}{4}+O(\mu^{-3}))$. Subtracting this from (14.9), and using computer algebra to refine the estimate further, gives us the answer we seek:

Theorem 6. *The expected value of the excess, when the number of edges is $\frac{1}{2}n(1+\mu n^{-1/3})$, approaches*

$$\frac{2}{3}\mu^3 + 1 + \frac{5}{24}\mu^{-3} + \frac{15}{16}\mu^{-6} + O(\mu^{-9}), \qquad (14.13)$$

for fixed $\mu \geq \delta > 0$ as $n \to \infty$. □

This method of calculation shows also that the variance will be $O(\mu^6)$ and the kth moment will be $O(\mu^{3k})$; each derivative of (14.6) can do no worse than multiply by μ^3, because of (14.7).

Incidentally, the $O(\mu^{-3})$ terms in all three equations of (14.12) turn out to be equal to $\frac{5}{48}\mu^{-3} + \frac{15}{32}\mu^{-6} + O(\mu^{-9})$, and this is no coincidence. We have, in fact,

$$\frac{\Gamma(\frac{5}{6})\Gamma(\frac{7}{6})}{\Gamma(\frac{1}{3})\Gamma(\frac{2}{3})} F\left(\frac{5}{6}, \frac{7}{6}; \frac{1}{3}, \frac{2}{3}; z^3\right) \sim \frac{\Gamma(\frac{7}{6})\Gamma(\frac{9}{6})}{\Gamma(\frac{2}{3})\Gamma(\frac{4}{3})} zF\left(\frac{7}{6}, \frac{3}{2}; \frac{2}{3}, \frac{4}{3}; z^3\right)$$

$$\sim \frac{\Gamma(\frac{3}{2})\Gamma(\frac{11}{6})}{\Gamma(\frac{4}{3})\Gamma(\frac{5}{3})} z^2F\left(\frac{3}{2}, \frac{11}{6}; \frac{4}{3}, \frac{5}{3}; z^3\right),$$

$$(14.14)$$

in the sense that all three functions have the same asymptotic series $\sum s_k z^{-3k}e^{z^3}$ as $z \to \infty$. This result follows because all three functions satisfy the same differential equation, and because their asymptotic behavior depends only on the differential equation except for a constant of proportionality. It is well known that the general hypergeometric functions $F(a_1,\ldots,a_m; b_1,\ldots,b_n; z)/(\Gamma(b_1)\ldots\Gamma(b_n))$ and $z^{1-b_1}F(a_1+1-b_1,\ldots,a_m+1-b_1; b_2+1-b_1,\ldots,b_n+1-b_1, 2-b_1; z)/ (\Gamma(b_2+1-b_1)\ldots\Gamma(b_n+1-b_1))$ both satisfy the differential equation $\vartheta(\vartheta+b_1-1)\ldots(\vartheta+b_n-1)F = z(\vartheta+a_1)\ldots(\vartheta+a_m)F$. In the case of (14.14), even more is true: The three asymptotically equivalent functions shown there can be written respectively as $\frac{1}{3}(G(z)+G(\omega z)+G(\omega^2 z))$, $\frac{1}{3}(G(z)+\omega^2 G(\omega z)+\omega G(\omega^2 z))$, $\frac{1}{3}(G(z)+\omega G(\omega z)+\omega^2 G(\omega^2 z))$, where

$$G(z) = \frac{1}{\sqrt{3}} \sum_{k\geq 0} \frac{\Gamma(3/2+k)z^k}{\Gamma(1/2+k/3)\,k!} \qquad (14.15)$$

and $\omega = e^{2\pi i/3}$.

15. Deficiency, Planarity, and Complexity

The calculations in the preceding section can be combined with the structure theory of Section 9 to yield the following general result.

Theorem 7. *Let $\overline{\overline{M}}$ be a reduced multigraph of excess r and deficiency d, that is, a reduced multigraph having $2r - d$ vertices and $3r - d$ edges. The probability that the complex part of a random graph or multigraph reduces to $\overline{\overline{M}}$ is asymptotically*

$$\frac{\sqrt{2\pi}\,\kappa(\overline{\overline{M}})}{(2r-d)!}\,A(3r + \tfrac{1}{2} - d, \mu)\,n^{-d/3} \qquad (15.1)$$

when there are $\frac{1}{2}n(1 + \mu n^{-1/3})$ edges and n vertices, $|\mu| = o(n^{1/12})$. Here $\kappa(\overline{\overline{M}})$ denotes the compensation factor (1.1), and $A(y, \mu)$ is defined in (10.2). The sum of (15.1) over all $\overline{\overline{M}}$ of deficiency 0 is 1. For each $d \geq 0$, the probability that a random multigraph has deficiency $\geq d$ is $O\big((1 + \mu^4)^d n^{-d/3}\big)$, uniformly in n and μ.

Proof. When $d = 0$, this theorem is a consequence of the corollary following (9.16), together with (13.17) and (14.1).

When $d > 0$, (15.1) is clear, but we need two auxiliary results of independent interest before we can prove the desired uniform estimate.

Lemma 4. *Let $E_{r(\geq d)}$ denote the generating function for all complex multigraphs of excess r whose deficiency is at least d. Then*

$$\frac{e_{rd}T}{(1-T)^{3r-d}} - \frac{(2r-d-1)e_{rd}T}{(1-T)^{3r-d-1}} \leq E_{r(\geq d)} \leq \frac{e_{rd}T}{(1-T)^{3r-d}}, \qquad (15.2)$$

where inequality between generating functions means that the coefficients of every power of z obey the stated relation.

Proof. The claim is trivial when $d = 2r - 1$; and it is true when $r = 1$, because $E_1 = \frac{5}{24}\,T(1-T)^{-3} - \frac{1}{12}\,T(1-T)^{-2}$. The lower bound is easily seen to be a lower bound on $e_{rd}T^{2r-d}/(1 - T)^{3r-d}$ itself.

The proof of the upper bound now proceeds by induction on r. Let

$$E'_r = \sum_{k=0}^{d-1} e_{rk}(1 + \zeta)^r \zeta^{2r-k} + e_{rd}\zeta(1 + \zeta)^{3r-d-1}, \qquad (15.3)$$

in the notation of Section 5. We want to prove that $E_r \leq E'_r$; it suffices to show, by (5.8), that

$$\big(r + (1-T(z))\vartheta\big)E'_r = \big(r + (1+\zeta)^{-1}\vartheta\big)E'_r \geq \tfrac{1}{2}\big(\tfrac{1}{2}\zeta(1+\zeta) + \vartheta\big)^2 E'_{r-1}, \qquad (15.4)$$

considering both sides as generating functions in powers of z. Proceeding as in (5.11) and (5.12) to form

$$A'_r = \left(\tfrac{1}{2}\,\zeta(1+\zeta)+\vartheta\right)E'_r\,, \qquad B'_r = \left(\tfrac{1}{2}\,\zeta(1+\zeta)+\vartheta\right)A'_r\,,$$

a bit of algebra shows that when $0 \le d \le 2r-3$ we have

$$\left(r+(1+\zeta)^{-1}\vartheta\right)E'_r - \tfrac{1}{2}B'_{r-1}$$

$$= \frac{1}{2}\zeta(1+\zeta)^{r+1}\Bigg(\sum_{k\ge 0}\zeta^{2r-d-2-k}\big((\alpha_k+\beta_k)e_{rd} - (\gamma_k+\delta_k+\epsilon_k)e_{(r-1)d}\big)$$

$$- (2r-1-d)^2 e_{(r-1)(d-1)}\zeta^{2r-d-2}\Bigg), \qquad (15.5)$$

where

$$\alpha_k = 2(3r-d)\binom{2r-d-2}{k+1}, \qquad \beta_k = 2(r+1)\binom{2r-d-2}{k};$$

$$\gamma_k = \left(3r-\tfrac{1}{2}-d\right)\left(3r-\tfrac{5}{2}-d\right)\binom{2r-d-3}{k+1},$$

$$\delta_k = \left(9r-\tfrac{13}{2}-3d\right)\binom{2r-d-3}{k}, \qquad \epsilon_k = \binom{2r-d-3}{k-1}.$$

Obviously $\beta_k e_{rd} \ge \epsilon_k e_{(r-1)d}$, since $e_{rd} \ge e_{(r-1)d}$. And the inequality $9r-\tfrac{13}{2}-3d \le (3r-\tfrac{1}{2}-d)(3r-\tfrac{5}{2}-d)$ for $0 \le d \le 2r-3$ yields

$$(\gamma_k+\delta_k)e_{(r-1)d} \le \left(3r-\tfrac{1}{2}-d\right)\left(3r-\tfrac{5}{2}-d\right)\binom{2r-d-2}{k+1}e_{(r-1)d} \le \alpha_k e_{rd}.$$

In fact, (5.11)–(5.13) imply that

$$2(3r-d)e_{rd} \ge \left(3r-\tfrac{1}{2}-d\right)\left(3r-\tfrac{5}{2}-d\right)e_{(r-1)d}$$
$$+ (2r-d)(2r-1-d)e_{(r-1)(d-1)};$$

so (15.5) is a polynomial in ζ with nonnegative coefficients, and thus a power series in z with nonnegative coefficients, proving (15.4). The case $d = 2r-2$ needs to be handled separately, but it offers no difficulty. $\square$

Lemma 5. *There exists a constant $\epsilon > 0$ such that, for every fixed $d \geq 0$, a random multigraph with n vertices and $m = \frac{n}{2}(1 + \mu n^{-1/3})$ edges has excess r and deficiency $\geq d$ with probability*

$$
\begin{cases}
O(\mu^{4d-3/2}n^{-d/3}e^{-\epsilon(r-\frac{2}{3}\mu^3)^2/\mu^3}), & \text{if } r \leq \mu^3, \\
O(n^{-d/3}e^{-\epsilon r}), & \text{if } r \geq \mu^3,
\end{cases}
$$

uniformly in n, r, and μ when $\mu \leq n^{1/12}$.

Proof. Let $p_{rd} = p_{rd}(n, \mu)$ be the stated probability. It suffices to prove the lemma when $\mu \geq 1$ and $r \geq 1$. For if $r = d = 0$, the result follows from Lemma 3; and $p_{0d} = 0$ when $d > 0$. On the other hand, if $\mu < 1$ we have $p_{rd}(n, \mu) \leq \sum_{j=r}^{\infty} p_{jd}(n, 1)$.

Using Lemma 4 and arguing as in the proof of Lemma 3, equation (10.11), we obtain

$$
p_{rd} = \frac{2^m m! \, n!}{n^{2m}} [z^n] \frac{U^{n-m+r}}{(n-m+r)!} e^V E_{r(\geq d)}
$$

$$
\leq \frac{2^m m! \, n!}{n^{2m}} [z^n] \frac{U^{n-m+r}}{(n-m+r)!} \frac{e_{rd} \, T}{(1-T)^{3r-d+1/2}}
$$

$$
= \frac{2^m m! \, n! \, e_{rd} \, e^n \, 2^{m-n-r}}{(n-m+r)! \, n^{2m} \, 2\pi i} \oint \left(\frac{z(2-z)}{1-z}\right)^r (1-z)^{d-2r+1/2} e^{nh(z)} dz,
$$

with $h(z)$ as in (10.12), and where the integral is taken around a circle $z = \rho e^{i\theta}$ with $0 < \rho < 1$. On this circle, both $|(2-z)/(1-z)|$ and $|1-z|^{-1}$ attain their maxima at $z = \rho$. Moreover, by (10.16) we have $\frac{d}{d\theta} \Re h = -\rho g(\theta) \sin \theta$, where $g(\theta) > ((2-\rho)^2-1)/9 > \frac{2}{9}(1-\rho)$; therefore

$$
\Re h(\rho e^{i\theta}) \leq h(\rho) + \frac{2}{9}(1-\rho)\rho(\cos\theta - 1) \leq h(\rho) - \frac{4}{9\pi^2}\rho(1-\rho)\theta^2, \text{ for } |\theta| \leq \pi.
$$

Now $p_{rd} = 0$ if $d \geq 2r$, because we are assuming that $r \geq 1$. Hence $d - 2r + 1/2 < 0$, and the contour integral including the factor $1/(2\pi i)$ is less than

$$
\frac{\rho}{2\pi} \left(\frac{\rho(2-\rho)}{1-\rho}\right)^r (1-\rho)^{d-2r+1/2} e^{nh(\rho)} \int_{-\pi}^{\pi} \exp\left(-\frac{4n\rho(1-\rho)}{9\pi^2}\theta^2\right) d\theta
$$

$$
< \frac{3}{4}\sqrt{\frac{\pi}{n}} \rho^{r+1/2}(2-\rho)^r (1-\rho)^{d-3r} e^{nh(\rho)}.
$$

In the following argument, unspecified positive constants will be denoted by ϵ_1, ϵ_2, ..., while positive numbers that may depend on d will be denoted by C_1, C_2, ... Let $\nu = n^{-1/3}$. If we apply (10.10) to the coefficient in front of the contour integral, and if we use the estimate

$$\frac{(n-m+r)!}{(n-m)!} > (n-m)^r = \left(\frac{n}{2}\right)^r (1 - \mu\nu)^r > \left(\frac{n}{2}\right)^r e^{-2\mu\nu r},$$

which is valid when $\mu\nu \leq \frac{1}{2}$, we obtain the upper bound

$$p_{rd} \leq C_1 e_{rd} n^{-r} \rho^r (2-\rho)^r (1-\rho)^{d-3r} e^{nh(\rho)-\mu^3/6+2\mu\nu r}, \qquad (15.6)$$

where ρ is any number between 0 and 1.

Suppose now that $r \leq 12\mu^3$, and set $\rho = 1 - \xi\mu\nu$, $r = \frac{2}{3}x\mu^3$. If $\xi = O(1)$, we have

$$nh(1 - \xi\mu\nu) = \frac{1}{3}\xi^3\mu^3 + \frac{1}{2}\xi^2\mu^3 + O(1)$$

as in (10.17). Therefore, since $\rho(2-\rho) < 1$,

$$p_{rd} \leq C_2 e_{rd} n^{-r} (\xi\mu\nu)^{d-3r} e^{\xi^3\mu^3/3+\xi^2\mu^3/2-\mu^3/6}$$

$$\leq C_3 3^r 2^{-r} r^{r+d-1/2} e^{-r} n^{-d/3} (\xi\mu)^{d-3r} e^{\xi^3\mu^3/3+\xi^2\mu^3/2-\mu^3/6}$$

$$= C_3 r^{d-1/2} n^{-d/3} (\xi\mu)^d e^{k(x,\xi)\mu^3/6},$$

by (7.16) and Stirling's formula, where

$$k(x,\xi) = 2\xi^3 + 3\xi^2 - 1 + 4x\ln\left(\frac{x}{e\xi^3}\right).$$

Given x between 0 and 18, we minimize $k(x,\xi)$ by letting ξ be the positive root of $\xi^3 + \xi^2 = 2x$; notice that this makes $\xi \leq 3$, justifying our assumption that $\xi = O(1)$. The minimum $k(x,\xi)$ satisfies

$$k(x,\xi) = \xi^2 - 1 + 2(\xi^3 + \xi^2)\ln\left(\frac{1+\xi^{-1}}{2}\right)$$

$$\leq \xi^2 - 1 + 2(\xi^3 + \xi^2)\frac{\xi^{-1} - 1}{2} = -(\xi+1)(\xi-1)^2.$$

We also have $|\xi - 1| \geq \epsilon_1|x - 1|$, hence $k(x,\xi) \leq -\epsilon_2(x-1)^2$. Our estimates have shown that

$$p_{rd} \leq C_5 r^{d-1/2} n^{-d/3} \mu^d e^{-\epsilon_2(x-1)^2\mu^3/6},$$

when $r = \frac{2}{3}x\mu^3 \leq 12\mu^3$, so the first half of the lemma has been proved.

When $r \geq 12\mu^3$, let us set $\rho = 1 - \eta$ and $r = yn$. In this case we will in fact prove the lemma for a much larger range of μ, assuming only that $\mu\nu \leq \delta$, when δ is a suitably small constant. If $\delta \leq \frac{1}{5}$ we can assume that $0 < y < \frac{3}{5}$, since m is at most $\frac{1+\delta}{2}n$ and since $p_{rd} = 0$ when $r \geq m$. Using (7.16) and (15.6) again, we find

$$p_{rd} \leq C_6 r^{d-1/2} \eta^d e^{nl(y,\eta) - \mu^3/6 + 2r\mu\nu},$$

where

$$l(y,\eta) = y \ln\left(\frac{3y(1-\eta^2)}{2e\eta^3}\right) + h(1-\eta)$$

$$= y \ln\left(\frac{3y(1-\eta^2)}{2e\eta^3}\right) - \eta - \ln(1-\eta) + \frac{1-\mu\nu}{2}\ln(1-\eta^2).$$

Given y, the minimum value of $l(y,\eta)$ occurs at the point $y = \eta^2(\eta + \mu\nu)/(3 - \eta^2)$. However, we do not need to find the exact minimum, in order to achieve the upper bound in the lemma; it will suffice to be close to the minimum when y is small. Therefore we choose η in such a way that the calculations will be relatively simple:

$$y = \frac{2\eta^3}{3(1-\eta^2)}. \tag{15.7}$$

With this choice, we always have $\eta < \frac{3}{4}$; and

$$l(y,\eta) = f(\eta) = -\frac{2\eta^3}{3(1-\eta^2)} - \eta - \ln(1-\eta) + \frac{1-\mu\nu}{2}\ln(1-\eta^2).$$

If we set $\eta = \mu\nu$, this function $f(\eta)$ reduces to

$$\sum_{k=1}^{\infty} \eta^{2k+1}\left(\frac{1}{2k} + \frac{1}{2k+1} - \frac{2}{3}\right) < \frac{\eta^3}{6} = \frac{\mu^3}{6n}.$$

On the other hand, the actual value of η must be larger than $2\mu\nu$, because $2\mu\nu$ is too small to satisfy (15.7):

$$\frac{2(2\mu\nu)^3}{3(1-(2\mu\nu)^2)} \leq \frac{16(\mu\nu)^3}{3(1-\frac{4}{25})} = \frac{400\mu^3}{63n} < \frac{r}{n} = y.$$

When $\eta > \mu\nu$ we have

$$f'(\eta) = -\frac{\eta(\eta^3 + 3\eta^2\mu\nu + 3(\eta - \mu\nu))}{3(1 - \eta^2)^2} < -\eta(\eta - \mu\nu) < -(\eta - \mu\nu)^2 \, ;$$

hence when η satisfies (15.7) we have

$$l(y, \eta) < \frac{\mu^3}{6n} - \frac{(\eta - \mu\nu)^3}{3} \leq \frac{\mu^3}{6n} - \epsilon_3\eta^3 \leq \frac{\mu^3}{6n} - \epsilon_4 y \, .$$

We have proved that

$$p_{rd} \leq C_7 r^d y^{d/3} e^{-\epsilon_4 r + 2r\mu\nu} \, ,$$

and this is at most $C_8 n^{-d/3} e^{-\epsilon_5 r}$ if δ is less than $\frac{1}{2}\epsilon_4$. □

Returning to the proof of Theorem 7, its final claim now follows for $\mu \leq n^{1/12}$ by summing the upper bounds of Lemma 5 over all values of r. The claim is trivial when $\mu > n^{1/12}$. □

As remarked earlier, the fact that (15.1) sums to 1 allows us to compute asymptotic probabilities of any collection of graphs or multigraphs obtained as a union over an infinite set of reduced multigraphs, as long as at least one multigraph in the set is clean (has deficiency zero). We simply sum the individual probabilities, neglecting unclean cases.

One corollary of Theorem 7 is the fact that a random graph with $\frac{1}{2}(n + \mu n^{2/3})$ edges is clean with probability $1 - o(1)$ whenever $\mu = o(n^{1/12})$. Stepanov proved this for $\mu \leq 0$ [36, Theorem 3] and conjectured that it would also hold for positive μ. His conjecture was proved for all fixed μ by Łuczak, Pittel, and Wierman [28].

Erdős and Rényi remarked in their pioneering paper [13, §8] that, if x is any real number, the probability that a graph with $\frac{1}{2}n + xn^{1/2}$ edges is nonplanar "has a positive lower limit, but we cannot calculate its value. It may even be 1, though this seems unlikely." They gave no proof that the limiting probability is positive, and their remark was embedded in a section of [13] that contains a technical error (see [27]); but a proof of their assertion was found later by Stepanov [36, Corollary 2 following (10)]. In the other direction, the fact that nonplanarity occurs with probability strictly less than 1 follows from the fact that a graph with $\frac{1}{2}n + o(n^{2/3})$ edges has excess 0 with probability $\sqrt{2/3}$, as observed in [14, Corollary 8].

We are now in a position to make a more precise estimate of the probability in question.

Theorem 8. *The probability that a graph with $\frac{1}{2}n + o(n^{2/3})$ edges is nonplanar approaches a limit ρ as $n \to \infty$, where*

$$0.000229 \le \rho \le 0.012926. \tag{15.8}$$

Proof. The condition $m = \frac{1}{2}n + o(n^{2/3})$ is equivalent to saying that $\mu = o(1)$ when $m = \frac{1}{2}n(1 + \mu n^{-1/3})$, so we can let $\mu = 0$ in the asymptotic formulas above. By Theorem 7, the constant ρ is the sum $\sum \sqrt{2\pi} A(3r + \frac{1}{2}, 0)\kappa(\overline{\overline{M}})/(2r)! = \sum \sqrt{\frac{2}{3}}\left(\frac{4}{3}\right)^r r! \,\kappa(\overline{\overline{M}})/(2r)!^2$ over all nonplanar, reduced, labeled, clean multigraphs $\overline{\overline{M}}$, where $r = r(\overline{\overline{M}})$ is the excess of $\overline{\overline{M}}$.

A clean multigraph cannot contain a subgraph that is homeomorphic to the complete graph K_5, namely, a subgraph that cancels to K_5, because K_5 has deficiency 5. Adding an edge to any multigraph increases the excess by 0 or 1 and increases the deficiency by 0, 1, or 2 (see Section 20 below); hence all subgraphs of a clean multigraph are clean. Indeed, this argument implies that a random graph with $\frac{1}{2}n + o(n^{2/3})$ edges has probability $O(n^{-5/3})$ of containing a K_5.

Therefore, if a sparse graph or multigraph is nonplanar, its nonplanarity comes almost surely from a subgraph that cancels to the complete bipartite graph $K_{3,3}$, which is clean and has excess 3.

One way to obtain bounds on ρ is to restrict consideration to reduced multigraphs whose components all have excess ≤ 3. If such a multigraph contains a $K_{3,3}$, it corresponds only to nonplanar graphs; if it does not, it corresponds only to planar graphs. The difference between the upper and lower bounds so obtained is the probability that a random graph of $\frac{1}{2}n + o(n^{2/3})$ edges has at least one component of excess ≥ 4, that is, that at least one component is more than tetracyclic.

The multigraph $K_{3,3}$ has compensation factor 1, because it is a graph; and its vertices can be labeled in $\frac{1}{2}\binom{6}{3} = 10$ different ways. Thus it contributes only $\frac{10}{6!} = \frac{1}{72}$ to the constant $c_3 = \frac{1105}{1152}$ that accounts for all clean connected multigraphs of excess 3.

Let $f_r = [z^r]\exp(c_1 z + c_2 z^2 + c_3 z^3)$ and $g_r = [z^r]\exp\big(c_1 z + c_2 z^2 + (c_3 - \frac{1}{72})z^3\big)$. Then the quantities

$$p = \sum_{r \ge 0} \sqrt{\frac{2}{3}}\left(\frac{4}{3}\right)^r f_r \frac{r!}{(2r)!} \qquad \text{and} \qquad q = \sum_{r \ge 0} \sqrt{\frac{2}{3}}\left(\frac{4}{3}\right)^r g_r \frac{r!}{(2r)!}$$

are respectively the probability that a sparse graph has all components of excess ≤ 3 and the probability that, moreover, no component cancels to $K_{3,3}$. These series converge rapidly and lead to the numerical bounds $p - q$ and $1 - q$ in (15.8). $\square$

It is interesting to study the expected number En_1 of vertices in complex components, as a function of μ, because it will be the expected number of vertices in the giant component when μ increases. We have $En_1 = \sum_r E(n_1|r)\Pr(\mathcal{E}_r)$. By (13.17) and the remarks preceding (13.13), each term in this sum can be approximated, to within relative error $O\big((1+\mu^4)n^{-1/3}\big)$, by $3r\sqrt{2\pi}\,e_r\,A\big(3r + \frac{5}{2}, \mu\big)n^{2/3}$. Let us assume for simplicity that μ is bounded. Then the proof of Lemma 5 is easily modified to show that the rth term of the sum is $O\big(n^{2/3}(r+1)e^{-\epsilon r}\big)$, uniformly in n and r. Thus, by dominated convergence, $En_1 = \big(f(\mu) + o(1)\big)n^{2/3}$, where

$$f(\mu) = \sum_{r \geq 0} 3r\sqrt{2\pi}\,e_r\,A\big(3r + \tfrac{5}{2}, \mu\big). \tag{15.9}$$

Equation (10.23) tells us that

$$\tfrac{1}{2}\mu^2 f(\mu) + f'(\mu) = \tfrac{1}{2}\sum_{r \geq 0} 3r\sqrt{2\pi}\,e_r\,A\big(3r + \tfrac{1}{2}, \mu\big) = \tfrac{3}{2}g(\mu), \tag{15.10}$$

where $g(\mu)$ is the expected value of r; we calculated $g(\mu)$ in the discussion leading up to Theorem 6. Thus, we obtain the estimate

$$f(\mu) = 2\mu - \mu^{-2} - \tfrac{27}{8}\mu^{-5} - \tfrac{495}{16}\mu^{-8} + O(\mu^{-11}), \tag{15.11}$$

for $\mu \geq \delta > 0$, by combining (15.9) with the asymptotic formula for $g(\mu)$ in (14.13).

We can express $f(\mu)$ in "closed hypergeometric form" by proceeding as in (14.9) and (14.10). The result is

$$
\begin{aligned}
f(\mu) = {}& -\frac{2^{-2/3}\pi}{3^{7/6}\,\Gamma\big(\frac{2}{3}\big)}\,e^{-\mu^3/6} + \mu - \frac{\mu}{4}e^{-\mu^3/6}\,F\Big(\frac{1}{3}; \frac{4}{3}; \frac{\mu^3}{6}\Big) \\
& + e^{-\mu^3/6}\Bigg(\frac{2^{1/3}\sqrt{\pi}}{3^{7/6}\,\Gamma\big(\frac{7}{6}\big)}\,F\Big(\frac{1}{2}, \frac{5}{6}; \frac{1}{3}, \frac{2}{3}; \frac{\mu^3}{6}\Big) \\
& \qquad\qquad - \frac{\mu}{2\sqrt{3}}\,F\Big(\frac{5}{6}, \frac{7}{6}; \frac{2}{3}, \frac{4}{3}; \frac{\mu^3}{6}\Big) \\
& \qquad\qquad + \frac{3^{1/6}\sqrt{\pi}\,\mu^2}{2^{7/3}\,\Gamma\big(\frac{5}{6}\big)}\,F\Big(\frac{7}{6}, \frac{3}{2}; \frac{4}{3}, \frac{5}{3}; \frac{\mu^3}{6}\Big)\Bigg). \tag{15.12}
\end{aligned}
$$

It is instructive to compare this expression with alternative formulas for the same quantity obtained in [28] by a different method:

$$
\begin{aligned}
f(\mu) &= \frac{1}{\sqrt{2\pi}}\int_0^\infty \Big(\sum_{r \geq 1} f_r x^{3r/2}\Big)e^{G(x,\mu)}\,dx \\
&= \mu + \frac{1}{\sqrt{2\pi}}\int_0^\infty \frac{1 - e^{G(x,\mu)}}{x^{3/2}}\,dx - \frac{1}{4}\int_0^\infty e^{G(x,\mu)}\,dx. \tag{15.13}
\end{aligned}
$$

Here $G(x, \mu) = \left((\mu - x)^3 - \mu^3\right)/6$, and $f_r n^{n + (3r - 1)/2}$ is Wright's asymptotic estimate [44] for the number of connected graphs with excess r.

16. Evolutionary Paths

Consider any graph or multigraph that evolves by starting out with isolated vertices and then by acquiring edges one at a time. Initially its excess is 0; then each new edge either preserves the current excess or increases it by 1. We observed in Section 4, following (4.7), that a new edge augments the excess if and only if both of its endpoints are currently in the cyclic part. We observed in Section 13 that many interesting statistics about random graphs can be usefully represented in terms of probabilities that are conditional on the graph having a given excess. Therefore it is natural to look more closely at the way a graph changes character as its excess grows.

Every evolution of a graph or multigraph traces a path from left to right in the following diagram, which shows the beginning of an infinite partial ordering of all possible configurations $[r_1, r_2, \ldots, r_q]$:

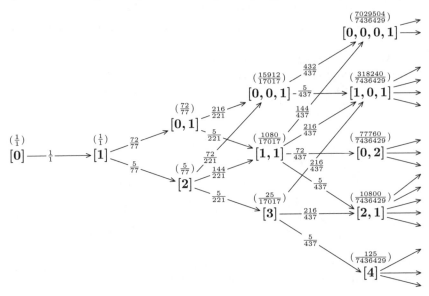

FIGURE 1. The evolution of complex components.   Each configuration $[r_1, r_2, \ldots, r_q]$ stands for a graph or multigraph with r_1 bicyclic components, r_2 tricyclic components, $\ldots$, r_q $(q+1)$-cyclic components. As a graph evolves, its excess $r_1 + 2r_2 + 3r_3 + \cdots$ increases in unit steps, and the configurations follow a path from left to right in this partial ordering.

When complex components begin to form, they follow a path in this diagram, with the indicated transition probabilities. The upper path is followed most frequently; on this path there is a unique complex component that will become the "giant." Parenthesized ratios are the probabilities of reaching a given configuration. At the moment the excess first reaches 2, the configuration must either be $[0, 1]$ (one tricyclic component) or $[2]$ (two bicyclic components). When the excess goes from 2 to 3, we go either from $[0, 1]$ to $[0, 0, 1]$ or $[1, 1]$, or from $[2]$ to $[0, 0, 1]$, $[1,1]$, or $[3]$; and so on. Each configuration $[r_1, r_2, \ldots, r_q]$ corresponds to a partition of the excess $r = r_1 + 2r_2 + \cdots + qr_q$. The fraction in parenthesis shown above each configuration in Figure 1 is the limiting probability $c_1^{r_1} c_2^{r_2} \ldots c_q^{r_q}/(r_1! \, r_2! \ldots r_q! \, e_r)$ that a random graph of excess r has configuration $[r_1, r_2, \ldots, r_q]$. This is the limiting probability that the infinite path traced out in the infinite extension of Figure 1 will pass through $[r_1, r_2, \ldots, r_q]$ during the evolution of a random graph or multigraph on a large number of vertices.

A random graph almost always acquires nearly $\frac{1}{2}n$ edges before taking the first step from $[0]$ to $[1]$ in Figure 1. Indeed, the uniform estimate (13.17), with $\mu = -n^{1/21}$, implies that the probability of excess r when $m = \frac{1}{2}n \exp(-n^{-2/7})$ is of order $n^{-r/7}$.

The fractions shown on arcs leading between configurations are transition probabilities, namely the limiting probabilities that a random graph of configuration $[r_1, r_2, \ldots, r_q]$ will go to another specified configuration when its excess next changes. For example, a random graph in configuration $[2]$, having two bicyclic components and no other complex components, will proceed next to configuration $[1,1]$ with probability $\frac{144}{221}$. These transition probabilities have a fairly simple characterization:

Theorem 9. *Let $r_1 + 2r_2 + \cdots + qr_q = r$ and $\delta_1 + 2\delta_2 + 3\delta_3 + \cdots = 1$. The asymptotic probability that a random graph or multigraph of configuration $[r_1, r_2, \ldots, r_q]$, having no acyclic components, will change to configuration $[r_1 + \delta_1, r_2 + \delta_2, \ldots, r_q + \delta_q, \delta_{q+1}, \ldots]$ when a random edge is added, can be computed as follows:*

Nonzero δ's	Probability
$\delta_1 = 1$,	$\frac{5}{4}/((3r + \frac{1}{2})(3r + \frac{5}{2}))$;
$\delta_j = -1$, $\delta_{j+1} = 1$,	$9j(j+1)r_j/((3r + \frac{1}{2})(3r + \frac{5}{2}))$;
$\delta_j = -2$, $\delta_{2j+1} = 1$,	$9j^2 r_j(r_j - 1)/((3r + \frac{1}{2})(3r + \frac{5}{2}))$;
$\delta_j = -1$, $\delta_k = -1$, $\delta_{j+k+1} = 1$, $j < k$,	$18j\,k\,r_j r_k/((3r + \frac{1}{2})(3r + \frac{5}{2}))$.

In all other cases, the probability is zero. The estimates are correct to within $O(n^{-1/2})$ when there are n vertices.

Proof. As usual, it is easiest to consider first the uniform multigraph process. We know that the generating function for the cyclic multigraphs under consideration is

$$S(z) = e^{V(z)} \frac{C_1(z)^{r_1}}{r_1!} \frac{C_2(z)^{r_2}}{r_2!} \cdots \frac{C_q(z)^{r_q}}{r_q!} ; \qquad (16.1)$$

the number of such multigraphs, weighted as usual by the compensation factor (1.1), is $[z^n] S(z)$. We also know from (3.4) that $V(z) = -\frac{1}{2} \ln(1 - T(z))$, hence

$$e^{V(z)} = \frac{1}{(1 - T(z))^{1/2}} .$$

We observed in Section 4 that the operator $\vartheta = z\frac{d}{dz}$ corresponds to "marking" or singling out a particular vertex. The function $\vartheta^2 S(z)$ can therefore be regarded as the generating function for multigraphs of configuration $[r_1, r_2, \ldots, r_q]$ together with an ordered pair of marked vertices $\langle x, y \rangle$. When $S(z)$ is a product $A(z)B(z)$, the familiar relation

$$\vartheta^2 \big(A(z)B(z)\big) = \big(\vartheta^2 A(z)\big)B(z) + 2\big(\vartheta A(z)\big)\big(\vartheta B(z)\big) + A(z)\big(\vartheta^2 B(z)\big) \tag{16.2}$$

has a natural combinatorial interpretation: The product $A(z)B(z)$ stands for ordered pairs of graphs, generated respectively by $A(z)$ and $B(z)$, with no edges between them; the first term $\big(\vartheta^2 A(z)\big)B(z)$ of (16.2) corresponds to cases when both of the marked vertices $\langle x, y \rangle$ are in the graph generated by $A(z)$; the last term corresponds to cases when both x and y belong to the $B(z)$ graph. The middle term $2\big(\vartheta A(z)\big)\big(\vartheta B(z)\big)$ corresponds to the cases where x is in A and y is in B or vice versa.

We can use this idea in connection with (16.1) to understand what happens when the graph gains a new edge. The coefficient of z^n in $\vartheta^2 S(z)$ represents all possibilities $\langle x, y \rangle$; we can divide this into cases by writing

$$\vartheta^2 S(z) = S(z)\left(\sum_{j=0}^{q} \frac{\vartheta^2 f_j(z)}{f_j(z)} + 2 \sum_{j=0}^{q-2} \sum_{k=j+1}^{q-1} \frac{\vartheta f_j(z)}{f_j(k)} \frac{\vartheta f_k(z)}{f_k(z)} \right) \qquad (16.3)$$

where $f_0(z) = e^{V(z)}$ and $f_j(z) = C_j(z)^{r_j}/r_j!$ for $j \geq 1$. A term like $S(z)\big(\vartheta^2 f_j(z)\big)/f_j(z)$, say, then corresponds to cases where x and y both belong to $(j+1)$-cyclic components.

Each of the factors $f_j(z)$ is a linear combination of powers of the quantity $\xi = 1 + \zeta = 1/(1 - T(z))$. For example, $f_0(z) = \xi^{1/2}$ and $f_1(z) = \frac{5}{24}\xi^3 - \frac{7}{24}\xi^2 + \frac{1}{12}\xi$, according to (3.4) and (11.3). Hence we can easily compute ϑf_j and $\vartheta^2 f_j$, using rule (4.5):

$$\vartheta(\xi^\alpha) = \alpha\xi^{\alpha+2} - \alpha\xi^{\alpha+1};$$

$$\vartheta^2(\xi^\alpha) = \alpha(\alpha+2)\xi^{\alpha+4} - \alpha(2\alpha+3)\xi^{\alpha+3} + \alpha(\alpha+1)\xi^{\alpha+2}. \quad (16.4)$$

The overall function $S(z)$ has the form $\xi^{3r+1/2}P(\xi^{-1})$ for some polynomial P, with $P(0) \neq 0$; hence the coefficient $[z^n]S(z)$ is equal to $t_n(3r + \frac{1}{2})P(0)(1 + O(n^{-1/2}))/n!$ by (3.8) and (3.9). It follows from (16.4) that $\vartheta^2 S(z) = \xi^{3r+9/2}Q(\xi^{-1})$ for some polynomial Q, where $Q(0) = (3r + \frac{1}{2})(3r + \frac{5}{2})P(0)$. Hence

$$n^2 = \frac{[z^n]\vartheta^2 S(z)}{[z^n]S(z)} = (3r + \tfrac{1}{2})(3r + \tfrac{5}{2})\frac{t_n(3r + \frac{9}{2})}{t_n(3r + \frac{1}{2})}\left(1 + O(n^{-1/2})\right).$$

$$(16.5)$$

The transition probabilities we wish to compute are the fractions of $(3r+\frac{1}{2})(3r+\frac{5}{2})$ that occur when ϑ^2 operates on individual factors of $S(z)$.

For example, consider first the term $S(z)(\vartheta^2 f_0(z))/f_0(z)$ of (16.3). This corresponds to the case where both x and y belong to a cyclic component (possibly the same one), thereby creating a new bicyclic component; thus it corresponds to having $\delta_1 = 1$ and all other $\delta_j = 0$. In this case $[z^n]S(z)(\vartheta^2 f_0(z))/f_0(z) \sim \frac{1}{2} \cdot \frac{5}{2}t_n(3r + \frac{9}{2})P(0)/n!$, and the latter is asymptotically $\frac{5}{4}/((3r + \frac{1}{2})(3r + \frac{5}{2}))$ of the total $[z^n]\vartheta^2 S(z)$.

The term $2S(z)(\vartheta f_0(z))(\vartheta f_j(z))/(f_0(z)f_j(z))$, similarly, gives the probability that a vertex from a cyclic component joins with a $(j+1)$-cyclic component; this case occurs with probability

$$2(\tfrac{1}{2})(3j\, r_j)/((3r + \tfrac{1}{2})(3r + \tfrac{5}{2})).$$

The net effect on components corresponds to $\delta_j = -1$, $\delta_{j+1} = +1$.

There is also another way to get $\delta_j = -1$ and $\delta_{j+1} = +1$, namely if both x and y belong to the same $(j+1)$-cyclic component. The probability of this case works out to be $(3j)(3j+2)r_j/((3r+\frac{1}{2})(3r+\frac{5}{2}))$; hence the total transition probability for $\delta_j = -1$ and $\delta_{j+1} = +1$ is $9j(j+1)r_j/((3r + \frac{1}{2})(3r + \frac{5}{2}))$ as stated in the theorem.

Notice that

$$\vartheta^2 C_j^{r_j} = r_j C_j^{r_j-1}(\vartheta^2 C_j) + r_j(r_j - 1)C_j^{r_j-2}(\vartheta C_j)^2. \quad (16.6)$$

We have just taken care of the first term; the second term corresponds to vertices x and y in distinct C_j's, when the new edge makes $\delta_j = -2$ and $\delta_{2j+1} = +1$. The probability comes to $9j^2 r_j(r_j - 1)/((3r + \frac{1}{2})(3r + \frac{5}{2}))$.

Finally, the term $2S(z)\big(\vartheta f_j(z)\big)\big(\vartheta f_k(z)\big)/\big(f_j(z)f_k(z)\big)$ of (16.3) represents a case that occurs with probability

$$2(3jr_j)(3kr_k)/((3r + \tfrac{1}{2})(3r + \tfrac{5}{2}))$$

and corresponds to $\delta_j = \delta_k = -1$, $\delta_{j+k+1} = +1$.

If we are working with the graph process instead of the multigraph process, we must use $\widehat{C}_j(z)$ instead of $C_j(z)$; but $f_0(z)$ is still essentially of degree $-1/2$ in ξ^{-1}, and $f_j(z)$ is still of degree $-3j$, so the asymptotic calculations work out as before.

However, in a random graph we must use the operator $\frac{1}{2}(\vartheta_z^2 - \vartheta_z) - \vartheta_w$ instead of ϑ_z^2, and we must work with bivariate generating functions, as discussed in Section 6. The bgf corresponding to (16.1) is almost univariate, however:

$$\widehat{F}(w, z) = w^r\, e^{\hat{V}(wz)}\, \frac{C_1(wz)^{r_1}}{r_1!}\, \frac{C_2(wz)^{r_2}}{r_2!} \cdots \frac{C_q(wz)^{r_q}}{r_q!}.$$

It is not difficult to see that the effect of ϑ_z^2 swamps the effects of ϑ_z and ϑ_w, asymptotically, so the multigraph analysis carries through. □

One amusing consequence of Theorem 9 is that we can use it to discover and prove formula (7.2) for the numbers e_r in a completely different way. The probability of reaching the configuration $[r]$, consisting of r bicyclic components and none of higher cyclic order, is $c_1^r/(r!\, e_r)$. The only way to reach this configuration, when $r > 0$, is from $[r - 1]$, and the transition probability is

$$\frac{\frac{5}{4}}{(3r - \tfrac{5}{2})(3r - \tfrac{1}{2})} = \frac{c_1^r/(r!\, e_r)}{c_1^{r-1}/((r-1)!\, e_{r-1})}.$$

Since $c_1 = 5/24$, we have $e_r = (6r - 5)(6r - 1)e_{r-1}/(24r)$, and (7.2) follows by induction. This indirect method is probably the simplest way to deduce the fact that Wright's constant is $1/(2\pi)$.

17. A Near-Markov Process

We proved in Theorem 9 that the transition probabilities shown in Figure 1 are the limiting probabilities, averaged over all multigraphs, that a multigraph reaching a particular state will take a particular step as its excess increases. But we did not prove that those transition probabilities are independent of past history. For all we know, the path taken to a particular configuration during the evolution of a random graph might strongly influence the probability distribution of its next leap forward. The next theorem addresses this question.

Theorem 10. *For any fixed R, an evolving random graph or multigraph almost surely carries out a random walk in the first R levels of the partial ordering shown in Figure 1, with transition probabilities that approach the limiting values derived in Theorem 9.*

Proof. As in previous proofs, it suffices to consider random multigraphs. We will show that the transition probabilities have the asymptotic behavior of Theorem 9 for all random multigraphs that remain clean — namely, for all multigraphs that reduce, under the pruning and cancelling algorithms of Section 9, to 3-regular multigraphs $\overline{\overline{M}}$ having $2r$ vertices and $3r$ edges, when the excess is $r \leq R$. We know from Theorem 7 that the multigraph will be clean with probability $1 - O\big((1 + \mu^4)n^{-1/3}\big)$; and we know from (13.17) that the probability of excess r becomes superpolynomially small as the number of edges passes $\frac{n}{2}$. So the excess will almost surely increase past any given value before a large multigraph becomes unclean. For example, if $\mu \to \infty$ with $\mu = o(n^{1/12})$, the probability of excess $\leq R$ approaches zero while the probability of remaining clean is $1 - o(1)$.

The proof for clean multigraphs is not as trivial as might be expected: Multigraphs that follow a given path to $[r_1, r_2, \ldots, r_q]$ in the partial ordering are *not* uniformly distributed, among all multigraphs whose complex parts are enumerated by the generating function

$$e^V (C_1^{r_1}/r_1!)(C_2^{r_2}/r_2!) \ldots (C_q^{r_q}/r_q!)$$

assumed in the proof of Theorem 9. Past history does affect the frequency of certain types of components. For example, a tricyclic component that prunes and cancels to $K_{3,3}$ cannot evolve along the path $[1] \to [2] \to [0,0,1]$; removing any edge of $K_{3,3}$ leaves a connected graph.

Let's try to clarify the situation by working an example. Consider the reduced multigraph

$$\text{(17.1)}$$

suppose we wish to compute the transition probabilities for multigraphs of excess 3 that prune and cancel to (17.1) after following the path $[1] \to [0,1] \to [0,0,1]$. The generating function for all such multigraphs, assuming that there are no acyclic components, would be equal to $\frac{1}{32} e^V T^6/(1 - T)^9$, if we did not specify the past history $[1] \to [0,1] \to [0,0,1]$; but it turns out to be only $\frac{8}{9}$ as much when we do prescribe the history. The reason is, intuitively, that (17.1) has 9 edges, and a multigraph with history $[1] \to [0,1] \to [0,0,1]$ can reduce to it only if the "middle" edge is not the last to be completed. The latter event happens with probability $\frac{8}{9}$.

A formal proof of the $\frac{8}{9}$ phenomenon can be given as follows. The generating function $e^V T^6 (1-T)^9$ expands to $e^V T^6 \sum_{n_1,n_2,\ldots,n_9 \geq 0} T^{n_1} T^{n_2} \ldots T^{n_9}$; the individual terms represent the insertion of $\langle n_1, \ldots, n_9 \rangle$ vertices into the nine edges of (17.1), after which a tree is sprouted at each vertex. The resulting multigraph will have n vertices and $m = n+3$ edges; there will be $6+n_1+\cdots+n_9$ root vertices and $9+n_1+\cdots+n_9$ "critical" edges on paths between root vertices. Suppose we color each critical edge with one of 9 colors, corresponding to the original edge of (17.1) from which it was subdivided. Then among the $m!$ permutations of edges that could generate any such multigraph, exactly $\frac{8}{9}$ have the property that the last critical edge has some color besides the "middle" color. (This follows by symmetry between $n_1, n_2, \ldots, n_9$.) Such permutations are precisely those for which the history will be $[1] \to [0,1] \to [0,0,1]$; hence we obtain (17.1) with exactly $\frac{8}{9}$ times its overall probability, given that history.

It turns out that there are 17 unlabeled clean, connected, reduced multigraphs of excess 3; and exactly 6 of them occur with weight $\frac{8}{9}$ when the past history is $[1] \to [0,1] \to [0,0,1]$. Those 6 occur with weight $\frac{1}{9}$ when the past history is $[1] \to [2] \to [0,0,1]$, and the other 11 do not occur at all in that case.

In general, given any $\overline{\overline{M}}$ that can arise for a given past history, there will be a fraction $\beta > 0$ such that each multigraph reducing to $\overline{\overline{M}}$ arises β times as often with the given history as it does overall. The reason is a slight generalization of the method by which we proved the $\frac{8}{9}$ phenomenon: Each permutation of colors of critical edges is equally likely to be the sequence of last appearances in a random permutation of $n_1 + n_2 + \cdots + n_{3r}$ critical edges, and such permutations determine the past history. The generating function for $\overline{\overline{M}}$ will then be a constant multiple of $e^V \big(T^{2r_1}/(1-T)^{3r_1}\big)\big(T^{2r_2}/(1-T)^{3r_2}\big) \ldots \big(T^{2r_q}/(1-T)^{3r_q}\big)$. Hence the asymptotic transition probabilities will be the same for every feasible $\overline{\overline{M}}$, exactly as calculated in Theorem 9. $\square$

18. An Emerging Giant

The classic papers of Erdős and Rényi [12, 13] tell us that an evolving graph almost surely develops a single giant component, which eventually is surrounded by only a few trees and later by only isolated vertices, until the entire graph becomes connected. Thus there will almost surely be a time when the graph reaches some configuration $[0, 0, \ldots, 0, 1]$ on the top line of Figure 1 and stays on that top line ever afterward.

Indeed, the most probable path in Figure 1 is the one that goes directly from $[1]$ to $[0, 1]$ to $[0, 0, 1]$ and so on, never leaving the top line. The first transition probability is $\frac{72}{77}$, the next is $\frac{216}{221}$, and subsequent steps are ever more likely to stay in line. In such cases we can see the "seed" around which the giant component is forming, before that component has become in any way gigantic. (The complex components of any given finite excess almost always have only $O(n^{2/3})$ vertices, a vanishingly small percentage of the total; each step at the beginning of Figure 1 occurs after adding about $n^{2/3}$ more edges.)

If we assume that the transition probabilities in Figure 1 are exact, the overall probability that an evolving graph adheres strictly to the top line — never having more than one complex component throughout its entire evolution — is

$$\prod_{r=1}^{\infty} \frac{r(r+1)}{(r+\frac{1}{6})(r+\frac{5}{6})} = \frac{\Gamma(\frac{7}{6})\Gamma(\frac{11}{6})}{\Gamma(1)\Gamma(2)} = \frac{5}{36}\Gamma\left(\frac{1}{6}\right)\Gamma\left(\frac{5}{6}\right) = \frac{5\pi}{18}. \qquad (18.1)$$

Numerically, this comes to 0.8726646, roughly 7 times out of every 8.

Is $\frac{5\pi}{18}$ the true limiting probability that an evolving graph or multigraph never acquires two simultaneous components of positive excess, throughout its evolution? We can at least prove that $\frac{5\pi}{18}$ is an upper bound. For we know from Theorem 10 that an evolving graph will hug the top line of Figure 1 for at least R steps with probability

$$\prod_{r=1}^{R} \frac{r(r+1)}{(r+\frac{1}{6})(r+\frac{5}{6})} = \frac{5\pi}{18} + O(R^{-1}) + O(n^{-1/3}) \qquad (18.2)$$

for any fixed R, as $n \to \infty$.

It is natural to conjecture that $\frac{5\pi}{18}$ is also a lower bound, because a large component tends to propagate itself as soon as it becomes large enough. Still, it is conceivable that a random graph might have a tendency to leave the top line briefly when it first becomes unclean. The transition probability for remaining on the top line becomes strictly less than $r(r+1)/((r+\frac{1}{6})(r+\frac{5}{6}))$ when the graph has a positive deficiency. For example, suppose the initial bicyclic component is already unclean; it will then correspond to the double self-loop of (9.15). We know from (13.16) that this case arises with probability $O(n^{-1/3})$. But if it does occur, the generating function for the complex part will be a constant multiple of $T/(1-T)^2$ instead of $T^2/(1-T)^3$, so the proof technique of Theorem 9 will yield a transition probability from [1] to $[0,1]$ of only $\frac{8}{9}$ instead of $\frac{72}{77}$. In general, when the deficiency is d, the asymptotic transition probability drops to

$$((r - \tfrac{d}{3})(r - \tfrac{d}{3} + 1))/((r - \tfrac{d}{3} + \tfrac{1}{6})(r - \tfrac{d}{3} + \tfrac{5}{6})) \,.$$

This probability estimate is, moreover, valid only when the excess is reasonably small as a function of n; otherwise the trees that sprout from the pruned multigraph $\overline{M}$ will not be large enough to assert their asymptotic behavior.

19. A Monotonicity Property

During the time when an evolving graph or multigraph stays clean, we can show that the asymptotic top-line transition probabilities $(r(r+1))/((r+\tfrac{1}{6})(r+\tfrac{5}{6}))$ are in fact *lower* bounds for the correct (non-asymptotic) probabilities. More precisely, the proof of Theorem 9 shows that the true transition probability is a ratio of expressions involving the tree polynomials $t_n(y)$, when there are n vertices in the cyclic part of the multigraph. We will prove that this ratio decreases monotonically to $(r(r+1))/((r+\tfrac{1}{6})(r+\tfrac{5}{6}))$ as $n \to \infty$.

First we need to prove an auxiliary result about tree polynomials that is interesting in its own right. Let us generalize the definition of $t_n(y)$ in (3.8) by introducing a new parameter $m \geq 0$:

$$\frac{T(z)^m}{(1-T(z))^y} = \sum_{n=0}^{\infty} t_{m,n}(y) \frac{z^n}{n!} \,. \tag{19.1}$$

Thus

$$t_{m,n}(y) = \sum_{j=0}^{m} \binom{m}{j}(-1)^j t_n(y-j) \tag{19.2}$$

is the mth backward difference of $t_n(y)$.

Lemma 6. *Let m be a nonnegative integer. For any fixed integer $n > m$ and arbitrary real $y > 0$, the ratio $t_{m,n+1}(y)/t_{m,n}(y)$ is an increasing function of y. Equivalently, for fixed $y > 0$ and any integer $n > m$, the ratio $t'_{m,n}(y)/t_{m,n}(y)$ is an increasing function of n.*

Proof. The two statements of the lemma are clearly equivalent, because $t_{m,n}(y)$ is positive when $y > 0$ and $n > m$.

Equation (2.12) of [24] states that

$$t_n(y) = n^{n-1} \sum_{k \geq 0} \frac{y^{\overline{k+1}}}{k!} \frac{(n-1)^{\underline{k}}}{n^k}, \tag{19.3}$$

where $x^{\overline{k}}$ means $x(x+1)\dots(x+k-1)$ and $x^{\underline{k}}$ means $x(x-1)\dots(x-k+1)$. Therefore, by (19.2),

$$t_{m,n}(y) = n^{n-1} \sum_{k \geq m-1} (k+1) \frac{y^{\overline{k+1-m}}}{(k+1-m)!} \frac{(n-1)^{\underline{k}}}{n^k}$$

$$= n^{n-m} \sum_{k=0}^{n-m} (k+m) \frac{y^{\overline{k}}}{k!} \frac{(n-1)^{\underline{k+m-1}}}{n^k}. \tag{19.4}$$

It follows that the inequality $t'_{m,n}(y)/t_{m,n}(y) < t'_{m,n+1}(y)/t_{m,n+1}(y)$ is equivalent to

$$\frac{\sum_{k=0}^{N} a_k \alpha_k}{\sum_{k=0}^{N} b_k \alpha_k} > \frac{\sum_{k=0}^{N} a_k \beta_k}{\sum_{k=0}^{N} b_k \beta_k}, \tag{19.5}$$

where $N = n + 1 - m$ and

$$a_k = (k+m) \frac{y^{\overline{k}}}{k!}, \qquad b_k = (k+m) \frac{d}{dy} \frac{y^{\overline{k}}}{k!}, \tag{19.6}$$

$$\alpha_k = (n-1)^{\underline{k+m-1}} n^{n-m-k}, \qquad \beta_k = n^{\underline{k+m-1}} (n+1)^{n+1-m-k}.$$

The following condition is sufficient to prove (19.5), assuming positive denominators:

$$\frac{a_0}{b_0} > \frac{a_1}{b_1} > \dots > \frac{a_N}{b_N} \qquad \text{and} \qquad \frac{\alpha_0}{\beta_0} > \frac{\alpha_1}{\beta_1} > \dots > \frac{\alpha_N}{\beta_N}. \tag{19.7}$$

For we have

$$\sum_{k=0}^{N} b_k \beta_k \sum_{j=0}^{N} a_j \alpha_j - \sum_{j=0}^{N} b_j \alpha_j \sum_{k=0}^{N} a_k \beta_k$$

$$= \sum_{0 \leq j < k \leq n} (b_k a_j - b_j a_k)(\beta_k \alpha_j - \beta_j \alpha_k) > 0. \tag{19.8}$$

(*Historical note:* Inequality (19.5) under condition (19.7) goes back at least to Seitz in 1936 [33]; see [29, Section 2.5, Theorem 4], where a supplementary condition is needed: The product of the denominators must be positive. In linearly ordered discrete probability space, the inequality is equivalent to saying that $E(f(X)g(X)) \geq E(f(X))E(g(X))$ whenever f and g are increasing functions of the random variable X. This inequality is, in turn, a special case of the celebrated FKG inequality [15], which applies to certain partially ordered probability spaces. The equality in (19.8), which reduces to Lagrange's identity when we set $a_k = \alpha_k$ and $b_k = \beta_k$, is the Binet–Cauchy identity for det AB when A is a matrix of size $2 \times n$ and B is $n \times 2$.)

And (19.7) is not difficult to verify, under the substitutions (19.6). We have

$$\frac{b_{k+1}}{a_{k+1}} = \frac{1}{y} + \frac{1}{y+1} + \cdots + \frac{1}{y+k} = \frac{b_k}{a_k} + \frac{1}{y+k};$$

$$\frac{\alpha_{k+1}}{\alpha_k} = \frac{n-k-m}{n} < \frac{n-k-m+1}{n+1} = \frac{\beta_{k+1}}{\beta_k}.$$

(When $m = 0$ we omit the terms for $k = 0$.)   □

Assume now that the cyclic part of a random multigraph contains n vertices. The "top line" transition probability from a single clean component of excess r to a single component of excess $r+1$ is $1 - p_{nr}$, where p_{nr} is the probability that a new bicyclic component will be formed. By the argument of Theorem 9,

$$p_{nr} = \frac{[z^n]\left(\vartheta^2 V(z)\right)S(z)}{[z^n]\,\vartheta^2\left(V(z)S(z)\right)}, \tag{19.9}$$

where $V(z) = 1/(1 - T(z))^{1/2}$ is the generating function for unicyclic components and $S(z) = T(z)^{2r}/(1-T(z))^{3r}$ is a prototypical generating function for clean components of excess r. We want to show that p_{nr} is an increasing function of n, since we want $1 - p_{nr}$ to be decreasing.

Let's work on a simpler problem first, showing that

$$q_{nr} = \frac{[z^n]\left(\vartheta\, A(z)\right)S(z)}{[z^n]\,\vartheta\left(A(z)S(z)\right)} \tag{19.10}$$

is an increasing function of n whenever

$$A(z) = \frac{T(z)^a}{(1 - T(z))^b}, \qquad b > \frac{3}{2}\,a. \tag{19.11}$$

Here a is a nonnegative integer; we will assume that $n \geq 2r + a$, so that the denominator of (19.10) is nonzero. We have

$$\vartheta A(z) = \frac{b\, T(z)^a}{\left(1 - T(z)\right)^{b+2}} - \frac{(b - a)\, T(z)^a}{\left(1 - T(z)\right)^{b+1}},$$

$$\vartheta\big(A(z)S(z)\big) = \frac{(3r + b)T(z)^{2r+a}}{\left(1 - T(z)\right)^{3r+b+2}} - \frac{(r + b - a)T(z)^{2r+a}}{\left(1 - T(z)\right)^{3r+b+1}};$$

hence

$$q_{nr} = \frac{b\, t_{2r+a,n}(3r + b + 2) - (b - a)\, t_{2r+a,n}(3r + b + 1)}{(3r + b)\, t_{2r+a,n}(3r + b + 2) - (r + b - a)\, t_{2r+a,n}(3r + b + 1)}$$

$$= \frac{b}{3r + b}\left(1 - \frac{r(2b - 3a)/(3rb + b^2)}{\dfrac{t_{2r+a,n}(3r + b + 2)}{t_{2r+a,n}(3r + b + 1)} - \dfrac{r + b - a}{3r + b}}\right).$$

Since the coefficients of $t_{2r+a,n}(y)$ are nonnegative, we have

$$t_{2r+a,n}(3r + b + 2)/t_{2r+a,n}(3r + b + 1) \geq 1 > (r + b - a)/(3r + b).$$

It follows that q_{nr} is increasing if and only if

$$\frac{t_{2r+a,n}(3r + b + 2)}{t_{2r+a,n}(3r + b + 1)} < \frac{t_{2r+a,n+1}(3r + b + 2)}{t_{2r+a,n+1}(3r + b + 1)}. \tag{19.12}$$

And (19.12) does hold, because $t_{2r+a,n+1}(y)/t_{2r+a,n}(y)$ is an increasing function of y by Lemma 6.

Incidentally, this argument also shows that q_{nr} is constant when $b = \frac{3}{2}a$ and decreasing when $0 < b < \frac{3}{2}a$.

Now to prove that p_{nr} is increasing, we can write

$$p_{nr} = \frac{[z^n]\left(\vartheta^2 V(z)\right)S(z)}{[z^n]\,\vartheta\big((\vartheta V(z))S(z)\big)}\, \frac{[z^n]\,\vartheta\big((\vartheta V(z))S(z)\big)}{[z^n]\,\vartheta^2\big(V(z)S(z)\big)}$$

$$= \frac{[z^n]\left(\vartheta^2 V(z)\right)S(z)}{[z^n]\,\vartheta\big((\vartheta V(z))S(z)\big)}\, \frac{[z^n]\left(\vartheta V(z)\right)S(z)}{[z^n]\,\vartheta\big(V(z)S(z)\big)}.$$

The first factor is of type q_{nr} if we put

$$A(z) = \vartheta V(z) = \frac{1}{2}T(z)/\left(1 - T(z)\right)^{5/2};$$

here $a = 1$ and $b = \frac{5}{2}$, so q_{nr} is increasing. The second factor is of type q_{nr} if we put $A(z) = V(z)$; here $a = 0$, $b = \frac{1}{2}$, and again q_{nr} is increasing. We have proved

Theorem 11. *The probability that a clean random multigraph of excess $r > 0$ will not acquire a new bicyclic component when its excess next changes is strictly greater than the limiting value*

$$\frac{r(r+1)}{(r+\frac{1}{6})(r+\frac{5}{6})} \cdot \quad \square \qquad (19.13)$$

Theorem 11 gives further support to the $\frac{5\pi}{18}$ conjecture of Section 18, because $\frac{5\pi}{18}$ was shown there to be an upper bound. If the top-line transition probability were always strictly greater than (19.13), we could establish $\frac{5\pi}{18}$ as a lower bound. However, Theorem 11 does not prove the conjecture, because the probability becomes smaller than (19.13) when a graph becomes unclean.

Incidentally, when the number of edges gets large, we may need asymptotic formulas for $t_n(y)$ that are valid when y goes to infinity with n. Formula (3.9) can be extended to

$$t_n(y) = \frac{\sqrt{2\pi}\, n^{n-1/2+y/2}}{2^{y/2}\, \Gamma(y/2)} \left(1 + O(y^{3/2} n^{-1/2})\right), \qquad (19.14)$$

uniformly for $1 \le y \le n^{1/3}$, using the proof technique of Lemma 3. Still larger values of y can be handled by using the saddle point method to derive the following general estimate:

$$t_{a\lambda n, n}(\lambda n + b) = \frac{n!\, e^{n\rho}\, \rho^{(a\lambda-1)n}\, \lambda^{(1-b)/2}}{2\sqrt{\pi n}\, (1-\rho)^{\lambda n}}$$
$$\times \left(1 + O(\sqrt{\lambda}) + O(1/\sqrt{\lambda n})\right), \qquad (19.15)$$

for fixed a and b as $\lambda \to 0$ and $\lambda n/(\log n)^2 \to \infty$, where

$$\rho = 1 + c\lambda - \sqrt{\lambda(1 + c^2\lambda)} = 1 - \sqrt{\lambda} + c\lambda - \frac{c^2}{2}\lambda^{3/2} + O(\lambda^{5/2}), \quad (19.6)$$

if $c = (1-a)/2$.

For example, to estimate $t_{2r,n}(3r)$ when $r = n^{1/2}$, we can use (19.15) with $a = \frac{2}{3}$, $b = 0$, and $\lambda = 3n^{-1/2}$. The complicated dependence on ρ can also be expressed as

$$\frac{e^{n\rho}\, \rho^{(a\lambda-1)n}}{(1-\rho)^{\lambda n}} = \exp\left(n\left(1 - \tfrac{1}{2}\lambda\ln\lambda + \tfrac{1}{2}\lambda + (\tfrac{1}{3} - a)\lambda^{3/2}\right.\right.$$
$$\left.\left. - \tfrac{1}{4}a^2\lambda^2 + O(\lambda^{5/2})\right)\right), \qquad (19.17)$$

which is sufficiently accurate if $\lambda \le n^{-1/4}$.

20. The Evolution of Uncleanness

We get further insight into the behavior of an evolving multigraph by studying how its reduced multigraph $\overline{\overline{M}}$ changes as the excess increases. Let's review the theory of Section 9 in light of what we have learned since then. The generating function for the cyclic part of all multigraphs having excess r and deficiency d is

$$E_{rd}(z) = e_{rd} \frac{T(z)^{2r-d}}{\left(1 - T(z)\right)^{3r-d+1/2}} . \tag{20.1}$$

We can interpret it as follows, ignoring the constant factor e_{rd} for a moment: There is a reduced multigraph $\overline{\overline{M}}$ having $\nu = 2r - d$ vertices and $\mu = 3r - d$ edges; each vertex has degree ≥ 3, where a self-loop is considered to add 2 to the degree. We can obtain all cyclic multigraphs M that reduce to $\overline{\overline{M}}$ by a two-step process. First we insert 0 or more vertices of degree 2 on each edge; and we also construct any desired number of cycles, as separate components. All of the newly constructed vertices, including the vertices in the cycles, have degree 2. This first step creates a set of multigraphs with the univariate generating function $z^{\nu}(1 - z)^{-\mu}(1 - z)^{-1/2}$, because each edge subdivision corresponds to $(1 - z)^{-1}$, and because the cycles are generated by

$$\exp(\tfrac{1}{2}z + \tfrac{1}{4}z^2 + \tfrac{1}{6}z^3 + \cdots) = (1 - z)^{-1/2}.$$

Now we proceed to step two, which sprouts a rooted tree from every vertex; this changes z to $T(z)$ in the generating function.

The excess increases by 1 when we add a new edge $\langle x, y \rangle$ to M. How does the new edge change $\overline{\overline{M}}$? A moment's thought shows that $\overline{\overline{M}}$ will gain 2, 1, or 0 vertices; this means the deficiency will either stay the same or it will increase by 1 or 2.

In fact there is a nice algebraic and quantitative way to understand what happens, in terms of the generating function. Again we consider a two-step process: First we choose a vertex x of M; this means we apply the marking operator ϑ to the generating function. There are three cases: The marked vertex either belongs to a tree attached to one of the ν special vertices of $\overline{\overline{M}}$, or it belongs to a tree attached to a vertex within one of the μ edges, or it belongs to a tree attached to a vertex in some cycle. We represent Case 1 by attaching a "half-edge" to the existing vertex; we represent Case 2 by introducing a new vertex into the split edge and attaching a half-edge to it; we represent Case 3 by introducing a new vertex with a self-loop and attaching a half-edge to it.

A half-edge is like an edge but it touches only one vertex. For example, if $\overline{\overline{M}}$ is the multigraph K_4, the symbolic representations of the three possible outcomes of step 1 are

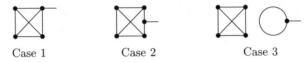

Case 1 　　　　 Case 2 　　　　 Case 3

Let's call this augmented multigraph $\overline{\overline{M}}'$.

A cyclic multigraph M' with a marked vertex can be reduced by attaching a half-edge to the marked vertex, then pruning all vertices of degree 1 and cancelling all vertices of degree 2. Conversely, the marked cyclic multigraphs that reduce to a given $\overline{\overline{M}}'$ are obtained by adding zero or more vertices to each edge (*including* the half edge), also adding cycles, then sprouting trees from each vertex. Thus the generating function for M' in Case 1 is

$$e_{rd}\,\frac{\nu\,T(z)^\nu}{\left(1-T(z)\right)^{\mu+3/2}}\,;\qquad(20.2)$$

the ν in the numerator accounts for the number of vertices that can be chosen, and the extra $\left(1-T(z)\right)$ in the denominator accounts for the new half-edge. The generating function for M' in Case 2 is

$$e_{rd}\,\frac{\mu\,T(z)^{\nu+1}}{\left(1-T(z)\right)^{\mu+5/2}}\,;\qquad(20.3)$$

now we have μ edges that can be split, and we include an additional $T(z)$ in the numerator for the new vertex and an additional $\left(1-T(z)\right)^2$ in the denominator for the new half-edge and the additional split edge. Finally, the generating function for M' in Case 3 is

$$e_{rd}\,\frac{\frac{1}{2}\,T(z)^{\nu+1}}{\left(1-T(z)\right)^{\mu+5/2}}\,;\qquad(20.4)$$

as in Case 2, the diagram has gained one vertex and two edges. The factor $\frac{1}{2}$ is due to the compensation factor κ of a self-loop.

If our calculations are correct, the sum $(20.2)+(20.3)+(20.4)$ should be the result of applying ϑ to the overall generating function (20.1). And sure enough,

$$\vartheta\,\frac{T(z)^\nu}{\left(1-T(z)\right)^{\mu+1/2}}=\frac{\nu\,T(z)^\nu}{\left(1-T(z)\right)^{\mu+3/2}}+\frac{\left(\mu+\frac{1}{2}\right)T(z)^{\nu+1}}{\left(1-T(z)\right)^{\mu+5/2}}\,;\qquad(20.5)$$

everything checks out fine.

The next step, choosing y, is the same, except that now we mark a vertex of M' and obtain M''. The transition from $\overline{\overline{M}}'$ to $\overline{\overline{M}}''$ leads again to three cases; we attach another half-edge and possibly split an existing edge or add a new self-loop. In particular, we might split the half-edge of $\overline{\overline{M}}'$. The change in the generating function is once again represented by (20.5), but this time ν and μ have to be adjusted to equal the number of vertices and edges of $\overline{\overline{M}}'$. The left term of (20.5) therefore becomes

$$\frac{\nu^2 T(z)^\nu}{\left(1 - T(z)\right)^{\mu+5/2}} + \frac{\nu(\mu + \frac{3}{2}) T(z)^{\nu+1}}{\left(1 - T(z)\right)^{\mu+7/2}}, \tag{20.6}$$

and the right term becomes

$$\frac{(\mu + \frac{1}{2})(\nu + 1) T(z)^{\nu+1}}{\left(1 - T(z)\right)^{\mu+7/2}} + \frac{(\mu + \frac{1}{2})(\mu + \frac{5}{2}) T(z)^{\nu+2}}{\left(1 - T(z)\right)^{\mu+9/2}}. \tag{20.7}$$

Notice that the first term of (20.5) corresponds to the case that the deficiency increases by 1 when x is chosen, while the second term corresponds to the case where the deficiency stays the same. Similarly, the first terms of (20.6) and (20.7) correspond to an increase in deficiency when y is chosen, after x has already been marked.

By looking at the coefficients of these generating functions we can see why the deficiency rarely increases unless the total number of vertices in the cyclic part is not much larger than ν. Suppose we change the generating function to

$$F(z, s) = \frac{T(z)^\nu}{\left(1 - s T(z)\right)^{\mu+1/2}};$$

then

$$\frac{[z^n] \frac{\partial}{\partial s} F(z, s)\big|_{s=1}}{[z^n] F(z, s)\big|_{s=1}}$$

will be the average number of tree-root vertices that appear within the edges of $\overline{\overline{M}}$. For fixed ν and μ as $n \to \infty$ this number is

$$\frac{[z^n] (\mu + \frac{1}{2}) T(z)^{\nu+1} \left(1 - T(z)\right)^{-\mu-3/2}}{[z^n] T(z)^\nu \left(1 - T(z)\right)^{-\mu-1/2}}$$

$$= \frac{(\mu + \frac{1}{2}) t_n(\mu + \frac{3}{2})}{t_n(\mu + \frac{1}{2})} \left(1 + O(n^{-1/2})\right), \tag{20.8}$$

which is approximately $\sqrt{\mu n}$ by (3.9), when μ is large. Thus, there are about $\sqrt{\mu n}$ tree roots, only ν of which will increase the deficiency when chosen; almost all choices of x and y will fall in trees that add new vertices to $\overline{\overline{M}}$ and $\overline{\overline{M}}'$.

If we replace one of the factors $T(z)$ in the numerator of the generating function by $\vartheta T(z) = T(z)/\big(1 - T(z)\big)$, we multiply the coefficient of z^n by the average size of a rooted tree; we find that each rooted tree contains approximately $\sqrt{n/\mu}$ vertices.

The number n in these calculations has been the number of vertices in the cyclic part of a multigraph, and the number μ is $3r$. Let's return to our other notational convention, where n is the total number of vertices in the evolving multigraph and $m = \frac{n}{2}(1 + \mu n^{-1/3})$ is the total number of edges. Recall that the average excess r grows as $\frac{2}{3}\mu^3$, for $\mu \le n^{1/12}$; the size of the cyclic part, similarly, has order $\mu n^{2/3}$. The probability that a random new edge falls in the cyclic part (and therefore increases the excess) is therefore of order $(\mu n^{2/3}/n)^2 = \mu^2 n^{-2/3}$; we must add about $n^{2/3}/\mu^2$ more edges before the excess increases. And when it does, the probability of choosing a "bad" x or y, making the new multigraph unclean, is the ratio of $2r$ to the *total* number of tree roots, which is of order

$$\frac{2r}{\sqrt{3r(\mu n^{2/3})}} \approx \frac{\mu^3}{\sqrt{\mu^4 n^{2/3}}} = \mu n^{-1/3}\,.$$

We will probably have to do $n^{1/3}/\mu$ augmentations of excess, adding $(n^{2/3}/\mu^2)(n^{1/3}/\mu) = n/\mu^3$ more edges, before we reach an unclean multigraph. That is why the multigraph tends to stay clean until $\mu = n^{1/12}$, as asserted in Theorem 7.

After x and y are chosen to form the endpoints of a new edge, a third step takes place: This new edge is merged or integrated with the other edges. Symbolically, the two half-edges for x and y are now spliced together. We can complete our study of how the generating function changes at the time of excess augmentation by considering this third and final step.

It is easiest to consider the *inverse* of the final step, namely the operation of marking an edge whose removal would *decrease* the excess. Such an edge must be in the complex part, not the acyclic or unicyclic part. The operator that corresponds to marking an arbitrary edge in a complex multigraph of excess r is $r + \vartheta$, because this multiplies the coefficient of z^n by $r + n$, the total number of edges. However, we also need to figure out the generating function for "insignificant" edges, edges whose removal would leave the excess unchanged. Such edges can be described by an ordered pair consisting of a rooted tree and a multigraph of excess r with a marked vertex; one end of the edge is attached to the marked vertex and the other end is attached to the root of the tree. Thus the appropriate operator for insignificant edges is $T(z)\vartheta$. Altogether then, the generating function that corresponds to marking a significant edge, given a family of complex multigraphs of excess r, is $r + \vartheta - T\vartheta$. We also should multiply this by two, because we assign an orientation to the edge with the ordered pair $\langle x, y \rangle$.

When the operator $2(r + \vartheta - T(z)\vartheta)$ is applied to a generating function of the form $T(z)^\nu / (1 - T(z))^\mu$, with $\mu = \nu + r$, we get

$$2(r + (1 - T(z))\vartheta)\frac{T(z)^\nu}{(1 - T(z))^\mu} = \frac{2((r + \nu)T(z)^\nu)}{(1 - T(z))^\mu} + \frac{2\mu T(z)^{\nu+1}}{(1 - T(z))^{\mu+1}}$$

$$= \frac{2\mu T(z)^\nu}{(1 - T(z))^{\mu+1}}. \qquad (20.9)$$

Therefore the inverse operation we seek, which merges an ordered $\langle x, y \rangle$ into the set of existing edges, takes

$$\frac{T(z)^\nu}{(1 - T(z))^{\mu+1}} \longmapsto \frac{1}{2\mu}\frac{T(z)^\nu}{(1 - T(z))^\mu}. \qquad (20.10)$$

For example, the first term of (20.6) will go into

$$\frac{\nu^2 T(z)^\nu}{2(\mu + 1)(1 - T(z))^{\mu+3/2}}.$$

(First we multiply by $(1 - T(z))^{1/2}$ to get rid of the unicyclic components, then we apply the inverse operation (20.10), then we put the unicyclic components back.)

Altogether we find that the generating function

$$T(z)^{2r-d} / (1 - T(z))^{3r-d+1/2}$$

for cyclic multigraphs of excess r and deficiency d makes the following contributions to the generating functions for cyclic multigraphs of excess $r + 1$ and deficiencies d, $d+1$, and $d+2$, according to (20.6), (20.7), and (20.10):

$$\frac{(6r-2d+5)(6r-2d+1)}{8(3r-d+3)}\frac{T(z)^{2r+2-d}}{(1-T(z))^{3r+3-d+1/2}}$$

$$+ \frac{(2r-d)(6r-2d+3) + (2r-d+1)(6r-2d+1)}{4(3r-d+2)}\frac{T(z)^{2r+1-d}}{(1-T(z))^{3r+2-d+1/2}}$$

$$+ \frac{(2r-d)^2}{2(3r-d+1)}\frac{T(z)^{2r-d}}{(1-T(z))^{3r+1-d+1/2}}. \qquad (20.11)$$

This formula is essentially the same as the recurrence relation for e_{rd} in equations (5.11)–(5.13).

We can illustrate the observations of this section by introducing another partial ordering analogous to Figure 1. Every evolving graph or multigraph traces a path in Figure 2, just as it does in Figure 1; but in Figure 2 the state (r, d) represents excess r and deficiency d. Fractions in brackets above each state are the coefficients e_{rd} of the generating function (5.10). Fractions on the arrows are not transition probabilities but rather the amounts by which each generating function coefficient affects the coefficients at the next level; these fractions are the coefficients in (20.11).

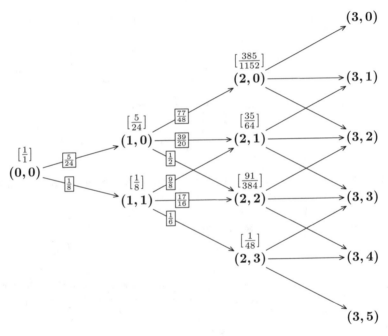

FIGURE 2. The evolution of deficiency. Each configuration (r, d) stands for a graph or multigraph whose complex part reduces to a multigraph with $2r - d$ vertices and $3r - d$ edges, when vertices of degrees 1 and 2 are eliminated. A graph or multigraph with deficiency 0 is called "clean"; the reduced multigraphs in such cases are 3-regular. When r is small, each unit increase in deficiency occurs with probability of order $n^{-1/3}$; therefore most random graphs stay clean until r is quite large.

21. Waiting for Uncleanness

We have seen that a graph almost surely stays clean while it has $\frac{1}{2}(n + \mu n^{2/3})$ edges, as long as μ is $o(n^{1/12})$. What happens when μ gets a bit larger? Another contour integral provides the answer; in this one, we rescale μ in preparation for the appearance of the giant component, but we allow μ to be small enough that there is a substantial overlap with the estimate (10.1) of Lemma 3.

Note: In the next few sections, the parameter μ will be $n^{-1/3}$ times as large as it was in our previous discussions.

Lemma 7. If $m = \frac{1}{2}(n + \mu n)$ and $r = \frac{2}{3}\mu^3 n + \rho\sqrt{\mu^3 n}$, we have

$$\frac{2^m m!\, n!\, e_r}{n^{2m}(n - m + r)!} \, [z^n] \, \frac{U(z)^{n-m+r}T(z)^{2r}}{\left(1 - T(z)\right)^{3r+y}}$$

$$= B(y, \mu, \rho, n) \exp\left(O\left((1 + |\rho|^3)\mu^{-3/2}n^{-1/2} + (1 + |\rho|)\mu^{5/2}n^{1/2}\right)\right),$$
(21.1)

where

$$B(y, \mu, \rho, n) = \sqrt{\frac{3}{20\pi n}} \, \mu^{-1-y} \exp\left(-\frac{2}{3}\mu^4 n - \frac{3}{20}\rho^2\right),$$
(21.2)

uniformly for $n^{-1/3}\log n \le \mu \le n^{-1/5}$, $|\rho| \le \frac{2}{3}\mu^{3/2}n^{1/2}$, and fixed y, as $n \to \infty$.

Proof. This is the sort of lemma for which computer algebra really pays off. We can begin by using Stirling's approximation to show that

$$\log\left(\frac{2^m m!\, n!\, e_r}{n^{2m}(n - m + r)!\, 2^{n-m+r}}\right) = -n + 3r\ln\mu - \frac{5}{6}\mu^3 n$$

$$- \frac{3}{2}\ln\mu + \frac{1}{2}\ln\frac{3}{2} + \frac{2}{3}\mu^4 n + \frac{3}{4}\rho^2$$

$$+ O\left((1 + |\rho|^3)\mu^{-3/2}n^{-1/2} + (1 + |\rho|)\mu^{5/2}n^{1/2}\right). \quad (21.3)$$

Now we express the remaining factor by using the trick of (10.11):

$$[z^n] \, \frac{(2U(z))^{n-m+r}T(z)^{2r}}{\left(1 - T(z)\right)^{3r+y}} = \frac{1}{2\pi i}\oint (1 - z)^{1-y}e^{g(z)}\,\frac{dz}{z}, \quad (21.4)$$

where

$$g(z) = nz + (3r - m)\ln z - 3r\ln(1 - z) + (n - m + r)\ln(2 - z). \quad (21.5)$$

As before we can show that the asymptotic value of the integral depends only on the behavior of the integrand near $z = 1$. This time we need not worry about a three-legged saddle point, because we are sufficiently far from the critical region near $\mu = 0$. A good path of integration turns out to be $z = 1 - \alpha + it\mu^{-1/2}n^{-1/2}$, where $\alpha = \mu - \frac{2}{3}\mu^2 + \frac{3}{5}\rho\mu^{-1/2}n^{-1/2}$. Indeed, some beautiful cancellation occurs in the most significant terms:

$$\begin{aligned}
g(1 &- \alpha + it\mu^{-1/2}n^{-1/2}) \\
&= g(1 - \alpha) - \tfrac{5}{2}t^2 + O\big((\mu^{5/2}n^{1/2} + \mu^{-3/2}n^{-1/2}\rho^2)t\big) \\
&\quad + O\big(((1 + |\rho|)\mu^{-3/2}n^{-1/2} + \mu)t^2\big),
\end{aligned} \tag{21.6}$$

when $|t| \le \log n$. The O bounds follow from the fact that the power series for $\log z$, $\log(1 - z)$, and $\log(2 - z)$ converge in the stated ranges.

The other factors of the integrand, besides $e^{g(z)}$, are

$$(1 - z)^{1-y}\frac{dz}{z} = \mu^{1-y}i\mu^{-1/2}n^{-1/2}\,dt\sum_{k=0}^{\infty}\mu^k\beta^{k+1-y},$$

where $\beta = (\alpha - it\mu^{-1/2}n^{-1/2})/\mu = 1 - \frac{2}{3}\mu + (\frac{3}{5}\rho - it)\mu^{-3/2}n^{-1/2}$. We can now write the integral as a factor independent of t times

$$\int e^{-5t^2/2}(1 + \gamma_1 t + \gamma_2 t^2 + \cdots)\,dt, \tag{21.7}$$

where the γ's are functions of μ and ρ, and the series is convergent for $|t| \le \log n$. The integrand is superpolynomially small when $t = \pm\log n$; hence we can bound the error terms for $|t| \le \log n$, then integrate from $-\infty$ to ∞, showing that (21.7) is

$$\sqrt{\frac{2\pi}{5}}\left(1 + O(\mu^{5/2}n^{1/2} + (1 + \rho^2)\mu^{-3/2}n^{-1/2})\right). \tag{21.8}$$

Finally we observe that the other factors nicely cancel the leading terms of (21.3); only (21.1) and (21.2) are left. The overall formula (21.1) has a weaker estimate than (21.8) because Stirling's approximation (21.3) is more sensitive to the value of ρ and because of the term $g(1 - \alpha)$. $\square$

Notice that Lemma 7 matches the first estimate of Lemma 5 in Section 15, which says that the asymptotic probability of excess r is like that for a normal distribution with mean $\frac{2}{3}\mu^3$ and variance of order μ^3, as long as $r = O(\mu^3)$. On the other hand, the extreme tails for larger values of r are not as small as they would be in a normal distribution; they decrease only as shown in the second estimate of Lemma 5. For example, with probability 100^{-m} all edges will join vertices in the first $n/10$ vertices; so there will be at least $0.9n$ isolated vertices, and the excess will be at least $m - n + 0.9n > 0.4n$.

Theorem 12. *The probability that a random multigraph with n vertices and $m = \frac{1}{2}(n + \mu n)$ edges is clean, when $0 \le \mu \le n^{-1/5}$, is*

$$\exp\left(-\tfrac{2}{3}\mu^4 n + O\left(\mu^{5/2} n^{1/2} \log n + \mu^{-3/2} n^{-1/2} (\log n)^3\right)\right). \tag{21.9}$$

Proof. The probability decreases as μ increases. Therefore we need to verify the result only for μ greater than $n^{-3/11}$ or so, when the error estimate $\mu^{-3/2} n^{-1/2} (\log n)^3$ does not swamp the main term $\exp(-\tfrac{2}{3}\mu^4 n) = 1 - \tfrac{2}{3}\mu^4 n + O(\mu^8 n^2)$.

Formula (21.1) is the probability that a random graph or multigraph with m edges is clean and has excess r, if we set $y = \frac{1}{2}$. That probability is superpolynomially small unless $|\rho| \le \log n$, because of the term $-\rho^2$ in the exponent. Extremely large values of ρ, not covered by the hypotheses of Lemma 7, are also negligible. Therefore we can sum over r by integrating over ρ from $-\log n$ to $+\log n$; and we can then extend the integral from $-\infty$ to ∞ without changing its asymptotic value. Hence the probability of cleanliness is

$$n^{-1/2}\sqrt{\frac{3}{20\pi}}\,\mu^{-3/2}\exp\left(-\tfrac{2}{3}\mu^4 n\right)\int_{-\infty}^{\infty} e^{-3\rho^2/20}\sqrt{\mu^3 n}\,d\rho = \exp\left(-\tfrac{2}{3}\mu^4 n\right),$$

plus the error term. Another nice bit of cancellation.   □

Corollary. *The average number of edges added to an evolving multigraph until it first becomes unclean is*

$$\tfrac{1}{2}n + \frac{3^{1/4}\Gamma\left(\tfrac{1}{4}\right)}{2^{13/4}}\,n^{3/4} + O(n^{8/11+\epsilon}), \tag{21.10}$$

and the standard deviation is of order $n^{3/4}$.

Proof. The stated average number is $\sum_{m\ge 0} p_m$, where p_m is the probability in the theorem. When $\mu \le 0$, the probability of uncleanliness is $O(n^{-1/3})$ by Theorem 7, so the sum for $0 \le m < \frac{1}{2}n$ is $\frac{1}{2}n - O(n^{2/3})$. When $0 \le \mu \le n^{-3/11}(\log n)^{6/11}$, the probability of uncleanliness is

$O(n^{-1/11}(\log n)^{24/11})$ by (21.9); after that the error is negligible in comparison with the integral

$$\frac{1}{2}n\int_0^\infty e^{-(2/3)\mu^4 n}\,d\mu = n^{3/4}\frac{1}{4}\left(\frac{3}{2}\right)^{1/4}\int_0^\infty e^{-u}u^{-3/4}\,du = cn^{3/4}\,,$$

where c is the coefficient of $n^{3/4}$ in (21.10). This proves (21.10).

The expected value of m^2 at the stopping time is $\sum_{m\geq 0}(2m+1)p_m$, and we need to be especially careful when evaluating this sum; the simple estimate $p_m = 1 - O(n^{-1/3})$ for $m \leq \frac{1}{2}n$ will not do, because it will obliterate significant terms by adding $O(n^{5/3})$. Appropriate accuracy is maintained by computing the expected value of $(m - \frac{1}{2}n)^2$, which is

$$\frac{n^2}{4} + \sum_{m\geq 0}(2m+1-n)p_m = \sum_{m=0}^{n/2}(n-2m)(1-p_m) + \sum_{m=n/2}^{\infty}(2m-n)p_m + O(n)\,.$$

We can show that the terms for $m \leq \frac{1}{2}n$ are now negligible, because the cleanliness probability p_m is bounded below by the probability that a multigraph with m edges has excess 0. Therefore $1 - p_m = O(n^2/(n-2m)^3)$ when $m \leq m_0 = \frac{1}{2}n - n^{2/3+\epsilon}$, by the remarks preceding (13.23); and

$$\sum_{m=0}^{n/2}(n-2m)(1-p_m) = \sum_{m=0}^{m_0}O\left(\frac{n^2}{(n-2m)^2}\right) + \sum_{m=m_0}^{n/2}O(n-2m)$$

$$= O(n^{4/3-\epsilon}) + O(n^{4/3+2\epsilon})\,.$$

The other terms can be approximated by

$$\sum_{m=n/2}^{\infty}(2m-n)p_m = \int_0^\infty \frac{n^2\mu}{2}e^{-(2/3)\mu^4 n}\,d\mu + O(n^{16/11+\epsilon})\,,$$

with an error estimate coming from the range $0 \leq \mu \leq n^{-3/11+\epsilon}$ as before. It follows that the variance is asymptotic to this integral minus the square of $((21.10) - \frac{1}{2}n)$, namely $(3^{1/2}\Gamma(\frac{1}{2})2^{-7/2} - c^2)n^{3/2}$.

Incidentally, the value of c is approximately 0.50155, and the standard deviation is approximately $0.1407n^{3/4}$. $\square$

Once a graph begins to get dirty, its deficiency rises rapidly. For fixed d we can estimate the probability of excess r and deficiency d by taking $y = \frac{1}{2} - d$ and multiplying (21.1) by $r^d/d!$, because of (7.16). The fact that (21.1) has $T(z)^{2r}$ in the numerator instead of $T(z)^{2r-d}$ is unimportant, since $T(z)^{2r} = T(z)^{2r-d} \sum \binom{d}{k} (T(z) - 1)^k$. We obtain a probability about $(\frac{2}{3}\mu^4 n)^d/d!$ times as large as before, but this is damped rapidly by the factor $\exp(-\frac{2}{3}\mu^4 n)$ when μ becomes greater than $n^{-1/4}$. We will look further at the growth of deficiency in Section 23.

22. A Closer Look

The structure theory of Section 20 gives us more detailed information about what happens when an evolving multigraph first changes from clean to unclean. We learned in that section that the process of adding a new edge $\langle x, y \rangle$ can be broken into three parts, namely the introduction of half-edges at x and y followed by the joining of those two edges. The deficiency can increase by 1 during each of the first two stages.

The probability that a clean graph becomes potentially deficient when a half-edge is attached to x is the probability that the image $\overline{\overline{M}}'$ of the half-edge after pruning and cancellation does not create a new vertex not in $\overline{M}$. According to the analysis of Section 20, the expected number of times this happens is

$$p_1(n) = \sum_m \frac{2^m m! \, n!}{n^{2m+1}} \, [w^m z^n] \, G_1(w, z), \tag{22.1}$$

where

$$G_1(w, z) = e^{U(wz)/w} \sum_{r \geq 0} e_r w^r \frac{2r \, T(wz)^{2r}}{\left(1 - T(wz)\right)^{3r+3/2}}$$

$$= \tfrac{5}{12} w^3 z^2 + \tfrac{5}{24} \left(2w^3 + 13w^4\right) z^3 + \cdots. \tag{22.2}$$

The factor $2r$ covers the deficient choices of x, as in the first term of (20.5).

Actually (22.2) is an overestimate, because some apparently bad choices of x are "false alarms." If the half-edge of x does not add a vertex to $\overline{\overline{M}}$, there's still a possibility that y will be chosen in the acyclic part; then the new edge $\langle x, y \rangle$ will not increase the excess and the multigraph will still be clean. The expected number of false alarms is

$$p_1'(n) = \sum_m \frac{2^m m! \, n!}{n^{2m+2}} \, [w^m z^n] \, \frac{T(wz)}{w} \, G_1(w, z). \tag{22.3}$$

The multigraph becomes unclean when y is chosen if the half-edge for y prunes and cancels to a reduced multigraph $\overline{\overline{M}}''$ having the same $2r + 1$ vertices as $\overline{\overline{M}}'$. This occurs with probability

$$p_2(n) = \sum_m \frac{2^m m! \, n!}{n^{2m+2}} [w^m z^n] \, G_2(w, z), \qquad (22.4)$$

$$G_2(w, z) = e^{U(wz)/w} \sum_{r \geq 0} e_r w^r \frac{(3r + \frac{1}{2})(2r + 1) \, T(wz)^{2r+1}}{\left(1 - T(wz)\right)^{3r+7/2}}$$

$$= \tfrac{1}{2} wz + \tfrac{1}{4}(2w + 9w^2) z^2 + \cdots . \qquad (22.5)$$

Consequently we must have

$$p_1(n) - p_1'(n) + p_2(n) = 1 \qquad (22.6)$$

for all n; this identity is a nontrivial property of the bivariate generating functions $G_1(w, z)$ and $G_2(w, z)$. When $n = 6$, for example, computer calculations show that

$$p_1(6) = \frac{10288260775}{22039921152} \approx 0.4668 \, ; \qquad p_1'(6) = \frac{38865625}{612220032} \approx 0.0635 \, ;$$

$$p_2(6) = \frac{13150822877}{22039921152} \approx 0.5967 \, .$$

We can use Lemma 7 to calculate the approximate values of these quantities when n is large, ignoring extreme terms not covered by that lemma:

$$p_1(n) \sim \frac{1}{n} \int_0^\infty \int_{-\infty}^\infty 2r B(\tfrac{3}{2}, \mu, \rho, n)(\mu^{3/2} n^{1/2} \, d\rho)(\tfrac{1}{2} n \, d\mu)$$

$$= 2^{-7/4} 3^{-1/4} \Gamma(\tfrac{3}{4}) n^{1/4} \, ; \qquad (22.7)$$

$$p_1'(n) = \sum_m \frac{2^{m-1}(m-1)! \, n!}{n^{2m}} [w^m z^n] \, T(wz) G_1(w, z)$$

$$\sim \frac{1}{2} \int_0^\infty m^{-1} \int_{-\infty}^\infty 2r B(\tfrac{3}{2}, \mu, \rho, n)(\mu^{3/2} n^{1/2} \, d\rho)(\tfrac{1}{2} n \, d\mu)$$

$$\sim 2^{-7/4} 3^{-1/4} \Gamma(\tfrac{3}{4}) n^{1/4} \, ; \qquad (22.8)$$

$$p_2(n) \sim \frac{1}{n^2} \int_{n^{-1/3}}^\infty \int_{-\infty}^\infty (3r + \tfrac{1}{2})(2r + 1) B(\tfrac{7}{2}, \mu, \rho, n)(\mu^{3/2} n^{1/2} \, d\rho)(\tfrac{1}{2} n \, d\mu)$$

$$\sim \frac{1}{2} \, . \qquad (22.9)$$

Notice that $p_1(n)$ and $p'_1(n)$ are unbounded, so they must be regarded as expected values (not probabilities). But $p_1(n) - p'_1(n)$ is the probability of a "true alarm." As we might have guessed, the transition from clean to unclean occurs about half the time when x is chosen, half the time when y is chosen.

23. Giant Growth

We know from the classical theory [13] that a giant component will emerge when the number of edges is $\frac{n}{2}(1 + \mu)$ for a positive constant μ. The classical theory deals with graphs, but the same phenomenon will occur with multigraphs, because random graphs are generated by the multigraph process if we discard self-loops and duplicate edges; discarded edges do not affect the size of components, and comparatively few edges are discarded until the graph has gotten rather dense (see [4]).

Instead of relying on the classical theory, we can also deduce the existence of a giant component by studying the generating function $G(w, z)$. The proof is indirect: First we count the vertices that lie in trees and unicyclic components, showing that there probably aren't too many of those. Then we show that it is improbable to have two distinct complex components.

The first part is easy, because there is a simple closed form for the expected number of vertices in trees. If we mark just the vertices in trees of size k, by differentiating the generating function

$$G(w, z) \exp\left(-k^{k-2} w^{k-1} z^k / k! + k^{k-2} w^{k-1} z^k s^k / k!\right)$$

with respect to s and setting $s = 1$, we see that the expected number of such vertices is just

$$\frac{2^m m! \, n!}{n^{2m}} [w^m z^n] \frac{k^{k-1}}{k!} w^{k-1} z^k G(w, z)$$

$$= \frac{2^m m! \, n! \, k^{k-1}}{n^{2m} k!} [w^{m-k+1} z^{n-k}] G(w, z)$$

$$= \frac{2^m m! \, n! \, k^{k-1}}{n^{2m} k!} \frac{(n-k)^{2(m-k+1)}}{2^{m-k+1}(m-k+1)! \, (n-k)!};$$

this expression can be written

$$\frac{k^{k-1}}{k!} \frac{2^{k-1} m^{\underline{k-1}} n^{\underline{k}}}{(n-k)^{2k-2}} \left(\frac{n-k}{n}\right)^{2m} \tag{23.1}$$

in terms of falling factorial powers $x^{\underline{k}} = x(x-1) \ldots (x-k+1)$.

Asymptotically, we have $n^{\underline{k}} = n^k\left(1 + O(k^2/n)\right)$ and $(n - k)^k = n^k\left(1 + O(k^2/n)\right)$ for all k; also $(1 - k/n)^n = e^{-k}\left(1 + O(k^2/n)\right)$ for $k \leq \sqrt{n}$ and $(1 - k/n)^n \leq e^{-k}$ for $k \leq n$. If μ is a nonzero constant, $\mu > -1$, and if $m = \frac{n}{2}(1 + \mu)$, expression (23.1) is

$$\frac{n}{1 + \mu} \frac{k^{k-1}}{k!}(1 + \mu)^k e^{-k(1+\mu)}\left(1 + O(\tfrac{k^2}{n})\right) \qquad (23.2)$$

for $k \leq \sqrt{n}$; and it is superpolynomially small when $k = \sqrt{n}$, because it is $O\left(\left((1+\mu)e^{-\mu}\right)^k k^{1/2}\right)$ and $(1+\mu)e^{-\mu} < 1$. It is also superpolynomially small when $k > \sqrt{n}$, because we will prove in Section 27 below that a continuous approximation of the quantity

$$e^k \sqrt{\frac{m - k}{m}} \sqrt{\frac{n - k}{n}} \frac{2^k m^{\underline{k}} n^{\underline{k}}}{(n - k)^{2k}}\left(\frac{n - k}{n}\right)^{2m} \qquad (23.3)$$

decreases when k increases.

Let σ be defined by the formula

$$(1 + \mu)e^{-\mu} = (1 - \sigma)e^{\sigma}, \qquad \sigma = \mu + O(\mu^2). \qquad (23.4)$$

Then σ is the quantity called $1 - x\left(\frac{1}{2}(1 + \mu)\right)$ in [13], and we have

$$\sum_{k \geq 1} \frac{k^{k-1}}{k!}\left((1 + \mu)e^{-(1+\mu)}\right)^k = \sum_{k \geq 1} \frac{k^{k-1}}{k!}\left((1 - \sigma)e^{-(1-\sigma)}\right)^k$$

$$= T\left((1 - \sigma)e^{-(1-\sigma)}\right) = 1 - \sigma,$$

when μ is positive. By summing (23.2) over all k, we conclude that the expected total number of vertices in trees is

$$\frac{1 - \sigma}{1 + \mu}n + O(\sigma^{-3}); \qquad (23.5)$$

the error term $O(\sigma^{-3})$ here comes from summing $\vartheta^2 T\left((1 - \sigma)e^{-(1-\sigma)}\right)$, which brings a factor of k^2 into each term.

For example, if $1 + \mu = \ln 4$, we have $1 - \sigma = \ln 2$, because $\frac{1}{4}\ln 4 = \frac{1}{2}\ln 2$. When the number of edges reaches $n \ln 2$ the expected number of vertices in trees will be $\frac{1}{2}n$. And in general when the number of edges reaches $\frac{n}{2x}\ln\frac{1}{1-x}$, the expected number of vertices in trees will be $(1 - x)n$, for $0 < x < 1$.

The expected number of vertices in unicyclic components can be found in a similar way, by differentiating

$$G(w, z)\, e^{-V(wz)+V(wzs)}$$

with respect to s and setting $s = 1$. The generating function is

$$\frac{1}{2} \frac{T(wz)}{(1 - T(wz))^2}\, G(w, z) = \big(\vartheta V(wz)\big)\, G(w, z)\,, \qquad (23.6)$$

and we have

$$\frac{T(z)}{(1 - T(z))^2} = \sum_{k \geq 1} \frac{k^k Q(k)}{k!}\, z^k \qquad (23.7)$$

by (3.12). The expected number of vertices belonging to unicyclic components of size k therefore can be expressed in closed form, analogous to (23.1):

$$\frac{1}{2} \frac{k^k Q(k)}{k!} \frac{2^m m!\, n!}{n^{2m}}\, [w^{m-k} z^{n-k}]\, G(w, z)$$

$$= \frac{1}{2} \frac{k^k Q(k)}{k!} \frac{2^k m^{\underline{k}} n^{\underline{k}}}{(n-k)^{2k}} \left(\frac{n-k}{n}\right)^{2m}. \qquad (23.8)$$

Summing over k, and breaking the sum into two parts $k \leq \sqrt{n}$ and $k > \sqrt{n}$ as above, now yields

$$\frac{1}{2} \sum_{k \geq 1} \frac{k^k Q(k)}{k!} \left((1 + \mu) e^{-(1+\mu)}\right)^k \left(1 + O\!\left(\frac{k^2}{n}\right)\right) = \frac{1 - \sigma}{2\sigma^2} + O(\sigma^{-6} n^{-1}).$$

$$(23.9)$$

(We will obtain sharper bounds in Section 27.)

We have assumed in this discussion that μ is a constant. But our relatively coarse asymptotic arguments are in fact valid if μ varies with n, provided that it is not too small. Relation (23.4) defines σ as an analytic function of μ,

$$\sigma = \mu - \frac{2}{3}\mu^2 + \frac{4}{9}\mu^3 - \frac{44}{135}\mu^4 + \frac{104}{405}\mu^5 - \frac{40}{189}\mu^6 + \frac{7648}{42525}\mu^7$$

$$- \frac{2848}{18225}\mu^8 + \frac{31712}{229635}\mu^9 - \frac{23429344}{189448875}\mu^{10} + \cdots, \qquad (23.10)$$

where the power series converges for $|\mu| < 1$. The quantity $\left((1+\mu) e^{-\mu}\right)^k$ is superpolynomially small for $k = \sqrt{n}$ if μ is at least, say, $n^{-1/4} \log n$.

We are therefore justified in using (23.5)+(23.9) as the expected number of vertices in non-complex components whenever $\mu \geq n^{-1/4} \log n$.

Suppose $\mu = n^{-1/4} \log n$. Then the expected number of vertices in unicyclic components is approximately $\frac{1}{2}\sigma^{-2} \sim \frac{1}{2}\mu^{-2} = \frac{1}{2}n^{1/2}(\log n)^{-2}$, and a similar argument proves that the expected value of the square of this number is approximately $\frac{5}{4}\sigma^{-4} \sim \frac{5}{4}n(\log n)^{-4}$. So the probability of choosing two vertices in unicyclic components is approximately $\frac{5}{4}n^{-1}(\log n)^{-4}$. This probability decreases steadily as m increases, but even if it stayed fixed we would have to add about $\frac{4}{5}n(\log n)^4$ more edges before hitting two unicyclic vertices, that is, before creating a new bicyclic component. By that time the expected number of vertices in trees and unicyclic components will be nearly zero, so the multigraph will almost surely contain no such vertices. Therefore, if there is only one complex component present when $\mu = n^{-1/4} \log n$, there will almost surely be only one complex component from that time on; it will become gigantic. (We will obtain sharper results in Section 27; see Lemma 9 and its corollary.)

Let's look more closely at what happens as the giant component develops. According to (23.5), it will have approximately

$$\left(1 - \frac{1-\sigma}{1+\mu}\right) n = \frac{\mu + \sigma}{1+\mu} n = 2\mu n + O(\mu^2 n) \qquad (23.11)$$

vertices when $m = \frac{n}{2}(1+\mu)$; this is substantially larger than the number $\frac{1}{2}\mu^{-2} + O(\mu^{-1})$ of unicyclic vertices. When m increases by 1, the value of μn increases by 2, so (23.11) increases by 4. Notice that (23.11) agrees with the leading term of (15.11).

We saw in Section 21 that the expected excess r is approximately $\frac{2}{3}\mu^3 n$ when $m = \frac{n}{2}(1 + \mu)$, at least for $0 \leq \mu \leq n^{-1/4}$. We will prove momentarily that this relationship continues to hold as long as μ remains $o(1)$; but before giving the proof, let's look at the situation heuristically. The probability that a new edge increases the excess is the probability that both of its endpoints lie in the cyclic part, namely $(2\mu)^2$. The change in r with respect to m is $(dr/d\mu)(d\mu/dm) = (2\mu^2 n)(2/n)$, and this too is $(2\mu)^2$. So the relation $r = \frac{2}{3}\mu^3 n$ is consistent with (23.11) when μ is not too large.

The expected value of the deficiency d turns out to be approximately $\frac{2}{3}\mu^4 n$, about μ times r. Heuristic justification comes from the considerations of Section 20: When a new edge $\langle x, y \rangle$ falls in the cyclic part, the probability that x is "bad" (in the sense that it increases the deficiency) will be the number of reduced vertices $2r - d$ divided by the square root

of $3r - d$ times the size of the complex part (see the remarks following
(20.8)). So it will be approximately $\frac{4}{3}\mu^3 n$ divided by $((2\mu^3 n)(2\mu n))^{1/2}$,
namely $\frac{2}{3}\mu$. The same holds for y. Hence the expected increase in d,
given that r increases, is $\frac{4}{3}\mu$. And the derivative of $\frac{2}{3}\mu^4 n$ with respect
to μ is indeed $\frac{4}{3}\mu$ times the derivative of $\frac{2}{3}\mu^3 n$.

In order to carry out a rigorous proof as μ increases from $n^{-1/4}$ to
$n^{-1/5}$ to $n^{-1/6}$ and so on, we need to track the full asymptotic spectrum
of the behavior of r and d, not using just the leading terms. It turns out
that r and d are approximately given by the following joint functions
of μ and σ, whose asymptotic series can be computed from (23.10):

$$r_\mu = \frac{\mu^2 - \sigma^2}{2(1 + \mu)}\, n\,; \tag{23.12}$$

$$d_\mu = \frac{3\mu^2 - 3\sigma^2 - \sigma(\mu + \sigma)^2}{2(1 + \mu)}\, n\,. \tag{23.13}$$

Notice that the numerators of both r_μ and d_μ are divisible by $(\mu + \sigma)n$,
so r_μ and d_μ are multiples of the formula $(\mu + \sigma)n/(1 + \mu)$ for giant
component size (23.11). The quantity $\mu + \sigma$ can also, incidentally, be
expressed as $\ln(1 + \mu) - \ln(1 - \sigma)$.

These values r_μ and d_μ also have a surprising relation to the confluent hypergeometric series $F(z) = F(1; 4; 4z)$ of (7.5). It is not difficult
to check that

$$F\big((\mu + \sigma)/4\big) = \frac{6e^{\mu + \sigma}}{(\mu + \sigma)^3}\left(\frac{2r_\mu - d_\mu}{n}\right)$$

$$= \frac{6(1 + \mu)}{(1 - \sigma)(\mu + \sigma)^3}\left(\frac{2r_\mu - d_\mu}{n}\right)\,; \tag{23.14}$$

$$\frac{\vartheta F\big((\mu + \sigma)/4\big)}{F\big((\mu + \sigma)/4\big)} = \frac{d_\mu}{2r_\mu - d_\mu}\,. \tag{23.15}$$

The quantities r_μ and d_μ are not the exact expected values of r
and d. Indeed, the exact expected values are rational numbers, when m
and n are integers, while σ is always irrational when μ is rational. But
we will prove that the distributions of r and d are approximately normal
with expectations r_μ and d_μ.

Before we can prove such a claim, we need to improve the estimate
of e_{rd} in (7.16), because that estimate was derived only for fixed d.

Lemma 8. *Let $F(z)$ be the function defined in (7.5). If $r \to \infty$ and if d varies in such a way that $d/r \to 0$, the polynomial $P_d(r) = [z^d] F(z)^{2r-d}$ satisfies*

$$P_d(r) = \frac{F(s)^{2r-d}}{s^d} \frac{(d/e)^d}{d!} \left(1 + O\left(\frac{d}{r}\right)\right), \qquad (23.16)$$

where s is the solution to $\vartheta F(s)/F(s) = d/(2r-d)$.

Proof.   We have

$$P_d(r) = \frac{1}{2\pi i} \oint \frac{F(z)^{2r-d}}{z^d} \frac{dz}{z} = \frac{1}{2\pi i} \oint e^{(2r-d)f(z)} \frac{dz}{z},$$

where $f(z) = \ln F(z) - (d/(2r-d)) \ln z$, integrated on the circle $|z| = s$. By hypothesis, $\vartheta f(s) = 0$. Using the expansion formula

$$f(se^t) = \sum_{k=0}^{n} \frac{t^k}{k!} \vartheta^k f(s) + \int_0^t \frac{x^n}{n!} \vartheta^{n+1} f(se^{t-x}) \, dx \qquad (23.17)$$

with $n = 2$ and $t = i\theta$, we obtain

$$f(se^{i\theta}) = f(s) - \tfrac{1}{2}\theta^2 \vartheta^2 f(s) + O(\theta^3 s)$$

because $|\vartheta^3 f(se^{i\theta})| = O(s)$. If $d \to \infty$, the contour integral is

$$\frac{1}{2\pi} \int_{-\pi}^{\pi} \exp\left((2r-d)\left(f(s) - \tfrac{1}{2}\theta^2 \vartheta^2 f(s) + O(\theta^3 s)\right)\right) d\theta$$

$$= \frac{1}{2\pi\sqrt{d}} \int_{-\pi\sqrt{d}}^{\pi\sqrt{d}} \exp\left((2r-d)f(s) - t^2/2 + O(t^2 d/r) + O(t^3 d^{-1/2})\right) dt$$

$$= \frac{F(s)^{2r-d}}{s^d \sqrt{2\pi d}} \left(1 + O(d/r) + O(d^{-1/2})\right), \qquad (23.18)$$

because $\vartheta^2 f(s) = s + O(s^2) = d/(2r-d) + O(d^2/r^2)$. The terms $O(t^2 d/r)$ and $O(t^3 d^{-1/2})$ can safely be moved out of the exponent because they are bounded when $|t| \le d^{1/6}$ and $|t| \le \sqrt{r/d}$. Larger values of $|t|$ are unimportant in the integral because of the factor $e^{-t^2/2}$, and because the relation

$$F(z) = 3 \int_0^1 (1-u)^2 e^{4zu} \, du \qquad (23.19)$$

implies that $|F(z)| \le F(\Re z)$; once the real part is sufficiently small, we can neglect the remaining part of the path.

Equation (23.18) does not match (23.16) perfectly, although it would be sufficient for the applications considered below. To derive the sharper estimate claimed in (23.16) when d is small, we can apply (23.17) to $f(z) - z$ instead of to $f(z)$, obtaining

$$f(se^{i\theta}) - se^{i\theta} = f(s) - s - i\theta s + O(\theta^2 s^2);$$

$$f(se^{i\theta}) = f(s) + s(e^{i\theta} - i\theta - 1) + O(\theta^2 s^2)$$

$$= f(s) + \frac{d}{2r - d}(e^{i\theta} - i\theta - 1) + O\left(\frac{\theta^2 d^2}{r^2}\right).$$

The contour integral without the O term can be evaluated exactly,

$$\frac{1}{2\pi} \int_{-\pi}^{\pi} \exp\left((2r - d)\left(f(s) + (e^{i\theta} - i\theta - 1)d/(2r - d)\right)\right) d\theta$$

$$= \frac{F(s)^{2r-d}}{s^d} \cdot \frac{1}{2\pi} \int_{-\pi}^{\pi} e^{(e^{i\theta} - 1)d} \, d\theta / e^{i\theta d}$$

$$= \frac{F(s)^{2r-d}}{s^d} [z^d] e^{(z-1)d} = \frac{F(s)^{2r-d}}{s^d} \frac{(d/e)^d}{d!}.$$

The O term contributes a relative error of $O(d/r)$, because we have

$$\int_{-\pi}^{\pi} \left|\exp\left((e^{i\theta} - 1 - i\theta)d\right)\right| \theta^2 \, d\theta = \int_{-\pi}^{\pi} e^{(\cos\theta - 1)d} \theta^2 \, d\theta$$

$$\leq \int_{-\pi}^{\pi} e^{-c\theta^2 d} \theta^2 \, d\theta = O(d^{-3/2}),$$

where $c = 2/\pi^2$. $\square$

Theorem 13. *The joint distribution of the excess r and deficiency d of a random multigraph with $m = \frac{n}{2}(1 + \mu)$ edges is approximately normal about the expected values r_μ and d_μ in (23.12) and (23.13), with zero covariance. More precisely, there exists $\epsilon > 0$ such that if*

$$r = r_\mu + \rho\sqrt{\mu^3 n}, \qquad d = d_\mu + \delta\sqrt{\mu^4 n}, \qquad (23.20)$$

the probability that a random multigraph has excess r and deficiency d is

$$\frac{3}{4\pi\sqrt{5}\,\mu^{7/2}n} \exp\left(-\frac{3}{20}\rho^2 - \frac{3}{4}\delta^2\right)$$

$$+ O\left((1 + |\rho| + |\delta|)^2 \mu^{1/2} + \frac{1 + |\rho|^3}{(\mu^3 n)^{1/2}} + \frac{1 + |\delta|^3}{(\mu^4 n)^{1/2}}\right), \qquad (23.21)$$

when $n^{-1/4} \leq \mu \leq \epsilon$ and $n \to \infty$, uniformly for $|\rho| \leq \frac{1}{2}\sqrt{\mu^3 n}$ and $|\delta| \leq \frac{1}{2}\sqrt{\mu^4 n}$.

Proof. Before proving formula (23.21), we can verify that its leading factor yields total probability 1 when integrated over all values of r and d near r_μ and d_μ: The integral over d gives a factor of $\sqrt{4\pi\mu^4 n/3}$, and the integral over r gives a factor of $\sqrt{20\pi\mu^3 n/3}$.

Let r and d be given by (23.20); the probability of excess r and deficiency d is then

$$\frac{2^m m!\, n!\, e_{rd}}{n^{2m}(n-m+r)!\, 2^{n-m+r}}\, [z^n]\, \frac{\left(2-T(z)\right)^{n-m+r}T(z)^{n-m+3r-d}}{\left(1-T(z)\right)^{3r-d+1/2}}. \quad (23.22)$$

We find the coefficient of z^n by evaluating a contour integral as in (10.11) and (21.4); it is

$$\frac{1}{2\pi i}\oint e^{g(z)}(1-z)^{1/2}\frac{dz}{z}, \quad (23.23)$$

$$g(z) = nz + (3r-d)\left(\ln z - \ln(1-z)\right) + r\ln(2-z)$$
$$- m\ln z + (n-m)\ln(2-z). \quad (23.24)$$

The key to Theorem 13 is the fact that, when $\rho=\delta=0$, there is a saddle point at $z=1-\sigma$:

$$\frac{g'(1-\sigma)}{n} = 1 + \frac{\mu+\sigma}{2(1+\mu)}\left(\frac{\sigma(\mu+\sigma)}{1-\sigma}+\frac{\sigma(\mu+\sigma)}{\sigma}-\frac{\mu-\sigma}{1+\sigma}\right)$$
$$-\frac{1+\mu}{2(1-\sigma)}-\frac{1-\mu}{2(1+\sigma)}=0. \quad (23.25)$$

Moreover, $g''(1-\sigma) = 5\mu n + O(\mu^2 n)$ in that case. If we integrate on the path $z = 1-\sigma+it/\sqrt{\mu n}$, as we did in Lemma 7 (Section 21), the logarithm of the result will be

$$g(1-\sigma) + \ln\frac{2^m m!\, n!\, e_{rd}}{n^{2m}(n-m+r)!\, 2^{n-m+r}}\sqrt{\frac{1}{10\pi n}}+O(\mu)+O(\mu^{-3/2}n^{-1/2}),$$

where $r=r_\mu$ and $d=d_\mu$. The relevant quantity s needed in Lemma 8 is

$$s = \frac{\mu+\sigma}{4} \quad (23.26)$$

because of (23.15). The evaluation of the stated logarithm is tedious, but it can be done in a reasonable amount of time with computer assistance, using some simplifications such as

$$3r - d = \frac{\sigma(\mu + \sigma)^2}{2(1 + \mu)}\, n\,, \qquad n - m + r = \frac{1 - \sigma^2}{2(1 + \mu)}\, n\,.$$

The term $\ln\big((6r - 2d)!/(3r - d)!\big)$ from (7.3) can be evaluated as $(3r - d)\ln(3r - d) - 3r + d + (6r - 2d)\ln 2 + \frac{1}{2}\ln 2 + O(\mu)$. It is not difficult to verify that the terms involving $n \ln n$ cancel. There are three terms involving $n \ln \mu$, namely $(3r - d)\ln \mu^2$, $-d \ln \mu$, and $-(2r - d)\ln \mu^3$, coming respectively from within expansions of $(3r - d)\ln(3r - d)$, $-d \ln s$, and $-(2r - d)\ln(2r - d)$; there are two other terms, $-(3r - d)\ln \sigma$ from within $g(1 - \sigma)$ and $+(3r - d)\ln \sigma$ from within $(3r - d)\ln(3r - d)$, which also cancel. The most difficult part of the computation is the sum of about 16 terms that are rational functions in μ and σ, times n; these too sum to zero, using relations (23.14). The net result is that the complicated logarithm sums to $\ln 3 - 2\ln 2 - \ln \pi - \frac{1}{2}\ln 5 - \frac{7}{2}\ln \mu - \ln n + O(\mu) + O(\mu^{-4}n^{-1})$; this proves the theorem when $\rho = \delta = 0$.

For the case of general ρ and δ the calculations are similar but even worse. We now choose the integration path

$$z = 1 - \sigma - \tfrac{3}{5}\,\rho/\sqrt{\mu n} + it/\sqrt{\mu n}\,; \qquad (23.27)$$

the first-order effects of ρ and δ then cancel out, and the second-order effects contribute $-\frac{3}{20}\rho^2 - \frac{3}{4}\delta^2$ to the logarithm of the result.   □

24. A Waiting Game

Now let's consider a little game. Start with an empty multigraph and add edges repeatedly at random until either (1) two different complex components are present; or (2) the multigraph is unclean. Case 1 represents the event "we have left the top line of Figure 1 before leaving the top line of Figure 2."

Let $G_0(w, z)$ be the bgf for all multigraphs such that the game has not yet stopped. Then

$$\sum_m \frac{2^m m!\, n!}{n^{2m}}\, [w^m z^n]\, G_0(w, z) \qquad (24.1)$$

is the expected running time of the game. We have

$$G_0(w, z) = e^{U(wz)/w} \sum_r w^r K_r(wz)\,, \qquad (24.2)$$

where $K_r(z)$ generates all clean cyclic multigraphs, weighted by the probability that they will arise as the cyclic part of a multigraph occurring during the game.

We learned in Section 17 how to compute weighting factors that account for the history of transitions in Figure 1 among clean multigraphs; and we learned more specifically in Section 20 how these coefficients arise as a multigraph gains random edges. In consequence, we can conclude that $K_r(z) = k_r T(z)^{2r}/(1 - T(z))^{3r+1/2}$, where $k_1 = e_1 = \frac{5}{24}$ and the later coefficients obey the rule

$$k_{r+1} = \frac{3}{2}r k_r \,. \tag{24.3}$$

Here's why: Given $k_r T^{2r}/(1 - T)^{3r+1/2}$, the generating function for a clean vertex x is

$$\tfrac{1}{2}k_r T^{2r+1}/(1 - T)^{3r+5/2} + 3rk_r T^{2r+1}/(1 - T)^{3r+5/2}\,, \tag{24.4}$$

where the first term corresponds to cases where x is in the unicyclic part. Similarly, given the generating function $\frac{1}{2}k_r T^{2r+1}/(1 - T)^{3r+5/2}$ after vertex x is chosen to be unicyclic, the generating function for a clean unicyclic y is

$$\left(\tfrac{5}{2}\right)\left(\tfrac{1}{2}\right)k_r T^{2r+2}/(1 - T)^{3r+9/2}\,; \tag{24.5}$$

here $\frac{5}{2} = 1 + 1 + \frac{1}{2}$, for choosing y on the half-edge to x, or on the self-loop attached to that half-edge, or in a different unicyclic component. We obtain a new bicyclic component if and only if both x and y were unicyclic. Therefore the generating function for cases where the game continues is

$$\left((3r + \tfrac{5}{2})(3r + \tfrac{1}{2}) - \tfrac{5}{4}\right)k_r T^{2r+2}/(1 - T)^{3r+9/2}\,.$$

As in (20.9) and (20.10), we multiply by $(1 - T)/(6r + 6)$ to account for merging $\langle x, y \rangle$ with the existing edges. This proves (24.3).

Equation (24.3) implies, of course, that

$$k_r = \frac{5}{36}\left(\frac{3}{2}\right)^r (r - 1)! \,. \tag{24.6}$$

Comparing this to the case $d = 0$ of (7.16), we have

$$k_r = \frac{5\pi}{18}\, e_r \left(1 + O(r^{-1})\right). \tag{24.7}$$

Therefore the similar calculations of Section 22, where we found that $p_1(n) - p_1'(n) = 1 - p_2(n) \sim \frac{1}{2}$, tell us that *the game will stop in Case* (2) *with probability* $\frac{5\pi}{18}$. This provides further evidence in support of the top-line conjecture that was made in Section 18.

We can now try to compute the expected time for the game to be completed, but it appears to be quite complicated. The contribution to (24.1) from a given m and r can be obtained by changing e_r to k_r in (13.17) when m and r are not too large; this means we want to evaluate

$$\sum_{k \geq 0} \sqrt{\frac{2\pi}{3}} \frac{\left(\frac{1}{2} 3^{2/3} \mu\right)^k}{k!} \left(\frac{1}{\Gamma(1/2 - 2k/3)} + \sum_{r \geq 1} \frac{5}{36} \left(\frac{1}{2}\right)^r \frac{(r-1)!}{\Gamma(r+1/2 - 2k/3)} \right) \tag{24.8}$$

in place of (14.1), representing $e^{\mu^3/6}$ times the probability that the game is still alive after m edges. The inner sum is known to be $\frac{5}{72}$ times

$$\sum_{r \geq 0} \left(\frac{1}{2}\right)^r \frac{r!}{\Gamma(r + 3/2 - 2k/3)}$$

$$= \frac{1}{\Gamma(3/2 - 2k/3)} F\left(1, 1; \frac{3}{2} - \frac{2k}{3}; \frac{1}{2}\right)$$

$$= \frac{1}{\Gamma(1/2 - 2k/3)} \left(\psi\left(\frac{3}{4} - \frac{k}{3}\right) - \psi\left(\frac{1}{4} - \frac{k}{3}\right) \right), \tag{24.9}$$

which has the value $\sqrt{\pi}$ when $k = 0$. (Here, as usual, $\psi(z) = \Gamma'(z)/\Gamma(z)$.) Further study of (24.8) should prove to be interesting.

25. Waiting Time in General

Bivariate generating functions provide a useful tool for studying the "first occurrences" of particular graphs or multigraphs, as shown in [14]. The special problems considered in that paper can be put into the following general framework.

Let S be any collection of multigraphs, with bgf $S(w, z)$. Suppose we wish to study the first time that an evolving multigraph on n vertices does not lie in S. If $[z^n] S(0, z) = 0$, the empty graph on n vertices is not in S, so the process never gets started. Otherwise, the probability that an evolving multigraph lies in S when it has $m - 1$ edges but not when it has m is

$$\frac{2^{m-1}(m-1)! \, n!}{n^{2m}} [w^m z^n] (w \vartheta_z^2 - 2\vartheta_w) S(w, z). \tag{25.1}$$

The proof is simple, by definition of the operators ϑ_z and ϑ_w, because the probability in question is

$$\frac{2^{m-1}(m-1)!\,n!}{n^{2(m-1)}}\,[w^{m-1}z^n]\,S(w,z) - \frac{2^m m!\,n!}{n^{2m}}\,[w^m z^n]\,S(w,z)\,.$$

For convenience we shall write

$$\nabla\, S(w,z) = (w\vartheta_z^2 - 2\vartheta_w)\,S(w,z)\,; \tag{25.2}$$

we call ∇S the bgf for "stopping configurations," while S itself is the bgf for "going configurations."

The operator Φ_n, introduced in [14], is

$$\Phi_n F(w,z) = \sum_{m=1}^{\infty} \frac{2^{m-1}(m-1)!\,n!}{n^{2m}}\,[w^m z^n]\,F(w,z)\,. \tag{25.3}$$

Equations (25.1)–(25.3) imply that $\Phi_n\nabla S(w,z)$ is the probability that a stopping configuration will be encountered when some edge is added to an initially empty multigraph. A similar operator

$$\widehat{\Phi}_n\,\widehat{F}(w,z) = \sum_{m=1}^{\infty} \frac{n!}{2m\binom{n(n-1)/2}{m}}\,[w^m z^n]\,\widehat{F}(w,z) \tag{25.4}$$

for graphs instead of multigraphs is considered in [14], but we will restrict consideration to multigraphs for simplicity. (As one might expect from Section 6, we should use the operator

$$\widehat{\nabla} = w(\vartheta_z^2 - \vartheta_z - 2\vartheta_w) - 2\vartheta_w \tag{25.5}$$

in place of ∇ when defining stopping configurations for the graph process.)

Several examples will help clarify these definitions and demonstrate their usefulness. Since the bgf $G(w,z)$ for all multigraphs satisfies $\vartheta_z^2 G = 2w^{-1}\vartheta_w G$, equation (4.2), we have $\nabla G = 0$; this, of course, is obvious, because there are no stopping configurations when all multigraphs are permitted.

Example 1. Let $S(w,z)$ be the bgf for all multigraphs having nothing but self-loops. Clearly $S(w,z) = e^{ze^{w/2}}$, because $ze^{w/2}$ is the bgf for a single vertex with nothing but self-loops. Formula (25.2) now tells us that

$$\nabla S(w,z) = wz^2 e^w e^{ze^{w/2}}\,, \tag{25.6}$$

because $\vartheta_z^2 S = z^2 e^w S + z e^{w/2} S$ and $\vartheta_w S = \frac{w}{2} z e^{w/2} S$. Thus, by (25.1), the probability that an evolving multigraph first fails to lie in S when it acquires the mth edge is

$$\frac{2^{m-1}(m-1)!\,n!}{n^{2m}}\,[w^m z^n]\,wz^2 e^w e^{ze^{w/2}}$$

$$= \frac{2^{m-1}(m-1)!\,n!}{n^{2m}}\,[w^m]\,\frac{we^w e^{(n-2)w/2}}{(n-2)!}$$

$$= \frac{2^{m-1}(m-1)!\,n(n-1)}{n^{2m}}\,[w^{m-1}]\,e^{nw/2} = \frac{n(n-1)}{n^{m+1}}\,.$$

And sure enough, $n^{1-m} - n^{-m}$ is obviously the probability that a random sequence of edges $\langle x_1, y_1 \rangle \ldots \langle x_m, y_m \rangle$ will have $x_1 = y_1, \ldots,$ $x_{m-1} = y_{m-1}$, $x_m \neq y_m$.

Example 2. Let $S(w, z)$ be the bgf for all acyclic multigraphs, namely $e^{U(w,z)} = e^{U(wz)/w}$. The formulas

$$\vartheta_z U = w^{-1} T, \quad \vartheta_z^2 U = w^{-1} T/(1-T), \quad \vartheta_w U = \tfrac{1}{2} w^{-1} T^2 \qquad (25.7)$$

were derived in Section 4, and we have

$$\vartheta_z^2 e^F = (\vartheta_z^2 F) e^F + (\vartheta_z F)^2 e^F \qquad (25.8)$$

for any $F = F(w, z)$; hence

$$\nabla e^U = \frac{T}{1-T}\,e^U\,. \qquad (25.9)$$

These are the stopping configurations that define the appearance of the first cycle in an evolving multigraph. The term $T^k e^U$ corresponds to a first cycle of length k; therefore if we replace T^k by kT^k and sum over all stopping times, we get an expression for the expected length of the first cycle,

$$\Phi_n \frac{T}{(1-T)^2}\,e^U\,. \qquad (25.10)$$

The first cycle was one of the main topics studied in [14], where it was shown that the expected length is proportional to $n^{1/6}$ although the standard deviation is proportional to $n^{1/4}$.

Example 3. Let $S(w, z) = U(w, z)$ be the bgf for unrooted trees. This is a perverse example, thrown in primarily because (25.7) gives us the information we need to calculate

$$\nabla U = \frac{T}{1 - T} - \frac{T^2}{w}$$

$$= wz + (-w + 2w^2)z^2 + (-2w^2 + \tfrac{9}{2}w^3)z^3 + \cdots . \quad (25.11)$$

What is the meaning of these negative coefficients?

The example does make sense, if we rephrase our interpretation of (25.1). The exact meaning of

$$\frac{2^{m-1}(m-1)!\, n!}{n^{2m}}\, [w^m z^n]\, \nabla S(w, z)$$

is, "the probability that an evolving multigraph leaves S when the mth edge is added, minus the probability that it enters S when the mth edge is added." In our example, $U(0, z) = z$; when there are two or more vertices, the empty multigraph is not a tree, but it can become one later. The bgf for becoming a tree is $w^{-1}T^2$, corresponding to an ordered pair of rooted trees with $m - 1$ edges. The bgf for adding a new edge $\langle x, y \rangle$ to a tree is $\sum_{k \geq 1} T^k$, where the term T^k corresponds to cases where x and y are at distance k. (Each appearance of $T = T(wz)$ includes an implicit edge touching the tree root, because w and z appear with equal powers in every term.)

Example 3 cautions us to interpret the operators ∇ and Φ_n a bit more carefully. In general, we have the identity

$$\Phi_n \nabla S(w, z) = n!\, [z^n]\, S(0, z) - \lim_{m \to \infty} \frac{2^m m!\, n!}{n^{2m}}\, [w^m z^n]\, S(w, z), \quad (25.12)$$

for any bgf $S(w, z)$ such that the limit exists, because

$$\Phi_n \nabla S(w, z) = \sum_{m=1}^{\infty} \left(\frac{2^{m-1}(m-1)!\, n!}{n^{2(m-1)}}\, [w^{m-1} z^n]\, S(w, z) \right.$$

$$\left. - \frac{2^m m!\, n!}{n^{2m}}\, [w^m z^n]\, S(w, z) \right).$$

A sufficient condition for the limit to exist is that the coefficients of $\nabla S(w, z)$ are nonnegative. A sufficient condition for the coefficients to be nonnegative is that $S(w, z)$ should represent a family of multigraphs S with the property that the deletion of any edge from a member of S yields another member of S.

Example 4. Let $S(w, z) = G(w, z) - C(w, z)$ be the bgf for all disconnected multigraphs. The stopping configurations now represent the first time an evolving multigraph becomes connected. Since $G(w, z) = e^{C(w,z)}$, we have

$$\vartheta_w C = \vartheta_w \ln G = (\vartheta_w G)/G\,;$$
$$\vartheta_z C = \vartheta_z \ln G = (\vartheta_z G)/G\,;$$
$$\vartheta_z^2 C = (\vartheta_z^2 G)/G - (\vartheta_z G)^2/G^2\,;$$

hence

$$\nabla S = \nabla G - \nabla C = w(\vartheta_z C)^2\,. \tag{25.13}$$

Of course! This is an edge that joins an ordered pair of vertices marked in distinct components.

Example 5. Let $S(w, z)$ be any bgf of the form

$$S(w, z) = e^{U(w,z)+V(w,z)} H(w, z)\,. \tag{25.14}$$

Then we can use (25.7) and (4.9) to compute

$$\nabla S = e^{U+V}\left((2T\vartheta_z - 2\vartheta_w + we^{-V}\vartheta_z^2 e^V)H\right)\,. \tag{25.15}$$

For example, when $S(w, z) = G(w, z)$, the left side of (25.15) is zero, and $H(w, z)$ is the bgf we have called $E(w, z)$. Equating the right side of (25.15) to zero gives the differential equation (5.1) that we originally used to compute $E(w, z)$.

In the special case $H(w, z) = 1$, the stopping configurations correspond to the first time an evolving multigraph acquires a bicyclic component, that is, the time when its excess changes from 0 to 1. This is another problem that was considered in [14], where it was shown that the expected number of unicyclic components present at the time of first excess is $\frac{1}{6} \ln n + O(1)$.

If we express H in terms of univariate generating functions,

$$H(w, z) = \sum_{r \geq 0} w^r H_r(wz)\,, \tag{25.16}$$

then (25.15) can be written

$$\nabla S = e^{U+V} \sum_{r \geq 1} w^r \nabla H_r(wz)\,, \tag{25.17}$$

where the univariate function $H_r(z)$ is related to (5.3):

$$\nabla H_r(z) = e^{-V}\vartheta^2 e^V H_{r-1}(z) - 2(r + (1 - T)\vartheta)H_r(z)\,. \tag{25.18}$$

Example 6. Specializing Example 5 further, let

$$S(w, z) = e^{U(w,z)+V(w,z)} \sum_{r=0}^{R} w^r E_r(wz), \qquad (25.19)$$

where R is any nonnegative integer. Then the stopping configurations ∇S represent the time when an evolving multigraph first acquires excess $R + 1$. Expression (25.18) becomes almost trivial because ∇H_r is zero for all $r \neq R + 1$; we have

$$\nabla S(w, z) = w^{R+1} e^{U(w,z)} \vartheta_z^2 e^{V(wz)} E_R(wz). \qquad (25.20)$$

This family S has the property that $\Phi_n \nabla S = 1$, by (25.12), because a multigraph surely acquires excess $R + 1$ at some time $m \leq n + R + 1$. We can write the identity $\Phi_n \nabla S = 1$ more explicitly, using our known formula for E_R, and using r in place of R: The formula

$$\Phi_n \left(w^{r+1} e^{U(wz)/w} \vartheta_z^2 \sum_{d=0}^{2r} e_{rd} \frac{T(wz)^{2r-d}}{(1 - T(wz))^{3r-d+1/2}} \right) = 1 \qquad (25.21)$$

holds for all $n \geq 1$ and $r \geq 0$. Moreover, we can write (25.20) in the form

$$\nabla S(w, z) = 2w^{R+1} e^{U(w,z)+V(wz)} \big(R + 1 + (1 - T)\vartheta_z\big) E_{R+1}(wz),$$

using (5.3). Setting $r = R + 1$ and applying (20.9) gives us another way to express (25.21),

$$\Phi_n \left(w^r e^{U(wz)/w} \sum_{d=0}^{2r} (6r-2d) e_{rd} \frac{T(wz)^{2r-d}}{(1 - T(wz))^{3r-d+3/2}} \right) = 1, \qquad (25.22)$$

for all $n \geq 1$ and $r \geq 1$.

For example, the case $r = 1$ of (25.22) is

$$\Phi_n \left(we^U \left(\frac{5}{4} \frac{T^2}{(1-T)^{9/2}} + \frac{1}{2} \frac{T}{(1-T)^{7/2}} \right) \right) = 1. \qquad (25.23)$$

The operator Φ_n is defined in (25.3) to be a sum over m, and the mth term of (25.23) is

$$\frac{2^{m-1}(m-1)! \, n!}{n^{2m}} [w^m z^n] \, we^{U(w,z)} f\big(T(wz)\big)$$

$$= \frac{1}{2m(n-m+1)} \frac{2^m m! \, n!}{n^{2m}(n-m)!} [z^n] \, U(z)^{n-m+1} f\big(T(z)\big)$$

$$= \frac{1}{4m(n-m+1)} \frac{2^m m! \, n!}{n^{2m}(n-m)!} [z^n] \, U(z)^{n-m} g\big(T(z)\big), \qquad (25.24)$$

where $f(T) = \frac{5}{4}T^2/(1-T)^{9/2} + \frac{1}{2}T/(1-T)^{7/2}$ and $g(T) = (2-T)Tf(T)$. We can write

$$g(T) = \frac{5/4}{(1-T)^{9/2}} - \frac{2}{(1-T)^{7/2}} - \frac{1/2}{(1-T)^{5/2}} + \frac{2}{(1-T)^{3/2}} - \frac{3/4}{(1-T)^{1/2}},$$

so we can evaluate (25.24) by summing five applications of formula (10.1). The value is negligibly small unless m is $\frac{1}{2}n + O(n^{2/3})$, hence the factor $4m(n - m + 1)$ can be assumed to equal $n^2 + O(n^{5/3})$. The five terms of g yield values of order $n^{4/3}$, n, $n^{2/3}$, $n^{1/3}$, and 1 respectively, according to (10.1); thus the leading term $\frac{5}{4}/(1-T)^{9/2}$ must be responsible for the major contribution to (25.23), and the mth term of the sum when $m = \frac{1}{2}n + \frac{1}{2}\mu n^{2/3}$ will be

$$\tfrac{5}{4}n^{-2/3}\sqrt{2\pi}\,A(\tfrac{9}{2},\mu) + O(n^{-1}).$$

Summing over m yields 1. Therefore it must be true that

$$\int_{-\infty}^{\infty} A\left(\frac{9}{2},\mu\right) d\mu = \frac{8/5}{\sqrt{2\pi}}.$$

This integral formula is not at all obvious from the definition of $A(y,\mu)$ in (10.2), and it would be interesting to find a direct proof.

The argument we have just given can be extended to arbitrary r, starting with (25.22), and it implies the following remarkable result:

$$\int_{-\infty}^{\infty} A\left(3r + \tfrac{3}{2},\mu\right) d\mu = \frac{1}{3re_r\sqrt{2\pi}}, \qquad \text{integer } r \geq 1. \qquad (25.25)$$

By (8.17) we can also write

$$\int_{-\infty}^{\infty} A\left(3r + \tfrac{3}{2},\mu\right) d\mu = \frac{1}{3}\left(\frac{2}{3}\right)^r \frac{\Gamma(r)\sqrt{2\pi}}{\Gamma(r+\frac{5}{6})\Gamma(r+\frac{1}{6})}, \qquad \text{integer } r \geq 1. \quad (25.26)$$

Let $M_{r,n} = \frac{1}{2}n + \frac{1}{2}U_{r,n}n^{2/3}$ be the number of edges when the excess first reaches r. We have just proved that

$$\Pr(M_{r,n} = m) = 6re_r\sqrt{2\pi}\,A\left(3r + \tfrac{3}{2},\mu\right)n^{-2/3} + O(n^{-1}). \qquad (25.27)$$

Hence $U_{r,n} \to U_r$ in distribution, where U_r has the density function

$$f_r(\mu) = 3re_r\sqrt{2\pi}\,A\left(3r + \tfrac{3}{2},\mu\right), \qquad -\infty < \mu < \infty. \qquad (25.28)$$

Combining this formula with (13.17), we have

$$\sqrt{2\pi}\, e_r A(3r + \tfrac{1}{2}, \mu) = \lim_{n\to\infty} \Pr(\mathcal{E}_r) = \lim_{n\to\infty} \Pr(M_{r,n} \le m < M_{r+1,n})$$

$$= \int_{-\infty}^{\mu} \left(f_r(u) - f_{r+1}(u) \right) du \,,$$

whence

$$\sqrt{2\pi}\, e_r\, A'(3r + \tfrac{1}{2}, \mu) = f_r(\mu) - f_{r+1}(\mu). \tag{25.29}$$

In fact, (25.29) can be derived also by setting $y = 3r + \tfrac{1}{2}$ in the formula

$$A'(y, \mu) = (y - \tfrac{1}{2})A(y + 1, \mu) - \tfrac{1}{2} y(y + 2)A(y + 4, \mu), \tag{25.30}$$

which is a consequence of (10.22) and (10.23). (Differentiation in these formulas is with respect to μ.)

26. Continuous Excess

Let $I(y)$ be the integral in (25.25) when the parameter r is not necessarily an integer:

$$I(y) = \int_{-\infty}^{\infty} A(y, \mu)\, d\mu\,. \tag{26.1}$$

It is natural to conjecture that formula (25.26) holds for y in general:

$$I(y) = \frac{2^{y/3}\sqrt{\pi}\,\Gamma(y/3 - 1/2)}{3^{y/3+1/2}\,\Gamma(y/3 + 1/3)\,\Gamma(y/3 - 1/3)}\,, \qquad y > \frac{3}{2}. \tag{26.2}$$

(The condition $y > \tfrac{3}{2}$ is necessary and sufficient for convergence of the integral, because of (10.3) and (10.4).) And indeed, this conjecture is true.

Theorem 14. *The integral* (26.1) *has the closed form* (26.2).

Proof. Let $I_0(y)$ be the right-hand side of (26.2); we wish to show that $I(y) = I_0(y)$. Clearly

$$I_0(y + 3) = \frac{2y - 3}{(y + 1)(y - 1)}\, I_0(y), \qquad y > \frac{3}{2}. \tag{26.3}$$

Since $\int_{-\infty}^{\infty} A'(y, \mu)\, d\mu = 0$ for $y > \tfrac{1}{2}$, by (10.3) and (10.4), we can integrate (25.30) and replace y by $y - 1$ to get the same recurrence for $I(y)$:

$$I(y + 3) = \frac{2y - 3}{(y + 1)(y - 1)}\, I(y), \qquad y > \frac{3}{2}. \tag{26.4}$$

Therefore $I(y)/I_0(y)$ is a periodic function, and we need only prove asymptotic equivalence $I(y) \sim I_0(y)$ as $y \to \infty$ in order to verify strict equality $I(y) = I_0(y)$ for all $y > \tfrac{3}{2}$.

The duplication and triplication formulas for the Gamma function provide us with an alternative expression for $I_0(y)$:

$$I_0(3y) = \left(\frac{9}{2}\right)^{y-1} \frac{\Gamma(2y-1)}{\Gamma(3y-1)} \sim \frac{1}{\sqrt{6}} \left(\frac{2e}{3y}\right)^y. \tag{26.5}$$

To show that $I(y)$ has the same asymptotic behavior, we break the integral into two parts,

$$I(y) = \int_{-\infty}^0 A(y,\mu)\,d\mu + \int_0^\infty A(y,\mu)\,d\mu = I_-(y) + I_+(y). \tag{26.6}$$

By definition (10.2) we have

$$I_+(y) = \frac{1}{3^{(y+1)/3}} \int_0^\infty \sum_{k\geq 0} \frac{e^{-\mu^3/6}\left(\frac{1}{2}3^{2/3}\mu\right)^k d\mu}{k!\,\Gamma((y+1-2k)/3)}; \tag{26.7}$$

we will show that the asymptotic value of $I_+(y)$ can be obtained by interchanging summation and integration, then estimating the resulting sum.

Let a_k be the kth term after integration,

$$\begin{aligned} a_k &= \int_0^\infty \frac{e^{-\mu^3/6}\left(\frac{1}{2}3^{2/3}\mu\right)^k d\mu}{k!\,\Gamma((y+1-2k)/3)} \\ &= \frac{2^{(1-2k)/3}3^{k-2/3}\,\Gamma((k+1)/3)}{k!\,\Gamma((y+1-2k)/3)}. \end{aligned} \tag{26.8}$$

If $a_k = 0$ then $a_{k+3} = 0$; otherwise we have

$$\frac{a_{k+3}}{a_k} = \frac{(2k+5-y)(2k+2-y)}{4(k+2)(k+3)}, \tag{26.9}$$

which is greater than 1 when $k < \frac{1}{4}y - \frac{17}{8} - \frac{5}{4}(y+\frac{3}{2})^{-1}$, less than 1 when k exceeds that value, and nonnegative except for one or two values of k near $\frac{1}{2}y$. So the largest terms a_k occur when k is near $\frac{1}{4}y$. If $y > 5$ and $k > y/2$, we have

$$\left|\frac{a_{k+3}}{a_k}\right| \leq \left(1 - \frac{y-5}{2k}\right)^2 \leq \left(e^{3/k}\right)^{(5-y)/3} \leq \left(\frac{k+3}{k}\right)^{(5-y)/3},$$

and it follows that $a_k = O(k^{(5-y)/3})$ as $k \to \infty$. Therefore $\sum |a_k|$ exists, and the interchange of summation and integration is justified, at least for large y.

Let $k = \frac{1}{4}y + x$, where $|x| \leq y^{1/2+\epsilon}$. Then Stirling's formula tells us that

$$\ln a_k = \frac{y+3}{3} \ln 2 + \frac{y-2}{3} \ln 3 - \frac{2y+3}{6} \ln y$$
$$+ \frac{y}{3} - \ln \sqrt{\pi} - \frac{8x^2}{3y} + O(y^{3\epsilon-1/2}). \qquad (26.10)$$

If $0 < \epsilon < \frac{1}{6}$, this estimate implies that the sum of all terms for $|x| > y^{1/2+\epsilon}$ is superpolynomially small in relation to the sum of terms for $|x| \leq y^{1/2+\epsilon}$; hence

$$\sum_{k=0}^{\infty} a_k \sim \frac{2^{(y+3)/3} 3^{(y-2)/3} e^{y/3}}{y^{(2y+3)/6} \sqrt{\pi}} \int_{-\infty}^{\infty} e^{-8x^2/(3y)} \, dx$$

and we have

$$I_+(y) = \frac{1}{3^{(y+1)/3}} \sum_{k=0}^{\infty} a_k \sim \frac{1}{\sqrt{6}} \left(\frac{2e}{y}\right)^{y/3} \sim I_0(y). \qquad (26.11)$$

The proof of (26.2) will therefore be complete if we can show that $I_-(y)/I_+(y) \to 0$ as $y \to \infty$. First, we can use (10.9) to show that

$$A(y, -\alpha) \leq \frac{1}{2\pi} \alpha^{1/2-y} \int_{-\infty}^{\infty} \left|1 + \frac{it}{\alpha^{3/2}}\right|^{1-y} e^{-t^2/2} \, dt \leq \frac{\alpha^{1/2-y}}{\sqrt{2\pi}} \, ;$$

therefore the first portion of $I_-(y)$ is quite small,

$$\int_{-\infty}^{-y^{1/3}} A(y, \mu) \, d\mu \leq \frac{1}{\sqrt{2\pi}} \int_{y^{1/3}}^{\infty} \alpha^{1/2-y} \, d\alpha = O(y^{-1/2-y/3}).$$

On the other hand when $-y^{1/3} \leq \mu \leq 0$ we can integrate (10.7) from $y^{1/3} - i\infty$ to $y^{1/3} + i\infty$, obtaining

$$A(y, \mu) \leq \frac{1}{2\pi} \int_{-\infty}^{\infty} |y^{1/3} + it|^{1-y} e^{\Re K(\mu, y^{1/3}+it)} \, dt$$
$$\leq \frac{1}{2\pi} y^{(1-y)/3} e^{K(\mu, y^{1/3})} \int_{-\infty}^{\infty} \exp\left(-\left(y^{1/3} + \frac{\mu}{2}\right)t^2\right) dt$$
$$= \frac{y^{(1-y)/3} \exp\left(y/3 + \mu(3y^{2/3} - \mu^2)/6\right)}{\sqrt{2\pi (2y^{1/3} + \mu)}} \leq \frac{y^{1/6}}{\sqrt{2\pi}} \left(\frac{e}{y}\right)^{y/3} \, ;$$

hence

$$\int_{-y^{1/3}}^{0} A(y,\mu)\,d\mu = O\left(y^{1/2}\left(\frac{e}{y}\right)^{y/3}\right)$$

and $I_-(y) \ll I_+(y)$ as desired. $\square$

Theorem 14 sheds further light on the results of [14], where the first cycle of a random multigraph was shown to have average length asymptotic to $\sqrt{\pi/2}\,I(2)\,n^{1/6}$. According to a lengthy numerical calculation sketched in the appendix to that paper, this coefficient was determined to be 2.0337, correct to four decimal places. "Substantially faster methods would need to be devised if we wanted to calculate K to, say, 100 decimal places." Sure enough, equation (26.2) now confirms that the exact value is

$$K = \frac{\pi^{1/2}\,\Gamma(1/3)}{2^{1/6}\,3^{2/3}}$$

$$= 2.03369\ 20140\ 63898\ 89186\ 17247\ 01028\ 49830\ 16693- \ ; \quad (26.12)$$

and we could obtain thousands of further digits without great difficulty.

Section 7 of [14] also proves implicitly that, if the random variables L and S are respectively the length of the first cycle and the size of the component containing that cycle, we have

$$\mathrm{E}_n\,L^k \ \sim \ \sqrt{\frac{\pi}{2}}\,k!\,I(k+1)\,n^{k/3-1/6}\ ; \qquad (26.13)$$

$$\mathrm{E}_n\,S^k \ \sim \ 2^{k-1/2}\,\Gamma\!\left(k+\tfrac{1}{2}\right)I(2k+1)\,n^{2k/3-1/6}. \qquad (26.14)$$

In particular, the variance of L is asymptotically $\sqrt{2\pi n}$; the asymptotic mean and variance of S are $\sqrt{\pi n/2}$ and $K n^{7/6}$, where K is the constant in (26.12). For graphs instead of multigraphs, these coefficients should all be multiplied by $e^{3/4}$.

Notice that $I(3) = 1$. Hence the function $A(3,\mu)$, which is expressible in terms of Airy series or Bessel functions $\bigl($see (10.32)$\bigr)$, defines a probability density.

Let V_y be a random variable with density function $A(y,\mu)/I(y)$, assuming that $y > \frac{3}{2}$. Then, by (10.22),

$$\mathrm{E}\,V_y = \int_{-\infty}^{\infty} \frac{\mu\,A(y,\mu)\,d\mu}{I(y)} = \frac{y\,I(y+2) - I(y-1)}{I(y)}$$

$$= \frac{(y-3)I(y-1)}{(y-2)I(y)}\,, \qquad (26.15)$$

if $y > \frac{5}{2}$. In particular, the variable U_r of (25.28), which is $V_{3r+3/2}$, has the mean value

$$\frac{(3r - 3/2)I(3r + 1/2)}{(3r - 1/2)I(3r + 3/2)} = \left(\frac{3}{4}\right)^{1/3} \frac{\Gamma(2r - 2/3)}{\Gamma(2r - 1)}. \tag{26.16}$$

This is the limit as $n \to \infty$ of $E\,U_{r,n}$, which represents the mean waiting time $\frac{1}{2}n + \frac{1}{2}E\,U_r n^{2/3}$ for a graph or multigraph to reach excess r. The values are 0.8113, 1.2621, 1.5191, 1.7104, 1.8666, 2.0002, 2.1181, 2.2241, 2.3209, 2.4102, when $1 \le r \le 10$.

Similarly, (10.23) implies that

$$E\,V_y^2 = \frac{I(y-2)}{I(y)} = \frac{(y-2)}{6^{1/3}} \frac{\Gamma\big((2y - 7)/3\big)}{\Gamma(2y/3 - 2)}, \quad \text{if } y > \tfrac{7}{2}. \tag{26.17}$$

Hence
$$E\,V_y = (y/2)^{1/3}\big(1 - \tfrac{7}{6}y^{-1} + O(y^{-2})\big),$$
$$E\,V_y^2 = (y/2)^{2/3}\big(1 - \tfrac{2}{3}y^{-1} + O(y^{-2})\big),$$

and we have
$$\mathrm{Var}\,V_y = \frac{5}{2^{2/3}\,3}\,y^{-1/3} + O(y^{-4/3}). \tag{26.18}$$

Let us now set $\mu = (y/2)^{1/3} + \sigma z$, where

$$\sigma^2 = \frac{5}{2^{2/3}\,3}\,y^{-1/3}. \tag{26.19}$$

An argument similar to the derivation of (26.11) proves that

$$\frac{A(y,\mu)}{I(y)} \sim \frac{1}{\sqrt{2\pi\sigma^2}}\,e^{-z^2/2}, \qquad z = O(1), \quad y \to \infty. \tag{26.20}$$

Therefore $(\mathrm{Var}\,V_y)^{-1/2}(V_y - E\,V_y)$ approaches the normal distribution $N(0,1)$ as $y \to \infty$. In particular, this establishes a kind of asymptotic normality of $U_{r,n}$ (and $M_{r,n}$), if we first let $n \to \infty$ and then $r \to \infty$.

27. Proof of the Top-Line Conjecture

We are almost ready to settle the conjecture that was made in Section 18, but first we should carry out the promised refinement of our estimates (23.5) and (23.9) for the sizes of the acyclic and unicyclic parts of a random multigraph.

The first step is to consider the quantity (23.3), when $m = \frac{1}{2}n(1+\mu)$ and $k = \kappa n$. If $k \geq m$ or $k \geq n$, expression (23.3) is zero; otherwise $0 \leq \kappa < \min\left(\frac{1+\mu}{2}, 1\right)$, and Stirling's approximation yields

$$e^k \sqrt{\frac{m-k}{m}} \sqrt{\frac{n-k}{n}} \frac{2^k \, m^{\frac{k}{}} n^{\frac{k}{}}}{(n-k)^{2k}} \left(\frac{n-k}{n}\right)^{2m}$$

$$= \exp\left(nf(\kappa, \mu) + O\left(\frac{1}{m-k}\right) + O\left(\frac{1}{n-k}\right)\right), \quad (27.1)$$

where

$$f(\kappa, \mu) = \frac{1+\mu}{2} \ln(1+\mu) - \frac{(1+\mu-2k)}{2} \ln(1+\mu-2\kappa)$$
$$+ (\mu - \kappa) \ln(1-\kappa) - \kappa. \quad (27.2)$$

Notice that

$$\frac{\partial f(\kappa, \mu)}{\partial \kappa} = \ln(1+\mu-2\kappa) - \ln(1-\kappa) + \frac{\kappa - \mu}{1-\kappa},$$

$$\frac{\partial^2 f(\kappa, \mu)}{\partial \kappa^2} = \frac{(1-\mu)(\mu-\kappa)}{(1+\mu-2\kappa)(1-\kappa)^2};$$

so both first and second derivatives vanish when $\kappa = \mu$. The first derivative is ≤ 0 when $\kappa = 0$; if $0 < \mu < 1$ it increases to zero when $\kappa = \mu$, then becomes negative; if $\mu \leq 0$ or $\mu \geq 1$ it decreases steadily. Thus $f(\kappa, \mu)$ is a decreasing function of κ, as claimed in Section 23.

We also have

$$\frac{\partial f(\kappa, \mu)}{\partial \mu} = \frac{\ln(1+\mu)}{2} - \frac{\ln(1+\mu-2\kappa)}{2} + \ln(1-\kappa);$$

this derivative decreases steadily, passing through zero when μ reaches $\kappa/(2-\kappa)$. Therefore we have

$$f(\kappa, \mu) \leq f\left(\kappa, \frac{\kappa}{2-\kappa}\right)$$

$$= -\left(\frac{\kappa^3}{24} + \frac{\kappa^4}{24} + \frac{11\kappa^5}{320} + \cdots + \frac{1-2j/2^j}{j(j-1)} \kappa^j + \cdots\right), \quad (27.3)$$

for all $\mu > 2\kappa - 1$. In particular, we can conclude that terms like (23.1) and (23.8) are superpolynomially small for all $k \geq n^{2/3+\epsilon}$, since they are $O\left(\exp(-n^{3\epsilon}/24)\right)$ when $k = n^{2/3+\epsilon}$.

Our next goal is to estimate the sum of (23.8) for $k \geq 1$ when $\mu \geq n^{-1/3}$. This sum $V(m,n)$ is the expected number of vertices in unicyclic components after m steps of the multigraph process. The formulas above allow us to write

$$V(m,n) = \sum_{k \leq n^{2/3+\epsilon}} \frac{1}{2} \frac{k^k Q(k)}{k! \, e^k} \sqrt{\frac{m}{m-k}} \sqrt{\frac{n}{n-k}} \, e^{nf(k/n,\mu)+O(n^{-1})}$$

$$+ O(e^{-n^\epsilon})$$

$$= \sum_{k \leq n^{2/3+\epsilon}} \frac{1}{2} \frac{k^k Q(k)}{k! \, e^k} \exp\left(k\big(\ln(1+\mu) - \mu\big) + \frac{\mu k^2}{2n} - \frac{k^3}{6n^2} \right.$$

$$\left. + O\left(\frac{\mu^2 k^2}{n} + \frac{k^4}{n^3} + \frac{k}{n} \right) \right) + O(e^{-n^\epsilon}). \qquad (27.4)$$

Let $\mu = \alpha n^{-1/3}$, so that α is the quantity we called μ in Sections 10–20 above. We will assume that $\alpha \geq 1$, and also that $\alpha \leq cn^{1/3}$ (hence $\mu \leq c$), where c is a sufficiently small constant. The terms of $V(m,n)$ are negligible for $k \geq n^{2/3+\epsilon}$, regardless of the value of μ; and when $n^{-1/3} \leq \mu \leq c$ we can in fact ignore all terms for $k > \alpha^\epsilon \mu^{-2}$. The reason is that

$$k\big(\ln(1+\mu) - \mu\big) + \frac{\mu k^2}{2n} - \frac{k^3}{6n^2} = \frac{-\mu^2 k}{2} \left(\frac{1}{4} + \frac{1}{3}\left(\frac{3}{2} - \frac{k}{\mu n} \right)^2 \right) + O(k\mu^3)$$

$$\leq \frac{-\mu^2 k}{8} (1 + O(\mu)) \leq \frac{-\mu^2 k}{100}$$

if we choose c small enough. The sum of $O(e^{-\mu^2 k/100})$ for $\alpha^\epsilon/\mu^2 < k < \infty$ is then $O(\mu^{-2} e^{-\alpha^\epsilon/100})$, which is dominated by the error bounds we will encounter below.

When $k \leq \alpha^\epsilon \mu^{-2}$, we have $\mu k^2/2n \leq \alpha^{2\epsilon-3}/2 \leq 1/2$. Therefore we are justified in moving terms out of the exponent in (27.4):

$$V(m,n) = \sum_{k \geq 1} \frac{1}{2} \frac{k^k Q(k)}{k!} \big((1+\mu)e^{-(1+\mu)}\big)^k \left(1 + \frac{\mu k^2}{2n} - \frac{k^3}{6n^2} \right.$$

$$\left. + O\left(\frac{\mu k^2}{2n} - \frac{k^3}{6n^2} \right)^2 + O\left(\frac{\mu^2 k^2}{n} + \frac{k^4}{n^3} + \frac{k}{n} \right) \right)$$

$$= \sum_{k \geq 1} \frac{1}{2} \frac{k^k Q(k)}{k!} \big((1-\sigma)e^{-(1-\sigma)}\big)^k \left(1 + \frac{\mu k^2}{2n} \right.$$

$$\left. + O(\alpha^{4\epsilon-6}) + O\left(\frac{\alpha^{2\epsilon-2}}{n^{1/3}} \right) \right). \qquad (27.5)$$

Here σ is the "shadow" of μ as in (23.4) and (23.10), and the error bounds are computed under the assumption $k \leq a^\epsilon/\mu^2$. The trick of (23.5) and (23.9) now applies, using (23.7), and we have

$$V(m,n) =$$

$$= \frac{1}{2}\left(1 + O(\alpha^{4\epsilon-6}) + O\left(\frac{\alpha^{2\epsilon-2}}{n^{1/3}}\right) + \frac{\mu}{2n}\vartheta^2\right)\frac{T\big((1-\sigma)e^{-(1-\sigma)}\big)}{\big(1 - T\big((1-\sigma)e^{-(1-\sigma)}\big)\big)^2}$$

$$= \frac{1}{2}\left(\left(1 + O(\alpha^{4\epsilon-6}) + O\left(\frac{\alpha^{2\epsilon-2}}{n^{1/3}}\right)\right)\frac{1-\sigma}{\sigma^2}\right.$$

$$\left. + \left(\frac{\mu}{2n}\frac{8 - 17\sigma + 11\sigma^2 - 2\sigma^3}{\sigma^6}\right)\right). \qquad (27.6)$$

If we had expanded the summand further, we would have obtained still more accuracy; therefore we are allowed to set $\epsilon = 0$ in (27.6). The term $O(\alpha^{-2}n^{-1/3})$ dominates $O(\alpha^{-6})$ when $\alpha \geq n^{1/12}$; it comes from both $O(\mu^2k^2/n)$ and $O(k/n)$ in (27.5).

We are assuming that μ is small, hence $\sigma = \mu\big(1 + O(\mu)\big)$. Thus (27.6) can be simplified to

$$V(m,n) = \left(\frac{1}{2\alpha^2} + \frac{2}{\alpha^5} + O\left(\frac{1}{\alpha^8}\right)\right)n^{2/3}\big(1 + O(\mu)\big),$$

and with an extension of the same approach we obtain an asymptotic expansion that begins

$$V(m,n) = \left(\frac{1}{2\alpha^2} + \frac{2}{\alpha^5} + \frac{20}{\alpha^8} + \frac{320}{\alpha^{11}} + O\left(\frac{1}{\alpha^{14}}\right)\right)n^{2/3}\big(1+O(\mu)\big). \quad (27.7)$$

This expansion is readily computed if we note that

$$\vartheta^k \frac{T(z)}{\big(1 - T(z)\big)^2} = \frac{2^k\,k!}{\big(1 - T(z)\big)^{2k+2}} + \cdots, \qquad (27.8)$$

where the remaining terms $a_{k1}/\big(1 - T(z)\big)^{2k+1} + a_{k2}/\big(1 - T(z)\big)^{2k} + \cdots$ are negligible when we replace $T(z)$ by $1 - \mu - O(\mu^2)$. The asymptotic series in (27.7) is obtained also from the integral

$$\frac{1}{4}\int_0^\infty e^{-\alpha^2 t/2 + \alpha t^2/2 - t^3/6}\,dt = \frac{1}{4}\int_0^\infty e^{(\alpha-t)^3/6 - \alpha^3/6}\,dt, \qquad (27.9)$$

because we can expand $e^{\alpha t^2/2 - t^3/6}$ into powers of t and use the formula

$$\int_0^\infty e^{-\alpha^2 t/2}t^k\,dt = \frac{2^{k+1}\,k!}{\alpha^{2k+2}}, \qquad (27.10)$$

which matches (27.8). The coefficients of (27.7) follow a simple pattern; for example, $320 = 16 \cdot 10 \cdot 4 / 2$. Thus we are led to conjecture the asymptotic series

$$\int_0^\infty e^{(\alpha-t)^3/6 - \alpha^3/6}\, dt \sim 2F\left(\tfrac{2}{3}, 1;\; ; 6/\alpha^3\right)/\alpha^2 \quad \text{as} \quad \alpha \to \infty; \quad (27.11)$$

the right-hand side here is a formal power series that diverges for all finite α. And indeed, this conjecture is true, as we will see momentarily.

A similar calculation allows us to estimate $U(m, n)$, the expected number of vertices in trees. The analog of (27.5) is

$$U(m, n) = \frac{n}{1 + \mu} \sum_{k \geq 1} \frac{k^{k-1}}{k!} \left((1 - \sigma) e^{-(1-\sigma)}\right)^k$$
$$\times \left(1 + \frac{\mu k^2}{2n} + O(k\alpha^{3\epsilon-4} n^{-2/3}) + O(k\alpha^\epsilon n^{-1})\right); \quad (27.5')$$

we leave a factor of k in the O terms because it will lead to a better final estimate. Then the analogs of (27.6)–(27.10) are

$$U(m, n) = \frac{n}{1 + \mu}\left(1 - \sigma + O(\alpha^{3\epsilon-5} n^{-1/3})\right.$$
$$\left. + O(\alpha^{\epsilon-1} n^{-2/3}) + \frac{\mu}{2n}\left(\frac{1-\sigma}{\sigma^3}\right)\right); \quad (27.6')$$

$$U(m, n) = n + \left(-2\alpha + \frac{1}{2\alpha^2} + \frac{11}{8\alpha^5} + \frac{175}{16\alpha^8} + \frac{19005}{128\alpha^{11}}\right.$$
$$\left. + O\left(\frac{1}{\alpha^{14}}\right)\right) n^{2/3} (1 + O(\mu)); \quad (27.7')$$

$$\vartheta^k T(z) = \frac{2^{k-1}\,\Gamma(k-1/2)}{\Gamma(1/2)} \frac{1}{(1-T(z))^{2k-1}} + \cdots, \quad k \geq 1; \quad (27.8')$$

$$\frac{1}{\sqrt{2\pi}} \int_0^\infty \frac{e^{-\alpha^2 t/2}(e^{\alpha t^2/2 - t^3/6} - 1)\, dt}{t^{3/2}}$$
$$= \frac{1}{\sqrt{2\pi}} \int_0^\infty \frac{e^{(\alpha-t)^3/6 - \alpha^3/6} - 1}{t^{3/2}}\, dt + \alpha; \quad (27.9')$$

$$\frac{1}{\sqrt{2\pi}} \int_0^\infty e^{-\alpha^2 t/2} t^{k-3/2}\, dt = \frac{2^{k-1}\,\Gamma(k-1/2)}{\sqrt{\pi}\,\alpha^{2k-1}}, \quad k \geq 1. \quad (27.10')$$

The asymptotic series (27.7) and (27.7′) for $\alpha \to \infty$ blend perfectly with the results obtained in [28] when α is any constant (positive, negative, or zero):

$$V(m,n) = \frac{1}{4}\left(\int_0^\infty e^{(\alpha-t)^3/6-\alpha^3/6}\,dt\right) n^{2/3} + O(n^{1/3});\qquad (27.12)$$

$$U(m,n) = n + \left(-\alpha + \frac{1}{\sqrt{2\pi}}\int_0^\infty \frac{e^{(\alpha-t)^3/6-\alpha^3/6}-1}{t^{3/2}}\,dt\right) n^{2/3}$$
$$+ O(n^{1/3}).\qquad (27.12')$$

These integrals are entire functions of α,

$$\int_0^\infty e^{(\alpha-t)^3/6-\alpha^3/6}\,dt = \frac{6^{1/3}\,\Gamma(1/3)}{3}\,e^{-\alpha^3/6} + \alpha\,F(1;\tfrac{4}{3};-\alpha^3/6);\ (27.13)$$

$$\int_0^\infty \frac{e^{(\alpha-t)^3/6-\alpha^3/6}-1}{t^{3/2}}\,dt = -e^{-\alpha^3/6}\Big(\big(6^{5/6}\Gamma(\tfrac{5}{6})/3\big)\,F(\tfrac{1}{2},\tfrac{5}{6};\tfrac{1}{3},\tfrac{2}{3};\alpha^3/6)$$
$$- \big(6^{1/2}\Gamma(\tfrac{1}{2})/6\big)\alpha F(\tfrac{5}{6},\tfrac{7}{6};\tfrac{2}{3},\tfrac{4}{3};\alpha^3/6)$$
$$+ \big(6^{1/6}\Gamma(\tfrac{1}{6})/8\big)\alpha^2 F(\tfrac{7}{6},\tfrac{3}{2};\tfrac{4}{3},\tfrac{5}{3};\alpha^3/6)\Big).\qquad (27.13')$$

Equation (27.13) is proved by observing that if $g(\alpha) = \int_0^\infty e^{(\alpha-t)^3/6}\,dt$, then $g'(\alpha) = \int_0^\infty (\alpha-t)^2 e^{(\alpha-t)^3/6}\,dt/2 = e^{\alpha^3/6}$. It implies (27.11) by well-known properties of confluent hypergeometric series. Equation (27.13′) is proved by setting

$$h(\alpha) = \int_0^\infty \big(e^{(\alpha-t)^3/6} - e^{\alpha^3/6}\big)\,t^{-3/2}\,dt = -\int_0^\infty (\alpha-t)^2\,e^{(\alpha-t)^3/6}t^{-1/2}\,dt$$

and showing that $h'''(\alpha) = \frac{1}{2}\alpha^2 h''(\alpha) + \frac{5}{2}\alpha h'(\alpha) + \frac{15}{8}\,h(\alpha)$, hence

$$[\alpha^{k+3}]\,h(\alpha) = [\alpha^k]\,h(\alpha)(k+\tfrac{3}{2})(k+\tfrac{5}{2})/\big(2(k+1)(k+2)(k+3)\big).$$

Recall that we enumerated $n - U(m,n) - V(m,n)$, the expected number of vertices in complex components, using a complementary approach in (15.13), by summing over the excess r.

Lemma 9. *Let V_{mn} be the number of vertices in unicyclic components of a random multigraph with m edges and n vertices. If $m = \frac{1}{2}n(1+\mu)$ and $\mu \geq n^{-1/3}$, the expected value of V_{mn}^l is $O(\mu^{-2l})$, for every fixed integer $l \geq 1$.*

Proof. Equation (27.7) proves this for $l = 1$ and $n^{-1/3} \leq \mu \leq c$, where c is some positive constant. A similar argument applies for arbitrary l, because the generating function $\vartheta^l e^{V(z)}$ is $e^{V(z)}/\big(1 - T(z)\big)^{2l}$ times a polynomial in $T(z)$; this means we are summing terms like (27.5), but with $Q(k)$ replaced by a semipolynomial in k of degree $l - \frac{1}{2}$. (See the proof of Theorem 3 in Section 8.) The analog of (27.6) will then be $O(\sigma^{-2l})$, which is $O(\mu^{-2l})$ if $\mu \leq c$. Incidentally, for this range of μ we will have

$$
\mathrm{E}\, V_{mn}^l = \frac{\Gamma(l+1/4)}{\Gamma(1/4)} \left(\frac{2}{\sigma^2}\right)^l \big(1 + O(\mu) + O(\mu^{-3}n^{-1})\big). \qquad (27.14)
$$

If $c \leq \mu \leq n^\epsilon$, with $\epsilon < \frac{1}{4}$, let $0 < \delta < 1 - \ln(1+c)/c$. Then each term in the analog of (27.4) with $k \leq n^{3/4}$ is

$$
O\big(k^{l-1} \exp\big(k(\ln(1 + \mu) - \mu) + O(\mu^2 k^2/n))\big)\big) = O\big(k^{l-1}\exp(-k\delta\mu)\big)
$$
$$
= O\big(k^{l-1}e^{-\delta\mu - \delta c k}\big).
$$

Hence $\mathrm{E}\, V_{mn}^l = O(e^{-\delta\mu})$.

Finally, if $\mu \geq n^\epsilon$ the value of $\mathrm{E}\, V_{mn}^l$ is superpolynomially small, for it is a sum of n terms each of which is bounded by a polynomial in m and n times $(1 - 1/n)^{2m}$, which is $O(\mu^d e^{-\mu})$ for some finite degree d. □

Corollary. *The probability that a random multigraph never acquires a new complex component after it has gained $m = \frac{1}{2}(n + \alpha n^{2/3}) > \frac{1}{2}n$ edges is $1 - O(\alpha^{-3})$.*

Proof. We may assume that $\alpha \geq 1$. A new complex component must be bicyclic. A multigraph gains a new bicyclic component if and only if the endpoints of a new edge both fall in unicyclic components. The probability that this occurs at time $m = \frac{1}{2}(n + \mu n)$ is $\mathrm{E}\, V_{mn}^2/n^2 = O(\mu^{-4}n^{-2})$, by the lemma. Summing for $m \geq \frac{1}{2}(n + \alpha n^{2/3})$ gives $O(\alpha^{-3})$ as an upper bound on the probability that at least one new bicyclic component appears after time $\frac{1}{2}(n + \alpha n^{2/3})$. □

Theorem 15. *The probability that an evolving graph or multigraph on n vertices never has more than one complex component throughout its evolution approaches $\frac{5\pi}{18}$ as $n \to \infty$.*

Proof. Let $\epsilon > 0$ be fixed. By the corollary just proved, there exists a number α, depending on ϵ but not on n, such that the probability of a random multigraph obtaining a new complex component after time $m = \frac{1}{2}(n + \alpha n^{2/3})$ is less than ϵ.

By Section 14 and the corollary of Section 13, there is a number R, independent of n, such that the probability of having excess $> R$ at this time m is less than ϵ. So the probability that a random multigraph leaves the top line after excess R is $< 2\epsilon$. (Either it reaches excess R before time m, or it leaves the top line after time m.)

But the probability that a random multigraph leaves the top line before excess R is $1 - \frac{5\pi}{18} + O(R^{-1}) + O(n^{-1/3})$, by (18.2). We may choose R sufficiently large that this $O(R^{-1})$ is less than ϵ; then we may choose n sufficiently large that the $O(n^{-1/3})$ is less than ϵ. The probability that a random multigraph leaves the top line for such n is therefore between $1 - \frac{5\pi}{18} - 2\epsilon$ and $1 - \frac{5\pi}{18} + 4\epsilon$.

For graphs, we note that an evolving graph may be constructed from an evolving multigraph by ignoring all new edges that would be loops or parallel to an existing edge. Since this reduction preserves or decreases both the excess and the number of complex components, it follows that if the graph leaves the top line after excess R, then the multigraph does too. Hence this event likewise has probability $< 2\epsilon$, and the proof is completed as for multigraphs. □

Theorem 16. *Given any set S of infinite paths in Figure 1, the probability that the evolution of a random multigraph follows a path in S converges as $n \to \infty$ to the corresponding probability for the Markov chain with the transition probabilities given in Theorem 9. Similarly, if the evolution of a random graph, which stops at excess $\binom{n}{2} - n$ when the complete graph is reached, is continued along the top line to an infinite path in Figure 1, then the probability that this path lies in S converges to the same limit.*

Proof. Given $\epsilon > 0$, let R be as in the preceding proof so that a random graph or multigraph leaves the top line after excess R with probability $< 2\epsilon$. We can also choose R large enough that $c_R > e_R(1 - \epsilon)$, by (8.7). Since c_R/e_R is the sum of all Markov transition probabilities for paths that intersect the top line at excess R, if we cut Figure 1 at excess R, the Markov probabilities for paths in S that do not have this property must sum to less than ϵ. When R is large enough, the sum of Markov probabilities for all paths that diverge from the top line after excess R is likewise less than ϵ, because it is $O(\sum_R^\infty r^{-2}) = O(R^{-1})$.

Let $P_n(S)$ be the probability that the evolution of a random graph or multigraph on n vertices follows a path in S, and let $P_\infty(S)$ denote the corresponding Markov probability. If S_R is the subset of S having all paths on the top line when the excess is $\geq R$, then we have $0 \leq P_n(S) - P_n(S_R) < 2\epsilon$ for all $n \leq \infty$. Similarly, if S'_R is the set of all paths that follow a path in S_R up to excess R, but afterwards are arbitrary, then $0 \leq P_n(S'_R) - P_n(S_R) < 2\epsilon$, for $n \leq \infty$. Finally, by Theorem 10, $|P_n(S'_R) - P_\infty(S'_R)| < \epsilon$ if n is large enough, and we have $|P_n(S) - P_\infty(S)| < 5\epsilon$. $\square$

Theorem 16 says that the evolutionary path, regarded as a random element of the set of all paths in Figure 1, converges in distribution to the Markov process. There are uncountably many paths, but the theorem needs no measurability restriction since the distributions for finite n and for the limit are concentrated on the countable set of paths that eventually follow the top line. Note that we cannot strengthen the statement for random graphs to deduce the limiting probability that the evolution follows a path in S until it stops at excess $\binom{n}{2} - n$; for example, if S is the set of all paths that do *not* eventually follow the top line, the Markov probability $P_\infty(S)$ is zero, while $P_n(S) = 1$ for all finite n.

Corollary. *The probability that an evolving graph or multigraph never has more than l complex components converges to a limit P_l.* $\square$

Closed form expressions for P_l might not exist when $l \geq 2$, but the values can be estimated from below using the following related probabilities:

Corollary. *The probability that an evolving graph or multigraph acquires exactly $l \geq 1$ new complex components during the evolution converges to*

$$p'_l = \Pr\left(\sum_{r=0}^\infty I_r = l\right) = \Pr\left(\sum_{r=1}^\infty I_r = l - 1\right), \qquad (27.15)$$

where $I_0, I_1, I_2, I_3, \ldots$ are independent Bernoulli distributed random variables with $\Pr(I_r = 1) = 1 - \Pr(I_r = 0) = 5/(6r + 1)(6r + 5)$.

In other words, the number of new complex components converges in distribution to $\sum_{r=0}^\infty I_r$.

Proof. Let $I_r = 1$ if the Markov process acquires a new bicyclic component when the excess goes from r to $r + 1$, and $I_r = 0$ otherwise; in particular $I_0 = 1$ always. By Theorem 9, $\Pr(I_r = 1) = 5/((6r + 1)(6r + 5))$ independently of the previous history, and thus the variables are independent. □

The probabilities p'_l have a surprisingly simple generating function: We have

$$p'_l = [z^l] \prod_{r=0}^{\infty} \left(1 + (z - 1)\frac{5}{(6r + 1)(6r + 5)}\right)$$

$$= [z^l] \prod_{r=0}^{\infty} \frac{\left(r + \frac{1}{2} + \frac{1}{6}\sqrt{9 - 5z}\right)\left(r + \frac{1}{2} - \frac{1}{6}\sqrt{9 - 5z}\right)}{\left(r + \frac{1}{6}\right)\left(r + \frac{5}{6}\right)}$$

$$= [z^l] \frac{\Gamma\left(\frac{1}{6}\right)\Gamma\left(\frac{5}{6}\right)}{\Gamma\left(\frac{1}{2} + \frac{1}{6}\sqrt{9 - 5z}\right)\Gamma\left(\frac{1}{2} - \frac{1}{6}\sqrt{9 - 5z}\right)}$$

$$= [z^l] \cos\left(\frac{\pi}{6}\sqrt{9 - 5z}\right) / \cos\frac{\pi}{3}. \tag{27.16}$$

Computing the coefficients of the Taylor series for $\cos\left(\frac{\pi}{6}\sqrt{9 - 5z}\right)$, we find that the numbers p'_l are rational polynomials in π:

$$p'_1 = \frac{5\pi}{18} \approx 0.87266\,;$$

$$p'_2 = \frac{50\pi}{6^4} \approx 0.12120\,;$$

$$p'_3 = \frac{500\pi}{6^6}\left(1 - \frac{\pi^2}{12}\right) \approx 0.00598\,;$$

$$p'_4 = \frac{6250\pi}{6^8}\left(1 - \frac{\pi^2}{10}\right) \approx 0.00015\,.$$

Let $P'_l = \sum_{j=1}^{l} p'_j$; numerically we have $P_2 > P'_2 \approx 0.99387$, $P_3 > P'_3 \approx 0.99985$, $P_4 > P'_4 \approx 0.999998$.

The number of new complex components is also studied in [19], where further results are given. The methods of [19] do not, however, seem to yield the sharp results obtainable with generating functions.

28. Empirical Data

Computer simulations of random multigraphs tend to confirm the theoretical results derived above. In this section we will discuss some of the statistics computed during 10,000 trials of the multigraph process on 20,000 vertices, so that readers can obtain a feel for the way in which random multigraphs actually evolve in practice. Several additional runs gave roughly the same results; only the first 10,000 outputs are recorded here, because we did not wish to introduce any subjective bias.

When a statistic is given in the form '$x \pm y$' below, x is the sample mean and y is the sample standard deviation divided by $\sqrt{10000}$. The sample standard deviation has been computed by taking the square root of an unbiased estimate of the variance. The "time" of an event is the number of edges present when that event occurred.

The first cycle was formed at time 6714 ± 30; this agrees reasonably well with the asymptotic formula $n/3$ found in [14, Corollary 3]. The size of the first unicyclic component was 177 ± 4. According to (26.14), the mean should be approximately $\sqrt{\pi n/2} \approx 177$; perhaps we were lucky.

The length of the first cycle was 9.5 ± 0.2; in fact, the histogram was

length =	1	2	3	4	5	6	7	≥ 8
actual =	3320	1398	798	558	434	318	244	2930
theoretical =	3333	1333	762	508	369	284	227	3183

The distribution has infinite mean, approximately $2.03n^{1/6} + O(n^{3/22})$, and its standard deviation is of order $n^{1/4}$ by (26.13), so the length of the first cycle should not be expected to be a robust statistic. Still, the empirical result is near $2.03n^{1/6} \approx 10.6$.

Several people have suggested in conversation that the "last cycle" ought to have the same statistical characteristics as the first. The last cycle is the cycle in the last unicyclic component that is present during a multigraph's evolution: After that component becomes or is absorbed into a component of higher complexity, no further unicycles exist, and no further unicycles are formed. (If two cycles disappear simultaneously when the edge $\langle x, y \rangle$ is added, we say that the cycle containing y was the last to go.) The manner in which the giant component swallows other structures is rather like the initial stages of evolution but in reverse: First the unicycles tend to go, then the larger trees, and finally only isolated vertices are left (see Bollobás [6, Sections VI.3 and VII.1]). A strong formulation of this symmetry principle was proved by Łuczak [25]; the phenomenon can be explained by the symmetry between $T(z)$ and $2 - T(z)$ in $U(z)$. However, the length of the last cycle has a distinctly different distribution from the length of the first cycle (see [20]). In these computer runs it had the following histogram:

length =	1	2	3	4	5	6	7	≥ 8
observed =	4347	1617	930	522	423	315	217	1629

with mean 5.7 ± 0.1.

The total number of unicyclic components formed during the entire evolution was

number =	1	2	3	4	5	6	7	≥ 8
observed =	444	1426	2241	2289	1709	1034	505	352

with mean 4.05 ± 0.02.

The excess of the multigraph changed from 0 to 1 at time 10324 ± 4. The number of unicyclic components present was about 2.7 just before this event, and about 1.5 just after. As soon as the excess became positive it began a steady rise:

excess	time	unicyclic size just before	unicyclic size just after	complex size just before	complex size just after
1	10324 ± 4	1558 ± 7	174 ± 3	0	1385 ± 7
2	10491 ± 3	284 ± 5	133 ± 2	1712 ± 7	1863 ± 6
3	10590 ± 3	179 ± 3	109 ± 2	2118 ± 6	2188 ± 6
4	10665 ± 3	133 ± 2	94 ± 2	2392 ± 6	2430 ± 6
5	10726 ± 2	109 ± 2	82 ± 2	2606 ± 6	2634 ± 6
6	10779 ± 2	92 ± 2	73 ± 1	2788 ± 5	2806 ± 5
7	10826 ± 2	81 ± 2	65 ± 1	2944 ± 5	2961 ± 5
8	10870 ± 2	71 ± 1	60 ± 1	3091 ± 5	3103 ± 5
9	10910 ± 2	65 ± 1	56 ± 1	3219 ± 5	3228 ± 5
10	10947 ± 2	60 ± 1	53 ± 1	3338 ± 5	3346 ± 5

The value of $n^{2/3}$ is approximately 737 when $n = 20000$, so each additional edge increases the parameter μ of Lemma 3 by approximately 0.0027. The value of μ when $m = 10947$ is approximately 2.57; then

$$\frac{2}{3}\mu^3 + 1 + \frac{5}{24}\mu^{-3} + \frac{15}{16}\mu^{-6} \approx 12.3,$$

so the excess is not quite keeping up with the expected value in Theorem 6. Similarly, formula (26.16) predicts that the excess will reach 1 when $m \approx 10299$, and 10 when $m \approx 10888$; random multigraphs for finite n seem to become complex a bit "late." It is interesting to note that the observed standard deviations kept decreasing as the excess increased, while the discrepancy from (26.16) kept increasing.

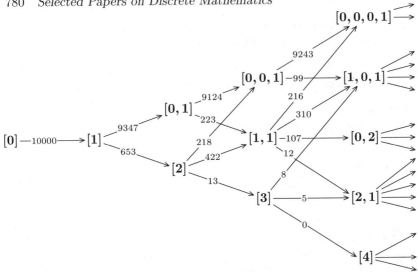

FIGURE 3. The number of times the paths in Figure 1 were actually traced, when 10000 random multigraphs on 20000 vertices were generated in experimental tests.

The random multigraphs followed paths in Figure 1 with the frequencies shown in Figure 3. When the excess changed from 9 to 10, the transition was from a single C_9 to C_{10} in 9762 cases, from (C_1, C_8) to (C_1, C_9) in 154 cases, from (C_1, C_8) to C_{10} in 29 cases, from (C_2, C_7) to (C_2, C_8) in 13 cases, from C_9 to (C_1, C_9) in 11 cases, from (C_2, C_7) to C_{10} in 8 cases, and in eleven less common ways in the remaining 23 cases. Altogether 8700 of the 10000 random multigraphs remained on the top line of Figure 1 throughout their evolution.

There comes a time when the giant component first succeeds in annihilating everything except isolated vertices, after which it remains the only component with edges. In these runs that time was 58399 ± 70. The number of isolated vertices still remaining was then 70.4 ± 0.4.

The multigraph finally became connected at time 104620 ± 127. The expected time for an evolving multigraph to have no isolated vertices is $\frac{1}{2}nH_n = \frac{1}{2}n\ln n + \frac{1}{2}\gamma n + \frac{1}{4} + O(n^{-1})$, which is approximately 104807 when $n = 20000$.

29. Open Problems

The topics discussed in this paper raise a host of interesting questions, and the answers to those questions will no doubt bring additional striking patterns to light.

But the reader may have noticed that this paper is already rather long. Therefore it seems wise to stop at this point, with the hope that researchers all over the world will enjoy exploring the tantalizing questions that remain.

For example, it would be interesting to find a basis for as many linear combinations of terms $w^r T^a/(1-T)^b$ as possible such that

$$\Phi_n \, w^r \, e^U \, T^a/(1-T)^b$$

has a known value, as in (25.22). We can find many linear combinations of such functions for which Φ_n gives 0, because $\Phi_n \nabla S$ is usually 0 or 1. Notice that

$$\frac{T^a}{(1-T)^{b+1}} = \frac{T^a}{(1-T)^b} + \frac{T^{a+1}}{(1-T)^{b+1}} \, ; \qquad (29.1)$$

hence terms of excess $r+1$ can be expressed as combinations of terms of excess r. Conversely, we can go from excess r to excess $r+1$, because

$$\frac{T^a}{(1-T)^b} = \frac{T^a}{(1-T)^{b+1}} - \frac{T^{a+1}}{(1-T)^{b+2}} + \frac{T^{a+2}}{(1-T)^{b+3}} - \cdots \qquad (29.2)$$

is an infinite series that always "converges" under application of Φ_n; all terms after a certain point are multiples of T^{n+1}, so they do not change the coefficient of z^n.

The stopping configuration machinery suggests many further problems of interest. For example, we should be able to deduce more about the nature of a random multigraph when its deficiency first exceeds a given number d.

The discussion in Section 23 characterizes the stochastic behavior of r and d when $\mu = o(1)$; what happens thereafter? Relations (23.12) and (23.13) may well continue to describe the approximate mean values of r and d as $\mu \to \infty$. The shadow point σ defined in (23.4) will approach 0, but it remains an analytic function of μ, and $1 - \sigma$ remains a saddle point of the contour integral for $[z^n] \, U^{n-m+r} T^{2r-d}/(1-T)^{3r-d+1/2}$.

The analytic function $T(z)$ has an interesting Riemann surface: There is a quadratic singularity at $z = e^{-1}$, and if we travel around that point we get to a second sheet in which there is a logarithmic singularity at $z = 0$. Winding around that logarithmic singularity takes us to infinitely many other sheets having no finite singularities besides 0. It may be possible to work out a theory under which contour integrals of importance in the study of random graphs could be evaluated by paths that pass through the point $1 + \mu$, which lies on the "wrong side" of the quadratic singularity of $T(z)$; $1 + \mu$ turns out to be a saddle point for several important generating functions.

Identity (8.15)–(8.16) suggests that the generating functions for random multigraphs might have interesting continued fraction forms. Such expressions could well be of special importance, because they often converge when power series do not.

The fact that the recurrence for the coefficients e_{rd} can be "solved" to yield (7.3)–(7.5) should prove to be a good challenge for computer systems that are now being constructed to solve recurrence relations automatically. The similar recurrence for the coefficients e'_{rd}, discussed in (7.24) and (7.25), will probably be an even greater challenge; at least, no simple derivation of (7.21) from (7.26) is known.

The solution to the recurrence for e_{rd} in Section 7 relies on the introduction of a "half excess" stage, in which the polynomials must be evaluated at integers plus $\frac{1}{2}$ although the recurrence in which they are used involves integers only. In Section 20 we found, similarly, that it was fruitful to break the process of adding an edge into stages in which "half-edges" were added. Perhaps the theory of fractional differentiation will be of value in future investigations. However, the operators $D^{1/2}$ and $\vartheta^{1/2}$ do not seem to transform the basic functions $T^a/(1 - T)^b$ very nicely.

Is there an equation (27.11′) analogous to (27.11)? There must be a reason why the coefficients of (27.7′) tend to have small prime factors.

We have seen numerous examples in which the multigraph process leads to formulas that are mathematically cleaner than the analogous formulas for the graph process. This suggests that an analogous theory be introduced in place of the alternative '$\mathbf{G}_{n,p}$' model of random graphs: Instead of saying that each edge is present with probability p, the multiplicity of each edge should be allowed to have a Poisson distribution with mean p. Readers are encouraged to experiment with such an approach.

Convergence to limiting distributions often appears to be monotonic. For example, the probability that an evolving multigraph on n vertices stays on the top line appears to be strictly decreasing as n increases. How could this be proved?

Our proof of the top-line probability in Theorem 15 was independent of the difficult analyses in Lemma 7 and Theorem 13 about the behavior of random multigraphs with more than $\frac{1}{2}(n + n^{2/3+\epsilon})$ edges; moreover, it did not use the stopping-configuration machinery of Sections 24–26, although that theory was in fact motivated by attempts to prove Theorem 15 in a sharper form via generating functions. The top-line phenomenon may perhaps be understood more deeply if we use a generating-function-based approach, and the following ideas may therefore prove to be useful. Let $S(w, z)$ be the bgf for all multigraphs that never leave the top line of Figure 1, where each multigraph is weighted by the probability of having a purely top-line history as discussed in Section 17. The discussion of Sections 19 and 20 shows that

$$S(w, z) = e^{U(w,z)+V(w,z)} H(w, z), \qquad (29.3)$$

where $H(w, z)$ satisfies a differential equation almost like the equation (5.1) that defines $E(w, z)$:

$$\frac{1}{w} (\vartheta_w - T\vartheta_z)H = \frac{1}{2} e^{-V} \vartheta_z^2 e^V H - \frac{1}{2} e^{-V} (\vartheta_z^2 e^V)(H - 1). \qquad (29.4)$$

The subtracted term $\frac{1}{2} e^{-V} (\vartheta_z^2 e^V)(H - 1)$ accounts for the forbidden case that a new edge marked by ϑ_z^2 lies entirely in the unicyclic part generated by e^V; a second complex component arises if and only if this happens. The correction applies to $H - 1$, not H, because the very first complex component does not violate the top-line condition.

Expressing $H(w, z)$ in the form (25.16), we have $H_1 = E_1$, but H_2 is smaller than E_2:

$$H_2 = \frac{5}{16} \frac{T^4}{(1 - T)^6} + \frac{25}{48} \frac{T^3}{(1 - T)^5} + \frac{11}{48} \frac{T^2}{(1 - T)^4} + \frac{1}{48} \frac{T}{(1 - T)^3}.$$

In general we can write

$$H_r = \sum h_{rd} \frac{T^{2r-d}}{(1 - T)^{3r-d}} \qquad (29.5)$$

for appropriate coefficients h_{rd}. The special case $\mu = \nu = 0$ of (20.7) tells us that

$$\vartheta^2 e^V = \vartheta^2 (1 - T)^{-1/2} = \frac{1}{2} T(1 - T)^{-7/2} + \frac{5}{4} T^2 (1 - T)^{-9/2}; \qquad (29.6)$$

therefore we can compute the coefficients h_{rd} by making a slight change to the rule for computing e_{rd} that is expressed in (20.11): Subtract 5 from the numerator of the first coefficient term in (20.11), and subtract 1 from the numerator of the second coefficient. The first coefficient now simplifies to

$$\frac{(6r - 2d + 5)(6r - 2d + 1) - 5}{8(3r - d + 3)} = \frac{3r - d}{2}.$$

In particular, when $d = 0$ we have $h_{(r+1)0} = \frac{3}{2}rh_{r0}$; hence h_{r0} is the number we called k_r in (24.3).

Equation (25.17) now gives us a useful expression for the stopping configurations,

$$\nabla S = e^{U(w,z)} \sum_{r \geq 2} w^r (\vartheta_z^2 e^V) H_{r-1}(wz)$$

$$= e^{U(w,z)} \sum_{r \geq 2} w^r \Big(\frac{1}{2} \frac{T(wz)}{(1 - T(wz))^{7/2}}$$

$$+ \frac{5}{4} \frac{T(wz)^2}{(1 - T(wz))^{9/2}} \Big) H_{r-1}(wz). \qquad (29.7)$$

The probability that an evolving multigraph on n vertices leaves the top line of Figure 1 is $\Phi_n \nabla S$.

For fixed r we can evaluate the contribution made to $\Phi_n \nabla S$ by the rth term of (29.7), to within $O(n^{-1/3})$, because the leading coefficient $h_{(r-1)0}$ controls the asymptotic behavior. Indeed, we know from (25.22) and the subsequent discussion that

$$\Phi_n \Big(w^r \frac{e^{U(w,z)} T(wz)^{2r}}{(1 - T(wz))^{3r+3/2}} \Big) = \frac{1}{6re_r} + O(n^{-1/3}) \qquad (29.8)$$

for all fixed r. Therefore when Φ_n is applied to the rth term of (29.7) we get

$$\Phi_n e^U w^r \Big(\frac{1}{2} \frac{T}{(1 - T)^{7/2}} + \frac{5}{4} \frac{T^2}{(1 - T)^{9/2}} \Big) H_{r-1}$$

$$= \frac{5k_{r-1}}{24re_r} + O(n^{-1/3}). \qquad (29.9)$$

When $r = 2$, the limit is $\frac{5}{77}$; when $r > 2$, (7.1) and (24.3) imply that

$$\frac{5k_{r-1}}{24re_r} = \Big(\frac{5k_{r-2}}{24(r-1)e_{r-1}} \Big) \Big(\frac{36(r-1)(r-2)}{(6r-1)(6r-5)} \Big).$$

It follows by induction that

$$\frac{5k_{r-1}}{24re_r} = \frac{5}{36(r-1)r} \prod_{k=1}^{r-1} \frac{k(k+1)}{(k + \frac{1}{6})(k + \frac{5}{6})}$$

$$= \prod_{k=1}^{r-2} \frac{k(k+1)}{(k + \frac{1}{6})(k + \frac{5}{6})} - \prod_{k=1}^{r-1} \frac{k(k+1)}{(k + \frac{1}{6})(k + \frac{5}{6})}.$$

So the sum over r is a telescoping series,

$$\sum_{r \geq 2} \frac{5k_{r-1}}{24re_r} = 1 - \prod_{k=1}^{\infty} \frac{k(k+1)}{(k+\frac{1}{6})(k+\frac{5}{6})} = 1 - \frac{5\pi}{18}. \qquad (29.10)$$

In other words, convergence to the top-line probability depends entirely on the sum over r of the error term in (29.9).

   The number of challenging and potentially fruitful questions that remain unanswered seems to be almost endless. But we shall close this list of research problems by stating what seems to be the single most important related area ripe for investigation at the present time. Wright [42] gave a procedure for computing the number of *strongly connected labeled digraphs* of excess r, analogous to his formulas for connected labeled undirected graphs. Random directed multigraphs are of great importance in computer applications, and it is shocking that so little attention has been given to their study so far. Karp [21] carried Wright's investigations further and discovered a beautiful theorem: A random digraph with $n(1 + \mu)$ directed arcs almost surely has a giant strong component of size $\sim \Theta(\mu)^2 n$, when $\Theta(\mu)$ is the factor such that an undirected graph with $\frac{1}{2}n(1 + \mu)$ edges almost surely has a giant component of size $\sim \Theta(\mu)n$. (The function $\Theta(\mu)$ is $(\mu + \sigma)/(1 + \mu)$, according to (23.11). Karp's investigation was based on $\mathbf{D}_{n,p}$, in which every directed arc is present with probability p, but a similar result surely holds for other models of random digraphs.) A complete analysis of the random *directed* multigraph process is clearly called for, preferably based on generating functions so that extensive quantitative information can be derived without difficulty.

   Here is a sketch of how such an investigation might begin. The *directed multigraph process* consists of adding directed arcs $x \to y$ repeatedly to an initially empty multiset of arcs on the vertices $\{1, 2, \ldots, n\}$, where x and y are independently and uniformly distributed between 1 and n. The *compensation factor* $\kappa(M)$ of a multidigraph M with m_{xy} arcs from x to y is $1 / \prod_{x=1}^{n} \prod_{y=1}^{n} m_{xy}!$; we can use it to compute bivariate generating functions as in (2.1). The bgf for all possible multidigraphs is $\sum_{n \geq 0} e^{n^2} w z^n / n! = G(2w, z)$.

   Let $\mathcal{A}$ be the family of all multidigraphs such that all vertices are reachable from vertex 1 via a directed path, and let $A(w, z)$ be the corresponding bgf. There is a nice relation between $A(w, z)$ and the bgf $C(w, z)$ for connected undirected multigraphs, (2.10): If $A(w, z) =$

$\sum_{n\geq 1} a_n(w)z^n/n!$, we have

$$\sum_{n\geq 1} a_n(w)\, e^{-n^2 w/2}\, \frac{z^n}{n!} \;=\; C(w,z)\,. \qquad (29.11)$$

This can be proved by replacing z by $ze^{-w/2}$ and noting that the function $C(w, ze^{-w/2})$ is the bgf for connected multigraphs without self-loops, and by showing that all members of $\mathcal{A}$ are obtainable from such connected multigraphs M by the following reversible construction: Define a linear ordering $\prec$ on the vertices $\{1, 2, \ldots, n\}$ by saying that $x \prec y$ if $d(x) < d(y)$ or $d(x) = d(y)$ and $x < y$, where $d(x)$ is the distance from 1 to x in M. Then define a multidigraph $D \in \mathcal{A}$ by arcs $x \to y$ whenever $x - y$ in M and $x \prec y$; include arbitrary additional arcs $x \to y$ for all pairs of vertices with $x \succeq y$. The construction is reversible because $d(x)$ is easily seen to be the distance from 1 to x in D, regardless of the choice of additional arcs. The additional arcs correspond to a multiplicative factor

$$e^{\binom{n+1}{2}w} = e^{n^2 w/2} (e^{w/2})^n$$

in an n-vertex multigraph, with one factor e^w for each of the $\binom{n+1}{2}$ vertex pairs $x \succeq y$.

Let $\mathcal{S}$ be the family of all strongly connected multidigraphs, and let $S(w, z) = s_1(w)z + s_2(w)z^2/2! + s_3(w)z^3/3! + \cdots$ be the corresponding bgf. A nontrivial identity discovered by Wright [40] implies that we can calculate the coefficients $s_n(w)$ by using the formula

$$\sum_{n\geq 1} s_n(w)\, e^{-n^2 w/2}\, \frac{z^{n-1}}{(n-1)!}\, \frac{G(w,z)}{G(w, ze^{-nw})} \;=\; C'(w,z)\,, \qquad (29.12)$$

where the prime in $C'(w, z)$ denotes differentiation with respect to z. Notice that our generating function $G(w, z)$ satisfies

$$G'(w, z) = e^{w/2}G(w, ze^w), \quad G''(w, z) = e^{2w}G(w, ze^{2w}),$$
$$\ldots, \quad G^{(n)}(w, z) = e^{n^2 w/2}G(w, ze^{nw}), \quad \ldots; \qquad (29.13)$$

thus the denominator $G(w, ze^{-nw})$ in (29.12) is essentially an n-fold integral of $G(w, z)$.

Wright [42] proved that the number of strongly connected digraphs with $n + r$ arcs on n vertices, disallowing self-loops and multiple arcs, is $n!$ times a polynomial in n of degree $3r - 1$, when $n > r > 0$. His proof can be adapted to multidigraphs, and everything becomes much simpler, just as formula (9.4) for multigraphs is simpler than formula (9.20) for graphs. The analogs of (2.11) and (3.4) are

$$S(w, z) = w^{-1}S_{-1}(wz) + S_0(wz) + wS_1(wz) + w^2 S_2(wz) + \cdots, \quad (29.14)$$

where

$$S_{-1}(z) = z, \quad\quad (29.15)$$
$$S_0(z) = -\ln(1 - z), \quad\quad (29.16)$$

and $S_r(z)$ for $r \geq 1$ can easily be shown to be $(1 - z)^{-3r}$ times a polynomial in z of degree $< 3r$. For example, the multidigraphs enumerated by $wS_1(wz)$ all arise by inserting ("uncancelling") vertices in the arcs of the reduced multidigraphs

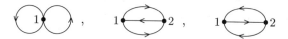

whose generating functions are respectively $\frac{1}{2}w^2 z$, $\frac{1}{4}w^3 z^2$, $\frac{1}{4}w^3 z^2$. The operation of uncancelling corresponds to replacing w by $w/(1 - wz)$, as in Lemma 1; so

$$wS_1(wz) = \frac{1}{2}w^2 z/(1 - wz)^2 + \frac{1}{2}w^3 z^2/(1 - wz)^3 = \frac{1}{2}w^2 z/(1 - wz)^3,$$

and $S_1(z) = \frac{1}{2}z/(1 - z)^3$.

In fact, the numerator of $S_r(z)$ turns out to have a surprisingly small degree. Computer calculations indicate that we can write

$$S_r(z) = \frac{s_{r0}z^{2r-1}}{(1 - z)^{3r}} + \frac{s_{r1}z^{2r-2}}{(1 - z)^{3r-1}} + \cdots + \frac{s_{r(2r-2)}z}{(1 - z)^{r+2}}, \quad (29.17)$$

a formula analogous to (8.4), at least when $r \leq 5$. The coefficients are

$d =$	0	1	2	3	4	5	6	7	8
$s_{1d} =$	$\frac{1}{2}$								
$s_{2d} =$	$\frac{17}{8}$	$\frac{13}{8}$	$\frac{1}{6}$						
$s_{3d} =$	$\frac{275}{12}$	$\frac{427}{12}$	$\frac{391}{24}$	$\frac{13}{6}$	$\frac{1}{24}$				
$s_{4d} =$	$\frac{26141}{64}$	$\frac{61231}{64}$	$\frac{51299}{64}$	$\frac{18473}{64}$	$\frac{6047}{144}$	$\frac{263}{144}$	$\frac{1}{120}$		
$s_{5d} =$	$\frac{1630711}{160}$	$\frac{1276481}{40}$	$\frac{3125933}{80}$	$\frac{2840093}{120}$	$\frac{3546283}{480}$	$\frac{6743}{6}$	$\frac{25307}{360}$	$\frac{43}{36}$	$\frac{1}{720}$

No reason why $S_r(z)$ should have the simple form (29.17) is apparent; this phenomenon cries out for explanation, if it is indeed true for all $r > 0$, and the explanation will probably lead to new theorems of interest. It can be shown that this conjecture is equivalent to the assertion that the sum of $(-1)^\nu \kappa/\nu!$, over all labeled, reduced, strongly connected multidigraphs of excess r, is zero; or in other words, if we choose a labeled, reduced, strongly connected multidigraph of excess r at random, with probabilities weighted in the natural way by the compensation factor κ, then the probability is $\frac{1}{2}$ that there will be an even number of vertices.

Is there a simple recurrence governing the leading coefficients s_{10}, s_{20}, s_{30}, ..., perhaps analogous to the relation we observed for ordinary connected components in (8.5)?

Acknowledgments

The authors wish to thank Prof. Richard Askey for helpful correspondence relating to this research.

This work was supported in part by the National Science Foundation under grant CCR-86-10181, and by Office of Naval Research contract N00014-87-K-0502.

References

[1] G. N. Bagaev, "Sluchaĭnye grafy so stepen'iu sviâznosti 2," *Diskretnyĭ Analiz* **22** (1973), 3–14.

[2] G. N. Bagaev and E. F. Dmitriev, "Perechislenie sviâznykh otmechennykh dvudol'nykh grafov," *Doklady Akademiiâ Nauk BSSR* **28** (1984), 1061–1063.

[3] Edward A. Bender, E. Rodney Canfield, and Brendan D. McKay, "The asymptotic number of labeled connected graphs with a given number of vertices and edges," *Random Structures and Algorithms* **1** (1990), 127–169.

[4] Béla Bollobás, "A probabilistic proof of an asymptotic formula for the number of labelled regular graphs," *European Journal of Combinatorics* **1** (1980), 311–316.

[5] Béla Bollobás, "The evolution of random graphs," *Transactions of the American Mathematical Society* **286** (1984), 257–274.

[6] Béla Bollobás, *Random Graphs* (London: Academic Press, 1985).

[7] B. Bollobás and A. Frieze, "On matchings and Hamiltonian cycles in random graphs," in *Random Graphs '83*, edited by Michał Karoński

and Andrzej Ruciński, *Annals of Discrete Mathematics* **28** (1985), 23–46.

[8] C. W. Borchardt, "Ueber eine der Interpolation entsprechende Darstellung der Eliminations-Resultante," *Journal für die reine und angewandte Mathematik* **57** (1860), 111–121.

[9] V. E. Britikov, "O strukture sluchaĭnogo grafa vblizi kriticheskoĭ tochki," *Diskretnaiā Matematika* **1**,3 (1989), 121–128. English translation, "On the random graph structure near the critical point," *Discrete Mathematics and Applications* **1**,3 (1991), 301–309.

[10] A. Cayley, "A theorem on trees," *Quarterly Journal of Pure and Applied Mathematics* **23** (1889), 376–378. Reprinted in his *Collected Mathematical Papers* **13**, 26–28.

[11] Gotthold Eisenstein, "Entwicklung von $\alpha^{\alpha^{\alpha^{\cdots}}}$," *Journal für die reine und angewandte Mathematik* **28** (1844), 49–52.

[12] P. Erdős and A. Rényi, "On random graphs I," *Publicationes Mathematicae (Debrecen)* **6** (1959), 290–297. Reprinted in *Paul Erdős: The Art of Counting* (Cambridge, Massachusetts: MIT Press, 1973), 561–568; and in *Selected Papers of Alfréd Rényi* (Budapest: Akadémiai Kiadó, 1976), 308–315.

[13] P. Erdős and A. Rényi, "On the evolution of random graphs," *Magyar Tudományos Akadémia Matematikai Kutató Intézetének Közleményei* **5** (1960), 17–61. Reprinted in *Paul Erdős: The Art of Counting* (Cambridge, Massachusetts: MIT Press, 1973), 574–618; and in *Selected Papers of Alfréd Rényi* (Budapest: Akadémiai Kiadó, 1976), 482–525.

[14] Philippe Flajolet, Donald E. Knuth, and Boris Pittel, "The first cycles in an evolving graph," *Discrete Mathematics* **75** (1989), 167–215. [Reprinted as Chapter 40 of the present volume.]

[15] C. M. Fortuin, P. W. Kasteleyn, and J. Ginibre, "Correlation inequalities on some partially ordered sets," *Communications in Mathematical Physics* **22** (1971), 89–103.

[16] I. P. Goulden and D. M. Jackson, *Combinatorial Enumeration* (New York: Wiley, 1983).

[17] Ronald L. Graham, Donald E. Knuth, and Oren Patashnik, *Concrete Mathematics* (Reading, Massachusetts: Addison–Wesley, 1989). [Some of the equations cited above were numbered incorrectly in the original printing of 1989.]

[18] Peter Henrici, *Applied and Computational Complex Analysis*, Volume 2 (New York: Wiley–Interscience, 1977).

[19] Svante Janson, "Multicyclic components in a random graph process," *Random Structures and Algorithms* **4** (1993), 71–84.

[20] Svante Janson and Tomasz Łuczak, "The size of the last cycle in the random graph process," *Abstracts of Papers Presented to the American Mathematical Society* **13** (1992), 354, abstract 875-05-131.

[21] Richard M. Karp, "The transitive closure of a random digraph," *Random Structures and Algorithms* **1** (1990), 73–93.

[22] Donald E. Knuth, "An analysis of optimum caching," *Journal of Algorithms* **6** (1985), 181–199. [Reprinted as Chapter 17 of *Selected Papers on Analysis of Algorithms*, CSLI Lecture Notes 102 (Stanford, California: Center for the Study of Language and Information, 2000), 235–255.]

[23] Donald E. Knuth, "Convolution polynomials," *The Mathematica Journal* **2**,4 (Fall 1992), 67–78. [Reprinted as Chapter 15 of the present volume.]

[24] Donald E. Knuth and Boris Pittel, "A recurrence related to trees," *Proceedings of the American Mathematical Society* **105** (1989), 335–349. [Reprinted as Chapter 39 of the present volume.]

[25] Tomasz Łuczak, "Component behavior near the critical point of the random graph process," *Random Structures and Algorithms* **1** (1990), 287–310.

[26] Tomasz Łuczak, "Cycles in a random graph near the critical point," *Random Structures and Algorithms* **2** (1991), 421–439.

[27] Tomasz Łuczak and John C. Wierman, "The chromatic number of random graphs at the double-jump threshold," *Combinatorica* **9** (1989), 39–49.

[28] Tomasz Łuczak, Boris Pittel, and John C. Wierman, "The structure of a random graph at the point of the phase transition," *Transactions of the American Mathematical Society* **341** (1994), 721–748.

[29] D. S. Mitrinović, *Analytic Inequalities* (Berlin: Springer–Verlag, 1970).

[30] S. Ramanujan, "Questions for solution, number 294," *Journal of the Indian Mathematical Society* **3** (1911), 128; **4** (1912), 151–152.

[31] Alfred Rényi, "Some remarks on the theory of trees," *Magyar Tudományos Akadémia Matematikai Kutató Intézetének Közleményei*

4 (1959), 73–85. Reprinted in *Selected Papers of Alfréd Rényi* **2**, 363–374.

[32] Robert James Riddell, Jr., *Contributions to the Theory of Condensation* (Ph.D. thesis, University of Michigan, 1951). The main results of this dissertation were published as R. J. Riddell, Jr., and G. E. Uhlenbeck, "On the theory of the virial development of the equation of state of monoatomic gases," *Journal of Chemical Physics* **21** (1953), 2056–2064.

[33] G. Seitz, "Une remarque aux inégalités," *Aktuarské Vědy* **6** (1936, 1937), 167–171.

[34] L. J. Slater, "Expansions of generalized Whittaker functions," *Proceedings of the Cambridge Philosophical Society* **50** (1954), 628–630.

[35] V. E. Stepanov, "Neskol'ko teorem otnositel'no sluchaĭnykh grafov," *Veroiatnostnye metody v diskretnoĭ matematike* (Karel'skiĭ filial Akademiiâ Nauk SSSR, Petrozavodsk, 1983), 90–92.

[36] V. E. Stepanov, "O nekotorykh osobennostiâkh stroeniiâ sluchaĭnogo grafa vblizi kriticheskoĭ tochki," *Teoriya Veroyatnostei i ee Primeneniiâ* **32** (1988), 633–657. English translation, "On some features of the structure of a random graph near a critical point," *Theory of Probability and Its Applications* **32** (1988), 573–594.

[37] J. J. Sylvester, "On the change of systems of independent variables," *Quarterly Journal of Pure and Applied Mathematics* **1** (1857), 42–56. Reprinted in his *Mathematical Papers* **2**, 65–85.

[38] V. A. Voblyĭ, "O koeffitsientakh Raĭta i Stepanova–Raĭta," *Matematicheskie Zametki* **42** (1987), 854–862. English translation, V. A. Voblyi, "Wright and Stepanov–Wright coefficients," *Mathematical Notes of the Academy of Sciences of the USSR* **42** (1987), 969–974.

[39] E. M. Wright, "A relationship between two sequences," *Proceedings of the London Mathematical Society* (3) **17** (1967), 296–304, 547–552.

[40] E. M. Wright, "The number of strong digraphs," *Bulletin of the London Mathematical Society* **3** (1971), 348–350.

[41] E. M. Wright, "The number of connected sparsely edged graphs," *Journal of Graph Theory* **1** (1977), 317–330.

[42] E. M. Wright, "Formulae for the number of sparsely-edged strong labelled digraphs," *Quarterly Journal of Mathematics*, Oxford (2), **28** (1977), 363–367.

[43] E. M. Wright, "The number of connected sparsely edged graphs. II. Smooth graphs and blocks," *Journal of Graph Theory* **2** (1978), 299–305.

[44] E. M. Wright, "The number of connected sparsely edged graphs. III. Asymptotic results," *Journal of Graph Theory* **4** (1980), 393–407.

[45] E. M. Wright, "The number of connected sparsely edged graphs. IV. Large nonseparable graphs," *Journal of Graph Theory* **7** (1983), 219–229.

[46] E. M. Wright, "Enumeration of smooth labelled graphs," *Proceedings of the Royal Society of Edinburgh* **91A** (1982), 205–212; **107A** (1987), 197.

Addendum

The empirical results reported in Section 28 of the original paper were faulty, due to a bug in my computer program of 1991; therefore I recomputed the relevant statistical data in 2003, while preparing the present book for publication. I thank Charles Stein for calling my attention to the fact that the former numbers were too much in conflict with the theory to be believable. Stein also pointed out some methods for obtaining much more reliable estimates of the expected first cycle length, based on an elegant way to use an idea that I had called "avoiding stupid moves" in my paper on estimating the size of a backtrack tree. (See §5 of Chapter 6 in *Selected Papers on Analysis of Algorithms*.) Those improvements were not, however, necessary to obtain the newly computed results.

A complementary way to study how components come together during random graph processes has been developed by David Aldous, "Brownian excursions, critical random graphs and the multiplicative coalescent," *The Annals of Probability* **25** (1997), 812–854.

Index

0^0, 22.

0–1 matrices, 360, 439–441, 446, 480, 482.

1-factor, 104; *see also* perfect matching.

3-dimensional matrices, 321.

$\Im$ (imaginary part), 598.

$\Re$ (real part), 598, 693.

$\alpha(G)$ (stability number), 129.

β-ary partitions, 543, 546.

ϑ operator for differentiation, 51, 251–252, 619, 655, 735.

   Taylor series analog, 752.

$\vartheta(G)$ (Lovász number), x, 123, 129.

$\kappa(G)$ (q-stability number), 129.

$\Lambda(A)$ (largest eigenvalue), 131.

π (circle ratio), as data source, 408–409, 587–588.

   mnemonic for, 87.

ϕ (golden ratio), 554.

$\chi(G)$ (chromatic number), 123, 130.

$\omega(G)$ (clique number), 123, 130.

Aardenne-Ehrenfest, Tatyana van, 207–208.

Abel, Niels Henrik, 222–223.

   binomial theorem, xi, 218, 222–224.

Abeles, Francine Forte, 120.

Abelian group, 6, 8.

   elementary, 308.

abl, 160.

Abramowitz, Milton, 39, 641.

absolute value, 20.

acyclic components, 652.

Aczél, János, 381, 385.

addition, 273.

adjacent vertices, 126.

Agarwal, Ratan Prakash, 501, 509.

Airy, George Biddle, function, 635, 638, 696–697, 767.

al-Uqlīdisī, Abū al-Ḥasan Aḥmad ibn Ibrāhīm, 88.

Alanen, Jack David, xiii, 277.

Albert, Abraham Adrian, 303, 306, 317–318, 336, 343, 348, 355.

Aldous, David John, 792.

Aleksandrov, Aleksandr Danilovich, 94, 103–104.

Algoet, Paul, 119.

ALGOL 60 language, 487–488.

ALGOL W language, 414, 416, 422.

Alizadeh, Farid, 172, 174.

almost surely, 588.

Alon, Noga, 438.

alternating sign matrices, 113.

André, Antoine Désiré, 438.

Andrews, George W. Eyre, 38, 39, 481, 497, 501, 508, 509.

annulus, 47.

anti-isomorphism, 331, 368.

anti-reflective functions, 64–67, 73–74.
antiblocker, 160.
antisymmetric function, 477.
APL language, 16, 24.
Appell, Paul Émile, 253.
Arbogast, Louis François Antoine, 570.
arc digraph, 203.
arcsin, 236, 711.
Arora, Sanjeev, 174.
Art of Computer Programming, 26, 124, 177, 224, 256, 464.
Askey, Richard Allen, 508, 509, 788.
assertions, 448–449.
associated Stirling number, 253.
associative inequality, 377.
associative law, 322–324.
asymmetrization, 404.
asymptotic growth, 248–253, 425–426, 526–534, 545–563.
asymptotic series, 36–37, 82, 638–640, 713, 772.
Atkin, Arthur Oliver Lonsdale, 480, 489, 563.
Austin, Thomas LeRoy, Jr., 217, 219.
automorphisms, 153, 307–308, 314, 350.
autotopisms, 316–319, 329, 333, 350.
Aztec diamond, 187.

b-ary sequence, 266.
Babbage, Charles, 40.
Bachmann, Paul Gustav Heinrich, 516, 523.
backtracking, 3–5, 487–488, 792.
backward difference, 64.
Bagaev, Gennadiĭ Nikolaevich, 653, 676, 788.
balanced codes, 433–438.

balanced digraph, 207.
Ball, Walter William Rouse, 9.
basic (m, t) set, 179.
basic hypergeometric series, 507–508.
basis of a vector space, 511–513.
basis set of a matroid, 409–410.
Bauer, Henry Raymond, III, 422.
Baumert, Leonard Daniel, 10.
Becker, Sheldon, 422.
Bell, Eric Temple, 253.
    numbers, 230, 235.
Bellman, Richard Ernest, 9.
Bender, Edward Anton, xiii, 464, 465, 481, 489, 788.
Bendix, Peter Bernard, 372.
Berge, Claude, 162, 403.
Bergmann, Ernest Eisenhardt, 438.
Bernoulli, Jacques (= Jakob = James), numbers, 23, 75, 79–82, 245.
    polynomials, 77–82.
    trials, 577.
Berry, Andrew Campbell, 577.
Bessel, Friedrich Wilhelm, functions, 635, 677, 697–698, 767.
Beta function, 605.
Bézout, Etienne, 113.
bgf, 591, 648.
bi-imaginary system, 274–276.
bicyclic components, 618–622, 652, 699, 761.
big bang, 646.
bijection, 210–213, 223, 433, 451–452, 456, 458, 466–470, 485–486, 510, 543–544.
    signed, 682.
binary number system, 85–88, 521, 528.
binary partitions, 525–526, 546.
binary semifields, 333, 337, 349, 355.

binary trees, 228, 377, 541–542.
Binet, Jacques Philippe Marie, 732.
binomial coefficient, 225.
  generalized, xii, 38–39, 511–524.
binomial power series, 81, 228, 236, 246, 249.
binomial theorem, 225.
  generalized, xi, 218, 222–224.
binomial type, 233–234.
bipartite graphs, x, 104, 110, 119, 187–188, 627.
Birch, Bryan John, 563.
Birkhoff, Garrett, 96, 103.
bits variables, 416.
bitwise operations, 416.
bivariate generating function, 591, 648–649.
Bixby, Robert Eugene, 423.
Blackburn, John E., 425, 427.
Blaha, Stephen, 472, 480.
Blaschke, Wilhelm, 114.
block designs, 7, 177.
Boethius, Anicius Manlius Torquatus Severinus, 86.
Bollobás, Béla, 174, 585, 606, 641, 644–645, 778, 788.
Borchardt, Carl Wilhelm, 191, 219, 652, 789.
Bose, Bella, 438.
Bose, Raj Chandra, xv, 9, 221.
bound variables, 51.
bracket notation, 45–59, 226, 590, 649.
brackets versus parentheses, 24.
Bratley, Paul, 480, 489.
Brent, Richard Peirce, 253.
Brill, John, 118, 119.
Brillhart, John David, 303.
Brioschi, Francesco, 119.
Britikov, Vladimir Evgen'evich, 708, 789.
British Museum, x, 34, 85–86.
Brown, Kimberly Ann Lewis, xiii.

Brown, Robert, motion, 792.
Bruck, Richard Hubert, 7, 9.
Bruhat, François, 481.
Bruijn, Nicolaas Govert de, 203, 207–208, 538, 545–546, 563, 641.
Brylawski, Thomas Henry, 422–423.
bubble sort, 571.
bumping, 449.
Burnside, William Snow, problem, ix.
Burroughs 220 computer, 333.
Burroughs B5000 computer, 534.
Burroughs Corporation, 534.
Buseman, Herbert, 94, 103.
Bussey, William Henry, 277, 303.

Cajori, Florian, 43.
California Institute of Technology, ix.
Cambridge University, 83.
cancellation law, 383.
cancelling a multidigraph, 787.
cancelling a multigraph, 679.
Canfield, Earl Rodney, 788.
Cantor, Georg Ferdinand Ludwig Philip, ternary set, 551.
Capelli, Alfredo, 256.
Carlitz, Leonard, 244, 254, 480, 508.
Carmichael, Robert Daniel, 516, 523.
carries, 515, 519–521.
Carter, John Lawrence, 13.
Cartesian product of graphs, 201.
Case Institute of Technology, 303.
Catalan, Eugène Charles, numbers, 80, 228–229, 236.
Cauchy, Augustin Louis, 22, 40, 464, 732.
  inequality, 124, 137–138, 162.
  integral formula, 593, 692.

Cayley, Arthur, 117–119, 209, 217, 219, 238, 254, 546, 563, 652, 789.
  graph, 201.
central factorial numbers, 71, 229, 237.
central groupoids, 357–375.
certificate, 139.
Chakravarti, Indra Mohan, 9.
Champernowne, David Gawen, 267, 270.
characteristic labeling, 127.
characteristic of a semifield, 308.
characteristic polynomial, 188, 195, 361.
Chaundy, Theodore William, 480.
checkers, 187.
Cheema, Mohindar Singh, 466, 480.
Child, James Mark, 88.
Chinese ring puzzle, 2.
Chou, Jim Chih-Chun, xiii.
Chow, Timothy Yi-Chung, 192.
chromatic number of a graph, 123, 524.
Churchhouse, Robert Francis, 304, 546, 563.
circles, 87.
circuit in a matroid, 410, 420.
circulant matrix, 151.
circular list, 416.
Ciucu, Mihai, 192.
classroom example, 357.
Clausen, Thomas, 23.
clean multigraph, 646, 743.
clique, 123.
clique labeling, 127.
closed subsets, 406.
closure, difunctional, 402.
  in a matroid, 406.
  reflexive, 401.
  transitive, 87, 393.
clutter, 413–414, 423.
codes, 5, 433–438.

Cohn, Martin, 10.
collineations, 314, 350.
coloring a graph, 123, 524.
column vectors, 124.
combinations, 86.
combinatorial geometries, 408, 425.
combinatorial matrices, x, 177–186.
common eigenvectors, 159, 186.
commutative law, 377–378.
commutative semifields, 334, 336, 345–355.
commutators, ix.
companion matrix, 296.
compatible marriages, 96.
compatible matrices, 157.
compatible partitions, 199.
compensation factor, 648, 678, 680, 785.
complement of a graph, 123.
complement of a relation, 393.
completely equidistributed sequence, 265–270.
completing the square, 90.
complex component, 644, 657.
complex multigraph, 654.
complex partition, 502–505.
complex variable theory, 633–640.
components, 570, 650.
  acyclic, 652.
  bicyclic, 618–622, 652, 699, 761.
  complex, 644, 657, 722, 780.
  giant, 589, 643–792, 785.
  sparse, 671.
  strong, 395–397, 785.
  trivial, 396.
  unicyclic, 619, 644, 652, 749, 778, 779.
  weak, 395–397, 404.
composition of functions, 55, 238.
composition of relations, 393.
computer science versus mathematics, x.

computers in mathematical
    research, 1–13, 345.
Comtet, Louis, 238, 254, 641.
Concrete Mathematics, 15, 24, 28,
    84, 564.
condensation method, 112.
cone, 133–134.
confluent hypergeometric series,
    638, 666, 773.
conjugates, 190–191.
connected multigraph, 650–651,
    761, 780.
connected relation, 395.
continued fraction, 677, 782.
continuous mathematics, xi, 15.
converse of a relation, 394.
convex combination, 95–96.
convex corner, 160.
convex function, 538–539.
convolution family, 225–227.
convolution matrix, 234, 238.
convolution minimization, 538.
convolution polynomials, 225–256,
    569.
Conway, John Horton, 304.
Coppersmith, Don, 438.
coproduct of graphs, 147.
Cordero Brana, Minerva, 344.
core of a multigraph, 678.
Corless, Robert Malcolm, 256.
coset, 359.
cost of a vector, 127.
cotrees, 191.
coupon collector's problem, 57.
covering relation, 180, 407.
Coxeter, Harold Scott Macdonald,
    9.
Cramer, Gabriel, 117.
Crapo, Henry Howland, 422–427.
Crelle, August Leopold, 20, 222.
Croft, William James, xiii.
Čulik, Karel, 191.

Cunningham, Allan Joseph
    Champneys, 303.
Curtis, Frank Edward, 375.
Cvetković, Dragoš Mladen, 189,
    192, 202.
cycle, 210.
    first, 585–642, 759, 767, 778, 792.
cyclic group, 295.
cyclic shift, 429.
cyclotomic number field, 190.
cyclotomic polynomials, 299, 521.

De Bruijn, Nicolaas Govert, 203,
    207–208, 538, 545–546, 563,
    641.
de Finetti, Bruno, 44.
decomposable matrix, 97.
deficiency of a multigraph, 646–647,
    714, 729, 735, 740, 753.
Delsarte, Philippe, 176.
Desargues, Gérard, 317.
    projective plane, 8–9, 333.
Désarménien, Jacques Robert
    Jeanésarménien, 105, 481–482.
Desnanot, Pierre, 112, 114, 120.
determinants, x, 110–114, 119.
Dickson, Leonard Eugene, 306,
    335–336, 341, 343, 519, 523.
difference sets, 7–8.
differential posets, 186.
differentiation, 50, 231, 656.
difunctional closure, 402.
digraphs (directed graphs), xii, 357.
    random, 785–788.
    strong components, 395–397, 785.
    strongly connected, 387–391,
        785–788.
    weak components, 395–397, 404.
Dijkstra, Edsger Wijbe, 58.
Dijkstra Debets, Maria C., 58.
direct cosum of graphs, 145.
direct sum of graphs, 143.
directed graphs, xii, 357.

directed multigraph process,
785–788.
Dirichlet, Johann Peter Gustav
Lejeune, integrals, 554.
discrete mathematics, x–xii, 15.
distributive laws, 306, 308, 309.
distributive quasifield, 306.
Dixon, Alfred Cardew, 53, 59.
Dmitriev, E. F., 676, 788.
Dodgson, Charles Lutwidge,
112–113, 120.
Dold, Albrecht, 192, 202.
Doob, Michael, 192, 202.
dot product, 45, 124.
of matrices, 133.
of tensors, 322.
doubly infinite series, 45–51.
doubly stochastic matrix, 95–98,
104.
Dowling, Thomas Allan, xv, 221.
Dress, Andreas Walter Martin,
119, 120.
Drew, John Henry, 375.
dual argument, 361–362.
dual feasible matrix, 165.
dual of a planar graph, 189.
dual of a projective plane, 331–332.
dual tableau, 456–457.
dual ternary operation, 312–313,
344.
duality principle, 133, 169.
Duckworth, Richard, 13.
Dulmage, Andrew Lloyd, 11.
duplication formula, 665, 699, 765.
Dwyer, Paul Sumner, 303.
dynamic programming, 5, 526,
537, 541.

Ear decomposition, 391.
echelon form, 512.
Eckmann, Beno, 192, 202.
edge automorphism, 680.
edge-symmetric graph, 153.

Edmonds, John Robert (= Jack),
405, 422.
Edwards, Anthony William
Fairbank, 62, 78–79, 83.
Egerváry, Jenő (= Eugen), 96, 103.
egf, 590.
Egorychev, Georgii Petrovich, 58,
89, 94, 103–104.
Ehresmann, Charles, 513.
eigenvalues, 92, 125, 155, 182, 188,
199.
eigenvectors, 125, 181, 185.
Eilers, Daniel Ralph, xiii.
Eisenstein, Ferdinand Gotthold
Max, 567, 582, 652, 789.
elementary Abelian group, 308.
Elkies, Noam David, 187, 191, 192.
elliptic theta function, 494, 571.
empirical data, 778–780, 792.
enlargements, 406, 408, 414–415.
enveloping series, 639–640.
equilibrium vector, 97.
equivalence relation, 395, 407.
equivalent hypercubes, 326.
equivalent quadratic forms, 90–91.
Erdélyi, Arthur, 37, 43, 431–432.
Erdős, Pál (= Paul), 546, 563,
585–588, 615, 618, 626, 628,
641, 643–644, 647, 719, 728,
789.
erections, 423.
Esseen, Carl-Gustav, 577.
Etchemendy, Max, xiii.
Etherington, Ivor Malcolm Haddon,
431–432.
Ettingshausen, Andreas von, 33–34,
40.
Euler, Leonhard, 5, 494, 498, 501,
502, 509, 546, 563, 710.
summation formula, 82, 572.
totient function $\varphi(n)$, 295, 521.
Eulerian circuits, 207–208.
Evans, T. A., 8, 11.

Evans, Trevor, 357, 359–360, 364, 371–372, 375.
evolution of complex components, 644, 722, 780.
evolution of deficiency, 740.
evolving graph, 585–792.
Ewing, John Harwood, 39, 44.
excess of a multigraph, 646–647, 651, 703, 722, 735, 740, 753, 761, 763, 779.
exp function in finite field, 278–279.
expansion formula for $f(se^t)$, 752.
exponent modulo p, 280.
exponential generating function, 590.
exponentiation, 88, 301–302.
Ezekiel, son of Buzi, 387.

Facets, 163, 166, 170.
factorial powers, 30–33, 36–38, 221, 228, 232–233, 256, 731.
factorials, 86.
Falikman, Dmitry Iechielovich, 104.
Fano, Robert Mano, 12.
Farkas, Imre, 134.
Faulhaber, Johann, 61–84.
    portrait, 63.
feasible matrix, 131.
Feller, Willibald (= Vilim = Willy = William), 577, 582.
Fenchel, Werner, 94, 103.
Fermat, Pierre de, little theorem, 281.
Fibonacci, Leonardo, sequence, 516, 520–523, 525–527, 553–554.
Fibonomial coefficients, 38, 515–516, 521.
field polynomial, 295.
Finetti, Bruno de, 44.
finite fields, xi, 277–304, 511.
finite Fourier transform, 150.
first cycle, 585–642, 759, 767, 778, 792.

Fischer, Ludwig Joseph, 10.
Fishburn, Peter Clingerman, 404.
FKG inequality, 732.
Flajolet, Philippe Patrick Michel, xii, xiii, 39, 40, 44, 58, 59, 570, 573, 582, 585, 789.
Fletcher, Raymond Russwald, III, 375.
floating-point arithmetic, 88.
floor function, xi, 257–264.
Flores, Ivan, 10.
flowchart, 273.
Foata, Dominique Cyprien, xiv, 105.
Fontaine des Bertins, Alexis, 113.
Ford, Lester Randolph, 44.
Ford, Lester Randolph, Jr., 268, 270.
forest, 622.
formal power series, 49.
Fortuin, C. M., 789.
four-color problem, 5.
Fourier, Jean Baptiste Joseph, 34, 40.
    transform, finite, 150.
fractional part, 257.
Franklin, Fabian, 481, 494–497, 502, 509, 510, 513.
Franklin, Joel Nick, 265, 270.
Fray, Robert Dutton, 515, 524.
Fredman, Michael Lawrence, xiii, 537, 556, 563.
free central groupoids, 371–373.
free erections, 423.
free relations, 397.
free trees, 209, 217, 565, 588–591.
Frenzen, Christopher Lee, 40.
Frieze, Alan Michael, 788.
Frobenius, Ferdinand Georg, 153, 202, 469, 480.
Furck, Sebastian, 63.

Gale, David, 441, 443.

Galois, Évariste, fields, xi, 277–304, 511.

Galtman, Amanda Carol, 176.

Galton, Francis, 563.

Gamma function, 665, 689, 696, 699, 710, 765.

gapless partition, 543.

Gardy, Danièle, 58.

Garey, Michael Randolph, 394, 402, 404.

Garsia, Adriano Mario, 24, 39, 40, 254.

Gauß (= Gauss), Johann Friderich Carl (= Carl Friedrich), 509, 510, 524, 710.

distribution, 577, 753, 768.

integers, 502.

q-nomial coefficients, xii, 38, 511–524.

gcd, 517, 522.

Gee, Tony, xiii.

generating function, 496, 504, 546, 553, 651–653.

bivariate, 591, 648–651.

generation of combinatorial patterns, 1.

combinations, 3.

n-tuples, 1.

partitions, 3.

perfect matchings, 106, 117.

permutations, 2–3, 13.

Gessel, Ira Martin, 35–36, 40, 53, 58, 59, 78–79, 83–84, 244, 253.

Giambelli, Giovanni Zeno, 120.

giant component, 589, 643–792.

giant strong component, 785.

Ginibre, Jean, 789.

Glaser, Anton, xiv, 41.

Glassey, Charles Roger, 382, 385.

Gnewt, Ursula Nom-de-Plume, 44.

Godsil, Christopher David, 192–195, 202.

going configurations, 622, 758.

Goldberg, Karl, 25.

golden ratio (ϕ), 523.

Goldman, Jay Robert, 513.

Golomb, Solomon Wolf, 10, 270.

Golub, Gene Howard, 202.

Golumbic, Martin Charles, 385.

Gomes Teixeira, Francisco, 236, 254.

Good, Irving John, 209, 215, 219, 248, 254, 267, 270.

Gordon, Basil, 466–467, 471, 475–476, 480–482.

Gouarné, René, 343.

Gould, Henry Wadsworth, 35, 38, 40, 219, 254.

Goulden, Ian Peter, 45, 59, 641, 789.

Goursat, Édouard Jean-Baptiste, 404.

Graham, Ronald Lewis, xiii, 28, 40, 84, 254, 393, 439–441, 443, 789.

Graham Harrison, Susan Lois, 422.

graph process, 647.

graph theory, x–xiii.

Grassman, Hermann Günther, 513.

Gray, Frank, binary code, 2.

greatest integer function, 257.

Greene, Daniel Hill, 41.

Grötschel, Martin, 123–124, 128, 174–175.

group, 6, 306.

group theory, ix.

Guilbaud, Georges Théodule, 403, 404.

Gupta, Hansraj, 35, 41.

Gutman, Ivan, 189, 192.

Hadamard, Jacques Salomon, 524.

matrix, 5.

Hagen, Johann George, 41.

Håland Knutson, Inger Johanne, xiii, 257, 264.

half-edge, 735–736, 745–746, 782.

Hall, Marshall, Jr., v, ix–x, xiii, 1, 7–10, 311, 343, 344, 372, 377.

Hall, Philip, 96, 103.

Hamel, Angele Marie, 121.

Hamilton, William Rowan, cycles, 788.

Hamming, Richard Wesley, code, 426–427.

Hankel, Hermann, contour, 637, 689, 691.

symbol, 677.

Harary, Frank, 203, 208, 219, 641.

Hardy, Godfrey Harold, 23, 41, 270, 480, 494, 509, 535.

harmonic numbers, 531, 574.

generalized, 82–83.

Harriot, Thomas, 34, 85–88.

Harris, Bernard, 254.

Hathaway, Arthur Stafford, 510.

Hawlitschek, Kurt, 84.

Hayashi, Harry Sumio, 10.

Haynsworth, Emilie Virginia, 25.

Heaviside, Oliver, 20, 42.

Heine, Heinrich Eduard, 508, 509.

Heller, Sidney, 546, 563.

Henrici, Peter Karl Eugen, 47, 49, 59, 790.

Hensel, Kurt Wilhelm Sebastian, 42.

Herz, Jean-Claude, 304.

Higgs, Denis Arthur, 425, 427.

Hindenburg, Carl Friedrich, 13, 41.

history of mathematics, 19–22, 29–35, 44, 61–64, 68–70, 85–88, 117–119, 502.

Hoare, Charles Antony Richard, xiv, 45, 422.

Hobby, John Douglas, xiii.

Hodge, William Vallance Douglas, 104.

Hoffman, Alan Jerome, 152, 373.

Holt Hopfenberg, Anatol Wolf, 391.

homogeneous coordinates, 8–9, 310–312, 344.

homomorphism, 369.

Householder, Alston Scott, 125, 175.

Houten, Lorne, 466–467, 471, 475–476, 480.

Hsu, Leetsch Charles, 254.

Hu, Te Chiang, 381–382, 385.

Huffman, David Albert, 385.

algebra, 377.

algorithm, xii, 377–379, 382–383.

Hughes, Daniel Richard, 11, 306, 341–343.

Hurwitz, Adolf, 222–223.

Hutchinson, G., 11.

Hwang, Frank Kwangming, 386.

hypercubes, 321–326.

hypergeometric series, 221, 635, 697–698, 710–713, 721.

basic, 507–508.

hyperoctahedral group, 680.

hyperplane, 134.

IBM 360/67 computer, 414.

IBM 7094 computer, 364.

ideal primes, 518, 523.

idempotent elements, 360, 368.

idempotent mappings, 230.

identity element, 306.

Ikehara, Shikao, 561, 564.

imaginary number system, 271–276.

imaginary part, 598.

increasing binary operation, 377.

independent set, 409–410, 420.

index function, 278–279.

index of a sequence, 486–487.

indexing polynomials, 279, 304.

induced functions, 256.

inequalities, 731–732.

input/output, 422.

Institut Mittag-Leffler, 174, 186.

integration, 656.

interior-point methods, 172.
interval arithmetic, 88.
interval notation, xi.
invariant relations, 448–449.
inverse of a relation, 394.
inversion digraph, 452–453.
involutions, 480, 494–495, 503, 508, 510.
irreducible polynomial modulo p, 279–280.
irreducible representation, 445.
isolated vertices, 588–589, 614, 704, 728, 778, 780.
isomorph elimination, 5.
isomorphism, 314, 365.
isotopism, 313–316, 328, 348.
Itai, Alon, 386.
iteration of a function, 239–241.
Iverson, Kenneth Eugene, xi, 41.
    convention, 15–25, 44, 90.

Jabotinsky, Eri, 234, 237–238, 240, 244, 254.
Jackson, David Martin Rhŷs, 45, 59, 641, 789.
Jacobi, Carl Gustav Jacob, 62, 76–77, 79, 83, 84, 106, 114, 117–118, 120, 303, 463, 478, 494, 501–502, 505–506, 509–510.
    logarithm, 283.
Jacquet, Philippe Pierre, 59.
Janson, Carl Svante, xiii, 253, 605, 631, 641, 643, 790.
Jarden, Dov, 516, 524.
Jaworski, Jerzy, 641.
Jesus Christ, 69.
Jiggs, B. H., 11.
Johns Hopkins University, 502.
Johnson, Diane Mary, 11.
Johnson, Selmer Martin, 2, 11, 13, 186.

Jordan, Camille, canonical form, 368–369.
Jordán, Károly (= Charles), 26, 34–35, 41, 253, 255.
Juhász, Ferenc, 172, 175.

k-distributed sequence, 266.
$\mathcal{K}$-structure, 211.
Kac, Mark, 23, 41.
Kahaner, David Kenneth, 255.
Kanerva, Lauri, xiii.
Kantor, William M., 355.
Karamata, Jovan, 44.
Karloff, Howard, 186.
Karoński, Michał, 641, 788.
Karp, Richard Manning, 382, 385, 644–645, 785, 790.
Kashin, Boris Sergeevich, 174, 175.
Kasteleyn, Pieter Willem, 789.
Kaucký, Josef, 254.
Kerber, Adalbert, 105.
kernel of a multigraph, 679.
kernel system, 181.
Killgrove, Raymond Bruce, 10, 11.
Kirschenhofer, Peter, 641.
Klamkin, Murray Seymour, 40.
Kleinfeld, Erwin, 9, 11, 306, 335, 341–343.
Kleitman, Daniel J (Isaiah Solomon), 385, 421, 429, 432.
Knopp, Konrad Hermann Theodor, 535.
Knowlton, Kenneth Charles, 439, 443.
Knuth, Donald Ervin, iv, xiii, 9, 11, 13, 38, 40, 41, 59, 84, 209, 219, 254, 255, 271, 275, 303, 306, 333, 343, 355, 372, 422, 423, 443, 464, 466, 481, 489, 524, 546, 564–566, 571, 578, 582–583, 642, 644, 653, 673, 789, 790.
Kolesova, Galina Ivanovna, 13.

Koniāgin, Sergeĭ Vladimirovich, 174, 175.
Kostka, Carl Franz Albert, 463.
Kou, Lawrence Tien-Yi, 385, 386.
Kramp, Christian, 30, 32–34, 36–37, 41, 42.
Krattenthaler, Christian, 482.
Krause, Karl Christian Friedrich, 10.
Kreweras, Germain, 192.
Krogdahl, Stein, 421.
Kronecker, Leopold, 19, 24, 42.
  delta, 19, 472.
  product of matrices, 146–147, 188, 192, 195.
Kruskal, Martin David, 575, 578, 583.
Kuczma, Marek, 241, 255.
kudology, 355.
Kuipers, Lauwerens, 264.
Kummer, Ernst Eduard, 515–518, 521, 524.
Kuperberg, Gregory John (= Grzegorz Jan), 187, 192.
Kuratowski, Kasimir, 394, 404.

L-classes, 368.
L-series, 49.
labeled objects, 590, 649.
Labelle, Gilbert, 256.
Lagarias, Jeffrey Clark, 104.
Lagrange (= de la Grange), Joseph Louis, Comte, 91, 112, 114, 732.
  interpolation, 240.
  inversion, 54, 59, 215, 241–244, 569, 595.
  multipliers, 99.
Laguerre, Edmond Nicolas, polynomials, 239.
Lah, Ivo, 255–256.
  numbers, 239, 244.
Laksov, Dan, 114, 120.

Lam, Clement Wing Hong, 13.
Lambek, Joachim, 404.
Landau, Edmund Georg Hermann, 561, 564.
Langford, Eric, 404.
language design, 416.
Laplace (= de la Place), Pierre Simon, Marquis de, 690.
Laplacian matrix of a digraph, 204.
Laplacian matrix of a graph, 189.
Larsen, Michael, 187, 192.
Las Vegas, 389–391.
Lascoux, Alain, 114, 116, 119, 120.
last cycle, 778–779.
latin squares, 4–7.
Laurent, Paul Mathieu Hermann, series, 49, 56, 58.
Lawler, Eugene Leighton, 422.
leaves, 377.
Lebesgue, Henri Léon, 561.
  measure, 262.
Leclerc, Bernard, 119, 120.
Leech, John, 372.
Leeuwen, Jan van, 382, 386.
left nucleus, 309.
Legendre (= Le Gendre), Adrien Marie, function, 710.
Lehmer, Derrick Henry, 3, 11, 61, 303.
Leibniz, Gottfried Wilhelm, Freiherr von, 15, 85.
Leiss, Ernst Ludwig, 438.
length of a vector, 124.
Lewis Brown, Kimberly Ann, xiii.
lexicographic (dictionary) order, 381, 447, 487.
Li, Chi-Kwong, 375.
Libri, Guglielmo Icilio Bruto Timoleone, 20–22, 42.
Lidl, Rudolf, 304.
linear functional, 346, 593–594.
linear operators, 233–234.
linear programming, 169.

linear recurrence, 516, 525.
linear transformations, 90.
linked list, 416.
Lint, Jacobus Hendricus van, 104.
literate programming, 422.
Littlewood, Dudley Ernest, 463,
    464, 477–478, 481.
Lloyd, Edward Keith, 104.
Logan, Benjamin Franklin (= Tex),
    Jr., 28, 35–37, 39, 42.
London, David, 101, 103.
long labels, 417, 422.
longest cycle, 627.
loop: semigroup with cancellation
    laws and identity element, 345.
loop: edge from a vertex to itself,
    648, 758–759, 786.
Lovász, László, 120, 123–124,
    128, 131, 152, 156–157, 165,
    172–175, 186, 192, 391.
    number $\vartheta(G)$, x, 123, 129, 176.
Lucas, François Édouard Anatole,
    515, 520, 524.
Lucchesi, Cláudio Leonardo, 391.
Luce, Robert Duncan, 391.
Łuczak, Tomasz Jan, xiii, 643, 719,
    778, 790.
Lund, Carsten, 174.

Macdonald, Ian Grant, 480, 481,
    489.
Maclagan-Wedderburn, Joseph
    Henry, 307, 343; *see also*
    Veblen–Wedderburn systems.
MacMahon, Percy Alexander, xii,
    447, 465–466, 468, 470, 476,
    481–484, 489, 491.
MACSYMA system, 613–614.
Mahler, Kurt, 535, 546, 564.
Mann, Henry Berthold, 8, 11.
mappings, 569–570, 575.
Marcus, Marvin, 99, 103, 192, 202.

Markov, Andrei Andreevich (the
    elder), chain, 775–776.
Markov, Andrei Andreevich (the
    younger), process, 646–647.
Martian horsemen, 602.
Martin, Monroe Harnish, 270.
martingales, 606.
Marx, Imanuel, 26, 42.
matchings, perfect, 119.
Mathematica, iv, 225–226, 231–232,
    235, 239, 241, 248, 253.
mathematics, concrete, 15, 24, 28.
mathematics and computer science,
    x.
matrices, 3-dimensional, 321.
    circulant, 151.
    combinatorial, x, 177–186.
    companion, 296.
    compatible, 157.
    convolution, 234, 238.
    decomposable, 97.
    dot product of, 133.
    doubly stochastic, 95–98, 104.
    dual feasible, 165.
    feasible, 131.
    Kronecker product of, 146–147,
        188, 192, 195.
    minimal, 98.
    minors of, 98.
    multiplication of, 322.
    n-dimensional, 321–326.
    nonsingular, 90, 324.
    of nonnegative integers, 446.
    of zeros and ones, 360, 439–441,
        446, 480, 482.
    orthogonal, 125, 196.
    permanent of, 92, 98, 104.
    permutation, 96.
    positive semidefinite, 132.
    symmetric, 89, 125, 447.
    trace of, 132, 189.
    transpose of, 321, 447, 455–457.
    triangular, 89.

matrix tree theorem, 189.
    oriented, 204, 212, 214.
matroids, xii, 405–427.
Maurolico, Francesco, 86.
McEliece, Robert James, 175.
McKay, Brendan Damien, 192–195,
    202, 788.
McKay, John Keith Stuart, 480,
    489.
McMechan, William Edgar, xiii.
medial law, 377–378, 381.
Meertens, Lambert Guillaume
    Louis Théodore, 676.
Mendelsohn, Nathan Saul, 11, 369.
merging, 539.
Mersenne, Marin, number, 303.
Mertens, Franz Carl Joseph, 118,
    120.
METAPOST, xiii.
Metzler, William Henry, 110, 120.
middle nucleus, 309.
Minc, Henrik, 101, 104, 192, 202.
Minding, Ferdinand, 114.
minimal matrix, 98.
minimization, 537–538.
minor of a matrix, 98.
minvolution, 538.
Mitrinović, Dragoslav Svetislav,
    790.
Mittag-Leffler, Magnus Gösta (=
    Gustaf), Institute, 174, 186.
mixed-radix number system, 519.
mnemonic for π, 87.
Möbius, August Ferdinand, 22,
    42, 524.
    function μ, 522.
modd, 280.
Monte Carlo method, 3–4.
Moon, John Wesley, 209, 219, 641.
Moore, Douglas Houston, 42.
Morrison, Emily, 40.
Morrison, Philip, 40.
Motwani, Rajeev, 174.

Motzkin, Theodor Samuel, xiii, 48,
    59, 239, 255, 393, 516, 524.
Muir, Thomas, 110, 120.
Mullin, Ronald, 244, 255.
multidigraphs, 785–788.
multigraph process, 647, 782.
multigraphs, 626, 647–648.
multiplication, 274.
multiset union, 381.

n-cube, 128.
n-simplex, 128.
Naegelsbach, Hans Eduard von,
    463, 478.
Napier, John, Laird of Merchiston,
    87.
Nash-Williams, Crispin St. John
    Alvah, 391.
National Science Foundation, 9, 39,
    103, 208, 218, 391, 403, 421,
    427, 432, 438, 508, 523, 534,
    563, 582, 641, 788.
natural central groupoid, 359, 362.
negative radix, 272–273, 275.
Netto, Otto Erwin Johannes
    Eugen, 42.
Newman, Morris, 25, 99, 103.
Newton, Isaac, 15.
    identities, 21, 189.
Nguyen, Hien Quang, 423.
Nicole, François, 32, 42.
Niederreiter, Harald Günther, 264,
    304.
Nielsen, Niels, 34–35, 39, 42.
non-Desarguesian projective planes,
    9, 310, 345.
nonassociative division rings, 306.
noncentral elements, 320.
nonperiodic words, 522.
nonsingular hypercubes, 324.
nonsingular matrices, 90, 324.
Norges Almenvitenskapelige
    Forskningsråd, 427.

normal distribution, 577, 753, 768.
normal numbers, 266–267.
Norman, Robert Zane, 203, 208.
Norton, Horace Wakeman, III, 7,
    11, 12.
notation, x–xi, 15, 34, 44–46, 56,
    221, 256, 311, 344.
NP-hard problem, x.
nucleus of a semifield, 309, 338,
    342.

Odlyzko, Andrew Michael, 39, 40,
    255, 438, 582.
Office of Naval Research, 103, 372,
    391, 403, 421, 427, 432, 508,
    523, 563, 582, 641, 788.
optimal trees, 541.
order ideals, 412, 414.
order-preserving binary operation,
    377.
order-preserving mappings, 35.
ordered trees, 209.
oriented subtrees, 204.
oriented trees, 209–210, 217.
orthogonal labelings, 126.
orthogonal latin squares, 4–5.
orthogonal matrices, 125, 196.
orthogonal vectors, 124.
Ostrowski, R. T., 11.
Overton, Michael Lockhart, 172,
    175.
Oxford University, 85.

P-partition, 485.
Pacault, Jean François, 404.
Padberg, Manfred William, 128,
    176.
Paige, Lowell J., 10, 11.
Palmer, Edgar Milan, 586, 641,
    642.
paradox, 46–49, 632–633.
parallel decoding, 434–435, 437.
parentheses versus brackets, 24.

Parker, Douglass Stott, Jr., 385,
    386.
Parker, Ernest Tilden, 5, 8, 11.
Parker, R. V., 35, 42.
partial Bell polynomial, 234.
partial differential equations, 117.
partial order, 35, 396, 484.
partial quotients, 564.
partition function, 534.
partition geometry, 426.
partitioned tensor products,
    193–202.
partitions, xii, 493–513.
    generation of, 3.
    into powers, 525–528, 543.
    of a set, 26–27, 230, 235,
        395–396, 407, 439.
    plane, xii, 119, 447–448, 465–485.
    solid, 483–491.
Patashnik, Oren, 40, 84, 254, 789.
Paterson, Michael Stewart, xiii,
    493.
Paule, Peter, 54, 58, 59.
Peleg, David, 404.
Pennington, William Barry, 546,
    564.
Pepper, Jon Vivian, 85, 88.
perfect graphs, 162–165, 168.
perfect matchings, 106, 117, 190,
    192.
permanent of a matrix, 92, 98, 104.
permutation matrix, 96.
permutation model, 587, 594, 647.
permutations, 26–27, 86, 429–432.
    generalized, 447.
    generation of, 2–3, 13.
    multiplication of, 6.
Perron, Oskar, 153.
Petersen, Julius Peter Christian,
    graph, 156, 168.
Petri, Carl Adam, net, 391.
Pfaff, Johann Friedrich, x, 40, 42,
    117, 120.

Pfaffian, 105–121.
phase transition, 589, 644–647, 688.
pi (π), as data source, 408–409, 587–588.
mnemonic for, 87.
Pickert, Günter, 343.
Pierce, William A., 7, 12.
Piff, Michael John, 425–427.
Pippenger, Nicholas John, 564.
Pitman, James William, 256.
Pittel, Boris Gershon, xiii, 59, 255, 565, 585, 628, 642–644, 653, 719, 789, 790.
planar graphs, 189, 628, 719–720.
plane partitions, xii, 119, 447–448, 465–485.
Plummer, Michael David, 120.
Pochhammer, Leo, 31, 42.
Poisson, Siméon Denis,
distribution, 577, 782.
process, 605–606, 641.
Pólya, György (= George), 493, 642.
polytope, 168.
Pope, David Alexander, 13.
population growth, 87.
Poritsky, Hillel, 42.
positive semidefinite matrix, 132.
postfix notation, 238.
poweroids, 234.
Pragel, Daniel Michael, 375.
pre-semifields, 308, 326, 346.
preferential arrangements, 238–239.
preservation of order, 377.
prime factors, 190.
primitive elements, 278.
primitive polynomials, 279.
modulo p, 304.
primitive roots, 278, 282.
principal isotopes, 314.
principle of optimality, 541.
Proctor, Robert Alan, 481.
Prodinger, Helmut, 641.

projective planes, ix, 1, 6–9, 13, 310.
coordinates for, 8–9, 310–312, 344.
proof of an algorithm, 448–449.
proper semifield, 307.
Propp, James Gary, 187, 192.
Prüfer, Ernst Paul Heinz, 210–211, 219, 223.
pruning a multigraph, 677.
psi function, 757.

q-nomial coefficients, xii, 38, 511–524.
QSTAB, 127, 161.
quadratic forms, 89.
quadratic programming, 133.
quater-imaginary number system, 271–274.
quaternary number system, 272.
queues, 383, 385.

R-classes, 368.
R-series, 49.
radix list sort, 421.
Raleigh, Sir Walter, 85.
Ramanujan Iyengar, Srinivasa, 506, 509–510, 535, 790.
function, 566, 569, 575, 654, 673.
Ramshaw, Lyle Harold, 119.
random graph process, 587, 647.
random graphs, xii–xiii, 52, 585–792.
random multigraph process, 587, 647.
random objects, 4, 406, 414.
random walk, 727.
randomness defined, 265.
Raney, George Neal, 209, 219, 223–224.
rank in a matroid, 411, 416.
rank of apparition, 518.
real part, 598, 693.
reciprocity theorem, 35–36.

recurrence relations, xii, 75–78, 106,
    525–583, 606–607, 696, 782.
reduced multigraph, 679.
referees, 119, 191, 355, 438.
reflective functions, 64–67, 73–74.
reflexive closure, 401.
regular digraphs, 207.
regular graphs, 152.
regularly divisible sequences, 517,
    522–523.
Reiffen, Barney, 13.
renewal arrays, 238.
Rényi, Alfréd, 569, 583, 585–588,
    615, 618, 626, 628, 641,
    643–644, 647, 654, 719, 728,
    789–791.
reversion of series, 241.
Riddell, Robert James, Jr., 650,
    791.
Riemann, Georg Friedrich
    Bernhard, 561, 781.
right nucleus, 309.
Riguet, Jacques, 404.
Riordan, John, 71, 84, 209, 215,
    219, 221–222, 229, 237, 255,
    569, 583.
Robbins, David Peter, 113, 121.
Roberts, Leigh, 423.
Robinson, D. W., 524.
Robinson, Gilbert de Beauregard,
    447, 456, 464.
Rodemich, Eugene Richard, 175.
Rødseth, Øystein Johan, 546, 564.
Rogers, Douglas George, 255.
Rogers, Leonard James, 497, 509.
Roman, Steven, 57–59, 244, 255.
rooted tree, 589–591, 652, 656.
Roscoe, Andrew William, xiv, 45.
Rosenstiehl, Pierre, 403, 404.
Rota, Gian-Carlo, 244, 255, 422,
    425–427, 481, 513.
Rothe, Heinrich August, 231.
round-robin tournament, 432.

Ruciński, Andrzej, 641, 789.
Rüdiger, Christian Friedrich, 13.
ruler function ρ, 528.
Rumsey, Howard Calvin, Jr., 113,
    121, 175.
Rutherford, Daniel Edwin, 464.
Ryser, Herbert John, 7, 9, 441, 443.

S..., 22, 40, 43.
Saalschütz, Louis, 53, 59, 119, 121.
Sachs, Horst, 192, 202.
saddle point method, 248, 598–599,
    615–616.
Sade, Albert, 7, 12.
Salmeri, Antonio, 26, 43.
Samuel, Isaac, 343.
Samuel, Pierre, 304.
Sandler, Reuben, 12, 306, 337, 343.
sandwich inequality, 123, 130.
Satterthwaite, Edwin Hallowell,
    Jr., 422.
Scheibner, Wilhelm, 118, 121.
Schensted, Craige Eugene (= Ea
    Ea), 447–449, 453, 455, 457,
    461–462, 464.
Schläfli, Ludwig, 33, 43.
Schlömilch, Oscar Xaver, 33, 43.
Schneider, Ivo, 83–84.
Schneider, Rolf, 94, 104.
Schoenfeld, Lowell, 254.
Schönhage, Arnold, 565–566, 578,
    583, 642.
Schrijver, Alexander, 104, 123, 128,
    157, 173–175.
Schröder, Friedrich Wilhelm Karl
    Ernst, 403, 404.
Schur, Issai, 115, 121, 463, 477–478,
    535.
Schützenberger, Marcel Paul, 450,
    456, 464.
Schwartz, Eugene Sidney, 383, 386.
Science Citation Index, 21, 94.
Scoins, Hubert Ian, 217, 219.

Scott, Dana Stewart, 12.
Seitz, G., 731, 791.
self-loops, 648, 758–759, 786.
semideterminants, 118.
semifields, 9, 305–355.
semipolynomials, 674.
  of a set, 26–27, 230, 235,
    395–396, 407, 439.
sets of lists, 239.
Shader, Leslie Elwin, 373, 375.
shadow point, 748, 781.
Shanks, Daniel Charles, 500, 509.
Shannon, Claude Elwood, Jr., 128,
  175, 176.
shears of a semifield plane,
  317–319, 350.
Siegel, Carl Ludwig, 546.
sign of a permutation, 105–108.
sign of a word, 106–107.
signed permutation, 680.
simple weak component, 396.
Singer, James, 8, 12.
singular value decomposition, 196.
sink vertex, 169.
skew symmetry, 105, 117.
Slater, Lucy Joan, 791.
Slaught, Herbert Ellsworth, 1.
smooth multigraph, 677.
solid partitions, 483–491.
Soria, Michèle, 570, 582.
sorting, 421.
Sosa, Christine, xiii.
source vertex, 169, 203.
spanning trees, 189–192.
sparse components, 671.
Sperner, Emanuel, 438.
  theorem, 433.
Sprugnoli, Renzo, 51.
spuds, 135.
STAB, 127, 161.
stable labeling, 127.
stable set, 127.
staircase, 79.

standard deviation, 778.
standard form of a hypercube, 328.
Stanford University, 15, 83.
Stanley, Richard Peter, 28, 35–36,
  39, 40, 43, 94, 104, 186, 187,
  190, 192, 464, 481–482, 513.
Stanton, Dennis Warren, 53, 59,
  186.
Staudt, Karl Georg Christian von,
  23.
Stedman, Fabian, 13.
Steffensen, Johan Frederik, 234,
  255.
  polynomials, 229.
Stegun, Irene Anne, 39, 641.
Stein, Charles Max, 792.
Steiner, Jacob, triple systems, 5.
Stembridge, John Reese, 114, 119,
  121.
Stenger, Allen, 509.
Stepanov, Vadim Evdokimovich,
  570, 583, 644–645, 676, 686,
  719, 791.
Stieltjes, Thomas Jan, constants,
  545.
Stirling, James, 25, 29–34, 37, 39,
  43.
  approximation, 37, 249, 426, 434,
    532, 571, 576, 581, 649, 667,
    692, 717, 741–742, 766, 769.
  cycle numbers, 25–39, 44, 87,
    235, 238–239, 243, 530–532.
  numbers, history, x, 29–35, 87.
  numbers, notation for, 15, 25–26,
    44.
  polynomials, 245.
  subset numbers, 25–39, 44,
    71–72, 230, 235, 238–239,
    243, 686.
Stockmal, Frank, 12.
stopping configurations, 590, 758.
Strehl, Karl Ernst Volker, 105.

strict plane partitions, 465,
474–476.
strictly enveloping series, 639–640.
strong Huffman algebra, 383–384.
strong components, 395–397, 785.
strong orthogonal labelings, 173.
strong product of graphs, 146.
strong zero, 25.
strongly connected digraphs,
387–391.
strongly connected labeled
digraphs, 785–788.
Subbarao, Mathukumalli Venkata,
497, 510.
subcubes, 322.
subspaces, 511–513.
Sudan, Madhu, 174.
Sudler, Culbreth, Jr., 466, 469,
481, 510.
summation, repeated, 66.
sums, manipulation of, 16–19.
sums of powers, 61–84.
superpolynomially small, 691.
SWAC computer, 8.
Swiercz, Stanisław, 13.
Swift, Jonathan Dean, 5, 10, 12,
280, 303.
switch operation, 375.
Sylvester, James Joseph, 91, 114,
481, 494, 502–505, 510, 511,
513, 652, 791.
symmetric functions, 477–479.
symmetric group, 445.
symmetric matrices, 89, 125, 447.
symmetrization, 404.
Szegedy, Márió, 174.
Szegő, Gábor, 642.

T-graph, 210.
tableau algebra, 461.
tableaux, generalized Young, xii,
445–475, 477.
tail-exchange method, 690, 709.

Tamaki, Jeanne Keiko, 383, 385,
386.
Tanner, Henry William Lloyd, 107,
109, 118, 121.
Tanner, Rosalind Cecilia Hildegard
Young, 87.
Tanturri, Alberto, 546, 564.
Tarjan, Robert Endre, 404.
Tauber, Alfred, 550, 561, 581.
Taylor, Brook, series, 574, 777.
using ϑ, 752.
with remainder, 639.
Teissier, Bernard, 94, 104.
tensor product of matrices, 195.
tensors, 321–322.
ternary operation, 310–312.
ternary ring, 313.
TEX, iv, 31.
TH, 127, 140, 161.
*The Art of Computer
Programming*, 26, 124, 177,
224, 256, 464.
theory of functions, 633–640.
theory versus practice, xi, 270.
theta function (elliptic), 494, 571.
Thiel, Larry Henry, 13.
Thiele, Thorvald Nicolai, 34.
Thimonier, Loÿs, 58.
Thorup, Anders, 114, 120.
three-legged saddle, 602, 628, 693,
742.
Tits, L., 62, 84.
toll booths, 389–390.
Tompkins, Charles Brown, 5, 11,
12.
top line, 722, 729, 755, 775.
topological sequence, 486–488.
topological sorting, 486.
Torelli, Gabriele, 116, 121.
Torgašev, Aleksandar, 192.
Toscano, Letterio, 256.
total order, 396, 400, 484, 649.
trace of a matrix, 132, 189.

transitive closure, 87, 393.
translation of a semifield plane,
    317–319, 350.
transpose of a matrix, 321, 447,
    455–457.
transpose of a projective plane,
    331.
transpose of a relation, 394.
transpose of a Veblen–Wedderburn
    system, 332, 344.
transversals, 5.
tree function, 229–230, 249,
    590–591, 781–782.
tree polynomials, 230, 233,
    566–575, 653, 730–731.
trees, 588–591, 644, 652, 704,
    747–748, 760, 778.
triangular matrix, 89.
triangularization, 512.
Tricomi, Francesco Giacomo
    Filippo, 37, 43.
triplication formula, 665, 710, 765.
trivial component, 396.
Trotter, Hale Freeman, 2, 12, 13.
Trudi, Nicolò, 463.
truncation in a matroid, 413.
Tucker, Alan Curtiss, 382, 385.
Turing, Alan Mathison, machine,
    21.
Tutte, William Thomas, 204–205,
    207, 208, 212, 214, 219, 405,
    422.
Tweedie, Charles, 34, 43.
twisted fields, 336.

Uhlenbeck, George Eugene, 791.
umbral calculus, 57, 222, 255.
understatement, 781.
unicyclic components, 619, 644,
    652, 749, 778, 779.
uniform model, 587, 593, 647.
uniformly distributed sequences,
    258, 263–270.

unique representation, 275–276.
University of North Carolina, 221.
University of South Alabama, 83.
UNIX, 31.
unrooted trees, 588–591, 644, 652,
    656, 704, 747–748, 760, 778.

Valid modification, 98.
van Aardenne-Ehrenfest, Tatyana,
    207–208.
van der Waerden, Bartel Leendert,
    x, 89, 98, 103–104.
Van Duren, K. D., 11.
van Leeuwen, Jan, 382, 386.
van Lint, Jacobus Hendricus, 104.
Van Loan, Charles Francis, 202.
Vandermonde, Alexandre
    Théophile, convolution, 225,
    229.
Veblen, Oswald, 9, 332, 344.
Veblen–Wedderburn systems, 9,
    332, 344.
vector space, 309, 405.
Veltmann, Wilhelm, 118, 121.
verbose labels, 422.
vertex-symmetric graph, 153.
Viennot, Gérard Xavier, 78–79, 84.
Vobly˘ı, Vitali˘ı Antonievich,
    676–677, 791.
von Ettingshausen, Andreas, 33–34,
    40.
von Staudt, Karl Georg Christian,
    23.

Waerden, Bartel Leendert van der,
    x, 89, 98, 103–104.
Walker, Robert John, 9, 10, 12,
    333, 344, 345, 355.
Wall, Donald Dines, 523, 524.
Ward, Mark Daniel, xiii.
Ward, Morgan, 58, 59, 253, 256.
Watson, Eric John, 304.
weak components, 395–397, 404.
weak direct product, 187.

weak nucleus, 338–342.
Wedderburn, Joseph Henry
    Maclagan, 307, 343; *see also*
    Veblen–Wedderburn systems.
Weichsel, Paul Morris, 192.
Weierstrass (= Weierstraß), Karl
    Theodor Wilhelm, 20, 43.
weighted Lovász number, 129.
weighted path length, 381, 385.
Welch, Lloyd Richard, 270.
well-ordered set, 484.
Wells, Mark Brimhall, 2, 13.
Welsh, Dominic James Anthony,
    425, 427.
Wene, Gregory Peter, 344.
Wenzel, Walter, 119–121.
Wermuth, Udo Wilhelm Emil, xiii.
White, Neil Lawrence, 423.
Whitney, Hassler, 405, 422.
Whittaker, Edmund Taylor, 791.
Wiener, Christian, 44.
Wiener, Norbert, 561, 564.
Wierman, John Charles, 719, 790.
Wilf, Herbert Saul, xiii, 39, 41, 44,
    83, 186, 515, 524.
Williamson, Stanley Gill, 186.
wires, identifying, 439.
Wirth, Niklaus Emil, 422.
Woodall, Herbert J., 303.
Wouk, Arthur, 274.
Wozencraft, John McReynolds, 13.
Wright, Edward Maitland, 23, 41,
    270, 480, 489, 494, 509, 510,
    620, 642, 653, 654, 675–677,
    680, 684, 686, 722, 726,
    785–786, 791–792.

Yale University, 303.
Yamabe, Hidehiko, 13.
Yee, Ae Ja, 510.
Young, Alfred, 445, 464.
    generalized tableaux, 445–475.

Zajączkowski, Władysław, 118, 121.

Zariski, Oscar (= Zaristky, Asher),
    304.
Zassenhaus, Hans Julius, 404.
zeta function, 545, 572.
Zolnowsky, John Edward, 503, 510.